Food Microbiology and Food Safety

Series Editor:
Michael P. Doyle

Food Microbiology and Food Safety Series

Food Microbiology and Food Safety publishes valuable, practical, and timely resources for professionals and researchers working on microbiological topics associated with foods, as well as food safety issues and problems.

For other titles published in this series, go to
http://www.springer.com/series/7131

Effie Tsakalidou • Konstantinos Papadimitriou
Editors

Stress Responses of Lactic Acid Bacteria

Springer

Editors
Effie Tsakalidou
Laboratory of Dairy Research
Department of Food Science
and Technology
Agricultural University of Athens
Iera Odos 75, 118 55 Athens
Greece
et@aua.gr

Konstantinos Papadimitriou
Laboratory of Dairy Research
Department of Food Science
and Technology
Agricultural University of Athens
Iera Odos 75, 118 55 Athens
Greece
kpapadimitriou@aua.gr

ISBN 978-1-4614-2975-3 ISBN 978-0-387-92771-8 (eBook)
DOI 10.1007/978-0-387-92771-8
Springer New York Dordrecht Heidelberg London

Printed on acid-free paper

Springer is part of Springer Science+Business Media (www.springer.com)

To the former, current and future members
of the Laboratory of Dairy Research
at the Agricultural University of Athens,
Greece.

Preface

All organisms will be exposed to stress during their lifetime. From single cells to multicellular organisms, they all are able to sense and monitor their environment and to trigger appropriate protective physiological mechanisms in case environmental conditions interfere with or threaten their homeostasis. Such adaptive responses are necessary for any organism to survive and persist under stress. It is well established today that bacteria are among the most resilient organisms known to man and they occupy highly divergent ecological niches, including some that would be prohibiting for most other forms of life. Their inherent adaptive nature is one of the reasons bacteria have served as important models for studying the stress responses of cellular organisms. For many years, research on *Escherichia coli* and *Bacillus subtilis*, the Gram-negative and Gram-positive paradigms, has dominated the field of bacterial stress physiology.

Lactic acid bacteria (LAB) constitute a heterogeneous group of bacteria that are renowned for the crucial role they play in the health of humans and animals. Several LAB species are eminent industrial microorganisms, since they are employed as starters for the majority of existing fermented foods. During fermentation, LAB acidify their environment by catabolizing carbohydrates into acidic end-products (primarily lactic acid) and thus are responsible for the bio-transformation of raw materials of plant or animal origin into foods with extended shelf-life. In parallel, their overall metabolic properties contribute to the development of the flavor, aroma and texture, i.e. the organoleptic characteristics of fermented foods. A number of LAB starter strains have a long history of harmless consumption and have been awarded the general regarded as safe (GRAS) status. Interestingly, surpassing the GRAS status, certain LAB exhibit probiotic properties and they exert additional health benefits to the consumer. Finally, LAB are also commensals and they can be found as part of the natural flora of mammals. In contrast to food-related and probiotic LAB, host-associated LAB include some of the most severe human and veterinary pathogens.

Due to their economic importance for the food industry and their health-related implications as probiotics or pathogens, LAB have attracted much attention, and their genetics, physiology and metabolism have been under rigorous investigation over the past decades. Among the different fields of LAB research, the exploration of their stress responses was realized early on to be fundamental. During food processing and storage, LAB reside under adverse environmental conditions designed to be bacteriostatic or bactericidal for food spoilage microorganisms and foodborne pathogens. Apparently, the ability of starter LAB to respond to and withstand such technological stresses is essential for an incessant fermentation process and can seriously influence the overall quality of the final product. Robustness towards technological stresses is also a prerequisite for probiotic LAB for the preparation of probiotic formulas and foods. However, during consumption, the key feature of probiotic strains is their aptitude to survive through the harsh environment of the gastrointestinal tract of the host so as to reach and colonize the intestine and exert their health-promoting effects. Furthermore, the pathogenic nature of certain LAB species has been clearly associated with their tolerance to environmental stresses. In contrast to starter and probiotic LAB, pathogenic LAB must cope with the host's innate immune response that imposes invading microorganisms to severe stress conditions and bactericidal substances.

The focus of this book is to present the current knowledge of the stress physiology of LAB. Even though the topic started to appear in the literature by the late 70s to early 80s, it was not until the 90s when the development of tools for the elaborate genetic manipulation of several LAB strains laid the background for the first detailed studies of their stress responses. Most importantly, over the past 10 years, the field has greatly benefited from advances in genomics, transcriptomics and proteomics, and there has been a burst of information about the stress behavior of LAB.

The book is organized into five different parts. Part I provides an introductory overview of the stress responses of LAB. Part II deals with the responses of LAB towards specific environmental stresses. As a response to different stress conditions, LAB coordinate the expression of gene and protein networks in an attempt to counteract perturbations caused by the specific environmental insult. Such adaptive responses of LAB are well elucidated. Some of the mechanisms that underpin stress adaptation are conserved among LAB, while others may be species- or even subspecies-dependent. In addition, induction of a stress response may confer protection against either a single environmental stress or multiple stresses. All these aspects are the essence of the stress behavior of LAB and are thoroughly discussed in this part of the book. In Part III, the stress responses of LAB in the context of species or genera are presented. Taking into account the large number of species and genera in the LAB group, research about their stress physiology has been uneven, with the majority of LAB being virtually unexplored. However, several members of the group have been extensively studied and the mechanisms of their stress resistance are reviewed here. Part IV focuses on the applications and the future aspects of LAB stress research. Information available today has shed new light on the responses of LAB to stress. Our overall perception of LAB stress behavior has significantly advanced and we have reached the point where targeted manipulation of their robustness is now feasible. Moreover, our knowledge of LAB stress responses has served as the basis for applied research in forefront applications. In many of these cases, there is a clear need to ensure survival of LAB when exposed to multiple stresses or to stresses that exceed the typical definition of the 'abiotic' stress (e.g., in the case of bacteriocins or bacteriophages). Finally, Part V highlights in a concluding summary the most important research challenges in the field that will significantly improve our conception of stress behavior in LAB.

It should be emphasized that in this book pathogenic LAB are treated equally with food-related and probiotic LAB and information about their stress adaptive mechanisms is included whenever available. It should also be mentioned that in Part III, a chapter is dedicated to the stress physiology of Bifidobacteria, which belong to *Actinobacteria*, but have been traditionally included in LAB *sensu lato*.

We would like to express our gratitude to all authors who contributed to this book. We sincerely appreciated their hard work, their patience during the editing process, as well as their determination to see this book completed. We believe that 'Stress Responses of Lactic Acid Bacteria' will serve as an essential guidebook to researchers in the field, industry professionals, and advanced students in the area.

Athens, Greece

Effie Tsakalidou
Konstantinos Papadimitriou

Acknowledgments

We sincerely thank all of the authors for their contribution to this book. We would also like to thank Eleni Mavrogonatou for her assistance in the editing process and all of our colleagues in the laboratory for their patience during the preparation of the book.

Contents

Part V [illegible]

Contributors

Helge K. Abicht Department Clinical Research, University of Bern, Murtenstrasse 35, 3010 Berne, Switzerland

Sikder M. Asaduzzaman Tissue Banking and Biomaterial Research Unit, Atomic Energy Research Establishment, Savar, GPO Box 3787, 1000 Dhaka, Bangladesh

Laboratory of Microbial Technology, Division of Applied Molecular Microbiology and Biomass Chemistry, Department of Bioscience and Biotechnology, Graduate School, Kyushu University, 6-10-1 Hakozaki, Higashi-ku, Fukuoka, 812-8581 Japan

Yanick Auffray Laboratoire de Microbiologie de l'Environnement, USC INRA 2017/EA UCBN956, Université de Caen, 14032 Caen Cedex, France

Rodolphe Barrangou Danisco USA Inc, 3329 Agriculture Drive, Madison, WI 53716, USA

Abdellah Benachour Laboratoire de Microbiologie de l'Environnement, Université de Caen, 14032, Caen Cedex, France

Peter A. Bron TI Food and Nutrition, Nieuwe Kanaal 9A, 6709 PA Wageningen, The Netherlands

NIZO Food Research, P.O. Box 20, 6710 BA Ede, The Netherlands

Kluyver Centre for Genomics of Industrial Fermentation, Julianalaan 67, 2628 BC Delft, The Netherlands

Aurélie Budin-Verneuil Laboratoire de Microbiologie de l'Environnement, Université de Caen, 14032 Caen Cedex, France

Vittorio Capozzi Department of Food Science, University of Foggia, Via Napoli 25, 71122 Foggia, Italy

Bénédicte Cesselin Unité Bactéries Lactiques et Pathogènes Opportunistes, INRA UR888, Jouy en Josas 78352, France

Jian Chen School of Biotechnology, Jiangnan University, Wuxi, Jiangsu 214122, China

State Key Laboratory of Food Science and Technology, Jiangnan Uiversity, Wuxi, Jiangsu 214122, China

Maria De Angelis Department of Biologia e Chimica Agro-Forestale ed Ambientale, University of Bari, Via G. Amendola 165/a, 70125 Bari, Italy

Clara G. de los Reyes-Gavilán Department of Microbiology and Biochemistry of Dairy Products, IPLA – CSIC, Ctra. Infiesto s/n, 33300 Villaviciosa, Asturias, Spain

Aurélie Derré-Bobillot Unité Bactéries Lactiques et Pathogènes Opportunistes, INRA UR888, Jouy en Josas 78352, France

Guocheng Du School of Biotechnology, Jiangnan University, Wuxi, Jiangsu 214122, China

The Key Laboratory of Industrial Biotechnology, Ministry of Education, Wuxi, Jiangsu 214122, China

State Key Laboratory of Food Science and Technology, Jiangan University, Wuxi, Jiangsu 214122, China

Annabelle Fernandez Unité Bactéries Lactiques et Pathogènes Opportunistes, INRA UR888, Jouy en Josas 78352, France

Daniela Fiocco Department of Biomedical Sciences, University of Foggia, Via L. Pinto, 71122 Foggia, Italy

Gerald F. Fitzgerald Department of Microbiology, University College Cork, Cork, Ireland

Alimentary Pharmabiotic Centre, Cork, Ireland

Petra Foerst Chair for Food Process Engineering and Dairy Technology, Centre of Life and Food Sciences, Technische Universität München, Weihenstephaner Berg 1, 85354 Freising-Weihenstephan, Germany

Charles M.A.P. Franz Max Rubner-Institute, Federal Research Institute for Nutrition and Food, Department of Safety and Quality of Fruit and Vegetables, Haid-und-Neu-Strasse 9, D-76131 Karlsruhe, Germany

Christel Garrigues Department of Physiology, Chr. Hansen A/S, 2970 Hφrsholm, Denmark

Philippe Gaudu Unité Bactéries Lactiques et Pathogènes Opportunistes, INRA UR888, Jouy en Josas 78352, France

Jean-Christophe Giard Laboratoire de Microbiologie de l'Environnement, Université de Caen, 14032 Caen Cedex, France

Marco Gobbetti Department of Biologia e Chimica Agro-Forestale ed Ambientale, University of Bari, Via G. Amendola 165/a, 70125 Bari, Italy

Alexandra Gruss Unité Bactéries Lactiques et Pathogènes Opportunistes, INRA UR888, Jouy en Josas 78352, France

Jean Guzzo Equipe de Recherche en Vigne et Vin, Institut Jules Guyot, Université de Bourgogne, BP 27877–21078, Dijon, France

Axel Hartke Laboratoire de Microbiologie de l'Environnement, Université de Caen, 14032 Caen Cedex, France

Colin Hill Department of Microbiology, University College Cork, Cork, Ireland

Alimentary Pharmabiotic Centre, Cork, Ireland

Wilhelm H. Holzapfel Handong Global University, School of Life Sciences, Pohang, Gyeongbuk 791-708, South Korea

Philippe Horvath Danisco France SAS, Dangé-Saint-Romain 86220, France

Jessica K. Kajfasz Center for Oral Biology, University of Rochester, School of Medicine and Dentistry, 601 Elmwood Avenue, P.O. Box 611, Rochester, NY 14580, USA

Michiel Kleerebezem TI Food and Nutrition, Nieuwe Kanaal 9A, 6709 PA Wageningen, The Netherlands

NIZO Food Research, P.O. Box 20, 6710 BA Ede, The Netherlands

Laboratory for Microbiology, Wageningen University, Dreijeplein 10, 6703 HB Wageningen, The Netherlands

Jan Kok Department of Molecular Genetics, Groningen Biomolecular Sciences and Biotechnology Institute, University of Groningen, Nijenborgh 7, 9747 AG Groningen, The Netherlands

Top Institute Food and Nutrition, Nieuwe Kanaal 9A, 6709 PA Wageningen, The Netherlands

The Netherlands Kluyver Centre for Genomics of Industrial Fermentations, NCSB, Julianalaan 67, 2628 BC Delft, The Netherlands

Oscar P. Kuipers Department of Molecular Genetics, Groningen Biomolecular Sciences and Biotechnology Institute, University of Groningen, Nijenborgh 7, 9747 AG Groningen, The Netherlands

Top Institute Food and Nutrition, Nieuwe Kanaal 9A, 6709 PA Wageningen, The Netherlands

The Netherlands Kluyver Centre for Genomics of Industrial Fermentations, NCSB, Julianalaan 67, 2628 BC Delft, The Netherlands

Ulrich Kulozik Chair for Food Process Engineering and Dairy Technology, Centre of Life and Food Sciences, Technische Universität München, Weihenstephaner Berg 1, 85354 Freising-Weihenstephan, Germany

Gilles Lamberet Unité Bactéries Lactiques et Pathogènes Opportunistes, INRA UR888, Jouy en Josas 78352, France

Delphine Lechardeur Unité Bactéries Lactiques et Pathogènes Opportunistes, INRA UR888, Jouy en Josas 78352, France

Claire Le Marrec Ecole Nationale Supérieure de Chimie Biologie Physique, Institut Polytechnique de Bordeaux, Unité de Recherche Œnologie - EA 4577 - USC 1219 INRA., Institut des Sciences de la Vigne et du Vin, 210 Chemin de Leysotte, CS50008 Villenave d'Ornon, France

José A. Lemos Center for Oral Biology and Department of Microbiology and Immunology, University of Rochester Medical Center, 601 Elmwood Avenue, Box 611, Rochester, NY 14642, USA

Stefano Mancini Department Clinical Research, University of Bern, Murtenstrasse 35, 3010 Berne, Switzerland

Abelardo Margolles Department of Microbiology and Biochemistry of Dairy Products, IPLA – CSIC, Ctra, Infiesto s/n, 33300 Villaviciosa, Asturias, Spain

Mélanie Mermod Department Clinical Research, University of Bern, Murtenstrasse 35, 3010 Berne, Switzerland

Konstantinos Papadimitriou Laboratory of Dairy Research, Department of Food Science and Technology, Agricultural University of Athens, Iera Odos 75, 118 55 Athens, Greece

Martin Bastian Pedersen Department of Physiology, Chr. Hansen A/S, 2970 Hϕrsholm, Denmark

Vianney Pichereau Laboratoire de Microbiologie de l'Environnement, Université de Caen, 14032 Caen Cedex, France

João P.C. Pinto Department of Molecular Genetics, Groningen Biomolecular Sciences and Biotechnology Institute, University of Groningen, Nijenborgh 7, 9747 AG Groningen, The Netherlands

Robert G. Quivey Jr. Center for Oral Biology, University of Rochester, School of Medicine and Dentistry, Rochester, NY 14580, USA

Department of Microbiology and Immunology, University of Rochester, School of Medicine and Dentistry, 601 Elmwood Avenue, P.O. Box 611, Rochester, NY 14580, USA

Alain Rincé Laboratoire de Microbiologie de l'Environnement, Université de Caen, 14032 Caen Cedex, France

R. Paul Ross Teagasc, Moorepark Food Research Centre, Fermoy, Co Cork, Ireland

Alimentary Pharmabiotic Centre, Cork, Ireland

Chalat Santivarangkna Chair for Food Process Engineering and Dairy Technology, Centre of Life and Food Sciences, Technische Universität München, Weihenstephaner Berg 1, 85354 Freising-Weihenstephan, Germany

Nicolas Sauvageot Laboratoire de Microbiologie de l'Environnement, Université de Caen, 14032 Caen Cedex, France

Kirsi Savijoki Department of Food and Environmental Sciences, University of Helsinki, P.O. Box 66, 00014 Helsinki, Finland

Marc Solioz Department Clinical Research, University of Bern, Murtenstrasse 35, 3010 Berne, Switzerland

Kenji Sonomoto Laboratory of Microbial Technology, Division of Applied Molecular Microbiology and Biomass Chemistry, Department of Bioscience and Biotechnology, Faculty of Agriculture, Graduate School, Kyushu University, 6-10-1 Hakozaki, Higashi-ku, Fukuoka 812-8581, Japan

Laboratory of Functional Food Design, Department of Functional Metabolic Design, Bio-Architecture Center, Kyushu University, 6-10-1 Hakozaki, Higashi-ku, Fukuoka 812-8581, Japan

Giuseppe Spano Department of Food Science, University of Foggia, Via Napoli 25, 71100 Foggia, Italy

Bioagromed University of Foggia, 71100 Foggia, Italy

Catherine Stanton Teagasc, Moorepark Food Research Centre, Fermoy, Co Cork, Ireland

Alimentary Pharmabiotic Centre, Cork, Ireland

Effie Tsakalidou Laboratory of Dairy Research, Department of Food Science and Technology, Agricultural University of Athens, Iera Odos 75, 118 55 Athens, Greece

Francesca Turroni Laboratory of Probiogenomics, Department of Genetics, Biology of Microorganisms, Anthropology, Evolution, University of Parma, Parco Area delle Scienze 11A, 43124 Parma, Italy

Aditya Upadrasta Teagasc, Moorepark Food Research Centre, Fermoy, Co Cork, Ireland

Department of Microbiology, University College Cork, Cork, Ireland

Hermien van Bokhorst-van de Veen TI Food and Nutrition, Nieuwe Kanaal 9A, 6709 PA Wageningen, The Netherlands

NIZO Food Research, P.O. Box 20, 6710 BA Ede, The Netherlands

Laboratory for Microbiology, Wageningen University, Dreijeplein 10, 6703 HB Wageningen, The Netherlands

Douwe van Sinderen Alimentary Pharmabiotic Centre and Department of Microbiology, Bioscience Institute, National University of Ireland, Western Road, Cork, Ireland

Pekka Varmanen Department of Food and Environmental Sciences, University of Helsinki, P.O. Box 66, 00014 Helsinki, Finland

Marco Ventura Laboratory of Probiogenomics, Department of Genetics, Biology of Microorganisms, Anthropology, Evolution, University of Parma, Parco Area delle Scienze 11A, 43124 Parma, Italy

Nicolas Verneuil Laboratoire de Microbiologie de l'Environnement, Université de Caen, 14032 Caen Cedex, France

Bart C. Weimer Department of Population Health and Reproduction, School of Veterinary Medicine, University of California, Davis, CA 95616, USA

Michiel Wels TI Food and Nutrition, Nieuwe Kanaal 9A, 6709 PA Wageningen, The Netherlands

Nizo food research, P.O. Box 20, 6710 BA Ede, The Netherlands

Center for Molecular and Biomolecular Informatics, Radboud University Nijmegen Medical Center, PO Box 9101, 6500 HB Nijmegen, The Netherlands

Chongde Wu School of Biotechnology, Jiangnan University, Wuxi, Jiangsu 214122, China

The Key Laboratory of Industrial Biotechnology, Ministry of Education, Wuxi, Jiangsu 214122, China

Feng Xue School of Biotechnology, Jiangnan University, Wuxi, Jiangsu 214122, China

The Key Laboratory of Industrial Biotechnology, Ministry of Education, Wuxi, Jiangsu 214122, China

Yuji Yamamoto Laboratory of Cellular Microbiology, School of Veterinary Medicine, Kitasato University, Towada 034-8628, Aomori, Japan

Juan Zhang School of Biotechnology, Jiangnan University, Wuxi, Jiangsu 214122, China

The Key Laboratory of Industrial Biotechnology, Ministry of Education, Wuxi, Jiangsu 214122, China

Aldert Zomer Alimentary Pharmabiotic Centre and Department of Microbiology, Bioscience Institute, National University of Ireland, Western Road, Cork, Ireland

Part I
Introduction

Chapter 1
The Importance of Understanding the Stress Physiology of Lactic Acid Bacteria

Charles M.A.P. Franz and Wilhelm H. Holzapfel

1.1 Introduction

Throughout human history, lactic acid bacteria (LAB) have had a significant impact on human culture, traditions, and well-being. In modern times, their economic importance has increased tremendously as a result of the industrialization of food bio-transformations. In particular, they have a key function in the development of the sensory and safety features of fermented food products. Thus, the reliability of starter strains is important in terms of quality and functional properties (relating to aroma and texture) and in terms of growth performance (fast growth, rapid acidification of the substrate, phage and bacteriocin resistance) and robustness of strains (during starter handling, storage and preservation by lyophilization, freezing, or spray-drying) (van de Guchte et al. 2002). Moreover, in the development of new applications such as probiotic foods, pharmaceutical preparations, and live vaccines, the need for robust LAB is even more important. Probiotics should survive the production and handling procedures as well as the environmental conditions encountered in the product they are added to, such as dairy-related products (e.g., fermented milks, frozen fermented dairy desserts, ice cream, and cheese). Furthermore, once they are consumed and encounter conditions in the gastrointestinal tract, they need to survive and be metabolically active under the range of stressful conditions typical of this environment. Thus, they should exhibit resistance against the autochthonous microbiota, demonstrate an ability to colonize the digestive or urogenital mucosa, and express specific probiotic functions under conditions

W.H. Holzapfel (✉)
Handong Global University, School of Life Sciences,
Pohang, Gyeongbuk 791–708, South Korea
e-mail: wilhelm@woodapple.net

E. Tsakalidou and K. Papadimitriou (eds.), *Stress Responses of Lactic Acid Bacteria*,
Food Microbiology and Food Safety, DOI 10.1007/978-0-387-92771-8_1,

that are unfavorable to growth (van de Guchte et al. 2002; Lorca and Font de Valdez 2009). Probiotic strains should therefore typically be acid resistant and bile tolerant, they should survive gastrointestinal conditions, and they should be able to bind to epithelial cells or the gastrointestinal mucus in order to colonize (van de Guchte et al. 2002). It should be noted, however, that, in practice, a drop in the viability of probiotic bacteria generally occurs (Dave and Shah 1997; Kailasapathy and Chin 2000; Ainsley Reid et al. 2007), typically due to the sensitivity of probiotic bacteria to heat treatments or to exposure to oxygen, hydrogen peroxide, bacteriocins of starter cultures, or acid environments (Dave and Shah 1998; Ainsley Reid et al. 2007). Thus, despite selection for robust strains, there is also still a need for the development of methods that protect the bacteria and increase their viability during processing and storage.

As a result of the association of LAB with fermented foods and with human and animal health, LAB enjoy increasing importance and consideration from scientists, industries, and consumers in modern society. In terms of total biomass, enormous quantities of LAB are being consumed in our daily diet. Traditionally, this is primarily related to fermented foods, but an increasing amount of LAB biomass can now also be allocated to functional foods, of which the probiotics comprise the largest and most rapidly growing segment of the market. The International Dairy Federation (IDF) reported the average annual consumption of fermented milk products to be 22 kg per capita in Europe (Mogensen et al. 2002), which amounts to around 8.5 billion kg of fermented milk per year. This figure does not take into account the LAB used in nondairy food fermentations (vegetables, meat, legumes, etc.) or probiotic products containing selected strains with beneficial health features. Thus, the actual amount of lactic acid bacterial biomass would be far greater (Hummel et al. 2007).

An improved understanding of the stress-response mechanisms of LAB is important, as it will provide a deeper insight into and an understanding of the adaptive responses and cross-protection to varying stresses. Therefore, the exploitation of LAB in industrial processes should become more strongly target-oriented and "streamlined." According to van de Guchte et al. (2002), the identification of crucial stress-related genes will reveal targets

- for specific modulation (to promote or limit growth),
- to develop tools to screen for tolerant or sensitive strains,
- to evaluate the fitness and level of adaptation of a culture.

Thus, future genome and transcriptome analyses can become a strong tool toward the selection and adaptation of LAB strains for industrial applications.

In general terms, the recovery ratio of a strain subsequent to a stress situation will depend on the prevailing environmental conditions. Moreover, the physiological state of microbial populations has a decisive influence on the quality of a fermented food, for example, depending on the maximum viability and physiological vigor of a starter strain. In contrast, this may also have an impact on the safety with regard to the survival and viability of pathogenic or spoilage bacteria.

1.2 Taxonomy of the Lactic Acid Bacteria

LAB form part of the Gram-positive bacteria with "low" (≤55 mol%) G+C in the DNA. They belong to class III (*Bacilli*) of the phylum *Firmicutes,* which comprises three classes, the other two being *Clostridia* (class I) and *Mollicutes* (class II). Within class III, LAB are represented in order II, the *Lactobacillales* (Garrity and Holt 2001). They may be considered as a "rapidly expanding" group of bacteria, presently with six families and about 40 genera. For some of these genera the family position has not been finalized definitively. Some new genera such as *Atopobacter* and *Bavariicoccus,* for example, appear to belong to the family Carnobacteriaceae. The wide range of the six families gives an impression of the diversity within LAB:

- Aerococcaceae (with seven genera)
- Carnobacteriaceae (with 16 genera)
- Enterococcaceae (with seven genera)
- Lactobacillaceae (with three genera)
- Leuconostoccaceae (with four genera)
- Streptococcaceae (with three genera).

Orla-Jensen (1942), in an early definition, described LAB as Gram-positive, non-motile, nonspore-forming, rod- or coccus-shaped organisms that ferment carbohydrates and higher alcohols to form mainly lactic acid. For several decades, the LAB taxonomy was based on this classical approach, involving morphological and physiological characteristics. However, due to advances in taxonomic considerations and molecular techniques, modern approaches no longer recognize such an unequivocal definition (Stiles and Holzapfel 1997; Axelsson 2004). Still, in present-day terms, a "typical" lactic acid bacterium is considered to be Gram-positive, nonspore-forming, catalase-negative, devoid of cytochromes, nonaerobic but aerotolerant, fastidious, acid-tolerant, and strictly fermentative, with lactic acid as the major end product of sugar fermentation (Klein et al. 1998; Axelsson 2004). LAB are phylogenetically quite diverse, and the genera generally associated with foods include *Lactobacillus, Lactococcus, Enterococcus, Streptococcus, Leuconostoc, Weissella, Oenococcus, Pediococcus, Tetragenococcus,* and *Carnobacterium.*

The recent availability of the complete genomes of representative LAB strains of all major families of the *Lactobacillales* enables an analysis of their evolutionary relationships (Makarova et al. 2006; Makarova and Koonin 2007). These investigations could show on the basis of a phylogenetic tree, reconstructed from concatenated alignments of the four subunits of the DNA-dependent RNA polymerase sequences, a division of *Lactobacillus* into three distinct groups, the first being comprised of *L. brevis, L. plantarum,* and *P. pentosaceus,* to which *L. salivarius* was basal. The second group consists of *L. gasseri, L. johnsonii, L. delbrueckii,* and *L. acidophilus*. An additional branch with *Leuconostoc mesenteroides* and *Oenococcus oeni* is wedged between the *L. brevis, P. pentosaceus,* and *L. plantarum* group and *L. salivarius;* thus, *L. salivarius* is also basal to this branch. The third group consists of *L. casei* and *L. sakei* and is basal to the *L. gasseri, L. johnsonii, L. delbrueckii,* and *L. acidophilus* groups (Makarova et al. 2006; Makarova and Koonin 2007).

Whole-genome comparisons of five *Lactobacillus* species (*L. salivarius, L. plantarum, L. acidophilus, L. johnsonii,* and *L. sakei*) that were completely sequenced showed that there is no extensive synteny of the genome sequences of these five species (Canchaya et al. 2006). The sequences with the best alignments were *L. johnsonii* and *L. acidophilus*, but alignments of these two species with the other three species (*L. plantarum, L. salivarius,* and *L. sakei*) showed much lower degrees of synteny at the interspecies level than observed in other species' genome comparisons with high- and low-G+C-content Gram-positive bacteria (Canchaya et al. 2006). Claesson et al. (2008) used the genomic data from 12 *Lactobacillus* strains to investigate whether a single, congruent phylogeny could be inferred. By reconstructing phylogenetic trees from concatenated sequences of 141 core proteins, as well as concatenated RNA polymerase subunit sequences, considerable incongruence was noticed, but it was still possible to distinguish four subgeneric groups, namely, Group A (*L. acidophilus, L. helveticus, L. delbrueckii* subsp. *bulgaricus, L. johnsonii,* and *L. gasseri*), Group B (*L. salivarius, L. plantarum, L. reuteri, L. brevis,* and *P. pentosaceus*), Group C (*L. sakei* and *L. casei*) and Group D (*Leuconostoc mesenteroides* and *O. oeni*). However, based on significantly different branching patterns within some groups and the availability of genomic data for too few members of the groups, three of the four groups could not confidently be identified as candidate novel genera within the current genus (Claesson et al. 2008). LAB commonly used as starters for fermented foods or as probiotics and their taxonomic status are shown in Table 1.1.

Although modern molecular biological methods and whole-genome analyses have provided a better understanding of, and a deeper insight into, the taxonomy of LAB, it appears that even more data are required to obtain a satisfactory picture of the phylogenetic relationships among LAB. In particular, the taxonomy of the largest genus, *Lactobacillus,* is far from satisfactory, as clearly indicated by the genomic data and the still widely accepted "artificial" division among obligately homofermentative, facultatively heterofermentative, and obligately heterofermentative species. Fortunately, most commercialized strains presently in use as starter cultures and probiotics are taxonomically well characterized and defined. This is an essential precondition for successful regulation and marketing.

1.3 Lactic Acid Bacterial Stress Responses

LAB are commonly regarded as fastidious, with complex growth requirements, and associated with "moderate" environmental conditions. However, when looking at LAB as a whole, they appear to be a highly heterogeneous group, showing wide biological diversity in terms of their physiology and their adaptation to ecological

Table 1.1 Lactic acid bacteria and starter cultures associated with fermented foods and products and the stresses they can encounter during production (Holzapfel, unpublished results)

		LAB genera			
Fermented product or probiotic	Stress encountered	*Lactobacillus (L.)*	*Lactococcus (Lc.)/ Streptococcus (S.)*	*Leuconostoc (Leuc.)/ Weissella/Oenococcus*	*Pediococcus/ Tetragenoccus*
Dairy products					
Hard cheese without eyes	Salt, acid, osmotic, oxidative		*Lc. lactis* subsp. *lactis*		
	High temperatures (some cheeses, yogurt)		*Lc. lactis* subsp. *cremoris*		
Cheese with small eyes	Cold temperatures (storage)		*Lc. lactis* subsp. *lactis* *Lc. lactis* subsp. *cremoris*	*Leuc. mesenteroides* subsp. *cremoris*	
Italian- and Swiss-type cheeses		*L. delbrueckii* subsp. *lactis* *L. delbrueckii* subsp. *bulgaricus* *L. helveticus*, *L. casei*	*S. thermophilus*		
Butter and buttermilk			*Lc. lactis* subsp. *lactis* *Lc. lactis* subsp. *cremoris*	*Leuc. mesenteroides* subsp. *cremoris*	
Yogurt		*L. delbrueckii* subsp. *bulgaricus* *L. acidophilus* (mild/probiotic yogurts)	*S. thermophilus*		
Kefir		*L. kefiri*, *L. parakefiri*, *L. kefiranofaciens*, *L. brevis*			

(continued)

Table 1.1 (continued)

Fermented product or probiotic	Stress encountered	LAB genera: *Lactobacillus (L.)*	*Lactococcus (Lc.)/ Streptococcus (S.)*	*Leuconostoc (Leuc.)/ Weissella/Oenococcus*	*Pediococcus/ Tetragenoccus*
Fermented meat products					
Raw fermented sausages (salami, cervelat, Teewurst, Landjäger, Chorizo, etc.), raw ham	Nitrite, salt (curing salts), osmotic, oxidative, cold temperatures, smoke, low water activity	*L. sakei, L. curvatus*			*P. acidilactici* *P. pentosaceus* *T. halophilus* (ham)
Fermented plant products					
Olives	Salt, osmotic, acid, cold temperatures	*L. pentosus, L. plantarum*		*Leuc. mesenteroides*	
Sauerkraut		*L. plantarum*		*Leuc. mesenteroides* *Leuc. fallax*	*P. acidilactici*
Pickles		*L. plantarum, L. brevis*		*Leuc. mesenteroides*	*P. cerevisiae*
Vegetables		*L. plantarum, L. fermentum*			*P. acidilactici* *P. pentosaceus*
Soy sauce	Salt, osmotic				*T. halophilus*

Fermented cereals			
Sourdough	Acid, CO_2, ethanol, oxidative	*L. sanfranciscensis, L. farciminis, L. fermentum, L. brevis, L. plantarum, L. amylovorus, L. reuteri, L. pontis, L. panis, L. alimentarius, L. rossiae*	*W. cibaria*
Alcoholic beverages			
Wine	Alcohol, acid, low temperatures		*O. oeni*
Rice wine		*L. sakei*	
Probiotics	In product: Acid, low temperatures, drying In gastrointestinal tract: Acid, bile, pancreatic juice	*L. acidophilus, L. amylovorus, L. crispatus, L. delbrueckii* subsp. *bulgaricus, L. fermentum, L. gallinarum, L. gasseri, L. helveticus, L. johnsonii, L. paracasei, L. plantarum, L. reuteri, L. rhamnosus, L. salivarius*	

conditions, some of which even range into being "extreme" conditions. Examples are given in Table 1.2, showing species such as *Lactobacillus suebicus,* which is able to grow and/or tolerate pH levels < 3.0, and others (e.g., some enterococci and carnobacteria) growing at pH values of 9.6. *Tetragenococcus muriaticus* and *T. halophilus* are adapted to high salt concentrations of 18% and even higher and are associated with

Table 1.2 Examples of growth/tolerance and the association of LAB with extreme conditions (Holzapfel, unpublished data)

Factor	Value	Organism	Substrate/habitat
Low pH	pH 2.8	*L. suebicus*	Fermenting apple/pear mash
	pH 3.2	*L. acetotolerans*	Rice vinegar
	pH ~ 3.0	*L. acidophilus* "group"	Stomach/upper duodenum
High pH	pH 9.6	*Enterococcus faecium*	*Bacillus* fermentation of soya
		Carnobacterium	Meat
Bile salts and pancreatic juice	Physiological concentrations	*L. acidophilus* "group"	Small intestines
		L. reuteri/ L. paracasei	
Salt (NaCl)	18–24%	*Tetragenococcus muriaticus*	Salted fermenting fish
	26.4% (tolerance)	*Carnobacterium viridans*	From vacuum-packed bologna
Low temperatures	0–2°C	*Leuconostoc gelidum*	Chill-stored vacuum-packed meats
		Some carnobacteria	
High temperatures	55°C	*L. delbrueckii* subsp. *delbrueckii*	Emmental-type cheese
	ca. 50°C	*Streptococcus thermophilus*	Italian-type hard cheeses
		Lactobacillus helveticus	
Nitrite	>150 ppm	Several LAB	Cured meat
Sorbic acid	>2 g/kg	Several LAB	Preserved juices, etc.
Hops resistance[a]		*Pediococcus damnosus*	Beer
		Pediococcus claussenii	
		L. brevis	
Ethanol	15%	*L. fructivorans*	Isolated from ketchup
	13%	*Oenococcus oeni*	Wine
Radiation resistance	γD_{10} = >1.0 kGy	*L. sakei*[b]	Radurized meat
Heat resistance	D_{65} = 20–30 min	*Weissella viridescens*	Processed meats
		Ent. faecalis	

[a]In relation to "bitter hop compounds" at concentrations ranging around 55 ppm of iso-α-acids
[b]Higher resistance during exponential growth than in stationary phase (Hastings et al. 1986)

salted fermenting fish and shellfish. Interestingly, *Carnobacterium viridans*, isolated from vacuum-packed bologna, was found to tolerate 26±4% (w/v) NaCl (saturated brine) for long periods at 4°C (Holley et al. 2002).

Stress can be defined as a change in the genome, the proteome, or the environment that results in a decrease in the growth rate or survival of a microorganism (Spano and Massa 2006; Sugimoto et al. 2008). Stress responses are extremely important for microorganisms, which experience continual changes in factors such as temperature, nutrient, and water availability or osmotic pressure in the environments in which they occur. The stressors, or stress factors, can be chemical, physical, or biological in nature. Some are of environmental origin (i.e., temperature, osmotic pressure, pH, ethanol concentration, available oxygen, presence of bile, antimicrobials), while others can be self-generated (acidity, starvation/low nutrient availability as a result of metabolism, generation of reactive oxygen species) (van de Guchte et al. 2002; Miyoshi et al. 2003; Spano and Massa 2006; Bruno-Bárcena et al. 2010).

Therefore, both the physiological status of the cells and environmental factors will affect the mechanism of response to stress. LAB have developed stress-sensing systems that detect these stresses and activate defenses that allow the bacteria to withstand harsh conditions or sudden environmental changes (van de Guchte et al. 2002; Spano and Massa 2006; Lorca and Font de Valdez 2009). The time taken to initiate the stress response is different for different types of stress. For example, bacteria respond to heat shock and osmotic shock quickly (in minutes) compared to cold shock (hours) (Wouters et al. 2001; Rosen and Ron 2002; Spano and Massa 2006). The activation of defenses against stress conditions depends on regulated gene expression. Although bacteria could theoretically have specific regulators for each stress, this would imply a tremendous genetic burden. Instead, regulators often control several genes and sometimes even other regulators (Van Bogelen et al. 1999) in integrated regulation systems. Bacterial stress responses therefore rely on the coordinated expression of genes that alter cellular processes such as cell division, DNA metabolism, housekeeping, membrane composition, metabolism, and transport (van de Guchte et al. 2002), acting in concert to increase the bacterial stress tolerance (Storz and Hengge-Aronis 2000; van de Guchte et al. 2002). The integration of these stress responses is accomplished by networks of regulators that allow the cell to react to various complex changes that affect cell survival and growth.

The stress-resistance systems can be divided into three classes: (1) specific, induced by a sublethal dose of the stress. This adaptive response is usually associated with the log phase of growth and involves the induction of specific groups of genes or regulons designed to cope with specific stress conditions; (2) general systems, where the adaptation to one stress condition can render cells resistant to other stress conditions; (3) stationary-phase-associated stress response, which involves the induction of numerous regulons designed to overcome several stress conditions. Unlike the adaptive response, the stationary-phase-associated response does not require any preexposure to stress for resistance development (Lorca and Font de Valdez 2009) and can be characterized as a general-type stress response (van de Guchte et al. 2002). Sometimes, as in the case for high temperature and acid stress, multiple mechanisms (both logarithmic and stationary-phase responses) can be present and are orchestrated in response to the stress challenge (van de Guchte et al.

2002; Azcárate-Peril et al. 2004). Cross-resistance, where one stress condition can render cells resistant to other stress conditions, is, in fact, a general theme among resistance systems in LAB but appears to vary among species (van de Guchte et al. 2002). An overview of the different stresses experienced by LAB, their reported cross-resistances, and the resistance mechanisms are shown in Table 1.3. From this table, it is also clear that certain effectors such as DnaK and GroEL play a central role in the global stress response.

A common regulatory mechanism in the stress response of bacteria involves the modification of sigma factors whose primary role is to bind core RNA polymerase-conferring promoter specificity (Haldenwang 1995). In *Escherichia coli* and other enteric bacteria, sigma S (RpoS) is a major global regulator for stress response in the cells. In *Bacillus* and *Listeria,* the gene most closely serving the function of *rpo*S is *sigB*, for regulation of the stress response (Völker et al. 1994; Becker et al. 1998) in these bacteria and also for the regulation of the virulence of *Staphylococcus aureus* (Bischoff et al. 2001). *B. subtilis* responds to environmental stress signals by producing over 40 general stress genes under the control of the σ^B transcription factor (Haldenwang 1995; Akbar et al. 1997). The σ^B transcription regulon includes the catalase gene *katE*, the gene encoding an osmoregulated proline transporter *opu*E, the *clpC* gene, which is similar to the ATPase subunits of ClpP-type proteases, the *gtaB* gene encoding a UDP-glucose pyrophosphorylase involved in trehalose synthesis, and several other genes (Abee and Wouters 1999). Among the differences between LAB species and *Bacillus*, one of the most striking is the absence of a σ^B ortholog, while several stress proteins (e.g., DnaK, GroEL, Clp) and their regulators (HrcA and CtsR) are conserved. In lactobacilli, the stress response is negatively regulated by these one-component system regulators HrcA and CtsR and by two-component signal transduction systems (Lorca and Font de Valdez 2009). A two-component regulatory system of *L. acidophilus* similar to the acid-related *lis*RK from *Listeria monocytogenes,* for example, was shown to be involved in the stress response to acid and ethanol (Azcárate-Peril et al. 2005). Thus, it is not yet clear how lactobacilli sense and respond to various stimuli and stresses from the environment, whether this is by global transcriptional regulators or by two-component regulatory systems, or perhaps both (Lorca and Font de Valdez 2009), and the unraveling of the occurrence and interaction of different stress responses remains an interesting challenge to which genomic and proteomic studies are contributing greatly (Lim et al. 2000; Gouesbet et al. 2002; Marceau et al. 2004; Xie et al. 2004; Azcárate-Peril et al. 2005; Pieterse et al. 2005; Denou et al. 2007; Stevens et al. 2010).

1.4 Industrial Importance of Understanding the Stress Response of Lactic Acid Bacteria

LAB play vital roles in food production and human health, while other Gram-positive pathogens, on the other hand, can cause diseases ranging from dental caries to potentially fatal gastrointestinal infections. While a stress response such

Table 1.3 Response mechanisms of lactic acid bacteria to various stresses encountered during processing or application and the major stress proteins or enzymes involved in the response (adapted from Van de Guchte et al. 2000; Azcárate-Peril et al. 2004; Lorca and Font de Valdez 2009; Spano and Massa 2006; Sugimoto et al. 2008; Walter et al. 2005; Bruno-Bárcena et al. 2010)

Stress response	Reported stress cross-resistance	Stress-related resistance mechanism	Involved stress-related proteins (enzymes)
Acid stress response	Heat, osmotic, oxidative (varies between species)	ATP-dependent expulsion of protons by proton pumps	F_0F_1-ATPAse
Two general types			K^+-ATPAse
During log growth phase (L-ATR; induced by nonlethal low pH)		Activation of arginine deiminase pathway – production of basic compounds (e.g., ammonia)	Arginine deiminase
In stationary phase, induction of general stress response			Urease
		Amino acid decarboxylation reactions and electrogenic transport	Ornithine/arginine/lysine decarboxylase
		Change in cell envelope composition	Lo18
		Repair of damaged proteins, DNA, and cell components	Ffh
		Incremental expression of regulators that promote minor or global responses	Heat shock chaperones and regulators (DnaK, GroEL; HrcA, CtsR)
		Induction of heat shock proteins	recA, AP endonuclease, UvrSystem
Oxidative stress response	Heat, acid, general stress resistance	Reducing intracellular environment	Glutathione peroxidase, glutathione reductase
		Prevention of reactive oxygen species formation	Thioredoxin, thioredoxin reductase
		Target protection	NADH oxidase
		Repair of oxidative damage	Catalase
			Pseudocatalase
			Superoxide dismutase
			Methionine sulfoxide reductase
			Mannose phosphotransferase system
			FLP (FNR-like proteins)
			RecA
			Phosphate ABC transporter

(continued)

Table 1.3 (continued)

Stress response	Reported stress cross-resistance	Stress-related resistance mechanism	Involved stress-related proteins (enzymes)
Cold stress response	Heat shock, freezing (cryotolerance)	Production of cold-induced proteins (CIPs) to maintain	CIPs involved in
Transient adaptive response, i.e., cold shock response		membrane fluidity	Sugar metabolism (Hpr, CcpA, β-PGM, β-phosphoglucomutase)
		DNA supercoiling	Chromosome structuring (HslA)
		Transcription and translation	Signal transduction (LlrC)
			Stress adaptation (OsmC)
			Proteolysis of misfolded proteins (ClpX ATPAse)
			Cold shock proteins (CSPs)
			CspA-CspG, vary in number according to species
Osmotic stress response	Heat shock	Exchange of compatible solutes to maintain osmotic balance	ATP-dependent uptake system (QacT) or ABC transporter (OpuA or BusA) (species-dependent) for uptake of glycine-betaine, carnitine, and proline during hyperosmotic stress conditions, efflux by unidentified channel protein
			General stress proteins (GroEL, GroES, DnaK)
			Proteases FtsH, HtrA
Heat stress response	Acid, oxidative, cold, osmotic, alcohol	Production of heat-inducible chaperones	Chaperone complexes DnaK-GrpE-DnaJ and GroES-GroEL
		Production of heat-inducible proteases	HtrA/DegP protease, FtsH/HflB protease, Clp proteases (ClpB, C, E, P, Q, X and Y), Lon protease
		Production of heat shock proteins	Trigger factor, HrcA, HSP10, HSP23 (ClpP), HSP26, HSP33, HSP40, DnaK/HSP70, GroEL/HSP60, HSP84, HSP85, HSP100

			Small heat shock proteins (sHSPs), e.g., Lo18, HSP18.5, HSP18.55, HSP19.3, HSP16.4, HSP20
Bile stress response	Heat	Metabolism of bile salts	Bile salt hydrolase (BSH)
		Adaptation to bile stress	DnaK, GroEL
		MDR efflux	MDR transporters
Nutrient starvation stress response	Heat, oxidative, ethanol, acid, osmotic	Modification of cell morphology	General stress proteins
		Regulation of metabolism	Proteins involved in carbon metabolism (triose phosphate isomerase, putative dihydroxy-acetone kinase, Gls24 protein)
		Amino acid (arginine) catabolism	Proteins involved in amino acid catabolism (carbamate kinase, putative glycine cleavage system, L-serine dehydrogenase)
Ethanol stress response	Heat, acid	Production of heat shock proteins	sHSPs: HSP18

as acid resistance is considered a functional property and hence a beneficial trait for LAB starter cultures or probiotics, such a property would assist the survival of pathogens such as *L. monocytogenes* in the stomach or in a macrophage phagosome and thus would be considered a virulence factor. The same also holds true for other survival traits that depend on stress responses, such as bile tolerance, tolerance to low or high temperatures, or resistance to other environmental stresses, that would be considered beneficial for starter cultures or probiotics but virulence factors for foodborne pathogens. The justification for considering the stress responses of industrially important LAB and probiotics (with the possible exemption of the enterococci) as beneficial traits lies in their safe use for thousands of years in food production, the absence of virulence factors, and a low incidence of association with human infections (Franz et al. 2010). To ensure that lactic starters or probiotic strains indeed survive in high numbers to guarantee their desired functional effects, two strategies can be adopted. First, their survival is enhanced by designing appropriate handling, storage, or production methods, for example, by using oxygen-impermeable containers or by the microencapsulation or incorporation of protective nutrients or buffering substances in the food product (Kailasapathy and Supriadi 1996; Ravula and Shah 1998; Shah 2000; Sultana et al. 2000). Second, stress-resistant strains are selected that will survive the conditions encountered from production to final application. A further option would also be a combination of the previous two, that is, choosing strains that are relatively stress-resistant and adjusting the handling, storage, and production methods to allow adequate survival. The beneficial interactions of LAB with food matrices and some examples of industrial uses of specific LAB species and the environmental stresses that impact these bacteria in their interaction with food or the human body are described next.

1.4.1 Beneficial Interactions of Lactic Acid Bacteria in Health and in Food Production

The diversity of physicochemical factors typical of a food system is probably as wide and diverse as the LAB taxa associated with these products, where they may be present either as spoilage organisms or as dominant fermentation microbiota, but also as natural contaminants (see Tables 1.1 and 1.2). LAB are generally a microbial group of major importance in the production of fermented foods, and this includes dairy, meat, vegetable, and cereal foods, mainly as a result of their acidification and associated preservation effects. The benefits of lactic fermentation may be summarized as follows:

- Preservation/safeguarding, by (a) competition and (b) metabolic activities involving metabolites such as lactic acid, acetic acid, alcohol, CO_2, H_2O_2, bacteriocins, etc.
- Quality improvement/enrichment of the diet (improvement of aroma, taste, and texture, etc.).

- Improvement of the digestibility, leading to a reduction in preparation time and in the energy required.
- Biological enrichment by the production or formation of essential amino acids, essential fatty acids, vitamins, etc.
- Detoxification during fermentation, for example, of patulin in some fruit and vegetable juices (Arici and Holzapfel, unpubl. results).
- Degradation of antinutritive factors, for example, of linamarin (a cyanogenic glucoside) during cassava fermentation.
- In addition to the traditional food fermentations typical of a region, culture, or country, industrial fermentations of raw materials such as milk, meat, vegetables, cereals, and fruits have become "internationalized." These industrial products of fermentation are being marketed in industrialized and many developing countries, thereby serving to enrich the diet. Many of these foods enjoy a positive image among consumers as health or well-being foods. Moreover, the quality and sensory attributes of a range of other food commodities such as cocoa (Camu et al. 2007; Kostinek et al. 2008) and coffee are essentially determined by a post-harvest fermentation process in which LAB play a vital role.

1.4.2 Environmental Stresses Encountered by Lactic Acid Bacteria

The LAB genera of major importance in food fermentations are first and foremost *Lactobacillus, Lactococcus, Leuconostoc, Weissella,* and *Pediococcus*, for all of which only particular species are of relevance. Representatives of other genera such as *Enterococcus, Oenococcus, Streptococcus, Tetragenococcus,* and *Carnobacterium* may play a role and survive under relatively extreme conditions (see Tables 1.1 and 1.2). Strains associated with food fermentations show particular abilities for surviving and competing either within a specific natural food ecosystem and/or under a set of food processing factors, while strains used as probiotics show abilities for colonizing and surviving the gastrointestinal system, as well as surviving and interacting with the host immune system. As a heterogeneous group of bacteria, LAB may grow in quite diverse environments and thus do not encounter identical stress conditions. As a consequence, these bacteria have developed stress responses that fit the specific constraints of a given substrate, environment, and/or process (e.g., meat, milk, or meat/milk fermentation). This has become particularly evident from the information gained from whole-genome sequence studies of probiotic LAB or LAB associated with milk and meat fermentations, which show the presence of specific genes that encode regulators capable of dealing with the relevant stress conditions of the respective niches. An overview of such different stresses is shown in Table 1.3. The stresses encountered and the way specific LAB species deal with these are the topic of this book, and studies in these areas will provide extremely valuable data for selecting successful strains and effective production processes.

References

Abee T, Wouters JA (1999) Microbial stress response in minimal processing. Int J Food Microbiol 50:65–91

Ainsley Reid A, Champagne CP, Gardner N, Fustier P, Vuillemard JC (2007) Survival in food systems of *Lactobacillus rhamnosus* R011 microentrapped in whey protein gel particles. J Food Sci 72:31–37

Akbar S, Min Kang C, Gaidenko TA, Price CW (1997) Modelling protein RsbR regulates environmental signalling in the general stress pathway of *Bacillus subtilis*. Mol Microbiol 24:567–578

Axelsson L (2004) Lactic acid bacteria: Classification and physiology. In: Salminen S, von Wright A, Ouwehand A (Eds.), Lactic acid bacteria, Marcel Dekker, New York, pp. 1–66

Azcárate-Peril MA, Altermann E, Hoover-Fitzula RL, Cano RJ, Klaenhammer TR (2004) Identification and inactivation of genetic loci involved with *Lactobacillus acidophilus* acid tolerance. Appl Environ Microbiol 70:5315–5322

Azcárate-Peril MA, McAuliffe O, Altermann E, Lick S, Russell WM, Klaenhammer TR (2005) Microarray analysis of a two-component regulatory system involved in acid resistance and proteolytic activity in *Lactobacillus acidophilus*. Appl Environ Microbiol 71:5794–5804

Becker LA, Cetin MS, Hutkins RW, Benson AK (1998) Identification of the gene encoding the alternative sigma factor σ^B from *Listeria monocytogenes* and its role in osmotolerance. J Bacteriol 180:4547–4554

Bischoff M, Entenza JM, Giachino P (2001) Influence of a functional sigB operon on the global regulators *sar* and *agr* in *Staphylococcus aureus*. J Bacteriol 183:5171–5179

Bruno-Bárcena JM, Azcárate-Peril MA, Hassan HM (2010) Role of antioxidant enzymes in bacterial resistance to organic acids. Appl Environ Microbiol 76:2747–2753

Camu N, De Winter T, Verbrugghe K, Cleenwerck I, Vandamme P, Takrama JS, Vancanneyt M, De Vuyst L (2007) Dynamics and biodiversity of populations of lactic acid bacteria and acetic acid bacteria involved in spontaneous heap fermentation of cocoa beans in Ghana. Appl Environ Microbiol 73:1809–1824

Canchaya C, Claesson MJ, Fitzgerald GF, van Sinderen D, O'Toole PW (2006) Diversity of the genus *Lactobacillus* revealed by comparative genomics of five species. Microbiology 152:3185–3196

Claesson MJ, van Sinderen D, O'Toole PW (2008) *Lactobacillus* phylogenomics – towards a reclassification of the genus. Int J Syst Evol Microbiol 58:2945–2954

Dave RI, Shah NP (1997) Viability of yoghurt and probiotic bacteria in yoghurts made from commercial starter cultures. Int Dairy J 7:31–41

Dave RI, Shah NP (1998) Ingredient supplementation effects on viability of probiotic bacteria in yoghurt. J Dairy Sci 81:2804–2816

Denou E, Berger B, Barretto C, Panoff JM, Arigoni F, Brussow H (2007) Gene expression of commensal *Lactobacillus johnsonii* strain NCC533 during in vitro growth and in murine gut. J Bacteriol 189:8109–8119

Franz CMAP, Cho GY, Holzapfel WH, Gálvez A (2010) Safety of lactic acid bacteria. In: Mozzi F, Vignolo G, Raya R (Eds.), Biotechnology of lactic acid bacteria: Novel applications. Wiley-Blackwell, Ames, IA, pp. 341–359

Garrity GM, Holt JG (2001) The road map to the manual. In: Boone DR, Castenholz RW (Eds.), Bergey`s manual of systematic bacteriology, Springer, New York, pp. 119–123

Gouesbet G, Jan G, Boyaval P (2002) Two-dimensional electrophoresis study of *Lactobacillus delbrueckii* subsp. *bulgaricus* thermotolerance. Appl Environ Microbiol 68:1055–1063

Haldenwang WG (1995) The sigma factors of *Bacillus subtilis*. Microbiol Rev 59:1–30

Hastings JW, Holzapfel WH, Niemand JG (1986) Radiation resistance of lactobacilli isolated from radurized meat relative to growth phase and substrate. Appl Environ Microbiol 52:898–901

Holley RA, Guan TZ, Peirson M, Yost CK (2002) *Carnobacterium viridans* sp. *nov.*, an alkaliphilic, facultative anaerobe isolated from refrigerated, vacuum-packed bologna sausage. Int J Syst Evol Microbiol 52:1881–1885

Hummel A, Holzapfel WH, Franz CMAP (2007) Characterisation and transfer of antibiotic resistance genes from enterococci isolated from food. Syst Appl Microbiol 30:1–7

Kailasapathy K, Chin J (2000) Survival and therapeutic potential of probiotic organisms with reference to *Lactobacillus acidophilus* and *Bifidobacterium* spp. Immunol Cell Biol 78:80–88

Kailasapathy K, Supriadi D (1996) Effect of whey protein concentrates on the survival of *L. acidophilus* in lactose hydrolysed yoghurt during refrigerated storage. Milchwissenschaft 51:565–568

Klein G, Pack A, Bonaparte C, Reuter, G (1998) Taxonomy and physiology of probiotic lactic acid bacteria. Int J Food Microbiol 41:103–125

Kostinek M, Ban-Koffi L, Ottah-Atikpo M, Teniola D, Schillinger U, Holzapfel WH, Franz CM (2008) Diversity of predominant lactic acid bacteria associated with cocoa fermentation in Nigeria. Curr Microbiol 56:306–314

Lim EM, Ehrlich SD, Maguin E (2000) Identification of stress inducible proteins in *Lactobacillus delbrueckii* subsp. *bulgaricus*. Electropohoresis 21:2557–2561

Lorca GL, Font de Valdez G (2009) *Lactobacillus* stress responses. In: Ljungh A, Waldström T (Eds), *Lactobacillus* molecular biology. Caister Academic Press, Norfolk, pp. 115–137

Makarova KS, Koonin EV (2007) Evolutionary genomics of lactic acid bacteria. J Bacteriol 189:1199–1208

Makarova K, Slesarev A, Wolf Y, Sorokin A, Mirkin B, Koonin E, Pavlov A, Pavlova N, Karamychev V, Polouchine N, Shakhova V, Grigoriev I, Lou Y, Rohksar D, Lucas S, Huang K, Goodstein DM, Hawkins T, Plengvidhya V, Welker D, Hughes J, Goh Y, Benson A, Baldwin K, Lee JH, Diaz-Muniz I, Dosti B, Smeianov V, Wechter W, Barabote R, Lorca G, Altermann E, Barrangou R, Ganesan B, Xie Y, Rawsthorne H, Tamir D, Parker C, Breidt F, Broadbent J, Hutkins R, O'Sullivan D, Steele J, Unlu G, Saier M, Klaenhammer T, Richardson P, Kozyavkin S, Weimer B, Mills D (2006) Comparative genomics of the lactic acid bacteria. Proc Natl Acad Sci USA 103:15611–15616

Marceau A, Zagorec M, Chaillou S, Mera T, Champomier-Verges MC (2004) Evidence for the involvement of at least six proteins for adaptation of *Lactobacillus sakei* to cold temperature and addition of NaCl. Appl Environ Microbiol 70:7260–7268

Miyoshi A, Rochat T, Gratadoux J-J, Le Loir Y, Oliveira SC, Langella P, Azevedo V (2003) Oxidative stress in *Lactococcus lactis*. Genet Mol Res 2:348–359

Mogensen G, Salminen S, O'Brien J, Ouwehand A, Holzapfel WH, Shortt C, Fonden R, Miller GD, Donohue D, Playne M, Crittenden R, Bianchi Salvadori B, Zink R (2002) Food microorganisms – health benefits, safety evaluation and strains with documented history of use in foods. Bull Int Dairy Fed 377:4–9

Orla-Jensen S (1942) The lactic acid bacteria, 2nd ed. Munksgaard, Copenhagen

Pieterse B, Leer RJ, Schuren FHJ, van der Werf MJ (2005) Unraveling the multiple effects of lactic acid stress on *Lactobacillus plantarum* by transcription profiling. Microbiology 151:3881–3894

Ravula RR, Shah NP (1998) Effect of acid casein hydrolysate and cysteine on the viability of yoghurt and probiotic bacteria in fermented frozen dairy desserts. Aust J Dairy Technol 53:175–179

Rosen R, Ron EZ (2002) Proteome analysis in the study of the bacterial heat shock response. Mass Spectrom Rev 21:244–265

Shah NP (2000) Probiotic bacteria: Selective enumeration and survival in dairy foods. J Dairy Sci 83:894–907

Spano G, Massa S (2006) Environmental stress response in wine lactic acid bacteria: beyond *Bacillus subtilis*. Crit Rev Microbiol 32:77–86

Stevens MJA, Molenaar D, de Jong A, de Vos WM, Kleerebezem M (2010) Involvement of the mannose phosphotransferase system of *Lactobacillus plantarum* WCFS1 in peroxide stress tolerance. Appl Environ Microbiol 76:3748–3752

Stiles ME, Holzapfel WH (1997) Lactic acid bacteria of foods and their current taxonomy. Int J Food Microbiol 36:1–29

Storz G, Hengge-Aronis R (Eds.) (2000) Bacterial stress responses. ASM Press, Washington, DC

Sugimoto S, Al-Mahin A, Sonomoto K (2008) Molecular chaperones in lactic acid bacteria: physiological consequences and biochemical properties. J Biosci Bioeng 106:324–336

Sultana K, Godward G, Reynolds N, Arumugaswamy R, Peiris P, Kailasapathy K (2000) Encapsulation of probiotic bacteria with alginate-starch and evaluation of survival in simulated gastrointestinal conditions and in yoghurt. Int J Food Microbiol 62:47–55

van de Guchte M, Serror P, Chervaux C, Smokvina T, Ehrlich SD, Maguin E (2002) Stress responses in lactic acid bacteria. Antonie van Leeuwenhoek 82:187–216

Van Bogelen RA, Greis KD, Blumenthal M, Tani TH, Matthews RG (1999) Mapping regulatory networks in microbial cells. Trends Microbiol 7:320–328

Völker U, Engelmann S, Maul B, Riethdorf S, Volker A, Schmid R, Mach H, Hecker M (1994) Analysis of the induction of general stress proteins of *Bacillus subtilis*. Microbiology 140:741–752

Walter J, Chagnaud P, Tannock GW, Loach DM, Dal Bello F, Jenkinson HF, Hammes WP, and Hertel C (2005) A high-molecular-mass surface protein (Lsp) and methionine sulfoxide reductase B (MsrB) contribute to the ecological performance of *Lactobacillus reuteri* in the murine gut. Appl Environ Microbiol 71:979–986

Wouters JA, Mailhes M, Rombouts FM, Kuipers OP, Abee T (2001) Cold shock proteins of *Lactococcus lactis* MG 1363 are involved in cryoprotection and in the production of cold induced proteins. Appl Environ Microbiol 65:5171–5178

Xie Y, Chou LS, Cutler A, Kuipers OP, Abee T (2004) DNA macroarray profiling of *Lactococcus lactis* subsp. *lactis* IL1403 gene expression during environmental stresses. Appl Environ Microbiol 70:6738–6747

Part II
Responses of Lactic Acid Bacteria Towards Specific Environmental Stresses

Chapter 2
Responses of Lactic Acid Bacteria to Acid Stress

Jessica K. Kajfasz and Robert G. Quivey Jr.

2.1 Introduction

The expanse of microbial diversity and its influence on humans and their environment has always been a central concept in microbiology. The concept has also always been one of increasing complexity as the abilities to sample particular environments, and to study the genomes of the bacteria populating these environments, grow at increasing rates. Indeed, advances in genomic sequencing have eliminated the need for bacteria to be grown in culture for genetic information to become catalogued and compared by informaticists. The result, so far, has been an enlarging appreciation for how many closely related organisms exist and how many exist in a given niche. The increase in species numbers suggests that microenvironments are abundant in nature, even in spaces thought to be small and homogeneous.

The rising accessibility of high-throughput DNA sequencing facilitates the acquisition of genomic information for bacterial species and their variants, allowing the comparison of organisms and their metabolic pathways in ways that were only imaginable a decade ago. In this way, the comparison of genomic and transcriptional data of microbial niches will inform us of the effects of the environment on microbes and will provide insights into how these environments have selected the related species that are now in existence.

In this chapter, a review of acid tolerance in lactic acid bacteria (LAB) will be presented. LAB exhibit astonishing diversity in the range of niches they occupy.

R.G. Quivey Jr. (✉)
Center for Oral Biology, University of Rochester, School of Medicine and Dentistry, Rochester, NY 14580, USA

Department of Microbiology and Immunology, University of Rochester, School of Medicine and Dentistry, 601 Elmwood Avenue, P.O. Box 611, Rochester, NY 14580, USA
e-mail: Robert_Quivey@urmc.rochester.edu

E. Tsakalidou and K. Papadimitriou (eds.), *Stress Responses of Lactic Acid Bacteria*, Food Microbiology and Food Safety, DOI 10.1007/978-0-387-92771-8_2,

Included among these bacteria are those used in the industrial-scale production of dairies, wine, and bread. Certain LAB are also prominent human pathogens. Indeed, the oral streptococci are among the most common infectious agents on our planet, and virtually all persons in industrialized nations harbor *Streptococcus mutans* in their mouths. *Streptococcus pneumoniae* is another common and potentially dangerous pathogen, as is *Streptococcus pyogenes*, the so-called flesh-eating bacterium. It is somewhat difficult then to reconcile these pathogens with benign LAB, like *Lactococcus lactis*, *Streptococcus thermophilus,* and all other species used as starters or adjuncts in food fermentations. The commonality of these bacteria is, of course, that they ferment sugars through the Embden–Meyerhof–Parnas pathway to organic acids, principally lactic acid.

The production of lactic acid from fermentable carbohydrates occurs at very rapid rates. For example, lactic acid production by *S. mutans* can lower pH values two to three orders of magnitude in pH in a few minutes. Given that pH values are scaled in logarithmic increments, the production of a 1,000-fold increase in lactic acid in less than 5 min is very impressive and clearly results in a major challenge for the organism's survival. LAB, as a group, inherently acidify their environment quite rapidly, self-imposing acid stress. There is a need, then, for these bacteria to respond to acidification to ensure physical and genetic integrity and, as discussed ahead, to rely on the involvement of multiple mechanisms to survive.

2.2 Acid Tolerance Contributes to a Substantial Competitive Advantage

It is clear that the ability to resist the inimical effects of environmental acidification would be positive for any given microbe. In the case of bacteria that grow in highly competitive environments among a substantial number of competitors, acid tolerance could likely be the paramount reason by which LAB dominate their niche. This is certainly true for the oral microbiome, which presently is estimated to consist of hundreds of species (Becker et al. 2002; Aas et al. 2008; Keijser et al. 2008), with as much as 40% represented by various species of streptococci or related LAB. These recent metagenomic data provide substantial evidence that acid tolerance is a key fitness attribute of the bacteria, permitting these organisms to survive periods of starvation and acidification and to outcompete other organisms during periods of relative abundance of available carbohydrates. A similar mechanism is also true for food-related LAB that, by acidifying the food matrix, prevail over food spoilage and pathogenic bacteria.

In the 1980s, results from studies with mixed cultures of oral bacteria, using chemostats to control conditions, showed clearly that pH values had a dramatic effect on the proportions of *S. mutans*, *Streptococcus sanguinis,* and *Lactobacillus casei* in cultures (McDermid et al. 1986; Bradshaw et al. 1989). *L. casei* was least affected by changes in the culture pH, whereas the population of *S. mutans* was reduced at neutral pH. In contrast, the population of *S. sanguinis* plummeted during low pH values, indicating its inherently impaired ability to compete in a situation where the pH was held at acidic levels (pH = 4.1) (McDermid et al. 1986). Subsequent work (Bradshaw et al. 1989), again in mixed-culture chemostats, showed equivalent results when pulses of glucose were fed to the cultures, allowing the metabolism of sugar to organic acids to reduce the external pH values. Again, *L. casei* and *S. mutans* dominated cultures, whereas *S. sanguinis* nearly disappeared from the cultures. The data showed that fermentation of sugars to acid end products was itself an effective means for oral LAB to compete with organisms incapable of growing in acidic conditions. Somewhat surprising, however, was the observation reported by the Marsh laboratory that, despite sharp drop-offs in the proportion in a number of the oral bacteria used in the cultures, none of the species in the mixed-culture experiments ever completely disappeared, indicating long-term survival strategies that are not yet well understood. Nevertheless, the in vitro work helped to set the stage for molecular approaches to gain an understanding of acid resistance in *S. mutans*.

2.3 The Role of F-ATPase in Acid-Tolerance

The data from mixed-culture experiments in the Marsh laboratory provided a context for exploring the mechanisms underlying the capacity of oral LAB to cope with acidification.

Initial studies focused on the ability of intact bacterial membranes to resist physical damage caused by acidic conditions. It was shown that oral bacteria exhibit distinct differences in the ability to resist membrane damage due to acidification, as measured by magnesium release (Bender et al. 1986). In those experiments, the oral streptococcal species (*S. mutans*, *S. sanguinis,* and *Streptococcus salivarius*) all showed a similar release of magnesium at pH values of approximately 4.0, whereas *L. casei* was considerably more resistant, showing damage at pH values of 3.0 and below. Potential variations in membrane composition that may relate to those differences remain unclear. However, the Bender et al. study also assessed proton movement across the membranes of the test bacteria. The authors showed that proton movement across membranes virtually ceased at pH 5.0, 7.0, and 6.0 for *S. mutans*, *S. sanguinis,* and *S. salivarius*, respectively (Bender et al. 1986). The addition of the ATPase inhibitor dicyclohexylcarbodiimide (DCCD) resulted in an increase in the proton permeability of all of the test organisms, indicating that proton movement was bidirectional across the membranes and that the outflow involved a proton-translocating ATPase (F_1F_0-ATPase). The purpose of that F-ATPase was to remove protons from the cytoplasm by pumping them through the membrane-bound subunit (subunit c), at the expense of the ATP that was produced during glycolysis. The result is a cytoplasm that is more alkaline than the external environment.

Subsequent studies with permeabilized membranes showed that the differences in proton movement were directly attributable to the amount of F-ATPase produced by the oral bacteria and to the specific activity of each enzyme (Bender et al. 1986; Bender and Marquis 1987). Moreover, the F-ATPases also exhibited different pH optima, such that the enzyme from *L. casei* not only produced the most ATPase in comparison to oral streptococci, but also had the lowest pH optimum (approximately 5.0) reported. It became clear that the F-ATPase activity in a given organism was directly related to its ability to move protons out of the cytoplasm and to maintain internal pH homeostasis. In effect, the more F-ATPase a LAB has, and the lower its pH optimum for the activity is, the more acid-resistant the organism will be. This tenet is important in the sense that the F-ATPases have an important role in the ability of LAB to compete with other organisms and that the enzyme provides a strong basal level of inherent acid tolerance (Sturr and Marquis 1992). The value of the proton pump to the oral streptococci is that acid-sensitive glycolytic enzymes, such as enolase and glyceraldehyde 3-phosphate dehydrogenase (GAPDH), are protected by the action of the ATPase. Hence, ATP production continues in situations in which the external pH is below the point at which growth can be sustained by the organism. The ability to produce ATP at low pH, which can be used metabolically or by the F-ATPase to pump protons out of the cell, is a key facet in the organism's ability to survive long-term acidic conditions and to compete with other organisms (Bender et al. 1986).

Later experiments invoked a new facet of ATPase biology in the oral streptococci: that production of the F-ATPase increases during growth at low pH values (Belli and Marquis 1991). Moreover, the results showed that a pH-dependent increase in F-ATPase activity was independent of growth rate, suggesting an independent mechanism of regulation for production of the ATPase that might be responsive to external pH values. By this point, it was understood that acid tolerance was dependent to a large degree on the activity of the F-ATPase, which uses the energy from ATP cleavage to remove protons from the cell cytoplasm. That ability is clearly dependent on the availability of ATP. The results from work carried out with *S. mutans* GS-5 showed that available sugars influence the amount of ATP available to the F-ATPase (Belli and Marquis 1994). In this work, it was shown that galactose-grown cells actually exhibited lower acid tolerance than cells grown on glucose. The conclusion was that galactose enters *S. mutans* by a proton motive force (PMF)-driven permease, contributing to a reduction in the ΔpH, whereas glucose enters cells via a phosphoenol pyruvate-dependent phosphotransferase system (PEP-PTS), preserves ΔpH, and results in a higher ATP yield (Belli and Marquis 1994).

The experiments described above presented a physiological framework for acid tolerance in oral streptococci in particular, but also for the LAB in general. The F-ATPase, a membrane-bound, proton-translocating pump, is a substantial component of acid tolerance. Subsequent experiments in the area were directed to an understanding of F-ATPase production at the genetic level and whether differences existed between *S. mutans*, a highly aciduric strain, and *S. sanguinis*, a relatively acid-sensitive strain of oral streptococci.

The F-ATPase from *Escherichia coli* has been well established as an eight-subunit holoenzyme, containing three membrane-bound or membrane-associated subunits, and five subunits in the cytoplasm that function as the ATPase, or ATP synthase (Walker et al. 1984). The available information for the catalytic domain of the enzyme, the beta subunit encoded by *atp*D, indicated a high level of conservation among known bacterial F-ATPase sequences at the time, in the mid-1990s. Experiments were undertaken to elucidate the means by which *S. mutans* regulated the production of its F-ATPase in response to external pH values. Using polymerase chain reaction (PCR) and highly degenerate primers to a conserved region of the beta-subunit gene, a fragment of the *S. mutans* F-ATPase ortholog for *atp*D was cloned and used to probe a library of *S. mutans* genomic fragments. The approach resulted in identifying the DNA encoding the entire operon of the F-ATPase in *S. mutans* (Smith et al. 1996). The operon was organized somewhat differently than that described earlier for *E. coli*. In the case of *S. mutans* (and subsequently, all LAB for which genomic sequence is now available), the gene order was *atp*EBFHAGDC. The *atp*CBF genes encode the membrane-bound portion of the enzyme, through which protons flow, moving through the channel by interaction with an aspartic acid residue on the c subunit (encoded by *atp*E). Somewhat surprisingly, the membrane-subunit genes occur transcriptionally prior to the catalytic domain genes *atp*HAGDC. This suggests the possibility that membrane pores are formed before they can be capped, and regulated, by the presence of the catalytic subunits resting in the cytoplasmic side of the enzyme and controlling proton flow. Because freely available membrane subunits have not been identified in any bacterium to date, it appears that F-ATPases are synthesized and assembled immediately into membranes. Mutation of the *S. mutans ffh* gene, encoding a membrane-protein assembly chaperone, has been shown to result in reduced F-ATPase activity and acid sensitivity (Gutierrez et al. 1999). The observation supports both the role of the F-ATPase in acid tolerance and also the concept that F-ATPase subunit assembly into membranes relies, at least in part, on a chaperone mechanism (Gutierrez et al. 1999; Hasona et al. 2005). In addition to the structural genes in the *S. mutans* operon, an upstream-DNA sequence was obtained that showed two things: the absence of an *atp*I gene, which is in the *E. coli* operon, and an intergenic sequence of approximately 260 base pairs, containing two sets of inverted DNA repeats. The presence of the repeated sequences suggested that a DNA-binding motif for the operon might exist. The sequence provided the impetus to explore the possibility of identifying a pH-dependent controlling mechanism for the *atp* operon.

Experiments were undertaken to determine whether the promoter-region DNA sequences for the *S. mutans atp* operon were involved in its regulation and whether that regulation was dependent on external pH. The availability of the *S. mutans* F-ATPase operon promoter facilitated the cloning of the *S. sanguinis* promoter, which was important to the discussion of acid-resistant *S. mutans* compared to acid-sensitive *S. sanguinis*. The question at that time was whether the promoters of the two operons were functionally similar and whether external pH values affected their respective transcription. *S. mutans* and *S. sanguinis atp*-promoter fusions to a chloramphenicol acetyl-transferase (CAT) gene were constructed and placed in the chromosomes of each organism and on low-copy plasmids. The experiment was also conducted with the *S. sanguinis atp*-promoter-CAT constructs in *S. mutans* and the *S. mutans atp*-promoter-CAT constructs in *S. sanguinis*. The results provided several clear insights. The first was that the *atp* promoters for both *S. mutans* and

S. sanguinis were upregulated during growth at pH 5.0 compared to cells grown at pH 7.0 (Kuhnert et al. 2004), in accordance with the enzyme activity levels for *S. mutans* reported earlier (Belli and Marquis 1991). The second observation was that the *S. mutans* and *S. sanguinis* promoters were each responsive to pH in the opposite genetic background, meaning that the *S. mutans atp* promoter functioned normally in *S. sanguinis* and that the *S. sanguinis* promoter functioned normally in the *S. mutans* background. The conclusions were that the transcriptional machinery of the oral streptococci, at least, was likely to be very similar. Indeed, in parallel work in *S. pneumoniae*, a virtually identical *atp* promoter sequence was identified during experiments that showed that the operon contained an "extended"-10 motif. Further, a disruption of the one-base extension reduced transcription of the operon (Martín-Galiano et al. 2001). Recent work involving *Lactobacillus plantarum* also shows that elevated transcription of the *atp* operon is correlated with acid resistance (Duary et al. 2010). The conclusions from the oral streptococci were also that the F-ATPase transcriptional machinery and catalytic domain sequences were so similar that identifying an inhibitor specific to the acid-tolerant strains of oral streptococci alone might be impossible. However, there were, in all bacteria, differences in the membrane-bound subunits to an extent, indicating that disruption of the membrane subunits, or their interaction with the membrane lipids, might be a viable approach to developing new therapeutic agents.

Two examples of F-ATPase membrane-bound subunit drugs have been established for *S. pneumoniae*. Quinine is an example of a compound that inhibits the proteolipid subunit (the c subunit, encoded by *atp*E) of the F-ATPase in *S. pneumoniae* (Munoz et al. 1996) and optochin binds to the *atp*C gene product, the epsilon membrane subunit (Fenoll et al. 1994, 1995). The observations regarding the conservation of transcriptional control in *S. mutans* and *S. sanguinis*, and the possibility of identifying membrane-specific changes, dependent on external pH values, led to the studies conducted with membrane fatty acids described next.

2.4 The Role of Membrane Fatty Acids in Acid Tolerance

2.4.1 Production of Unsaturated Membrane Fatty Acids

In *S. mutans*, growth under acidic conditions resulted in a substantial increase in the proportion of unsaturated membrane fatty acids compared to growth at neutral pH values as well as a rise in the proportion of the longer-chain, unsaturated fatty acids $C_{18:1}$ and $C_{20:1}$ (UFAs) (Quivey et al. 2000; Fozo and Quivey 2004a). The shift to UFAs was shown to occur rapidly, measurably within 20 min, following the addition of glucose to a culture growing at pH 7.0, after which the pH began to fall immediately (Fozo and Quivey 2004a). Parallel work with *S. pneumoniae* revealed the existence of a *trans*-2, *cis*-3 isomerase activity that was likely responsible for the formation of UFAs in that organism (Marrakchi et al. 2002). The isomerase was named FabM in *S. pneumoniae* (Marrakchi et al. 2002), and the gene encoding the enzyme was subsequently identified in *S. mutans* (Fozo and Quivey 2004b). The location of the *fab*M gene is immediately upstream of the fatty acid biosynthetic gene cluster in both *S. mutans* and *S. pneumoniae*. However, it has apparently been difficult to isolate a mutation in the *fab*M gene in *S. pneumoniae*, whereas insertional mutants in *fab*M and deletions have been constructed and used to evaluate the loss of FabM in *S. mutans*. In the absence of the *fab*M gene, *S. mutans* contains only trace amounts of UFAs, compared to the approximately 60% of membrane fatty acids during growth at pH 5.0. Thus, the data indicate that FabM is likely the sole mechanism by which UFAs are formed in streptococci. Indeed, a small but representative sampling of other oral LAB showed that the acid-resistant organisms *Streptococcus gordonii, S. salivarius*, and *L. casei* all exhibited increases in UFAs as a function of growth at low pH values, whereas the more well-established acid-sensitive *S. sanguinis* did not (Fozo et al. 2004).

The loss of FabM activity in *S. mutans* resulted in sensitivity to growth under normal acidic conditions (pH 5.0) and to extreme sensitivity to pH 3.5 in acid-challenge experiments. Moreover, the ability to produce acid was reduced substantially, indicating that either glycolytic enzymes or sugar transport mechanisms were inhibited by the loss of UFAs in the organism's membrane. Subsequent studies showed that loss of the UFAs actually resulted in reduced ability of the *fab*M mutant strain to maintain the normal ΔpH, indicating that acidic conditions were likely affecting enolase and GAPDH, which were known to be acid-sensitive (Belli et al. 1995). The transport of sugar through PEP-PTS was also diminished, which suggested a membrane–protein interaction involving UFAs via the main glucose-transporter EII_{man} (Fozo and Quivey 2004b). In addition, the F-ATPase operon was transcriptionally elevated in the *fab*M mutant, supporting the concept that membrane UFAs participate in protecting *S. mutans* from internal acidification. The genetic and physiological ramifications of the *fab*M mutation could be largely alleviated by growing the mutant strain in the presence of $C_{18:1}$ or $C_{20:1}$ fatty acids added exogenously, showing that the fatty acids are readily transported into the organism and incorporated into membranes (Fozo and Quivey 2004b).

The role of *fab*M in the pathogenesis of *S. mutans* was tested using a well-established rat model for dental caries (known popularly as cavities). The results showed very clearly that loss of the FabM activity reduced the number of caries by approximately 30% and the extent of the damage of caries by nearly 90% (Fozo et al. 2007). These data represented one of the first reports linking bacterial acid resistance, attributable to a single biochemical activity, to the ability to cause disease. Certainly, the concept is not difficult to convey, but in this case, the linkage was experimentally established and the presence of membrane UFAs in *S. mutans* was directly linked to its ability to initiate and promote the development of oral disease.

The regulation of the production of UFAs in *S. pneumoniae* is linked to the regulation of fatty acid biosynthesis itself by the FabT repressor. The *fab*T gene is located immediately upstream of the fatty acid biosynthetic gene cluster (*fab*) in *S. pneumoniae* and *S. mutans*. A mutation of *fab*T resulted in higher levels of saturated fatty acids and a greater sensitivity to acidic growth conditions, illustrating that UFAs also play a role in acid tolerance for *S. pneumoniae* (Lu and Rock 2006). Of great interest was the subsequent finding that FabT repressor, an MarR family member containing a typical ligand-binding domain, was enhanced in the binding to its target motif by the presence of acyl-acyl carrier protein (Jerga and Rock 2009). While the interaction of FabT occurred with a range of fatty acid (acyl) lengths, the strongest interaction was with $C_{18:1}$, *cis*-vaccenoyl-acyl-carrier protein. *S. mutans fab* genes are also regulated by FabT, though acyl-acyl carrier protein-binding partners have not yet been established. The loss of FabT in *S. mutans* deregulates the saturated/unsaturated ratio of membrane fatty acids, as shown in *S. pneumoniae*, indicating a common mechanism of controlling the balance of the two major forms of fatty acids in their membranes. The external growth pH and its relationship to pathogenesis are not generally considered for *S. pneumoniae*. However, existing data for the loss of *fab*T and the reduction in the proportion of membrane UFAs suggests that membrane biology is likely an important component of the *S. pneumoniae* pathogenic repertoire as well.

2.4.2 Production of Cyclopropane Fatty Acids in Response to Acidification

The production of cyclopropane-containing fatty acids (CFAs) by LAB has been known for a considerable period of time. Indeed, these molecules are referred to as lactobacillic acid (C19cyc11) and dehydrosterculic acid (C19cyc9). The production of CFAs occurs via the transfer of a methyl group from S-adenosyl methionine to fatty acids via the action of CFA synthase.

CFA synthase is induced during growth at pH 5.0 in *L. lactis* MG1363 (Budin-Verneuil et al. 2005a). It was found that the *cfa* gene was repressed in a *rel*A mutant strain, suggesting a connection of CFA production with amino acid starvation via the stringent response. Briefly, organisms starved for amino acids will elicit the production of ppGpp and (p)ppGpp, referred to historically as bacterial alarmones, via the RelA protein, which is associated with the ribosome. The effects of the stringent response are global for bacterial growth and metabolism, though reports of the effects on alarmone production vary widely in bacteria, and a universally shared model for the stringent response may or may not exist (Potrykus and Cashel 2008; Traxler et al. 2008). Recent work with *L. casei* ATCC334 indicates that acid-adapted cells contain elevated levels of saturated fatty acids and CFAs at the expense of UFAs (Broadbent et al. 2010). Transcriptional profiling of acid-adapted *L. casei*, compared to control cells, indicated the possibility that CFA production could be part of the stringent response in *L. casei*. The production of CFAs also occurs in *Lactobacillus helveticus* and *Lactobacillus sanfranciscensis*. *L. sanfranciscensis* appears to have evolved a strategy of producing shorter-membrane fatty acids during stress to maintain its membrane fluidity (Montanari et al. 2010). These observations fit the notion that carbon flow, in at least some LAB, is affected by the stringent response. The participation of RelA in the starvation-response of *S. mutans* has been established (Lemos et al. 2004), though the organism does produce significant CFAs under conditions reported so far. In this case, the strongest effects of the mutation appeared to be in the greater acid resistance of the *rel*A mutant strain, following growth in biofilms, compared to its parent strains. The conclusion, by those authors, was that the stringent pathway may involve quorum sensing, but effects on fatty acid metabolism have not yet been reported.

The increased acid resistance of a *rel*A strain of *L. lactis* MG1363 was correlated with constitutively high levels of GAPDH activity, suggesting a higher throughput of carbon through glycolysis, though that was in fact not the case, and the authors posited that the improved acid survival was due to slower throughput of glycolysis (Mercade et al. 2006). This apparent conundrum will have to be resolved over time, but it might be anticipated that slower growth affords protection from metabolic products that could be lethal to cells. For example, if electron movement through oxidases and dehydrogenases becomes decoupled from protonation, one result might be that free radical formation could be reduced in cells growing at low metabolic rates. The slow growth rates themselves suggest that fewer targets for damage are available, whether it is in planktonic cultures or in biofilms.

2.5 Metabolic Responses to Acid Stress in Lactic Acid Bacteria: Protection from Acid-Induced Damage

In addition to the membrane-oriented mechanisms of acid resistance in LAB, the microbes also respond with several mechanisms to deal with the effects of acid-mediated damage and to ameliorate falling internal pH values. The triggering events for the participation, or induction, of the acid-stress responses in LAB are not well understood at present, though it seems very likely that the interruption of carbon-flow normal metabolic channels, namely, the Embden–Meyerhof–Parnas pathway, would be a primary signal that homeostasis in the cells is threatened.

What is well understood at present is that the timeframe of inducing a response to acidification is relatively fast and that in just a few minutes, acid response can be observed by an increased ammonia and carbon dioxide production, as well as an increased production of proteins or compounds related to intracellular repair processes. In the following sections, brief descriptions of the established mechanisms for LAB defenses against the effects of acidification are presented.

In addition to the membrane-associated events occurring during the acid response, LAB also produce or, more correctly, release compounds that act to buffer acidic conditions inside cells. The releases of ammonia and lactic acid plus carbon dioxide, via the arginine deiminase (ADI) system and malolactic fermentation (MLF), respectively, have become metabolic hallmarks of the LAB. Indeed, the wine industry, in particular, benefits from MLF by *Oenococcus oeni*. These mechanisms are widely conserved in LAB, and available data strongly support their role in the acid resistance of these bacteria.

2.5.1 Arginine and Agmatine Deiminase Systems, Urease

The release of ammonia, which acts to raise pH values, appears to be a well-distributed mechanism of acid resistance among LAB. The most common sources of ammonia are arginine and the less well-known amino acid agmatine. The enzymatic mechanism of ammonia release involves the ADI system, which relies on the activity of three enzymes: arginine deiminase (AD, encoded by the *arc*A gene), ornithine carbamoyltransferase (*arc*B), and carbamate kinase (*arc*C). Biochemically, arginine is taken up by cells via an arginine-specific porter, as part of an arginine-ornithine antiporter system. First described in experiments involving the antiporter system of *Enterococcus hirae*, [arginine_{in}/ornthine_{out}], it was recognized that arginine was required for induction of the system and that the antiporter system was also found in *S. sanguinis* and *Streptococcus milleri* (Poolman et al. 1987). Once arginine has been acquired by LAB, the compound is cleaved by AD to yield citrulline and ammonia. The citrulline is further deaminated and cleaved to the products ornithine and carbamoyl phosphate by ornithine transcarbamylase (OTC). The carbamoyl phosphate is then cleaved to carbon dioxide and ammonia by carbamate kinase (CK), with an accompanying substrate-level phosphorylation of ADP to ATP. Thus, the system delivers two moles of ammonia for each arginine transported into cells plus an ATP, in an energy-forming series of reactions.

Following description of the ADI pathway (Poolman et al. 1987), reports quickly appeared describing the role of ammonia release in the protection of *S. sanguinis* and *Streptococcus rattus* from acidification. Pertinent to the composition of the oral microflora, the enzymes of the ADI system were shown to function at surprisingly low pH values, as approximately 10% of full activity still remained in *S. rattus* at pH values below 2.6, which is two pH units below the level needed for growth (Marquis et al. 1987; Casiano-Colon and Marquis 1988). These data explain, in part, how *S. sanguinis* could survive at low abundance in the low pH environments of the mixed-culture experiments described above (McKee et al. 1985; McDermid et al. 1986; Bradshaw et al. 1989).

The genes encoding the ADI system, referred to as *arc*ABC (*arc*A encoding AD, *arc*B encoding OTC, and *arc*C encoding CK), were cloned initially from *S. sanguinis* NCTC10904 and were shown to be active in *E. coli* to the extent that they complemented mutant strains defective in OTC and CK, though more effectively following the addition of arginine to the culture medium (Burne et al. 1989). The authors also provided the first evidence that the genes function as a transcriptional unit. Subsequent work with other LAB revealed the broad distribution of the pathway including strains involved with wine production (Tonon and Lonvaud-Funel 2000; Arena et al. 2002) and meat fermentation (Champomier Verges et al. 1999). A number of reports in the literature indicate that the production of the ADI system enzymes is affected by environmental conditions, including salinity in the case of the sourdough LAB, *Lactobacillus fermentum*, suggesting that NaCl-mediated stress induces AD-catalyzed conversion of arginine to ornithine (Vrancken et al. 2009a, b). The operon in *L. lactis* was induced during growth at low pH (Budin-Verneuil et al. 2006), again supporting a broad role for the system as a mitigator of low pH in LAB.

Despite the common occurrence of the ADI system in LAB, it is still not entirely clear that AD is important for acid resistance in all organisms. For example, *Lactobacillus sakei* contains an intact *arc* operon, including *arc*R, the regulatory protein. However, the loss of *arc*R did not change growth of the organism, such that the role of the operon in *L. sakei* is not yet clear (Zuniga et al. 2002), though it is clear that the utilization of arginine is beneficial to the organism (Champomier Verges et al. 1999).

In addition to its role in ameliorating low pH to protect most LAB species, the ADI system also has potential health consequences for humans, both positive and negative. In the case of oral health, there are significant benefits from the presence of ammonia-releasing activity by dental plaque bacteria (Nascimento et al. 2009). In studies involving caries-free and caries-active subjects, the authors showed that there was a strong negative correlation between high levels of *S. mutans* and low levels of ammonia released by AD systems and urease activity (see ahead). The data indicated a rise in the environment pH, via ammonia, that was likely to reduce the acid-mediated erosion of dental enamel. However, in the case of the wine industry, the production of carbamate, via the CK reaction of the ADI system, raises the concern that ethanol and carbamate could combine to form the potentially carcinogenic compound ethyl carbamate (more commonly known as urethane). The *arc* genes associated with heterofermentative wine bacteria have been correlated with trace amounts of ethyl carbamate (Araque et al. 2009).

In addition to the ADI system, many species of LAB have other abilities to produce ammonia for acid protection, by using agmatine and urea as substrates. Agmatine deiminase is broadly found in lactic acid species, including species as diverse as *S. mutans* (Griswold et al. 2004) and *Lactobacillus brevis* (Lucas et al. 2007). The urease system of streptococci has been extensively investigated in *S. salivarius* and in the yogurt-producing *S. thermophilus* (Pernoud et al. 2004; Mora et al. 2005).

The agmatine deiminase (AgDI) system appears to be distributed through the *mutans* streptococci (Griswold et al. 2009). The biochemical pathway, similar to that of the ADI system, yields ammonia, carbon dioxide, and ATP, while agmatine is converted to putrescine. Ammonia is produced by the system at low external pH values, indicating that the AgDI system could assist in acid tolerance (Griswold et al. 2004).

The genetic organization of the AgDI pathway operon is *agu*BDAC. A putative regulator of the AgDI system has been identified upstream of the *agu* operon at SMU.261c of the *S. mutans* UA159 genome (Ajdic et al. 2002). Inactivation of the gene resulted in substantial reduction of the deiminase activity and abolished the ability of agmatine to induce the system (Griswold et al. 2006). Like the ADI system of other LAB, the AgDI pathway is induced in the presence of exogenous agmatine and is repressed via the carbon catabolite system that acts through the CcpA protein. In addition to agmatine, the induction of the AgDI system in *S. mutans* requires the presence of the AguR transcriptional activator and low pH, indicating the likelihood that the system is part of the acid-stress response in this organism at least (Liu et al. 2009). The system also responds to multiple two-component regulatory systems (TCSs), including the VicRK, ComDE, and CiaRH systems (Liu and Burne 2009). All of these TCSs have been implicated in acid response in *S. mutans* (see ahead), though a clear picture of how acid stress is communicated to the TCSs has not yet emerged in complete form. A screen of nearly 200 *L. brevis* strains revealed that approximately three dozen strains contained the AgDI genes, including the agmatine/putrescine antiporter (Lucas et al. 2007). The presence of putrescine in wine, attributed to *O. oeni* strains, has also been reported (Mangani et al. 2005)

Urea-degrading systems are also present in some species of LAB, including *S. thermophilus* (Pernoud et al. 2004; Mora et al. 2005) and *S. salivarius*. The production of ammonia via urease is far more complex than that of the ADI and AgDI pathways described above. The urease-encoding genetic system in *S. salivarius* begins with *ure*I, followed by *ure*ABC (structural genes) and *ure*EFGD (accessory genes) (Chen et al. 1998). In addition, the enzyme requires a nickel-uptake system to provide nickel as a cofactor for the urease, designated *ure*M, *ure*Q, and *ure*O, respectively, which encode an Ni(2+)-specific ATP-binding cassette transporter (Chen and Burne 2003). The enzyme was shown to be elevated in biofilms, where the pH is typically low, indicating its potential usefulness to ameliorate dental plaque pH values (Li et al. 2000). Experiments with multiorganism chemostats showed clearly that ammonia production, via urease-producing streptococci, would influence the proportion of species in the culture, providing additional support of the concept that the urease might be of benefit to oral health care (Shu et al. 2003). Subsequent work with caries-active and caries-free patients indicated a strong correlation between measurable urease levels in the mouth and caries status (Shu et al. 2007; Nascimento et al. 2009). In experiments to determine whether ammonia-producing oral bacteria could limit low pH in situ, it was found that plaque bacteria had difficulty in neutralizing sugar given as a bolus, but that over time the organisms may have the ability to modulate plaque pH values and plaque composition (Toro et al. 2010).

2.5.2 Malolactic Acid Fermentation

The ability of LAB to utilize L-malate as a source of carbon dioxide via the MLF pathway has been known for many years. The release of carbon dioxide from L-malic acid, by the malolactic enzyme, results in the formation of a lactic acid molecule and a molecule of carbon dioxide, which can be converted to bicarbonate via carbonic anhydrase (Pilone and Kunkee 1970). Typically, there is an antiporter system for secreting the lactic acid and taking in a molecule of L-malate. In bioenergetic terms, a single negative-charge compound is secreted in exchange for a dianionic compound, thereby increasing the PMF available for the cell to take up other nutrients. Because the carbon dioxide is captured, in part, by carbonic anhydase, there is buffering capacity available to the cell. Moreover, the organoleptic properties of malic acid, a tart substance found in grape must (and in substantial abundance in apples), is converted to lactic acid, which conveys a softer taste, usually described as buttery, to wines. Because the MLF pathway is found in many LAB, including those unrelated to the wine industry, it has become apparent that the reaction is related to protective buffering for the host bacterium, at a minimum.

Even though the capacity to carry out MLF is broadly distributed among LAB, clear differences exist between strains, particularly with regard to the ability to survive at low pH values and in the presence of alcohol formed during a primary fermentation of sugar. For example, several reports showed that MLF reactions in *O. oeni* function following primary fermentation and at pH values of 3.5, whereas other LAB, present in wine samples with *O. oeni,* do not survive at that pH (Lafon-Lafourcade et al. 1983; Davis et al. 1986).

MLF gene clusters include a regulatory protein, MleR; the malolactic synthase protein responsible for decarboxylation of L-malate, MleS; and the MleP protein, which acts as the porter for L-malate. The *mle*R gene in *L. lactis* has been reported as required for MLF (Renault et al. 1989). In contrast, it appears that the presence of *mle*R is probably essential for full activity of the *mle* genetic cluster to function in *S. mutans*, but not required for measurable amounts of activity (Lemme et al. 2010; Sheng et al. 2010).

The *mle*S gene, encoding the malate-cleaving enzyme of the pathway, was initially reported from *L. plantarum* (Ansanay et al. 1993), quickly followed by reports from other LAB, for example, *L. lactis* (Denayrolles et al. 1994) and *O. oeni* (Labarre et al. 1996), continuing to the present day (Lemme et al. 2010; Sheng et al. 2010). In all cases, the malolactic acid gene clusters include an *mle*R gene, encoding a DNA-binding regulatory protein, which regulates transcription of the *mle*S and *mle*P genes. In most cases, the presence of L-malate is necessary to fully induce the system, though there has been a report that the MLF system is not inducible by the presence of L-malate in *O. oeni* (Labarre et al. 1996). The MleR regulators belong, in general, to the DNA-binding proteins of the LysR family, containing an allosteric domain for binding L-malate.

The malate/lactate exchange is now known to generate membrane potential that is accomplished by the MleP, the malate porter. Once inside LAB, malate is decarboxylated by MleS, the malolactic enzyme, to release lactic acid and a carbon dioxide molecule. The net effect of the transport and enzymatic reactions is the formation of a pH gradient inside alkaline. The gradient contributes to the PMF available to the organism for taking up additional nutrients as part of what has been referred to as a secondary PMF system (Poolman et al. 1987, 1991; Renault et al. 1988; Cox and Henick-Kling 1989; Salema et al. 1994, 1996; Lolkema et al. 1995; Konings et al. 1997).

With the acquisition of genomic information and the physiological characterization of a number of MLF systems, the MleP porter is presently considered to be a member of the 3-hydroxycarboxylate family of porters that also includes the citrate carriers. These proteins do exhibit substrate flexibility in the molecules that are transported, though the main requirement is that they adhere to the general chemical structure of R1R2COHCOOH, which includes glycerol to citrate (Bandell et al. 1997).

More recently, work in MLF has been extended to the oral streptococci, with the question being whether malolactic acid provides protection from acidification in the oral cavity. Here, again, it was shown that L-malate is capable of inducing the system and that the MLF system functioned optimally at pH values of 4.0 in *S. mutans* UA159. Malate itself does not support growth of *S. mutans* but does protect the organism from acid challenge. Additionally, malate utilization contributes to the maintenance of ATP pools, indicating that alkali formation by MLF could be a major piece of the acid-resistance repertoire of *S. mutans* (Sheng and Marquis 2007). Subsequent work showed that, like other LAB, the MleR regulator controls transcription of the *mle*SP genes but that maximal induction required malate and low pH and that, interestingly, MleR would bind to specific sites, depending on the presence or absence of L-malate (Lemme et al. 2010). The loss of the *mle*P, *mle*S, and, to a lesser extent, *mle*R, resulted in reduced ability of *S. mutans* to survive acidic challenge (Sheng et al. 2010). The availability of the mutant strains of *S. mutans* in *mle*P, S, and R also permitted the authors to show that MLF acted additively with the ADI system to protect *S. sanguinis*, but did not interact additively with the urease system of *S. salivarius* (Sheng et al. 2010).

2.5.3 Additional Acid-Protective Decarboxylation Reactions That Are Beneficial to Lactic Acid Bacteria

In addition to the release of carbon dioxide from malic acid, citrate metabolism by LAB has also been found to be of benefit to them during growth in acidic conditions. A recent report has indicated that citrate stimulated the growth rate of *L. lactis* CRL264 (biovar *diacetylactis*) during growth at low pH (Sánchez et al. 2008). ^{13}C-driven NMR studies strongly implicated the presence of a citrate/lactate antiporter in the organism (Sánchez et al. 2008). These data suggest a mechanism of generating a PMF similar to that described for the malolactic acid system. These authors argued that the effects of citrate were due less to alkalinization of the cells during growth at low pH than to the reduction of ATP expenditure toward removing protons from the cell cytoplasm. The two points are important and suggest that additional research would be useful. Clearly, alkalinization of the organism's cytoplasm by the release of carbon dioxide would benefit potentially acid-sensitive enzymes of the glycolytic pathway, providing a mechanism to protect ATP production and a higher net yield of ATP formation. The acid-inducible nature of citrate lyase reported for *L. lactis* CRL264 (biovar *diacetylactis*) again indicates an acid-inducible mechanism for carbon dioxide release and the removal of organic anions via an antiporter system that assists in generating PMF (Martín et al. 2004).

Decarboxylation systems involving specific antiporters are a well-established mechanism for mitigating internal acidification (Small and Waterman 1998). The availability of genomic information and a high degree of amino acid conservation for these enzymes permit the precise study of these mechanisms. In the case of the probiotic LAB, *Lactobacillus acidophilus* NCFM, bioinformatics was used to identify open-reading frames with homology to known decarboxylation-antiporter systems. Genes identified as a result of the analysis included a glutamate γ-aminobutyrate antiporter, an ornithine decarboxylase, an amino acid permease, and a putative transcriptional regulator, in an analogous arrangement to many malolactic gene clusters. Mutation of each of the four genes resulted in acid-sensitive phenotypes indicating their importance in protection of the organism from internal acidification, though acid adaptation itself was not abolished in any of the strains (Azcarate-Peril et al. 2004).

Building on those results, the analysis of a putative glutamate-utilization operon in *S. mutans* showed that exogenous glutamate would facilitate the increased production of lactic acid (Krastel et al. 2010). The loss of the operon *gln*QHMP resulted in the inability to accumulate glutamate and in a slow-growth phenotype. However, the *gln*-deficient mutant strain survived exposure to pH 3.5 better than its parental strain UA159, suggesting that additional, compensatory acid-adaptive mechanisms are available to the organism, and probably to most LAB.

2.5.4 The Role of Molecular Chaperones, Clp and HtrA Proteases, and Heat-Shock Proteins

The Clp proteases function in bacteria as processing enzymes that act either to assist in correct folding or to cleave and degrade misfolded proteins. The number of Clp proteins in a given bacterium varies. For example, there are five Clp-related proteins in *S. mutans;* of these, the ClpL chaperone and ClpXP protease affect the expression of virulence attributes associated with *S. mutans,* including sensitivity to growth at low pH (Kajfasz et al. 2009). However, strains deleted for *clp*P and *clp*X were less sensitive to killing by acid, which led to a genomic search for suppressor genes, culminating in the identification of the *spx*A and B genes, which appear also to play a role in regulating not only the Clp chaperones, but also responses to the environment, including responsiveness to acid and oxidative conditions. Previous work with *clp*C and *clp*P had shown that they were induced during growth at low pH, indicating the need for either the removal of damaged proteins or the necessity of acting as chaperones for stabilizing the proper folding of proteins (Lemos and Burne 2002). It was also shown that in an *S. mutans clp*P mutant strain, the heat-shock proteins GroESL and DnaK were upregulated, indicating the likelihood that the cells were stressed, presumably due to the accumulation of damaged proteins (Lemos and Burne 2002). The involvement of the heat-shock proteins with the Clp proteins, pertinent to acid-mediated stress, is probably a common feature across LAB, if not bacteria in general, as reported for *S. mutans* (Jayaraman et al. 1997), *Enterococcus faecalis* (Laport et al. 2004, 2006), *Lactobacillus bulgaricus* (Fernandez et al. 2008), and *Lactobacillus johnsonii* (Walker et al. 1999). Subsequent work with a *clp*P mutant strain of *S. mutans* showed that approximately 100 gene transcripts were either up- or downregulated in a *clp*P deletion, indicating its global effect on the *S. mutans* transcriptome (Zhang et al. 2009; Chattoraj et al. 2010). The involvement of Clp proteins in stress regulation has not been limited to oral streptococci. In *O. oeni*, ClpX is an early phase of growth-related protein, and ClpP and ClpL appear to be involved at multiple levels of stress regulation (Guzzo et al. 2000; Beltramo et al. 2004). Consistent with these reports, ClpL is involved in early response to acidification in *Lactobacillus reuteri* (Wall et al. 2007). In addition to the Clp systems, an extracellular protease, HtrA, has been implicated in multiple stress responses, including acid-mediated stress in *S. mutans* (Biswas and Biswas 2005; Kang et al. 2010) and *L. lactis* (Morello et al. 2008). The enzyme is known in other Gram-positive bacteria and is also referred to as DegP, a serine protease (Pallen and Wren 1997; Ehrmann and Clausen 2004; Ingmer and Brondsted 2009). In the case of *S. mutans*, HtrA apparently can degrade, at least, the extracellular glucosyltransferases, thereby affecting in vitro biofilm formation (Biswas and Biswas 2005). Remodeling of cell surface proteins affected by oxidative agents or acid attack could be useful for LAB, though how these types of stress are communicated to the HtrA synthesis machinery remains to be established.

2.5.5 Other Protective Mechanisms: Overlap of the Acid-Stress Response with Additional Stress Responses

The response of LAB to acid stress also appears to overlap, at least partially, with what are generally thought of as oxidative stress-resistant mechanisms. For example, glutathione has proven to be protective for *L. lactis* subsp. *cremoris* at pH values of 4.0 and 2.5 (Zhang et al. 2007). The use of glutathione was also responsible for increased yields of *L. salivarius*, the probiotic organism, during growth at a pH value of 4.0 (Lee et al. 2010). The mechanism by which glutathione is able to protect these organisms at low pH is not yet clear, though the reports indicate that thiols are at risk during growth in acidic conditions by LAB.

In addition, DNA-repair mechanisms, often characterized as responsive to oxidative stress, have also been shown to be involved in acid-stress responses in LAB. For example, the *uvr*A excision nuclease, which has been classically categorized as a participant in the nucleotide excision repair (NER) pathway, is part of the acid-response of LAB members *L. helveticus* CNBL1156 (Cappa et al. 2005) and *S. mutans* strain JH1005 (Hanna et al. 2001). DNA damage, as a result of acidification of the cytoplasm, is not surprising. What is less understood presently is how the formation of DNA damage, by acid attack on either the nucleotide or phosphate bonds, is communicated to the cells such that DNA-repair genes are induced. In *S. mutans* GS-5, it has been shown that the acid repertoire includes an *rec*A-independent, acid-growth-inducible mechanism of DNA repair, which may include NER or base excision pathways (Quivey et al. 1995).

2.6 Sensing Acid Stress and Signaling

2.6.1 The Stringent Response and Early Signaling Events for Acid-Stress Response

Up to this point, the discussion of acid stress has been largely devoted to what is known about how LAB respond to acidification, compared to how cells recognize that the environment has become acidic.

The guanidine nucleotide pools of bacteria typically reflect starvation, which can be considered as any disruption of carbon flow through the Embden–Meyerhoff–Parnas glycolytic pathway. The formation of (p)ppGpp, guanosine penta- or tetra-phosphate, occurs by the activity of the RelA pppGpp synthase, or related enzymes, when the translation of mRNA stalls on ribosomes due to a shortage of amino acids. Because (p)ppGpp is at the interface of translation, carbon flow, and phosphate pools, the molecule is an ideal location to potentially affect many cellular changes. Not surprisingly, the production of proteins in many species of bacteria is affected by the loss of RelA or its equivalents (Magnusson et al. 2005). Relatively little work has been reported in this area from the LAB, though it is known, for example, that *L. lactis* responds very quickly to acidification (Rallu et al. 1996) involving pppGpp (Rallu et al. 2000). Further, in proteomic experiments with *L. lactis* mutant strains defective in *rel*A, *gua*A (the GMP synthase), and *pst*S (phosphate transporter) compared to the parent strain (MG1363), numerous proteins were deregulated, as shown by two-dimensional (2D) gel electrophoresis coupled with mass spectrometric identification of proteins (Budin-Verneuil et al. 2007). However, only six proteins overlapped the mutant strains: RecA (the DNA-repair-recombination protein), pyruvate carboxylase, CTP synthase, glutamyl tRNA synthetase, R30S ribosomal protein S1, and the subunit of DNA polymerase DnaN, suggesting that these proteins play a role in a constitutive acid response. Similarly, in global transcriptional studies conducted with an *rel*A mutant strain of *S. mutans*, approximately 50 transcripts were shown to be involved in the stringent response attributable to the RelP- and RelQ-mediated, RelA-independent stress response, indicating that the three mechanisms of producing (p)ppGpp in *S. mutans* are all involved in sensing stress (Nascimento et al. 2009). Moreover, RelA is known to affect acid resistance in *S. mutans* grown in biofilms and the LuxS protein, a participant in AI-2-mediated quorum sensing, may be involved (Lemos et al. 2004).

2.6.2 Two-Component Systems and Their Involvement in the Acid Responses of Lactic Acid Bacteria

Similar to the use of global transcriptional profiling to characterize acid-stress responses of LAB, and most other species of culturable bacteria, the exploration of two-component signal-sensing and -response circuits in LAB is an area of intense investigation, with a large and growing literature. The interested reader should refer to recent reviews on the general concepts of TCSs and their role in regulation of the vast breadth of bacterial responses to external stimuli (Gao and Stock 2009; Krell et al. 2010). In the case of this discussion, substantial interest exists regarding the question of how LAB recognize falling pH values and transduce that information into a protective response.

A number of studies have been published implicating TCSs in the control of acid-stress responses in bacteria and in LAB specifically. Two general examples of the situation are included here. In the case of *S. mutans*, multiple TCSs affect the sensitivity to acidification. In one of the earliest reports, Li et al. (2002) used existing information from the HK03 and RR03 protein sequences of *S. pneumoniae* as the basis for BLAST searching the *S. mutans* UA159 genomic sequence (Ajdic et al. 2002), which resulted in the identification of a putative TCS designated *hk11* and *rr11* in *S. mutans* NG8. Mutation of the *hk11* gene, encoding the histidine kinase of the system, resulted in an acid-sensitive phenotype. A more exhaustive study, in which all 14 of the recognizable TCS pairs in *S. mutans* (Levesque et al. 2007), showed that three systems – TCS-2, referred to as a homolog of the CiaRH systems in other bacteria, TCS-3, similar to the ScnRK-like systems, and TCS-9 – were all involved at some level in acid resistance. Later studies showed that the VicRK TCS of *S. mutans* was also involved in acid resistance of *S. mutans* and that loss of the *vicK* gene affects 89 transcripts in the microarray analysis of the *vic*K mutant strain (Senadheera et al. 2009). Additional reports, also involving mutation of all 14 sensor/kinase pairs of TCSs in *S. mutans,* provided information on the role of the systems in the stress response (Biswas et al. 2008; Kawada-Matsuo et al. 2009), though the clearest message to emerge from these types of studies is that the potential for overlapping responses among multiple TCSs appears to be substantial. Studies have already begun to uncover information that TCS expression has to be interpreted with care, in that an overlap in at least some of the sensing does indeed occur (Chong et al. 2008). As information continues to develop regarding DNA-binding motifs for the response-regulator partner of TCSs (Senadheera et al. 2005), the ability to distinguish between the direct and indirect effects of TCSs in regulating the acid response will continue to improve.

While relatively understudied, with respect to the effects of low pH on virulence, a recent study with Group B streptococcus (GBS), strain V/R 2603, has been published, reporting the global transcriptional analysis of the organism following growth at pH values of 7.0 and 5.5 (Santi et al. 2009). Like many studies involving global gene transcription, the authors found that over 300 genes were upregulated during growth at a pH value of 5.5, compared to growth at pH 7.0, and that 61 genes were downregulated during growth at pH 5.5. The genes expressed during acidic growth involved all of the major metabolic and stress-responsive pathways. In this study, focus was given to pH-responsive genes known to be controlled by the CsrRS TCS. The results showed that approximately 90% of the downregulated genes and nearly 60% of the upregulated genes were CsrRS-dependent. The implication of the study was that movement of GBS from the vaginal tract of the mother (acidic) to the lungs of infants might signal the upregulation of virulence factors and initiate the invasive phenotype of GBS, including the surface protein BibA, a GBS-vaccine candidate, that was pH-responsive (Santi et al. 2009).

2.7 Global Analysis of Acid-Stress Responses in Lactic Acid Bacteria: Proteomic Approaches

Over the past 10 years, a large number of bacterial genome sequences have become available, for LAB and many other organisms. The acquisition of genomic information continues to grow at a rapid rate, which facilitates comparisons of strains and also provides maps for aligning transcriptional data arising from microarray analyses, proteomic determinations, and, more recently, RNA-seq experiments (Wang et al. 2009), a method for transcriptional analysis that does not require the use of microarray chips.

The initial proteomic mapping of LAB was conducted using 2D gel electrophoresis to separate proteins, which were later identified using Edman degradation or, more recently, matrix-assisted laser desorbtion-ionization (MALDI) mass spectrometry. The resulting amino acid sequences were then compared to existing protein databases. Now, MALDI data are directly aligned by BLAST searches of the more extensive genomic databases.

The types of proteomic questions that have been addressed in LAB are in several categories: a comparison of expressed proteins during growth at varying pH levels; a comparison of expressed proteins from cells at different points in their growth curves; and a comparison of expressed proteomes from highly similar but different strains or species of LAB.

In one of the first 2D gel studies with LAB, using radiolabeled cells of *L. bulgaricus*, a comparison was made of proteins produced by cells grown at pH values of 6.0 and 4.75 (Lim et al. 2000). The results showed several proteins that were prominently increased in abundance in the extracts of the pH 4.75-grown cells. N-terminal sequencing of three proteins revealed them to be GroES, GroEL, and DnaK, all of which are well-known proteins that function as molecular chaperones during stressful conditions. In a similar fashion, comparisons of radiolabeled cultures of *S. mutans* grown under a variety of stress conditions, including acidic conditions, oxidative stress, and variation in temperature, showed a variety of protein patterns, including a small number of unidentified proteins that were present in all conditions, suggesting their involvement as general stress proteins (Svensater et al. 2000). Later studies compared extracts of *S. mutans* cells from planktonic and biofilm cultures. Again, the expressed protein profiles differed depending on culture conditions, which supports the concept that LAB are versatile in their response to their environment (Svensater et al. 2001; Welin et al. 2004; Welin-Neilands and Svensäter 2007). During this period of time, many LAB were examined by 2D gel electrophoresis. The stationary-phase acid tolerance of *L. acidophilus* CRL639 was characterized by comparison of cells grown at pH 4.5 and 6.0, which resulted in the identification of a variety of proteins that were induced during growth at the lower pH value (Lorca and Valdez 2001). The difficulty these investigators had at the time was that they lacked high-throughput mass spectrometric capability and the genomic maps that were in the future; thus, the identities and function of acid-induced proteins were restricted to reporting the molecular mass and estimating the isoelectric points.

As technology advanced, the use of MALDI-based technologies dramatically increased the ability to identify and characterize larger numbers of proteins in 2D gel separations and to construct metabolic maps of proteins that might be involved with acid responses. Certainly, the numbers of proteins shown to be involved in acid adaptation grew in magnitude from handfuls of proteins to dozens of proteins. For example, in a set of papers with *S. mutans*, proteins representing entire metabolic pathways were reconstructed as either upregulated or downregulated during growth at neutral and acidic conditions (Len et al. 2004a, b). Likewise, during this period, work with *L. lactis* MG1363 revealed the identity of 80 proteins in a comparison of the acid response of the organism during growth in two differing media. Of the identified proteins, 10 were shown to be upregulated in both media. These particular proteins fit into the existing concept of an acid response that includes arginine metabolism (ArgS), the ADI pathway (ArcB), the F-ATPase (AtpA), CFA synthase, and oxidative stress responsiveness via superoxide dismutase (SodA) (Budin-Verneuil et al. 2005b). More recent studies with *L. reuteri* have focused on the protein profile of the organism growing at specific pH values, to simulate the effect of moving through the human alimentary canal and its varying pH values. Results from those studies have shown that proteins involved with acid-stress responses in that organism, at least, are represented in essentially every major metabolic category of the organism, again supporting the notion of bacteria that have evolved to be highly flexible with respect to external pH values (Lee et al. 2008; Lee and Pi 2010).

Among the most detailed studies to date has been the proteomic characterization of *L. bulgaricus* (Fernandez et al. 2008). In this study, the transcription of specific genes was estimated and combined with proteomic data to provide a picture of carbon flow during growth at acidic pH values. Quantitative PCR was used to estimate the abundance of RNA transcripts from specific genes, and those data were coupled with proteomic data arising from mass spectrometric analysis of proteins separated by 2D gels. The authors of this elegant study were then able to reconstruct carbon-flow pathways showing that during acidic growth, carbon was diverted into fatty acid synthesis, which would affect membrane fluidity (Fernandez et al. 2008). The results from this study are probably extensible throughout LAB. In addition to their importance for food LAB, the data also provide insight into the acid-responsive membrane changes of membranes in pathogens, *S. mutans* (Fozo and Quivey 2004a, b; Fozo et al. 2007) and *S. pneumoniae* (Marrakchi et al. 2002; Jerga and Rock 2009).

2.8 Global Analysis of Acid-Stress Responses in Lactic Acid Bacteria: Transcriptional Approaches

As with most other bacterial species, the last decade has seen an intense focus on expression profiling of LAB genomes and how acid stress (among other stimuli) affects specific stress-response pathways. The information to be gained from these studies is related to the potential ability to manipulate pathways during the production of foods, such as cheeses and yogurt, or to curtail infections from the pathogenic LAB. The approach of many studies is to obtain data for gene expression from bacteria grown in various culture conditions.

In a recent study of stress responses in *L. lactics* subsp. *lactis* IL1403, macroarray analysis was used to characterize the response of the bacterium to heat, acid, and osmotic stress. Of a genomic subset of 375 genes, between 13 and 18% of the genes were responsive to one or more external stimuli, which is consistent with observations in many other bacterial systems, including LAB. The data from this particular study provided insight into differences between the IL1403 strain and strain MG1363, such as the variation in the methionine-metabolizing genes that was shown to differ between the strains. Of potential importance to the dairy industry, the report showed that peptide transporters were repressed in stress-inducing conditions, which, as the authors state, would have implications in cheese-ripening processes (Xie et al. 2004).

Transcriptional studies using RNA isolated from *L. plantarum*, grown in a comprehensive variety of culture conditions, have shown that microarray-based profiling can reveal changes in major metabolic pathways, with respect to growth in acidic conditions (Pieterse et al. 2005). The study took the interesting approach that the lactate/lactic acid ratio would have effects on the organism's transcriptome as an indicator of toxicity for the organism, with concomitant effects on gene expression. The large amount of data arising from these studies has demonstrated the value of using multiple conditions to evaluate the acid-stress response but also showed significant shifts in the transcription of genes relating to carbon flow, which could benefit manipulation of the organism in future commercial application (Pieterse et al. 2005).

In the case of *S. pneumoniae,* though not always thought of as an acid-resistant organism but certainly capable of eliciting a sustained acid resistance (Martín-Galiano et al. 2001), stationary-phase exposure to pH 4.4 induces relatively high levels of acid resistance (Martín-Galiano et al. 2005). Subsequent experiments were conducted to assess the differences in global gene expression in cells growing in cultures exposed to acidic conditions for short periods of time, compared to those from acid-adapted cultures grown at pH 6.0. The results of these studies revealed 126 genes whose expression was affected by external pH and that of those genes, nearly 60 were specific for acid adaptation and 37 genes were involved with growth in the stationary phase, or maintenance genes. Another 30 genes were reportedly found in both sets of conditions. Of the pH-responsive group of genes, chaperones and transporters appeared to be significant and, as with other studies of this kind, there was substantial overlap with genes thought or known to be involved with oxidative and osmotic stress. The magnitude of expression changes was somewhat smaller for genes in this study, compared to those of the 10- and 100-fold magnitude shown in studies with the oral streptococci *Streptococcus sobrinus* (Nascimento et al. 2004), *S. mutans* (Len et al. 2004b), and *Streptococcus oralis* (Wilkins et al. 2001), perhaps reflecting the larger variation of external pH values that the oral streptococci may be responsive to (Martín-Galiano et al. 2005). Regulatory motifs connecting some of the acid-regulated genes were reported, and these provide a highly useful tool in order to develop acid-response maps, similar to those described for *S. mutans* (Len et al. 2004a) and *L. bulgaricus* (Fernandez et al. 2008).

The combination of transcriptional studies with assays for metabolites (or metabolomics) is of substantial importance in studies going forward. The reason is that transcriptomes are very important measures of mRNA in a single snapshot. However, translational efficiencies of mRNAs vary widely in bacteria, and the regulation of translation, particularly during stress-inducing conditions, may have an enormous impact on the proteome of bacteria and on the levels of metabolites. For example, work published by Even et al. (2003) using *L. lactis* subsp. *cremoris* MG1363 showed that the regulation of translation superseded transcriptional regulation in acidic growth conditions. The authors of this study showed that changes in the abundance of glycolytic enzymes were not correlated with changes in transcript levels, indicating that metabolites, as a function of enzyme levels, were dominant in regulating carbon flow. Their data also indicated that total RNA levels fell during growth at low pH, providing support to the concept that translation in *L. lactis* is optimal during growth in acidic conditions (Even et al. 2003).

One major obstacle for the transcriptional profiling of bacteria has been the lack of appropriate microarray chips, now somewhat resolved by the development of high-throughput sequencing approaches for transcriptional profiling (Wang et al. 2009). However, the technique of comparative genomic hybridization has been used to study changes in mRNA levels of *S. sobrinus*. In this type of approach, microarrays produced for the *S. mutans* genome were used to evaluate the transcriptional output of *S. sobrinus*, an organism also capable of causing dental caries. While the interspecies approach is not ideal, the study did take advantage of the possibility that acid resistance may be highly conserved between the species. In addition, mass spectrometric analysis was applied to extracts of *S. sobrinus* to provide supporting data for the array experiments. Among the results presented by the authors of the study, MLF genes were substantially upregulated during acid growth, as well as energy-metabolism genes, including F-ATPase operon genes (Martinez et al. 2010).

2.9 Population Biology of the Lactic Acid Bacteria Acid Response

Two facets of the acid-stress response in LAB that have not received as much attention as they should have, and probably will have in the coming decade, are the following: the temporal regulation of acid-stress responses and the possibility that subpopulations of a given organism in growing cultures, or microenvironments, may elicit different levels of the response. There is abundant literature from studies conducted with LAB describing early logarithmic, mid-logarithmic, and stationary-phase expression of acid-stress responses. Difficulty exists in developing an entirely integrated view of how LAB deploy their acid-stress responses, which is due to species and strain differences, on the one hand, and to variation in media composition, culture conditions, and sampling effects on the other hand. As discussed above, the combined use of transcriptional profiling and proteomics is illustrative of how changes in acid-stress-responsive genes can be followed over time (Len et al. 2004a, b; Fernandez et al. 2008). Additional methods for observing the effects on individual cells, and for identifying specific cells in mixtures, are also developing. Perhaps not surprisingly, imaging methodologies, sometimes combined with other approaches, have been reported following variation in the expression of acid resistance (or tolerance). Recent reports for *Streptococcus macedonicus* have shown that arbitrarily primed PCR amplification of genomes, coupled with Fourier-transform infrared spectroscopy, can be used to track changes in the organism, over time, in response to acidic conditions or specific treatment regimes (Papadimitriou et al. 2008). Conceptually, the notion that LAB elicit a fluid response to acid stress is probably universally accepted. Providing evidence for that plasticity in populations of bacteria has been somewhat longer in arriving. These authors have shown that the acid response (or acid-tolerance response, ATR) is indeed plastic. Further, cells in differing phases of their responses can be separated using flow cytometry and cell sorting (Papadimitriou et al. 2007). The results from these two reports provided further evidence of multiple participants, including changes in F-ATPase and fatty acid enzyme expression, in the acid response of *S. macedonicus*, but perhaps more importantly, they suggested that populations of cells can be subdivided for elucidation of their stress responses. The use of additional imaging technologies has permitted the examination of bacteria located in three-dimensional space in biofilms and the effects of carbohydrates on three-dimensional spacing in matrices of extracellular polysaccharide. Confocal microscopy has shown that biofilm development differs when *S. mutans* utilizes sucrose to form biofilms, compared to those formed during growth on starch or combinations of starch and glucose (Klein et al. 2009). Laser-near-infrared Raman spectroscopy has been used to show that highly related oral streptococci, namely, *S. mutans* UA159 and *S. sanguinis* NCTC10904, can be identified and enumerated in biofilms by relying on the deconvolution of Raman spectra of the two organisms (Zhu et al. 2004, 2007; Beier et al. 2010), very similar to the approaches used with *S. macedonicus* and the resolution of *L. acidophilus* from *S. thermophilus* (Oberreuter et al. 2000) by FT-IR.

The ability to evaluate the effects of epigenetic events, such as acidification, and how LAB interact with other flora in situ will likely be the focus of work yet to come. The results from those experiments shall provide much-needed insights into metabolic networks of bacteria and how those can be used to explain disease-related processes such as the substantial acid resistance of oral LAB growing in biofilms (McNeill and Hamilton 2004; Welin et al. 2004; Welin-Neilands and Svensäter 2007) or the useful ways in which modification of the acid response might be used to influence the organoleptic and nutritional properties of foods.

2.10 Concluding Remarks

In the foregoing sections, the discussion was directed to a broad summary of the mechanisms that are now at least partly understood concerning the ability of LAB to protect themselves from their own metabolism. Reference was made in part to approaches for understanding the effects of genes on a more global scale. Transcriptional and proteomic approaches have come into vogue in the last decade, as technologies have enabled scientists to learn how environmental conditions affect bacterial responses, and what the loss of any given gene or genes will have on that global response. Initially, genome-wide transcriptional studies were limited to those organisms for which a microarray of probes was available, either commercially or locally produced. More recently, the development of high-throughput cDNA sequencing, referred to as RNA-Seq, promises to permit the transcriptional profiling of any bacterium from any test condition (Wang et al. 2009). The continued development of automated protein separation-mass spectrometry will further speed the acquisition of information related to the production of stress responses in all bacteria, in addition to the lactic streptococci. The culmination of these technologies, and the continued expansion of genomic information for species and strains, will facilitate experiments in which communities of bacteria, and their mutant strains, are evaluated for their responses to stress and the production of attractive organoleptic properties or virulence attributes.

References

Aas JA, Griffen AL, Dardis SR, Lee AM, Olsen I, Dewhirst FE, Leys EJ, Paster BJ (2008) Bacteria of dental caries in primary and permanent teeth in children and young adults. J Clin Microbiol 46:1407–1417

Ajdic D, McShan WM, McLaughlin RE, Savic G, Chang J, Carson MB, Primeaux C, Tian R, Kenton S, Jia H, Lin S, Qian Y, Li S, Zhu H, Najar F, Lai H, White J, Roe BA, Ferretti JJ (2002) Genome sequence of *Streptococcus mutans* UA159, a cariogenic dental pathogen. Proc Natl Acad Sci USA 99:14434–14439

Ansanay V, Dequin S, Blondin B, Barre P (1993) Cloning, sequence and expression of the gene encoding the malolactic enzyme from *Lactococcus lactis*. FEBS Lett 332:74–80

Araque I, Gil J, Carrete R, Bordons A, Reguant C (2009) Detection of arc genes related with the ethyl carbamate precursors in wine lactic acid bacteria. J Agric Food Chem 57:1841–1847

Arena ME, Manca de Nadra MC, Munoz R (2002) The arginine deiminase pathway in the wine lactic acid bacterium *Lactobacillus hilgardii* X1B: structural and functional study of the *arc*ABC genes. Gene 301:61–66

Azcarate-Peril MA, Altermann E, Hoover-Fitzula RL, Cano RJ, Klaenhammer TR (2004) Identification and inactivation of genetic loci involved with *Lactobacillus acidophilus* acid tolerance. Appl Environ Microbiol 70:5315–5322

Bandell M, Ansanay V, Rachidi N, Dequin S, Lolkema JS (1997) Membrane potential-generating malate (MleP) and citrate (CitP) transporters of lactic acid bacteria are homologous proteins. Substrate specificity of the 2-hydroxycarboxylate transporter family. J Biol Chem 272:18140–18146

Becker MR, Paster BJ, Leys EJ, Moeschberger ML, Kenyon SG, Galvin JL, Boches SK, Dewhirst FE, Griffen AL (2002) Molecular analysis of bacterial species associated with childhood caries. J Clin Microbiol 40:1001–1009

Beier BD, Quivey RGJ, Berger AJ (2010) Identification of different bacterial species in biofilms using confocal raman microscopy. J Biomed Opt 15(6):066001

Belli WA, Marquis RE (1991) Adaptation of *Streptococcus mutans* and *Enterococcus hirae* to acid stress in continuous culture. Appl Environ Microbiol 57:1134–1138

Belli WA, Marquis RE (1994) Catabolite modification of acid tolerance of *Streptococcus mutans* GS-5. Oral Microbiol Immunol 9:29–34

Belli WA, Buckley DH, Marquis RE (1995) Weak acid effects and fluoride inhibition of glycolysis by *Streptococcus mutans* GS-5. Can J Microbiol 41:785–791

Beltramo C, Grandvalet C, Pierre F, Guzzo J (2004) Evidence for multiple levels of regulation of *Oenococcus oeni clp*P-*clp*L locus expression in response to stress. J Bacteriol 186:2200–2205

Bender GR, Sutton SV, Marquis RE (1986) Acid tolerance, proton permeabilities, and membrane ATPases of oral streptococci. Infect Immun 53:331–338

Bender GR, Marquis RE (1987) Membrane ATPases and acid tolerance of *Actinomyces viscosus* and *Lactobacillus casei*. Appl Environ Microbiol 53:2124–2128

Biswas S, Biswas I (2005) Role of HtrA in surface protein expression and biofilm formation by *Streptococcus mutans*. Infect Immun 73:6923–6934

Biswas I, Drake L, Erkina D, Biswas S (2008) Involvement of sensor kinases in the stress tolerance response of *Streptococcus mutans*. J Bacteriol 190:68–77

Bradshaw DJ, McKee AS, Marsh PD (1989) Effects of carbohydrate pulses and pH on population shifts within oral microbial communities *in vitro*. J Dent Res 68:1298–1302

Broadbent JR, Larsen RL, Deibel V, Steele JL (2010) Physiological and transcriptional response of *Lactobacillus casei* ATCC 334 to acid stress. J Bacteriol 192:2445–2458

Budin-Verneuil A, Maguin E, Auffray Y, Ehrlich SD, Pichereau V (2005a) Transcriptional analysis of the cyclopropane fatty acid synthase gene of *Lactococcus lactis* MG1363 at low pH. FEMS Microbiol Lett 250:189–194

Budin-Verneuil A, Pichereau V, Auffray Y, Ehrlich DS, Maguin E (2005b) Proteomic characterization of the acid tolerance response in *Lactococcus lactis* MG1363. Proteomics 5:4794–4807

Budin-Verneuil A, Maguin E, Auffray Y, Ehrlich DS, Pichereau V (2006) Genetic structure and transcriptional analysis of the arginine deiminase (ADI) cluster in *Lactococcus lactis* MG1363. Can J Microbiol 52:617–622

Budin-Verneuil A, Pichereau V, Auffray Y, Ehrlich D, Maguin E (2007) Proteome phenotyping of acid stress-resistant mutants of *Lactococcus lactis* MG1363. Proteomics 7:2038–2046

Burne RA, Parsons DT, Marquis RE (1989) Cloning and expression in *Escherichia coli* of the genes of the arginine deiminase system of *Streptococcus sanguis* NCTC 10904. Infect Immun 57:3540–3548

Cappa F, Cattivelli D, Cocconcelli PS (2005) The *uvr*A gene is involved in oxidative and acid stress responses in *Lactobacillus helveticus* CNBL1156. Res Microbiol 156:1039–1047

Casiano-Colon A, Marquis RE (1988) Role of the arginine deiminase system in protecting oral bacteria and an enzymatic basis for acid tolerance. Appl Environ Microbiol 54:1318–1324

Champomier Verges MC, Zuniga M, Morel-Deville F, Perez-Martinez G, Zagorec M, Ehrlich SD (1999) Relationships between arginine degradation, pH and survival in *Lactobacillus sakei*. FEMS Microbiol Lett 180:297–304

Chattoraj P, Banerjee A, Biswas S, Biswas I (2010) ClpP of *Streptococcus mutans* differentially regulates expression of genomic islands, mutacin production, and antibiotic tolerance. J Bacteriol 192:1312–1323

Chen YY, Burne RA (2003) Identification and characterization of the nickel uptake system for urease biogenesis in *Streptococcus salivarius* 57.I. J Bacteriol 185:6773–6779

Chen YY, Weaver CA, Mendelsohn DR, Burne RA (1998) Transcriptional regulation of the *Streptococcus salivarius* 57.I urease operon. J Bacteriol 180:5769–5775

Chong P, Drake L, Biswas I (2008) LiaS regulates virulence factor expression in *Streptococcus mutans*. Infect Immun 76:3093–3099

Cox DJ, Henick-Kling T (1989) Chemiosmotic energy from malolactic fermentation. J Bacteriol 171:5750–5752

Davis CR, Wibowo DJ, Lee TH, Fleet GH (1986) Growth and metabolism of lactic acid bacteria during and after malolactic fermentation of wines at different pH. Appl Environ Microbiol 51:539–545

Denayrolles M, Aigle M, Lonvaud-Funel A (1994) Cloning and sequence analysis of the gene encoding *Lactococcus lactis* malolactic enzyme: relationships with malic enzymes. FEMS Microbiol Lett 116:79–86

Duary RK, Batish VK, Grover S (2010) Expression of the atpD gene in probiotic *Lactobacillus plantarum* strains under in vitro acidic conditions using RT-qPCR. Res Microbiol 161:399–405

Ehrmann M, Clausen T (2004) Proteolysis as a regulatory mechanism. Annu Rev Genet 38:709–724

Even S, Lindley ND, Cocaign-Bousquet M (2003) Transcriptional, translational and metabolic regulation of glycolysis in *Lactococcus lactis* subsp. *cremoris* MG 1363 grown in continuous acidic cultures. Microbiology (Reading, Engl) 149:1935–1944

Fenoll A, Munoz R, Garcia E, de la Campa AG (1994) Molecular basis of the optochin-sensitive phenotype of pneumococcus: characterization of the genes encoding the F0 complex of the *Streptococcus pneumoniae* and *Streptococcus oralis* H(+)-ATPases. Mol Microbiol 12:587–598

Fenoll A, Munoz R, Garcia E, de la Campa AG (1995) Optochin sensitivity is encoded by the *atp*C gene of the *Streptococcus pneumonia atp* (F0F1 H(+)-ATPase) operon. Dev Biol Stand 85:287–291

Fernandez A, Ogawa J, Penaud S, Boudebbouze S, Ehrlich D, van de Guchte M, Maguin E (2008) Rerouting of pyruvate metabolism during acid adaptation in *Lactobacillus bulgaricus*. Proteomics 8:3154–3163

Fozo EM, Quivey RG Jr (2004a) Shifts in the membrane fatty acid profile of *Streptococcus mutans* enhance survival in acidic environments. Appl Environ Microbiol 70:929–936

Fozo EM, Quivey RG Jr (2004b) The *fab*M gene product of *Streptococcus mutans* is responsible for the synthesis of monounsaturated fatty acids and is necessary for survival at low pH. J Bacteriol 186:4152–4158

Fozo EM, Kajfasz JK, Quivey RG Jr (2004) Low pH-induced membrane fatty acid alterations in oral bacteria. FEMS Microbiol Lett 238:291–295

Fozo EM, Scott-Anne K, Koo H, Quivey RG Jr (2007) Role of unsaturated fatty acid biosynthesis in virulence of *Streptococcus mutans*. Infect Immun 75:1537–1539

Gao R, Stock AM (2009) Biological insights from structures of two-component proteins. Annu Rev Microbiol 63:133–154

Griswold AR, Chen YY, Burne RA (2004) Analysis of an agmatine deiminase gene cluster in *Streptococcus mutans* UA159. J Bacteriol 186:1902–1904

Griswold AR, Jameson-Lee M, Burne RA (2006) Regulation and physiologic significance of the agmatine deiminase system of *Streptococcus mutans* UA159. J Bacteriol 188:834–841

Griswold AR, Nascimento MM, Burne RA (2009) Distribution, regulation and role of the agmatine deiminase system in *mutans* streptococci. Oral Microbiol Immunol 24:79–82

Gutierrez JA, Crowley PJ, Cvitkovitch DG, Brady LJ, Hamilton IR, Hillman JD, Bleiweis AS (1999) *Streptococcus mutans ffh*, a gene encoding a homologue of the 54 kDa subunit of the signal recognition particle, is involved in resistance to acid stress. Microbiology 145(Pt 2):357–366

Guzzo J, Jobin MP, Delmas F, Fortier LC, Garmyn D, Tourdot-Maréchal R, Lee B, Diviès C (2000) Regulation of stress response in *Oenococcus oeni* as a function of environmental changes and growth phase. Int J Food Microbiol 55:27–31

Hanna MN, Ferguson RJ, Li YH, Cvitkovitch DG (2001) *uvr*A is an acid-inducible gene involved in the adaptive response to low pH in *Streptococcus mutans*. J Bacteriol 183:5964–5973

Hasona A, Crowley PJ, Levesque CM, Mair RW, Cvitkovitch DG, Bleiweis AS, Brady LJ (2005) Streptococcal viability and diminished stress tolerance in mutants lacking the signal recognition particle pathway or YidC2. Proc Natl Acad Sci USA 102:17466–17471

Ingmer H, Brondsted L (2009) Proteases in bacterial pathogenesis. Res Microbiol 160:704–710

Jayaraman GC, Penders JE, Burne RA (1997) Transcriptional analysis of the *Streptococcus mutans* *hrc*A, *grp*E and *dna*K genes and regulation of expression in response to heat shock and environmental acidification. Mol Microbiol 25:329–341

Jerga A, Rock CO (2009) Acyl-acyl carrier protein regulates transcription of fatty acid biosynthetic genes via the FabT repressor in *Streptococcus pneumoniae*. J Biol Chem 284: 15364–15368

Kajfasz JK, Martinez AR, Rivera-Ramos I, Abranches J, Koo H, Quivey RG Jr, Lemos JA (2009) Role of Clp proteins in expression of virulence properties of *Streptococcus mutans*. J Bacteriol 191:2060–2068

Kang K-h, Lee J-S, Yoo M, Jin I (2010) The influence of HtrA expression on the growth of *Streptococcus mutans* during acid stress. Mol Cells 29:297–304

Kawada-Matsuo M, Shibata Y, Yamashita Y (2009) Role of two component signaling response regulators in acid tolerance of *Streptococcus mutans*. Oral Microbiol Immunol 24:173–176

Keijser BJ, Zaura E, Huse SM, van der Vossen JM, Schuren FH, Montijn RC, ten Cate JM, Crielaard W (2008) Pyrosequencing analysis of the oral microflora of healthy adults. J Dent Res 87:1016–1020

Klein MI, Duarte S, Xiao J, Mitra S, Foster TH, Koo H (2009) Structural and molecular basis of the role of starch and sucrose in *Streptococcus mutans* biofilm development. Appl Environ Microbiol 75:837–841

Konings WN, Lolkema JS, Bolhuis H, van Veen HW, Poolman B, Driessen AJ (1997) The role of transport processes in survival of lactic acid bacteria. Energy transduction and multidrug resistance. Antonie van Leeuwenhoek 71:117–128

Krastel K, Senadheera DB, Mair R, Downey JS, Goodman SD, Cvitkovitch DG (2010) Characterization of a glutamate transporter operon, *gln*QHMP, in *Streptococcus mutans* and its role in acid tolerance. J Bacteriol 192:984–993

Krell T, Lacal J, Busch A, Silva-Jimenez H, Guazzaroni ME, Ramos JL (2010) Bacterial sensor kinases: diversity in the recognition of environmental signals. Annu Rev Microbiol 64:539–559

Kuhnert WL, Zheng G, Faustoferri RC, Quivey RG Jr (2004) The F-ATPase operon promoter of *Streptococcus mutans* is transcriptionally regulated in response to external pH. J Bacteriol 186:8524–8528

Labarre C, Divies C, Guzzo J (1996) Genetic organization of the mle locus and identification of a *mle*R-like gene from *Leuconostoc oenos*. Appl Environ Microbiol 62:4493–4498

Lafon-Lafourcade S, Carre E, Ribereau-Gayon P (1983) Occurrence of lactic acid bacteria during the different stages of vinification and conservation of wines. Appl Environ Microbiol 46:874–880

Laport MS, Lemos JA, Bastos Md Mdo C, Burne RA, Giambiagi-De Marval M (2004) Transcriptional analysis of the groE and dnaK heat-shock operons of *Enterococcus faecalis*. Res Microbiol 155:252–258

Laport MS, Dos Santos LL, Lemos JA, do Carmo FBM, Burne RA, Giambiagi-Demarval M (2006) Organization of heat shock *dna*K and *gro*E operons of the nosocomial pathogen *Enterococcus faecium*. Res Microbiol 157:162–168

Lee K, Lee H-G, Pi K, Choi Y-J (2008) The effect of low pH on protein expression by the probiotic bacterium *Lactobacillus reuteri*. Proteomics 8:1624–1630

Lee K, Pi K (2010) Effect of transient acid stress on the proteome of intestinal probiotic *Lactobacillus reuteri*. Biochemistry (Mosc) 75:460–465

Lee K, Pi K, Kim EB, Rho B-S, Kang S-K, Lee HG, Choi Y-J (2010) Glutathione-mediated response to acid stress in the probiotic bacterium, *Lactobacillus salivarius*. Biotechnol Lett 32:969–972

Lemme A, Sztajer H, Wagner-Döbler I (2010) Characterization of *mle*R, a positive regulator of malolactic fermentation and part of the acid tolerance response in *Streptococcus mutans*. BMC Microbiol 10:58

Lemos JA, Burne RA (2002) Regulation and physiological significance of ClpC and ClpP in *Streptococcus mutans*. J Bacteriol 184:6357–6366

Lemos JA, Brown TA Jr, Burne RA (2004) Effects of RelA on key virulence properties of planktonic and biofilm populations of *Streptococcus mutans*. Infect Immun 72:1431–1440

Len ACL, Harty DWS, Jacques NA (2004a) Proteome analysis of *Streptococcus mutans* metabolic phenotype during acid tolerance. Microbiology (Reading, Engl) 150:1353–1366

Len ACL, Harty DWS, Jacques NA (2004b) Stress-responsive proteins are upregulated in *Streptococcus mutans* during acid tolerance. Microbiology (Reading, Engl) 150:1339–1351

Levesque CM, Mair RW, Perry JA, Lau PC, Li YH, Cvitkovitch DG (2007) Systemic inactivation and phenotypic characterization of two-component systems in expression of *Streptococcus mutans* virulence properties. Lett Appl Microbiol 45:398–404

Li YH, Chen YY, Burne RA (2000) Regulation of urease gene expression by *Streptococcus salivarius* growing in biofilms. Environ Microbiol 2:169–177

Li YH, Lau PC, Tang N, Svensater G, Ellen RP, Cvitkovitch DG (2002) Novel two-component regulatory system involved in biofilm formation and acid resistance in *Streptococcus mutans*. J Bacteriol 184:6333–6342

Lim EM, Ehrlich SD, Maguin E (2000) Identification of stress-inducible proteins in *Lactobacillus delbrueckii* subsp. *bulgaricus*. Electrophoresis 21:2557–2561

Liu Y, Burne RA (2009) Multiple two-component systems of *Streptococcus mutans* regulate agmatine deiminase gene expression and stress tolerance. J Bacteriol 191:7363–7366

Liu Y, Zeng L, Burne RA (2009) AguR is required for induction of the *Streptococcus mutans* agmatine deiminase system by low pH and agmatine. Appl Environ Microbiol 75: 2629–2637

Lolkema JS, Poolman B, Konings WN (1995) Role of scalar protons in metabolic energy generation in lactic acid bacteria. J Bioenerg Biomembr 27:467–473

Lorca GL, Valdez GF (2001) A low-pH-inducible, stationary-phase acid tolerance response in *Lactobacillus acidophilus* CRL 639. Curr Microbiol 42:21–25

Lu YJ, Rock CO (2006) Transcriptional regulation of fatty acid biosynthesis in *Streptococcus pneumoniae*. Mol Microbiol 59:551–566

Lucas PM, Blancato VS, Claisse O, Magni C, Lolkema JS, Lonvaud-Funel A (2007) Agmatine deiminase pathway genes in *Lactobacillus brevis* are linked to the tyrosine decarboxylation operon in a putative acid resistance locus. Microbiology 153:2221–2230

Magnusson LU, Farewell A, Nystrom T (2005) ppGpp: A global regulator in *Escherichia coli*. Trends Microbiol 13:236–242

Mangani S, Guerrini S, Granchi L, Vincenzini M (2005) Putrescine accumulation in wine: role of *Oenococcus oeni*. Curr Microbiol 51:6–10

Marquis RE, Bender GR, Murray DR, Wong A (1987) Arginine deiminase system and bacterial adaptation to acid environments. Appl Environ Microbiol 53:198–200

Marrakchi H, Choi KH, Rock CO (2002) A new mechanism for anaerobic unsaturated fatty acid formation in *Streptococcus pneumoniae*. J Biol Chem 277:44809–44816

Martín MG, Sender PD, Peirú S, de Mendoza D, Magni C (2004) Acid-inducible transcription of the operon encoding the citrate lyase complex of *Lactococcus lactis* Biovar *diacetylactis* CRL264. J Bacteriol 186:5649–5660

Martín-Galiano AJ, Ferrandiz MJ, de la Campa AG (2001) The promoter of the operon encoding the F0F1 ATPase of *Streptococcus pneumoniae* is inducible by pH. Mol Microbiol 41:1327–1338

Martín-Galiano AJ, Overweg K, Ferrándiz MJ, Reuter M, Wells JM, de la Campa AG (2005) Transcriptional analysis of the acid tolerance response in *Streptococcus pneumoniae*. Microbiology (Reading, Engl) 151:3935–3946

Martinez AR, Abranches J, Kajfasz JK, Lemos JA (2010) Characterization of the *Streptococcus sobrinus* acid-stress response by interspecies microarrays and proteomics. Mol Oral Microbiol 25:331–342

McDermid AS, McKee AS, Ellwood DC, Marsh PD (1986) The effect of lowering the pH on the composition and metabolism of a community of nine oral bacteria grown in a chemostat. J Gen Microbiol 132:1205–1214

McKee AS, McDermid AS, Ellwood DC, Marsh PD (1985) The establishment of reproducible, complex communities of oral bacteria in the chemostat using defined inocula. J Appl Bacteriol 59:263–275

McNeill K, Hamilton IR (2004) Effect of acid stress on the physiology of biofilm cells of *Streptococcus mutans*. Microbiology (Reading, Engl) 150:735–742

Mercade M, Cocaign-Bousquet M, Loubière P (2006) Glyceraldehyde-3-phosphate dehydrogenase regulation in *Lactococcus lactis* ssp. *cremoris* MG1363 or relA mutant at low pH. J Appl Microbiol 100:1364–1372

Montanari C, Sado Kamdem SL, Serrazanetti DI, Etoa F-X, Guerzoni ME (2010) Synthesis of cyclopropane fatty acids in *Lactobacillus helveticus* and *Lactobacillus sanfranciscensis* and their cellular fatty acids changes following short term acid and cold stresses. Food Microbiol 27:493–502

Mora D, Monnet C, Parini C, Guglielmetti S, Mariani A, Pintus P, Molinari F, Daffonchio D, Manachini PL (2005) Urease biogenesis in *Streptococcus thermophilus*. Res Microbiol 156:897–903

Morello E, Bermudez-Humaran LG, Llull D, Sole V, Miraglio N, Langella P, Poquet I (2008) *Lactococcus lactis*, an efficient cell factory for recombinant protein production and secretion. J Mol Microbiol Biotechnol 14:48–58

Munoz R, Garcia E, De la Campa AG (1996) Quinine specifically inhibits the proteolipid subunit of the F0F1 H+-ATPase of *Streptococcus pneumoniae*. J Bacteriol 178:2455–2458

Nascimento MM, Lemos JA, Abranches J, Goncalves RB, Burne RA (2004) Adaptive acid tolerance response of *Streptococcus sobrinus*. J Bacteriol 186:6383–6390

Nascimento MM, Gordan VV, Garvan CW, Browngardt CM, Burne RA (2009) Correlations of oral bacterial arginine and urea catabolism with caries experience. Oral Microbiol Immunol 24:89–95

Oberreuter H, Mertens F, Seiler H, Scherer S (2000) Quantification of micro-organisms in binary mixed populations by Fourier transform infrared (FT-IR) spectroscopy. Lett Appl Microbiol 30:85–89

Pallen MJ, Wren BW (1997) The HtrA family of serine proteases. Mol Microbiol 26:209–221

Papadimitriou K, Pratsinis H, Nebe-von-Caron G, Kletsas D, Tsakalidou E (2007) Acid tolerance of *Streptococcus macedonicus* as assessed by flow cytometry and single-cell sorting. Appl Environ Microbiol 73:465–476

Papadimitriou K, Boutou E, Zoumpopoulou G, Tarantilis PA, Polissiou M, Vorgias CE, Tsakalidou E (2008) RNA arbitrarily primed PCR and Fourier transform infrared spectroscopy reveal plasticity in the acid tolerance response of *Streptococcus macedonicus*. Appl Environ Microbiol 74:6068–6076

Pernoud S, Fremaux C, Sepulchre A, Corrieu G, Monnet C (2004) Effect of the metabolism of urea on the acidifying activity of *Streptococcus thermophilus*. J Dairy Sci 87:550–555

Pieterse B, Leer RJ, Schuren FHJ, van der Werf MJ (2005) Unravelling the multiple effects of lactic acid stress on *Lactobacillus plantarum* by transcription profiling. Microbiology (Reading, Engl) 151:3881–3894

Pilone GJ, Kunkee RE (1970) Carbonic acid from decarboxylation by "malic" enzyme in lactic acid bacteria. J Bacteriol 103:404–409

Poolman B, Driessen AJ, Konings WN (1987) Regulation of arginine-ornithine exchange and the arginine deiminase pathway in *Streptococcus lactis*. J Bacteriol 169:5597–5604

Poolman B, Molenaar D, Smid EJ, Ubbink T, Abee T, Renault PP, Konings WN (1991) Malolactic fermentation: electrogenic malate uptake and malate/lactate antiport generate metabolic energy. J Bacteriol 173:6030–6037

Potrykus K, Cashel M (2008) (p)ppGpp: Still magical? Annu Rev Microbiol 62:35–51

Quivey RG Jr, Faustoferri RC, Clancy KA, Marquis RE (1995) Acid adaptation in *Streptococcus mutans* UA159 alleviates sensitization to environmental stress due to RecA deficiency. FEMS Microbiol Lett 126:257–261

Quivey RG Jr, Faustoferri R, Monahan K, Marquis R (2000) Shifts in membrane fatty acid profiles associated with acid adaptation of *Streptococcus mutans*. FEMS Microbiol Lett 189:89–92

Rallu F, Gruss A, Maguin E (1996) *Lactococcus lactis* and stress. Antonie van Leeuwenhoek 70:243–251

Rallu F, Gruss A, Ehrlich SD, Maguin E (2000) Acid- and multistress-resistant mutants of *Lactococcus lactis*: identification of intracellular stress signals. Mol Microbiol 35:517–528

Renault P, Gaillardin C, Heslot H (1988) Role of malolactic fermentation in lactic acid bacteria. Biochimie 70:375–379

Renault P, Gaillardin C, Heslot H (1989) Product of the *Lactococcus lactis* gene required for malolactic fermentation is homologous to a family of positive regulators. J Bacteriol 171:3108–3114

Salema M, Poolman B, Lolkema JS, Dias MC, Konings WN (1994) Uniport of monoanionic L-malate in membrane vesicles from *Leuconostoc oenos*. Eur J Biochem 225:289–295

Salema M, Capucho I, Poolman B, San Romao MV, Dias MC (1996) *In vitro* reassembly of the malolactic fermentation pathway of *Leuconostoc oenos* (*Oenococcus oeni*). J Bacteriol 178:5537–5539

Sánchez C, Neves AR, Cavalheiro J, dos Santos MM, García-Quintáns N, López P, Santos H (2008) Contribution of citrate metabolism to the growth of *Lactococcus lactis* CRL264 at low pH. Appl Environ Microbiol 74:1136–1144

Santi I, Grifantini R, Jiang S-M, Brettoni C, Grandi G, Wessels MR, Soriani M (2009) CsrRS regulates group B *Streptococcus* virulence gene expression in response to environmental pH: a new perspective on vaccine development. J Bacteriol 191:5387–5397

Senadheera D, Krastel K, Mair R, Persadmehr A, Abranches J, Burne RA, Cvitkovitch DG (2009) Inactivation of VicK affects acid production and acid survival of *Streptococcus mutans*. J Bacteriol 191:6415–6424

Senadheera MD, Guggenheim B, Spatafora GA, Huang YC, Choi J, Hung DC, Treglown JS, Goodman SD, Ellen RP, Cvitkovitch DG (2005) A VicRK signal transduction system in *Streptococcus mutans* affects *gtf*BCD, *gbp*B, and *ftf* expression, biofilm formation, and genetic competence development. J Bacteriol 187:4064–4076

Sheng J, Marquis RE (2007) Malolactic fermentation by *Streptococcus mutans*. FEMS Microbiol Lett 272:196–201

Sheng J, Baldeck JD, Nguyen PT, Quivey RG Jr, Marquis RE (2010) Alkali production associated with malolactic fermentation by oral streptococci and protection against acid, oxidative, or starvation damage. Can J Microbiol 56:539–547

Shu M, Browngardt CM, Chen YY, Burne RA (2003) Role of urease enzymes in stability of a 10-species oral biofilm consortium cultivated in a constant-depth film fermenter. Infect Immun 71:7188–7192

Shu M, Morou-Bermudez E, Suarez-Perez E, Rivera-Miranda C, Browngardt CM, Chen YY, Magnusson I, Burne RA (2007) The relationship between dental caries status and dental plaque urease activity. Oral Microbiol Immunol 22:61–66

Small PL, Waterman SR (1998) Acid stress, anaerobiosis and gadCB: lessons from *Lactococcus lactis* and *Escherichia coli*. Trends Microbiol 6:214–216

Smith AJ, Quivey RG Jr, Faustoferri RC (1996) Cloning and nucleotide sequence analysis of the *Streptococcus mutans* membrane-bound, proton-translocating ATPase operon. Gene 183:87–96

Sturr MG, Marquis RE (1992) Comparative acid tolerances and inhibitor sensitivities of isolated F-ATPases of oral lactic acid bacteria. Appl Environ Microbiol 58:2287–2291

Svensater G, Sjogreen B, Hamilton IR (2000) Multiple stress responses in *Streptococcus mutans* and the induction of general and stress-specific proteins. Microbiology 146(Pt 1):107–117

Svensater G, Welin J, Wilkins JC, Beighton D, Hamilton IR (2001) Protein expression by planktonic and biofilm cells of *Streptococcus mutans*. FEMS Microbiol Lett 205:139–146

Tonon T, Lonvaud-Funel A (2000) Metabolism of arginine and its positive effect on growth and revival of *Oenococcus oeni*. J Appl Microbiol 89:526–531

Toro E, Nascimento MM, Suarez-Perez E, Burne RA, Elias-Boneta A, Morou-Bermudez E (2010) The effect of sucrose on plaque and saliva urease levels *in vivo*. Arch Oral Biol 55:249–254

Traxler MF, Summers SM, Nguyen HT, Zacharia VM, Hightower GA, Smith JT, Conway T (2008) The global, ppGpp-mediated stringent response to amino acid starvation in *Escherichia coli*. Mol Microbiol 68:1128–1148

Vrancken G, Rimaux T, Weckx S, De Vuyst L, Leroy F (2009a) Environmental pH determines citrulline and ornithine release through the arginine deiminase pathway in *Lactobacillus fermentum* IMDO 130101. Int J Food Microbiol 135:216–222

Vrancken G, Rimaux T, Wouters D, Leroy F, De Vuyst L (2009b) The arginine deiminase pathway of *Lactobacillus fermentum* IMDO 130101 responds to growth under stress conditions of both temperature and salt. Food Microbiol 26:720–727

Walker DC, Girgis HS, Klaenhammer TR (1999) The *gro*ESL chaperone operon of *Lactobacillus johnsonii*. Appl Environ Microbiol 65:3033–3041

Walker JE, Saraste M, Gay NJ (1984) The *unc* operon. Nucleotide sequence, regulation and structure of ATP-synthase. Biochim Biophys Acta 768:164–200

Wall T, Båth K, Britton RA, Jonsson H, Versalovic J, Roos S (2007) The early response to acid shock in *Lactobacillus reuteri* involves the ClpL chaperone and a putative cell wall-altering esterase. Appl Environ Microbiol 73:3924–3935

Wang Z, Gerstein M, Snyder M (2009) RNA-Seq: a revolutionary tool for transcriptomics. Nat Rev Genet 10:57–63

Welin J, Wilkins JC, Beighton D, Svensater G (2004) Protein expression by *Streptococcus mutans* during initial stage of biofilm formation. Appl Environ Microbiol 70:3736–3741

Welin-Neilands J, Svensäter G (2007) Acid tolerance of biofilm cells of *Streptococcus mutans*. Appl Environ Microbiol 73:5633–5638

Wilkins JC, Homer KA, Beighton D (2001) Altered protein expression of *Streptococcus oralis* cultured at low pH revealed by two-dimensional gel electrophoresis. Appl Environ Microbiol 67:3396–3405

Xie Y, Chou L-S, Cutler A, Weimer B (2004) DNA macroarray profiling of *Lactococcus lactis* subsp. *lactis* IL1403 gene expression during environmental stresses. Appl Environ Microbiol 70:6738–6747

Zhang J, Fu R-Y, Hugenholtz J, Li Y, Chen J (2007) Glutathione protects *Lactococcus lactis* against acid stress. Appl Environ Microbiol 73:5268–5275

Zhang J, Banerjee A, Biswas I (2009) Transcription of *clp*P is enhanced by a unique tandem repeat sequence in *Streptococcus mutans*. J Bacteriol 191:1056–1065

Zhu Q, Quivey RG Jr, Berger AJ (2004) Measurement of bacterial concentration fractions in polymicrobial mixtures by Raman microspectroscopy. J Biomed Opt 9:1182–1186

Zhu Q, Quivey RG Jr, Berger AJ (2007) Raman spectroscopic measurement of relative concentrations in mixtures of oral bacteria. Appl Spectrosc 61:1233–1237

Zuniga M, Miralles Md Mdel C, Perez-Martinez G (2002) The Product of *arc*R, the sixth gene of the *arc* operon of *Lactobacillus sakei*, is essential for expression of the arginine deiminase pathway. Appl Environ Microbiol 68:6051–6058

Chapter 3
Responses of Lactic Acid Bacteria to Heat Stress

Pekka Varmanen and Kirsi Savijoki

3.1 Introduction

Lactic acid bacteria (LAB), like other free-living microbes, are often exposed to environmental stress conditions, which include limitations in nutrient supply, sudden changes in osmolarity, and up- or downshifts in temperature. Changes in temperature are probably the most common stress with which bacteria and other organisms are confronted in the natural world. Living cells from all kingdoms respond to a sudden increase in temperature by rapid changes in gene expression resulting in elevated levels of a set of proteins called *heat-shock proteins* (HSPs). Under normal conditions, HSPs assist in protein folding, assembly, transport, and degradation, and under stress these functions become especially important. The heat-shock response is one of the best-characterized physiological responses of the cell. The primary structure of most HSPs appears to be highly conserved during evolution, indicating that the HSP function has been conserved among diverse organisms. The unfolding and subsequent aggregation of denatured proteins are the hallmark cellular consequences of heat shock. Thus, it is not surprising that the two most common classes of HSPs are molecular chaperones and energy-dependent proteases (Georgopoulos and Welch 1993; Gottesman et al. 1997; Hendrick and Hartl 1993).

Despite the conservation of HSPs in different bacteria, the heat-shock response illustrates the amazing diversity of bacterial gene regulation, and recent studies have revealed a variety of *hsp* gene expression regulatory mechanisms in diverse bacteria including LAB. In bacteria, the regulation of gene expression occurs primarily in response to environmental changes and at the level of transcription. The model

P. Varmanen (✉)
Department of Food and Environmental Sciences, University of Helsinki,
P.O. Box 66, 00014 Helsinki, Finland
e-mail: pekka.varmanen@helsinki.fi

E. Tsakalidou and K. Papadimitriou (eds.), *Stress Responses of Lactic Acid Bacteria*,
Food Microbiology and Food Safety, DOI 10.1007/978-0-387-92771-8_3,

organisms *Escherichia coli* and *Bacillus subtilis* have served as paradigms of bacterial stress-response regulation. Intensive research that started long before the "genomic era" has already resulted in detailed molecular-level information regarding the regulation of *hsp* gene expression in these species (Narberhaus 1999).

3.2 Physiological Roles of Heat-Shock Proteins

The major HSPs, which include the classical chaperones DnaK, GroEL, and GroES, as well as the Clp family of proteins, play an indispensable role in protein quality control in both stressed and unstressed bacteria. Many HSPs are expressed in significant amounts in cells maintained under normal growth conditions and are essential for cellular growth at all physiologically relevant temperatures. The DnaK and GroEL proteins, for example, participate in protein folding, protein translocation, and possibly higher-order protein assembly (Georgopoulos and Welch 1993). The importance of the DnaK chaperone complex for growth and heat-shock-response regulation in *Lactococcus lactis* has been shown by the construction and physiological characterization of *dnaK* mutant strains (Koch et al. 1998).

The three-dimensional structure of cellular proteins is sensitive to small increases in temperature, and the misfolding and subsequent aggregation of unfolded proteins are the primary problems for cells subjected to heat stress. To combat the deleterious effects of protein denaturation, the cell induces the synthesis of chaperones and proteases to either refold or degrade damaged proteins (Georgopoulos and Welch 1993). In the last decade, results have explicitly demonstrated that ClpP proteolytic complexes play an essential role in disposing heat-damaged proteins in Gram-positive bacteria, including LAB (Frees et al. 2007). In fact, *L. lactis* was among the first organisms where this activity was observed. For many bacteria, including *L. lactis*, the role of ClpP-mediated proteolysis was first demonstrated indirectly by showing that bacterial mutants lacking ClpP were restricted for growth at high temperatures (Frees and Ingmer 1999; Frees et al. 2007). More direct evidence was provided by the observation that the inactivation of ClpP decreased the in vivo degradation of nonnative model substrates (puromycyl peptides) to less than 30% of the wild-type level (Frees and Ingmer 1999). As many stress conditions lead to protein denaturation, ClpP may generally dispose of the cells' damaged proteins during stress, explaining why the absence of ClpP hampers growth under many different conditions in most Gram-positive bacteria (Frees et al. 2007).

To recognize and degrade nonnative proteins, ClpP associates with a ClpATPase partner. In LAB, the partners of ClpP remain to be identified. However, the current knowledge suggests that ClpC is the principal ClpATPase responsible for recruiting ClpP to degrade nonnative proteins in many Gram-positive bacteria (Krüger et al. 2000; Frees et al. 2004, 2007; Kock et al. 2004). In Bacillus, a *clpX* mutant was reported to be heat-sensitive, and accordingly ClpX proved to be important (however, less important than ClpC) for the degradation of nonnative proteins in vivo

(Gerth et al. 1998; Krüger et al. 2000). In lactococci and streptococci, ClpX is essential, but the molecular basis for this observation is unknown (Robertson et al. 2003; Savijoki, unpublished results).

3.3 Heat-Shock Regulation

3.3.1 Regulation in the Model Organisms

Using calibrated Western analysis, Gerth et al. (2004) estimated that a cell in an exponentially growing *B. subtilis* culture contains approximately 1,200 ClpP tetradecamers, 1,400 ClpX hexamers, 250 ClpC hexamers, and 100 ClpE hexamers. These amounts were calculated to correspond to 0.44%, 0.5, 0.4, and 0.02% of the total soluble protein for ClpP, ClpX, ClpC, and ClpE, respectively (Gerth et al. 2004). It was calculated that in response to heat shock, the numbers increase maximally up to 2,500 ClpP tetradecamers, 1,000 ClpC hexamers, and 450 ClpE hexamers, whereas the number of hexameric ClpX does not increase. In *E. coli*, the expression of *clpP*, *clpB*, and *clpX* as well as all other major HSP encoding genes are controlled by the general HSP sigma factor σ^{32}. A single regulon arrangement of heat-shock genes appears to be the exception rather than the rule, and in many bacteria (including *B. subtilis*) heat-shock genes are part of several regulons. For this bacterium, at least five different heat-shock regulons can be distinguished: Class I genes, encoding the classical chaperones DnaK, GroES and GroEL, are controlled by the HrcA repressor, which recognizes the highly conserved CIRCE (controlling inverted repeat of chaperone expression) operator sequence (TTAGCACTC-N9-GAGTGCTAA); class II genes encode general stress proteins, and their expression is dependent on the σ^B sigma factor; class III heat-shock genes are regulated by CtsR (Class three stress gene Repressor), which recognizes a tandem heptanucleotide direct repeat (A/GGTCAAANANA/GGTCAAA); class IV genes are those that are not controlled by HrcA, σ^B, or CtsR; and class V genes are regulated by the two-component system CssRS (Darmon et al. 2002; Schumann et al. 2002). The tendency of HrcA to aggregate in vivo and in vitro is considered to be a fundamental aspect of its autoregulation (Mogk et al. 1997; Wilson et al. 2005). According to the present model, HrcA is released from ribosomes in an inactive form. To become an active repressor able to bind to its target operator sequences, it must interact with the GroEL chaperone. GroEL is titrated by nonnative proteins arising as a result of heat shock, driving the equilibrium toward inactive HrcA (Mogk et al. 1997; Wilson et al. 2005).

In *B. subtilis*, the *clpP* and *clpC* operons are preceded by two functional promoters, one recognized by the vegetative RNA polymerase σ^A and the other by the alternative sigma factor σ^B (Krüger et al. 1996; Gerth et al. 1998). The regulation of the stress-inducible *clpP* and *clpC* operons as well as *clpE* depends predominantly on the transcriptional repressor CtsR, which recognizes a heptanucleotide repeat A/GGTCAAA/T located on the σ^A promoter regions of target genes (Krüger and

Hecker 1998; Derré et al. 1999). Interestingly, the stability of CtsR appears to be modulated by three distinct Clp proteases. Under nonstress conditions, CtsR is a substrate for ClpXP (Derré et al. 2000), while following heat shock, a fine-tuned degradation of CtsR is mediated by the sequential activity of ClpEP and ClpCP (Krüger et al. 2001; Miethke et al. 2006). In addition, all gene products of the *clpC* operon, which consists of *ctsR*, *mcsA*, *mcsB*, and *clpC*, play a regulatory role in controlling the expression of class III genes. Detailed studies have revealed the function of McsA as an activator of McsB, which, by its kinase activity, phosphorylates CtsR and inhibits its DNA-binding activity (Krüger et al. 2001; Kirstein et al. 2005). ClpC, on the other hand, is an inhibitor of the kinase activity of McsB, and McsA is necessary for the heat-shock–induced degradation of CtsR (Kirstein et al. 2005). In the present model, McsA, McsB, and ClpC under nonstress conditions form a ternary complex where the kinase activity of McsB is inhibited (Kirstein et al. 2005; Kirstein and Turgay 2005). When denatured proteins accumulate under heat stress, ClpC binds to unfolded proteins and is titrated away from the ternary complex, and McsA becomes capable of stimulating the kinase activity of McsB (Kirstein and Turgay 2005). As a result, a new ternary complex consisting of phosphorylated McsA, McsB, and CtsR is formed. The formation of this complex and the phosphorylation of CtsR prevent CtsR from binding to its target promoters, leading to derepression of the CtsR regulon (Kirstein et al. 2005; Miethke et al. 2006). More details of this model are discussed by Kirstein and Turgay (2005).

In addition to regulation at the level of transcriptional initiation, cellular levels of Clp proteins are fine-tuned by changes in the stability of transcripts and proteins (Gerth et al. 2004). Another mechanism of controlling Clp protein activity has been shown to depend on the adaptor protein MecA, which is crucial for the major chaperone activity of ClpC (Schlothauer et al. 2003). The adaptor-controlled oligomerization of ClpC is considered to represent a new mechanism of regulating AAA+ (ATPases associated with diverse cellular activities) protein activity (Kirstein et al. 2006). Of all the *clp* genes studied, the regulation of the *clpX* transcription is the least well known. It has, however, been shown that the heat-shock–induced transcription of *clpX* in *B. subtilis* is driven from a σ^A-dependent promoter, and unlike the other *clp* genes, it is not a member of the CtsR regulon (Gerth et al. 2004).

3.3.2 Regulation of the Heat-Shock Response in Lactic Acid Bacteria

Until recently, relatively little was known about the molecular mechanisms regulating the stress response and the expression of *clp* genes in Gram-positive bacteria other than *B. subtilis*. Derré et al. (1999) reported the presence of CtsR-encoding genes and conserved binding sites in the upstream regions of several *clp* genes in a wide range of Gram-positive bacteria including LAB. Later, CtsR was characterized and shown to repress *clp* gene expression in *L. lactis* (Varmanen et al. 2000), *Oenococcus oeni* (Grandvalet et al. 2005), *Streptococcus thermophilus* (Zotta et al.

2009), and *Lactobacillus plantarum* (Fiocco et al. 2010). In *Streptococcus salivarius*, *clpP* is under dual regulation by CtsR and HrcA (Chastanet and Msadek 2003), and sequence analyses indicate that in several members of *Streptococcaceae* (including *L. lactis*), the HrcA and CtsR regulons partially overlap (Chastanet and Msadek 2003; Chastanet et al. 2003; Grandvalet et al. 2005). A conserved CIRCE element was identified overlapping the promoter region of *clpL* in *Lactobacillus rhamnosus,* indicating HrcA-dependent regulation (Suokko et al. 2005). To our knowledge, the attempts to inactivate *hrcA* in LAB or other bacteria have not been successful, which has hampered functional studies. Furthermore, HrcA proteins are somewhat difficult to purify in biologically active form because they tend to aggregate, hindering in vitro analyses of HrcA DNA-binding activity and specificity. A His-tagged HrcA protein of *Lactobacillus gasseri* was recently overexpressed in *E. coli* and purified in nonnative form from inclusion bodies under denaturing conditions (Suokko et al. 2008). This 6xHis-HrcA protein was reactivated by removing the denaturing agent (8 M urea) with extensive dialysis in the presence of nonspecific DNA. This protein was used in an electrophoretic mobility shift assay, which revealed a direct interaction between HrcA and the promoter of the *clpL* gene. Intriguingly, in *O. oeni* the widely conserved CIRCE-HrcA regulatory circuit appears to be absent, and the CtsR regulon contains genes encoding the classical chaperone proteins DnaK, GroEL, and GroES, in addition to the *clp* genes (Grandvalet et al. 2005). Genome sequences indicate a different arrangement in *Lactobacillus bulgaricus*, *Lactobacillus johnsonii,* and *Lactobacillus acidophilus*, which appear to be devoid of CtsR, and the *clp* genes are most probably regulated by HrcA in these bacteria (van de Guchte et al. 2006). Fiocco et al. (2009) reported recently that the *L. plantarum ftsH* gene, encoding a membrane-bound metalloprotease, is a novel member of the CtsR regulon. Further characterization of the CtsR regulon in *L. plantarum* revealed a direct role of CtsR in the regulation of several *clp* genes as well as *hsp1* (encoding one of the three small HSP20 family HSPs) (Fiocco et al. 2010). Thus, while the expression of the classical chaperones and the Clp proteins is quite commonly regulated by a repressor protein (either HrcA or CtsR), there appears to be limited conservation in the organization of these regulons among *firmicutes*.

At present, the mechanisms regulating the activity of CtsR and/or HrcA, which control *clp* gene expression, are poorly characterized in bacteria other than *B. subtilis*. In *L. lactis*, both the CtsR and the HrcA regulons are upregulated under conditions where misfolded proteins accumulate (Frees et al. 2001). ClpP-mediated proteolysis appears not to be a decisive step in the deactivation of CtsR, because genes of the CtsR regulon are derepressed in *Staphylococcus aureus*, *Streptococcus pneumoniae,* and *L. lactis* strains lacking ClpP (Frees and Ingmer 1999; Frees et al. 2001, 2004; Robertson et al. 2002). The MscA and MscB proteins and the *ctsR-mcsA-mcsB-clpC* operon structure are conserved in Gram-positive bacteria, including *Listeria monocytogenes*, *Bacillus anthracis*, *Clostridium acetobutylicum*, *Clostridium difficile*, and *S. aureus* (Krüger et al. 2001). Thus, it is tempting to speculate that in these bacteria, the products of the operon are involved in regulating the function and deactivation of CtsR in response to the accumulation of misfolded proteins. On the other hand, genome mining of the bacterial sequences available at http://www.ncbi.nlm.nih.gov/BLAST/

indicates that members of the *Lactobacillales* order, including *Enterococcus*, *Lactobacillus*, *Lactococcus*, *Leuconostoc*, *Streptococcus*, and *Oenococcus*, are devoid of McsA and MscB and that the modulation of CtsR activity is thus arranged differently in these genera. The *clpE* mutant derivatives of distantly related *L. lactis* and *B. subtilis* have strikingly similar phenotypes, including the prolonged derepression of CtsR-dependent gene expression after heat shock (Varmanen et al. 2003; Miethke et al. 2006), indicating a conserved role of ClpEP in the regulation of CtsR stability. Proteomic analysis of an *rgg* mutant strain of *Streptococcus pyogenes* revealed increased expression of ClpL, ClpP, and ClpE, suggesting a role for Rgg in regulating the synthesis or stability of these proteins (Chaussee et al. 2004). Furthermore, the inactivation of *covR*, encoding another regulator that coordinates virulence factor expression in *S. pyogenes*, was reported to increase the amount of *clpL* and *clpE* transcripts in cells at stationary phase (Graham et al. 2002).

3.4 Heat-Shock Responses in *Bifidobacteria* and Other *Actinobacteria*

Bifidobacteria are often considered to be members of LAB, although this classification is not unanimously accepted. *Bifidobacteria* belong to *Actinobacteria* (high G+C Gram-positive bacteria) and are not related to other LAB. As expected, there are considerable differences in the heat-shock response and its regulation between these groups of bacteria. Analogously to other cells, *Bifidobacteria* induce a specific set of proteins in response to heat shock that protect the cell from the deleterious effects of unfolded and/or misfolded proteins. As in other bacteria, several of these protective proteins act as molecular chaperones or as proteases (Zomer et al. 2009). Recent analysis of the *Bifidobacterium breve* UCC2003 genome sequence revealed the presence of genes encoding the chaperones GroEL, GroES, DnaK, GrpE, DnaJ1, and DnaJ2, as well as the *clp*-family genes *clpB*, *clpC*, *clpP1*, and *clpP2* (Ventura et al. 2006). Three heat-shock–related transcriptional regulators have been identified from *B. breve*: HrcA, HspR, and ClgR (Ventura et al. 2005a, b, c). In addition to *B. breve*, the transcriptional activator ClgR (*clp* gene regulator) has also been studied in other *Actinobacteria*, including *Corynebacterium glutamicum* (Engels et al. 2004) and *Streptomyces lividans* (Bellier and Mazodier 2004). It appears that in *Actinobacteria*, ClgR plays an analogous role to that of CtsR in low-G+C Gram-positive bacteria. In *C. glutanicum*, the ClgR regulon includes *clpC*, *clpP1*, and *clpP2*, as well as genes predicted to be involved in the repair of DNA damage after UV exposure (Engels et al. 2005). In this organism, the stability of ClgR increases in *clpC* and *clpP1P2* mutant strains, indicating ClpCP-dependent degradation (Engels et al. 2004, 2005). In *B. breve*, the ClgR-regulated expression of ClpC and ClpP1P2 is not induced under severe heat stress but under moderate heat-shock conditions (Ventura et al. 2005a). Among *Actinobacteria*, the regulatory mechanisms involved in *clp* gene expression have been most extensively investigated in *S. lividans*. In this species, five *clp* genes have been identified, and they are organized as one monocistronic transcriptional unit (*clpP5*) and two bicistronic operons (*clpP1P2* and *clpP3P4*). None of the *clp* genes are heat-induced in this organism; however, in *B. breve*, the expression of *clpB* is controlled by the repressor HspR, which was shown to bind the promoter region of *clpB* (Ventura et al. 2005a, b, c). The HspR regulon in *B. breve* has been reported to include *dnaK*, *grpE*, *dnaJ1*, *hspR*, *clgR*, *Bbr_1182*, *recA*, and *recX* (Zomer et al. 2009). With the aid of transcriptomics, DNA–protein interaction studies and reporter fusions, Zomer et al. (2009) shed light on the interactive regulatory mechanisms controlling stress responses in *B. breve*. According to this study, HspR controls the SOS response and the ClgR regulon, which in turn regulates and is regulated by HrcA in this member of *Actinobacteria* (Zomer et al. 2009).

3.5 HtrA, FtsH, sHSPs, RecA, and Other Important Factors

In addition to the major HSPs such as the classical chaperones DnaK, GroEL, and GroES, and the Clp family of proteins, several other players are involved in bacterial adaptation to sudden rises in temperature. This was demonstrated by DNA microarray analysis of the metabolic genes of *L. lactis* subsp. *lactis* IL1403, which revealed that almost 10% of the 375 genes analyzed were upregulated following heat shock (Xie et al. 2004). While the heat-shock–induced expression of central metabolic genes of LAB is likely to affect, for example, fermentation processes and flavor compound generation in cheese (Xie et al. 2004), the role of metabolic genes in heat tolerance is not clear yet. The genes in LAB with established roles in the heat-shock response include *htrA*, which encodes a housekeeping surface protease also known as DegP. HtrA is a widely conserved protein that combines chaperone and proteolytic activities in a single protein. In *L. lactis*, an *htrA* mutant strain showed a thermosensitive phenotype and HtrA appears to play a fundamental role in the degradation of abnormal exported proteins (Poquet et al. 2000). The heat-shock induction of *htrA* has also been shown in *Lactobacillus helveticus* (Smeds et al. 1998) and *Bifidobacterium longum* (Savijoki et al. 2005). However, the regulatory mechanism behind the temperature-dependent expression of *htrA* in LAB and *Bifidobacteria* remains to be elucidated.

FtsH is a membrane-bound HSP with dual chaperone–protease activity. The FtsH protein includes transmembrane segments located in the N-terminal part and a main cytosolic region consisting of an AAA^+ protein and a Zn^{2+} metalloprotease (Ito and Akiyama 2005). The functional role of FtsH in protein quality control under heat shock has been established in various bacteria, including *L. lactis* (Nilsson et al. 1994) and *L. plantarum* (Fiocco et al. 2009). By contrast, much less is known about the regulatory mechanisms controlling *ftsH* expression in Gram-positive bacteria, and with the exception of *L. plantarum* (Fiocco et al. 2009), they remain to be studied in LAB also.

The members of the small heat-shock protein (sHSP) family function as holding chaperones that assist the protein-folding process by stabilizing unfolded or partially folded proteins without actively promoting their remodeling (Han et al. 2008). The sHSP family is widely conserved, and members have been identified and characterized in LAB and *Bifidobacteria*, including *L. acidophilus* (Altermann et al. 2005), *L. bulgaricus* (van de Guchte et al. 2006), *L. johnsonii* (Pridmore et al. 2004), *L. plantarum* (Kleerebezem et al. 2003). and *B. breve* (Ventura et al. 2007).

The RecA protein plays a pivotal role in homologous recombination, DNA repair, SOS repair, and mutagenesis and is well conserved in LAB. However, in species such as *L. lactis*, RecA seems to affect the expression of genes associated with heat and oxygen stress (Duwat et al. 1995a, b). Consistent with this result, a link between the heat shock and DNA-damage responses in *L. lactis* was drawn from the findings that disruption of the *L. lactis recA* gene results in temperature sensitivity (Duwat et al. 1995a, b) and that this sensitivity can be suppressed by disruption of a gene, *trmA* (*spxA2*) (Duwat et al. 1999), that also suppresses the temperature sensitivity of a *clpP* mutation (Frees et al. 2001). SpxA2 was shown to negatively control proteolysis in *L. lactis* (Frees et al. 2001). In *B. subtilis*, Spx was shown to function as a regulator through interference with the activator-stimulated transcription that seems to be controlled by the Clp-mediated proteolysis of Spx (Nakano et al. 2002, 2003). Furthermore, the specific DNA-damage response regulated by RecA-LexA seems to be substituted by a RecA-HdiR- (heat shock and DNA damage-induced regulator) regulated SOS response that has been shown to coordinate cells' responses to both mitomycin C and heat shock in *L. lactis* strain MG1363 (Savijoki et al. 2003). Among other LAB, the role of RecA in the adaptation to heat shock has been established for another member of *Streptococcaceae*, *S. thermophilus* (Giliberti et al. 2002). Furthermore, it appears that the RecA-HdiR system is restricted to the *Streptococcaceae* family (Savijoki et al. 2003; Varhimo et al. 2007). Other LAB, like most bacteria, are equipped with the conserved RecA-LexA system.

3.6 Concluding Remarks

When exposed to a sudden upshift in temperature causing protein denaturation, living cells increase the expression of HSPs, including both chaperones and proteases, as a defense strategy to ensure survival. While the HSP function has been conserved among diverse organisms including LAB, the heat-shock response demonstrates the remarkable diversity of bacterial gene regulation. It has become clear that in order to understand the cellular regulatory mechanisms behind the cascade of events that reprogram the gene expression and engage HSPs in decisions to repair or degrade damaged proteins in LAB, we cannot entirely rely on previous work on model organisms like *E. coli* or even *B. subtilis*. Molecular-level information on stress-response regulation in different LAB has been increasing rapidly in recent years, and it appears that even inside the group of LAB, a variety of *hsp* gene expression regulatory mechanisms are present. Thus, it is expected that the intensive research on the diverse members of LAB will still provide us with some new and exciting information on heat-shock response.

References

Altermann E, Russell WM, Azcarate-Peril MA, Barrangou R, Buck BL, McAuliffe O, Souther N, Dobson A, Duong T, Callanan M, Lick S, Hamrick A, Cano R, Klaenhammer TR (2005) Complete genome sequence of the probiotic lactic acid bacterium *Lactobacillus acidophilus* NCFM. Proc Natl Acad Sci USA 102:3906–3912

Bellier A, Mazodier P (2004) ClgR, a novel regulator of *clp* and *lon* expression in *Streptomyces*. J Bacteriol 186:3238–3248

Chastanet A, Msadek T (2003) ClpP of *Streptococcus salivarius* is a novel member of the dually regulated class of stress response genes in gram-positive bacteria. J Bacteriol 185:683–687

Chastanet A, Fert J, Msadek T (2003) Comparative genomics reveal novel heat shock regulatory mechanisms in *Staphylococcus aureus* and other Gram-positive bacteria. Mol Microbiol 47:1061–1073

Chaussee MA, Callegari EA, Chaussee MS (2004) Rgg regulates growth phase-dependent expression of proteins associated with secondary metabolism and stress in *Streptococcus pyogenes*. J Bacteriol 186:7091–7099

Darmon E, Noone D, Masson A, Bron S, Kuipers OP, Devine KM, van Dijl JM (2002) A novel class of heat and secretion stress-responsive genes is controlled by the autoregulated CssRS two-component system of *Bacillus subtilis*. J Bacteriol 184:5661–5671

Derré I, Rapoport G, Msadek T (1999) CtsR, a novel regulator of stress and heat shock response, controls *clp* and molecular chaperone gene expression in Gram-positive bacteria. Mol Microbiol 31:117–132

Derré I, Rapoport G, Msadek T (2000) The CtsR regulator of stress response is active as a dimer and specifically degraded in vivo at 37°C. Mol Microbiol 38:335–347

Duwat P, Ehrlich SD, Gruss A (1995a) The *recA* gene of *Lactococcus lactis*: characterization and involvement in oxidative and thermal stress. Mol Microbiol 17:1121–1131

Duwat P, Sourice S, Ehrlich SD, Gruss A (1995b) *recA* Gene involvement in oxidative and thermal stress in *Lactococcus lactis*. Dev Biol Stand 85:455–467

Duwat P, Ehrlich SD, Gruss A (1999) Effects of metabolic flux on stress response pathways in *Lactococcus lactis*. Mol Microbiol 31:845–858

Engels S, Schweitzer JE, Ludwig C, Bott M, Schaffer S (2004) *clpC* and *clpP1P2* gene expression in *Corynebacterium glutamicum* is controlled by a regulatory network involving the transcriptional regulators ClgR and HspR as well as the ECF sigma factor sigmaH. Mol Microbiol 52:285–302

Engels S, Ludwig C, Schweitzer JE, Mack C, Bott M, Schaffer S (2005) The transcriptional activator ClgR controls transcription of genes involved in proteolysis and DNA repair in *Corynebacterium glutamicum*. Mol Microbiol 57:576–591

Fiocco D, Collins M, Muscariello L, Hols P, Kleerebezem M, Msadek T, Spano G (2009) The *Lactobacillus plantarum ftsH* gene is a novel member of the CtsR stress response regulon. J Bacteriol 191:1688–1694

Fiocco D, Capozzi V, Collins M, Gallone A, Hols P, Guzzo J, Weidmann S, Rieu A, Msadek T, Spano G (2010) Characterization of the CtsR stress response regulon in *Lactobacillus plantarum*. J Bacteriol 192:896–900

Frees D, Ingmer H (1999) ClpP participates in the degradation of misfolded protein in *Lactococcus lactis*. Mol Microbiol 31:79–87

Frees D, Varmanen P, Ingmer H (2001) Inactivation of a gene that is highly conserved in Gram-positive bacteria stimulates degradation of non-native proteins and concomitantly increases stress tolerance in *Lactococcus lactis*. Mol Microbiol 41:93–103

Frees D, Chastanet A, Qazi S, Sørensen K, Hill P, Msadek T, Ingmer H (2004) Clp ATPases are required for stress tolerance, intracellular replication and biofilm formation in *Staphylococcus aureus*. Mol Microbiol 54:1445–1462

Frees D, Savijoki K, Varmanen P, Ingmer H (2007) Clp ATPases and ClpP proteolytic complexes regulate vital biological processes in low GC, Gram-positive bacteria. Mol Microbiol 63:1285–1295

Georgopoulos C, Welch WJ (1993) Role of the major heat shock proteins as molecular chaperones. Annu Rev Cell Biol 9:601–634

Gerth U, Krüger E, Derre I, Msadek T, Hecker M (1998) Stress induction of the *Bacillus subtilis clpP* gene encoding a homologue of the proteolytic component of the Clp protease and the involvement of ClpP and ClpX in stress tolerance. Mol Microbiol 28:787–802

Gerth U, Kirstein J, Mostertz J, Waldminghaus T, Miethke M, Kock H, Hecker M (2004) Fine-tuning in regulation of Clp protein content in *Bacillus subtilis*. J Bacteriol 186:179–191

Giliberti G, Naclerio G, Martirani L, Ricca E, De Felice M (2002) Alteration of cell morphology and viability in a *recA* mutant of *Streptococcus thermophilus* upon induction of heat shock and nutrient starvation. Gene 295:1–6

Gottesman S, Wickner S, Maurizi MR (1997) Protein quality control: triage by chaperones and proteases. Genes Dev 11:815–823

Graham MR, Smoot LM, Migliaccio CA, Virtaneva K, Sturdevant DE, Porcella SF, Federle MJ, Adams GJ, Scott JR, Musser JM (2002) Virulence control in group A *Streptococcus* by a two-component gene regulatory system: global expression profiling and in vivo infection modeling. Proc Natl Acad Sci USA 99:13855–13860

Grandvalet C, Coucheney F, Beltramo C, Guzzo J (2005) CtsR is the master regulator of stress response gene expression in *Oenococcus oeni*. J Bacteriol 187:5614–5623

Han MJ, Yun H, Lee SY (2008) Microbial small heat shock proteins and their use in biotechnology. Biotechnol Adv 26:591–609

Hendrick JP, Hartl F-U (1993) Molecular chaperone functions of heat-shock proteins. Annu Rev Biochem 62:349–384

Ito K, Akiyama Y (2005) Cellular functions, mechanism of action, and regulation of FtsH protease. Annu Rev Microbiol 59:211–231

Kirstein J, Turgay K (2005) A new tyrosine phosphorylation mechanism involved in signal transduction in *Bacillus subtilis*. J Mol Microbiol Biotechnol 9:182–188

Kirstein J, Zuhlke D, Gerth U, Turgay K, Hecker M (2005) A tyrosine kinase and its activator control the activity of the CtsR heat shock repressor in *B. subtilis*. EMBO J 24:3435–3445

Kirstein J, Schlothauer T, Dougan DA, Lilie H, Tischendorf G, Mogk A, Bukau B, Turgay K (2006) Adaptor protein controlled oligomerization activates the AAA+protein ClpC. EMBO J 25:1481–1491

Kleerebezem M, Boekhorst J, van Kranenburg R, Molenaar D, Kuipers OP, Leer R, Tarchini R, Peters SA, Sandbrink HM, Fiers MW, Stiekema W, Lankhorst RM, Bron PA, Hoffer SM, Groot MN, Kerkhoven R, de Vries M, Ursing B, de Vos WM, Siezen RJ (2003) Complete genome sequence of *Lactobacillus plantarum* WCFS1. Proc Natl Acad Sci USA 100:1990–1995

Koch B, Kilstrup M, Vogensen FK, Hammer K (1998) Induced levels of heat shock proteins in a *dnaK* mutant of *Lactococcus lactis*. J Bacteriol 180:3873–3881

Kock H, Gerth U, Hecker M (2004) The ClpP peptidase is the major determinant of bulk protein turnover in *Bacillus subtilis*. J Bacteriol 186:5856–5864

Krüger E, Hecker M (1998) The first gene of the *Bacillus subtilis clpC* operon, *ctsR*, encodes a negative regulator of its own operon and other class III heat shock genes. J Bacteriol 180:6681–6688

Krüger E, Msadek T, Hecker M (1996) Alternate promoters direct stress-induced transcription of the *Bacillus subtilis clpC* operon. Mol Microbiol 20:713–723

Krüger E, Witt E, Ohlmeier S, Hanschke R, Hecker M (2000) The clp proteases of *Bacillus subtilis* are directly involved in degradation of misfolded proteins. J Bacteriol 182:3259–3265

Krüger E, Zühlke D, Witt E, Ludwig H, Hecker M (2001) Clp-mediated proteolysis in Gram-positive bacteria is autoregulated by the stability of a repressor. EMBO J 20:852–863

Miethke M, Hecker M, Gerth U (2006) Involvement of *Bacillus subtilis* ClpE in CtsR degradation and protein quality control. J Bacteriol 188:4610–4619

Mogk A, Homuth G, Scholz C, Kim L, Schmid FX, Schumann W (1997) The GroE chaperonin machine is a major modulator of the CIRCE heat shock regulon of *Bacillus subtilis*. EMBO J 16:4579–4590

Nakano S, Zheng G, Nakano MM, Zuber P (2002) Multiple pathways of Spx (YjbD) proteolysis in *Bacillus subtilis*. J Bacteriol 184:3664–3670

Nakano S, Nakano MM, Zhang Y, Leelakriangsak M, Zuber P (2003) A regulatory protein that interferes with activator-stimulated transcription in bacteria. Proc Natl Acad Sci USA 100:4233–4238

Narberhaus F (1999) Negative regulation of bacterial heat shock genes. Mol Microbiol 31:1–8

Nilsson D, Lauridsen AA, Tomoyasu T, Ogura T (1994) A *Lactococcus lactis* gene encodes a membrane protein with putative ATPase activity that is homologous to the essential *Escherichia coli ftsH* gene product. Microbiology 140:2601–2610

Poquet I, Saint V, Seznec E, Simoes N, Bolotin A, Gruss A (2000) HtrA is the unique surface housekeeping protease in *Lactococcus lactis* and is required for natural protein processing. Mol Microbiol 35:1042–1051

Pridmore RD, Berger B, Desiere F, Vilanova D, Barretto C, Pittet AC, Zwahlen MC, Rouvet M, Altermann E, Barrangou R, Mollet B, Mercenier A, Klaenhammer T, Arigoni F, Schell MA (2004) The genome sequence of the probiotic intestinal bacterium *Lactobacillus johnsonii* NCC 533. Proc Natl Acad Sci USA 101:2512–2517

Robertson GT, Ng WL, Foley J, Gilmour R, Winkler ME (2002) Global transcriptional analysis of clpP mutations of type 2 *Streptococcus pneumoniae* and their effects on physiology and virulence. J Bacteriol 184:3508–3520

Robertson GT, Ng WL, Gilmour R, Winkler ME (2003) Essentiality of *clpX,* but not *clpP*, *clpL*, *clpC*, or *clpE*, in *Streptococcus pneumoniae* R6. J Bacteriol 185:2961–2966

Savijoki K, Ingmer H, Frees D, Vogensen FK, Palva A, Varmanen P (2003) Heat and DNA damage induction of the LexA-like regulator HdiR from *Lactococcus lactis* is mediated by RecA and ClpP. Mol Microbiol 50:609–621

Savijoki K, Suokko A, Palva A, Valmu L, Kalkkinen N, Varmanen P (2005) Effect of heat-shock and bile salts on protein synthesis of *Bifidobacterium longum* revealed by [^{35}S]methionine labelling and two-dimensional gel electrophoresis. FEMS Microbiol Lett 248:207–215

Schumann W, Hecker M, Msadek T (2002) Regulation and function of heat-inducible genes in *Bacillus subtilis*. In: Sonenshein AL, Hoch JA, Losick R (Eds.), *Bacillus subtilis* and its closest relatives: From genes to cells. ASM Press, Washington, DC, pp. 359–368

Schlothauer T, Mogk A, Dougan DA, Bukau B, Turgay K (2003) MecA, an adaptor protein necessary for ClpC chaperone activity. Proc Natl Acad Sci USA 100:2306–2311

Smeds A, Varmanen P, Palva A (1998) Molecular characterization of a stress-inducible gene from *Lactobacillus helveticus*. J Bacteriol 180:6148–6153

Suokko A, Savijoki K, Malinen E, Palva A, Varmanen P (2005) Characterization of a mobile *clpL* gene from *Lactobacillus rhamnosus*. Appl Environ Microbiol 71:2061–2069

Suokko A, Poutanen M, Savijoki K, Kalkkinen N, Varmanen P (2008) ClpL is essential for induction of thermotolerance and is potentially part of the HrcA regulon in *Lactobacillus gasseri*. Proteomics 8:1029–1041

van de Guchte M, Penaud S, Grimaldi C, Barbe V, Bryson K, Nicolas P, Robert C, Oztas S, Mangenot S, Couloux A, Loux V, Dervyn R, Bossy R, Bolotin A, Batto JM, Walunas T, Gibrat JF, Bessieres P, Weissenbach J, Ehrlich SD, Maguin E (2006) The complete genome sequence of *Lactobacillus bulgaricus* reveals extensive and ongoing reductive evolution. Proc Natl Acad Sci USA 103:9274–9279

Varhimo E, Savijoki K, Jalava J, Kuipers OP, Varmanen P (2007) Identification of a novel streptococcal gene cassette mediating SOS mutagenesis in *Streptococcus uberis*. J Bacteriol 189:5210–5222

Varmanen P, Ingmer H, Vogensen FK (2000) *ctsR* of *Lactococcus lactis* encodes a negative regulator of *clp* gene expression. Microbiology 146:1447–1455

Varmanen P, Vogensen FK, Hammer K, Palva A, Ingmer H (2003) ClpE from *Lactococcus lactis* promotes repression of CtsR-dependent gene expression. J Bacteriol 185:5117–5124

Ventura M, Zhang Z, Cronin M, Canchaya C, Kenny JG, Fitzgerald GF, van Sinderen D (2005a) The ClgR protein regulates transcription of the *clpP* operon in *Bifidobacterium breve* UCC 2003. J Bacteriol 187:8411–8426

Ventura M, Fitzgerald GF, van Sinderen D (2005b) Genetic and transcriptional organization of the *clpC* locus in *Bifidobacterium breve* UCC 2003. Appl Environ Microbiol 71:6282–6291

Ventura M, Kenny JG, Zhang Z, Fitzgerald GF, van Sinderen D (2005c) The *clpB* gene of *Bifidobacterium breve* UCC 2003: transcriptional analysis and first insights into stress induction. Microbiology 151:2861–2872

Ventura M, Canchaya C, Zhang Z, Bernini V, Fitzgerald GF, van Sinderen D (2006) How high G+C Gram-positive bacteria and in particular *Bifidobacteria* cope with heat stress: protein players and regulators. FEMS Microbiol Rev 30:734–759

Ventura M, Canchaya C, Zhang Z, Fitzgerald GF, van Sinderen D (2007) Molecular characterization of *hsp20*, encoding a small heat shock protein of *Bifidobacterium breve* UCC2003. Appl Environ Microbiol 73:4695–703

Wilson AC, Wu CC, Yates JR III, Tan M (2005) Chlamydial GroEL autoregulates its own expression through direct interactions with the HrcA repressor protein. J Bacteriol 187:7535–7342

Xie Y, Chou LS, Cutler A, Weimer B (2004) DNA Macroarray profiling of *Lactococcus lactis* subsp. *lactis* IL1403 gene expression during environmental stresses. Appl Environ Microbiol 70:6738–6747

Zomer A, Fernandez M, Kearney B, Fitzgerald GF, Ventura M, van Sinderen D (2009) An interactive regulatory network controls stress response in *Bifidobacterium breve* UCC2003. J Bacteriol 191:7039–7049

Zotta T, Asterinou K, Rossano R, Ricciardi A, Varcamonti M, Parente E (2009) Effect of inactivation of stress response regulators on the growth and survival of *Streptococcus thermophilus* Sfi39. Int J Food Microbiol 129:211–220

Chapter 4
Responses of Lactic Acid Bacteria to Osmotic Stress

Claire Le Marrec

4.1 Introduction

Lactic acid bacteria (LAB) are an important group of Gram-positive bacteria that are extensively exploited in various food production industries. Their usefulness also extends to other various biotechnical applications, including the production of antimicrobial substances, enzymes, ethanol, and lactic acid, and other newer applications (e.g., probiotic food production, live vaccines). LAB are widespread in nature, being resident in animals, plants, and fruits. LAB are also normal inhabitants of the oral cavity and the digestive tract in humans. Some species/strains are opportunistic pathogens and have been reported to be associated with an overwhelming number of human infections.

In industrial processes, during human infection, and in nature, LAB are frequently exposed to adverse environmental conditions. Among the challenges posed by an ever-changing environment, osmotic stress is a prominent constraint that can produce a decrease in growth rate or survival and affect metabolic activities. In natural ecosystems, LAB must cope with changes in water availability as a result of rain, flooding, or dehydration. In an industrial environment, osmotic constraint represents one of the major stresses encountered by LAB during cheese production and ripening, meat fermentation, and the yogurt-making process. Salt, an osmotically active agent, can reach concentrations of up to 2.8% in certain types of cheese. Some food products can achieve higher salt concentrations. Hence, *Tetragenococcus halophila*, used for brewing of Japanese sauces, faces concentrations of up to 18%.

Starter cultures are increasingly being used today in concentrated forms for direct inoculation to the food matrix. The formulation and preservation of these cultures

C. Le Marrec (✉)
Ecole Nationale Supérieure de Chimie Biologie Physique, Institut Polytechnique de Bordeaux, Unité de Recherche Œnologie - EA 4577 - USC 1219 INRA., Institut des Sciences de la Vigne et du Vin, 210 Chemin de Leysotte, CS50008 Villenave d'Ornon, France
e-mail: clehenaff@enscbp.fr

E. Tsakalidou and K. Papadimitriou (eds.), *Stress Responses of Lactic Acid Bacteria*, Food Microbiology and Food Safety, DOI 10.1007/978-0-387-92771-8_4,

impose specific constraints, including the osmotic imbalance occurring during freezing and dehydration. As a consequence, the need to increase the viability and stability of LAB in frozen, freeze-dried, and air-dried forms has driven significant research in both the dairy and food industries during recent years. The relief of osmotic stress is also of general importance to probiotic bacteria. The continuing expansion of interest in these bacteria has led to an increase in manufactured functional foods and medicines containing LAB. Strains that belong to certain species of *Lactobacillus* are extensively used in yogurts, dietary adjuncts, and other health-related products. Probiotics also include strains of the genus *Bifidobacterium*. Although phylogenetically unrelated to the other LAB, the genus is often included in this group on the grounds of similarities in its biochemistry, physiology, and ecology. Given their intestinal origin, these sensitive bacteria face enormous challenges to be in a highly viable state, not only throughout processing and storage, but also during gastrointestinal transit to the site of action in the human gut. To survive and proliferate within the gastrointestinal tract, probiotics must tolerate several environmental hurdles, including the elevated osmolarity in the upper small intestine. Osmotic stress is also prominent during human infection by pathogenic LAB and is provoked by the release of perspiration in skin infections.

Developing adaptive strategies to cope with osmotic stress is therefore essential for LAB to pose their functional characteristics in food fermentation or to deal with human defense systems. The study of osmoregulation – the adaptation of cells to changes in the external osmotic pressure – is thus central for our basic understanding of their important industrial and medical aspects.

4.2 Variations of Turgor: Accumulation and Release of Osmotically Active Solutes

4.2.1 Fundamental Principles

Bacterial cells accumulate solutes in their cytoplasm to concentrations far higher than those required for the metabolism of the cell to ensure that the direction of water flow during growth is into the cell. Consequently, all growing bacterial cells exhibit a high, outwardly directed turgor pressure that places the membrane in close proximity to the expanding peptidoglycan wall. The maintenance of constant positive turgor is generally considered the driving force for cell expansion, growth, and division. In general, Gram-positive bacteria maintain a higher turgor pressure (approximately 20 bars) than Gram-negative strains (3–5 bars) (Whatmore and Reed 1990; Ingraham and Marr 1996; Csonka and Epstein 1996).

Changes in extracellular water activity (the amount of water available to react) have direct consequences on the water activity of the cytoplasm and immediately trigger fluxes of water along the osmotic gradient. Bacterial cell envelopes are generally considered to exhibit a high permeability to water but form an effective barrier for most solutes. By diffusion over the semipermeable cell membrane, water can enter and leave the cell until equilibrium is established between internal and external osmotic concentrations. Osmotic concentrations are typically expressed as either milliosmoles/kilogram (mOsm/kg) of solvent, referred to as osmolality, or milliosmoles/liter (mOsm/L) of solution, referred to as osmolarity. Water flow can result either in swelling and bursting of the cell in hypotonic environments or in loss of turgor, plasmolysis, and dehydration under hypertonic conditions. A much accelerated water transit is achieved by diffusion through water-selective channels embedded in the membrane, the so-called aquaporins. Such channels mediate water fluxes in both directions in response to sudden osmotic up- or downshifts. Aquaporins belong to the ubiquitous major intrinsic protein (MIP) family of transporters. This family also includes glycerol facilitators and aquaglyceroporins, which allow the passage of several small molecules, including glycerol and other polyols, dihydroxyacetone, CO_2, urea, and ammonium. The functional characterization of Gla_{Llac} from *Lactococcus lactis* has demonstrated a role in both glycerol and water efflux (Froger et al. 2001). The synthesis of aquaporins has been shown to be induced after the transfer of certain bacteria to hyperosmotic media, suggesting a role in osmotic shock protection. The recent completion of the genome sequencing of various LAB has enabled the identification of putative aquaporins (Lorca et al. 2007). Their physiological role in osmoregulation remains to be fully explored.

Instead of a passive volume regulation, microorganisms have developed efficient and rapid countermeasures to avoid experiencing detrimental conditions. Bacteria control turgor by actively modulating the pool of osmotically active solutes in their cytoplasm, thereby allowing the water content to be adjusted by osmosis. The mechanisms directly involved in the recovery of turgor represent one of the most studied aspects of the response of LAB to osmotic stress. In addition to their contribution to the osmotic balance with the extracellular environment, these solutes enhance the stability of enzymes and preserve the integrity of biological membranes (Welsh 2000).

4.2.2 Growth and Survival Under Hyperosmotic Stress

The survival and growth of LAB upon exposure to osmotic stress has been generally examined in media where the water activity was decreased by the addition of salt. The physiological response to suboptimal salt concentrations has been investigated in various lactobacilli, *L. lactis*, *T. halophila*, *Pediococcus pentosaceus*, *Oenococcus oeni*, and the opportunistic pathogen *Enterococcus faecalis*. *T. halophila*, formerly known as *Pediococcus halophilus*, shows the greatest tolerance. Based on 16S rDNA sequence studies, this moderately halophilic bacterium shows a close phylogenetic relationship to enterococci and lactobacilli. Unlike these genera, the bacterium can tolerate high salt concentrations (up to 3.2 M NaCl) in MRS medium (Robert et al. 2000). In the same rich medium, the detrimental concentration of NaCl inhibiting growth is 2.5 M for *P. pentosaceus*, 1.5 M for *Lactobacillus plantarum*, *Carnobacterium piscicola*, and *Leuconostoc mesenteroides* (Baliarda et al. 2003a), and 1.4 M for *E. faecalis* in BHI (Pichereau et al. 1999). Other LAB are less tolerant of NaCl. A concentration of 0.4 M NaCl was shown to reduce the growth rate of *L. lactis* to 25–50% compared with the unstressed rate (Kilstrup et al. 1997). Hutkins et al. (1987) observed that different strains of *Lactobacillus acidophilus* and *Lactobacillus bulgaricus* could tolerate no more than 0.3–0.6 M NaCl.

During industrial use, LAB are also likely to experience osmotic stress with sugars (Sunny-Roberts and Knorr 2008). Glaasker et al. (1996a, 1998a) compared the effects on the growth of *L. plantarum* by raising the medium's osmolarity with high concentrations of salts and isoosmotic concentrations of lactose and sucrose. Hyperosmotic conditions imposed by sugar stress were much less detrimental and were only transient, because cells were able to equilibrate the extracellular and intracellular concentrations of lactose and sucrose. Similar observations have also been reported in *Lactobacillus rhamnosus* and *L. lactis* (Prasad et al. 2003; Molina-Höppner et al. 2004). The uptake of sugars was suggested to occur by facilitated diffusion via system(s) with a very low affinity for the substrates. On the minute time scale, the sugars caused osmotic stress. However, on a longer time scale, the external and internal sugars equilibrated. Consequently, growth inhibition did not occur at medium osmolalities at which equiosmolar salt concentrations were already inhibitory.

The response to an abrupt increase in external osmolality (or upshock) has been largely explored for many LAB. In particular, tremendous efforts have been made in recent years in order to understand the mechanisms leading to cell inactivation during the drastic variations in osmotic pressure encountered due to drying or freezing of LAB, both processes being similar with regard to low-water stress (Santivarangkna et al. 2008a). During dehydration, the rate at which water is removed from cells is a determinant of cell viability (Poirier et al. 1997; Mille et al. 2004). Cell death occurring after a rapid osmotic shock elicited by glycerol was not linearly related to stress intensity as observed for *L. bulgaricus* and *L. plantarum*. Cell viability collapsed at a critical threshold of osmotic pressure, differing according to the species. Hence, under similar conditions, *L. bulgaricus* displays a higher sensitivity to dehydration compared to *L. plantarum* (Mille et al. 2005).

As observed for other challenges, most studies on the osmotic stress response of LAB have been conducted with one or a few strains in a given species. If there is a body of proofs that osmotolerance is species-specific, additional intraspecies characteristics have to be considered. As an example, unexpected levels of halotolerance have been reported for *L. acidophilus* IFO3532 and *L. lactis* subsp. *lactis* isolates obtained from the intestinal tract of fish (Hutkins et al. 1987; Itoi et al. 2008). In the *L. lactis* species, strains from the subsp. *cremoris* are more sensitive to osmotic stress than subsp. *lactis* strains (Obis et al. 2001).

4.2.3 Strategies to Restore Turgor

The first response in *Escherichia coli* and other Gram-negative bacteria is a massive uptake of K^+ (and its counter-ion glutamate). This mechanism is followed by a dramatic increase in the cytoplasmic concentration (either by synthesis and/or by uptake) of neutral osmoprotective compounds, representing the secondary response. These compounds can be accumulated to molar levels without negative effects and are compatible with macromolar structure and function, the so-called compatible solutes. Perturbating inorganic ions are therefore replaced by compatible osmolytes, glycine betaine (GB, *N,N,N*-trimethylammonioacetate) being one of the most widely accumulated osmolytes in nature, as well as the most growth-stimulating osmoprotectant for bacteria. Such biphasic response is not observed among LAB. Even though potassium is essential for growth in media of varied salinities, it is not considered a key molecule for the osmotic adaptation of *L. plantarum* and *T. halophila* since (1) the apparent intracellular K^+ concentration did not increase in correlation with increasing external NaCl concentrations and (2) the accumulation of GB did not significantly affect the steady-state intracellular concentrations of potassium (Glaasker et al. 1996b, 1998b; Poirier et al. 1998; Robert et al. 2000). Thus, potassium plays a relatively small role in the achievement of osmotic balance in LAB, and the intracellular amassing of organic osmolytes is considered so far to be the core of the hyperosmotic stress response in this family. LAB use dedicated transporters to scavenge these molecules, and transport is tightly regulated by the osmolarity of the medium.

4.2.4 Osmoprotectants Identified in Lactic Acid Bacteria

HPLC and natural-abundance (13)C-nuclear magnetic resonance have been widely used to probe and identify the intracellular solute content that might be accumulated in salt-stressed cultures. When grown in hypertonic-rich medium, GB was significantly accumulated in the cytoplasm of *L. acidophilus* (Hutkins et al. 1987), *L. plantarum* (Kets and de Bont 1994), *E. faecalis* (Pichereau et al. 1999), *T. halophila* (Robert et al. 2000), and *P. pentosaceus* (Baliarda et al. 2003a). Peaks of carnitine (CA) were also identified in *L. plantarum* and *T. halophila*. LAB are polyauxotrophic bacteria, routinely cultivated in media supplemented with complex nutrients that potentially contain osmoprotective compounds: GB, choline, and CA. Clear interpretations of their role and mechanisms of accumulation (synthesis/uptake) were obtained by growing cells in hypertonic chemically defined media (CDM), lacking compounds known to act as bacterial osmoprotectants. Large differences in bacterial osmotolerance levels in rich and defined media were observed. As an example, in MRS medium and CDM, *T. halophila* could tolerate up to 3.2 and 1.6 M NaCl, respectively (Robert et al. 2000). In addition, organic osmolytes (such as GB and CA) could not be detected in cytoplasmic pools of cells cultured in hypertonic CDM, unless the molecules were provided in the medium (Robert et al. 2000). These observations have supported the general idea that LAB have limited biosynthetic capacities and rely on the presence of osmoprotectants in the surroundings.

The accumulation of various exogenously provided compounds has been shown to trigger growth restoration and also to increase the limit of tolerance to hyperosmolarity of LAB in CDM (Table 4.1). Bacteria can use several compounds,

Table 4.1 Presence (+) or absence (–) of growth-stimulatory properties (increased growth rate and maximal OD) of different osmoprotectants on LAB at high osmolarity

	(NaCl and corresponding reduction of growth yield)				
Additives	*P. pentosaceus* ATCC 33316[a] (0.8 M=80%)	*T. halophila* ATCC 33315[a, b] (2 M=70%)	*L. lactis*[c, d] (0.4 M=70%)	*L. plantarum*[a, e, f, g] (0.4 M=58%)[a]	*E. faecalis*[h] (0.75 M=90%)
L-Carnitine	+	+[b]	+[c], –[d]	+[e]	+
Choline	+	+[b]	–[c, d]	+[f], –[g]	nd
DMSA	+	+[a]	+[c]	nd	+
DMSP	–	+[a]	+[c]	nd	+
Ectoine	–	+[a]	–[d]	–[a]	–
Glycine betaine	+	+[b]	+[c]	+[e, f]	+
Pipecolic acid	–	–[a]	–	nd	–
Proline	+	–[a]	–[c], +[d]	+[f]	–
Sarcosine	–	–[a]	nd	–[e]	nd
Taurine	–	–[a]	–[d]	nd	nd

DMSA dimethylsulfonioacetate; *DMSP* dimethylsulfoniopropionate; *nd* not determined
[a]Baliarda et al. (2003b) (*L. plantarum* NCIMB8826)
[b]Robert et al. (2000)
[c]Uguen et al. (1999) (*L. lactis* ADRIA 85 LO 30)
[d]Obis et al. (1999) (*L. lactis* NCDO763)
[e]Kets et al. (1994)
[f]Kets et al. (1997) (*L. plantarum* P743)
[g]Glaasker et al. (1996b) (*L. plantarum* ATCC14917)
[h]Pichereau et al. (1999)

and as observed in other families, the wider spectrum is exhibited by the most osmotolerant LAB, *T. halophila*. Quaternary ammonium compounds are preferred, with GB being the most effective compound in reducing the inhibitory effects of NaCl. A noticeable exception is *O. oeni*, since no exogenously provided compounds of this family could restore growth of the bacterium under hyperosmotic conditions (Le Marrec et al. 2007). The GB structural analogs 2-dimethylsulfonioacetate (DMSA) and 3-dimethylsulfoniopropionate (DMSP) were also highly osmoprotective, while CA protected LAB to a lesser extent. No stereospecific effect of the compound was observed since both D- and L-CA configurations showed similar results (Kets et al. 1994; Pichereau et al. 1999; Baliarda et al. 2003a). The least efficient quaternary ammonium compound for *T. halophila* is choline (Robert et al. 2000), which is also protective for *P. pentosaceus* (Baliarda et al. 2003a). Depending on which strain was used, choline has also been reported to have osmoprotective effects in *L. plantarum* (Kets et al. 1997). In this model, choline was demonstrated to be osmoprotective per se. In contrast, the molecule had to be further converted into betaine to display osmoprotective properties in *T. halophila*. The conversion occurred under aerobic growth conditions, and the choline–GB pathway was suggested to be activated and/or induced by elevated salinity (Robert et al. 2000). No effect of choline on the growth parameters of two *L. lactis* strains grown under hyperosmotic stress has been reported (Obis et al. 1999; Uguen et al. 1999).

Apart from the uptake of quaternary compounds, osmotic stress involves qualitative and quantitative modifications in the intracellular pools of amino acids in LAB, which are widely influenced by the composition of the growth medium. The intracellular concentrations of a few amino acids were specifically affected under conditions of osmotic imbalance in CDM: proline, glutamate, and alanine in *L. plantarum* (Glaasker et al. 1998b); proline, glutamate and aspartate in *L. lactis* and *T. halophila* (Molenaar et al. 1993; Robert et al. 2000). Exogenously provided proline was observed to be efficient in the growth recovery of *P. pentosaceus* and *L. lactis* in hypertonic CDM, and its accumulation increased with osmolarity (Baliarda et al. 2003a; Obis et al. 1999). Growth inhibition by the osmotic stress was relieved by the addition of exogenous aspartate in *O. oeni*, and not by alanine (Le Marrec et al. 2007). The real extent of the contribution of amino acids in osmoregulation has been assessed in *L. plantarum*. The accumulation of amino acids under osmotic stress by strain ATCC14917 was reduced or suppressed in the presence of trimethyl ammonium compounds. In the case of glutamate, transport was observed to be regulated at the level of protein synthesis. As this process is slow, glutamate was not considered a crucial actor in the response of *L. plantarum*. In addition, compared to GB, the increase in amino acids could only partly compensate for the increase in external osmolarity, diminishing the role of amino acids in osmoregulation in this bacterium. A role of an amino acid derivative as a potent osmoprotectant has been unambiguously demonstrated in *T. halophila*. The tetrahydropyrimidine ectoine, one of the most widely synthetized osmoprotectants in the microbial world, is scavenged by *T. halophila* from the medium for osmoregulatory purposes. Unlike other moderate halophilic bacteria, *T. halophila* is not able to produce this potent osmoprotectant (Robert et al. 2000).

Di- and tripeptides have also been suggested to act as osmoprotectants in LAB. Growth stimulation in hyperosmotic medium has been reported in *Lactobacillus zeae* (Piuri et al. 2003) and *O. oeni* (Le Marrec et al. 2007). Due to the fact that osmotic stress reduces DNA supercoiling in bacteria, Piuri et al. (2003) observed that the presence of small peptides resulted in the reestablishment of plasmid DNA supercoiling in hyperosmotic medium, indicating that small peptides were used for osmotic purposes in *L. zeae*. The cell envelope-associated proteinase PrtP and the PepX peptidase also increased activity in high-osmolarity medium, suggesting a contribution of the whole proteolytic system peptide supply in the osmotic adaptation of *L. zeae* to high osmolarity.

As observed for other Gram-positive bacteria, unstressed LAB tend to scavenge osmotically active solutes in standard rich and minimal laboratory media (Glaasker et al. 1996b; Robert et al. 2000). In contrast, Gram-negative bacteria do not amass such solutes unless they are subjected to hyperosmotic conditions. This difference may be related to the fact that Gram-positive bacteria need to maintain a higher turgor than Gram-negative strains.

4.2.5 Sources of Osmoprotectants

Osmoprotectants efficient among LAB have a wide distribution in natural environments and food products. Betaine is present at high concentrations in meat, bovine milk whey, sugar beets, and other foods of plant origin. Typically, concentrations range from 0.3 to 0.5 nmol/mg fresh weight in meats, which is sufficient for it to act as a potential osmoprotectant. CA is predicted to be the most important osmolyte in foods of animal origin. The levels in meat range from 0.2 to 1.0 nmol/mg fresh weight, though in vegetable matter CA is present at a concentration approximately tenfold lower than this (Smith 1996). CA is also found in milk. Nonesterified (free) CA represents the major individual component of the CA pool (50–60%) in bovine milk, and the concentrations in skim milk are in the range of 120–140 nmol/g (Smith 1996). There are also likely to be substantial pools of free amino acids in many foods. Proline has been described in casein, where its content ranges from 10 to 35%. In addition, degradative enzymes that liberate peptides and amino acids from macromolecules make them available for bacterial cells.

4.3 Hyperosmotic Stress: Uptake of Osmotically Active Solutes and its Regulation

The ability of a bacterium to use a specific osmoprotectant appears highly dependent on the performance and selectivity of its osmoprotectant uptake system(s). LAB have evolved highly efficient osmodependent transport systems dedicated to solute uptake, and the knowledge reported in the literature mainly refers to the accumulation of GB by *L. plantarum* and *L. lactis*. In these bacteria, accumulation is energy-dependent and GB uptake is mediated via a single transport system. In contrast, transport studies and cross-competition experiments have suggested that the moderate halophilic bacterium *T. halophila* uses two systems. These transporters will be described in terms of substrate specificity and mechanisms of osmotic regulation.

4.3.1 Lactococcus lactis

A *L. lactis* mutant deficient in GB/proline transport has been isolated through transposon mutagenesis. Examination of the regions surrounding the transposon integration site has led to the identification of two genes, *busAA* and *busAB*, grouped in an operon, named *busA*, the major betaine uptake system in *L. lactis* (Obis et al. 1999). The designation OpuA (osmoprotectant uptake) has also been used by Bouvier et al. (2000) and van der Heide and Poolman (2000a). BusA/OpuA has a high affinity for GB (K_m = 1.7 μM). It is also responsible for the osmodependent proline transport observed in *L. lactis*. However, the affinity of the transporter for proline is low: A 1,000-fold excess of proline (50 mM) over betaine in a competition experiment did not inhibit betaine uptake (Obis et al. 1999). BusA/OpuA belongs to the ATP-binding cassette (ABC) superfamily of transporters: The *busAA*/*opuAA* gene codes for the ATP-binding protein, and *busAB*/*opuAB* endodes a 573-residue polypeptide that presents two striking features: (1) a fusion between the regions encoding the transmembrane domain and the substrate-binding domain (SBD), and (2) a swapping of the SBD subdomains when compared to the *Bacillus subtilis* betaine-binding protein OpuAC. BusAA and BusAB components are sufficient to observe a transport activity after reconstruction in a proteoliposomal system in the presence of an ATP-regenerating system in the vesicle lumen. Two molecules of ATP are hydrolyzed per molecule of GB translocated, demonstrating that in the functional state, the OpuA transporter is composed of two chimeric substrate-binding/translocator proteins and two ATPase subunits (Patzlaff et al. 2003). The two SBDs of OpuA interact in a cooperative manner in the translocation process by stimulating either the docking of the SBDs onto the translocator or the delivery of GB to the translocator (Biemans-Oldehinkel and Poolman 2003).

Obis et al. (2001) reported on the osmosensitive phenotype of *L. lactis* subsp. *cremoris* strains, distinguishing them from the *lactis* isolates. The phenotype could be associated with a very low activity of the BusA system, or a low expression, or the absence of the operon. A relaxation of selection pressure during the thorough cultivation of *cremoris* strains in milk (a rich and stable medium) might have resulted in the loss of important functions for the cell survival outside the dairy environment.

The BusA-dependent uptake of osmoprotectants is stimulated at elevated osmolarity through two additive mechanisms. When reconstituted in a proteoliposomal system, OpuA is activated by increased concentrations of luminal ions. Activation is instantaneous both in vivo and in vitro and only requires threshold levels of ionic osmolytes, irrespective of the electrolyte composition of the medium. The threshold for activation by ions depends on the lipid content of the membrane, indicating that the signal is transduced to the transporter by critical interactions of a protein domain with membrane lipids (van der Heide and Poolman 2000b; Guillot et al. 2000). By systematically varying the acyl chain lengths, the number and position of the *cis* or *trans* double bond, and the lipid headgroups, it could be shown that only the charge of the lipid headgroups influences the osmotic activation profile of the transporter. The lipid-dependent ion sensing by OpuA is mediated by a regulatory domain consisting of two cystathionine beta-synthase (CBS) domains in tandem at the C-terminal end of the ATPase subunit OpuAA. With two ATPase subunits per functional unit, a total of four CBS domains is present in OpuA. At physiologically relevant fractions of anionic lipids and low ionic strength, the transporter is inactive but can be "switched on" rapidly by increasing the ionic strength. The activation–deactivation process is entirely reversible. The C-terminal tail of the tandem CBS in OpuA is highly charged, with ten acidic and two basic residues in a stretch of 15 amino acids. This anionic tail modulates the activity of the CBS domains (Biemans-Oldehinkel et al. 2006; Mahmood et al. 2009). As a conclusion, BusA/OpuA not only acts as osmoregulator but also functions as an osmosensor for *L. lactis*.

The regulation of transport through OpuA/BusA also occurs at the level of gene expression. The *busA* operon transcription is enhanced by an osmotic upshock, resulting in increased amounts of the BusA transporter (Obis et al. 1999; Bouvier et al. 2000; Xie et al. 2004). The gene located upstream of *busA* encodes the BusR protein. This protein represses the transcription of *busA* by binding to a site overlapping the *busA* promoter. The osmotic induction of *busA* is due to the relief of the repression of the *busA* promoter by the regulator protein BusR (Romeo et al. 2003). The binding of BusR to the *busA* promoter region has been demonstrated to be affected by ionic strength in vitro (Romeo et al. 2007). Therefore, it appears that cytoplasmic ionic strength exerts a dual effect on the transport of osmoprotectant molecules in *L. lactis*, increasing both the amount and the specific activity of BusA.

Although no osmoprotective properties have been demonstrated for choline, a putative choline/GB ABC transporter named ChoQS is present in the genome of *L. lactis*. The *choS* gene is induced by osmotic upshock, while both genes are induced during acid stress (Xie et al. 2004). The spectrum and contribution of this system during osmotic stress are unknown.

4.3.2 *Lactobacillus plantarum*

The accumulation of quaternary ammonium compounds in *L. plantarum* ATCC14917 is mediated by a single transport system with a high affinity for GB (K_m = 18 μM) and CA and a low affinity for proline (K_m = 950 μM). The transport system, termed QacT (for quaternary ammonium compounds transporter), is described as a binding-protein-dependent system, and uptake is driven by ATP (Glaasker et al. 1998b). The QacT system is expressed semiconstitutively; proline, GB, and the osmolality of the medium all fail to have a large effect on the expression. Upon osmotic upshift the system is activated, and kinetic analysis has suggested a dual mode of regulation. Hence, QacT is activated through a diminished inhibition by *trans* substrate as well as through a turgor-related increase in the activity. Studies by Glaasker et al. (1998b) did not provide any indication for more than one major uptake system for quaternary ammonium compounds and proline in *L. plantarum* ATCC14917. The published genomes of both *L. plantarum* ATCC14917 and WCFS1 reveal two ABC systems: a four-component system, related to OpuC from *Listeria monocytogenes*, and a ChoQS transporter.

4.3.3 Other Lactic Acid Bacteria

Uptake systems responsible for the accumulation of GB and related molecules by *T. halophila* are not induced and are not activated, or partly activated, by elevated salinity. Evidence for a regulation by *trans* inhibition has not been provided (Robert et al. 2000). CA and GB are transported via two different uptake routes: A CA transport system takes up CA, GB, and choline, and a GB transport system takes up GB only. Accordingly, a gene coding for a permease specific for GB has been cloned in *T. halophila* through the functional complementation of an *E. coli* mutant defective in GB uptake (Baliarda et al. 2003b). The single-component transporter ButA (betaine uptake transporter) is a member of the betaine choline carnitine transporter (BCCT) family that includes a variety of previously characterized compatible solute transporters, such as OpuD from *B. subtilis* and EctP and BetP from *Corynebacterium glutamicum*. These proteins are specific for compounds containing a quaternary nitrogen atom and form a separate subfamily in the sodium/solute/symporter/superfamily (SSSS) (Jung 2002). In order to identify the second transporter present in *T. halophila*, a PCR-based technique that amplifies an internal fragment in putative *opuAA* genes was developed. This was achieved through the use of degenerate primers targeting DNA sequences specifying highly conserved amino acids motifs in the ATPase (the so-called WalkerA signature, responsible for ATP binding). An amplicon with the expected size was detected in *T. halophila* and *Tetragenococcus muriaticus*. The partially deduced protein sequence was found to be related to OpuA/BusA from *B. subtilis* and *L. lactis* (Le Marrec et al. 2007).

The recent availability of complete genome sequences in the LAB family provides a tool for the identification of potential transport systems dedicated to the uptake of quaternary ammonium compounds. One interesting point is the absence of *butA*-related sequences in complete genomes. The moderate halophilic *T. halophila* is therefore the sole LAB using a permease to scavenge osmoprotectants

(Robert et al. 2000). Lorca et al. (2007) support this observation. These authors analyzed the genomes of various Gram-positive bacteria (including nine LAB) for their complement of genes encoding transport proteins and found that SSSS was not represented in the low-G + C Gram-positive bacteria analyzed. In contrast, a number of orthologs of the ABC-encoding genes described above (*busA/opuA*, *opuC*, *choQS*) are present in the genomes of LAB. Their number and type are variable, and further investigations are required to understand how they contribute to the stress response. *Streptococcus mutans* has a *busA/opuA* operon preceded by a *busR* regulator, and a second operon related to *opuC* from *L. monocytogenes*. Both ATPase-encoding genes (*opuAA* and *opuCA*) are induced by an osmotic upshift (Abranches et al. 2006). Baliarda et al. (2003b) have suggested the existence of a common transporter for GB, DMSA, CA, choline, and proline operating in *P. pentosaceus*. The genome of strain ATCC25745 shows the presence of genes homologous to those reported in *L. plantarum*. Interestingly, a large plasmid from *Lactobacillus salivarius* has been shown to encode a ChoQS-related system, which contributed to the resistance to high salt concentrations when expressed in *L. lactis* (Flynn 2001; Fang et al. 2008). As shown in Table 4.1, taurine has not been systematically tested as a putative osmoprotectant in LAB. Strikingly, TauT family transporters specific for this compound are found in various LAB (Lorca et al. 2007). These systems may therefore, in addition to allowing utilization of taurine, provide defense against osmotic stress.

It is worth mentioning that a few LAB adapted to specific niches appear devoid of obvious compatible solute-uptake systems. This has been observed in *O. oeni*, the LAB driving the malolactic fermentation of wines. Due to its adaptation to its particular environment, this LAB may have followed a divergent evolutionary pathway, especially concerning the defense against hyperosmolarity, through loss-of-function events (Le Marrec et al. 2007).

4.3.4 Uptake of Osmotically Active Solutes Under Sugar Stress

Betaine does not confer osmotic tolerance to *L. plantarum* or *L. lactis* when the bacteria are challenged with various nonionic solutes such as sucrose, lactose, or sorbitol. When stressed with nonionic solutes, analysis of the cytoplasmic contents shows the presence of sugars (Glaasker et al. 1998a; Obis et al. 1999; Molina-Höppner et al. 2004). When present in the medium, GB is accumulated in *L. plantarum*. However, the levels of betaine are significantly lower in the presence of lactose or sucrose than with isoosmotic concentrations of salt. The initial upshock elicited by the addition of electrolytes and nonelectrolytes was demonstrated to enhance the rate of GB uptake to the same extent. However, the facilitated influx of sugars diminished the osmotic gradient in time. Thus, the osmostatic conditions were restored and the net rate of GB uptake ceased in sugar-stressed cells (Glaasker et al. 1998a).

4.4 Protection Against Extreme Turgor: Release of Osmoprotectants

Hypoosmotic shocks lead to a rapid and massive influx of water into the cell, increasing turgor and requiring the bacterium to quickly reduce its intracellular solute pool in order to avoid cell lysis. In *Lactobacillus casei* subsp. *rhamnosus*, an efflux of 90% of the accumulated proline occurred within 1 min upon osmotic downshock (Jewell and Kashket 1991). In similar conditions, an overall exit of a limited number of solutes is observed in *L. plantarum* cells, including GB, proline, and some glutamates, while the pools of the other amino acids remain unaffected (Glaasker et al. 1996b). An efflux of GB consists of a rapid initial phase and a slower second phase. Mutants with an impaired GB and proline uptake were demonstrated to behave similarly to the wild-type strain with regard to the rapid and slow components of efflux when an osmotic downshock was applied. These observations established the independence of the QacT solute importer and the efflux systems. The rapid efflux system had a half-life of less than 2 s and was unaffected by the metabolic status of the cells. The use of chloropromazine, a cationic amphipath that increases the open probability of mechanosensitive channels (MSCs), triggered a rapid GB efflux that mimicked the efflux elicited by an osmotic downshock. The rapid system was therefore established as being regulated via membrane stretch or torsion, following an on/off mechanism. MSCs open in response to the increased membrane tension effected by the rapid increase in cell volume, and the rapid release of solutes eliminates the driving force for water entry. Such channels serve therefore as safety valves for the release of solutes during hypoosmotic shock.

The slower phase of GB efflux observed in *L. plantarum* was suggested as serving to fine-tune turgor pressure and was suggested to be protein-mediated and inhibited upon osmotic upshock. It is interesting to note that the unidirectional rate of uptake through QacT was lowered upon osmotic downshock. In *L. plantarum*, osmostasis is therefore maintained through positive and negative regulation of both GB uptake and efflux (Glaasker et al. 1996a).

The contribution of MSCs in protection against extreme turgor has also been demonstrated in *L. lactis* (Folgering et al. 2005). This LAB possesses two such channel proteins, a small MSC (MscS) and a large-conductance MSC (MscL), encoded by the *mscL* and *yncB* genes, respectively. Both MscS and MscL form functional MSCs when expressed in *E. coli* or *L. lactis*. However, *L. lactis* IL1403 does not seem to possess detectable levels of functional MscS, although the *yncB* gene is transcribed. Contrary to the *busA/opuA* genes (dedicated to GB influx), the expression of the *mscL* and *yncB* genes in *L. lactis* IL1403 is not influenced by the osmolality of the medium. Membrane prepared from an *mscL* disruption *L. lactis* mutant did not show any MSC activity, irrespective of whether cells had been grown on a low- or high-osmolarity medium. In addition, the *mscL* mutant was also significantly compromised in its survival upon osmotic downshifts. MscL is therefore the primary, if not the only, MSC used by *L. lactis* through which a major portion of compatible solutes such as GB is released upon osmotic downshift. Interestingly, Folgering et al. (2005)

also observed that the glucose fermentation capacity of the *mscL L. lactis* mutant decreased by 60% after osmotic downshift. Strikingly, the decrease compared well with the observed decrease in the cell viability of the mutant strain.

The presence of *mscL* and *mscS* orthologs has been identified in the genome data of LAB (Lorca et al. 2007). Their numbers and types were observed to vary. For example, *L. casei*, *Lactobacillus delbrueckii*, and *L. cremoris* have two conductance MSC proteins, one from the MscL family and one from the MscS family. In contrast, *L. mesenteroides* lacks an MscS channel homolog.

To get a complete overview of the mechanisms involved in the regulation of hypoosmotic shocks among LAB, it is relevant here to cite an interesting study by Velamakanni et al. (2009) indicating a novel role for the LmrA multidrug ABC transporter from *L. lactis* during periods of disturbed ionic balance. The intervening action of LmrA and its homolog OmrA from *O. oeni* during an osmotic upshock with NaCl had been previously suggested by Bourdineaud et al. (2004). LmrA was demonstrated to transport NaCl by a secondary active mechanism and could mediate apparent H^+-Na^+-Cl^- symport in *L. lactis* strain NZ9000. Remarkably, LmrA activity significantly enhanced the survival of high-salt adapted lactococcal cells during downshifts in the extracellular ion concentration.

4.5 Compatible Solutes and Their Beneficiary Effects

4.5.1 Nature of the Protective Effects

Osmotically active compounds described in the LAB family serve a dual function in osmoregulating cells. In addition to their role in maintaining cell turgor, they also display significant protective effects on biomolecules in vitro during stressful conditions. In various bacteria, the accumulation of GB, CA, and proline provides cell protection against salinity and other stress factors, such as elevated temperature (Caldas et al. 1999). A DNA macroarray profiling of *L. lactis* gene expression during environmental stresses recently showed that heat shock changed the expression of the *busAA* and *busAB* genes dedicated to GB uptake (Xie et al. 2004). This was consistent with a previous observation that GB transport activity in *L. lactis* subsp. *cremoris* NCDO763 increases with a high growth temperature (Guillot et al. 2000). These findings can be attributed to the thermoprotectant function of GB (Caldas et al. 1999). The beneficiary effects of GB have recently been illustrated in probiotic LAB, where the cloning and expression of the listerial betaine-uptake system BetL were found to significantly enhance acid and salt tolerance in *Bifidobacterium breve* UCC2003 and osmo-, cryo-, chill-, and baro-tolerance in *L. salivarius* UCC118 (Sheehan et al. 2006, 2007). The accumulation of compatible solutes has also been involved in the baroprotective effect of NaCl in *L. lactis* (Molina-Höppner et al. 2004). Another important role of compatible solutes is to counteract the damaging effects of water loss due to drying. Exogenously provided GB has been shown to protect

L. plantarum during drying processes (Kets et al. 1996). Increased resistance to spray- and freeze-drying was also reported for the *L. salivarius betL*+ strain described above (Sheehan et al. 2006).

4.5.2 Stabilization of Macromolecules

Betaine, proline, and DMSP protect soluble and peripherally membrane-bound proteins and other cytoplasmic labile macromolecular structures from the potentially inhibitory effects of changes in intracellular ionic strength and water availability. Research on protein stabilization by compatible solutes has led to the development of some theories concerning solute/protein interactions reviewed by Kurz (2008). The four most outstanding among them discuss preferential interaction, water replacement, water density fractions, and osmophobic effects as the mechanisms of solute/protein interactions. The latter model predicts the existence of a thermodynamically unfavorable interaction with the protein backbone, which is therefore preferentially excluded from a compatible solute solution. This leads to stabilization of the conformation of the backbone and the prevention of protein denaturation (Liu and Bolen 1995). The mechanism is balanced by a thermodynamically favorable interaction of the osmolyte with the protein side chains. In addition to their activity as chemical chaperones that directly control protein stability, GB and proline may also indirectly regulate protein homeostasis in the cell by controlling the activity of molecular chaperones (Diamant et al. 2001). Beneficial interactions of compatible solutes with nucleic acids and protein–nucleic acid complexes also exist in bacterial cells, although they are less documented than the corresponding interactions of compatible solutes with proteins (Kurz 2008). Their existence is supported by the theory proposed by Cayley and Record (2004) suggesting that growth-stimulating effects of osmoprotectants result from their efficiency at increasing the amount of free (bulk) water in cells, decreasing the concentration of biopolymers and hence the level of crowding.

The predominantly protective roles of sugars are recognized during the preparation process of dried LAB starters (Santivarangkna et al. 2008b). Their influence on the viability of dried starter culture starts from the beginning (cultivation) to the end (storage) of the process. As seen before, some LAB react to osmotic upshock with sugars by the equilibration of intracellular and extracellular sugar concentrations. These compounds are therefore present on both sides of the membrane and are in contact with internal cytosolic proteins. The salt-induced accumulation of trehalose in *L. plantarum* (Kets et al. 1994) or of oligosaccharides in *L. rhamnosus* (Prasad et al. 2003) has been observed. In the latter case, the sugar/protein ratio was higher in stressed cells compared to standard conditions. Lower levels of glycerol and higher levels of modified oligosaccharides were also observed in osmotically stressed cells. Electrospray spectrometry analysis revealed that they corresponded to saccharides modified with glycerol. This modification was suggested to increase the number of hydroxyl groups available in the molecule for interaction with macromolecules (via hydrogen bonding) under conditions of dehydration.

4.5.3 *Effect of Betaine on the Cytoplasmic Sodium Level in T. halophila*

An interesting beneficial effect of compatible solutes has been observed in *T. halophila*, which uses betaines as osmotic as well as salt stabilizers (Robert et al. 2000). During growth under salt stress in a defined medium without osmoprotectants, *T. halophila* shows optimal growth at 0.4–0.8 M NaCl. At these osmolarities, the lower intracellular concentration of Na^+ was observed. In contrast, higher Na^+ concentrations were measured in the absence of NaCl, and above 1 M. It was therefore suggested that the bacterium was unable to control its intracellular Na^+ level on both sides of the optimal growth salinities, explaining the inhibition of growth. Interestingly, GB and CA were accumulated from the medium, not only at high salinity but also in media lacking NaCl. Over a wide range of salinities, GB and CA maintained intracellular Na^+ concentration close to the level observed in the optimal growth conditions and in the absence of betaines. The possible mechanisms of stimulation of Na^+ exclusion in the presence of betaines have not been further elucidated.

4.6 Other Cellular Reactions to Osmotic/Salt Stress

Significant advances have been made toward having a global picture of the osmostress response of LAB on a genomic and cellular scale by analyzing the proteome and the transcriptome of cells during salt stress. New osmoregulated genes have been detected, completing the information regarding the osmoadaptation strategies displayed by LAB. Interactions between osmotic and other types of stress responses have also been explored. Relevant contributions to understanding the variations in the composition of the bacterial cell envelope during hyperosmotic stress have recently been made and are supposed to represent additional osmoregulatory mechanisms.

4.6.1 *Stress Proteins*

The main proteins that are affected during salt adaptation have been characterized with the use of two-dimensional electrophoresis (2-DE). In addition to modulating the intracellular solutes during hyperosmotic conditions, LAB rapidly alter the expression of a sizable number of genes that are not directly involved in osmoprotectant uptake. However, few of these proteins are specifically induced by salt stress: twelve out of 96 in *E. faecalis* (Pichereau et al. 1999) and 10 out of 52 in *S. mutans* (Svensäter et al. 2000). Most correspond to general stress-response proteins (GSPs). Proteins of this category are cross-induced by various stress conditions and are therefore not specific to osmotic/salt stress. Hence, a high similarity in the high-hydrostatic-pressure proteome of *Lactobacillus sanfranciscensis* was found with NaCl-stressed cells, with 11 overlapping proteins (Hörmann et al. 2006). All 12 salt-induced proteins in *L. lactis* MG1363, as identified by 2-DE, were shown to belong to the group of proteins induced by heat, showing a complete overlapping between the heat and salt stress responses. GSPs induced during hyperosmotic stress include molecular chaperones and proteases, which represent a major stress-response system in LAB dedicated to protein quality control. This system is also involved in the adaptation to other stresses, including temperature, pH, alcohol, and organic acids. Transcriptional and proteomic studies have shown that the DnaK, GroEL, and GroES chaperones are engaged in the response to salt stress in *L. lactis* (Kilstrup et al. 1997; Xie et al. 2004). Hyperosmotic conditions (sucrose and NaCl) also induced the transcription of *dnaK* in *E. faecalis* (Flahaut et al. 1996). Similar observations were reported during adaptation to high-salt conditions in *T. halophila* (Fukuda et al. 2002) and various lactobacilli (Prasad et al. 2003; De Angelis and Gobbetti 2004). Other GSPs are induced in *E. faecalis* and *Lactobacillus sakei* when submitted to a salt treatment. Disruption of the corresponding genes did not result in significant modification of the cell growth in the presence of salt (Giard et al. 2000; Marceau et al. 2004). Alternatively, small heat-shock proteins as well as proteases (HtrA, FtsH) are commonly induced by multiple stresses, including osmotic pressure, and are therefore also expected to play an important role in resistance to these conditions (Smeds et al. 1998; Bourdineaud et al. 2003).

4.6.2 Metabolic Activities Under Salt Stress

Expression changes observed after a shift to high salt also aim at adjusting cellular metabolism. Using 2-DE, rapid changes in the synthesis of some glycolytic enzymes have been observed when *L. rhamnosus* is upshocked with 0.6 M NaCl (Prasad et al. 2003). The 6-phosphate-fructokinase, which catalyzes the conversion of fructose-6-phosphate into fructose-1.6-biphosphate, was underexpressed when *L. sakei* was grown in the presence of 4% NaCl, suggesting that the flux through the glycolytic pathway decreases in stressed cells (Marceau et al. 2004). The overall effect of salt stress on *L. lactis* gene expression was recently measured in microarray experiments. Salt stress (30 min, 4% NaCl) had a profound impact on a large number of cellular functions, with 68 genes being significantly up- or downregulated by hyperosmotic stress. Apart from the *busA* and *dnaK*, genes induced after salt stress treatment were involved in nucleotide salvage and glutamate and lysine biosynthesis. Genes involved in fatty acid biosynthesis, citrate and malate fermentation, and the arginine deiminase pathway were repressed. In addition, the gene expression of related proteolytic enzymes and the β-glucoside-specific PTS system were also repressed. Xie et al. (2004) also studied gene expression during heat and acid stresses and reported that a number of stress responses were common in all the stress treatments, such as the repression of four transporter genes and the induction of two nucleotide kinase genes. Lastly, osmotic as well as heat stresses were shown to induce stringent response in *L. lactis*. In particular, the differential expression of GTP pyrophosphokinase (*relA*) under osmotic stress was observed. Since ten genes had similar gene expression patterns under both heat and osmotic treatment, Xie et al. (2004) suggested that some of this regulation might belong to the general stress mechanism induced by the stringent response.

It is important to consider that the addition of ionic solutes (NaCl) in the medium is considered to provide the stresses of high osmotic strength and high concentrations of ions. Microarray experiments using isotonic concentrations of NaCl and sucrose have recently been conducted in *E. coli*. Importantly, the study demonstrated that about two thirds of the genes were similarly regulated in response to NaCl and sucrose treatments, while the remainder displayed ionic stress-specific regulation (Shabala et al. 2009). Findings in this study support the idea that different osmotic conditions induce different adaptive responses, at least on a short time scale.

4.6.3 Variations in the Cell Wall and Membrane Compositions

An important aspect of the osmoadaptation of bacteria is the variation in their envelope compositions as a function of external osmolality. Growth at high osmolality influences the lipid composition of bacterial membranes. As depicted earlier, such variations (and especially those modifying the charge of the lipid headgroups) influence the osmotic activation profile of the osmoprotectant transporter operating in *L. lactis* through lipid–protein interactions. Modifications are also likely to influence membrane permeability. *L. bulgaricus* cells grown in MRS showed a higher decrease of volume after a given exposure to an osmotic upshock elicited by NaCl, in comparison with cells grown at high osmolality in the presence of polyethylene glycol (PEG) or NaCl. Growth in the presence of PEG induced a lower change in water permeability compared to the presence of salt. The permeation response was correlated with the physical properties of the membrane. Cells grown in the presence of PEG possessed a more rigid structure compared to the growth in standard MRS, which was linked to an increase in the saturated/unsaturated ratio for the total pools of bacterial lipids (Tymczyszyn et al. 2005). A similar trend was established in *L. zeae*. However, the induced slight decrease in the fluidity of the membrane was suggested to be compensated by other variations, including an increase in cyclic fatty acids (Machado et al. 2004). In *L. lactis*, the main modification in membrane fatty acid composition in response to high osmolality was the increase of cyclopropane fatty acid, whereas the unsaturated/saturated ratio remains unchanged (Guillot et al. 2000). Modifications of the membrane properties in *L. bulgaricus* were also related to an increase in the sugar content of the whole pool of lipids (Tymczyszyn et al. 2005). The response of *L. casei* to hyperosmotic conditions did not result in significant differences in the glycolipid/phospholipid ratio (Machado et al. 2004). However, the individual glycolipid and phospholipid compositions showed some significant variations. In particular, a small increase in the glycolipids involved in the formation of the lipoteichoic acid was observed and may be related to the increased cell surface hydrophobicity observed during osmotic stress in this LAB. A significant increase in some phospholipids, including cardiolipin, was observed. This membrane component localizes in the poles and septa of different bacteria, such as *E. coli*, *B. subtilis*, and also *Streptococcus pyogenes*, and may serve to recruit proteins to these specific cellular locations. Cardiolipin-varying content is proposed to be a key element of osmotic adaptation by bacteria (Romantsov et al. 2009).

LAB also have the ability to respond to osmotic stress by altering the nature of their cell wall. Modifications triggered during growth under hypertonic conditions have major impacts during the industrial applications of LAB, as they may modify the sensitivity to lysis (Piuri et al. 2005; Koch et al. 2007). This phenomenon is essential for the development of the texture and flavor of fermented products or for the release of heterologous proteins in recombinant DNA technology. Another applied aspect involving lysis of LAB is the tolerance to lytic enzymes as a requirement for probiotic strains during oral administration. The retardation of the growth of *L. casei* in high salt could be associated with an increase in the cellular size and modifications in the cell envelope as visualized by transmission electronic microscopy (Piuri et al. 2005). In addition, stressed cells were more sensitive to antibiotics that target the peptidoglycan and to the presence of lytic enzymes as well. These modifications could be correlated with a decrease in peptidoglycan cross-linking, involving penicillin-binding proteins (Piuri et al. 2005). In *L. lactis*, a short osmotic stress (30 min with 4% NaCl) resulted in the induction of the *murF* and *murG* genes involved in peptidoglycan biosynthesis (Xie et al. 2004). These observations are also reminiscent of data by Le Breton et al. (2002) reporting the screening of a library of insertional mutants and the isolation of an *E. faecalis* NaCl-sensitive mutant. The mutation affected a gene potentially involved in peptidoglycan biosynthesis pathways.

4.7 Concluding Remarks

For a graded reaction to osmotic shifts, cells must be able to (1) sense stimuli related to osmotic stress, (2) transduce corresponding signals to systems that properly respond, (3) activate transport or enzymatic functions, or (4) change gene expression profiles. Characterizing the signal transduction pathways originating from osmotic challenge and leading to efficient cell response warrants further investigation. LAB deal with environmental stress with the lack of a battery of sigma factors that are found in other Gram-positive bacteria. Recently, two-component systems (TCSs) have been shown to play a role in stress sensing and signaling in the LAB family (El-Sharoud 2005), and their role probably expands to osmotic protection. Six TCS systems have been identified in *L. lactis* MG1363, including KinD/LlrD, which was induced when cells entered the stationary phase. Since a mutant was impaired in its ability to survive an osmotic upshift, KinD/LlrD has therefore been suggested to play a role in the response to osmotic or salt stress (O'Connell-Motherway et al. 2000). A link has also been established between the essential VicRK system from *S. mutans* and the response to osmotic stress. VicR inactivation altered the transcription of 18 genes, including the osmoprotectant transporter OpuA, which was upregulated (Liu et al. 2006).

Major contributions to our understanding of the osmostress response of LAB have concerned a few models, including *L. lactis*. We lack a comprehensive appreciation of the mechanisms that allow other LAB to contend with osmotic stress, in particular LAB having different life environments. Genomics and functional genome analyses combined with classical biochemistry will contribute to comparing the adaptive responses displayed by LAB and to understanding their diversity. Such studies will unambiguously provide an important insight into selecting and producing starters with enhanced viability and robustness during industrial processes.

Acknowledgments Dr. Marguerite Dols and Prof. Mohamed Jebbar are gratefully acknowledged for their helpful comments as well as Cate Evans for her help in preparing the manuscript.

References

Abranches J, Lemos JA, Burne RA (2006) Osmotic responses of *Streptococcus mutans* UA159. FEMS Microbiol Lett 255:240–246

Baliarda A, Robert H, Jebbar M, Blanco C, Deschamps A, Le Marrec C (2003a) Potential osmoprotectants for the lactic acid bacteria *Pediococcus pentosaceus* and *Tetragenococcus halophila*. Int J Food Microbiol 84:13–20

Baliarda A, Robert H, Jebbar M, Blanco C, Le Marrec C (2003b) Isolation and characterization of ButA, a secondary glycine betaine transport system operating in *Tetragenococcus halophila*. Curr Microbiol 47:347–351

Biemans-Oldehinkel E, Poolman B (2003) On the role of the two extracytoplasmic substrate-binding domains in the ABC transporter OpuA. EMBO J 22:5983–5993

Biemans-Oldehinkel E, Mahmood NA, Poolman B (2006) A sensor for intracellular ionic strength. Proc Natl Acad Sci USA 103:10624–10629

Bourdineaud JP, Nehmé B, Tesse S, Lonvaud-Funel A (2003) The *ftsH* gene of the wine bacterium *Oenococcus oeni* is involved in protection against environmental stress. Appl Environ Microbiol 69:2512–2520

Bourdineaud JP, Nehmé B, Tesse S, Lonvaud-Funel A (2004) A bacterial gene homologous to ABC transporters protects *Oenococcus oeni* from ethanol and other stress factors in wine. Int J Food Microbiol 92:1–14

Bouvier J, Bordes P, Romeo Y, Fourçans A, Bouvier I, Gutierrez C (2000) Characterization of OpuA, a glycine-betaine uptake system of *Lactococcus lactis*. Mol Microbiol Biotechnol 2:199–205

Caldas T, Demont-Caulet N, Ghazi A, Richarme G (1999) Thermoprotection by glycine betaine and choline. Microbiology 145:2543–2548

Cayley S, Record MT (2004) Large changes in cytoplasmic biopolymer concentration with osmolality indicate that macromolecular crowding may regulate protein-DNA interactions and growth rate in osmotically stressed *Escherichia coli* K-12. J Mol Recog 17:488–496

Csonka LN, Epstein W (1996) Osmoregulation. In: Neidhardt FC, Curtiss III R, Ingraham JL, Lin ECC, Low KB, Magasanik B, Reznikoff WS, Riley M, Schaechter M, Umbarger HE (Eds.),

Escherichia coli and *Salmonella*: cellular and molecular biology, 2nd ed. ASM Press, Washington, DC, pp.1210–1223

De Angelis M, Gobbetti M (2004) Environmental stress responses in *Lactobacillus*: a review. Proteomics 4:106–122

Diamant S, Elaihu N, Rosenthal D, Goloubinoff P (2001) Chemical chaperones regulate molecular chaperones *in vitro* and in cells under combined salt and heat stresses. J Biol Chem 276:39586–39591

El-Sharoud WM (2005) Two-component signal transduction systems as key players in stress response of lactic acid bacteria. Science Progress 88:203–228

Fang F, Flynn S, Li Y, Claesson MJ, van Pijkeren JP, Collins JK, van Sinderen D, O'Toole PW (2008) Characterization of endogenous plasmids from *Lactobacillus salivarius* UCC118. Appl Environ Microbiol 74:3216–3228

Flahaut S, Benachour A, Giard JC, Boutibonnes P, Auffray Y (1996) Defence against lethal treatments and *de novo* protein synthesis induced by NaCl in *Enterococcus faecalis* ATCC19433. Arch Microbiol 165:317–324

Flynn S (2001) Molecular characterisation of bacteriocin producing genes and plasmid encoded functions of the probiotic strain *Lactobacillus salivarius* subsp. *salivarius* UCC118. PhD thesis. University College Cork, Ireland

Folgering JH, Moe PC, Schuurman-Wolters GK, Blount P, Poolman B (2005) *Lactococcus lactis* uses MscL as its principal mechanosensitive channel. J Biol Chem 280:8784–8792

Froger A, Rolland JP, Bron P, Lagrée V, Le Cahérec F, Deschamps S, Hubert JF, Pellerin I, Thomas D, Delamarche C (2001) Functional characterization of a microbial aquaglyceroporin. Microbiology 147:1129–1135

Fukuda D, Watanabe M, Sonezaki S, Sugimoto S, Sonomoto K, Ishizaki A (2002) Molecular characterization and regulatory analysis of *dnaK* operon of halophilic lactic acid bacterium *Tetragenococcus halophila.* J Biosci Bioeng 93:388–394

Giard JC, Rincé A, Capiaux H, Auffray Y, Hartke A (2000) Inactivation of the stress- and starvation-inducible *gls24* operon has a pleiotrophic effect on cell morphology, stress sensitivity and gene expression in *Enterococcus faecalis*. J Bacteriol 182:4512–4520

Glaasker E, Konings WN, Poolman B (1996a) Glycine betaine fluxes in *Lactobacillus plantarum* during osmostasis and hyper- and hypo-osmotic shock. J Biol Chem 271:10060–10065

Glaasker E, Konings WN, Poolman B (1996b) Osmotic regulation of intracellular solute pools in *Lactobacillus plantarum*. J Bacteriol 178:575–582

Glaasker E, Tjan FS, Ter Steeg PF, Konings WN, Poolman B (1998a) Physiological response of *Lactobacillus plantarum* to salt and nonelectrolyte stress. J Bacteriol 180:4718–4723

Glaasker E, Heuberger EH, Konings WN, Poolman B (1998b) Mechanism of osmotic activation of the quaternary ammonium compound transporter (QacT) of *Lactobacillus plantarum*. J Bacteriol 180:5540–5546

Guillot A, Obis D, Mistou MY (2000) Fatty acid membrane composition and activation of glycine-betaine transport in *Lactococcus lactis* subjected to osmotic stress. Int J Food Microbiol 55:47–51

Hörmann S, Scheyhing C, Behr J, Pavlovic M, Ehrmann M, Vogel RF (2006) Comparative proteome approach to characterize the high-pressure stress response of *Lactobacillus sanfranciscensis* DSM 20451(T). Proteomics 6:1878–1885

Hutkins RW, Ellefson WL, Kashket ER (1987) Betaine transport imparts osmotolerance on a strain of *Lactobacillus acidophilus*. Appl Environ Microbiol 53:2275–2281

Ingraham JL, Marr AG (1996) Effects of temperature, pressure, pH, and osmotic stress on growth. In: Neidhardt FC, Curtiss III R, Ingraham JL, Lin ECC, Low KB, Magasanik B, Reznikoff WS, Riley M, Schaechter M, Umbarger HE (Eds.), *Escherichia coli* and *Salmonella*: cellular and molecular biology, 2nd ed. ASM Press, Washington, DC, pp. 1570–1578

Itoi S, Abe T, Washio S, Ikuno E, Kanomata Y, Sugita H (2008) Isolation of halotolerant *Lactococcus lactis* subsp. *lactis* from intestinal tract of coastal fish. Int J Food Microbiol 21:116–121

Jewell JB, Kashket ER (1991) Osmotically regulated transport of proline by *Lactobacillus acidophilus* IFO 3532. Appl Environ Microbiol 57:2829–2833

Jung H (2002) The sodium/substrate symporter family: structural and functional features. FEBS Lett 529:73–77

Kets EP, de Bont JAM (1994) Protective effects of betaine on survival of *Lactobacillus plantarum* subjected to drying. FEMS Microbiol Lett 116:251–256

Kets EP, Galinski EA, de Bont JAM (1994) Carnitine: a novel compatible solute in *Lactobacillus plantarum*. Arch Microbiol 162:243–248

Kets EP, Teunissen P, de Bont JAM (1996) Effect of compatible solutes on survival of lactic acid bacteria subjected to drying. Appl Environ Microbiol 62:259–261

Kets EP, Groot M, Galinski EA, de Bont JAM (1997) Choline and acetylcholine: novel cationic osmolytes in *Lactobacillus plantarum*. Appl Microbiol Biotechnol 48:94–98

Kilstrup M, Jacobsen S, Hammer K, Voensen FK (1997) Induction of heat shock proteins Dank, GroEL, and GroES by salt stress in *Lactococcus lactis*. Appl Environ Microbiol 63:1826–1837

Koch S, Oberson G, Eugster-Meier E, Meile L, Lacroix C (2007) Osmotic stress induced by salt increases cell yield, autolytic activity, and survival of lyophilization of *Lactobacillus delbrueckii* subsp. *lactis*. Int J Food Microbiol 117:36–42

Kurz M (2008) Compatible solute influence on nucleic acids: many questions but few answers. Saline Systems 4:6

Le Breton Y, Pichereau V, Flahaut S, Auffray Y, Rincé A (2002) Identification of new genes related to osmotic adaptation in *Enterococcus faecalis*. Sci Alim 22:87–96

Le Marrec C, Bon E, Lonvaud-Funel A (2007) Tolerance to high osmolality of the lactic acid bacterium *Oenococcus oeni* and identification of potential osmoprotectants. Int J Food Microbiol 115:335–342

Liu Y, Bolen DW (1995) The peptide backbone plays a dominant role in protein stabilization by naturally occurring osmolytes. Biochemistry 34:12884–12891

Liu M, Hanks TS, Zhang J, McClure MJ, Siemsen DW, Elser J, Quinn MT, Lei B (2006) Defects in *ex vivo* and *in vivo* growth and sensitivity to osmotic stress of group A *Streptococcus* caused by interruption of response regulator *vicR*. Microbiol 152:967–978

Lorca GL, Barabote RD, Zlotopolski V, Tran C, Winnen B, Hvorup RN, Stonestrom AJ, Nguyen E, Huang LW, Kim DS, Saier MH Jr (2007) Transport capabilities of eleven Gram-positive bacteria: comparative genomic analyses. Biochim Biophys Acta 1768:1342–1366

Machado MC, Lopez CS, Heras H, Rivas EA (2004) Osmotic response in *Lactobacillus casei* ATCC 393: biochemical and biophysical characteristics of membrane. Arch Biochem Biophys 422:61–70

Mahmood NA, Biemans-Oldehinkel E, Poolman B (2009) Engineering of ion sensing by the cystathionine beta-synthase module of the ABC transporter OpuA. J Biol Chem 284: 14368–14376

Marceau A, Zagorec M, Chaillou S, Méra T, Champomier-Vergès MC (2004) Evidence for involvement of at least six proteins in adaptation of *Lactobacillus sakei* to cold temperature and addition of NaCl. Appl Environ Microbiol 70:7260–7268

Mille Y, Obert JP, Beney L, Gervais P (2004) New drying process for lactic bacteria based on their dehydration behavior in liquid medium. Biotechnol Bioeng 88:71–76

Mille Y, Beney L, Gervais P (2005) Compared tolerance to osmotic stress in various microorganisms: towards a survival prediction test. Biotechnol Bioeng 92:479–484

Molenaar D, Hagting A, Alkema H, Driessen AJ, Konings WN (1993) Characteristics and osmoregulatory roles of uptake systems for proline and glycine betaine in *Lactococcus lactis*. J Bacteriol 175:5438–5444

Molina-Höppner A, Doster W, Vogel RF, Gänzle MG (2004) Protective effect of sucrose and sodium chloride for *Lactococcus lactis* during sublethal and lethal high-pressure treatments. Appl Environ Microbiol 70:2013–2020

Obis D, Guillot A, Gripon JC, Renault P, Bolotin A, Mistou MY (1999) Genetic and biochemical characterization of a high-affinity betaine uptake system (BusA) in *Lactococcus lactis* reveals a new functional organization within bacterial ABC transporters. J Bacteriol 181:6238–6246

Obis D, Guillot A, Mistou MY (2001) Tolerance to high osmolality of *Lactococcus lactis* subsp. *lactis* and *cremoris* is related to the activity of a betaine transport system. FEMS Microbiol Lett 202:39–44

O'Connell-Motherway M, van Sinderen D, Morel-Deville F, Fitzgerald GF, Ehrlich SD, Morel P (2000) Six putative two-component regulatory systems isolated from *Lactococcus lactis* subsp. *cremoris* MG1363. Microbiology 146:935–947

Patzlaff JS, van der Heide T, Poolman B (2003) The ATP/substrate stoichiometry of the ATP-binding cassette (ABC) transporter OpuA. J Biol Chem 278:29546–29551

Pichereau V, Bourot S, Flahaut S, Blanco C, Auffray Y, Bernard T (1999) The osmoprotectant glycine betaine inhibits salt-induced cross-tolerance towards lethal treatment in *Enterococcus faecalis*. Microbiology 145:427–435

Piuri M, Sanchez-Rivas C, Ruzal SM (2003) Adaptation to high salt in *Lactobacillus*: role of peptides and proteolytic enzymes. J Appl Microbiol 95:372–379

Piuri M, Sanchez-Rivas C, Ruzal SM (2005) Cell wall modifications during osmotic stress in *Lactobacillus casei*. J Appl Microbiol 98:84–95

Poirier I, Marechal PJ, Gervais P (1997) Effects of kinetics of water potential variation on bacterial viability. J Appl Microbiol 82:101–106

Poirier I, Marechal PJ, Gervais P (1998) *Escherichia coli* and *Lactobacillus plantarum* responses to osmotic stress. Appl Microbiol Biotechnol 50:704–709

Prasad J, McJarrow P, Gopal P (2003) Heat and osmotic stress responses of probiotic *Lactobacillus rhamnosus* HN001 (DR20) in relation to viability after drying. Appl Environ Microbiol 69:917–925

Robert H, Le Marrec C, Blanco C, Jebbar M (2000) Glycine betaine, carnitine, and choline enhance salinity tolerance and prevent the accumulation of sodium to a level inhibiting growth of *Tetragenococcus halophila*. Appl Environ Microbiol 66:509–517

Romantsov T, Guan Z, Wood JM (2009) Cardiolipin and the osmotic stress responses of bacteria. Biochim Biophys Acta 1788:2092–2100

Romeo Y, Obis D, Bouvier J, Guillot A, Fourçans A, Bouvier I, Gutierrez C, Mistou MY (2003) Osmoregulation in *Lactococcus lactis*: BusR, a transcriptional repressor of the glycine betaine uptake system BusA. Mol Microbiol 47:1135–1147

Romeo Y, Bouvier J, Gutierrez C (2007) Osmotic regulation of transcription in *Lactococcus lactis:* ionic strength-dependent binding of the BusR repressor to the *busA* promoter. FEBS Lett 581:3387–3390

Santivarangkna C, Kulozik U, Foerst P (2008a) Inactivation mechanisms of lactic acid starter cultures preserved by drying processes. J Appl Microbiol 105:1–13

Santivarangkna C, Higl B, Foerst P (2008b) Protection mechanism of sugars during stages of preparation process during dried lactic acid starter cultures. Food Microbiol 25:429–441

Shabala L, Bowman J, Brown J, Ross T, McMeekin T, Shaabala S (2009) Ion transport and osmotic adjustment in response to ionic and non-ionic osmotic. Environ Microbiol 11:137–148

Sheehan VM, Sleator RD, Fitzgerald GF, Hill C (2006) Heterologous expression of BetL, a betaine uptake system, enhances the stress tolerance of *Lactobacillus salivarius* UCC118. Appl Environ Microbiol 72:2170–2177

Sheehan VM, Sleator RD, Hill C, Fitzgerald GF (2007) Improving gastric transit, gastrointestinal persistence and therapeutic efficacy of the probiotic strain *Bifidobacterium breve* UCC2003. Microbiology 153:3563–3571

Smeds A, Varmanen P, Palva A (1998) Molecular characterization of a stress-inducible gene from *Lactobacillus helveticus*. J Bacteriol 180:6148–6153

Smith LT (1996) Role of osmolytes in adaptation of osmotically stressed and chill-stressed *Listeria monocytogenes* grown in liquid media and on processed meat surfaces. Appl Environ Microbiol 62:3088–3093

Sunny-Roberts EO, Knorr D (2008) Evaluation of the response of *Lactobacillus rhamnosus* VTT E-97800 to sucrose-induced osmotic stress. Food Microbiol 25:183–189

Svensäter G, Sjögreen B, Hamilton IR (2000) Multiple stress responses in *Streptococcus mutans* and the induction of general and stress-specific proteins. Microbiology 146:107–117

Tymczyszyn EE, Gómez-Zavaglia A, Disalvo EA (2005) Influence of the growth at high osmolality on the lipid composition, water permeability and osmotic response of *Lactobacillus bulgaricus*. Arch Biochem Biophys 443:66–73

Uguen P, Hamelin, J, Le Pennec JP, Blanco C (1999) Influence of osmolarity and the presence of an osmoprotectant on *Lactococcus lactis* growth and bacteriocin production. Appl Environ Microbiol 65:291–293

van Der Heide T, Poolman B (2000a) Glycine betaine transport in *Lactococcus lactis* is osmotically regulated at the level of expression and translocation activity. J Bacteriol 182:203–206

van der Heide T, Poolman B (2000b) Osmoregulated ABC-transport system of *Lactococcus lactis* senses water stress *via* changes in the physical state of the membrane. Proc Natl Acad Sci USA 97:7102–7106

Velamakanni S, Lau CH, Gutmann DA, Venter H, Barrera NP, Seeger MA, Woebking B, Matak-Vinkovic D, Balakrishnan L, Yao Y, U EC, Shilling RA, Robinson CV, Thorn P, van Veen HW (2009) A multidrug ABC transporter with a taste for salt. PLoS One 4:6137

Welsh DT (2000) Ecological significance of compatible solute accumulation by micro-organisms: from single cells to global climate. FEMS Microbiol Rev 24:263–290

Whatmore AM, Reed RH (1990) Determination of turgor pressure in *Bacillus subtilis*: possible role for K+in turgor regulation. J Gen Microbiol 136:2521–2526

Xie Y, Chou LS, Cutler A, Weimer B (2004) DNA macroarray profiling of *Lactococcus lactis* subsp. *lactis* IL1403 gene expression during environmental stresses. Appl Environ Microbiol 70:6738–6747

Chapter 5
Responses of Lactic Acid Bacteria to Cold Stress

Vittorio Capozzi, Daniela Fiocco, and Giuseppe Spano

5.1 Introduction

Lactic acid bacteria (LAB) constitute a heterogeneous group of bacteria that are traditionally used to produce fermented foods. The industrialization of food transformations increased the economical importance of LAB because they play a crucial role in the development of the organoleptic and hygienic quality of fermented products. However, the strains selected for industrial purpose should tolerate the adverse conditions encountered in industrial processes, either during starter handling and storage (cryoconservation, freeze-drying, or spray-drying) or throughout food processing. Indeed, during the industrial production of starter cultures, bacteria are subjected to different types of stress, which result in a loss of viability and metabolic activity, membrane damage, and morphological cell changes (Stead and Park 2000; Ouvry et al. 2002; Dumont et al. 2004; Saarela et al. 2005). These stresses arise from the bacterial environment (thermic, acid, nutritional, and osmotic stresses) or from the starter production process (mechanical, hypothermic, osmotic, and oxidative stresses). Among them, cold stress is of particular interest because it concerns all kinds of cells when they are subjected to a stabilization process, it is required for starter or probiotic handling and production or for culture collections.

To more effectively withstand adverse conditions, bacteria have evolved stress-sensing systems and defense mechanisms, which allow them to survive harsh conditions and sudden environmental changes. The time taken to initiate the stress response is different according to the stress type. For example, bacteria respond to heat shock

G. Spano (✉)
Department of Food Science, University of Foggia, Via Napoli 25, 71100 Foggia, Italy

Bioagromed University of Foggia, 71100 Foggia, Italy
e-mail: g.spano@unifg.it

E. Tsakalidou and K. Papadimitriou (eds.), *Stress Responses of Lactic Acid Bacteria*, Food Microbiology and Food Safety, DOI 10.1007/978-0-387-92771-8_5,

and osmotic shock quickly (in minutes) compared to cold shock (in hours) (Rosen and Ron 2002). Temperature is one of the major stresses that all living organisms have to face, and among the stress-induced responses of bacteria that have been studied, the heat-shock response has been described in great detail (van de Guchte et al. 2002; Chastanet et al. 2003; Spano and Massa 2006; Fiocco et al. 2009). Although the regulatory mechanisms of the heat-shock response are sometimes unique in LAB and not well understood yet, heat shock usually causes well-defined damage to the cells, including the unfolding or denaturation of proteins and modification of membrane fluidification. Cells respond quickly by producing a subset of proteins, heat-shock proteins (HSPs), or molecular chaperones that assist in protein folding.

In contrast, cold shock does not cause such a well-defined cellular damage. Exposure to low temperatures is a frequent event that is encountered by different bacterial species in various situations. Food-related bacteria such as LAB are particularly and repeatedly exposed to low temperatures during food handling and storage. This exposure to cold temperatures has increased over the last few decades because of the extended use of refrigeration and freezing in food preservation. It is currently appreciated that most bacterial species can respond to cold shock by the transient induction of arrays of specific proteins, termed cold-induced proteins (CIPs), and the repression of other proteins synthesized during active growth or upon exposure to other stressful conditions such as heat shock. Such a response is presumed to aid cells in overcoming the physiological stress generated by cold shock.

The negative influence of exposure to cold stress stems mainly from the physical effect of low temperature on the cell structures and enzymatic reactions.

Cold-shock response is classically exhibited when an exponentially growing culture is shifted from its optimum growth temperature to a lower temperature. In most bacteria, such as *Bacillus subtilis*, a temperature downshift causes a transient cell growth arrest, during which the general protein synthesis is severely inhibited. However, under these conditions the synthesis of CIPs is triggered. Eventually, the synthesis of these proteins decreases, cells become acclimated to the low temperature, and growth resumes (Jones et al. 1987). The effect of cold shock may be observed at multiple levels: (1) a decrease in the membrane fluidity affecting the membrane-associated functions such as active transport and protein secretion; (2) a stabilization of RNA and DNA secondary structures, leading to a reduced efficiency of mRNA translation and transcription; (3) slow or inefficient folding of some proteins; and (4) the adaptation of ribosomes to cold in order to function properly at low temperatures (Phadtare 2004). Moreover, cold shock drastically disturbs the metabolism of bacterial cells; a fine network between metabolic regulation and cold response has recently been suggested in *Lactobacillus casei* (Beaufilis et al. 2007).

The response to freezing stress is often passive and leads to a decrease in viability and metabolic activity that is associated with cryoinjury (van de Guchte et al. 2002). In contrast, cold but positive temperatures lead to two kinds of adaptive responses. First, a subset of CIPs, called cold-shock proteins (CSPs), are synthesized. Second, changes in the membrane fatty acid composition, such as an increase in unsaturated and cyclic fatty acids contents, permit a modification of the membrane fluidity (Phadtare 2004). These adaptive responses generate transient physiological modifications of the cell proteome pattern and of the membrane's properties, allowing bacterial cells to better face further stress.

The study of cold-shock response in LAB is now in the limelight because of its commercial implications. Understanding the basis of the cold-stress response provides important insight into molecular adaptation to cold and aids to improve the selection, storage, and performance of current industrial strains.

5.2 Cold Sensors

A proper response to temperature changes require a precise perception of fluctuations related to this crucial environmental physical property. Bacteria cope with cold stress by different transiently complex molecular mechanisms. This response requires elements by which the prokaryotic cell converts cold-stress physical stimuli into another kind of signal, thus representing the molecular basis of cold thermodetection and signal transduction. DNA, RNA, proteins, and lipids may all have a role in "sensing" a thermal downshift, which may influence gene expression (cold-shock stimulon) at the transcriptional, translational, or posttranslational level. The lowest-limit temperature for bacterial growth is established to be 12°C, as a consequence of the physical properties of the inner and external cellular aqueous system (Margesin et al. 2007). In contrast to cold-loving bacteria that are usually able to deal with a great reduction in temperature, the growth and metabolism of LAB are drastically influenced even when a slow decrease in temperature occurs (van de Guchte et al. 2002). Therefore, an efficient sensing of temperature decreases allows bacteria to respond quickly and effectively in order to prevent cell damage (Digel et al. 2008).

According to Rivals et al. (2007), the hypothesis of a unique cellular thermostat, at least in some LAB, should be rejected in favor of multiple cell-stress-sensing mechanisms. In the model Gram-negative and -positive bacteria, *Escherichia coli* and *B. subtilis*, respectively, the cold-shock stimulon seems to be controlled by at least three different levels of temperature perception systems: DNA, ribosome, and membrane (Weber and Marahiel 2002) or DNA, RNA, and proteins (Schumann 2009).

An important example of a DNA-based cold-sensing mechanism is represented by H-NS, a nucleoid-associated DNA-binding protein of enteric bacteria (Eriksson et al. 2002). By binding to DNA, and affecting the level of supercoiling and DNA condensation, H-NS can silence transcription. H-NS is known to be responsible for the cold repression of bacterial genes. A number of H-NS homologs have been found in gammaproteobacteria, but not in bacteria that are phylogenetically distantly related to proteobacteria, such as Gram-positive bacteria (Tendeng and Bertin 2003). A nucleoid-associated protein, Lsr2, has recently been identified as the first H-NS-like protein in Gram-positive bacteria (Gordon et al. 2008). In vitro biochemical studies showed that Lsr2 is a DNA-bridging protein, while genetic complementation experiments showed that Lsr2 is functionally homologous to H-NS (Gordon et al. 2008). This study opens new perspectives on the analysis of Gram-positive cold sensors.

Histone-like proteins were recently identified in several bacteria (Dixon-Fyle and Caro 1999; Dorman and Deighan 2003) and their essentiality for bacterial viability and normal growth has been reported (Liu et al. 2008). Genes encoding histone-like HU (heat-unstable) proteins (*hlbA* and *hlbB*) showing extensive similarity to other bacterial homologs were also identified in LAB such as *Lactobacillus delbrueckii* subsp. *bulgaricus* (Chouayekh et al. 2009).

In *B. subtilis*, an increase in the negative supercoiling of DNA has been observed after cold shock without a significant alteration of gyrase subunit *gyrA* and *gyrB* mRNA levels, even under conditions that block de novo protein synthesis (Aguilar et al. 1999). It is well recognized that DNA topology significantly modulates the relative location of −35 and −10 promoter regions. Therefore, the mechanism described in *B. subtilis* might represent a general mechanism of gene expression control upon cold shock for organisms distantly related to *B. subtilis* (Aguilar et al. 1999).

Basically, two classes of RNA sensors can be distinguished: *cis*-acting RNA elements that regulate the expression of genes located "on the same side as", and *trans*-acting small, noncoding RNAs that function by base-pairing with complementary mRNA sequences encoded elsewhere in the genome (Schumann 2009). This type of thermosensor was first discovered in Gram-negative bacteria. In *E. coli*, binding at low temperature of the small DsrA RNA to the *rpoS* leader mRNA enhances its translation (Sledjeski et al. 1996). The *cspA* gene, encoding a major CSP of *E. coli*, is efficiently transcribed at all temperatures; however, translation of the *cspA* mRNA is hampered because of its extreme instability at 37°C. In contrast, upon cold shock, *cspA* transcript becomes stabilized (Fang et al. 1997). Interestingly, the 5′ end of the *cspA* mRNA is involved in its own cold stabilization (Mitta et al. 1997), suggesting that the RNA secondary structure may be a factor of the differential temperature-dependent stabilization of mRNA. In addition, PNPase, an exoribonuclease polynucleotide phosphorylase involved in mRNA processing and degradation in bacteria (Nierlich and Murakawa 1996), has been identified as another member of the CSP family, given its strong induction at low temperature (Jones et al. 1987). The *pnp* gene was found to be autogenously regulated, during cold shock, at the level of transcriptional termination (Zangrossi et al. 2000; Schumann 2009).

Even ribosomes may sense cold and temperature shifts. Intriguingly, in *B. subtilis*, a cold shock-like response can be induced at 37°C by the addition of sublethal amounts of chloramphenicol as translation inhibitor, suggesting a direct correlation between ribosome blocking and cold-shock response (Graumann et al. 1997). This finding suggests that the translation apparatus might have a role in the temperature perception system (Weber and Marahiel 2002). Nevertheless, the biochemical nature of the sensor still remains unclear, whether it is related to intrinsic properties of the ribosome or mRNAs, or both.

There is a number of structurally and functionally diverse protein-based thermosensors, including transcriptional regulators, chemosensory proteins, chaperones, and proteases (Jin et al. 1993; Lee et al. 1993; Smirnova et al. 2008; Klinkert and Narberhaus 2009). However, the two-component regulatory systems (2CRSs), frequently used by bacteria to adapt cellular functions to changes in environmental parameters, may play an important role as thermosensors. Practically, a 2CRS transforms external signals into an internal cytoplasmic cascade that, in turn, will activate an adequate cellular response. Typical 2CRSs consist of a membrane-bound sensor histidine kinase that senses environmental stimuli and a cytoplasmic response regulator that affects gene expression upon the phosphorylation of an aspartic acid (Aguilar et al. 2001). For example, the transcription of the *des* gene, coding for a Δ5-lipid desaturase, is induced by cold and is regulated by a two-component system in *B. subtilis* (Aguilar et al. 2001). The system is composed of a sensor kinase, DesK, and a response regulator, DesR, responsible for the cold induction of the *des* gene. The histidine kinase, DesK, is able to sense low temperatures by autophosphorylation and transduction of a phosphate group to the response regulator DesR. In this case, the 2CRS indirectly affects the fatty acid desaturation level in order to maintain an adequate membrane fluidity even at low temperatures. Recently, the detection and characterization of new 2CRSs have been greatly facilitated in several LAB strains by the completion of genome sequencing. However, the influence of 2CRSs on the physiology, growth, and survival of LAB exposed to environmental stress should also be investigated under industrial conditions, including cold storage temperatures and lyophilization (El-Sharoud 2005).

The cell envelope is usually the first cellular compartment to sense changes in the external temperature. Exposure to cold stress drastically alters cytoplasmic membrane properties. Could bacteria utilize changes in membrane fluidity and constitution as a bio-thermometer? As already reported, in *B. subtilis* DesK sits embedded in the cytoplasmic membrane, it acts as sensor-kinase, and its membrane domain is the temperature-sensing element (Hunger et al. 2004). It appears that the combination of membrane physical state and protein conformation is able to sense the temperature and to translate this sensing event into proper gene expression. Thus, it is reasonable to suppose that membranes might be part of a cold-sensing complex in which the final sensor is a protein (Delmas et al. 2001; Weber and Marahiel 2002).

5.3 Cold-Stress Proteins

The most strongly induced proteins during cold stress include a family of closely related low-molecular-weight proteins termed CSPs. These proteins share a high degree of sequence identity (45%) and have been found in multiple copies (from two to nine) in many Gram-positive and Gram-negative bacteria (Graumann and Marahiel 1998, 1999; Wouters et al. 2000; Phadtare 2004). CSPs have been identified in almost all psychrotrophic, mesophilic, thermophilic, and hyperthermophilic bacteria examined so far, and their presence in *Thermotoga* and *Aquifex* indicates an ancient origin (Phadtare et al. 2003). CSPs are believed to function as transcriptional and translational regulators. They also act as molecular chaperones because they nonspecifically bind single-stranded nucleic acids and destabilize their secondary structures at low temperature (Gualerzi et al. 2003; Zeeb et al. 2006).

In general, CSPs are involved in establishing a "new" cell balance, which in turn facilitates survival at low temperature. It is postulated that the CSPs also enhance microbial tolerance to freezing due to cryoprotectant properties (Willimsky et al. 1992). Indeed, it was found that the cell viability of some bacterial strains increased after being subjected to a cold shock prior to freezing. This implies that the cold shock induces cryotolerance (Kim and Dunn 1997).

Studies on CSPs have mainly focused on those produced by meso- and thermophiles. Three-dimensional structures have been determined for four CSPs from mesophilic, thermophilic, and hyperthermophilic bacteria, namely, *E. coli Ec*-CspA (Feng et al. 1998), *B. subtilis Bs*-CspB (Schindelin et al. 1993), *Bacillus caldolyticus Bc*-Csp (Mueller et al. 2000), and *Thermotoga maritima Tm*-Csp (Kremer et al. 2001). Recently, the 3D structure of the CspA-like protein from the psychrotrophic bacterium *Psychrobacter* sp. B6 was resolved, too (Kaufman-Szymczyk et al. 2009). All of these small CSPs exhibit a very similar conformation and contain five antiparallel β-strands folded into a β-barrel structure.

These proteins also share a high sequence homology (43%) with the Y-box factors, which are a family of eukaryotic nucleic acid-binding proteins. In such proteins, the domain involved in the nucleic acid binding is referred to as the *cold-shock domain*. This domain preferentially binds to the so-called Y-box, a nucleotide sequence element found in the promoter region of mammalian major histocompatibility complex class II genes. The Y-box is characterized by the highly conserved sequence ATTGG (Phadtare 2004). This sequence has also been shown to exist in the promoter regions of at least two cold-shock genes, *hns*, encoding the nucleoid protein H-NS, and *gyrA*, encoding a subunit of DNA gyrase. It has been demonstrated that CspA binds to the ATTGG element in the promoter region of *gyrA* in *E. coli* (Panoff et al. 1998). This sequence seems to be characteristic of several cold-shock genes in *E. coli* (*recA*, *pnp*) (Jones et al. 1992a). It has also been demonstrated that CspB can bind to single-stranded DNA that contains the ATTGG element as well as the complementary CCAAT sequence (Panoff et al. 1998). Therefore, it has been suggested that CspA and CspB could act as transcriptional enhancers to cold-shock genes by recognizing the putative ATTGG sequence (Phadtare 2004).

The presence of CSPs in a cell is also determined by the stability of the proteins. The CSPs of *B. subtilis* undergo very rapid folding and unfolding transitions, and they exhibit low conformational stability in solution. These CSPs are rapidly degraded by proteases in vitro, but are protected against proteolysis by binding to RNA. Moreover, the overproduction of *E. coli* CspA leads to increased levels of three CIPs (Jones et al. 1992b), and the heterologous expression of *B. subtilis* CspB in *E. coli* results in a reduction of cellular growth and in the production of several proteins that resemble the cold-shock response (Graumann and Marahiel 1997).

For *B. subtilis* strains in which *csp* genes have been deleted, compensatory effects of the remaining CSPs have been reported (Graumann et al. 1997), and a similar response is suggested for *Lactococcus lactis* (Wouters et al. 2000). Interestingly, multiple deletion analysis showed that at least one functional CSP is required for cell viability in *B. subtilis*, indicating that CSPs play an important role not only during cold-shock adaptation but also during active growth under physiologic temperatures (Graumann et al. 1997). Several homologous copies of *csp* genes have been identified in the majority of known LAB genomes (Table 5.1). However, only a few

Table 5.1 Genes encoding cold-shock proteins as identified in some of the sequenced genomes of lactic acid bacteria (LAB)

Organism	Size (Mbp)	Growth (15°C)	No. of *csp* genes	Locus tag
Lactobacillus acidophilus NCFM	2	–	1	LBA0818
Lactobacillus brevis ATCC367	2.35	+	1	LVIS_1440
Lactobacillus casei ATCC334	2.93	+	3	LSEI_0636; LSEI_1114; LSEI_1281
Lactobacillus casei BL23	3.1	+	3	LCABL_06950; LCABL_12740; LCABL_15120
Lactobacillus delbrueckii subsp. *bulgaricus* ATCC11842	1.9	–	2	Ldb0749; Ldb2117
Lactobacillus delbrueckii subsp. *bulgaricus* ATCC BAA-365	1.9	–	2	LBUL_1957; LBUL_0682
Lactobacillus fermentum IFO3956	2.1	–	2	LAF_1542; LAF_0583
Lactobacillus gasseri ATCC33323	1.9	–	1	LGAS_1195
Lactobacillus helveticus DPC4571	2.1	–	2	lhv_0576; lhv_0863
Lactobacillus johnsonii NCC 533	2	+	1	LJ_0980b
Lactobacillus plantarum JDM1	3.2	+	3	JDM1_0028; JDM1_0826; JDM1_0954
Lactobacillus plantarum WCFS1	3.34	+	3	lp_0031; lp_0997; lp_1160
Lactobacillus reuteri DSM20016	2	–	2	Lreu_0599; Lreu_1505
Lactobacillus reuteri JCM1112	2	–	2	LAR_0579; LAR_1414
Lactobacillus rhamnosus GG	3	+	3	LGG_00608; LGG_01072; LGG_01292
Lactobacillus rhamnosus Lc705	3.06	+	3	LC705_00587; LC705_01149; LC705_01310
Lactobacillus sakei subsp. *sakei* 23K	1.9	+	4	LCA_0768; LCA_0946; LCA_1152_b; LCA_1561
Lactobacillus salivarius UCC118	2.1	–	2	LSL_1069; LSL_1480
Lactococcus lactis subsp. *cremoris* MG1363	2.5	+	6	llmg_0180; llmg_1238; llmg_1255; llmg_1256; llmg_1846; llmg_1847
Lactococcus lactis subsp. *cremoris* SK11	2.56	+	3	LACR_0174; LACR_0755; LACR_0756

(continued)

Table 5.1 (continued)

Organism	Size (Mbp)	Growth (15°C)	No. of *csp* genes	Locus tag
Lactococcus lactis subsp. *lactis* Il1403	2.4	+	2	L172505; L117090
Leuconostoc citreum KM20	1.9	+	2	LCK_01134; LCK_01552
Leuconostoc mesenteroides subsp. *mesenteroides* ATCC8293	2.04	+	2	LEUM_0703; LEUM_1878
Oenococcus oeni PSU-1	1.8	+	1	OEOE_1376
Pediococcus pentosaceus ATCC25745	1.8	+	2	PEPE_1178; PEPE_1445
Streptococcus thermophilus CNRZ1066	1.8	–	2	str0837; str0838
Streptococcus thermophilus LMD-9	1.91	–	2	STER_0879; STER_0880
Streptococcus thermophilus LMG18311	1.8	–	2	stu0837; stu0838

For each microbial species, the genome size, the number of *csp* genes, and the genome locus tags are reported

of them have been studied in detail. The *L. lactis* MG1363 chromosome was found to contain two pairs of tandemly located, cold-inducible *csp* genes (*cspA-cspB* and *cspC-cspD*), a single, constitutively expressed *cspE* gene, and a putative cold-shock gene *cspD2*. Upon cold shock of *L. lactis* IL1403 by transfer from 30 to 15°C, a tenfold induction of *cspB*-directed galactosidase activity is observed (Wouters et al. 2000). Similar cold-induced expression has been reported for *E. coli cspA*, and data revealed that the transient induction of *E. coli* CspA occurs at the level of both transcription and mRNA stabilization (Goldenberg et al. 1996; Fang et al. 1997).

Furthermore, CIPs mRNAs have been reported to be still translated under cold-shock conditions because of the presence of a so-called downstream box, which enhances the ability to form the translation initiation complex with nonadapted ribosomes at low temperatures (Mitta et al. 1997).

Three *csp* genes (*cspL*, *cspC*, and *cspP*) were recently identified in *Lactobacillus plantarum* strain NC8. Interestingly, the overproduction of each CSP causes distinct phenotypic effects on *L. plantarum* (Derzelle et al. 2002, 2003). CspL overproduction transiently alleviated the cold-shock impairment of growth, while CspP overproduction enhanced the cryotolerance. Most notably, *L. plantarum* cells contain a substantial amount of the *cspC* transcript during early exponential growth at 27°C, and CspC overproduction improves growth adaptation at optimal temperatures (Derzelle et al. 2003). Therefore, CSPs may have a significant role in fermentation carried out either at low (CspL) or optimal temperatures (CspC). A common interesting feature of the noncold-induced *csp* genes of *L. plantarum* (*cspP* and *cspC*) and *L. lactis* (*cspE*) consists of a longer 5-UTR compared to their cold-induced counterparts. This difference might account for the higher instability of the transcripts and correlate with the expression level of the *csp* genes (Wouters et al. 1999a).

5.4 Cold and Heat Stress: Talking for Survival

Although cold and heat shock are normally considered independent processes, recent evidence points to the occurrence of an interlock between them in some microorganisms. In *B. subtilis*, low temperatures induce a stress response that is characterized by the strong repression of major cellular metabolic activities, whereas only a limited number of processes essential for cold adaptation are induced. In particular, the repression of a wide range of heat-shock genes has been documented, indicating that cold- and heat-shock genes are often regulated antagonistically. Transcriptional analyses indicated that several heat-shock genes, such as *grpE*, *dnaK*, *dnaJ*, *yqeT*, and *yqeU*, as well as *groEL*, *groES*, and *clpP*, were repressed three to sevenfold after cold stress (Beckering et al. 2002). In addition, a proteomic approach showed that the HSPs GroES and ClpP were repressed upon cold shock (Graumann et al. 1996). Therefore, it seems reasonable that *B. subtilis* has no need for HSPs at low temperatures and hence can save valuable resources by repressing the corresponding genes (Beckering et al. 2002).

In contrast, physiological and molecular experiments indicate a cross-talk between cold and heat stresses in LAB. Possible correlations between the heat- and cold-shock regulons were suggested by (1) the observation of an increased level of *Leuconostoc mesenteroides* homologs of the HSPs DnaK and GroEL upon cold shock (Salotra et al. 1995), (2) the cold induction of a group of small heat-shock genes in *L. plantarum* (Spano et al. 2004, 2005), and (3) a slight protection from freezing of *Lactobacillus johnsonii* (Walker et al. 1999) and *L. lactis* subsp. *lactis* (Broadbent and Lin 1999) cells upon induction of the heat-shock response.

Conversely, heat shock did not improve the cryotolerance of *L. lactis* subsp. *cremoris* strain MG1363 (Wouters et al. 1999a), suggesting a strain specificity of this phenomenon. However, an unidentified cold-inducible 45-kDa protein of *L. lactis* (Wouters et al. 1999a) was proposed to correspond to the heat-inducible ClpX ATPase (Skinner and Trempy 2001), a member of the large family of closely related ATPases found in both prokaryotic and eukaryotic cells, having multiple regulatory functions, including a general chaperone activity and the ability to enable the Clp proteases to recognize their substrates. Thus, a ClpX function may be to promote the proteolysis of misfolded proteins after both cold or heat shock. Recently, the synthesis of a Clp ATP-dependent protease (ClpL) was observed to be induced during both cold and heat stress in *Streptococcus thermophilus*, a moderate thermophilic LAB widely used in dairy fermentation (Varcamonti et al. 2006). Although the induction of ClpL protein in *S. thermophilus* was apparently essential for stress tolerance, it does not account by itself for the molecular relationships between cold and heat stress (Varcamonti et al. 2006).

Among mesophilic LAB, *L. plantarum* is a flexible and versatile species that is encountered in a variety of environmental niches, including fermented beverage, meat, and many vegetable or plant fermentations in which abiotic stresses are common. Moreover, *L. plantarum* is frequently encountered as a natural inhabitant of the human gastrointestinal tract, and a selected strain, *L. plantarum* 299v, is marketed as a probiotic that may confer various beneficial health effects to the consumer

(Ahrné et al. 1998). The ecological flexibility of *L. plantarum* matches the observation that this species has one of the largest known genomes among LAB (Kleerebezem et al. 2003; Molenaar et al. 2005). *L. plantarum* encodes genes for a number of stress-related proteins, including three small HSPs (sHSPs) that were recently characterized (Spano et al. 2004, 2005), and three highly homologous CSPs (CspL, CspC, CspP) (Derzelle et al. 2002, 2003). Based on their structure and on the regulatory elements found in the 5′ flanking regions, it was proposed that the small heat-shock (*shs*) genes identified so far in *L. plantarum* belong to class I (*hsp 18.5* and *hsp 19.3*) and class II (*hsp 18.55*) of the heat-shock gene family according to the *B. subtilis* nomenclature of heat response (Hecker et al. 1996). Using a quantitative real-time PCR approach (qRT-PCR), we analyzed the expression of *L. plantarum* genes involved in stress tolerance, including not only heat but also salt, ethanol, sulfite, and bile stresses. Genes such as *ctsR*, *ftsH*, *hsp33*, *dnaK*, and *shs* were analyzed during short or medium cold-induced stresses. We observed that although the majority of them were cold-repressed, a gene coding for an sHSP was strongly induced (Fig. 5.1) (unpubl. results). The induction of heat-responsive genes at low temperatures suggests a cross-talk between the heat- and cold-stress responses in LAB, and we cannot exclude that heat- and cold-stress genes may act in a coordinated fashion to help protect cells against the effects of low temperatures. For instance, although protein unfolding is much more evident during heat stress, it may also be identified in cold-damaged cells (Weber and Marahiel 2002). Recently, an interlink between cold and acid tolerance has also been evidenced in *L. delbrueckii*, on which an enhanced freeze-tolerance was observed after acidification at pH 5.25 for 30 min at the end of the fermentation process (Streit et al. 2007, 2008).

Taken together, these observations suggest that signal transduction machinery and transcription factors that mediate the genetic response to stressful conditions

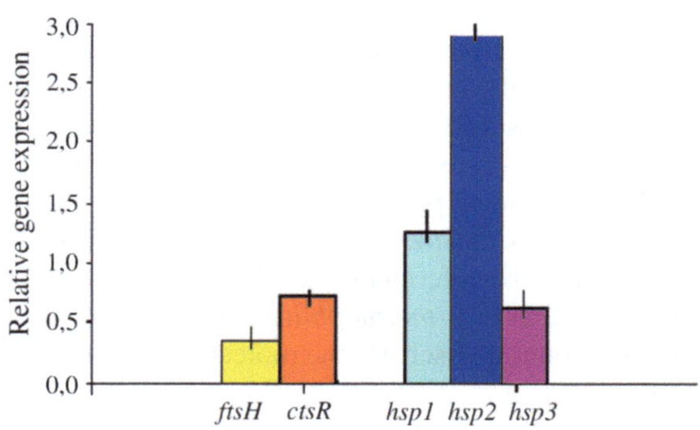

Fig. 5.1 Cold-shock induction of heat stress-responsive genes in *Lactobacillus plantarum*. Relative mRNA levels of *L. plantarum ftsH*, *ctsR*, and small heat-shock gene family (*hsp1*, *hsp2*, and *hsp3*) in response to cold stress (10°C) applied for 10 min as determined by quantitative reverse-transcription real-time PCR (qRT-PCR). *ldhD* was used as the internal control gene. Data shown are means ±SD of three independent experiments

appear to be shared by different stress-regulative networks, although the phenomenon of interlinked stress response is still poorly understood and needs to be further investigated (Chattopadhyay 2008).

5.5 Membrane Integrity and Cold Stress

Tolerance to cold stress is more than CSP production. Adaptive responses to cold stress in LAB vary among bacterial species and in relation to stress conditions. However, in addition to the induction of a specific set of proteins, a crucial response entails major changes in membrane fatty acid composition.

The performance of microorganisms under stress conditions requires the maintenance of the main functions of the cell membrane to control ionic permeability and to regulate the exchange of solutes between the cell and the external environment. The barrier properties of the cytoplasmic membrane are of special importance for the energy transduction systems of the bacterial cell and are known to depend critically upon the physical state of lipid bilayers, which, in turn, is affected by changes in the external temperature. Indeed, it is widely acknowledged that normal cell function requires the membrane lipid bilayers to be in a liquid-crystalline state, as it is at physiological temperatures. At lower temperatures, lipid bilayers undergo a reversible change of state from a disordered to an ordered array of fatty acid chains. The cellular adaptive mechanisms induced by cold shock consist of an increase in the unsaturated fatty acid content of membrane phospholipids, which leads to a decrease in the solid-to-fluid transition temperature and, thus, to an increase in membrane fluidity. Consequently, the ratio between unsaturated and saturated fatty acids (U/S) is inversely correlated with the growth temperature (Suutari and Laakso 1992). In addition, some specific fatty acids play an important role in stress response (Li et al. 2009). Fernandez Murga et al. (2000) observed an increase in C16:0 and C18:2 fatty acids in *Lactobacillus acidophilus* grown at low temperature (25°C). The C18:1 fatty acid concentration increased in response to low temperatures in *L. plantarum* (Russell et al. 1995), to acid pH in *S. thermophilus* (Beal et al. 2001), and to osmotic stress in *L. lactis* (Guillot et al. 2000). In contrast, the C18:1 fraction decreased in response to freezing in lactic streptococci and to spray-drying in *L. acidophilus* (Brennan et al. 1986). A high cycC19:0 concentration favored the cryotolerance in *L. bulgaricus*, *Lactobacillus helveticus*, and *L. acidophilus* (Gómez Zavaglia et al. 2000).

As it acts on the membrane fluidity, cold shock can also counterbalance the deleterious effect of other abiotic stresses. This feature has recently been observed in the wine starter *Oenococcus oeni* (Chu-Ky et al. 2005). The authors analyzed the effects of combined cold, acid, and ethanol on the membrane physical state and on the survival of *O. oeni*. Membrane stiffening was induced by ethanol and acid shock and correlated with total cell death. In contrast, *O. oeni* cells recovered their viability when sequentially subjected first to cold (8°C) and then to ethanol and acid shock (Chu-Ky et al. 2005). These results suggest a positive short-term effect of combined cold, acid, and ethanol shocks on the membrane fluidity and viability of *O. oeni*. In addition, this finding suggests new useful practical approaches for dealing with wine starters.

5.6 Freezing and Cryoprotection: An Industrial Issue

The preservation of LAB used in food and pharmaceutical industries for their contribution to aroma and texture, as well as for their probiotic role, is generally obtained by freezing or freeze-drying. Such techniques preserve the viability and main technological attributes of the bacteria: acidification activity, aroma production, texture formation, and probiotic properties (Fonseca et al. 2003). However, some strains are particularly sensitive to freezing and freeze-drying, both of which generally lower the performance of starter cultures. This sensitivity is related to the deterioration of the cell physiological state, caused by various stresses appearing during starter production and preservation (Muldrew and McGann 1990; Dumont et al. 2004).

Cold stress, which takes place during the cooling and freezing steps and throughout the frozen storage, is the main cause of loss of bacterial activity. Moreover, additional stresses, such as those deriving from an oxidative and/or hyperosmotic environment, may occur during freeze-thaw treatments (Muldrew and McGann 1990; Stead and Park 2000).

After the transfer of a bacterial cell population from the optimum temperature to a low temperature above 0°C, some bacterial strains develop an increased ability to survive at extremely low temperatures, a phenomenon known as *cryotolerance*. This adaptation to a freeze-thaw challenge depends on the temperature and the duration of cold preincubation as well as on the initial bacterial cell concentration. Cell cryopreservation requires specific optimization for each type of microorganism, and each type of cell has its own protocol for freezing. Numerous researchers have attempted to develop methods that permit a 100% preservation of freeze-thawing of diverse cellular specimens, but some microorganisms are still reluctant to preservation by freezing (Dumont et al. 2004).

Factors affecting LAB survival during freezing–thawing cycles include different factors such as growth media composition, growth phase, fermentation process, and cold temperatures (Bâati et al. 2000; Streit et al. 2007; Siaterlis et al. 2009). Different methods are presently proposed to preserve the quality of lactic and probiotic starters. Among these, the use of cryoprotectants (also called *chemical chaperones*) such as betaine, proline, glycerol, and trehalose (Panoff et al. 2000; Fonseca et al. 2003; Siaterlis et al. 2009) is considered one of the most effective. These molecules improve cell preservation by minimizing the cell water content and/or supporting the vitrification occurrence and finally by protecting the cell constitutive macromolecules (Adam et al. 1994; Dumont et al. 2003, 2004).

The use of adequate freezing and storage conditions (Foschino et al. 1996; Dumont et al. 2004) and the selection of more resistant strains (Monnet et al. 2003) are also useful tools in order to improve freezing tolerance. According to several authors, adequate attention should be paid to specific environmental conditions of fermentation, such as pH value, temperature, and centrifugation steps (Fernandez Murga et al. 1998; Palmfeldt and Hahn-Hagerdal 2000; Beal et al. 2001; Saarela et al. 2005; Wang et al. 2005a; Streit et al. 2007). For instance, the pH and temperature of fermentation have been reported to strongly influence the tolerance of *L. acidophilus* to freezing. In addition, cooling temperature and the cold acclimation of LAB such as *L. acidophilus* and *L. lactis* subsp. *diacetylactis* resulted in significantly improved cell viability after successive freezings and thawings (De Urraza and De Antoni 1997; Kim and Dunn 1997; Broadbent and Lin 1999; Lee 2004; Panoff et al. 1995; Wang et al. 2005b). Nevertheless, the intrinsic characteristics of the bacterial strain are usually crucial (Fonseca et al. 2001).

The resistance of bacterial cells to freezing might also be improved by genetic engineering. For example, the overproduction of the CSPs CspB and CspE has been shown to increase the survival of *L. lactis* after four freeze–thaw cycles of a ten- and fivefold factor, respectively (Wouters et al. 2000). Moreover, the overexpression of sHSPs in *L. plantarum* enable transformed cells to tolerate heat, solvent, and, importantly, cold stress (Fiocco et al. 2007).

5.7 Concluding Remarks

In order to respond to stressful conditions, microorganisms have evolved multiple adaptive mechanisms. Indeed, external signals can elicit dramatic changes in the expression pattern of a variety of stress-related genes, encoding proteins thought to improve adaptation to the changing environment. The continuously growing interest in probiotic bacteria has led to an increase in manufactured functional foods and medicines containing these microorganisms. As a consequence, the performances of "useful" bacteria are deeply analyzed in order to improve their viability during industrial processing and storage handling. The dissection of all those mechanisms involved in tolerance to abiotic stress will surely improve the storage and performance of industrial strains (Corcoran et al. 2008). In *B. subtilis*, the model Gram-positive bacteria, several mechanisms involved in cold stress tolerance have been elucidated, while in LAB several gaps still exist (Fig. 5.2).

Food-related bacteria are particularly and repeatedly exposed to low temperatures during food handling and storage (Wouters et al. 2000). A sudden downshift in the growth temperature (cold shock) triggers a drastic reprogramming of bacterial gene expression to allow cell survival under the new unfavorable conditions (Giuliodori et al. 2005). However, tolerance to cold goes beyond a specific response. An interlink between stresses is likely to exist, and cross-resistance or cell-to-cell communication might affect the survival of bacteria in food products (Vorob'eva 2004).

Freezing is commonly used to preserve the viability of LAB for a long time while maintaining their technological properties (acidification activity, organoleptic and preservative properties). This approach is of considerable interest for research

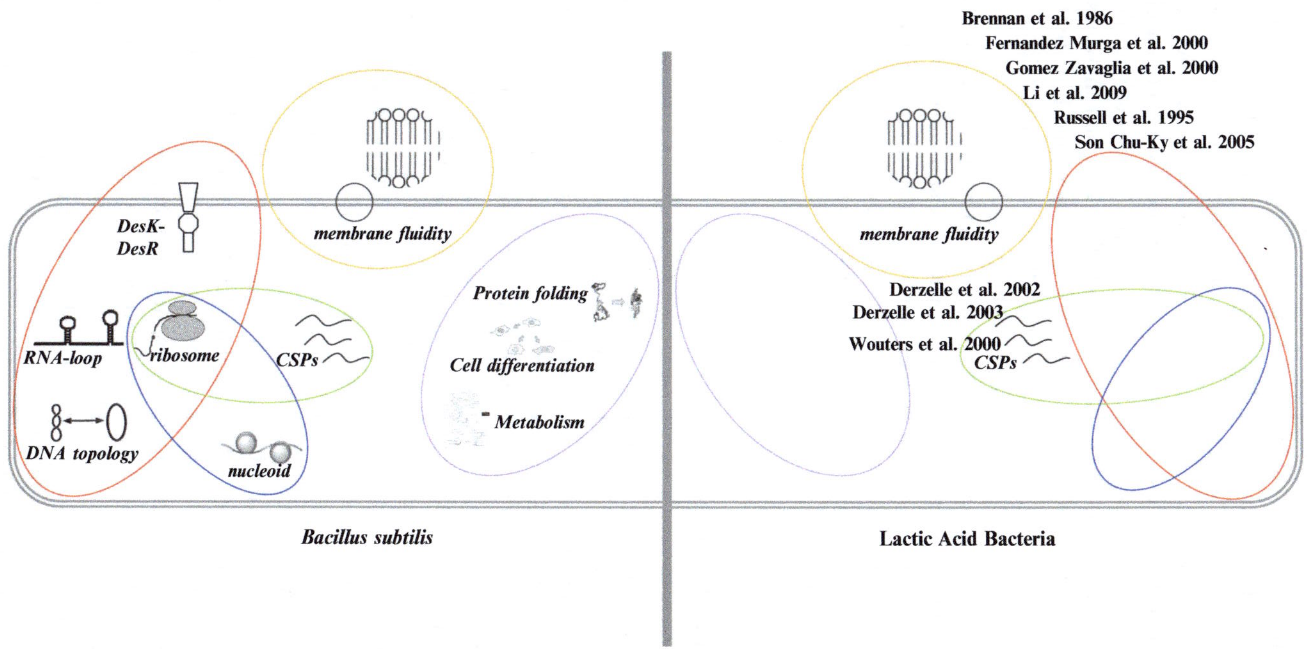

Fig. 5.2 Schematic representation of cold-shock response in *Bacillus subtilis*, the model microorganism for Gram-positive bacteria, compared with the corresponding state-of-the-art concerning lactic acid bacteria. The cold-shock response of *B. subtilis* is illustrated as reviewed by Weber and Marahiel (2002), distinguishing biological modifications related to temperature perception and signal transduction (*red*), membrane adaptation (*yellow*), CSPs and translation apparatus (*green*), nucleoid structure and transcription (*blue*), metabolism, protein folding, and cell differentiation (*violet*)

and industry as these bacteria are widely used for food processing and may be employed for the elaboration of pharmaceutical products (Corthier and Renault 1999). Thermophilic LAB exhibit different survival rates during freezing and frozen storage, depending on the processing conditions and, importantly, on the specific features of the strains tested. Therefore, a full knowledge of the strain physiology, the opportunity to dress a strain using its own fermentation process, and the use of cryoprotectants may facilitate bacteria storage and handling at freezing temperatures.

Understanding how cells respond to cold stress is also useful for the analysis of the biofilm formation in Gram-positive bacteria. In the human pathogens *Listeria monocytogenes* and *L. lactis* subsp. *cremoris*, cold shock was found to influence the biofilm-forming ability and the biofilm association with pathogens (Kives et al. 2005; Keskinen et al. 2008).

Finally, microbial biotechnology may progress using cold-regulated promoters. Qing et al. (2004) applied the unique features of the *cspA* gene of *E. coli* to develop a series of expression vectors, termed pCold vectors, that drive the high expression of cloned genes upon induction by cold shock (Hayashi and Kojima 2008). It will certainly be interesting to use or develop similar vectors in order to engineer LAB for the production of micro-bio-factories (MBF).

References

Adam MM, Rana KJ, McAndrew BJ (1994) Effect of cryoprotectants on activity of selected enzymes in fish embryos. Cryobiol 32:92–104

Aguilar PS, Lopez P, de Mendoza D (1999) Transcriptional control of the low-temperature-inducible *des* gene, encoding the Δ5 desaturase of *Bacillus subtilis*. J Bacteriol 181:7028–7033

Aguilar PS, Hernandez-Arriaga AM, Cybulski LE, Erazo AC, De Mendoza D (2001) Molecular basis of thermosensing: a two-component signal transduction thermometer in *Bacillus subtilis*. EMBO J 20:1681–1691

Ahrné S, Nobaek S, Jeppsson B, Adlerberth I, Wold AE, Molin G (1998) The normal *Lactobacillus* flora of healthy human rectal and oral mucosa. J Appl Microbiol 85:88–94

Bâati L, Fabre-Gea C, Auriol D, Blanc PJ (2000) Study of the cryotolerance of *Lactobacillus acidophilus*: effect of culture and freezing conditions on the viability and cellular protein levels. Int J Food Microbiol 59:241–247

Beal C, Fonseca F, Corrieu G (2001) Resistance to freezing and frozen storage of *Streptococcus thermophilus* is related to membrane fatty acid composition. J Dairy Sci 84:2347–2356

Beckering C, Steil CL, Weber MWH, Volker U, Marahiel MA (2002) Genome wide transcriptional analysis of the cold shock response in *Bacillus subtilis*. J Bacteriol 184:6395–6402

Beaufilis S, Sauvageot N, Mazé A, Laplace JM, Auffray Y, Deutscher J, Hartke A (2007) The cold shock response of *Lactobacillus casei*: relation between HPr phosphorylation and resistance to freeze/thaw cycles. J Mol Microbiol Biotechnol 13:65–75

Brennan M, Wanismail B, Johnson MC, Ray B (1986) Cellular damage in dried *Lactobacillus acidophilus*. J Food Prot 49:47–53

Broadbent JR, Lin C (1999) Effect of heat shock or cold shock treatment on the resistance of *Lactococcus lactis* to freezing and lyophilization. Cryobiology 39:88–102

Chastanet A, Ferrè T, Msadek T (2003) Comparative genomics reveal novel heat shock regulatory mechanisms in *Staphylococcus aureus* and other Gram-positive bacteria. Mol Microbiol 47:1061–1073

Chattopadhyay MK (2008) Cryotolerance in bacteria: interlink with adaptation to other stress factors. Trends Microbiol 16:455–460

Chouayekh H, Serror P, Boudebbouze S, Maguin E (2009) Highly efficient production of the staphylococcal nuclease reporter in *Lactobacillus bulgaricus* governed by the promoter of the *hlbA* gene. FEMS Microbiol Lett 293:232–239

Chu-Ky S, Tourdot-Marechal R, Marechal PA, Guzzo J (2005) Combined cold, acid, ethanol shocks in *Oenococcus oeni*: effects on membrane fluidity and cell viability. Biochim Biophys Acta 1717:118–124

Corcoran BM, Stanton C, Fitzgerald G, Ross RP (2008) Life under stress: the probiotic stress response and how it may be manipulated. Curr Pharm Des 14:1382–1399

Corthier G, Renault P (1999) Future directions for research on biotherapeutic agents: contribution of genetic approaches on lactic acid bacteria. In Elmer GW, McFarland L, Surawicz C (Eds.), Biotherapeutic agents and infectious diseases. Humana Press, Totowa, NJ, pp. 269–304

Delmas F, Pierre F, Coucheney F, Divies C, Guzzo J (2001) Biochemical and physiological studies of the small heat shock protein Lo18 from the lactic acid bacterium *Oenococcus oeni*. J Mol Microbiol Biotechnol 3:601–610

De Urraza P, De Antoni G (1997) Induced cryotolerance of *Lactobacillus delbrueckii* subsp. *bulgaricus* LBB by preincubation at suboptimal temperatures with fermentable sugar. Cryobiology 35:159–164

Derzelle S, Hallet B, Ferain T, Delcour J, Hols P (2002) Cold shock induction of the *cspL* gene in *Lactobacillus plantarum* involves transcriptional regulation. J Bacteriol 184:5518–5523

Derzelle S, Hallet B, Ferain T, Delcour J, Hols P (2003) Improved adaptation to cold shock, stationary-phase, and freezing stresses in *Lactobacillus plantarum* overproducing cold-shock proteins. Appl Environ Microbiol 69:4285–4290

Digel I, Kayser P, Artmann GM (2008) Molecular processes in biological thermosensation, J Biophys 2008:602870

Dixon-Fyle SM, Caro L (1999) Characterization *in vitro* and *in vivo* of a new HU family protein from *Streptococcus thermophilus* ST11. Plasmid 42:159–173

Dorman CJ, Deighan P (2003) Regulation of gene expression by histone-like proteins in bacteria. Curr Opin Genet Dev 13:179–184

Dumont F, Marechal PA, Gervais P (2003) Influence of cooling rate on *Saccharomyces cerevisiae* destruction during freezing: unexpected viability at ultra-rapid cooling rates. Cryobiology 46:33–42

Dumont F, Marechal PA, Gervais P (2004) Cell size and water permeability as determining factors for cell viability after freezing at different cooling rates. Appl Env Microbiol 70:268–272

El-Sharoud WM (2005) Two-component signal transduction systems as key players in stress responses of lactic acid bacteria. Sci Prog 88:203–228

Eriksson S, Hurme R, Rhen M (2002) Low-temperature sensors in bacteria. Phil Trans R Soc Lond 357:887–893

Fang L, Jiang W, Bae W, Inouye M (1997) Promoter-independent cold-shock induction of *cspA* and its derepression at 37°C by mRNA stabilization. Mol Microbiol 23:355–364

Feng W, Tejero R, Zimmerman DE, Inouye M, Montelione GT (1998) Solution NMR structure and backbone dynamics of the major cold-shock protein (CspA) from *Escherichia coli*: evidence for conformational dynamics in the single-stranded RNA-binding site. Biochem 37: 10881–10896

Fernandez Murga ML, Pesce de Ruiz Holgado A, Font de Valdez G (1998) Survival rate and enzyme activities of *Lactobacillus acidophilus* following frozen storage. Cryobiology 36:315–319

Fernandez Murga ML, Cabrera GM, De Valdez GF, Disalvo A, Seldes AM (2000) Influence of growth temperature on cryotolerance and lipid composition of *Lactobacillus acidophilus*. J Appl Microbiol 88:342–348

Fiocco D, Capozzi V, Goffin P, Hols P, Spano G (2007) Improved adaptation to heat, cold and solvent tolerance in *Lactobacillus plantarum*. Appl Microbiol Biotech 77:909–915

Fiocco D, Collins M, Muscariello L, Hols P, Kleerebezem M, Msadek T, Spano G (2009) The *Lactobacillus plantarum ftsH* gene is a novel member of the CtsR stress response regulon. J Bacteriol 191:1688–1694

Fonseca F, Béal C, Corrieu G (2001) Operating conditions that affect the resistance of lactic acid bacteria to freezing and frozen storage. Cryobiology 43:189–198

Fonseca F, Béal C, Mihoub F, Marin M, Corrieu G (2003) Improvement of cryopreservation of *Lactobacillus delbrueckii* subsp. *bulgaricus* CFL1 with additives displaying different protective effects. Int Dairy J 13:917–926

Foschino R, Fiori E, Galli A (1996) Survival and residual activity of *Lactobacillus acidophilus* frozen cultures under different conditions. J Dairy Res 63:295–303

Giuliodori AM, Gualerzi CO, Soto S, Vila J, Tavio MM (2005) Review on bacterial stress topics. Ann NY Acad Sci 1113:95–104

Goldenberg D, Azar I, Oppenheim A (1996) Differential stability of the *cspA* gene in the cold-shock response of *Escherichia coli*. Mol Microbiol 19:241–248

Gómez Zavaglia A, Disalvo EA, De Antoni GL (2000) Fatty acid composition and freeze-thaw resistance in lactobacilli. J Dairy Res 67:241–247

Gordon BR, Imperial R, Wang L, Navarre WW, Liu J (2008) Lsr2 of Mycobacterium represents a novel class of H-NS-like proteins. J Bacteriol 190:7052–7509

Graumann P, Marahiel MA (1997) Effects of heterologous expression of CspB, the major cold shock protein of *Bacillus subtilis*, on protein synthesis in *E. coli*. Mol Gen Genet 253:745–752

Graumann P, Marahiel MA (1998) A superfamily of proteins that contain the cold-shock domain. Trends Biochem Sci 23:286–290

Graumann PL, Marahiel MA (1999) Cold shock response in *Bacillus subtilis*. J Mol Microbiol Biotechnol 1:203–209

Graumann P, Schroder K, Schmid R, Marahiel MA (1996) Cold shock stress-induced proteins in *Bacillus subtilis*. J Bacteriol 178:4611–4619

Graumann P, Wendrich TM, Weber MHW, Schröder K, Marahiel MA (1997) A family of cold shock proteins in *Bacillus subtilis* is essential for cellular growth and for efficient protein synthesis at optimal and low temperatures. Mol Microbiol 25:741–756

Gualerzi CO, Giuliodori AM, Pon CL (2003) Transcriptional and post-transcriptional control of cold-shock genes. J Mol Biol 331:527–539

Guillot A, Obis D, Mistou MY (2000) Fatty acid composition and activation of glycine-betaine transport in *Lactococcus lactis* subjected to osmotic stress. Int J Food Microbiol 55:47–51

Hayashi K, Kojima C (2008) pCold-GST vector: a novel cold-shock vector containing GST tag for soluble protein production. Protein Expr Purif 62:120–127

Hecker M, Schumann W, Volker U (1996) Heat shock and general stress response in *Bacillus subtilis*. Mol Microbiol 19:417–428

Hunger K, Beckering CL, Marahiel MA (2004) Genetic evidence for the temperature-sensing ability of the membrane domain of the *Bacillus subtilis* histidine kinase DesK. FEMS Microbiol Lett 230:41–46

Jin S, Song YN, Deng WY, Gordon MP, Nester EW (1993) The regulatory VirA protein of *Agrobacterium tumefaciens* does not function at elevated temperatures. J Bacteriol 175:6830–6835

Jones PG, Van Bogelen RA, Neidhardt FC (1987) Induction of proteins in response to low temperature in *Escherichia coli*. J Bacteriol 169:2092–2095

Jones PG, Krah R, Tafuri SR, Wolffe AP (1992a) DNA gyrase, CS7.4, and the cold shock response in *Escherichia coli*. J Bacteriol 174:5798–5802

Jones PG, Cashel M, Glaser G, Neidhardt FC (1992b) Function of a relaxed-like state following temperature downshifts in *Escherichia coli*. J Bacteriol 174:3903–3914

Kaufman-Szymczyk A, Wojtasik A, Parniewsky P, Bialkowska A, Tkaczuk K, Turkiewicz M (2009) Identification of the *csp* gene and molecular modelling of the CspA-like protein from antarctic soil-dwelling psychrotrophic bacterium *Psychrobacter* sp. B6. Acta Biochim Polonica 56:63–69

Keskinen LA, Todd EC, Ryser ET (2008) Impact of bacterial stress and biofilm-forming ability on transfer of surface-dried *Listeria monocytogenes* during slicing of delicatessen meats. Int J Food Microbiol 127:298–304

Kim WS, Dunn NW (1997) Identification of a cold shock gene in lactic acid bacteria and the effect of cold shock on cryotolerance. Curr Microbiol 35:59–63

Kives J, Guadarrama D, Orgaz B, Rivera-Sen A, Vazquez J, San Jose C (2005) Interactions in biofilms *of Lactococcus lactis* ssp. *cremoris* and *Pseudomonas fluorescens* cultured in cold UHT Milk. J Dairy Sci 88:4165–4171

Kleerebezem M, Boekhorst J, van Kranenburg R, Molenaar D, Kuipers OP, Leer R, Tarchini R, Peters SA, Sandbrink HM, Fiers MW, Stiekema W, Lankhorst RM, Bron PA, Hoffer SM, Groot MN, Kerkhoven R, de Vries M, Ursing B, de Vos WM, Siezen RJ (2003) Complete genome sequence of *Lactobacillus plantarum* WCFS1. Proc Natl Acad Sci USA 100:1990–1995

Klinkert B, Narberhaus F (2009) Microbial thermosensors. Cell Mol Life Sci 66:2661–2676

Kremer W, Schuler B, Harrieder S, Geyer M, Gronwald W, Welker C, Jaenicke R, Kalbitzer HR (2001) Solution NMR structure of the cold-shock protein from the hyperthermophilic bacterium *Thermotoga maritima*. Eur J Biochem 268:2527–2539

Lee HS, Berger DK, Kustu S (1993). Activity of purified NIFA, a transcriptional activator of nitrogen fixation genes. Proc Natl Acad Sci USA 90:2266–2270

Lee K (2004) Cold shock response in *Lactococcus lactis* ssp. *diacetylactis*: a comparison of the protection generated by brief pre-treatment at less severe temperatures. Process Biochem 39:2233–2239

Li C, Jia-Liang Z, Yu-Tang W, Xu H, Ning L (2009) Synthesis of cyclopropane fatty acid and its effect on freeze-drying survival of *Lactobacillus bulgaricus* L2 at different growth conditions. World J Microbiol Biotechnol 25:1659–1665

Liu D, Yumoto H, Murakami K, Hirota K, Ono T, Nagamune H, Kayama S, Matsuo T, Miyake Y (2008) The essentiality and involvement of *Streptococcus intermedius* histone-like DNA-binding protein in bacterial viability and normal growth. Mol Microbiol 68:1268–1282

Margesin R, Neuner G, Storey KB (2007) Cold-loving microbes, plants, and animals-fundamental and applied aspects. Naturwissenschaften 94:77–99

Mitta M, Fang L, Inouye M (1997) Deletion analysis of *cspA* of *Escherichia coli*: requirement of the AT-rich UP element for cspA transcription and the downstream box in the coding region for its cold shock induction. Mol Microbiol 26:321–335

Molenaar D, Bringel F, Schuren FH, de Vos WM, Siezen RJ, Kleerebezem M (2005) Exploring *Lactobacillus plantarum* genome diversity by using microarrays. Microbiology 187:6119–6127

Monnet C, Béal C, Corrieu G (2003) Improvement of the resistance of *Lactobacillus delbrueckii* ssp. *bulgaricus* to freezing by natural selection. J Dairy Sci 86:3048–3053

Mueller U, Perl D, Schmid FX, Heinemann U (2000) Thermal stability and atomic-resolution crystal structure of the *Bacillus caldolyticus* cold shock protein. J Mol Biol 297:975–988

Muldrew K, McGann LE (1990) Mechanisms of intracellular ice formation. Biophys J 57:525–532

Nierlich DP, Murakawa GJ (1996) The decay of bacterial messenger RNA. Prog Nucleic Acid Res Mol Biol 52:153–216

Ouvry A, Wache Y, Tourdot-Maréchal R, Diviès C, Cachon R (2002) Effects of oxidoreduction potential combined with acetic acid, NaCl and temperature on the growth, acidification, and membrane properties of *Lactobacillus plantarum*. FEMS Microbiol Lett 214:257–261

Palmfeldt J, Hahn-Hagerdal B (2000) Influence of culture pH on survival of *Lactobacillus reuteri* subjected to freeze-drying. Int J Food Microbiol 55:235–238

Panoff JM, Thammavongs B, Laplace J, Hartke A, Boutibonnes P, Auffray Y (1995) Cryotolerance and cold adaptation in *Lactococcus lactis* subsp. *lactis* IL 1403. Cryobiology 32:516–520

Panoff JM, Bouachanh T, Micheline G, Boutibonnes P (1998) Cold stress response in mesophilic bacteria. Cryobiology 36:75–83

Panoff JM, Bouachanh T, Micheline G (2000) Cryoprotectants lead to phenotypic adaptation to freeze–thaw in *Lactobacillus delbrueckii* ssp. *bulgaricus* CIP 101027T. Cryobiology 40:264–269

Phadtare S (2004) Recent developments in bacterial cold-shock response. Curr Issues Mol Biol 6:125–136

Phadtare S, Hwang J, Severinov K, Inouye M (2003) CspB and CspL, thermostable cold-shock proteins from *Thermotoga maritima*. Genes Cells 8:801–810

Qing G, Ma LC, Khorchid A, Swapna GV, Mal TK, Takayama MM, Xia B, Phadtare S, Ke H, Acton T, Montelione GT, Ikura M, Inouye M (2004) Cold-shock induced high-yield protein production in *Escherichia coli*. Nat Biotechnol 22:877–882

Rivals JP, Beal C, Thammavongs B, Gueguen M, Panoff JM (2007) Cryotolerance of *Lactobacillus delbrueckii* subsp. *bulgaricus* CFL1 is modified by acquisition of antibiotic resistance. Cryobiology 55:19–26

Rosen R, Ron EZ (2002) Proteome analysis in the study of the bacterial heat shock response. Mass Spectr Rev 21:244–265

Russell NJ, Evans RI, ter Steeg PF, Hellemons J, Verheul A, Abee T (1995) Membranes as a target for stress adaptation. Int J Food Microbiol 28:255–261

Saarela M, Virkajärki I, Alakomi HL, Mattila-Sandholm T, Vaari A, Suomalainen T, Mättö J (2005) Influence of fermentation time, cryoprotectant and neutralization of cell concentrate on freeze-drying survival, storage stability, and acid and bile exposure of *Bifidobacterium animalis* ssp. *lactis* cells produced without milk-based ingredients. J Appl Microbiol 99:1330–1339

Salotra P, Singh DK, Seal KP, Krishna N, Jaffe H, Bhatnagar R (1995) Expression of DnaK and GroEL homologs in *Leuconostoc mesenteroides* in response to heat shock, cold shock or chemical stress. FEMS Microbiol Lett 131:57–62

Schindelin H, Marahiel MA, Heinemann U (1993) Universal nucleic acid-binding domain revealed by crystal structure of the *B. subtilis* major cold-shock protein. Nature 364:164–168

Schumann W (2009) Temperature sensors of eubacteria. Adv Appl Microbiol 67:213–256

Siaterlis A, Deepika G, Charalampopoulos D (2009) Effect of culture medium and cryoprotectants on the growth and survival of probiotic lactobacilli during freeze drying. Lett Appl Microbiol 48:295–301

Skinner MM, Trempy JE (2001) Expression of *clpX*, an ATPase subunit of the Clp protease, is heat and cold shock inducible in *Lactococcus lactis*. J Dairy Sci 84:1783–1785

Sledjeski DD, Gupta A, Gottesman S (1996) The small RNA, DsrA, is essential for the low temperature expression of RpoS during exponential growth in *Escherichia coli*. EMBO J 15:3993–4000

Smirnova AV, Braun Y, Ullrich MS (2008) Site-directed mutagenesis of the temperature-sensing histidine protein kinase CorS from *Pseudomonas syringae*. FEMS Microbiol Lett 283:231–238

Spano G, Massa S (2006) Environmental stress response in wine lactic acid bacteria: beyond *Bacillus subtilis*. Crit Rev Microbiol 32:77–86

Spano G, Capozzi V, Vernile A, Massa S (2004) Cloning, molecular characterization and expression analysis of two small heat shock genes isolated from wine *Lactobacillus plantarum*. J Appl Microbiol 97:774–782

Spano G, Beneduce L, Perrotta C, Massa S (2005) Cloning and characterization of the *hsp* 18.55 gene, a new member of the small heat shock genes family isolated from wine *Lactobacillus plantarum*. Res Microbiol 156:219–224

Stead D, Park SF (2000) Roles of Fe superoxide dismutase and catalase in resistance of *Campylobacter coli* to freeze–thaw stress. Appl Environ Microbiol 66:3110–3112

Streit F, Corrieu G, Béal C (2007) Acidification of fermented broth improves cryotolerance of *Lactobacillus delbrueckii* subsp. *bulgaricus* CFL1. J Biotech 128:659–667

Streit F, Delettre J, Corrieu G, Béal C (2008) Acid adaptation of *Lactobacillus delbrueckii* subsp. *bulgaricus* induces physiological responses at membrane and cytosolic levels that improves cryotolerance. J Appl Microbiol 105:1071–1080

Suutari M, Laakso S (1992) Changes in fatty acid branching and unsaturation of *Streptomyces griseus* and *Brevibacterium fermentans* as a response to growth temperature. Appl Environ Microbiol 58:2338–2340

Tendeng C, Bertin PN (2003) H-NS in Gram-negative bacteria: a family of multifaceted proteins. Trends Microbiol 11:511–518

van de Guchte M, Serror P, Chervaux C, Smokvina T, Ehrlich SD, Maguin E (2002) Stress responses in lactic acid bacteria. Antonie van Leeuwenhoek 82:187–216

Varcamonti M, Arsenijevic S, Martirani L, Fusco D, Naclerio G, De Felice M (2006) Expression of the heat shock gene *clpL* of *Streptococcus thermophilus* is induced by both heat and cold shock. Microbial Cells Factories 5:1–6

Vorob'eva LI (2004) Stressors, stress reactions, and survival of bacteria (a review). Prikl Biokhim Mikrobiol 40:261–269

Walker DC, Girgis HS, Klaenhammer TR (1999) The *groESL* chaperone operon of *Lactobacillus johnsonii*. Appl Environ Microbiol 65:3033–3041

Wang Y, Corrieu G, Béal C (2005a) Fermentation pH and temperature influence the cryotolerance of *Lactobacillus acidophilus* RD758. J Dairy Sci 88:21–29

Wang Y, Delettre J, Guillot A, Corrieu G, Béal C (2005b) Influence of cooling temperature and duration on cold adaptation of *Lactobacillus acidophilus* RD758. Cryobiology 50:294– 307

Weber MH, Marahiel MA (2002) Coping with the cold: the cold shock response in the Gram-positive soil bacterium *Bacillus subtilis*. Philos Trans R Soc Lond B Biol Sci 357:895–907

Willimsky G, Bang H, Fischer G, Marahiel MA (1992) Characterization of *cspB*, a *Bacillus subtilis* inducible cold shock gene affecting cell viability at low temperatures. J Bacteriol 174:6326–6335

Wouters JA, Jeynov B, Rombouts FM, de Vos WM, Kuipers OP, Abee T (1999a) Analysis of the role of 7 kDa cold- shock proteins of *Lactococcus lactis* MG1363 in cryoprotection. Microbiology 145:3185–3194

Wouters JA, Rombouts FM, Kuipers OP, De Vos WM, Abee T (2000) The role of cold-shock proteins in low-temperature adaptation of food-related bacteria. System Appl Microbiol 23:165–173

Zangrossi S, Briani F, Ghisotti D, Regonesi ME, Tortora P, Deho G (2000) Transcriptional and post-transcriptional control of polynucleotide phosphorylase during cold acclimation in *Escherichia coli*. Mol Microbiol 36:1470–1480

Zeeb M, Klaas EAM, Weininger U, Löw C, Sticht H, Balbach J (2006) Recognition of T-rich single-stranded DNA by the cold shock protein Bs-CspB in solution. Nucleic Acids Res 34:4561–4571

Chapter 6
Responses of Lactic Acid Bacteria to Oxidative Stress

Bénédicte Cesselin, Aurélie Derré-Bobillot, Annabelle Fernandez, Gilles Lamberet, Delphine Lechardeur, Yuji Yamamoto, Martin Bastian Pedersen, Christel Garrigues, Alexandra Gruss, and Philippe Gaudu

6.1 Introduction

Lactic acid bacteria that are generally recognized as safe (we used the term LAB_{GRAS} when discussing dairy bacteria because they also belong to the large family of bacteria producing lactic acid; some of them are pathogens) are widely used in dairy and other food fermentations and are therefore naturally present in the human gastrointestinal tract (GIT). Named for their capacity to degrade lactose or other carbohydrates into lactic acid, LAB_{GRAS} are used to promote milk coagulation and to contribute to the complex flavors, tastes, and textures of dairy products, for example, through the production of diacetyl and small peptides. They also may secrete small antimicrobial peptides (bacteriocins), which can limit the development of food contaminants like *Staphylococcus aureus* and *Listeria monocytogenes* and further improve food quality (Corr et al. 2007; Galvez et al. 2007). The "safe" and even "beneficial" status of LAB_{GRAS} is attested to by their widespread consumption in foods. In the last two decades, researchers aimed at developing new applications of LAB_{GRAS} in the area of biotechnology. For example, overproduction systems were developed in LAB_{GRAS} for large-scale recombinant protein purification and for protein delivery. The systems make use of natural promoters that respond to diverse molecules for induction, such as nisin (Mierau and Kleerebezem 2005), acid (Madsen et al. 2005), or metals (Morello et al. 2008). Additionally, the overproduced proteins can be targeted to the cytoplasm or directed into the medium (although high-efficiency secretion remains a bottleneck). The production of immunomodulators (interleukin) or enzymes was successfully reported in LAB_{GRAS} and might be useful for vaccine strategies to limit infection or in the treatment of cancer (Bermudez-Humaran et al. 2008).

P. Gaudu (✉)
Unité Bactéries Lactiques et Pathogènes Opportunistes, INRA UR888,
Jouy en Josas 78352, France
e-mail: philippe.gaudu@jouy.inra.fr

E. Tsakalidou and K. Papadimitriou (eds.), *Stress Responses of Lactic Acid Bacteria*,
Food Microbiology and Food Safety, DOI 10.1007/978-0-387-92771-8_6,

Environmental pressures, either in industrial fermentations or in the host, can have marked consequences on bacterial gene expression, growth, and survival. This may limit the uses and applications of LAB_{GRAS}. Major stress factors are acid, salt, heat, and oxygen, which are relevant to bacterial starter preparation, to dairy fermentations, and to LAB_{GRAS} life in the GIT, where cells are exposed to nutritional competition with other bacteria of the microbiota. To survive, bacteria have, as a nearly unique option, detection of the stress, which would lead to reprogramming of the gene expression and a consequent physiological adaptation to the new environment. LAB_{GRAS} that are used industrially, or are present in the GIT, grow via sugar fermentation. In view of the importance of sugars in the LAB_{GRAS} environment, this chapter focuses on aspects of oxidative stress mainly in *Lactococcus lactis* related to carbon metabolism. More recently, in addition to growth under fermentation, several LAB, including *L. lactis*, have been shown to undergo a respiration metabolism when heme, or heme and menaquinone, is available in the medium; heme is needed to activate a terminal cytochrome oxidase, while menaquinones serve as electron transfer molecules in the electron transport chain (Poole and Cook 2000).

6.2 Oxygen Species

6.2.1 *Oxygen*

Oxygen is a biradical molecule ($^{\cdot}$O-O$^{\cdot}$), because of its small size, it can easily access the catalytic site of proteins. Published analysis has established that O_2 reacts with certain metals, in particular with iron (present in heme-dependent cytochrome oxidase, for instance), and with activated amino acids, like the glycyl radical in some enzymes. In contrast to cytochrome oxidases that use O_2 as a substrate, pyruvate formate lyase (PFL), an anaerobic enzyme bearing a glycyl radical, is extremely sensitive to O_2 (Zhang et al. 2001). This reaction leads to the cleavage of the protein into two fragments and furthermore to irreversible inactivation of activity (Melchiorsen et al. 2000). However, reduction of O_2 generates products that are far more toxic than O_2 itself, such as superoxide anion radicals ($O_2^{\cdot-}$), hydrogen peroxide (H_2O_2), and hydroxyl radicals ($HO^{\cdot}$) (Imlay 2008). These molecules are physiologically more important in the oxidative stress than O_2, since for example *Escherichia coli* cells that are aerobically grown maintain concentrations of $O_2^{\cdot-}$ and H_2O_2 at 0.0001 and 0.1 μM, respectively, compared to an intracellular O_2 concentration of 210 μM (Imlay and Fridovich 1991; Gonzalez-Flecha and Demple 1995).

6.2.2 *Superoxide*

$O_2^{\cdot-}$ is the product of a one-electron reduction of O_2 (Imlay 2008). It is generated as a consequence of the oxidation of dihydroflavins ($FADH_2$, $FMNH_2$, Riboflavin H_2) or quinols (demethylmenaquinol, menaquinol, ubiquinol) (Huycke et al. 2001;

Rezaiki et al. 2008). The reason is that the radical forms of flavins and quinones are relatively stable, which then facilitates the one-electron reduction. Thus, quinones belonging to the respiratory chain, and flavoproteins were often supposed to be the major generators of $O_2^{\cdot-}$ in bacteria, including *Enterococcus faecalis*, *E. coli*, and *L. lactis* (Imlay and Fridovich 1991; Huycke et al. 2001; Korshunov and Imlay 2006; Rezaiki et al. 2008). Inactivation of the quinone biosynthesis pathway or NADH dehydrogenases of the respiratory chain significantly reduced $O_2^{\cdot-}$ production in *E. coli* (Korshunov and Imlay 2006), while cyanide, an inhibitor of cytochrome oxidases, strongly led to an increase in $O_2^{\cdot-}$ and H_2O_2 levels, as reflected by the overproduction of superoxide dismutase (SOD) and catalase (Hassan and Fridovich 1979).

$O_2^{\cdot-}$ is not considered a strong oxidant or reducer, and the presence of SOD remained an enigma for a long period. The construction of a strain deficient in SOD activity in *E. coli* finally helped to reveal the mechanism of $O_2^{\cdot-}$ toxicity (Farr et al. 1986). This mutant does not grow in chemical-defined medium in the presence of air and is strongly sensitive to $O_2^{\cdot}$-producing agents like paraquat (a catalyzer of $O_2^{\cdot-}$) in rich medium. The phenotypes result from oxidation of certain dehydratases containing particular [4Fe-4S] clusters, such as aconitase A or fumarase A and B (Imlay 2008). This [4Fe-4S] cluster is linked to proteins via only three irons, while the last one binds to the substrate (citrate and fumarate for aconitase and fumarase, respectively). This labile iron is attacked by $O_2^{\cdot-}$, leading to destruction of the cofactor and concomitant enzyme inactivation. This inactivation also coincides with increased free iron in cells according to the reaction [4Fe-4S] $+ O_2^{\cdot-} + 2H^+ \rightarrow [3Fe-4S] + Fe + H_2O_2$. It was also demonstrated in *E. coli* that $O_2^{\cdot-}$ reacted with an activated amino acid (tyrosyl radical) in proteins, as exemplified by aerobic deoxyribonucleotide di-phosphate reductase, a key enzyme of DNA synthesis (Gaudu et al. 1996).

In an acidic environment, the protonation of $O_2^{\cdot-}$ leads to a more reactive species, hydroperoxyl radical (HOO$^{\cdot}$). Due to the absence of charge, HOO$^{\cdot}$ can pass through the membrane and then return to its anionic state in the cytoplasm, as the intracellular environment is less acidic than the outside. Thus, bacteria that encounter acidic environments, for instance, the stomach or fermented products, might be particularly exposed to oxidative stress. In *L. lactis* MG1363, SOD was overproduced about fourfold after exposure of cells to acid stress (Budin-Verneuil et al. 2005). Thus, the presence of SOD should provide an advantage for colonization and survival in an acidic environment. Note that *Lactobacillus plantarum* species do not have SOD; they compensate for this absence by incorporating high concentrations of manganese (Archibald and Fridovich 1981a, b).

6.2.3 Hydrogen Peroxide and the Hydroxyl Radical

H_2O_2 is a neutral molecule that crosses the membrane barrier of the cell but is not reactive by itself. It is issued from two-electron reduction by flavoproteins, quinols, and metabolic enzymes (e.g., NADH peroxidase, pyruvate oxidase, and lactate oxidase) present in several LAB_{GRAS}, and in particular in lactobacilli (Lorquet et al. 2004; Goffin et al. 2006; Barre et al. 2007). Additionally, SOD or manganese produce this molecule from $O_2^{\cdot-}$. In some lactobacillus species, H_2O_2 can reach millimolar concentrations in some environments, which is high enough to stop growth (Ocana et al. 1999). As LAB have peroxidases and/or catalases (Kono and Fridovich 1983; Frankenberg et al. 2002) to eliminate peroxide, the high levels of peroxide they produce suggest that these enzymes might require specific conditions for activity and/or expression. For instance, many catalases contain heme as a cofactor in prokaryotes, although a manganese-dependent catalase has been reported in *L. plantarum* (Kono and Fridovich 1983). In the case of catalases using heme as a cofactor, *E. faecalis* would need to scavenge it from their environment, as it lacks the complete biosynthesis pathway like other LAB (Kono and Fridovich 1983; Duwat et al. 2001; Frankenberg et al. 2002). Peroxidase activity is mainly assured by the AhpF-AhpC complex. AhpF (also named Nox-1) is a dehydrogenase that drives electrons from NADH to AhpC in *Streptococcus mutans* (Higuchi et al. 1999, 2000). Furthermore, AhpC catalyzes the reduction of peroxide into water or alcohol via the oxidation of its cysteine residues. So a depletion of NADH due to the lactate dehydrogenase activity (or H_2O_2-producing NADH oxidase) probably limits the activity of AhpF-AhpC. Besides the aforementioned activity, AhpC can switch to a chaperone function as it forms oligomeric complexes as described in *Helicobacter pylori* and thus might stabilize the protein folding during stress conditions (Chuang et al. 2006). Interestingly, we recently found that AhpC is also a heme-binding protein, which is needed for full respiration activity of the opportunist pathogen *Streptococcus agalactiae* (Lechardeur et al. 2010).

H_2O_2 is activated by ultraviolet light or in the presence of metals and is converted into $HO^{\cdot}$, the most reactive species of O_2 byproducts. Nonheme iron was estimated to be around 10 μg/g of cell dried weight in *L. lactis* MG1363 (Gostick et al. 1999), under some conditions. While similar measurements are unavailable for LAB_{GRAS}, it is likely that iron is present, as since (1) some housekeeping enzymes require iron, such as dihydroorate reductase (Rowland et al. 2000), (2) LAB_{GRAS} carry genes for potential iron transporters, (3) LAB_{GRAS} genome analysis revealed genes coding for enzymes involved in iron utilization (Gostick et al. 1999, Thibessard et al. 2004), and (4) an iron chelator (like Desferal) protected DNA integrity when *L. lactis* was exposed to O_2 (Duwat et al. 1995; Rezaiki et al. 2004).

$HO^{\cdot}$ reacts at its site of production and oxidizes all biomolecules present in its vicinity. The steady-state concentration of this molecule is thus considered very low compared to $O_2^{\cdot-}$ and H_2O_2. In the case of DNA, the damage incurred suggested that iron (or other redox metals) is probably sequestered in the phosphate backbone of the DNA (Ouameur et al. 2005). To strengthen this, the family of Dps/Dpr proteins, which are used in iron storage, prevented $HO^{\cdot}$ formation and further protected the DNA (Yamamoto et al. 2002; 2004). Interestingly, Dps from *E. coli* and *L. lactis* bind to DNA (Martinez and Kolter 1997; Stillman et al. 2005), probably highlighting the importance of this class of proteins in DNA integrity in bacteria.

6.3 Target and Defense: Glyceraldehyde-3-Phosphate Dehydrogenase, Pyruvate Formate Lyase, and Respiration as General Barriers Against Reactive Oxygen Species

Besides some proteins that protect DNA from damage, others have recently been characterized for their sensitivity to oxidative stress in *L. lactis*, including glyceraldehyde-3-phosphate dehydrogenase (GAPDH) and PFL.

6.3.1 Glyceraldehyde-3-Phosphate Dehydrogenase

GAPDH, called GapB in *L. lactis* MG1363, catalyzes the conversion of glyceraldyde-3-phosphate to glycerate-1,3-biphosphate in glycolysis. A decrease in *gapB* gene expression resulted in a dramatic reduction in growth, suggesting that this enzyme, not surprisingly, is essential in lactococci (Solem et al. 2003). The *gapB* gene is highly expressed, and GapB is among the most abundant proteins in the *L. lactis* glycolytic pathway. It is readily detected on 2D gels, and appears as two spots with different isoelectric points but the same molecular weight, indicating that GapB is not cleaved, in contrast to PFL, when *L. lactis* cells are exposed to O_2 (Melchiorsen et al. 2000). The relative levels of these two spots were observed to change in *L. lactis* under two conditions: One is in a thioredoxin reductase mutant (*trxB1*), and the other is in respiration metabolism. This led us to consider that these two forms were related to oxidative stress (Vido et al. 2004, 2005). Using mass spectrometry, the two spots were analyzed in the *trxB1* background, and the sole difference observed was oxidation of the catalytic cysteine. The thiol group of this amino acid was converted into the sulfinic (SO_2H) or sulfonic (SO_3H) form, resulting in irreversible oxidation. To date, no reducing system to reverse this oxidation has been described in prokaryotes, while the reduction of sulfenic acid (SOH) and sulfoxide methionine has been reported. For instance, methionine sulfoxide reductase restores oxidized methionine in proteins (Boschi-Muller et al. 2008). Thus, to avoid cysteine oxidation in GapB, enzymatic reducing systems like thioredoxin-thioredoxin reductase systems are required to eliminate reactive oxygen species (ROS) before they attack the proteins. The large amount of GapB in *L. lactis* MG1363 may also be a means to avoid a decrease in the glycolytic flux occurring through oxidation of GapB, as a threefold lower GapB still maintained a fully active glycolysis (Solem et al. 2003). Moreover, the growth defect of the *trxB1* mutant is alleviated by the presence of metabolites like glutathione, cysteine, and pyruvate, indicating that these small molecules can protect GapB against ROS (Vido et al. 2005). *L. lactis* contains a second GAPDH, GapA. Like GapB, GapA is also present as two isoforms on 2D gels, indicating that this protein is also subject to oxidation (Willemoes et al. 2002). GapA is not overproduced in the *trxB1* mutant, suggesting that this paralog does not compensate for the *gapB* inactivation and may not be involved in the stress response (Vido et al. 2005).

6.3.2 Pyruvate Formate Lyase

Under anaerobiosis, PFL catalyzes the conversion of pyruvate into acetyl-CoA. During the shift from anaerobiosis to aerobiosis, PFL is cleaved into two fragments, resulting in its irreversible inactivation. One strategy to avoid the cleavage is to reduce the glycyl radical into glycine by the PFL deactivase (Melchiorsen et al. 2000). Otherwise, the induction of an alternative metabolic pathway is essential to maintain the production of acetyl-CoA under aeration. In this latter growth condition, genes encoding the pyruvate dehydrogenase complex (PdhABCD, PDHc) are overexpressed (Jensen et al. 2001). This complex also catalyzes the conversion of pyruvate into acetyl-CoA, but unlike PFL, the second end product of this reaction is NADH. As the lactate dehydrogenase recycles NADH into NAD but also consumes pyruvate, it cannot assist PDHc in its activity. Thus, other NADH oxidoreductases, such as the NADH:H_2O-forming oxidase (NOX), are required to recycle NAD independently of pyruvate. NOX catalyzes the four-electron reduction of O_2 into water. In some bacteria, this flavoprotein plays an important role in growth under aeration. It contributes both to the reduction of ROS formation through O_2 scavenging activity (Yamamoto et al. 2006) and to the production of acetyl-CoA coupled with PDHc. In *L. lactis*, an NADH:H_2O-forming oxidase overproduction produced a large amount of acetate, and acetoin, at the expense of lactate, indicating that this enzyme contributes significantly to the conversion of pyruvate into acetate (Lopez de Felipe et al. 1998; Hoefnagel et al. 2002), but the effect of NOX deletion on *L. lactis* growth or metabolic pathways remains unknown. In contrast, other *Streptococcaceae* may have less redundancy in this function. It is notable that NOX inactivation can be compensated for in respiring LAB by the NADH oxidase activity as part of the electron transport chain (Yamamoto et al. 2006). Interestingly, an *S. agalactiae nox-2* mutant failed to grow under aeration due to acetyl-CoA deficiency (its production requires NAD, which is produced less in the NOX mutant). As an acetyl-CoA is a precursor of fatty acids, the addition of lipids or serum restored the *S. agalactiae* growth defect (Yamamoto et al. 2006).

6.3.3 Respiration as a General Barrier Against Reactive Oxygen Species

LAB_{GRAS} species like *L. lactis* are traditionally considered to be obligate fermentative bacteria because even in aerobiosis they use sugar degradation for substrate-level phosphorylation, that is ATP production. However, experimental studies revealed that *L. lactis*, *E. faecalis*, and other LAB are capable of activating a heme-dependent cytochrome oxidase and of establishing a complete respiration chain (Winstedt et al. 2000; Duwat et al. 2001). In *L. lactis*, the respiratory chain comprises (1) an NADH dehydrogenase (likely to be encoded by *noxA* and/or *noxB*), which drives electrons to menaquinone (Gaudu et al. 2002; Brooijmans et al. 2009a; Tachon et al. 2009), (2) menaquinones (synthesized in *L. lactis* and *E. faecalis*, but lacking in other respiration-competent LAB), which transfer the electrons to the terminal acceptor (Huycke et al. 2001; Rezaiki et al. 2008; Brooijmans et al. 2009a; Tachon et al. 2009), and (3) heme-dependent cytochrome oxidase CydAB, the terminal electron acceptor, which finally reduces O_2 into water (Duwat et al. 2001) (Fig. 6.1). The heme biosynthesis pathway is incomplete in all LAB that have been sequenced. Some bacteria, like *L. lactis* and *E. faecalis*, encode a ferrochelatase (called HemH), which catalyzes the incorporation of iron in protoporphyrin-IX (PPIX), to generate heme (Duwat et al. 2001).

An unexpected finding for respiration-grown *L. lactis* was that the cells survive much better at 4°C than under fermentation once they attain the stationary phase (Duwat et al. 2001). For instance, after 2 weeks of storage, a 10^6-fold increase in the viable population was observed for respiration-grown cells compared to fermentation-grown cells. This effect is due to at least two factors: One concerns the higher pH of respiration cultures, due to less organic acid accumulation. Nevertheless, fermentative cells resuspended in medium buffered at pH 7.0 did not survive as well as cells cultured under respiration conditions, indicating that the pH difference is only part of the answer. The second and primary reason for the improved survival is therefore attributed to the O_2 scavenging properties of CydAB (Rezaiki et al. 2004). This class of oxidases is generally reported to have more affinity for O_2 than other

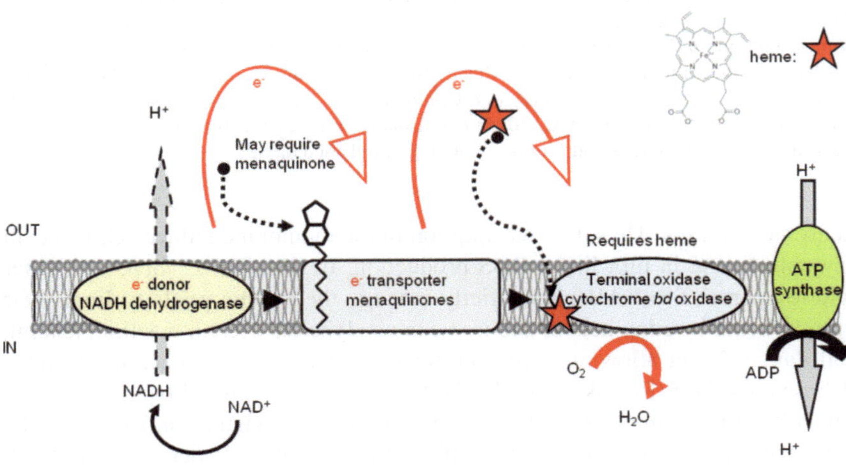

Fig. 6.1 A respiratory chain in LAB. Heme uptake probably involves a specific transporter(s), including the system encoded by the *fhu* operon in *L. lactis* (not schematized) (Gaudu et al. 2003). The mechanism of incorporation into cytochrome oxidase, CydAB, is still unknown. Electrons from NADH are transferred to CydAB via NADH dehydrogenase (electron donor) and menaquinone (electron transfer intermediate). Genes for menaquinone synthesis are present only in some of the LABs (see Fig. 6.2). The activated respiratory chain reduces O_2 to water and concomitantly extrudes protons from the cytoplasm. ATPase makes use of a pH gradient to produce extra ATP (Koebmann et al. 2008). To avoid a toxic heme overload, putative efflux systems, including YgfBA, are produced (not schematized) (Pedersen et al. 2008)

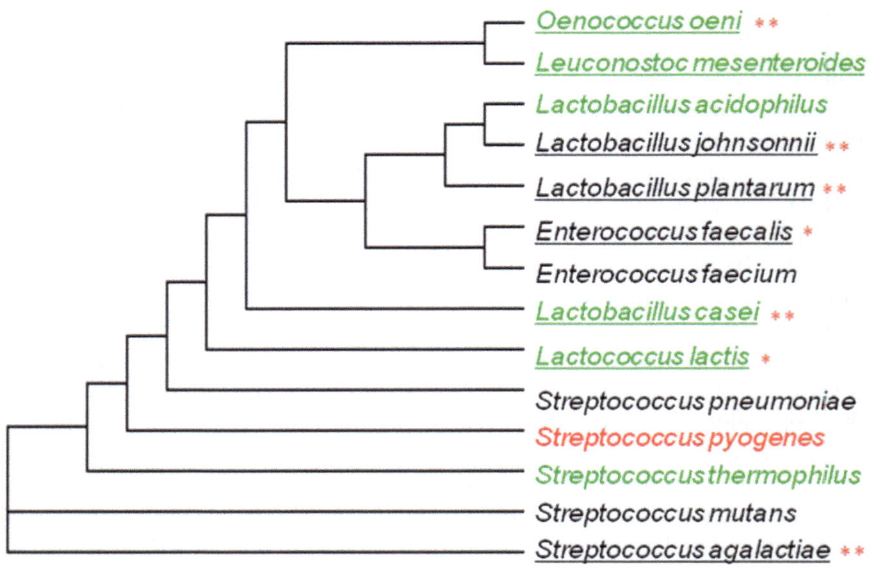

Fig. 6.2 Distribution of respiration capacity in related LAB_{GRAS} and other LAB. Related lactic acid bacteria (based on similarities of the conserved *recA* Orfs using ClustalW alignment) are presented. Bacteria are classified as food (*green*), commensal or opportunist pathogens (*black*), or pathogens (*red*). All these LAB lack heme biosynthesis genes. Among them, those encoding *cydABCD* genes are underlined. Those that require only heme to activate respiration metabolism are shown with one *red star*. Those that require heme and a menaquinone to activate respiration metabolism are shown with two *red stars*. Respiration metabolism was demonstrated experimentally for *L. lactis*, *E. faecalis*, *S. agalactiae*, *Lb. plantarum*, *L. mesenteroides* (Duwat et al. 2001; Huycke et al. 2001; Yamamoto et al. 2005; Brooijmans et al. 2009a; b); for *L. mesenteroides* (C. Foucaud, Alexandra Gruss and Philippe Gaudu, unpublished data)

cytochrome oxidases. Thus, O_2 consumption in the membrane reduces O_2 levels in the cytoplasm and in turn limits ROS production. In addition, *L. lactis* and other respiration-permissive LAB must strictly manage their heme pools to limit their reactivity (heme is redox-active, and iron can catalyze the Fenton reaction, leading to cell damage), yet allowing sufficient heme availability to activate the CydAB cytochrome oxidase (Rezaiki et al. 2004). Recently, we identified factors involved in intracellular heme management: AhpC was found to bind heme, and the YgfCBA proteins were found to be involved in efficient heme efflux (Pedersen et al. 2008; Lechardeur et al. 2010).

Several LAB encode the *cydABCD* genes, which are needed to produce the cytochrome bd oxidase (*cydCD* encodes functions implicated in cytochrome oxidase assembly) (Huycke et al. 2001; Yamamoto et al. 2005; Brooijmans et al. 2009a, b). These are summarized in Fig. 6.1. We consider this feature to be a reliable indication to identify which LAB have the potential to undergo a respiratory metabolism when heme, and in some cases a menaquinone source, is provided. For instance, *S. agalactiae*, an opportunistic human and bovine pathogen and a lactic acid bacterium, activates its respiratory chain once heme and menaquinone are supplied in aerobic cultures (Yamamoto et al. 2005). More recently, the respiration capacity of *cydABCD*-encoding organisms was also validated in *L. plantarum*, giving similar phenotypes as above (Brooijmans et al. 2009b). When heme (or PPIX) is supplied to *L. lactis* under aerobic conditions, carbon metabolism is dramatically altered: Acetate and acetoin are produced in substantial amounts at the expense of lactic acid (Kaneko et al. 1990; Duwat et al. 2001; Pedersen et al. 2008). This change is possibly regulated through the NADH/NAD^+ ratio through changes in gene expression in enzymes around pyruvate metabolism, as indicated for both transcriptome and proteome analysis (Vido et al. 2004; Pedersen et al. 2008). Note that respiratory chain activity expulses protons to the outside of the cells, which is taken up by the ATPase to produce ATP (Blank et al. 2001; Koebmann et al. 2008). Under the energetically favorable respiration conditions (i.e., in aerobic growth with heme), *L. lactis* displays a gain of biomass compared to fermentation growth (Duwat et al. 2001; Pedersen et al. 2008). Unexpectedly, in contrast to *L. lactis*, *L. plantarum* displayed an additional respiratory metabolism with nitrate as the electron acceptor. This nitrate respiration also required heme and menaquinone for activity (Brooijmans et al. 2009b).

6.4 Regulation of Oxidative Stress Response: Evidence and Models

Although several stress-defense proteins, in addition to the respiratory chain, have already been identified in LAB_{GRAS} (e.g., SOD, thioredoxin reductase, peroxidase, glutathione reductase, and RecA; see van de Guchte et al. (2002); and Hols et al. (2005) for a review) the mechanisms involved in their regulation remain widely unknown. Their genes are often induced under air or oxidative stress, as observed in other species like *Bacillus subtilis* or *E. coli*, suggesting that some strategies are common among bacteria. This hypothesis is supported by the presence of several potential regulators in LAB that are homologs to those characterized in *B. subtilis*, for instance. To date, only a few regulators have been identified in LAB_{GRAS}, and they can be classified in two categories: (1) transcriptional regulators that specifically recognize a motif in the promoter region of stress-response genes, to facilitate the recruitment of RNA polymerase, (2) proteins that modulate the activities of RNA polymerase.

6.4.1 Two-Component Systems

Among the six two-component systems (TCSs) identified in *L. lactis* MG1363, only two were found to be involved in oxidative stress: LlrF (DNA-binding regulator) and LlkinF (sensor) (O'Connell-Motherway et al. 2000). The inactivation of LlrF exhibited a greater sensitivity to peroxide: After 20 min of exposure to 4 mM of H_2O_2, only 9% of the mutant population remained viable compared to

70% in the control. The genes that are under control of this TCS, as well as the detected stress signal, remain to be identified in *L. lactis*. In *S. mutans*, ScnRK, a TCS, appears to regulate several genes related to stress, including *tpx*, encoding a thiol peroxidase (Chen et al. 2008). As this is also present in *L. lactis, tpx* might be controlled by LlrF/LlkinF.

6.4.2 PerR

Genome analysis in *L. lactis* revealed the presence of several homologs of *B. subtilis* PerR and OhrR, which are involved in the peroxide stress response. In *B. subtilis*, PerR is related to the ferric-uptake repressor family of proteins (Lee and Helmann 2006). It is a dimer and contains two metal ions per subunit: Zinc is coordinated to the protein via four cysteine residues, while ferrous iron or manganese ion is likely linked via histidines (H37, H91, and H93) and aspartic acids (D85 and D104). The $PerR_{Zn/Fe}$ form reacted via iron with peroxide by Fenton reaction, and furthermore, $HO^{\cdot}$ oxidized histidine at position 37 or 91 in the peptide sequence (Lee and Helmann 2006). This oxidation modified the affinity of PerR for the test promoter region of *mrgA* (encoding a Dps-like protein). In contrast, no oxidation occurred with the $PerR_{Zn/Mn}$ form, indicating that PerR is likely to use iron and zinc as metal cofactors in vivo. The *L. lactis* MG1363 homolog of PerR is named Fur (llmg_1023). In *E. faecalis*, a PerR-like regulator has been identified and found to contribute in oxidative stress response (Verneuil et al. 2005). Unexpectedly, in contrast to PerR, a new regulator HypR, belonging to the LysR family, has been shown to respond to peroxide stress in *E. faecalis*. Nevertheless, the roles of HypR versus PerR in stress regulation are not yet clearly understood in this bacterium (Verneuil et al. 2004).

6.4.3 OhrR

B. subtilis OhrR protein is a homodimer that belongs to the multiple antibiotic resistance (MarR) family. It regulates the expression of the *ohr* genes encoding thiol peroxidase, which reduces hydroperoxide to the corresponding alcohols. OhrR responds to hydroperoxide stress via the oxidation of its unique cysteine residue, located at position 15 in the polypeptide (Fuangthong and Helmann 2002). During oxidative stress, cysteine is oxidized to sulfenic acid (RSH $\rightarrow$ RSOH) or to more oxidized forms. The oxidized protein does not bind to the promoter region of *ohrA*, encoding an organic peroxide resistance protein. A closer homolog of OhrR is present in *L. lactis* IL1403 (RmaJ), but seems to be absent in MG1363. This indicates that the latter cannot cope with organic peroxide stress, or that the signaling pathway differs from OhrR.

6.4.4 Rex

Unlike OhrR and PerR, which detect stress by the oxidation of an amino acid, the Rex protein governs the expression of the cytochrome oxidase gene (*cydAB*) in *B. subtilis* and *Streptomyces ceolicolor* via the pool of NADH (Wang et al. 2008). This protein is a homodimer containing two domains: The N-terminal domain binds to the promoter region of target genes such as *cydABCD*, while the C-terminal recognizes the ligand, NADH. One model is that when cells reach the stationary phase (or when the O_2 tension decreases due to high cell density), the NADH pool increases, leading to its binding to Rex. The Rex-NADH-DNA complex becomes unstable, and thus the repression of genes controlled by Rex is off. In *L. lactis* MG1363, the homolog of Rex is llmg_1514. However, *E. faecalis* contains in its genome two ORFs similar to Rex, suggesting that one might respond to other signals like NADPH instead of NADH, for instance.

6.4.5 Factors Modulating RNA Polymerase

Modification of RNA polymerase activity is reportedly involved in the regulation of stress responses in *L. lactis*. Two candidates have been identified recently in this bacterium: ppGpp and a new class of proteins, Spx.

6.4.5.1 ppGpp

ppGpp is produced during the stringent response, and probably in all conditions that lead to growth arrest (Magnusson et al. 2005). Cross-linking experiments indicated that it interacted with the two β-subunits of RNA polymerase, close to the catalytic site (Chatterji et al. 1998). In *L. lactis*, inactivation of the high-affinity phosphate transporter, encoded by the *pst* locus, increased resistance to multiple stresses as reported for an *relA* mutant. The latter was shown to produce more ppGpp than the parent strain, indicating that the multiresistance observed in these mutants was correlated to ppGpp levels (Rallu et al. 2000). We developed a strategy to reveal *L. lactis* factors responding to oxidative stress, in which we disabled part of the oxidative stress response by growing cells in the presence of a reducing agent, dithiothreitol (DTT) (Cesselin et al. 2009). This molecule did not allow the production of SOD, AhpC, and TrxB1, confirming the use of a redox control in their respective gene expression. Unexpectedly, mutants isolated as resistant to stress in the presence of DTT were also inactivated in the *pst* locus. *Pst* mutants displayed an accrued resistance to copper, zinc, and tellurite (Turner et al. 2007; Cesselin et al. 2009). Copper and zinc bind to cysteine residues in proteins and, in the case of copper, also catalyze the Fenton reaction (Stadtman and Oliver 1991; Barre et al. 2007; Cesselin et al. 2009; Macomber and Imlay 2009). These observations suggest a probable link between ppGpp levels and metal homeostasis, as reported in *E. coli*. In *E. coli*, a metal limitation (iron) stimulated ppGpp production (Vinella et al. 2005). Note that two FNR-like proteins, containing redox cysteine residues, are involved in stress sensitivity and were shown to modulate zinc homeostasis (Gostick et al. 1999). These results all seem to point to the important role of metals in bacterial stress responses.

6.4.5.2 Spx

Spx proteins interact with the C-terminal domain of the RNA polymerase α-subunit in vitro and are thus a modulator of the transcriptional regulator – RNA polymerase interaction (Zuber 2004). Spx proteins are ubiquitous in Gram-positive bacteria, and the numbers of paralogs differ according to the species. For instance, *L. lactis* contains seven paralogs: The SpxB protein is involved in cell wall stress response (Veiga et al. 2007). TrmA (SpxA) inactivation leads to temperature resistance and compensates for defects due to *clpP* or *recA* inactivation (Duwat et al. 1999; Frees et al. 2001; Veiga et al. 2007). Functions of other Spx paralogs are under investigation in our laboratory (S. Kulakauskas, INRA). Interestingly, five of seven Spx proteins contain a CXXC motif in the N-terminal domain. This suggests that they might be under redox regulation. As this motif – present also in peroxiredoxin proteins such as thioredoxin – is highly conserved, it is tempting to propose that it might form disulfide bonds under certain conditions and thus might modulate Spx activity and its interactions with RNA polymerase. However, the work of You *et al.* suggests that this may not be the case (You et al. 2008). The addition of DTT, an agent that reverses disulfide bonds, had no effect on oxidized Spx after treatment of *B. subtilis* by paraquat. This observation suggests that cysteines of Spx proteins were probably oxidized in an irreversible form as observed in GapB (Vido et al. 2005). Moreover, the total absence of cysteine in the two other Spx proteins in *L. lactis* (llmg_1130 and llmg_1155) reveals a strong diversity in the activities of these proteins.

6.5 Concluding Remarks

In *E. coli*, $O_2^{\cdot -}$ reacts with particular iron-sulfur clusters present in some dehydratases (Keyer and Imlay 1996). As of now, no dehydratase containing the [4Fe-4S] cluster with a labile iron has been reported in LAB_{GRAS}, suggesting that the toxicity of $O_2^{\cdot -}$ might pass through a different mechanism of what was elucidated in the *E. coli* model. There are, nevertheless, *L. lactis* Orfs involved in amino acid synthesis, corresponding to dehydratases (like IlvD and SdaA), which likely involve Fe-S clusters. As the roles and requirements of iron in LAB_{GRAS} have not yet been firmly established, many lines of research remain to be explored to understand the toxicity of ROS in LAB_{GRAS}.

Similarly, the identification of stress-response regulators and signaling pathways in LAB_{GRAS} is incomplete, and most signals and targets are still unknown. Deeper characterizations of targets, for example, of the two-component LlrF-LlkinF system and of the roles of different Spx proteins in oxidative stress in LAB_{GRAS} are likely to bring information that cannot be transposed from studies of aerobic model bacteria. The impact of central metabolism on stress responses is still preliminary in LAB; for instance, the stress response under respiration conditions may be radically different from that under fermentation conditions. Global and modeling approaches will be valuable in obtaining an integral view on the stressed cell and should complement more in-depth studies of specific factors, which have led to some of the fundamental findings reported here.

Acknowledgments We thank members of the UBLO laboratory for stimulating discussions on the dynamics between bacteria and their environment. As the subjects we cover concern a broad scientific area, we apologize for any oversights in failing to mention topics related to the present review. Work reported from our laboratory in this review benefited from funding by the French Research Ministry (the ANR "StrepRespire" project) and Chr. Hansen A/S (Denmark).

References

Archibald FS, Fridovich I (1981a) Manganese and defenses against oxygen toxicity in *Lactobacillus plantarum*. J Bacteriol 145:442–451

Archibald FS, Fridovich I (1981b) Manganese, superoxide dismutase, and oxygen tolerance in some lactic acid bacteria. J Bacteriol 146:928–936

Barre O, Mourlane F, Solioz M (2007) Copper induction of lactate oxidase of *Lactococcus lactis*: a novel metal stress response. J Bacteriol 189:5947–5954

Bermudez-Humaran LG, Cortes-Perez NG, Ah-Leung S, Lefevre F, Yang G, Pang Q, Wu C, Zeng Y, Adel-Patient K, Langella P (2008) Current prophylactic and therapeutic uses of a recombinant *Lactococcus lactis* strain secreting biologically active interleukin-12. J Mol Microbiol Biotechnol 14:80–89

Blank LM, Koebmann BJ, Michelsen O, Nielsen LK, Jensen PR (2001) Hemin reconstitutes proton extrusion in an H(+)-ATPase-negative mutant of *Lactococcus lactis*. J Bacteriol 183:6707–6709

Boschi-Muller S, Gand A, Branlant G (2008) The methionine sulfoxide reductases: Catalysis and substrate specificities. Arch Biochem Biophys 474:266–273

Brooijmans R, Smit B, Santos F, van Riel J, de Vos WM, Hugenholtz J (2009a) Heme and menaquinone induced electron transport in lactic acid bacteria. Microb Cell Fact 8:28

Brooijmans RJ, de Vos WM, Hugenholtz J (2009b) *Lactobacillus plantarum* WCFS1 electron transport chains. Appl Environ Microbiol 75:3580–3585

Budin-Verneuil A, Pichereau V, Auffray Y, Ehrlich DS, Maguin E (2005) Proteomic characterization of the acid tolerance response in *Lactococcus lactis* MG1363. Proteomics 5:4794–4807

Cesselin B, Ali D, Gratadoux JJ, Gaudu P, Duwat P, Gruss A, El Karoui M (2009) Inactivation of the *Lactococcus lactis* high-affinity phosphate transporter confers oxygen and thiol resistance and alters metal homeostasis. Microbiology 155:2274–2281

Chatterji D, Fujita N, Ishihama A (1998) The mediator for stringent control, ppGpp, binds to the beta-subunit of Escherichia coli RNA polymerase. Genes Cells 3:279–287

Chen PM, Chen HC, Ho CT, Jung CJ, Lien HT, Chen JY, Chia JS (2008) The two-component system ScnRK of *Streptococcus mutans* affects hydrogen peroxide resistance and murine macrophage killing. Microbes Infect 10:293–301

Chuang MH, Wu MS, Lo WL, Lin JT, Wong CH, Chiou SH (2006) The antioxidant protein alkylhydroperoxide reductase of *Helicobacter pylori* switches from a peroxide reductase to a molecular chaperone function. Proc Natl Acad Sci USA 103:2552–2557

Corr SC, Li Y, Riedel CU, O'Toole PW, Hill C, Gahan CG (2007) Bacteriocin production as a mechanism for the antiinfective activity of *Lactobacillus salivarius* UCC118. Proc Natl Acad Sci USA 104:7617–7621

Duwat P, Ehrlich SD, Gruss A (1995) The *recA* gene of *Lactococcus lactis*: characterization and involvement in oxidative and thermal stress. Mol Microbiol 17:1121–1131

Duwat P, Ehrlich SD, Gruss A (1999) Effects of metabolic flux on stress response pathways in *Lactococcus lactis*. Mol Microbiol 31:845–858

Duwat P, Sourice S, Cesselin B, Lamberet G, Vido K, Gaudu P, Le Loir Y, Violet F, Loubiere P, Gruss A (2001) Respiration capacity of the fermenting bacterium *Lactococcus lactis* and its positive effects on growth and survival. J Bacteriol 183:4509–4516

Farr SB, D'Ari R, Touati D (1986) Oxygen-dependent mutagenesis in *Escherichia coli* lacking superoxide dismutase. Proc Natl Acad Sci USA 83:8268–8272

Frankenberg L, Brugna M, Hederstedt L (2002) *Enterococcus faecalis* heme-dependent catalase. J Bacteriol 184:6351–6356

Frees D, Varmanen P, Ingmer H (2001) Inactivation of a gene that is highly conserved in Gram-positive bacteria stimulates degradation of non-native proteins and concomitantly increases stress tolerance in *Lactococcus lactis*. Mol Microbiol 41:93–103

Fuangthong M, Helmann JD (2002) The OhrR repressor senses organic hydroperoxides by reversible formation of a cysteine-sulfenic acid derivative. Proc Natl Acad Sci USA 99:6690–6695

Galvez A, Abriouel H, Lopez RL, Ben Omar N (2007) Bacteriocin-based strategies for food biopreservation. Int J Food Microbiol 120:51–70

Gaudu P, Lamberet G, Poncet S, Gruss A (2003) CcpA regulation of aerobic and respiration growth in *Lactococcus lactis*. Mol Microbiol 50:183–192

Gaudu P, Niviere V, Petillot Y, Kauppi B, Fontecave M (1996) The irreversible inactivation of ribonucleotide reductase from *Escherichia coli* by superoxide radicals. FEBS Lett 387:137–140

Gaudu P, Vido K, Cesselin B, Kulakauskas S, Tremblay J, Rezaiki L, Lamberret G, Sourice S, Duwat P, Gruss A (2002) Respiration capacity and consequences in *Lactococcus lactis*. Antonie Van Leeuwenhoek 82:263–269

Goffin P, Muscariello L, Lorquet F, Stukkens A, Prozzi D, Sacco M, Kleerebezem M, Hols P (2006) Involvement of pyruvate oxidase activity and acetate production in the survival of *Lactobacillus plantarum* during the stationary phase of aerobic growth. Appl Environ Microbiol 72:7933–7940

Gonzalez-Flecha B, Demple B (1995) Metabolic sources of hydrogen peroxide in aerobically growing *Escherichia coli*. J Biol Chem 270:13681–23687

Gostick DO, Griffin HG, Shearman CA, Scott C, Green J, Gasson MJ, Guest JR (1999) Two operons that encode FNR-like proteins in *Lactococcus lactis*. Mol Microbiol 31:1523–1535

Hassan HM, Fridovich I (1979) Intracellular production of superoxide radical and of hydrogen peroxide by redox active compounds. Arch Biochem Biophys 196:385–395

Higuchi M, Yamamoto Y, Kamio Y (2000) Molecular biology of oxygen tolerance in lactic acid bacteria: functions of NADH oxidases and Dpr in oxidative stress. J Biosc Bioeng 90:484–493

Higuchi M, Yamamoto Y, Poole LB, Shimada M, Sato Y, Takahashi N, Kamio Y (1999) Functions of two types of NADH oxidases in energy metabolism and oxidative stress of *Streptococcus mutans*. J Bacteriol 181:5940–5947

Hoefnagel MH, Starrenburg MJ, Martens DE, Hugenholtz J, Kleerebezem M, Van Swam II, Bongers R, Westerhoff HV, Snoep JL (2002) Metabolic engineering of lactic acid bacteria, the combined approach: kinetic modelling, metabolic control and experimental analysis. Microbiology 148:1003–1013

Hols P, Hancy F, Fontaine L, Grossiord B, Prozzi D, Leblond-Bourget N, Decaris B, Bolotin A, Delorme C, Dusko Ehrlich S, Guedon E, Monnet V, Renault P, Kleerebezem M (2005) New insights in the molecular biology and physiology of *Streptococcus thermophilus* revealed by comparative genomics. FEMS Microbiol Rev 29:435–463

Huycke MM, Moore D, Joyce W, Wise P, Shepard L, Kotake Y, Gilmore MS (2001) Extracellular superoxide production by *Enterococcus faecalis* requires demethylmenaquinone and is attenuated by functional terminal quinol oxidases. Mol Microbiol 42:729–740

Imlay JA (2008) Cellular defenses against superoxide and hydrogen peroxide. Annu Rev Biochem 77:755–776

Imlay JA, Fridovich I (1991) Assay of metabolic superoxide production in *Escherichia coli.* J Biol Chem 266:6957–6965

Jensen NB, Melchiorsen CR, Jokumsen KV, Villadsen J (2001) Metabolic behavior of *Lactococcus lactis* MG1363 in microaerobic continuous cultivation at a low dilution rate. Appl Environ Microbiol 67:2677–2682

Kaneko T, Takahashi M, Suzuki H (1990) Acetoin Fermentation by Citrate-Positive *Lactococcus lactis* subsp. *lactis* 3022 Grown Aerobically in the Presence of Hemin or Cu. Appl Environ Microbiol 56:2644–2649

Keyer K, Imlay JA (1996) Superoxide accelerates DNA damage by elevating free-iron levels. Proc Natl Acad Sci USA 93:13635–13640

Koebmann B, Blank LM, Solem C, Petranovic D, Nielsen LK, Jensen PR (2008) Increased biomass yield of *Lactococcus lactis* during energetically limited growth and respiratory conditions. Biotechnol Appl Biochem 50:25–33

Kono Y, Fridovich I (1983) Isolation and characterization of the pseudocatalase of *Lactobacillus plantarum.* J Biol Chem 258:6015–6019

Korshunov S, Imlay JA (2006) Detection and quantification of superoxide formed within the periplasm of *Escherichia coli.* J Bacteriol 188:6326–6334

Lechardeur D, Fernandez A, Robert B, Gaudu P, Trieu-Cuot P, Lamberet G, Gruss A (2010) The 2-Cys peroxiredoxin alkyl hydroperoxide reductase c binds heme and participates in its intracellular availability in *Streptococcus agalactiae.* J Biol Chem 285:16032–16041

Lee JW, Helmann JD (2006) The PerR transcription factor senses H2O2 by metal-catalysed histidine oxidation. Nature 440:363–367

Lopez de Felipe F, Kleerebezem M, de Vos WM, Hugenholtz J (1998) Cofactor engineering: a novel approach to metabolic engineering in *Lactococcus lactis* by controlled expression of NADH oxidase. J Bacteriol 180:3804–3808

Lorquet F, Goffin P, Muscariello L, Baudry JB, Ladero V, Sacco M, Kleerebezem M, Hols P (2004) Characterization and functional analysis of the *poxB* gene, which encodes pyruvate oxidase in *Lactobacillus plantarum.* J Bacteriol 186:3749–3759

Macomber L, Imlay JA (2009) The iron-sulfur clusters of dehydratases are primary intracellular targets of copper toxicity. Proc Natl Acad Sci USA 106:8344–8349

Madsen SM, Hindre T, Le Pennec JP, Israelsen H, Dufour A (2005) Two acid-inducible promoters from *Lactococcus lactis* require the cis-acting ACiD-box and the transcription regulator *RcfB.* Mol Microbiol 56:735–746

Magnusson LU, Farewell A, Nystrom T (2005) ppGpp: a global regulator in *Escherichia coli.* Trends Microbiol 13:236–242

Martinez A, Kolter R (1997) Protection of DNA during oxidative stress by the nonspecific DNA-binding protein Dps. J Bacteriol 179:5188–5194

Melchiorsen CR, Jokumsen KV, Villadsen J, Johnsen MG, Israelsen H, Arnau J (2000) Synthesis and posttranslational regulation of pyruvate formate-lyase in *Lactococcus lactis.* J Bacteriol 182:4783–4788

Mierau I, Kleerebezem M (2005) 10 years of the nisin-controlled gene expression system (NICE) in *Lactococcus lactis.* Appl Microbiol Biotechnol 68:705–717

Morello E, Bermudez-Humaran LG, Llull D, Sole V, Miraglio N, Langella P, Poquet I (2008) *Lactococcus lactis*, an efficient cell factory for recombinant protein production and secretion. J Mol Microbiol Biotechnol 14:48–58

O'Connell-Motherway M, van Sinderen D, Morel-Deville F, Fitzgerald GF, Ehrlich SD, Morel P (2000) Six putative two-component regulatory systems isolated from *Lactococcus lactis* subsp. *cremoris* MG1363. Microbiology 146:935–947

Ocana VS, Pesce de Ruiz Holgado AA, Nader-Macias ME (1999) Selection of vaginal H2O2-generating Lactobacillus species for probiotic use. Curr Microbiol 38:279–284
Ouameur AA, Arakawa H, Ahmad R, Naoui M, Tajmir-Riahi HA (2005) A Comparative study of Fe(II) and Fe(III) interactions with DNA duplex: major and minor grooves bindings. DNA Cell Biol 24:394–401
Pedersen MB, Garrigues C, Tuphile K, Brun C, Vido K, Bennedsen M, Mollgaard H, Gaudu P, Gruss A (2008) Impact of aeration and heme-activated respiration on *Lactococcus lactis* gene expression: identification of a heme-responsive operon. J Bacteriol 190:4903–4911
Poole RK, Cook GM (2000) Redundancy of aerobic respiratory chains in bacteria? Routes, reasons and regulation. Adv Microb Physiol 43:165–224
Rallu F, Gruss A, Ehrlich SD, Maguin E (2000) Acid- and multistress-resistant mutants of *Lactococcus lactis*: identification of intracellular stress signals. Mol Microbiol 35:517–528
Rezaiki L, Cesselin B, Yamamoto Y, Vido K, van West E, Gaudu P, Gruss A (2004) Respiration metabolism reduces oxidative and acid stress to improve long-term survival of *Lactococcus lactis*. Mol Microbiol 53:1331–1342
Rezaiki L, Lamberet G, Derre A, Gruss A, Gaudu P (2008) *Lactococcus lactis* produces short-chain quinones that cross-feed Group B Streptococcus to activate respiration growth. Mol Microbiol 67:947–957
Rowland P NS, Jensen KF, Larsen S (2000) Structure of dihydroorotate dehydrogenase B: electron transfer between two flavin groups bridged by an iron-sulphur cluster. Structure 8:1227–35
Solem C, Koebmann BJ, Jensen PR (2003) Glyceraldehyde-3-phosphate dehydrogenase has no control over glycolytic flux in *Lactococcus lactis* MG1363. J Bacteriol 185:1564–1571
Stadtman ER, Oliver CN (1991) Metal-catalyzed oxidation of proteins. Physiological consequences. J Biol Chem 266:2005–2008
Stillman TJ, Upadhyay M, Norte VA, Sedelnikova SE, Carradus M, Tzokov S, Bullough PA, Shearman CA, Gasson MJ, Williams CH, Artymiuk PJ, Green J (2005) The crystal structures of *Lactococcus lactis* MG1363 Dps proteins reveal the presence of an N-terminal helix that is required for DNA binding. Mol Microbiol 57:1101–1112
Tachon S, Michelon D, Chambellon E, Cantonnet M, Mezange C, Henno L, Cachon R, Yvon M (2009) Experimental conditions affect the site of tetrazolium violet reduction in the electron transport chain of *Lactococcus lactis*. Microbiology 155:2941–2948
Thibessard A, Borges F, Fernandez A, Gintz B, Decaris B, Leblond-Bourget N (2004) Identification of *Streptococcus thermophilus* CNRZ368 genes involved in defense against superoxide stress. Appl Environ Microbiol 70:2220–2229
Turner MS, Tan YP, Giffard PM (2007) Inactivation of an iron transporter in *Lactococcus lactis* results in resistance to tellurite and oxidative stress. Appl Environ Microbiol 73:6144–6149
van de Guchte M, Serror P, Chervaux C, Smokvina T, Ehrlich SD, Maguin E (2002) Stress responses in lactic acid bacteria. Antonie Van Leeuwenhoek 82:187–216
Veiga P, Bulbarela-Sampieri C, Furlan S, Maisons A, Chapot-Chartier MP, Erkelenz M, Mervelet P, Noirot P, Frees D, Kuipers OP, Kok J, Gruss A, Buist G, Kulakauskas S (2007) SpxB regulates O-acetylation-dependent resistance of *Lactococcus lactis* peptidoglycan to hydrolysis. J Biol Chem 282:19342–19354
Verneuil N, Rince A, Sanguinetti M, Posteraro B, Fadda G, Auffray Y, Hartke A, Giard JC (2005) Contribution of a PerR-like regulator to the oxidative-stress response and virulence of *Enterococcus faecalis*. Microbiology 151:3997–4004
Verneuil N, Sanguinetti M, Le Breton Y, Posteraro B, Fadda G, Auffray Y, Hartke A, Giard JC (2004) Effects of the *Enterococcus faecalis hypR* gene encoding a new transcriptional regulator on oxidative stress response and intracellular survival within macrophages. Infect Immun 72:4424–4431
Vido K, Diemer H, Van Dorsselaer A, Leize E, Juillard V, Gruss A, Gaudu P (2005) Roles of thioredoxin reductase during the aerobic life of *Lactococcus lactis*. J Bacteriol 187:601–610
Vido K, Le Bars D, Mistou MY, Anglade P, Gruss A, Gaudu P (2004) Proteome analyses of heme-dependent respiration in *Lactococcus lactis*: involvement of the proteolytic system. J Bacteriol 186:1648–1657

Vinella D, Albrecht C, Cashel M, D'Ari R (2005) Iron limitation induces SpoT-dependent accumulation of ppGpp in *Escherichia coli*. Mol Microbiol 56:958–970

Wang E, Bauer MC, Rogstam A, Linse S, Logan DT, von Wachenfeldt C (2008) Structure and functional properties of the *Bacillus subtilis* transcriptional repressor Rex. Mol Microbiol 69:466–478

Willemoes M, Kilstrup M, Roepstorff P, Hammer K (2002) Proteome analysis of a *Lactococcus lactis* strain overexpressing *gapA* suggests that the gene product is an auxiliary glyceraldehyde 3-phosphate dehydrogenase. Proteomics 2:1041–1046

Winstedt L, Frankenberg L, Hederstedt L, von Wachenfeldt C (2000) *Enterococcus faecalis* V583 contains a cytochrome bd-type respiratory oxidase. J Bacteriol 182:3863–3866

Yamamoto Y, Fukui K, Koujin N, Ohya H, Kimura K, Kamio Y (2004) Regulation of the intracellular free iron pool by Dpr provides oxygen tolerance to *Streptococcus mutans*. J Bacteriol 186:5997–6002

Yamamoto Y, Pargade V, Lamberet G, Gaudu P, Thomas F, Texereau J, Gruss A, Trieu-Cuot P, Poyart C (2006) The Group B *Streptococcus* NADH oxidase Nox-2 is involved in fatty acid biosynthesis during aerobic growth and contributes to virulence. Mol Microbiol 62:772–785

Yamamoto Y, Poole LB, Hantgan RR, Kamio Y (2002) An iron-binding protein, Dpr, from *Streptococcus mutans* prevents iron-dependent hydroxyl radical formation in vitro. J Bacteriol 184:2931–2939

Yamamoto Y, Poyart C, Trieu-Cuot P, Lamberet G, Gruss A, Gaudu P (2005) Respiration metabolism of Group B Streptococcus is activated by environmental haem and quinone and contributes to virulence. Mol Microbiol 56:525–534

You C, Sekowska A, Francetic O, Martin-Verstraete I, Wang Y, Danchin A (2008) Spx mediates oxidative stress regulation of the methionine sulfoxide reductases operon in *Bacillus subtilis*. BMC Microbiol 8:128

Zhang W, Wong KK, Magliozzo RS, Kozarich JW (2001) Inactivation of pyruvate formate-lyase by dioxygen: defining the mechanistic interplay of glycine 734 and cysteine 419 by rapid freeze-quench EPR. Biochemistry 40:4123–4130

Zuber P (2004) Spx-RNA polymerase interaction and global transcriptional control during oxidative stress. J Bacteriol 186:1911–1918

Chapter 7
Responses of Lactic Acid Bacteria to Starvation

Bart C. Weimer

7.1 Introduction

Since the first description of viable but nonculturable (NC) organisms by Colwell and Grimes (2000), many microbes have been recognized to enter an NC state in the later stages of the stationary phase (Oliver 2010). Since then, most of the focus has been directed to pathogenic bacteria and their ability to become NC, to persist in the environment, and to initiate infections. Though it is well recognized that most bacteria in the environment are NC, this physiological state was not clearly demonstrated in lactic acid bacteria (LAB) until recently (Stuart et al. 1999; Ganesan et al. 2007). The realization that microbes in many habitats are largely NC is pushing microbiologists increasingly to utilize culture-independent measurement tools. Initially, gene chips were employed (Brodie et al. 2007). More recently, a metagenomic assessment for the definition of the community membership and the functional genes has been used; high-throughput DNA/RNA sequencing technologies are rapidly evolving to provide extensive genome coverage at a low cost from many samples. The realization that common food-borne pathogens are routinely NC is fueling an extensive debate about the impact on public health via water carriage and pathogen survival coupled to host infection, especially for enteric microbes and mycobacteria.

Raw milk also carries NC spoilage and possible pathogens that are heat-resistant and metabolically active (Gunasekera et al. 2002). Infection studies using NC *Salmonella enterica* sv. Typhimurium found that only actively dividing cells produced an infection (Passerat et al. 2009). While the vast majority of organisms associated with foods are capable of achieving the NC state, this chapter will focus on LAB, the NC state, and the metabolic capabilities associated with persistence of NC cells.

B.C. Weimer (✉)
Department of Population Health and Reproduction, School of Veterinary Medicine, University of California, Davis, CA 95616, USA
e-mail: bcweimer@ucdavis.edu

E. Tsakalidou and K. Papadimitriou (eds.), *Stress Responses of Lactic Acid Bacteria*, Food Microbiology and Food Safety, DOI 10.1007/978-0-387-92771-8_7,

7.2 Methods to Detect Nonculturable Cells

Traditional plating methods rely on the ability of the microbe to form colonies on a plate during cell division. This is the basis for virtually all methods to assess bacterial populations in regulatory situations. As cells become stressed, the ability to form colonies diminishes, yet they persist in broth. This phenotype demands that detection be based upon culture-independent methods, which are now becoming readily available. Specific dyes can be used to differentiate live and dead cells, relying on the ability of some to cross the membrane, while others do not unless the membrane is compromised (Cenciarini-Borde et al. 2009). Alternatively, dyes coupled to flow cytometry can be used to test and sort subpopulations that have different membrane potentials as well (Papadimitriou et al. 2006). The use of cellular DNA dyes enables flow cytometry to be used to assess the proportion of the population that is NC in the laboratory (Papadimitriou et al. 2007). However, these approaches have not achieved widespread use in food systems largely due to the interfering food particles, but they clearly demonstrate that subpopulations occur in LAB with differences in membrane integrity and culturability.

Genomics and culture-independent methods using metagenomics and PCR are gaining popularity to study NC populations directly from complex samples that can assess horizontal gene transfer and genetic diversity (Parnell et al. 2010). A concern of the food industry is the extraction of DNA from dead cells, hence the necessity for using membrane and DNA-binding dyes to differentiate live and dead cells. While it is possible that DNA remains stable to the environment during incubation, its stability can be compromised in fermented dairy products, where many nucleases are present (Weimer, unpubl. data). DNA extraction methods are now widely available in order to obtain high-quality DNA that is intact and of sufficient quality for PCR and sequencing. The use of DNA-complexing dyes, like propidium monoazide (PMA), a DNA-binding dye that is photo-cross-linked to DNA if the membrane is compromised, limits qPCR detection to intact, live cells (Nocker et al. 2006; Nocker and Camper 2009). It has been suggested that combining qPCR and PMA is an appropriate method to monitor disinfection (Nocker et al. 2007). While this is an interesting approach, it is unclear if it is applicable to industrial or food fermentations where biofilms occur. Recently, PMA and qPCR were used to determine the live/dead ratio of *Lactobacillus acidophilus* in yogurt to find that probiotic organisms survive very well during storage (Tabasco et al. 2009).

Alternatively, researchers are extracting RNA as a measure of cell viability to exploit the greater instability of RNA compared to DNA, and to gain access to information about gene expression, cell viability, and bacterial metabolism during product storage. Gene expression performed directly from complex samples and foods, especially dairy products containing NC LAB, is assessed routinely to determine changes during stress and storage (Weimer et al. 2004; Xie et al. 2004). This approach enables the detection of the gene expression of membrane proteins and transport systems during storage in food products that are informative of the community structure, cell state, and metabolism.

7.3 Resuscitation of Nonculturable Cells

Skeptics for the existence of the NC state were largely driven by the inability to "wake up" the cells after they became NC. This initial criticism is waning, but the method to induce a reculturable state from cells after they become NC remains largely unknown. While some Gram-negative proteobacteria (enterobacteria) and a few *actinobacteria* (e.g., micrococci, mycobacteria) have been resuscitated, relatively few molecules are known to induce the recovery of NC microbes. Resuscitation-promoting factors (*rpf* locus; Rpf), small proteins with muralytic activity excreted into the medium, initially identified in *Micrococcus luteus*, are now being discovered in many *actinobacteria*, including homologs in bifidobacteria and mycobacteria. These molecules are able to induce regrowth (Mukamolova et al. 1998; Kana and Mizrahi 2010). Work is progressing to understand the exact unresolved relationship between the predicted biochemical activity of Rpfs (i.e., the cleavage of the beta-1,4 glycosidic bond in the glycan backbone of peptidoglycan) and the resuscitation mechanisms (Kana and Mizrahi 2010). These diverse lytic transglycosylases are not well conserved and form a number of families. The commonality is cleavage of the glycosidic linkage between *N*-actetylmuramoyl and *N*-acetylglucosaminyl residues to remodel the peptidoglycan with the hypothesis that it will lead to regrowth (Scheurwater et al. 2008). Additionally, a new class of 2-nitrophenylthiocyanates was recently found that also regulates NC cells but does not influence actively dividing cells (Demina et al. 2009).

At this point, there is not a clear definition or mechanism of exactly how peptidoglycan remodeling and turnover occur in the large number of organisms that become NC. It appears that peptidoglycan hydrolases produce products that exert additional signal transduction-like activities through Ser/Thr protein kinases. Additionally, quorum-sensing molecules appear to be responsible for cellular communication during the NC state in biofilms for Gram-negative organisms, yet the specific molecules for Gram-positive organisms that do not respond to homoserine lactones have yet to be described (Oliver 2010). Lactococci and lactobacilli do not respond to homoserine lactones during normal growth or during the NC state, indicating that other small molecules act as signaling compounds for NC resuscitation (Stuart et al. 1999).

While carbohydrate exhaustion ushers LAB into the NC state, resuscitation has not been demonstrated. Many LAB, such as lactobacilli, lactococci, bifidobacteria, brevibacteria, and oenococci, enter the NC state, but they have yet to be resuscitated. Examination of the LAB genomes for the molecules described in other bacteria has yielded homologs that are not well conserved, suggesting that additional mechanisms for resuscitation in LAB are yet to be described. It appears that these genera contain distant homologs of Rfps as well as other uncharacterized genes in lactococci that may mediate resuscitation (Ravagnani et al. 2005) (Figs. 7.1 and 7.2). Interestingly, the operon downstream of *yuaE* that contains YuaC and LACR2197 is unique to lactococci and is not found in any other microbe.

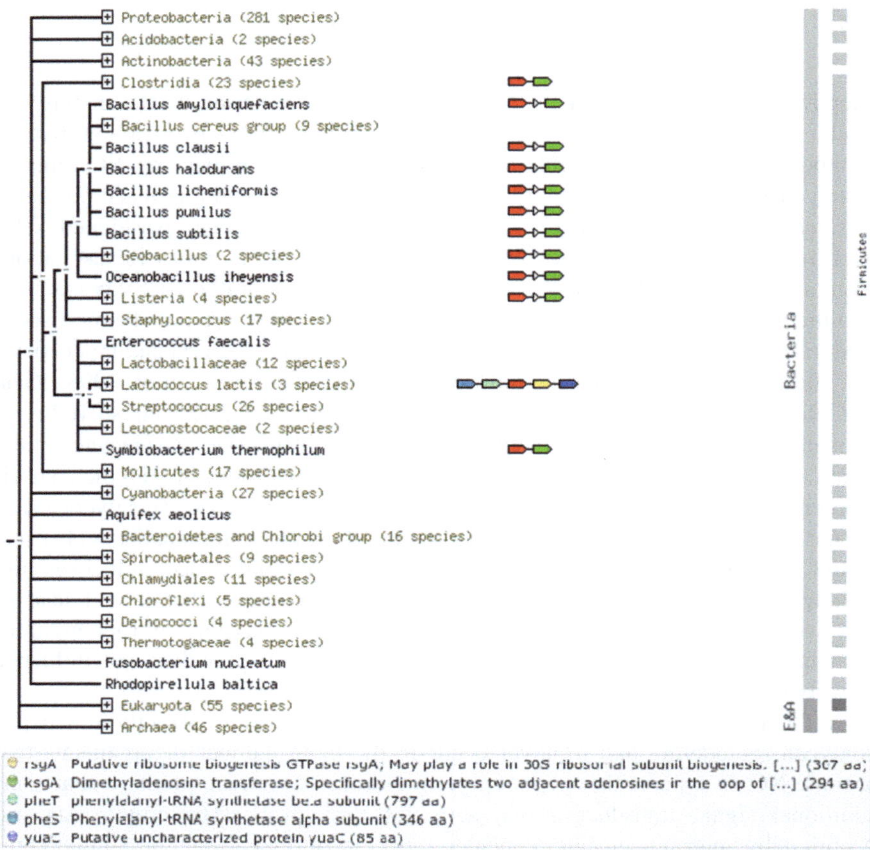

Fig. 7.1 Gene neighborhood of *yuaE*, a distant homolog of Rfp in *Lactococcus lactis* subsp. *cremoris* SK11. STRING was used to determine the predicted interacting proteins and the genome neighborhood. The structure in lactococci seems to be completely unique as compared to the other organisms with this protein

Yet, none of these homologs in these organisms has been demonstrated to lead to resuscitation. Some clues are appearing with the realization that the cells survive longer after salt stress in the presence of glycine/betaine (Weimer 2010) and after oxidative stress with the transport of glutathione (Zhang et al. 2010) and that galactose accumulation is coupled with the activity of autolysins in lactococci (*acmA*) (Steen et al. 2008). Interestingly, both glutathione and galactose accumulation led to changes in sugar metabolism and cell morphology and to resistance to autolysis. Lastly, the deacetylation of peptidoglycan in *Lactococcus lactis* via the regulation of peptidoglycan *N*-acetylglucosamine deacetylase (*pdgA* aka *xynD*) reduced autolysis due to *acmA* activity (Meyrand et al. 2007). Presumably, this was due to a modified charge on the cell wall and reduced interaction between AcmA and acetylglucosamine. Additional work is needed to fully determine the role of autolysins in

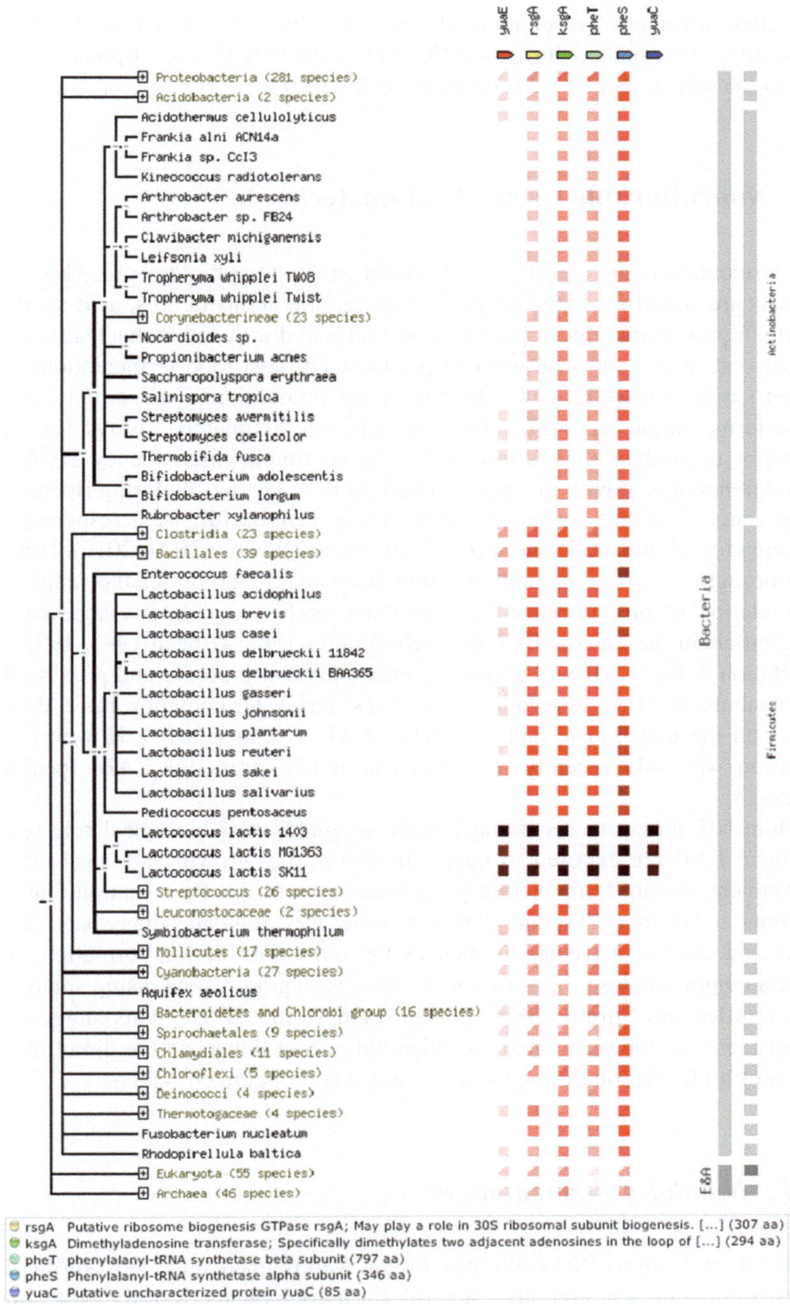

Fig. 7.2 Predicted association partners of *yuaE*, an Rpf-like homolog, from *Lactococcus lactis* subsp. *cremoris* SK11. STRING was used to determine the predicted interacting proteins and find other homologs

cell wall turnover and resuscitation. This is especially true for LAB, since *acmA* was not found to be regulated during the NC state, indicating that other processes are at play on the cell surface of LAB (Ganesan et al. 2007).

7.4 Nonculturable Lactic Acid Bacteria

Food fermentations are of specific importance in the stress biology of LAB since the products are stored for extended periods in variable temperatures, acid conditions (usually highly acidic), redox potential, and carbohydrate content that lead to unique cellular responses and metabolic end products. In most food fermentations, no or limited simple sugars remain after the initial production. For example, in hard cheese fermentation, lactose is exhausted within approximately 30 days, leaving the organisms to produce flavors from either fat or protein (McSweeney 2007). This observation, coupled with the apparent cell death of lactococci during ripening, led to the conclusion that nonstarter LAB, mostly lactobacilli, were responsible for accumulation of metabolic end products in cheese (Weimer 2007, 2010). However, the thorough list of compounds assembled by many authors cannot explain the accumulation of protein-related end products such as aromatic compounds and branched-chain fatty acids (BCFAs) (Schormüller 1968; Urbach 1993, 1995). The expectation is that LAB will produce compounds that preserve and provide flavor via metabolism. The combination of stresses during fermentation has a dramatic impact on the metabolism and survival of LAB (Weimer 2010). However, sugar starvation was only recently recognized as a key stress for LAB, specifically lactococci.

While NC states are becoming largely accepted, their biological role remains unknown. LAB are exposed to many stresses during growth and survival in the environment, during fermentation processes, and with host association that likely contribute to NC entry (Weimer 2007). Abiotic stress regulates many bacterial processes and cellular activities as well as the induction biofilm formation, where bacteria commonly lose the ability to divide and begin to signal using small molecules (Keller and Surette 2006; Straight and Kolter 2009). This suggests that perhaps some of the compounds so prized in cheese flavor are the direct result of LAB biofilm formation during the aging of LAB fermented products.

7.4.1 Nondairy Fermentations

In meat fermentations, NC cell types are not well defined since most of the flavor comes from lactic acid and other sugar-derived compounds from the large amount of added sugar. In sake and beer fermentations, lactobacilli become NC during aging and are related to spoilage (Koji et al. 2006). Using different media and cell-staining dyes, colonies and cell morphology were found to change during extended storage (Koji et al. 2006; Suzuki et al. 2008). While it seems likely that lactobacilli have the ability to become NC in beer, additional work is needed to demonstrate the relevant mechanisms.

7.4.2 Dairy Fermentations

During dairy fermentations where lactococci are used to initiate lactose fermentation, a number of groups noted that lactococci enter a different metabolic state during subsequent cheese aging, associated with amino acid metabolism and bacterial survival (Thomas and Batt 1968; Kunji et al. 1993; Rallu et al. 1996; Stuart et al. 1999). Leveraging previous observations and the ability of lactococci to utilize Arg for energy (Thomas and Batt 1968; Kunji et al. 1993), Chou demonstrated that a balance among lactose use, pH, and Arg consumption regulates pH and energy during growth and carbon starvation (Chou and Weimer 2001). It is widely believed that lactococci are alive only in the early steps of cheese making, where they produce lactic acid from lactose. The cheese-making protocol is designed to control the production of acid using heat and salt (O'Keeffe et al. 1976; Arora et al. 1995; Crow et al. 1995). Until recently, these two stresses, along with bacteriophages, were thought to solely slow metabolism and ultimately lead to autolysis, releasing intracellular enzymes that produced flavor compounds during ripening. While some amount of the population does lyse and contributes to flavor production (Crow et al. 1995), complete lysis leads to no flavor production (V. Crow and R. Holland, personal communication), indicating that metabolism by lactococci is critical for substrate conversion to flavor compounds in cheese.

While the growth curves using plate counts lead to the conclusion that lactococci lyse, changes during cheese ripening assist the conversion of amino acids to new compounds at concentrations that cannot be accounted for solely by cell lysis. This conclusion led to the assumption that apparent declines in the plate count were the result of widespread autolysis, and lactococci were simply "bags of enzymes" to release intracellular metabolic enzymes that are active during cheese ripening (Wilkinson et al. 1994; Ostlie et al. 1995; Buist et al. 1998). However, these conclusions presume that intracellular enzymes are active at the salt concentration (4–5% in the moisture), pH (5.0–5.3), and temperature (~5–15°C) found in aging cheese to produce the compounds associated with ripening.

Stuart et al. (1999) and Ganesan et al. (2007) demonstrated that carbohydrate starvation leads to the NC state within 12–48 h in lactococci. During short-term starvation (2 weeks), lactococci exhibited varying abilities to exhaust lactose, which delayed entry into the NC state, and a few strains never achieved lactose starvation (Stuart et al. 1999). Once the cells became NC, they continued to exhaust Arg and Met but produced Ser from the initial steps of glycolysis. Lactococci continue to remain metabolically active and play a causative role in cheese flavor in the NC state, as Ganesan estimates that only 0.001–0.1% of the total population undergoes autolysis (Ganesan et al. 2006, 2007).

Linking the inconsistencies among enzyme activities in the harsh conditions of cheese, the increase in amino acids during ripening, and the production of lipids that are not possible from the lipolysis of milk led to the exploration of the ability of lactococci to survive carbon stress via amino acid metabolism as a mechanism to explain flavor production by NC cells. Weimer's group found that when lactococci are allowed to completely deplete lactose, they lose the ability to divide but continue to regulate the proteome to proceed with Arg metabolism through arginine deiminase (Stuart et al. 1999; Chou and Weimer 2001). This leads to a subpopulation of lactococci that is undetected by bacterial plate counts, causing an apparent reduction in the total population during cheese ripening, leading to the conclusion that the starter culture dies. DNA-binding and membrane dyes verified the existence of these two different populations: those cells that can grow on plates and those that become NC (Stuart et al. 1999; Ganesan et al. 2006, 2007). The NC population consists of three different types of cells: (1) cells with intact cell walls, which remain capable of amino acid and peptide transport and metabolism; (2) cells that have compromised membranes; and (3) cells that are truly dead (Ganesan et al. 2006, 2007). The presence of intact cells reinforces the importance of transport systems that provide alternate substrates such as amino acids for catabolism during sugar starvation and cheese ripening to produce flavor-related compounds, which point to many amino acids, including Met, Arg, aromatics, and branched-chain amino acids (BCAAs). It was clearly established that amino acid utilization was induced in response to carbohydrate starvation (Stuart et al. 1999; Chou and Weimer 2001; Ganesan and Weimer 2004; Ganesan et al. 2007). This condition is also induced in the cheese matrix during aging (Fig. 7.3) (Xie et al. 2004). These findings indicate that lactococci are much more resistant to stress than previously appreciated. However, additional work is needed to completely demonstrate the specific mechanisms of stress resistance and the role of NC populations that contribute to metabolism and flavor production during cheese ripening.

In lactococci, NC cells were induced after carbohydrate starvation (Stuart et al. 1999; Chou and Weimer 2001; Ganesan and Weimer 2004; Ganesan et al. 2007). The machinery for replication was significantly repressed during entry into the NC state (Ganesan et al. 2007). For example, 8 of 10 cytoskeletal genes (*fts*) were repressed after entry into the NC state, while the genes needed to put Ala on the cell wall were significantly induced about 2.5-fold. These two processes suggest that the

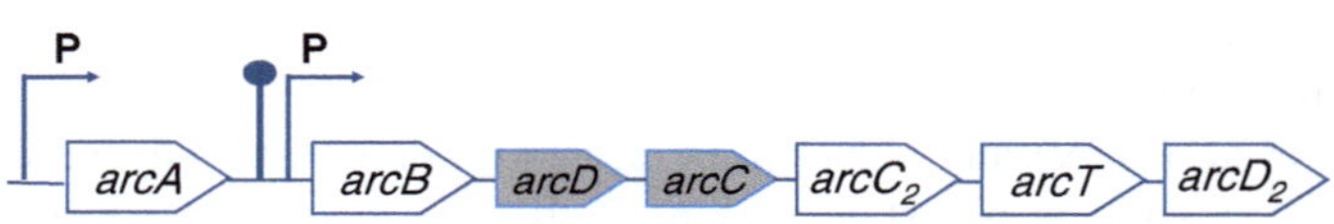

Age	ADI	OTC	A/O port	CK	End Products
Initial	+	+	+	+	ATP, CO_2, 2x NH_3
24 h	+	+		+	ATP, CO_2, 2x NH_3
45 d		+	+	+	ATP, CO_2, NH_3 – Orn transport

Fig. 7.3 Arginine utilization in Cheddar cheese during ripening using gene expression arrays (Xie et al. 2004). *ADI* (*arcA*) is arginine deiminase; *OTC* (*arcB*) is ornithine transcarbamylase; *CK* (*arcC*, *arcC*$_2$) is carbamate kinase; *A/O port* (*arcD* and *arcD*$_2$) is arginine/ornithine antiporter. *Plus* signs indicate the presence of activity in curd

cytoskeleton was no longer forming and that the cell wall was becoming resistant to hydrolysis by capping charged side groups making the cell wall less susceptible to hydrolysis by neuraminidases. None of the *acm* genes were expressed or regulated during the transition from active growth to the NC state, nor were holins from the three prophage families (pi148, pi251, and pi306) in *Lactococcus cremoris* SK11. Further, the carbon starvation protein (*cstA*) that signals broad carbon starvation was not induced during NC induction, suggesting that the cell was not starving for carbon. Presumably, amino acid metabolism was taking care of the carbon and nitrogen needs as further observed by induction of amino acid catabolic pathways and production of organic acids that lead to energy and redox control. These five changes alone can explain lactococci NC states. How the cell remains energized via metabolism needs to be elucidated, but amino acid catabolism is a starting point.

7.5 Metabolism and Nonculturable Cells

Lactococci are metabolically limited: They lack a functional electron transport system, contain relatively few metabolic pathways, depend on proteolysis for many amino acids, and have inducible complex carbohydrate metabolism systems (Makarova et al. 2006; Makarova and Koonin 2007).

Recently, a number of regulators (e.g., *ccpA*, *cstA*, and *codY*) were implicated in the transition to the stationary phase and forward to an NC state in lactococci. Cells lose the ability to form colonies on agar, but remain metabolically active, by the repression of *ftsZ*, which initiates cell division by forming the Z-ring. During this transition, Arg and BCAA catabolism increases, while sugar transport via the phosphotransferase system (PTS) is repressed due to the depletion of phosphoenolpyruvate (PEP) and the repression of HPr. This physiological state leads to the induction of new metabolic capabilities to produce BCFAs from BCAAs, which are only produced during carbon starvation in lactococci.

Previous studies focused on the growth and proliferation of lactococci due to the importance of phage resistance and acid production during the initial milk fermentation. The lack of consistency in flavor and the need for improved flavor in low-fat cheese reinvigorated interest in lactococcal metabolism, which began to lead to the bridge between the long list of compounds found in cheese and the metabolic routes for production (Gao et al. 1997; Weimer et al. 1997). Among the amino acid metabolic pathways, the arginine deiminase system in lactococci and lactobacilli appears to enable ATP production and pH regulation, and to directly link the switch from sugar to amino acid metabolism in the NC state (Stuart et al. 1999; Chou and Weimer 2001; Weimer et al. 2004; Xie et al. 2004). However, other amino acids, including aromatic, sulfur-containing, and BCAAs, can produce compounds directly impacting flavor (Weimer 2007, 2010). Interestingly, some flavor compounds in cheese are also now being recognized as signaling molecules in biofilms of Gram-positive bacteria (Keller and Surette 2006; Straight and Kolter 2009). These compounds were identified to be produced from amino acid metabolism during the stationary phase (aromatic), and some are induced only after sugar exhaustion and entry into the NC state (BCFAs) in LAB (Weimer 2007, 2010). Additionally, furanones are also potential signaling molecules in LAB (Vannini et al. 2007), but their exact role for signaling and production remains to be examined.

7.5.1 Amino Acid Metabolism Induced During Nonculturable State

During sugar starvation, LAB switch to the catabolism of amino acids via a committed step using aminotransferases (ATases), but their physiological role in NC metabolism and survival during cheese ripening remains to be fully characterized (Weimer et al. 2004, 2007; Weimer 2007). The lactococcal genome contains 8–12 ATase genes, among which only four are enzymatically characterized and most of which can utilize two to four substrates with varying affinity (Bolotin et al. 2001; Makarova et al. 2006). The aromatic, branched chain, sulfur, and aspartic acid ATases, with overlapping substrate specificities, are regulated by pH, amino acid concentration, and sugar source or entry in the NC state (Fig. 7.4). During the NC state, the PEP-PTS

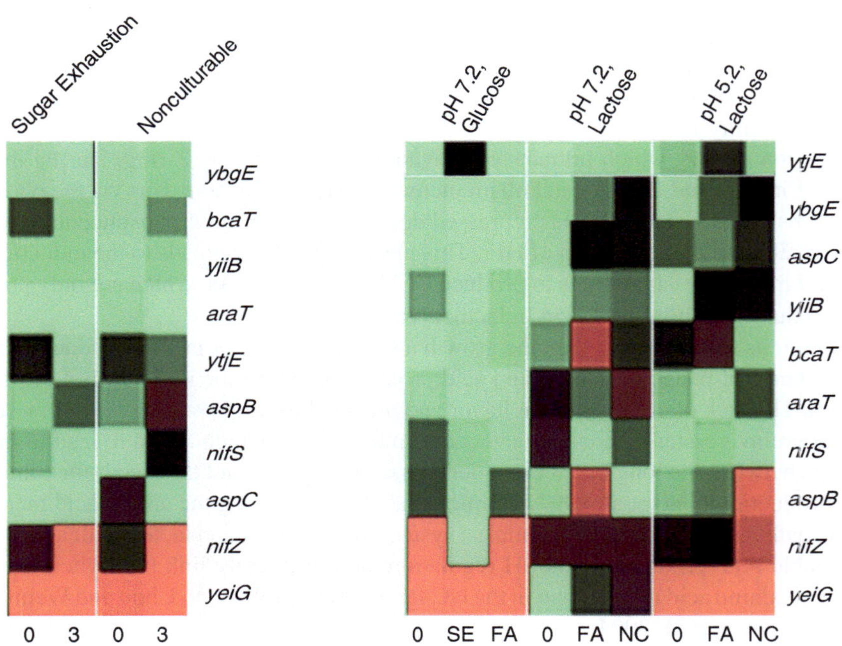

Fig. 7.4 Gene expression profile of lactococcal aminotransferases during sugar exhaustion (SE), entry to NC (NC), and branched-chain fatty acid production (FA). The numbers 0 and 3 indicate time (h). *Green* indicates repression, *black* is median expression, and *red* indicates induction. Data were collected during carbohydrate starvation using chemically defined media (Ganesan et al. 2004, 2007)

sugar transporters, the peptide transporters (*opp* and *opt* operons), and most aminopeptidases are repressed, while glycosyl hydrolases and the *dtp* operon, as well as the amino acid transporters, are induced (Ganesan et al. 2007). Interestingly, NC cells induce an entirely new set of amino acid metabolism machinery that begins with new transporters and ATases (Fig. 7.4) (Ganesan et al. 2007).

Amino acid to volatile fatty acids (VFAs) catabolic pathways are identified in LAB and cheese-associated genera (Ganesan et al. 2004, 2006; Ganesan and Weimer 2004). Interestingly, LAB accumulate VFAs, while brevibacteria cycle between production and consumption during NC incubation with BCAAs after carbohydrate exhaustion. A global transcriptional regulator that is unique to bacilliacae and found in LAB, *codY*, senses intracellular levels of BCAAs and aids in the catabolism of amino acids (Ganesan et al. 2004, 2006; Ganesan and Weimer 2004). Carbohydrate starvation induces BCFA production from BCAAs as a method to produce ATP during the NC state (Ganesan et al. 2006). It was noticed that substrate-level phosphorylation was used via a complex CoA-dependent multistep pathway to produce ATP. This approach produces sufficient ATP to enable cells to transport peptides during carbon and nitrogen starvation (Fig. 7.5).

Amino acid catabolism generates ATP via substrate-level phosphorylation and modulates the redox potential by the regeneration of NADH from NAD^+ under anaerobic conditions. This metabolism generates fatty acids only during carbon

Medium	Time (min)	pH	OD_{600}	Total Peptides in medium (peak area)
Basal medium	0	7.20±0.00	0.25±0.04	422.20±38.96
	120	6.70±0.00	0.33±0.03	423.05±13.32
Peptide-rich medium (casitone)	0	7.20±0.00	0.25±0.04	1223.85±34.18
	120	6.65±0.07	0.33±0.03	1324.30±32.59
Peptide hydrolyzed medium	0	7.20±0.00	0.23±0.04	889.05±28.95
	120	5.35±0.07	0.80±0.37	696.25±27.11

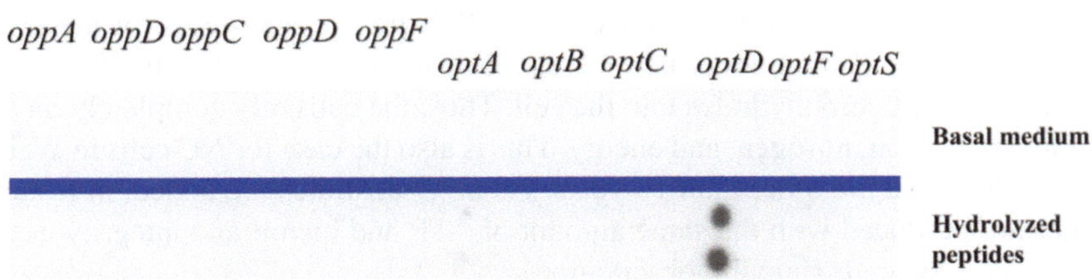

Fig. 7.5 Peptide transport during NC in *Lactococcus lactis* IL1403 with and without nitrogen starvation. The cells were incubated at 30°C and the peptide transport was measured with and without additional hydrolysis. Acidification was a measure of cell growth and metabolism. The peptide amount was measured using an HPLC to determine a reduction during the incubation. The expression of the peptide transport system was determined using gene expression arrays to find that NC cells express the *opt* operon and repress the *opp* operon that is used during log-phase growth

starvation that is initiated at the onset of the NC state. Considering that these pathways are not active in the presence of sugar, and that lactococci lose sugar catabolic traits via plasmids, the evidence toward amino acid catabolism to VFAs is only more convincing, as lactococci will lose glycolytic traits in the absence of sugars and continue with amino acid metabolism (Stuart et al. 1999; Ganesan et al. 2006).

While Arg is cometabolized with lactose, its catabolism is important only for the onset of starvation, as it is catabolized immediately in lactose limitation or lower pH (Chou and Weimer 2001), and Arg residues are present in a limited number in casein in comparison to BCAAs. It is clear that BCFA formation requires carbon starvation and accompanies ATP stability along with the control of nitrogen metabolism via Arg and Glu or 2-oxoglutamate. Interestingly, during fatty acid production, portions of the nitrogen regulation system of Glu, Gln, and 2-oxoglutamate were repressed, while other portions (i.e., Arg) were induced due to regulation by *ccpA* and *codY*. Further, many of the aminopeptidases were repressed at the NC state. These observations suggest that nitrogen metabolism is still tightly regulated during NC so that alternate catabolic shunts are used to control energy and redox. However, since LAB lack *rpoN* and shift between *codY* and *codZ* during NC, it is clear that additional study is needed to truly define the role of carbon regulation and the resulting impact on metabolism during the NC state in LAB.

7.5.2 *Polysaccharide Metabolism Induced During Nonculturable State*

The *ccpA* regulator that binds to the CRE element is also important for the carbon starvation response during polysaccharide utilization. Conservatively, there are approximately 1,000 DNA-binding regions in LAB for this regulator if one or two mismatches are allowed. The CRE box regulates sugar catabolism broadly – ribose, mannitol, galactose, maltose, gluconate, and the genes needed to degrade complex carbohydrates (alpha- and beta-glucosidases). This regulatory pair also regulates citrate metabolism in part. Interestingly, during the NC state, complex sugar metabolic enzymes were induced. However, the PTS transport systems for these sugars were repressed, leaving only permeases, or in some cases no transport systems, to bring complex carbohydrates into the cell. Thus, the cells rely completely on amino acids for carbon, nitrogen, and energy. This is also the case for NC cells in Weimer's laboratory that are older than 10 years. As new substrates are added, new metabolites are produced with the same amount of ATP and membrane integrity as newly produced NC cells (unpubl. observation).

7.6 Concluding Remarks

Sugar starvation induces the NC state in lactobacilli and lactococci, which is common in other bacteria that are found in the environment and associated with animals. The mechanisms underlying the NC state are intensively studied in many pathogens due to the public health implications. Methods to detect NC cells include membrane dyes and culture-independent genetic approaches. With the wide-scale availability of genomes and genomic tools, the study of NC cells is now tractable. The resuscitation of NC cells is possible in a few microbes and is becoming more common with the discovery of specific small molecules and peptides that are quorum-sensing molecules. Some of these molecules are flavor-active but have not been shown to resuscitate LAB. Due to the small size of the genome of LAB, many of the regulatory or quorum-sensing molecules are not produced. However, the accumulation of compatible molecules enables cellular stability, but no specific molecule or metabolic route has been identified to rescue LAB from the NC state. Stuart et al. (1998) determined that lactococci lose the ability to produce colonies after carbohydrate exhaustion, which is accompanied by the production of Met and Ser into the medium – just as observed during cheese ripening. Ganesan et al. (2006) extended this study using gene expression to confirm Stuart's work and further demonstrated that the cells continue to transcribe RNA even without the ability to form colonies. This observation was used to demonstrate BCFA production from amino acids using gene expression arrays. The authors also demonstrated that this cellular state can last as long as 3 years without the induction of the genes needed for cellular lysis. They proposed that a percentage between 0.1 and 0.001% of the starter culture lysed during this state. Hence, the culture appears to die because it cannot form colonies, yet the cells are intact and continue to metabolize peptides and amino acids to end products that impact flavor. The full impact of this metabolic state remains to be clarified in fermented foods. The biological significance of this ability remains to be demonstrated. However, it is clear that the regulation of carbon and nitrogen flow is still highly regulated in NC cells that are more than 10 years old.

References

Arora G, Cormier F, Lee B (1995) Analysis of odor-active volatiles in Cheddar cheese headspace by multidimensional GC/MS/sniffing. J Agric Food Chem 43:748–752

Bolotin A, Wincker P, Mauger S, Jaillon O, Malarme K, Weissenbach J, Ehrlich SD, Sorokin A (2001) The complete genome sequence of the lactic acid bacterium *Lactococcus lactis* subsp. *lactis* IL1403. Genome Res 11:731–753

Brodie EL, DeSantis TZ, Parker JP, Zubietta IX, Piceno YM, Andersen GL (2007) Urban aerosols harbor diverse and dynamic bacterial populations. Proc Natl Acad Sci USA 104:299–304

Buist G, Venema G, Kok J (1998) Autolysis of *Lactococcus lactis* is influenced by proteolysis. J Bacteriol 180:5947–5953

Cenciarini-Borde C, Courtois S, La Scola B (2009) Nucleic acids as viability markers for bacteria detection using molecular tools. Future Microbiol 4:45–64

Chou L-S, Weimer BC (2001) Interaction between glucose and arginine on the acid production in *Lactococcus lactis* ssp. *lactis*. Int Dairy J 11:253–258

Colwell RR, Grimes DJ (2000) Nonculturable microorganisms in the environment. ASM Press, Washington

Crow VL, Coolbear T, Gopal PK, Martley FG, McKay LL, Riepe H (1995) The role of autolysis of lactic acid bacteria in the ripening of cheese. Int Dairy J 5:855–875

Demina GR, Makarov VA, Nikitushkin VD, Ryabova OB, Vostroknutova GN, Salina EG, Shleeva MO, Goncharenko AV, Kaprelyants AS (2009) Finding of the low molecular weight inhibitors of resuscitation promoting factor enzymatic and resuscitation activity. PLoS One 4:e8174

Ganesan B, Weimer BC (2004) Role of aminotransferase IlvE in production of branched-chain fatty acids by *Lactococcus lactis* subsp. *lactis*. Appl Environ Microbiol 70:638–641

Ganesan B, Seefeldt K, Weimer BC (2004) Fatty acid production from amino acids and alpha-keto acids by *Brevibacterium linens* BL2. Appl Environ Microbiol 70:6385–6393

Ganesan B, Dobrowolski P, Weimer BC (2006) Identification of the leucine-to-2-methylbutyric acid catabolic pathway of *Lactococcus lactis*. Appl Environ Microbiol 72:4264–4273

Ganesan B, Stuart MR, Weimer BC (2007) Carbohydrate starvation causes a metabolically active but nonculturable state in *Lactococcus lactis*. Appl Environ Microbiol 73:2498–2512

Gao S, Oh D–H, Broadbent J, Johnson M, Weimer B, Steele J (1997) Aromatic amino acid catabolism by lactococci. Lait 77:371

Gunasekera TS, Sorensen A, Attfield PV, Sorensen SJ, Veal DA (2002) Inducible gene expression by nonculturable bacteria in milk after pasteurization. Appl Environ Microbiol 68:1988–1993

Kana BD, Mizrahi V (2010) Resuscitation-promoting factors as lytic enzymes for bacterial growth and signaling. FEMS Immunol Med Microbiol 58:39–50

Keller L, Surette MG (2006) Communication in bacteria: an ecological and evolutionary perspective. Nat Rev Microbiol 4:249–258

Koji S, Kazumaru I, Shizuka A, Hidetoshi K, Yasushi K (2006) Induction of viable but nonculturable state in beer spoilage lactic acid bacteria. J Inst Brewing 112:295–301

Kunji ERS, Ubbink T, Matin A, Poolman B, Konings WN (1993) Physiological responses of *Lactococcus lactis* ML3 to alternating conditions of growth and starvation. Arch Microbiol 159:372–379

Makarova KS, Koonin EV (2007) Evolutionary genomics of lactic acid bacteria. J Bacteriol 189:1199–1208

Makarova K, Slesarev A, Wolf Y, Sorokin A, Mirkin B, Koonin E, Pavlov A, Pavlova N, Karamychev V, Polouchine N, Shakhova V, Grigoriev I, Lou Y, Rohksar D, Lucas S, Huang K, Goodstein DM, Hawkins T, Plengvidhya V, Welker D, Hughes J, Goh Y, Benson A, Baldwin K, Lee JH, Díaz-Muñiz I, Dosti B, Smeianov V, Wechter W, Barabote R, Lorca G, Altermann E, Barrangou R, Ganesan B, Xie Y, Rawsthorne H, Tamir D, Parker C, Breidt F, Broadbent J, Hutkins R, O'Sullivan D, Steele J, Unlu G, Saier M, Klaenhammer T, Richardson P, Kozyavkin S, Weimer B, Mills D (2006) Comparative genomics of the lactic acid bacteria. Proc Natl Acad Sci USA 103:15611–15616

McSweeney P (2007) Cheese manufacture and ripening and their influence on cheese flavour. In: Weimer B (Ed.), Improving the flavor of cheese. Woodhead Publishing, Boca Raton, p 600

Meyrand M, Boughammoura A, Courtin P, Mezange C, Guillot A, Chapot-Chartier MP (2007) Peptidoglycan N-acetylglucosamine deacetylation decreases autolysis in *Lactococcus lactis*. Microbiology 153:3275–3285

Mukamolova GV, Kaprelyants AS, Young DI, Young M, Kell DB (1998) A bacterial cytokine. Proc Natl Acad Sci USA 95:8916–8921

Nocker A, Camper AK (2009) Novel approaches toward preferential detection of viable cells using nucleic acid amplification techniques. FEMS Microbiol Lett 291:137–142

Nocker A, Cheung CY, Camper AK (2006) Comparison of propidium monoazide with ethidium monoazide for differentiation of live vs. dead bacteria by selective removal of DNA from dead cells. J Microbiol Methods 67:310–320

Nocker A, Sossa KE, Camper AK (2007) Molecular monitoring of disinfection efficacy using propidium monoazide in combination with quantitative PCR. J Microbiol Methods 70:252–260

O'Keeffe RB, Fox PF, Daly C (1976) Contribution of rennet and starter proteases to proteolysis in cheddar cheese. J Dairy Res 43:97–107

Oliver JD (2010) Recent findings on the viable but nonculturable state in pathogenic bacteria. FEMS Microbiol Rev 34:415–425

Ostlie HM, Vegarud G, Langsrud T (1995) Autolysis of lactococci: detection of lytic enzymes by polyacrylamide gel electrophoresis and characterization in buffer systems. Appl Environ Microbiol 61:3598–3603

Papadimitriou K, Pratsinis H, Nebe-von-Caron G, Kletsas D, Tsakalidou E (2006) Rapid assessment of the physiological status of *Streptococcus macedonicus* by flow cytometry and fluorescence probes. Int J Food Microbiol 111:197–205

Papadimitriou K, Pratsinis H, Nebe-von-Caron G, Kletsas D, Tsakalidou E (2007) Acid tolerance of *Streptococcus macedonicus* as assessed by flow cytometry and single-cell sorting. Appl Environ Microbiol 73:465–476

Parnell JJ, Rompato G, Latta LC 4th, Pfrender ME, Van Nostrand JD, He Z, Zhou J, Andersen G, Champine P, Ganesan B, Weimer BC (2010) Functional biogeography as evidence of gene transfer in hypersaline microbial communities. PLoS One 5:e12919

Passerat J, Got P, Dukan S, Monfort P (2009) Respective roles of culturable and viable-but-nonculturable cells in the heterogeneity of *Salmonella enterica* serovar typhimurium invasiveness. Appl Environ Microbiol 75:5179–5185

Rallu F, Gruss A, Maguin E (1996) *Lactococcus lactis* and stress. Antonie von Leeuwenhoek 70:243–251

Ravagnani A, Finan CL, Young M (2005) A novel firmicute protein family related to the actinobacterial resuscitation-promoting factors by non-orthologous domain displacement. BMC Genom 6:39

Scheurwater E, Reid CW, Clarke AJ (2008) Lytic transglycosylases: bacterial space-making autolysins. Int J Biochem Cell Biol 40:586–591

Schormüller J (1968) The chemistry and biochemistry of cheese ripening. Adv Food Res 16:231

Steen A, Buist G, Kramer NE, Jalving R, Benus GF, Venema G, Kuipers OP, Kok J (2008) Reduced lysis upon growth of *Lactococcus lactis* on galactose is a consequence of decreased binding of the autolysin AcmA. Appl Environ Microbiol 74:4671–4679

Straight PD, Kolter R (2009) Interspecies chemical communication in bacterial development. Annu Rev Microbiol 63:99–118

Stuart M, Chou L–S, Weimer BC (1998) Influence of carbohydrate starvation on the culturability and amino acid utilization of *Lactococcus lactis* ssp. *lactis*. Appl Environ Microbiol 65:665–673

Stuart MR, Chou LS, Weimer BC (1999a) Influence of carbohydrate starvation and arginine on culturability and amino acid utilization of *Lactococcus lactis* subsp. *lactis*. Appl Environ Microbiol 65:665–673

Suzuki K, Asano S, Iijima K, Kitamoto K (2008) Sake and beer spoilage lactic acid bacteria – a review. J Inst Brewing 114:209–223

Tabasco R, Garcia-Cayuela T, Pelaez C, Requena T (2009) *Lactobacillus acidophilus* La-5 increases lactacin B production when it senses live target bacteria. Int J Food Microbiol 132:109–116

Thomas TD, Batt RD (1968) Survival of *Streptococcus lactis* in starvation conditions. J Gen Microbiol 50:367–382

Thomas TD, Batt RD (1969) Synthesis of protein and ribonucleic acid by starved *Streptococcus lactis* in relation to survival. J Gen Microbiol 58:363–369

Urbach G (1993) Relations between cheese flavour and chemical composition. Int Dairy J 3:389–422

Urbach G (1995) Contribution of lactic acid bacteria to flavour compound formation in dairy products. Int Dairy J 5:877–903

Vannini L, Ndagijimana M, Saracino P, Vernocchi P, Corsetti A, Vallicelli M, Cappa F, Cocconcelli PS, Guerzoni ME (2007) New signaling molecules in some Gram-positive and Gram-negative bacteria. Int J Food Microbiol 120:25–33

Weimer BC (2007) Improving the flavor of cheese. CRC Press/Woodhead, Boca Raton

Weimer BC (2010) Lactic acid bacteria: Physiology and stress response. In: Fuquay JW, Fox PF, McSweeney P (Eds.), Encyclopedia of dairy science, 2nd ed. Elsevier/Academic Press, London

Weimer BC, Brennand C, Broadbent J, Jaegi J, Johnson M, Milani F, Steele J, Sisson D (1997) Influence of flavor adjunct bacteria on the flavor and texture of 60% reduced fat cheddar cheese. Lait 77:383

Weimer BC, Xie Y, Chou L-S, Cutler A (2004) Gene expression arrays in food. In: Barredo J-L (Ed.), Microbial Products and Biotransformation. Humana Press, Totowa

Weimer B, Ganesan B, Rajan S (2007) Biotechnology of flavor production in dairy products. In: Belanger F (Ed.), Biotechnology of flavor production. Blackwell Publishing, Oxford

Wilkinson MG, Guinee TP, O'Callaghan DM, Fox PF (1994) Autolysis and proteolysis in different strains of starter bacteria during cheddar cheese ripening. J Dairy Res 61:249–262
Xie Y, Chou LS, Cutler A, Weimer B (2004) DNA Macroarray profiling of *Lactococcus lactis* subsp. *lactis* IL1403 gene expression during environmental stresses. Appl Environ Microbiol 70:6738–6747
Zhang Y, Zhu Y, Mao S, Li Y (2010) Proteomic analyses to reveal the protective role of glutathione in resistance of *Lactococcus lactis* to osmotic stress. Appl Environ Microbiol 76:3177–3186

Additional Resources

Biocyc (biocyc.org): a collection of over 160 pathway databases for metabolic reconstruction of specific organisms.
GOLD (www.genomesonline.org): provides current information about genome sequencing projects.
KEGG (www.genome.ad.jp/kegg/): a suite of databases and software to simulate the metabolism of cells from their genome information.
Metacyc (metacyc.org): a nonredundant metabolic encyclopedia of all known metabolic pathways.
National Center Biotechnology Information (www.ncbi.nlm.nih.gov): a genetic and bioinformatics resource within the National Institutes of Health that hosts GenBank files of genome sequences for public access.
The Joint Genome Institute (www.jgi.doe.gov): a genome sequencing facility hosted by the U.S. Department of Energy that provides public access to draft and finished genomes.
The Sanger Institute (www.sanger.ac.uk): a genome sequencing facility hosted by the Wellcome Trust Foundation that provides open-source tools for genome analysis.

Chapter 8
Responses of Lactic Acid Bacteria to Cell Envelope Stresses

João P.C. Pinto, Oscar P. Kuipers, and Jan Kok

8.1 Introduction

The existence of an enclosure that delimits cells from their environment is central in the biology of all living organisms. The complexity of these surfaces has evolved enormously, from putatively simple amphiphile-stabilized interfaces to the current broad range of multilayered structures enveloping modern organisms (Maurer et al. 2009; Panno 2004). All bacteria – and lactic acid bacteria (LAB) are no exceptions in this respect – have to endure in their natural habitats changes in external environmental parameters and a ferocious struggle with other bacteria, bacteriophages, and hosts, all of which could affect the structure of their cell envelopes. The ability of cells to respond and adapt to such aggressions, and to maintain the integrity of the cell envelope, defines the viability of the cell. Bacteria have developed sophisticated strategies to monitor and respond to these stresses. Here we review the state-of-the-art knowledge of the cell envelope stress responses of LAB, complementing it with findings from other bacteria. For more information about these responses, outside the scope of this review, and more broadly for Gram-positive or Gram-negative bacteria, see, for example, Jordan et al. (2008) or MacRitchie et al. (2008), respectively. It should be emphasized that many forms of stress are bound to activate a response that acts on the cell envelope. First, because the cell envelope is the first line of defense and a sensitive structure of the cell, and second, simply statistically, because

J. Kok (✉)
Department of Molecular Genetics, Groningen Biomolecular Sciences and Biotechnology Institute, University of Groningen, Nijenborgh 7, 9747 AG Groningen, The Netherlands

Top Institute Food and Nutrition, Nieuwe Kanaal 9A, 6709 PA Wageningen, The Netherlands

The Netherlands Kluyver Centre for Genomics of Industrial Fermentations, NCSB, Julianalaan 67, 2628 BC Delft, The Netherlands
e-mail: jan.kok@rug.nl

E. Tsakalidou and K. Papadimitriou (eds.), *Stress Responses of Lactic Acid Bacteria*, Food Microbiology and Food Safety, DOI 10.1007/978-0-387-92771-8_8,

more than 20% of all open reading frames (ORFs) in a given genome are predicted to code integral membrane proteins (Wallin and Heijne 1998). Here we only review responses that result from damage inflicted specifically on the cell envelope.

8.2 The Cell Envelope of Lactic Acid Bacteria

The structure of the cell envelope of eubacteria comes in two major forms: one that surrounds the Gram-positive bacteria and that of Gram-negatives (Gram 1884). The LAB clade encompasses a broad range of microorganisms that possess the cell envelope arrangement typical of Gram-positive bacteria. Still, the differences in cell envelope architecture and composition between members of this group of bacteria are quite significant, greatly influencing key features such as bacteria–host interactions for the pathogenic LAB (Roberts 1996), or texture formation for the ones used in the dairy industry (Cerning 1990). Cell envelope variations have been used to classify LAB taxonomically [LAB are divided into three morphological groups: cocci, bacilli, and bifidi (Delcour et al. 1999); for reviews on shape determination, see, for example, Carballido-López and Formstone (2007) and Osborn and Rothfield (2007)] or serologically (Lancefield 1933).

Generically, the cell envelope of LAB is composed of a cytoplasmic membrane, spaced from the pepdidoglycan sacculus by a periplasm. The cytoplasmic membrane contains proteins and other elements such as glycolipids, while the sacculus is usually decorated with proteins, teichoic acids (that can reach the cytoplasmic membrane), polysaccharides, and, often, a paracrystalline S-layer of proteins (Fig. 8.1). Each of the cell envelope elements is briefly described ahead. For more extensive reviews on the cell envelope of LAB, see, for example, Konings (2002) and Delcour et al. (1999).

8.2.1 The Cytoplasmic Membrane

The cytoplasmic membrane is essentially composed of a phospholipid bilayer. Its hydrophobic nature makes it a permeability barrier, blocking the diffusion of water and water-soluble molecules. The cytoplasmic membrane is thought to be a very active and crowded region of the cell, with a significant degree of differentiation, both functionally and structurally (Engelman 2005). It is also rich in fatty acids, of which the composition is modulated according to temperature, to keep the membrane optimally fluid (Shivaji and Prakash 2010).

The existence of a membrane barrier can also be viewed as a bottleneck to the cell when the goal is to translocate molecules in and out of the cell. To do so takes considerable amounts of resources and frequently requires ingenious mechanisms for transport. The insertion of membrane proteins into, or the translocation of secreted proteins across, the cytoplasmic membrane requires complex translocation/secretion machineries (Driessen and Nouwen 2008; Mandon et al. 2009).

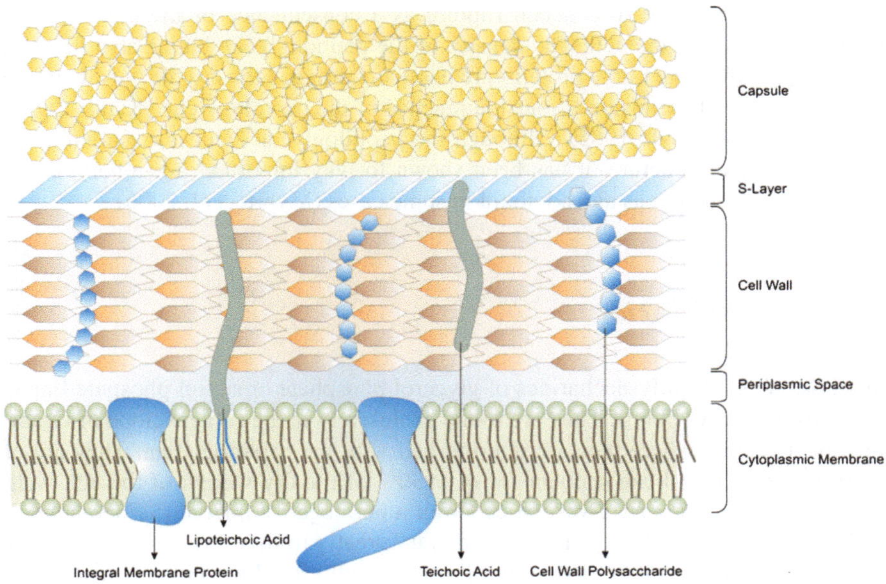

Fig. 8.1 A simplified scheme of the typical organization of the cell envelope of LAB. The cytoplasmic membrane contains (integral) membrane proteins. The cell wall is essentially composed of pepdidoglycan intermeshed with teichoic acids, lipoteichoic acids (which are lipid-anchored to the membrane), and cell wall polysaccharides. Some LAB may also contain a proteinaceous S-layer and/or a capsule made of polysaccharides. Pili and flagella, which are not depicted in this figure, can also be found in some LAB

In fact, the complexity of these export systems partially accounts for the limited success in producing a recombinant integral membrane or secreted proteins (Bowie 2005; Grisshammer 2006). Interestingly, LAB, and *Lactococcus lactis* in particular, seem to have a number of properties that might make them better suited than some other protein production hosts, even when the goal is to produce eukaryotic nonglycosylated membrane proteins (Morello et al. 2008; Kunji et al. 2005).

8.2.2 The Sacculus

The peptidoglycan (also known as murein) sacculus of LAB has the typical Gram-positive thick, multilayered structure, composed of inextensible sugar chains cross-linked by flexible peptide bridges, which is responsible for the withholding of the crystal violet dyes during the Gram staining assay (Gram 1884; Delcour et al. 1999). The glycan threads are made of the two alternating amino sugars *N*-acetylglucosamine and *N*-acetylmuramic acid, connected by cross-links of peptide chains of usually five alternating L- and D-amino acids (Delcour et al. 1999). This tough structure, resembling an exoskeleton, is required not only to shield cells from mechanical aggressions but also to reduce the risk of cell rupture as a consequence of the high internal osmotic pressure in the cell. The density of the peptidoglycan lattice also disables free protein diffusion. Large proteins are passively dragged toward the outside of the cell at the pace of the inside-to-outside renewal of the peptidoglycan (Kemper et al. 1993; Koch and Doyle 1985). Although the peptidoglycan layer is regarded as an inactive, passive structure, it has been shown to play important pro-inflammatory roles and can function as an endotoxin in some cases (Myhre et al. 2006).

8.2.3 Teichoic Acids

Teichoic acids are polysaccharides of glycerol phosphate or ribitol phospate linked via phosphodiester bonds. They are found exclusively in the cell envelope of Gram-positive bacteria, where they are intermeshed with the peptidoglycan. They are of vital importance and can account for over 50% of the dry weight of the cell wall. Their polyanionic nature gives the cell wall a net negative charge (Swoboda et al. 2010; Delcour et al. 1999). Teichoic acids vary greatly in structure and composition, depending on the bacterial strain and growth conditions. Lipoteichoic acids are a special class of teichoic acids that are lipid-anchored to the cytoplasmic membrane. They regulate autolysis and help hold together the two main functional structures of the cell envelope: the sacculus and the cytoplasmic membrane (Swoboda et al. 2010; Delcour et al. 1999).

8.2.4 The S-Layer

Surface layers (S-layers), composed of proteins or glycoproteins, are usually single-molecule-thick, planar, crystalline lattices that are commonly found in all prokaryotes. They are associated to the external side of the peptidoglycan in Gram-positive bacteria, where they add protection to low-pH environments and lytic enzymes (Engelhardt 2007; Sleytr and Beveridge 1999). S-layers have also been described in LAB, such as in lactobacilli, where the protein elements are usually smaller than those found in S-layers of other bacteria and have a predicted higher pI value (Avall-Jääskeläinen and Palva 2005).

8.2.5 Cell Wall Polysaccharides and Capsular Polysaccharides

Cell wall polysaccharides, also known as neutral polysaccharides (as opposed to the anionic polysaccharide teichoic acid), are complex molecules that vary in the nature of the sugar monomers (rhamnose is usually found as a constituent in LAB) and in structure (Delcour et al. 1999). The capsule is a thick structure typically composed of polysaccharides and is situated outside the cell wall. It is a virulence factor in pathogenic bacteria: It gives them a slimy coat preventing phagocytosis (Roberts 1996). Recently, a pellicle of polysaccharides (distinct from a thick capsule) was found to exist outside the cell wall of *L. lactis* (Chapot-Chartier et al. 2010). This pellicle also has the ability to disable phagocytosis and putatively assists in the separation of daughter cells at the septum. It should be underlined that differences in the polysaccharides of the cell wall and capsule are often at the basis of the different bacterial serotypes (Lancefield 1933; Roberts 1996).

8.2.6 Other Elements of the Cell Envelope

Although not extensively studied, pili and flagella are also found in LAB. Pili are filamentous structures that extend from the cell envelope and are central in the adhesion of bacteria to, for instance, host tissues (Sillanpää et al. 2009; Telford et al. 2006; Mandlik et al. 2008). The presence of human mucus-binding proteins in the pili of *Lactobacillus rhamnosus* GG explains the enhanced interaction of this widely used probiotic strain with host tissue during colonization of the gut (Kankainen et al. 2009; Walker 2009).

LAB are generally regarded as nonmotile, but some lactobacilli are motile due to the presence of flagella (Nielsen et al. 2007; Irisawa and Okada 2009). Flagella are involved in both cellular motility and protein export, including the secretion of virulence factors. They cross the entire cell envelope and, in Gram-positive bacteria, contain two basal body rings that act as mechanical bearings, one in the peptidoglycan and one in the cytoplasmic membrane (Macnab 2003).

8.3 Cell Envelope Stresses

LAB are faced with many different sorts and sources of stressors in their diverse natural habitats and in the various industrial processes in which they are employed. Several of these affect the integrity of the cell envelope and are presented in the following section. The responses that LAB developed to counteract these forms of aggression are presented in Sect. 8.4.

8.3.1 Chemical Stresses

The cell wall is a major target of many antibiotics that either inhibit an enzymatic step of its biosynthesis or capture one of its precursors (Silver 2003). Antibiotics are secondary metabolites that give their producers a competitive edge by inhibiting the growth of competitors. On the other hand, the evolution of these molecules has been paralleled in those targeted competitors by the development of stress responses and efficient resistance mechanisms.

The synthesis of the peptidoglycan represents a particularly critical point in this microbiological warfare. It is the target of several classes of antibiotics, such as the β-lactam penicillin and its analogs. This class of antibiotics disables the activity of a group of transpeptidases collectively known as penicillin-binding proteins (PBPs). The β-lactams do so by mimicking the terminus of the pentapeptide side chain of the peptidoglycan precursor and covalently binding to and blocking the active site of the PBPs. Thus, the β-lactam antibiotics inhibit the peptidoglycan chains from being cross-linked (Zapun et al. 2008; Sauvage et al. 2008). The resistance to these antibiotics is generally not driven by a cellular response but is the result of selective pressure. In β-lactam–resistant strains, the antibiotic-targeted PBP has often been replaced by a low-affinity derivative, through horizontal gene transfer, changing the relative expression profile of the ones already coded in the genome, or a mutation of an existing PBP (Chambers 1999; Zapun et al. 2008). Alternative resistance mechanisms include producing β-lactamases that inactivate the antibiotic (Fontana et al. 1990; Massova and Mobashery 1998) or extrusion of the antibiotic with efflux pumps (Poole 2007).

Another important step in peptidoglycan biosynthesis concerns the transport of the pentapeptide disaccharide precursors across the cytoplasmic membrane. Undecaprenol, a lipid-soluble molecule, functions as a carrier molecule by binding to the peptidoglycan precursor synthesized in the cytoplasm, thus producing a molecule designated lipid II, and then translocating it to the periplasm. The physical properties of lipid II allow the molecule to move across the cytoplasmic membrane, but the rate at which this happens in vivo suggests that this movement is enzyme-catalyzed (van Dam et al. 2007). This process, collectively known as the lipid II cycle due to the importance of the intermediate, is essential and serves as the target of several lantibiotics and vancomycin. For a more in-depth review on lipid II, see, for example, de Kruijff et al. (2008).

In total, 8 of the 14 conserved enzymatic steps in the biosynthesis of peptidoglycan have been shown to be targets of antibiotics (Silver 2006) and, despite the appearance of ever more resistant strains, cell wall synthesis remains a primary target for the development of novel antibiotics (Van Bambeke et al. 2008; Silver 2006).

LAB exhibit antimicrobial activity as they produce weak organic acids. In addition, many species and strains synthesize bacteriocins (Cotter et al. 2005). These antimicrobial peptides are classified into three major groups: Class I is for the lantibiotics (small, heat-stable peptides containing lanthione rings); class II includes small heat-stable peptides (containing an N-terminal consensus sequence); and class III is for the large, heat-labile proteins. Class I and II bacteriocins are pore-forming molecules that cause permeabilization of the membrane, while the ones from class III are generally murein hydrolases (De Vuyst and Leroy 2007).

Due to its greater exposure, the cell envelope is a prime target for many other biocides, although most do not exclusively harm that structure. Ortho-phtalaldehyde (OPA) and glutaraldehyde (GTA) are two nonoxidizing, highly reactive di-aldehyde molecules that are used as disinfectants and primarily affect the cell surface by cross-linking outer proteins of bacterial cells (Walsh et al. 1999). Sodium hypochlorite, commonly known as bleach, and other oxidizing agents can affect the cytoplasmic membrane and the cell wall to an extent that induces a loss of structure and function, and consequently cell lysis and death. Many organic compounds and detergents are able to denature proteins and destabilize the cytoplasmic membrane, leading to its disruption. For a comprehensive review on biocides and their cellular targets in bacteria, see Maillard (2002).

As a result of their metabolism, and also while passing through the gastrointestinal tract, LAB have to withstand acidic environments and have developed responses to cope with acid stress (see Chap. 2 for LAB responses to acid stress).

8.3.2 *Physical Stresses*

LAB may face adverse situations that result from changes in external physical parameters. These invariably affect the cell envelope, the first line of defense of bacteria, although they are not commonly referred to as cell envelope responses. Changes in temperature influence the fluidity of the cytoplasmic membrane (see Chap. 3 and 5 of this book for reviews on the responses of LAB to heat shock and cold shock). One main structural function of the cell envelope is to counteract the high internal osmotic pressure. LAB have response mechanisms to execute that function and maintain the essential balance between internal and external osmotic pressures when moisture and solute concentrations vary (see Chap. 4 for LAB responses to osmotic stress). Industrial applications of LAB usually force cells to endure mechanical and shear stresses, which result in damage inflicted on the cell envelope.

Applications of LAB, such as *L. lactis*, in the production of recombinant secreted and integral membrane proteins (Morello et al. 2008; Kunji et al. 2005) partially sequester the translocation/secretion machinery and may lead to a physical crowding of proteins in the cytoplasmic membrane (Marreddy et al. 2010; Pinto et al. 2010).

8.4 Cell Envelope–Stress Responses

Bacteria have developed mechanisms to directly monitor agents and parameters of stress, or indirectly by assessing the integrity of the cell envelope. Inputs for the response that lay downstream of the aggression, such as common indirect effects, allow the bacteria to produce broad responses, enabling the cells to resist a wide range of stresses. Nevertheless, these nonspecific responses may not always result in an increased resistance to specific sources of stress. Many forms of resistance, notably to antibiotics, are not a product of a real-time cellular response strategy, but rather an adaptation that results from natural/artificial selection (Summers 2006; Martínez 2008). Also, a clear distinction should be made between an immediate response and a nonpermanent (mutation-independent) cellular adaptation to a different situation. For example, when treated with increasing amounts of the bacteriocin nisin, *L. lactis* cells can become 75 times more resistant to the antimicrobial peptide than the unadapted strain (Kramer et al. 2006). This adaptation is not permanent, and cells regain nisin sensitivity once the pressure is removed. Although the underlying response mechanisms are not known, DNA microarray data indicate that this nisin resistance results from a broad rearrangement of gene expression-modulating cell wall thickness/density (*galE* and *pbp2A*) and charge (*dltD*), acidity outside the membrane (*arcAC1C2DT2*), membrane fluidity (*fabDG1G2Z1Z2*), and possibly the capacity to export nisin from the cells (Kramer et al. 2006).

Responses to cell envelope stress can be divided into two main categories: those mediated by two-component systems (TCSs) (Mascher 2006) and those mediated by extracytoplasmic function (ECF) sigma factors (Helmann 2002). The sensor-regulator devices of these two classes are analogous in their architecture, and both consist of two proteins: a membrane-anchored sensor (the histidine kinase or anti-sigma factor) that becomes active only under stress conditions and activates a cytoplasmic transcriptional regulator (the response regulator or ECF). Contrary to ECF sigma factors, TCSs are widespread and conserved in LAB (Mascher 2006). They are best classified according to the genes they regulate, that is, whether they aim to generally maintain the integrity of the cell envelope or to activate specific detoxification modules. Another type of stress response relies on one-component systems. They are putatively widely distributed in prokaryotes but have not been extensively studied. All these classes of stress responses and their presence in LAB are described ahead.

8.4.1 Extracytoplasmic Function Sigma Factors

ECF sigma factors are members of the sigma-70 family, which recognize the conserved "AAC" motif in the −35 region of a promoter and are usually cotranscribed with their cognate antisigma factor (Helmann 2002). Although most bacilli have many ECF sigma factors, they seem absent in most LAB (Jordan et al. 2008). To the best of our knowledge, only one ECF (SigV from *Enterococcus faecalis*) has been characterized and shown to play a role in the cell envelope–stress response of a LAB. SigV, which had been previously associated with a response to heat, acid, and ethanol stress (Benachour et al. 2005), also plays a central role in the lysozyme resistance and virulence of this bacterium (Le Jeune et al. 2010). This unusual resistance of *E. feacalis* seems to be specific to lysozymes and is not extrapolated to other antimicrobials such as nisin. The mechanism of resistance is still not known but is independent of OatA (an O-acetyl transferase) and DltA (involved in the D-alanylation of lipoteicoic acids), two proteins that also contribute to the resistance of this bacterium to lysozyme. This mechanism of resistance increases the virulence of *E. feacalis* since a *sigV* mutant displays a reduced potential to colonize host tissues (Le Jeune et al. 2010).

8.4.2 Two-Component Systems That Maintain the Integrity of the Cell Envelope

Some of the TCSs, like ECF sigma factors, respond to a wide range of cell envelope stresses and activate the expression of genes that maintain general aspects of cell envelope integrity. They are frequently associated with other (general) stress responses, growth, and competence development, supporting the idea that they have a more general role in maintaining the general homeostasis of the cell.

The TCS LiaRS is present in all LAB except the lactobacilli (Jordan et al. 2008). It invariably responds to perturbations in the integrity of the cell envelope, but the genes it regulates vary greatly among different species, perhaps representing a divergent evolution that resulted from different sources or susceptibilities to ecological aggressions. The *liaRS* genes are usually preceded by *liaF*, which encodes a strong inhibitor of LiaR (Jordan et al. 2006).

In *Bacillus subtilis*, where it was originally characterized, LiaRS is strongly induced by vancomycin and bacitracin, but only the promoters of *liaI* and *yhcY* were found to be regulated by LiaR (Jordan et al. 2006). LiaRS-dependent gene expression is repressed in growing cells by the transition state regulator AbrB. This repression is alleviated by the master regulator of sporulation, Spo0A, as cells enter the transition phase of growth (Jordan et al. 2007). In bacilli, *liaFRS* is always part of a larger operon, either *liaIHGFRS* (e.g., in *B. subtilis*) or *liaIHFRS* (e.g., in *Bacillus cereus*). Interestingly, the deletion of none of the *lia* genes modifies the sensitivity to the known inducers of LiaRS in *B. subtilis* (Jordan et al. 2008).

In *Staphylococcus aureus*, VraSR (another widely studied homolog of LiaRS) is strongly induced by a wide range of cell wall antibiotics; a *vraSR* deletion strain shows increased susceptibility to these inducers (Kuroda et al. 2003; Pietiäinen et al. 2009).

In LAB, LiaRS-like TCS mediated responses have been characterized in *L. lactis* (Martínez et al. 2007), *Streptococcus mutans* (Suntharalingam et al. 2009), *Streptococcus pneumoniae* (Eldholm et al. 2010), and *E. faecalis* (Hancock and Perego 2004). In all of these organisms only *liaFRS* are conserved, not the related, more extended operon as found in bacilli. The *liaRS* regulon is larger in LAB than in *B. subtilis*, indicating a different and more central role of LiaRS in the response against cell envelope stresses (Jordan et al. 2008). In all cases, LiaRS-like TCSs are positively autoregulated and respond to cell wall antibiotics that interfere with the lipid II cycle.

In *L. lactis*, the LiaRS-like TCS CesSR affects sensitivity to salts and osmotic pressure (O'Connell-Motherway et al. 2000) and is induced by vancomycin, bacitracin, and the two LAB bacteriocins Lcn972 and plantaricin C (Martínez et al. 2007).

The disruption of CesR only results in an increased susceptibility to those inducers that interfere with the lipid II cycle: bacitracin and plantaricin C. The *L. lactis* *ΔcesR* mutant is also more sensitive to nisin, despite the inability of this antimicrobial peptide to induce CesSR (Martínez et al. 2007). Interestingly, a nisin-adapted/resistant *L. lactis* strain, obtained by growing cells in the presence of increasing amounts of the antimicrobial, overexpresses CesSR and its regulon (Kramer et al. 2006). This strain is 75 times more resistant to nisin and is also 3 times more resistant to Lcn972, although a similar decreased susceptibility to Lcn972 was also found in the *cesR* deletion strain (Martínez et al. 2007). Most CesSR-regulated genes encode putative membrane proteins, supporting the idea that the response is specific to stresses that operate at the level of the cellular envelope (Martínez et al. 2007).

Notably, *lmrA* and *rmaB*, coding a multidrug ABC transporter and a transcriptional regulator of the MarR family, respectively, are members of the CesSR regulon. Llmg2163 [a putative transcriptional regulator, as it contains a PspC domain; Brissette et al. (1991)], Llmg2194, and Llmg0165 have protective roles in *L. lactis* even against unrelated and more general perturbations like temperature, pH, and salt stresses. The overexpression of *llmg2163* and *llmg2164* in *L. lactis* increased the resistance of the organism to Lcn972 (Roces et al. 2009). One of the most prominent targets of CesSR is SpxB, a protein that, by binding to RNA polymerase, induces the expression of *oatA* (Veiga et al. 2007). The activity of the encoded peptidoglycan *O*-acetylase OatA influences the level of peptidoglycan acetylation and therefore the sensitivity to peptidoglycan hydrolysis. The CesSR-dependent expression of *spxB* thus efficiently responds to cell envelope stress evoked by treating *L. lactis* cells with hen egg white lysozyme (Veiga et al. 2007). CesSR was also induced when *L. lactis* was forced to overproduce recombinant membrane proteins (Marreddy et al. 2010; Pinto et al. 2010). The response may result from an overcrowding of the cytoplasmic membrane with membrane proteins or may simply be caused by an overuse of the translocation/secretion machinery, as the same CesSR response was observed during the production of the secreted protein AmyQ (Pinto et al. unpublished data). This response might mimic normal periods in growth when the production of endogenous membrane proteins is high, since members of the translocation machinery, such as FtsH and OxaA2, are also regulated by CesSR. Disrupting *cesSR* hampers the growth of *L. lactis* when cells are induced to produce membrane proteins. On the other hand, the cooverproduction of CesSR resulted in a more than threefold improvement in the production yield of membrane proteins (Pinto et al. 2010).

In *S. pneumoniae,* LiaFSR has been shown to be activated by vancomycin (Haas et al. 2005), murein hydrolases, and lipid II–interacting antibiotics (Eldholm et al. 2010). In this organism, fratricide (by which competent cells lyze noncompetent sister cells) was shown to be assisted by the secretion of cell wall hydrolases such as CbpD, LytA, and LytC (Eldholm et al. 2009). The lysis of competent cells is partially prevented through a response coordinated by LiaFSR, after sensing the damage inflicted to the cell wall. Among the genes that LiaR regulates, PcpC (Spr0351) and Spr0810 are particularly important in preventing damage from lysis (Eldholm et al. 2010).

In *S. mutans*, LiaS regulates the levels of GbpC, a cell surface–associated protein that facilitates biofilm formation in environments such as the human oral cavity (Chong et al. 2008). Additionally, LiaSR was shown to regulate 174 genes, some of which are involved in membrane protein synthesis and peptidoglycan biosynthesis (Perry et al. 2008). Lipid II cycle–interfering antibiotics and other chemicals that disrupt the integrity of the cytoplasmic membrane activate LiaSR, and a disruption of *liaSR* makes *S. mutans* more susceptible to those compounds (Suntharalingam et al. 2009).

In *E. faecalis,* the inactivation of the LiaR homolog RR03 (EF2911) increased the susceptibility of this organism to bacitracin (Hancock and Perego 2004). Another TCS from *E. faecalis*, *croRS* (for "ceftriaxone resistance"; or RR05-HK05) is required for β-lactam resistance, and its disruption causes a 4,000-fold increase in the sensitivity to the cephalosporin ceftriaxone (Comenge et al. 2003). *croRS* is induced by narrow and broad-spectrum cephalosporins, imipenem, ampicillin, oxacillin, amdinocillin, and inhibitors of peptidoglycan synthesis such as phosphomycin, D-cycloserine, vancomycin, moenomycin, ramoplanin, and bacitracin (Comenge et al. 2003). CroRS regulates *glnQHMP* (Le Breton et al. 2007) and *salB* (previously *sagA*), which encodes a putative cell wall–attached protein that is important in cell division and in resistance to multiple stresses (Muller et al. 2006; Le Breton et al. 2003).

The CiaRH TCS of *S. pneumoniae*, initially identified in a screen for cefotaxime-resistant mutants (Guenzi et al. 1994), is activated by vancomycin and penicillin (Haas et al. 2005; Rogers et al. 2007) and is involved in virulence (Ibrahim et al. 2004), competence, and resistance to lysis (Dagkessamanskaia et al. 2004). The expression of the *dlt* operon and *htrA* is controlled by CiaRH in *S. pneumoniae* (Sebert et al. 2002, 2005). In *L. lactis*, the expression of *htrA* is triggered during the overproduction and secretion of recombinant proteins (Marreddy et al. 2010), while in *S. pneumoniae*, HtrA affects the activity of pneumocin MN, a two-peptide bacteriocin (Dawid et al. 2009). In *S. mutans*, the inactivation of *ciaH* diminishes the production of mutacin and affects processes such as competence development and biofilm formation (Qi et al. 2004).

8.4.3 Two-Component Systems That Activate Specific Detoxification Modules

Some TCSs regulate the transcription of genes that directly confer resistance to the antimicrobial that activates the TCS. These TCSs and the resistance element are usually encoded by neighboring genes and are typically found in contexts of high genetic mobility like plasmids or transposable elements (Weigel et al. 2003).

BceRS, a nonautoregulated TCS, specifically responds to bacitracin. In *B. subtilis*, BceS is not sufficient and requires BceAB, the corresponding bacitracin ABC transporter, to sense bacitracin itself (Bernard et al. 2007). MbrABCD from

S. mutans, a BceRSAB-like system, is also responsible for bacitracin resistance, and mutations in any of the respective genes lead to a more than 100-fold increased susceptibility to bacitracin (Tsuda et al. 2002).

VanRS, a TCS that is found in vancomycin-resitant enterococci, such as some *E. faecalis* and *Enterococcus faecium* isolates, senses vancomycin and triggers the expression of genes conferring resistance to the antibiotic and, depending on the genes it regulates, teicoplanin (Hong et al. 2008). Invariably, VanR activates the transcription of *vanHAX*, which encode enzymes that reprogram the synthesis of the cell wall by changing the terminal amino acid residues of the precursors from D-alanyl-D-alanine to D-alanyl-D-lactate or D-alanyl-D-serine (Bugg et al. 1991; Reynolds and Courvalin 2005).

8.4.4 One-Component Systems

Despite being less well characterized than TCSs, signal transduction systems consisting of only a single protein containing both a sensory and a DNA-binding domain are now thought to be more widespread in bacteria than TCSs (Ulrich et al. 2005). Their involvement in the response to cell envelope stresses has started to be revealed.

An example of such a one-component system is found in *E. faecalis*, in which a unique form of acquired bacitracin resistance is mediated by BcrAB, an ABC transporter that putatively exports bacitracin (Manson et al. 2004; Matos et al. 2009). The expression of the *bcrABD* operon is regulated by BcrR, a membrane-bound bacitracin sensor and DNA-binding protein (Manson et al. 2004; Gauntlett et al. 2008). This sensor/regulator seems to be active and induces the transcription of *bcrABD* only in the presence of bacitracin (Manson et al. 2004).

To the best of our knowledge, only two other one-component systems have to date been implicated in the resistance to cell envelope stresses. PrkC modulates antimicrobial resistance in *E. faecalis* (Kristich et al. 2007), and PknB modulates biofilm formation and resistance toward the envelope stress caused by H_2O_2 in *S. mutans* (Zhu and Kreth 2010). Both PrkC and PknB are one-component signaling proteins with a eukaryotic-type Ser/Thr kinase domain.

8.5 Concluding Remarks

The bacterial cell envelope is a complex structure, and many strategies are required to maintain its integrity and the optimal parameters of its physiology. Although a full characterization of the responses of LAB to aggressions affecting cell envelope integrity or functioning is imperative in order to understand, for example, bacterium–host interactions and the development of multidrug-resistant bacteria, this particular aspect of the biology of LAB remains largely unexplored. Also, LAB are evolutionarily sufficiently distinct from other well-studied model organisms to disable extrapolating, even remotely, the regulons of homologous response systems. The most striking particularity in LAB, when comparing them to other *firmicutes*, is the almost complete absence of stress-responsive sigma factors. LAB seem to have counterbalanced this apparent handicap in responding to cell envelope stresses by using two-component gene regulatory systems instead. Also, largely neglected, one-component systems are widely distributed in prokaryotes, and knowledge about their possible roles in the response to cell envelope stresses might prove essential. Most forms of cell envelope–stress responses in LAB are unspecific and aim at maintaining the integrity of the cell envelope. The underlining regulatory systems are often linked to other stress responses and to other important regulatory circuits, such as those involved in growth and competence. This indicates that cell envelope–stress responses have a broader role in maintaining the general fitness of the bacterial cell.

References

Avall-Jääskeläinen S, Palva A (2005) Lactobacillus surface layers and their applications. FEMS Microbiol Rev 29:511–529

Benachour A, Muller C, Dabrowski-Coton M, Le Breton Y, Giard J, Rincé A, Auffray Y, Hartke A (2005) The *Enterococcus faecalis* sigV protein is an extracytoplasmic function sigma factor contributing to survival following heat, acid, and ethanol treatments. J Bacteriol 187:1022–1035

Bernard R, Guiseppi A, Chippaux M, Foglino M, Denizot F (2007) Resistance to bacitracin in *Bacillus subtilis:* unexpected requirement of the BceAB ABC transporter in the control of expression of its own structural genes. J Bacteriol 189:8636–8642

Bowie JU (2005) Solving the membrane protein folding problem. Nature 438:581–589

Brissette JL, Weiner L, Ripmaster TL, Model P (1991) Characterization and sequence of the *Escherichia coli* stress-induced psp operon. J Mol Biol 220:35–48

Bugg TD, Wright GD, Dutka-Malen S, Arthur M, Courvalin P, Walsh CT (1991) Molecular basis for vancomycin resistance in *Enterococcus faecium* BM4147: biosynthesis of a depsipeptide peptidoglycan precursor by vancomycin resistance proteins VanH and VanA. Biochemistry 30:10408–10415

Carballido-López R, Formstone A (2007) Shape determination in *Bacillus subtilis*. Curr Opin Microbiol 10:611–616

Cerning J (1990) Exocellular polysaccharides produced by lactic acid bacteria. FEMS Microbiol Rev 7:113–130

Chambers HF (1999) Penicillin-binding protein-mediated resistance in pneumococci and staphylococci. J Infect Dis 179 Suppl 2:S353–S359

Chapot-Chartier M, Vinogradov E, Sadovskaya I, Andre G, Mistou M, Trieu-Cuot P, Furlan S, Bidnenko E, Courtin P, Pechoux C, Hols P, Dufrene YF, Kulakauskas S (2010) The cell surface of *Lactococcus lactis* is covered by a protective polysaccharide pellicle. J Biol Chem 285:10464–10471

Chong P, Drake L, Biswas I (2008) LiaS regulates virulence factor expression in *Streptococcus mutans*. Infect Immun 76:3093–3099

Comenge Y, Quintiliani R, Li L, Dubost L, Brouard J, Hugonnet J, Arthur M (2003) The CroRS two-component regulatory system is required for intrinsic beta-lactam resistance in *Enterococcus faecalis*. J Bacteriol 185:7184–7192

Cotter PD, Hill C, Ross RP (2005) Bacteriocins: developing innate immunity for food. Nat Rev Microbiol 3:777–788

Dagkessamanskaia A, Moscoso M, Hénard V, Guiral S, Overweg K, Reuter M, Martin B, Wells J, Claverys J (2004) Interconnection of competence, stress and CiaR regulons in *Streptococcus pneumoniae*: competence triggers stationary phase autolysis of *ciaR* mutant cells. Mol Microbiol 51:1071–1086

Dawid S, Sebert ME, Weiser JN (2009) Bacteriocin activity of *Streptococcus pneumoniae* is controlled by the serine protease HtrA via posttranscriptional regulation. J Bacteriol 191:1509–1518

de Kruijff B, van Dam V, Breukink E (2008) Lipid II: a central component in bacterial cell wall synthesis and a target for antibiotics. Prostaglandins Leukot Essent Fatty Acids 79:117–121

De Vuyst L, Leroy F (2007) Bacteriocins from lactic acid bacteria: production, purification, and food applications. J Mol Microbiol Biotechnol 13:194–199

Delcour J, Ferain T, Deghorain M, Palumbo E, Hols P (1999) The biosynthesis and functionality of the cell-wall of lactic acid bacteria. Antonie van Leeuwenhoek 76:159–184

Driessen AJM, Nouwen N (2008) Protein translocation across the bacterial cytoplasmic membrane. Annu Rev Biochem 77:643–667

Eldholm V, Johnsborg O, Haugen K, Ohnstad HS, Håvarstein LS (2009) Fratricide in *Streptococcus pneumoniae*: contributions and role of the cell wall hydrolases CbpD, LytA and LytC. Microbiology (Reading, Engl) 155:2223–2234

Eldholm V, Gutt B, Johnsborg O, Brückner R, Maurer P, Hakenbeck R, Mascher T, Håvarstein LS (2010) The pneumococcal cell envelope stress-sensing system LiaFSR is activated by murein hydrolases and lipid II-interacting antibiotics. J Bacteriol 192:1761–1773

Engelhardt H (2007) Are S-layers exoskeletons? The basic function of protein surface layers revisited. J Struct Biol 160:115–124

Engelman DM (2005) Membranes are more mosaic than fluid. Nature 438:578–580

Fontana R, Canepari P, Lleò MM, Satta G (1990) Mechanisms of resistance of enterococci to beta-lactam antibiotics. Eur J Clin Microbiol Infect Dis 9:103–105

Gauntlett JC, Gebhard S, Keis S, Manson JM, Pos KM, Cook GM (2008) Molecular analysis of BcrR, a membrane-bound bacitracin sensor and DNA-binding protein from *Enterococcus faecalis*. J Biol Chem 283:8591–8600

Gram H (1884) Über die isolierte Färbung der Schizomyceten in Schnitt- und Trockenpräparaten. Fortschritte der Medizin 2:185–189

Grisshammer R (2006) Understanding recombinant expression of membrane proteins. Curr Opin Biotechnol 17:337–340

Guenzi E, Gasc AM, Sicard MA, Hakenbeck R (1994) A two-component signal-transducing system is involved in competence and penicillin susceptibility in laboratory mutants of *Streptococcus pneumoniae*. Mol Microbiol 12:505–515

Haas W, Kaushal D, Sublett J, Obert C, Tuomanen EI (2005) Vancomycin stress response in a sensitive and a tolerant strain of *Streptococcus pneumoniae*. J Bacteriol 187:8205–8210

Hancock LE, Perego M (2004) Systematic inactivation and phenotypic characterization of two-component signal transduction systems of *Enterococcus faecalis* V583. J Bacteriol 186:7951–7958

Helmann JD (2002) The extracytoplasmic function (ECF) sigma factors. Adv Microb Physiol 46:47–110

Hong H, Hutchings MI, Buttner MJ (2008) Vancomycin resistance VanS/VanR two-component systems. Adv Exp Med Biol 631:200–213

Ibrahim YM, Kerr AR, McCluskey J, Mitchell TJ (2004) Control of virulence by the two-component system CiaR/H is mediated via HtrA, a major virulence factor of *Streptococcus pneumoniae*. J Bacteriol 186:5258–5266

Irisawa T, Okada S (2009) *Lactobacillus sucicola* sp. *nov.*, a motile lactic acid bacterium isolated from oak tree (*Quercus* sp.) sap. Int J Syst Evol Microbiol 59:2662–2665

Jordan S, Junker A, Helmann JD, Mascher T (2006) Regulation of LiaRS-dependent gene expression in *Bacillus subtilis*: identification of inhibitor proteins, regulator binding sites, and target genes of a conserved cell envelope stress-sensing two-component system. J Bacteriol 188:5153–5166

Jordan S, Rietkotter E, Strauch MA, Kalamorz F, Butcher BG, Helmann JD, Mascher T (2007) LiaRS-dependent gene expression is embedded in transition state regulation in *Bacillus subtilis*. Microbiology 153:2530–2540

Jordan S, Hutchings MI, Mascher T (2008) Cell envelope stress response in Gram-positive bacteria. FEMS Microbiol Rev 32:107–146

Kankainen M, Paulin L, Tynkkynen S, von Ossowski I, Reunanen J, Partanen P, Satokari R, Vesterlund S, Hendrickx APA, Lebeer S, De Keersmaecker SCJ, Vanderleyden J, Hämäläinen T, Laukkanen S, Salovuori N, Ritari J, Alatalo E, Korpela R, Mattila-Sandholm T, Lassig A, Hatakka K, Kinnunen KT, Karjalainen H, Saxelin M, Laakso K, Surakka A, Palva A, Salusjärvi T, Auvinen P, de Vos WM (2009) Comparative genomic analysis of *Lactobacillus rhamnosus* GG reveals pili containing a human- mucus binding protein. Proc Natl Acad Sci USA 106:17193–17198

Kemper MA, Urrutia MM, Beveridge TJ, Koch AL, Doyle RJ (1993) Proton motive force may regulate cell wall-associated enzymes of *Bacillus subtilis*. J Bacteriol 175:5690–5696

Koch AL, Doyle RJ (1985) Inside-to-outside growth and turnover of the wall of Gram-positive rods. J Theor Biol 117:137–157

Konings WN (2002) The cell membrane and the struggle for life of lactic acid bacteria. Antonie van Leeuwenhoek 82:3–27

Kramer NE, van Hijum SAFT, Knol J, Kok J, Kuipers OP (2006) Transcriptome analysis reveals mechanisms by which *Lactococcus lactis* acquires nisin resistance. Antimicrob Agents Chemother 50:1753–1761

Kristich CJ, Wells CL, Dunny GM (2007) A eukaryotic-type Ser/Thr kinase in *Enterococcus faecalis* mediates antimicrobial resistance and intestinal persistence. Proc Natl Acad Sci USA 104:3508–3513

Kunji ERS, Chan KW, Slotboom DJ, Floyd S, O'Connor R, Monné M (2005) Eukaryotic membrane protein overproduction in *Lactococcus lactis*. Curr Opin Biotechnol 16:546–551

Kuroda M, Kuroda H, Oshima T, Takeuchi F, Mori H, Hiramatsu K (2003) Two-component system VraSR positively modulates the regulation of cell-wall biosynthesis pathway in *Staphylococcus aureus*. Mol Microbiol 49:807–821

Lancefield RC (1933) A serological differentiation of human and other groups of hemolytic streptococci. J Exp Med 57:571–595

Le Breton Y, Boël G, Benachour A, Prévost H, Auffray Y, Rincé A (2003) Molecular characterization of *Enterococcus faecalis* two-component signal transduction pathways related to environmental stresses. Environ Microbiol 5:329–337

Le Breton Y, Muller C, Auffray Y, Rincé A (2007) New insights into the *Enterococcus faecalis* CroRS two-component system obtained using a differential-display random arbitrarily primed PCR approach. Appl Environ Microbiol 73:3738–3741

Le Jeune A, Torelli R, Sanguinetti M, Giard J, Hartke A, Auffray Y, Benachour A (2010) The extracytoplasmic function sigma factor SigV plays a key role in the original model of lysozyme resistance and virulence of *Enterococcus faecalis*. PLoS One 5: e9658

Macnab RM (2003) How bacteria assemble flagella. Annu Rev Microbiol 57:77–100

MacRitchie DM, Buelow DR, Price NL, Raivio TL (2008) Two-component signaling and gram negative envelope stress response systems. Adv Exp Med Biol 631:80–110

Maillard J (2002) Bacterial target sites for biocide action. J Appl Microbiol 92 Suppl:16S–27S

Mandlik A, Swierczynski A, Das A, Ton-That H (2008) Pili in Gram-positive bacteria: assembly, involvement in colonization and biofilm development. Trends Microbiol 16:33–40

Mandon EC, Trueman SF, Gilmore R (2009) Translocation of proteins through the Sec61 and SecYEG channels. Curr Opin Cell Biol 21:501–507

Manson JM, Keis S, Smith JMB, Cook GM (2004) Acquired bacitracin resistance in *Enterococcus faecalis* is mediated by an ABC transporter and a novel regulatory protein, BcrR. Antimicrob Agents Chemother 48:3743–3748

Marreddy RKR, Pinto JPC, Wolters JC, Geertsma ER, Fusetti F, Permentier HP, Kuipers OP, Kok J, Poolman B (2010) The response of *Lactococcus lactis* to membrane protein production. *Submitted*

Martínez B, Zomer AL, Rodríguez A, Kok J, Kuipers OP (2007) Cell envelope stress induced by the bacteriocin Lcn972 is sensed by the Lactococcal two-component system CesSR. Mol Microbiol 64:473–486

Martínez JL (2008) Antibiotics and antibiotic resistance genes in natural environments. Science 321:365–367

Mascher T (2006) Intramembrane-sensing histidine kinases: a new family of cell envelope stress sensors in Firmicutes bacteria. FEMS Microbiol Lett 264:133–144

Massova I, Mobashery S (1998) Kinship and diversification of bacterial penicillin-binding proteins and beta-lactamases. Antimicrob Agents Chemother 42:1–17

Matos R, Pinto VV, Ruivo M, Lopes MDFS (2009) Study on the dissemination of the *bcrABDR* cluster in *Enterococcus* spp. reveals that the BcrAB transporter is sufficient to confer high-level bacitracin resistance. Int J Antimicrob Agents 34:142–147

Maurer SE, Deamer DW, Boncella JM, Monnard P (2009) Chemical evolution of amphiphiles: glycerol monoacyl derivatives stabilize plausible prebiotic membranes. Astrobiology 9:979–987

Morello E, Bermúdez-Humarán LG, Llull D, Solé V, Miraglio N, Langella P, Poquet I (2008) *Lactococcus lactis*, an efficient cell factory for recombinant protein production and secretion. J Mol Microbiol Biotechnol 14:48–58

Muller C, Le Breton Y, Morin T, Benachour A, Auffray Y, Rincé A (2006) The response regulator CroR modulates expression of the secreted stress-induced SalB protein in *Enterococcus faecalis*. J Bacteriol 188:2636–2645

Myhre AE, Aasen AO, Thiemermann C, Wang JE (2006) Peptidoglycan – an endotoxin in its own right? Shock 25:227–235

Nielsen DS, Schillinger U, Franz CMAP, Bresciani J, Amoa-Awua W, Holzapfel WH, Jakobsen M (2007) *Lactobacillus ghanensis* sp. *nov*., a motile lactic acid bacterium isolated from Ghanaian cocoa fermentations. Int J Syst Evol Microbiol 57:1468–1472

O'Connell-Motherway M, van Sinderen D, Morel-Deville F, Fitzgerald GF, Ehrlich SD, Morel P (2000) Six putative two-component regulatory systems isolated from *Lactococcus lactis* subsp. *cremoris* MG1363. Microbiology 146 (Pt 4):935–947

Osborn MJ, Rothfield L (2007) Cell shape determination in *Escherichia coli*. Curr Opin Microbiol 10:606–610

Panno J (2004) The cell: evolution of the first organism. Facts on File, New York

Perry JA, Lévesque CM, Suntharaligam P, Mair RW, Bu M, Cline RT, Peterson SN, Cvitkovitch DG (2008) Involvement of *Streptococcus mutans* regulator RR11 in oxidative stress response during biofilm growth and in the development of genetic competence. Lett Appl Microbiol 47:439–444

Pietiäinen M, François P, Hyyryläinen H, Tangomo M, Sass V, Sahl H, Schrenzel J, Kontinen VP (2009) Transcriptome analysis of the responses of *Staphylococcus aureus* to antimicrobial peptides and characterization of the roles of *vraDE* and *vraSR* in antimicrobial resistance. BMC Genomics 10:429

Pinto JPC, Kuipers OP, Marreddy RKR, Poolman B, Kok J (2010) Efficient overproduction of membrane proteins in *Lactococcus lactis* requires the cell envelope stress sensor/regulator couple CesSR. *Submitted*

Poole K (2007) Efflux pumps as antimicrobial resistance mechanisms. Ann Med 39:162–176

Qi F, Merritt J, Lux R, Shi W (2004) Inactivation of the *ciaH* Gene in *Streptococcus mutans* diminishes mutacin production and competence development, alters sucrose-dependent biofilm formation, and reduces stress tolerance. Infect Immun 72:4895–4899

Reynolds PE, Courvalin P (2005) Vancomycin resistance in enterococci due to synthesis of precursors terminating in D-alanyl-D-serine. Antimicrob Agents Chemother 49:21–25

Roberts IS (1996) The biochemistry and genetics of capsular polysaccharide production in bacteria. Annu Rev Microbiol 50:285–315

Roces C, Campelo AB, Veiga P, Pinto JPC, Rodríguez A, Martínez B (2009) Contribution of the CesR-regulated genes *llmg0169* and *llmg2164-2163* to *Lactococcus lactis* fitness. Int J Food Microbiol 133:279–285

Rogers PD, Liu TT, Barker KS, Hilliard GM, English BK, Thornton J, Swiatlo E, McDaniel LS (2007) Gene expression profiling of the response of *Streptococcus pneumoniae* to penicillin. J Antimicrob Chemother 59:616–626

Sauvage E, Kerff F, Terrak M, Ayala JA, Charlier P (2008) The penicillin-binding proteins: structure and role in peptidoglycan biosynthesis. FEMS Microbiol Rev 32:234–258

Sebert ME, Palmer LM, Rosenberg M, Weiser JN (2002) Microarray-based identification of *htrA*, a *Streptococcus pneumoniae* gene that is regulated by the CiaRH two-component system and contributes to nasopharyngeal colonization. Infect Immun 70:4059–4067

Sebert ME, Patel KP, Plotnick M, Weiser JN (2005) Pneumococcal HtrA protease mediates inhibition of competence by the CiaRH two-component signaling system. J Bacteriol 187:3969–3979

Shivaji S, Prakash JSS (2010) How do bacteria sense and respond to low temperature? Arch Microbiol 192:85–95

Sillanpää J, Prakash VP, Nallapareddy SR, Murray BE (2009) Distribution of genes encoding MSCRAMMs and pili in clinical and natural populations of *Enterococcus faecium*. J Clin Microbiol 47:896–901

Silver LL (2003) Novel inhibitors of bacterial cell wall synthesis. Curr Opin Microbiol 6:431–438

Silver LL (2006) Does the cell wall of bacteria remain a viable source of targets for novel antibiotics? Biochem Pharmacol 71:996–1005

Sleytr UB, Beveridge TJ (1999) Bacterial S-layers. Trends Microbiol 7:253–260

Summers AO (2006) Genetic linkage and horizontal gene transfer, the roots of the antibiotic multiresistance problem. Anim Biotechnol 17:125–135

Suntharalingam P, Senadheera MD, Mair RW, Lévesque CM, Cvitkovitch DG (2009) The LiaFSR system regulates cell envelope stress response in *Streptococcus mutans*. J Bacteriol 191:2973–2984

Swoboda JG, Campbell J, Meredith TC, Walker S (2010) Wall teichoic acid function, biosynthesis, and inhibition. Chembiochem 11:35–45

Telford JL, Barocchi MA, Margarit I, Rappuoli R, Grandi G (2006) Pili in Gram-positive pathogens. Nat Rev Microbiol 4:509–519

Tsuda H, Yamashita Y, Shibata Y, Nakano Y, Koga T (2002) Genes involved in bacitracin resistance in *Streptococcus mutans*. Antimicrob Agents Chemother 46:3756–3764

Ulrich LE, Koonin EV, Zhulin IB (2005) One-component systems dominate signal transduction in prokaryotes. Trends Microbiol 13:52–56

Van Bambeke F, Mingeot-Leclercq M, Struelens MJ, Tulkens PM (2008) The bacterial envelope as a target for novel anti-MRSA antibiotics. Trends Pharmacol Sci 29:124–134

van Dam V, Sijbrandi R, Kol M, Swiezewska E, de Kruijff B, Breukink E (2007) Transmembrane transport of peptidoglycan precursors across model and bacterial membranes. Mol Microbiol 64:1105–1114

Veiga P, Bulbarela-Sampieri C, Furlan S, Maisons A, Chapot-Chartier M, Erkelenz M, Mervelet P, Noirot P, Frees D, Kuipers OP, Kok J, Gruss A, Buist G, Kulakauskas S (2007) SpxB regulates O-acetylation-dependent resistance of *Lactococcus lactis* peptidoglycan to hydrolysis. J Biol Chem 282:19342–19354

Walker A (2009) Genome watch: probiotics stick it to the man. Nat Rev Microbiol 7:843

Wallin E, Heijne GV (1998) Genome-wide analysis of integral membrane proteins from eubacterial, archaean, and eukaryotic organisms. Protein Sci 7:1029–1038

Walsh SE, Maillard JY, Simons C, Russell AD (1999) Studies on the mechanisms of the antibacterial action of ortho-phthalaldehyde. J Appl Microbiol 87:702–710

Weigel LM, Clewell DB, Gill SR, Clark NC, McDougal LK, Flannagan SE, Kolonay JF, Shetty J, Killgore GE, Tenover FC (2003) Genetic analysis of a high-level vancomycin-resistant isolate of *Staphylococcus aureus*. Science 302:1569–1571

Zapun A, Contreras-Martel C, Vernet T (2008) Penicillin-binding proteins and beta-lactam resistance. FEMS Microbiol Rev 32:361–385

Zhu L, Kreth J (2010) Role of *Streptococcus mutans* eukaryotic-type serine/threonine protein kinase in interspecies interactions with *Streptococcus sanguinis*. Arch Oral Biol 55:385–390

Chapter 9
Responses of Lactic Acid Bacteria to Heavy Metal Stress

Marc Solioz, Mélanie Mermod, Helge K. Abicht, and Stefano Mancini

9.1 Introduction

Lactic acid bacteria (LAB) belong to the order *Lactobacillales* and produce lactic acid as a result of carbohydrate fermentation. They are widely used in the production of fermented food, such as yogurt (streptococci and lactobacilli), cheeses (lactococci), sauerkraut (*Leuconostoc*), wine (oenococci), or cured sausages like Salami (pediococci, lactococci). They are heterotrophic and generally have complex nutritional requirements because they lack many biosynthetic capabilities. Because of this, LAB are generally abundant only in environments where these requirements can be provided, such as animal oral cavities and intestines (e.g., enterococci), plant leaves (*Lactobacillus, Leuconostoc*), decaying plant or animal matter, feces, and compost, etc.

LAB are used in the food industry for several reasons. Their growth lowers both the carbohydrate content of the foods that they ferment and the pH due to lactic acid production. This is often accompanied by the secretion of bacteriocins, such as nisin. Bacteriocins are proteinaceous toxins that inhibit the growth of similar or closely related bacterial strains. The combined action of low pH and bacteriocins is very important in food preservation to inhibit efficiently the growth of competing bacteria, including the most common human pathogens (Galvez et al. 2007). This bestows prolonged shelf lives on these foods. The acidity also changes the texture of the foods due to the precipitation of some proteins. In addition, the biochemical conversions involved in growth greatly enrich the flavor of fermented food. The acidic ambient generated by the secreted lactic acid can lead to the solubilization of complexed metal ions. For example, in traditional cheese making, the cells are challenged by copper released from the copper kettles (Kiermeier and Kyrein 1971).

M. Solioz (✉)
Department Clinical Research, University of Bern, Murtenstrasse 35, 3010 Berne, Switzerland
e-mail: marc.solioz@ikp.unibe.ch

E. Tsakalidou and K. Papadimitriou (eds.), *Stress Responses of Lactic Acid Bacteria*,
Food Microbiology and Food Safety, DOI 10.1007/978-0-387-92771-8_9,

Though this process is important for flavor development (Steffen et al. 2009), it also puts stress on the bacteria.

A distinction should be made between metal ions that are required by LAB for certain enzyme functions and are thus vital and metals that are only toxic without a benefit for life. Of the trace metals known to function in biochemical processes, iron, zinc, and magnesium are probably used by all bacteria, whereas nickel, cobalt, selenium, and molybdenum are only used by some. No function for copper or selenium has been identified in any member of the *Lactobacillales*, and only a few organisms of this order have an apparent requirement for nickel or cobalt (Table 9.1).

Some trace metals, like iron, selenium, molybdenum, manganese, and copper, are often scarce in the environment, and it can be assumed that cells requiring these metals possess corresponding uptake systems. However, few of these have been characterized to date. For essential metals like copper, nickel, cobalt, and zinc, which can occur in widely different bioavailable concentrations in the environment, bacteria must have homeostatic control mechanisms that can deal with excess as well as with deficiency. For purely toxic metals without a known function in biology, like lead, silver, or cadmium, specialized defense mechanisms have evolved in many bacterial species (a biological function for cadmium has been described in a marine diatom, but this may represent an exceptional case (Lane and Morel 2000)). Unfortunately, the knowledge of metal homeostasis and defense against metal stress by LAB is still very limited. Of all the biologically relevant metals, copper by far has received the widest attention. Copper homeostasis and the response to copper stress have been studied in detail in *Lactococcus lactis* and *Enterococcus hirae* and will be a major focus of this chapter. Stress responses to other metals, which have received little interest in LAB, will also be discussed for related bacteria such as *Bacillus subtilis*, to the extent that such work could be relevant to LAB on the basis of the known gene complements. Vanadium, molybdenum, and tungsten, which serve as cofactors in a variety of bacterial enzymes, will not be discussed because they are generally rare in the environment and have not received any attention in LAB.

9.2 Metal Toxicity Mechanisms

Several reactive oxygen species (ROS) and one thiyl radical ($RS^{\bullet}$) can be formed in cells and can exert toxicity by modifying biomolecules (see Miyoshi et al. (2003) for review). Metal ions can catalyze some of the reactions that lead to their formation, which is one of the underlying mechanisms of metal-induced stress. Superoxide radicals ($O_2^{\bullet-}$) are formed when oxygen takes up one electron. It is a product of "leaks" in the mitochondrial electron transport chain, but it can also be produced by macrophages in the "oxidative burst," which is an important bactericidal action by these cells. $O_2^{\bullet-}$ can be directly toxic, for example, by oxidizing and displacing iron from Fe-S clusters (Fig. 9.1), whereby the released iron can catalyze additional toxic reactions (see ahead). Alternatively, superoxide can be converted by SOD to

Table 9.1 Use of selected metals by lactic acid bacteria

Species	Genomes	Cu users (Ridge et al. 2008)	Ni users (Zhang et al. 2009)	Co users (Zhang et al. 2009)	Mo users (Zhang and Gladyshev 2008)	Se users (Zhang et al. 2008)
Enterococci	2	0	0	1	1	0
Lactobacilli	10	0	0	7	2	0
Lactococci	1	0	0	0	0	0
Leuconostoc	1	0	0	0	0	0
Oenococcus	1	0	0	0	0	0
Pediococcus	1	0	0	0	0	0
Streptococcus	6	0	1	2	0	0

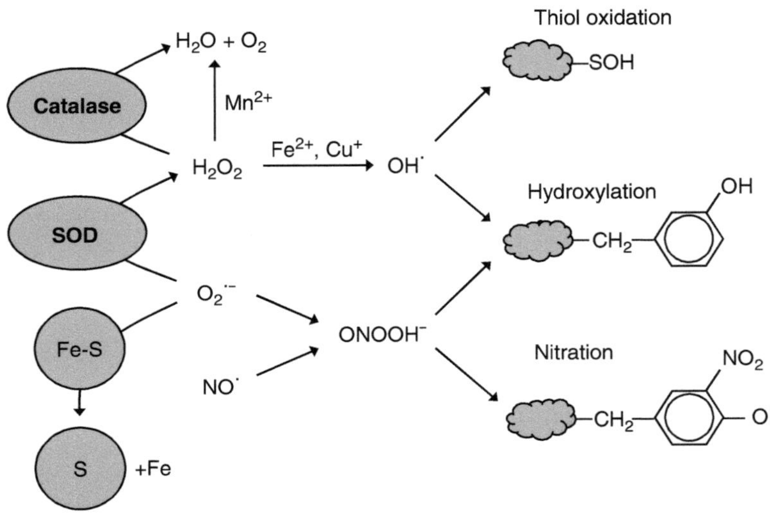

Fig. 9.1 Major oxidative damage mechanisms and their coupling to redox-active metals. Superoxide ($O_2^{\bullet-}$) produced by physiological reactions can attack iron-sulfur centers of enzymes and cause loss of the iron from the reactive center. For detoxification, superoxide is converted by superoxide dismutase (SOD) to hydrogen peroxide (H_2O_2). Hydrogen peroxide can be dismutated to water and oxygen either in a nonenzymatic reaction with Mn^{2+} or by catalase, but can also undergo Fenton chemistry catalyzed by Fe^{2+} or Cu^{+}, resulting in highly toxic hydroxyl radicals ($OH^{\bullet}$). These can lead to thiol oxidation and hydroxylation of cellular constituents. Superoxide also can react with nitrous oxide radicals ($NO^{\bullet}$) to form reactive peroxynitrite ($ONOOH^{-}$), which can nitrate or hydroxylate cellular components

less toxic hydrogen peroxide (H_2O_2). However, the combination of iron, H_2O_2, and superoxide leads to the generation of hydroxyl radicals ($OH^{\bullet}$) by a combination of the Fenton reaction (9.2) and the Haber–Weiss reaction (9.3):

$$Fe^{3+} + O_2^{\bullet-} \rightarrow Fe^{2+} + O_2 \qquad 9.1$$

$$Fe^{2+} + H_2O_2 \rightarrow Fe^{3+} + OH^{-} + OH^{\bullet} \qquad 9.2$$

$$O_2^{\bullet-} + H_2O_2 \rightarrow OH^{\bullet} + OH^{-} + O_2 \qquad 9.3$$

$OH^{\bullet}$ are very reactive and damage biomolecules by hydroxylation. In the overall ROS scheme, iron, copper, and other redox-active metal ions exert their effects by stimulating the Fenton reaction.

H_2O_2 is mainly produced by enzymatic reactions, such as the dehydrogenation of NAD(P)H. Because most lactococci do not possess catalase, H_2O_2 can reach levels of 1–2% (Rochat et al. 2006) and may thus be considered not very toxic per se. It can leave the cells by diffusion through the membrane. H_2O_2 is also produced by the dismutation of $O_2^{\bullet-}$ by superoxide dismutase (SOD). The combination of SOD and catalase provides an efficient antioxidant mechanism.

Nitric oxide is naturally formed in activated macrophages and endothelial cells and is considered as an active agent in several pathologies based on inflammation and organ reperfusion and may also play an important role in atherosclerosis. It has poor oxidizing power and is even antioxidant under physiological concentrations (up to 100 nM). It does, however, react rapidly with oxygen to yield nitrogen dioxide ($NO_2^{\bullet}$), which in turn may react with $NO^{\bullet}$ to yield nitrogen trioxide, N_2O_3. The rapid reaction of $O_2^{\bullet-}$ with $NO^{\bullet}$ gives the extremely reactive peroxynitrite, $ONOO^-$, which mediates oxidation, nitrozation, and nitration reactions.

Aliphatic thiols, RSH, are contained in living organisms in high concentrations. Typical levels of intracellular glutathione are 5–10 mM, and a similar level of RSH is provided by cysteines in proteins. RSH can be oxidized in the presence of redox-active metal ions like iron or copper ions according to reactions (9.4) and (9.5):

$$RSH + Cu^{2+} \rightarrow RS^{\bullet} + Cu^{+} + H^{+} \qquad 9.4$$

$$2Cu^{+} + 2H^{+} + O_2 \rightarrow 2\,Cu^{2+} + H_2O_2 \qquad 9.5$$

The thiyl radicals have strong reactivity toward oxygen (9.6):

$$RS^{\bullet} + O_2 \rightarrow RSO_2^{\bullet} \qquad 9.6$$

Furthermore, thiyl radicals are able to oxidize NADH to $NAD^{\bullet}$, as well as ascorbic acid, and to generate various free radicals such as $OH^{\bullet}$ and $O_2^{\bullet-}$. There can also be thiol depletion by reaction (9.7) in cyclic combination with reaction (9.7):

$$2\,Cu^{2+} + 2\,RSH \rightarrow 2\,Cu^{+} + RSSR + 2H^{+} \qquad 9.7$$

While lipid and protein damage by the above mechanism has been demonstrated in vitro in many studies, recent findings suggest that alternative mechanisms of metal toxicity may be responsible for the primary toxic effects of copper, iron, and related metals in vivo. First, the discovery that free copper or iron in the cell is extremely low or even nonexistent makes Fenton chemistry and sulfhydryl depletion very unlikely mechanisms (Changela et al. 2003). Second, most *Lactobacillales* are rather tolerant to H_2O_2. For example *L. lactis* IL1403, described in some detail below, generates H_2O_2 by NADH dehydrogenation but does not possess catalase for H_2O_2 removal (Bolotin et al. 2001; Marty-Teysset et al. 2000; Rochat et al. 2006). Third, Macomber et al. recently showed that copper-loaded *Escherichia coli* was less sensitive to killing by H_2O_2 than cells grown without copper. Also, copper decreased the rate of H_2O_2-induced DNA damage. High intracellular copper levels even impaired iron-mediated oxidative killing by H_2O_2 (Macomber et al. 2007).

Based on these observations, the authors suggested that copper exerts its toxicity by mechanisms other than oxidative stress.

A novel mechanism of copper toxicity was indeed recently demonstrated. It could be shown in vivo as well as in vitro that copper specifically damaged the iron-sulfur clusters of isopropylmalate dehydratase of *E. coli* (Macomber and Imlay 2009). This enzyme of the branched-chain amino acid biosynthesis pathway contains an iron-sulfur cluster from which the iron can be displaced by copper in the absence of oxygen. Copper efflux systems, chelation by glutathione, and cluster repair by assembly systems all enhance the resistance of cells to this type of copper toxicity. To establish whether this mechanism is a general route of copper toxicity in bacteria, including LAB, will require further investigation.

9.3 Response to Copper and Silver

9.3.1 Copper as a Bioelement

In the primordial, anaerobic world, copper was in the Cu(I) state in the form of water-insoluble sulfides under neutral pH conditions and was only bioavailable in the acidic waters near hydrothermal vents. The emergence of an oxygen-containing atmosphere by the action of oxygen-evolving microorganisms, probably cyanobacteria, less than 3×10^9 years ago was a dramatic event for most living organisms (Kasting and Siefert 2002). Most of them adapted to the new conditions by acquiring an oxidative metabolism. The "old" enzymes involved in anaerobic metabolism were designed to operate in the lower portion of the redox spectrum. The arrival of dioxygen created the need for a new redox active metal that could attain higher redox potentials. The oxidation of insoluble Cu(I) led to soluble and thus widely bioavailable Cu(II), which was ideally suited to exploit the oxidizing power of dioxygen (Crichton and Pierre 2001). Copper therefore is a modern bioelement (Kaim and Rall 1996). Concomitant with the arrival of oxygen, multicellular organisms developed.

Because of copper's ability to cycle between Cu^{2+} and Cu^{+} at biologically relevant redox potentials, it has become a cofactor for over 30 known enzymes in higher organisms (Karlin 1993). Prominent examples are lysyl oxidase, involved in the cross-linking of collagen; tyrosinase, required for melanin synthesis; dopamine β-hydroxylase of the catecholamine pathway; cytochrome *c* oxidase as a terminal electron acceptor of the respiratory chain; and SOD, required for defense against oxidative damage. Another class of copper proteins, such as plastocyanins or azurins, acts as electron carriers. Depending on the type of coordination of the copper to the protein, the redox potential can vary over the range of +200 to +800 mV. Concomitant with the lower complexity of bacteria, only ten cuproenzymes have so far been characterized in microbes (Table 9.2).

Strikingly, none of the sequenced *Lactobacillales* appears to be copper users based on bioinformatics analysis of known copper enzymes. The definition of

Table 9.2 Known bacterial copper-containing enzymes

Enzyme	Function	References
Cytochrome *c* oxidase	Terminal oxidase	Cavet et al. (2003)
NADH dehydrogenase-2	Electron transport, Cu reduction	Rapisarda et al. (2002); Rodriguez-Montelongo et al. (2006)
Nitrosocyanin, cuproredoxin-like	Electron transfer, other?	Arciero et al. (2002)
Plastocyanins	Electron transfer	Cavet et al. (2003)
Cu-containing nitrite reductases	Nitrite reduction	Ellis et al. (2007)
Tyrosinase	Phenol oxidation, melanin synthesis	Lopez-Serrano et al. (2004); Tsai and Lee (1998)
Cu amine oxidases	Oxidation of primary amines	Brazeau et al. (2004)
Particulate methane monooxygenase	Methane oxidation	Chan et al. (2004)
Cu,Zn-superoxide dismutase (cuprein)	Defense during infection?	Battistoni (2003)
Cu-containing laccase	Polyphenol oxidase	Hullo et al. (2001)

"users" is obviously based on the currently known bacterial cuproenzymes summarized in Table 9.2. However, not all functions of copper in LAB are known. It was observed, for example, that *Lactococcus lactis* subsp. *lactis* 3022 produced more biomass when grown aerobically with hemin and copper (Kaneko et al. 1990). The activity of diacetyl synthase was greatly stimulated by the addition of hemin or copper, and the activity of NAD-dependent diacetyl reductase was very high. Pyruvate formed via glycolysis was converted to diacetyl, which in turn was converted to acetoin by the NAD-dependent diacetyl reductase to reoxidize NADH. This suggests that hemin or copper stimulates acetyl coenzyme A formation from pyruvate, but the nature of this mechanism remains unknown. At any rate, some bacteria make extensive use of copper as a bioelement, while others, like the *Lactobacillales*, use it for only a few functions, if at all. It might be speculated that there is a connection between the small average genome size of *Lactobacillales* of only 2.3 Mb and those of copper-using Gram-positive organisms, with an average genome size of 3 Mb (Ridge et al. 2008).

Recently, an unexpected link between copper and molybdenum cofactor (MOCO) synthesis was discovered. Plant Cnx1G, a domain of the Cnx1GE protein, catalyzes the adenylation of molybdopterin. Cnx1G-bound molybdopterin was found to have copper bound to the molybdopterin dithiolate sulfurs (Kuper et al. 2004). The function of this bound copper is presently unknown, but copper might play a role in protecting the molybdopterin dithiolate from oxidation and/or in presenting a suitable leaving group for molybdenum insertion (Schwarz and Mendel 2006). It remains currently unclear if the binding of copper to molybdopterin is an essential step in MOCO synthesis, but if so, this pathway generates a copper requirement in addition to those considered in Table 9.1 (Zhang and Gladyshev 2008). If one looks across the bacterial phyla, a cooccurrence of copper use and MOCO synthesis strongly prevails. In the *Lactobacillales*, only three of 22 sequenced organisms appear to be capable of MOCO synthesis. Conceivably, this could create a need for copper for these apparent copper nonusers. In fact, the makeup of the copper homeostasis machinery of *E. hirae* strongly argues for a copper requirement in this organism (see ahead).

9.3.2 Copper Homeostasis in Enterococcus hirae

In LAB, copper homeostasis has most extensively been studied in *E. hirae,* and this system has served as a model for metal homeostasis in general (Solioz and Stoyanov 2003). The core element is an operon that consists of the four genes *copY*, *copZ*, *copA*, and *copB*. *CopZ* encodes a copper chaperone, *copY* encodes a copper-responsive repressor, and *copA* and *copB* encode copper transporting ATPases (Fig. 9.2). The *cop* operon enables *E. hirae* to grow in up to 8 mM copper and under copper-limiting conditions. The function of the four Cop proteins will be described next.

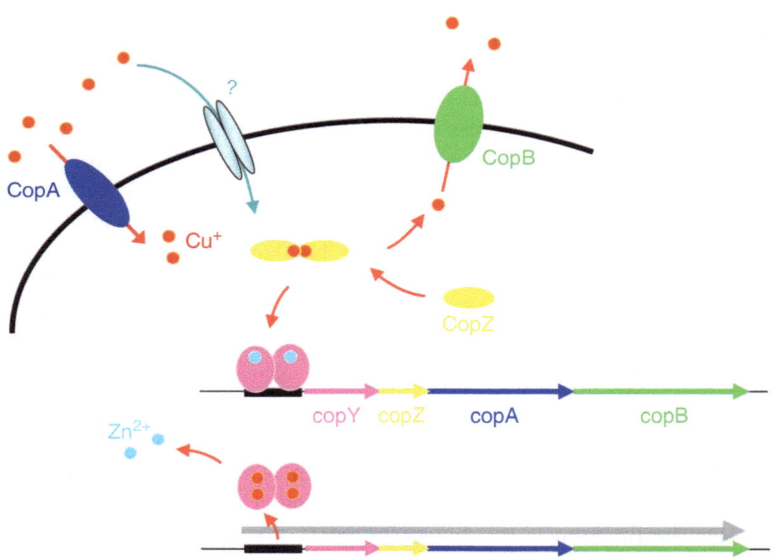

Fig. 9.2 Illustration of copper homeostasis in *E. hirae*. Copper enters the cell via *CopA* or by nonspecific leakage. Excess cytoplasmic copper binds to CopZ, which can then donate Cu^+ to CopB for export and to the CopY repressor to induce the *cop* operon. In low-copper conditions, CopY dimers in the zinc form are bound to the *cop*-boxes in front of the *cop* operon. When CopZ donates Cu^+ to CopY, one Zn^{2+} per CopY monomer is replaced by two Cu^+, with the concomitant release of CopY from the promoter and induction of transcription of the downstream genes

9.3.2.1 The CopZ Copper Chaperone

The identification of copper chaperones marked the emergence of a new concept in the handling of metal ions by cells, namely, the escorting of the metal by a protein to prevent nonspecific, damaging interactions. There is a range of different copper chaperones in eukaryotes to deliver copper to cytochrome *c* oxidase, SOD, or copper ATPases (Kim et al. 2008), but only two types of copper chaperones have so far been described in bacteria. Sco-like chaperones deliver copper to cytochrome oxidase; they appear to be absent in LAB. CopZ-like chaperones, on the other hand, transport copper to ATPases and transcriptional regulators and are ubiquitous in LAB (Huffman and O'Halloran 2001). Interestingly, no CopZ-like copper chaperones have been described in *Actinobacteria* or in *E. coli* and related organisms and it remains unknown how copper is escorted in the cytoplasm of those organisms.

CopZ of *E. hirae* is an 8-kDa protein, and the structure of CopZ and that of other CopZ-like proteins have been solved (see Davis and O'Halloran (2008) for review). They all share the same $\beta\alpha\beta\beta\alpha\beta$ ferredoxin-like structure, with two exposed cysteines of a CxxC motif located in a loop between $\beta1$ and $\alpha1$. There is still uncertainty as to how Cu^{+} is complexed by the chaperone in vivo. In principle, Cu^{+} can bind to the CxxC motif in a near-linear S–Cu–S bonding. However, X-ray structures of Hah1, the human CopZ-like copper chaperone, have revealed structures where a single Hg^{2+} or a Cu^{+} ion is complexed by the four cysteines of two chaperones in a dimeric arrangement (Rosenzweig 2001). Cu^{+}–CopZ of *E. hirae,* on the other hand, appeared to be dimeric in solution, with triagonally bound copper to be the most likely structure (Fig. 9.3) (Wimmer et al. 1999). The prevalence of homodimeric Cu^{+}-CopZ was also demonstrated by biochemical and light-spectroscopic techniques (Kihlken et al. 2002, 2008). A three-coordinate metal center is also supported by EXAFS measurements of Cu^{+}–thiol bonds (Pufahl et al. 1997; Wimmer et al. 1999). Glutathione was shown to inhibit dimer formation in vitro and could, in principle, be a ligand to monomeric Cu^{+}-CopZ inside the cell, where glutathione concentrations are high. It is also conceivable that there is an equilibrium among monomeric, dimeric, and even trimeric CopZ in the cell, but this will be difficult to assess.

CopZ of *E. hirae* was shown by surface plasmon resonance to interact with the CopA copper ATPase and the CopY repressor (Multhaup et al. 2001; Portmann et al. 2004). It is assumed that Cu^{+} imported by CopA is transferred to the CopZ

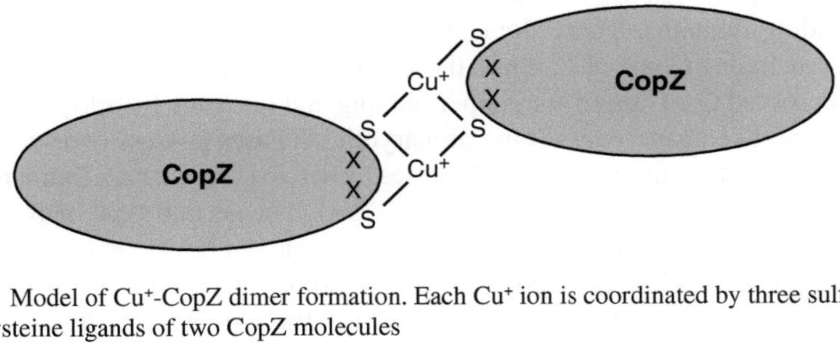

Fig. 9.3 Model of Cu^{+}-CopZ dimer formation. Each Cu^{+} ion is coordinated by three sulfur atoms of the cysteine ligands of two CopZ molecules

copper chaperone, which subsequently delivers copper to the CopY repressor for induction of the *cop* operon (discussed ahead) or to other sites requiring copper. An interaction of CopZ with the CopB copper-exporting ATPase has also been shown (unpubl. observation), suggesting that CopZ, in addition, has a role in copper export from the cell. CopZ interaction with the copper-exporting ATPase was also demonstrated in *B. subtilis* (Radford et al. 2003). In eukaryotes, the primary function of CopZ-like copper chaperones (Hah1, Atx1, Atox1) is in fact the delivery of copper to copper ATPases (Huffman and O'Halloran 2001).

9.3.2.2 Copper ATPases

The two *E. hirae* copper ATPases mark the discovery of ATP-driven copper transport across cell membranes in 1992 (Odermatt et al. 1992). Before that time, there was no concept and no serious discussion of how copper could cross cell membranes. According to the current model, CopA serves in the uptake of copper when copper is limiting, while CopB serves in copper extrusion under conditions of copper excess (Odermatt et al. 1994; Solioz and Odermatt 1995) (the nomenclature is confusing: Copper export is accomplished by CopB in *E. hirae*, but by enzymes called "CopA" in most other bacteria).

Copper ATPases belong to the superfamily of P-type ATPases, classically represented by eukaryotic Ca- and NaK-ATPases. The most prominent feature of this family of pumps is the formation of an acylphosphate intermediate (hence the name P-type ATPases) whereby the γ-phosphate of ATP phosphorylates the aspartic acid residue in the conserved motif DKTGT during the reaction cycle (Pedersen and Carafoli 1987). Detailed structures of the calcium ATPase of the sarcoplasmatic reticulum have given considerable insight into the working of such ATP-driven ion pumps (Toyoshima et al. 2003; Toyoshima and Mizutani 2004). Copper-transporting ATPases are a subgroup of the P-type ATPases. They have been termed heavy metal ATPases, or CPx-type ATPases due to a conserved intramembranous CPC or CPH motif (Lutsenko and Kaplan 1995; Solioz and Vulpe 1996), or P1B-type ATPases, based on more systematic phylogeny (Lutsenko and Kaplan 1995). CPx-type ATPases are widespread in nature and have been found to catalyze the transport of a range of transition and heavy metal ions, including Cu^{+}, Ag^{+}, Co^{2+}, Zn^{2+}, Cd^{2+}, Hg^{2+}, and Pb^{2+} (Axelsen and Palmgren 1998).

Another typical feature of CPx-type ATPases is the presence of N-terminal metal-binding domains. These domains can be of two kinds. In most CPx-type ATPases, including CopA of *E. hirae*, these domains consist of a CopZ-like module with a conserved CxxC motif for copper binding and the same βαββαβ fold exhibited by CopZ-like chaperones. Some prokaryotic ATPases possess two such CopZ-like domains, and eukaryotic copper ATPases have two (yeast) or six (humans) such domains (Hanson et al. 2001; Solioz et al. 1994). The second type of N-terminal metal-binding domain is found in CopB of *E. hirae* and a few poorly characterized bacterial CPx-type ATPases. It consists of a histidine- and methionine-rich region. Similar repeat structures were also found in two *Pseudomonas syringae* proteins, which were demonstrated to be periplasmic copper-binding proteins (Cha and Cooksey 1991).

The function of the N-terminal metal-binding domains of heavy metal ATPases remains unclear. Copper transfer from chaperones to the N-terminal metal-binding domains of CPx-type ATPases is now well documented, but it has never been shown that this copper can actually be transported across the membrane. Rather, it has been suggested that the N-terminus regulates the activity of the ATPase by domain–domain interaction (Arguello and Gonzalez-Guerrero 2008). Copper transport may thus require a separate copper-donation event by the chaperone to the membrane region of the ATPases (Gonzalez-Guerrero and Arguello 2008). In *B. subtilis*, the copper export pump CopA features two N-terminal CopZ-like copper-binding domains. It was shown that these motifs play a role in the dimerization of CopA, which could constitute a regulatory mechanism of the ATPase (Singleton et al. 2008; Singleton and Le Brun 2009).

ATP-driven copper transport from the cytoplasm to the extracytoplasmic space, catalyzed by copper ATPases, has been extensively studied and appears to take place in all bacterial species. Both Cu^+ and Ag^+ export by CopB of *E. hirae* have been directly demonstrated with radioisotopes in membrane vesicle and in whole cells loaded with silver (Odermatt et al. 1994; Solioz and Odermatt 1995). Copper-importing ATPases, on the other hand, have only been described in *E. hirae* (CopA), *Synechocystis* sp. (CtaA), and *B. subtilis* (YcnJ) (Chillappagari et al. 2009; Odermatt et al. 1994; Tottey et al. 2001). While the role of *E. hirae* CopA in cell physiology is still unclear, the CtaA of *Synechocystis* sp. imports copper for plastocyanin, a copper-containing thylakoid protein that functions in the photosynthetic electron transport chain.

Cyanobacteria (e.g., *Synechocystis* sp.) are the one bacterial group that has a known demand for cytoplasmic copper for the synthesis of copper-containing, thylakoid-localized plastocyanin and cytochrome oxidase (Tottey et al. 2005). In other organisms, the cuproenzymes are localized at the cytoplasmic membrane or in the periplasm, and copper loading of these proteins could take place in the periplasmic space. In many bacteria, including LAB, no intracellular copper requirements are known at all. The copper homeostatic machinery of these organisms may thus have the sole purpose of keeping copper out. Nevertheless, specific copper importers that are expressed under copper-limiting conditions have been described in *E. hirae* and *B. subtilis* (Chillappagari et al. 2009; Wunderli-Ye and Solioz 2001). Energy-dependent copper uptake has, however, not been directly demonstrated, neither by copper ATPases nor by alternative mechanisms such as with chalkophores (copper "siderophores") (Balasubramanian and Rosenzweig 2008; Kim et al. 2004) or as copper-substrate complexes through substrate transporters. In this light, copper import into the cytoplasm of LAB still needs rigorous experimental confirmation.

9.3.2.3 Regulation of Copper Homeostatic Genes

CopY of *E. hirae* is a copper-responsive transcriptional regulator. It responds to excessive copper in the cytoplasm by derepressing the *cop* operon. In LAB, two

Table 9.3 Copper-responsive regulators of *Actinobacteria*, *Firmicutes*, and Proteobacteria

Organisms	CopY-type	CsoR-type	CueR-type
Actinobacteria	0	43	0
Firmicutes			
Bacillales	3	30	7
Clostridia	0	25	0
Lactobacillales	39	3	0
Proteobacteria	0	54	215

types of copper-responsive transcriptional regulators have been identified by bioinformatics analysis of sequenced genomes, namely, CopY-type and CsoR-type regulators (Table 9.3). CopY-like repressors are the principal copper-responsive regulators of LAB and have experimentally been associated with gene regulation in *E. hirae* (Strausak and Solioz 1997), *Enterococcus faecium* (Hasman et al. 2006), *L. lactis* IL1403 (Magnani et al. 2008), *Streptococcus mutans* (Portmann et al. 2006; Vats and Lee 2001), and *Streptococcus gordonii* (Mitrakul et al. 2004).

CsoR-type regulators have only recently been described (Liu et al. 2007), although their occurrence is more widespread in the prokaryotic world than that of CopY-type repressors. In LAB, CsoR-related proteins occur only in a minority of the sequenced species, and no biochemical studies are as yet available. Finally, CueR-type regulators, which regulate copper homeostatic genes in *E. coli* (Outten et al. 2000), only occur in a few species of the *Bacillales*, but not in LAB. CueR-type regulators are thus the primary copper-responsive regulators of Gram-negative bacteria.

CopY has a bipartite structure: The N-terminus interacts with DNA, while the C-terminus interacts with zinc or copper. The N-terminus shows extensive sequence similarity to BlaI, MecI, and PenI, repressors that are involved in the regulation of β-lactamase in Gram-positive bacteria (Fig. 9.4a) (Garcia-Castellanos et al. 2004; Himeno et al. 1986; Van Melckebeke et al. 2003; Wittman and Wong 1988). The structure of the N-terminus of CopR of *L. lactis*, a CopY-homolog, has been solved by solution NMR (Cantini et al. 2009) and is in fact nearly superimposable on the structure of BlaI of *Bacillus licheniformis* (Fig. 9.4b). The C-terminus of CopY exhibits sequence similarity to the yeast copper-inducible repressors AMT1, ACE1, and Mac, and to the β-domain of metallothioneins (Bird 2008). All these proteins feature the consensus motif $CxCx_{4-5}CxC$. In newly synthesized CopY, this site is occupied by a single Zn^{2+}, which is coordinated by four sulfur atoms in a tetrahedral fashion (Cobine et al. 2002b).

At low ambient copper concentrations, CopY is present as a Zn(II)-containing homodimer and is bound to the operator-promoter region of the operon (Strausak and Solioz 1997). The CopY-dimer-binding sites feature the so called *cop*-box of consensus TACAnnTGTA, a motif that is widely conserved in the *Lactobacillales*. The DNA–CopY interaction has been assessed in quantitative terms by surface plasmon resonance analysis (Portmann et al. 2006). It was found that the CopY-type repressors of *L. lactis, E. hirae*, or *S. mutans* had very similar affinities for *cop*-boxes (Portmann et al. 2004). Interestingly, the β-lactamase regulators BlaI, MecI,

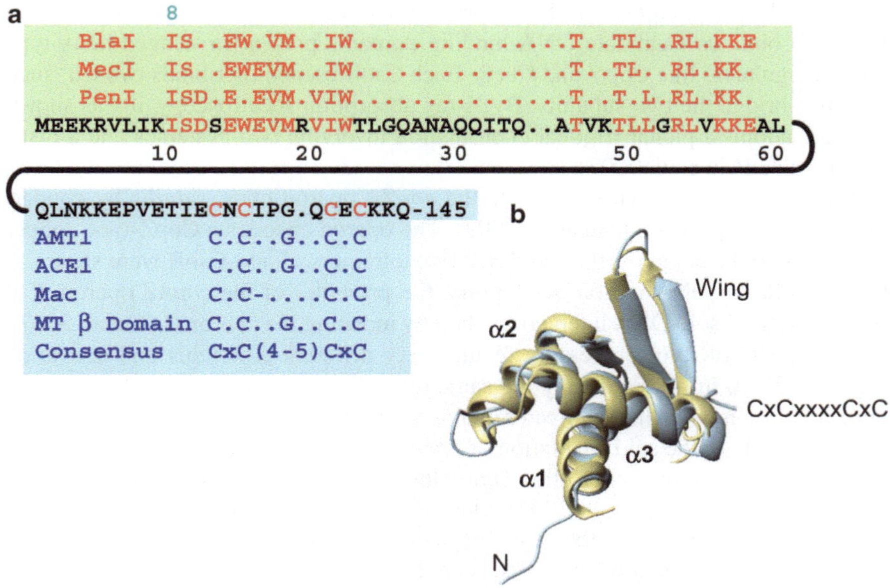

Fig. 9.4 (**a**) Alignment of the protein sequence of CopY of *E. hirae* with those of β-lactamase regulators in the N-terminal region and fungal transcriptional regulators and metallothionein in the C-terminal region. (**b**) Overlay of the N-terminal DNA-binding domain of *L. lactis* CopR (*blue*) and the BlaI β-lactamase regulator of *B. licheniformis* (*gold*)

and PenI, which feature an N-terminal DNA-binding domain essentially identical to that of CopY-like repressors, also recognize a "*cop*-box" (Sharma et al. 1998); the possible consequences of this have not been investigated.

Under low-copper conditions, a CopY dimer is bound to the *cop*-box and prevents transcription. When media copper is raised, two Cu^{+}-CopZ donate the copper ion to one CopY monomer. This displaces the bound Zn(II), and CopY is released from the DNA as Cu_2^{+}CopY, allowing transcription to proceed (Fig. 9.2). Protein–protein interaction between CopZ and CopY could be demonstrated by surface plasmon resonance spectroscopy (Multhaup et al. 2001), and the overall induction mechanism of CopY by copper and CopZ is experimentally well supported (Cobine et al. 1999, 2002a–c). At high intracellular copper levels, CopZ is degraded through a proteolytic pathway, conceivably because high levels of Cu^{+}-CopZ may be toxic to the cell (Lu and Solioz 2001). Following release from the DNA, Cu_2^{+}-CopY is probably also proteolytically degraded (unpubl. observation).

CsoR-type repressors have so far only been studied in *Mycobacterium tuberculosis* and *B. subtilis*. However, it can be assumed that CsoR-type repressors work similarly in LAB and will thus be briefly discussed. CsoR from *M. tuberculosis* represents the founding member of this new class of prokaryotic Cu(I) regulators, and its structure has recently been solved (Liu et al. 2007). CsoR is tetrameric, with two monomers each forming a stable homodimer that adopts an antiparallel four-helix bundle architecture. This represents a novel DNA-binding fold because it lacks the obvious candidate DNA-binding domains present in winged-helix-type metalloregulators like CopY and CueR. Each CsoR homodimer binds two Cu^+ such that they bridge the two subunits. By X-ray absorption spectroscopy, it was shown that Cu^+ adopts a planar trigonal coordination involving two cysteines and a histidine residue (Liu et al. 2007).

CsoR has been shown to regulate the *copZA* operon of *B. subtilis* by copper-dependent derepression (Ma et al. 2009). The operon encodes a CopZ-type copper chaperone and a copper efflux ATPase. Two tetramers of apo-CsoR were shown to bind to a 30-bp DNA region overlapping the promoter of the *copAZ* operon. Cu^+ weakened the CsoR-DNA interaction, thereby inducing the operon (Liu et al. 2007). CopY- and CsoR-type repressors feature very different structures and activation mechanisms to fulfill essentially the same role. Why such diverse mechanisms for gene regulation by copper evolved remains an interesting open question.

In the study of the acid adaptation of *Lactobacillus bulgaricus*, it was found that, among a range of three dozen other genes, three CPx-type ATPases were induced by low-pH stress (Penaud et al. 2006). One of these ATPases resembles CopB of *E. hirae,* and it appears likely that it serves in copper extrusion. Acidic conditions can lead to an increase in ambient copper concentrations through the release of bound copper and the induction of copper-exporting ATPases by acid stress makes physiological sense. No *cop*-boxes were present in any of the promoters of the *L. bulgaricus* CPx-type ATPases and the induction mechanism by low pH remains unknown. Acid-sensitive mutants in the unrelated microorganisms *Rhizobium leguminosarum* bv. *viciae* and *Sinorhizobim meliloti* were similarly found to have disrupted *actP* genes that encode CPx-type ATPases (Reeve et al. 2002). These mutant strains were also more sensitive to ambient copper. Copper induced the expression of the wild-type *actP* genes and low pH enhanced the induction two to threefold. Downstream of the ATPase genes of both organisms are genes encoding MerR-type transcriptional regulators, termed HmrR, which apparently regulate the expression of the ActP ATPases. In *E. coli*, a MerR-type transcriptional regulator, CueR, is responsible for the copper-induced transcription of the CopA copper-exporting ATPase and the periplasmic CueO copper oxidase (Outten et al. 2000, 2001). The acid induction of copper ATPases may be a more general phenomenon. Unfortunately, the acid induction of the copper homeostatic genes of *E. coli*, *E. hirae,* and *L. lactis* has not been addressed so far.

9.3.3 Copper Homeostasis in Lactococcus lactis

In *L. lactis* IL1403, the copper-inducible *copRZA* operon encodes the CopR repressor, a CopY-type repressor, the CopZ copper chaperone, and the CopA copper ATPase (Fig. 9.5). The latter exhibits 45% sequence identity to CopA of *E. hirae*. In contrast to *E. hirae* CopA, *L. lactis* CopA has been shown to be a copper-exporting ATPase (Magnani et al. 2008). The CopR repressor of *L. lactis* regulates the CopR

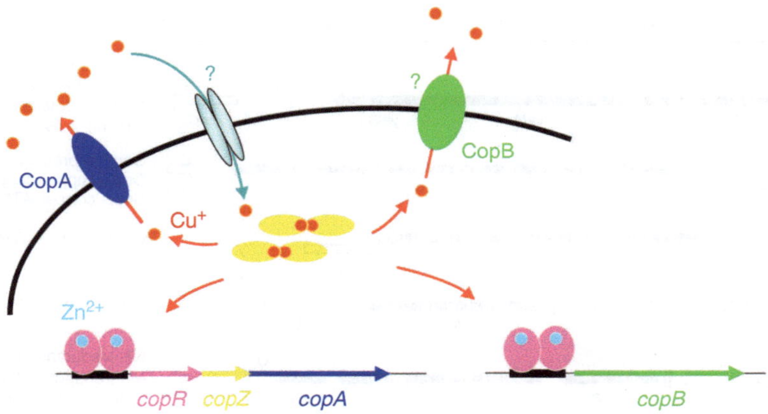

Fig. 9.5 Illustration of copper homeostasis in *L. lactis*. How copper enters the cell is unknown. Excess cytoplasmic copper binds to CopZ, which can then donate Cu^{+} to either the copper ATPases for export or the CopR repressor to induce transcription. In low-copper conditions, a CopR dimer in the zinc form is bound to the *cop*-box in front of the *copRZA* operon and the *copB* gene. When CopZ donates Cu^{+} to CopR, one Zn^{2+} per CopR monomer is replaced by two Cu^{+}, with the concomitant release of CopR from the promoter and induction of transcription of the downstream genes. CopA then accomplishes copper export from the cytoplasm. The function of CopB is unknown

regulon in a fashion analogous to CopY of *E. hirae*. The CopZ-like copper chaperone, finally, can be assumed to function in intracellular copper routing (Arnesano et al. 2002; Cobine et al. 2002b).

A second putative copper ATPase in *L. lactis* is encoded by the unlinked, monocistronic *copB* gene, which is also under the control of CopR. CopB features a histidine-rich N-terminus and shares 55% sequence identity with *E. hirae* CopB. However, no function could so far be assigned to this enzyme. It is notable that *E. hirae* CopB is encoded by the *copYZAB* operon, while CopB of *L. lactis* is encoded by a monocistronic gene. Whether these different gene organizations in *L. lactis* and *E. hirae* are a consequence of functional differences remains an open question.

9.3.3.1 Global Responses to Copper Stress by *Lactococcus lactis*

The CopR repressors of *L. lactis* also recognize the TACAnnTGTA consensus motif, the *cop*-box (Portmann et al. 2006). By performing a genome-wide search for *cop*-boxes in *L. lactis* IL1403, Barré and coworkers found 28 genes whose operator regions harbor the *cop*-box. Seven of these *cop*-boxes were shown to interact with CopR in a copper-responsive manner in vitro. The genes and operons associated with these *cop*-boxes were collectively termed *CopR regulon*. It encompasses a total of 14 genes, organized into four operons and two monocistronic genes (Fig. 9.6) (Magnani et al. 2008). What is the function of these additional copper-regulated genes in copper homeostasis? Only for *lctO*, an NAD-independent, flavin-containing lactate oxidase, has a function been proposed (Barré et al. 2007). Since LctO requires

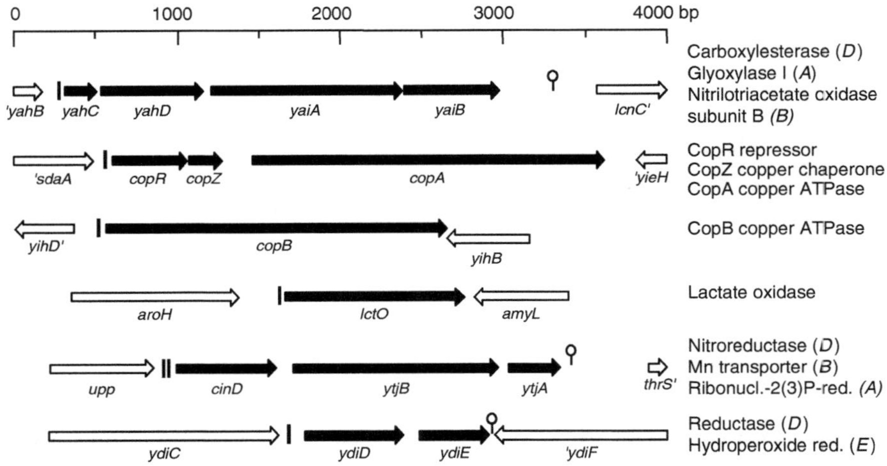

Fig. 9.6 CopR regulon of *L. lactis* IL1403. The genes in full color are regulated by CopR. *Vertical lines* indicate the location of *cop*-boxes and the *lariats* those of ρ-independent transcriptional terminators. The predicted functions of the genes are indicated on the *right*. Genes are drawn to the scale indicated in base pairs (bp) along the *top* of the figure

oxygen to convert lactate to pyruvate, it could serve in the elimination of molecular oxygen under copper stress, thereby attenuating the formation of reactive oxygen radicals (Barré et al. 2007). Similarly, an oxygen-consuming NADH oxidase has been proposed to be involved in the defense against oxidative stress in *Lactobacillus delbrueckii* subsp. *bulgaricus* by removing oxygen and thereby preventing the generation of H_2O_2 and its reaction products (Marty-Teysset et al. 2000).

9.3.4 Response to Silver

Silver has no known biological role and is highly toxic to microorganisms. In fact, silver-impregnated materials are starting to be employed to create aseptic surfaces or odorless clothing (Sondi and Salopek-Sondi 2004). Silver is not redox active like other toxic metals but remains in the Ag^+ form. Silver has a very high affinity to thiolates and binds avidly to sites normally occupied by Cu(I). Intracellular copper is always in the form of Cu(I), due to the reducing environment of the cytoplasm, and any site normally occupied by copper can be taken over by silver. In this way, silver can activate the copper-responsive repressors like CopY of *E. hirae* or CopR of *L. lactis*, which has experimentally been verified (Odermatt and Solioz 1995). It has also been shown that silver can be a substrate for copper-transporting ATPases of bacteria, fish, and mammals (Bury et al. 1999; Hanson et al. 2001; Kanamaru et al. 1994; Stoyanov et al. 2003). The copper efflux ATPase of *E. hirae*, CopB, was shown to pump Ag^+ with the same affinity and velocity as Cu^+ (Solioz and Odermatt 1995). Ag^+ also binds to copper chaperones in a fashion analogous to that of copper

(Kihlken et al. 2008; Narindrasorasak et al. 2004). Silver is thus a Cu(I) mimetic, and it can be assumed that all copper-resistance systems can also handle silver. However, due to the higher toxicity of silver, organisms can generally tolerate much less silver than copper.

A plasmid-born silver-resistance system has been isolated from silver-resistant *Salmonella* sp. (Gupta et al. 1999). The resistance determinant encodes a periplasmic silver-specific-binding protein plus two apparently parallel efflux pumps: a CPx-type ATPase, SilP, and a membrane potential-dependent cation/proton antiporter. The *sil* determinants are regulated by a two-component sensor kinase-response regulator system. Due to the similarity of Ag^{+} and Cu^{+}, it would be expected that the Sil system can also handle copper, but this was apparently not tested.

9.4 Response to Other Heavy Metals

Relatively few studies have been conducted on the response of LAB to heavy metals other than copper. We will therefore also discuss some of the key findings made in other bacterial species to the extent that they could be relevant to LAB. Mercury resistance, which has received considerable attention in many bacterial species, has not been addressed to any significant extent in LAB. The interested reader is referred to the excellent review on bacterial mercury resistance by Barkay et al. (2003).

9.4.1 Response to Iron

In air, Fe^{2+} is rapidly oxidized to Fe^{3+}, which forms hydroxides that are barely soluble at neutral pH. For this reason, bacteria generally have to deal with iron limitation rather than with iron excess. Hence, bacteria have developed a range of strategies to acquire iron from the environment. For one, they produce high-affinity chelators (siderophores) that can solubilize Fe^{3+}. In turn, corresponding ferrisiderophore-uptake systems take up the iron-siderophore complexes to cover the cellular demand for iron (Neilands 1995). It has been proposed that lactobacilli do not require iron for growth, based on the growth in iron-deficient media and other observations (Imbert and Blondeau 1998; Weinberg 1997). However, the genomes of LAB do contain genes that are predicted to have roles in iron acquisition. Also, recent work in this laboratory has identified HemN as an iron-requiring protein involved in heme metabolism. Conceivably, the iron requirement of LAB is conditional: Iron may only be required for aerobic growth, which also requires a supply of exogenous heme (Brooijmans et al. 2007).

LAB grown in the presence of oxygen produce damaging ROS, such as H_2O_2, $OH^{\bullet}$, or $O_2^{\bullet-}$. The hydroxyl and superoxide radicals, rather than H_2O_2, represent the ROS causing toxicity for LAB. H_2O_2 is membrane-permeable and can be accumulated in significant amounts by LAB. Many species, including *L. lactis* IL1403, do not possess catalase for the removal of H_2O_2 (Marty-Teysset et al. 2000; Rochat et al. 2006). However, in the presence of iron, H_2O_2 can be converted to highly reactive $OH^{\bullet}$ by a Fenton-type reaction. Therefore, intracellular iron levels may contribute significantly to the impact of high-H_2O_2 levels on cell survival.

Since the importance of iron for the growth of LAB has been discounted, little work has been performed on iron homeostasis. Here, findings from related Gram-positive bacteria, such as *B. subtilis* or *Lactobacillus plantarum,* will also be discussed since they may be extrapolated to LAB and may serve as a starting point for further investigations. In *B. subtilis*, iron homeostasis has been investigated in some detail. In these cells, the ferric uptake regulator (Fur) represses genes involved in iron uptake. Fur is a dimeric DNA-binding protein with one structural Zn^{2+} ion per monomer and possesses a regulatory Fe^{2+}-binding site (Bsat and Helmann 1999; Kehres et al. 2000). Iron starvation induced by the treatment of cultures with the iron chelator 2,2′-dipyridyl induces the Fur regulon, encompassing 20 operons with 39 genes. The same set of genes is also induced in *fur*-deletion mutants, supporting the nature of the Fur regulon (Baichoo et al. 2002). *L. lactis* IL1403 possesses a Fur-like protein of similar size to *B. subtilis* Fur (128 vs. 132 amino acids) and with 28% sequence identity, but experimental evidence for a function of this protein in iron homeostasis is not available.

The analysis of Fur-regulated genes in *B. subtilis* has led to the identification of various iron-uptake pathways that may also be present in *L. lactis*, such as FeuB (Accession: ABX75613, 328 amino acids, 38% sequence identity to *B. subtilis* FeuB, 334 amino acids). In *B. subtilis*, there is a range of iron-uptake systems: FeuBC is believed to take up the siderophores enterobactin and corynebactin, the latter being the siderophore produced by *B. subtilis*. Except for the YebLMN elemental iron-uptake system (related to the yeast FTS3 system), the iron transporters identified in *B. subtilis* belong to the ABC transporter family (Andrews et al. 2003; Moore and Helmann 2005). Four ABC transporters for the uptake of ferric citrate, corynebactin, and hydroxamate-type siderophores appear to be present in *B. subtilis* (FeuBC, YfiZ/YfhA, FhuBG, and YfmDE). Of all these iron acquisition proteins, *L. lactis* appears to possess only FeuB. This suggests, on the one hand, that there is a need for iron uptake by this LAB but, on the other hand, indicates a very low, maybe even nonessential, demand for iron.

An ABC transporter, MtsABC, involved in iron and zinc uptake has also been described for *Streptococcus pyogenes* (Janulczyk et al. 1999). The isolated protein exhibited high-affinity binding of Zn(II), Fe(III), and Cu(II) in vitro. An *mtsABC* mutant showed lower iron and zinc uptake but was not affected in its growth. In the light of these observations and of what is known about copper homeostasis, it appears unlikely that MtsABC acts as a copper importer in *S. pyogenes*. Convincing evidence of an ABC-type copper importer in any prokaryote has yet to be produced. All sequenced LAB genomes encode two or more ABC-type transporters, but the function of most of these has not yet been experimentally addressed.

Recently, an iron homeostatic gene, *mntH*, was identified in *Lactococcus lactis* MG1363, based on the resistance of tellurite (TeO_3^{2-}) and oxidative stress (Turner et al. 2007). Tellurite exerts oxidative stress by superoxide formation, which accompanies its reduction in the cytoplasm (Perez et al. 2007). The tellurite-resistant strain with a nonfunctional *mntH* gene exhibited greatly increased survival after 24 h of aerated growth, compared to the wild type. MntH is a member of the family of natural resistance-associated macrophage proteins (Nramp) (Richer et al. 2003). Members of this family have been shown to serve in Mn^{2+} and Fe^{2+} uptake (Kehres et al. 2000; Makui et al. 2000). The *mntH* mutant strain exhibited reduced iron uptake, suggesting that MntH serves in Fe^{2+} uptake. This observation does not, however, rule out that MntH also has a role in manganese acquisition (Turner et al. 2007). A strain deleted in MntH was still respiration-competent when supplied with heme or protoporphyrin IX, indicating that iron is still taken up. However, excess iron may be taken up by the wild type via MntH, and this iron could participate in oxygen-dependent toxicity in *L. lactis*.

9.4.2 Response to Zinc

Zinc is an essential metal ion but can be toxic if in excess. It plays a vital role as a cofactor for more than 300 enzymes, such as SOD, alcohol dehydrogenase, and DNA-binding proteins. It also functions as a structural scaffold for RNA polymerase, tRNA synthases, and approximately 40 additional proteins (Coleman 1998; Dunn et al. 2003; Outten and O'Halloran 2001; Sun and Plapp 1992; Vallee and Falchuk 1993). Additionally, zinc can also function as an antioxidant by protecting sulfhydryl groups of proteins from the attack of reactive free radical species and by antagonizing free radical formation by competing with redox-active transition metals like copper and iron (Powell 2000). In line with this, a mutant of *L. lactis* deficient in the low- and high-affinity zinc-uptake system was found to be more sensitive to H_2O_2 (Scott et al. 2000). In *B. subtilis*, H_2O_2 induces the Fur-like PerR repressor, which controls the expression of a dedicated zinc-uptake system, ZosA, in addition to catalase and some other genes (Gaballa and Helmann 2002). A *zosA* mutant exhibited significantly lower resistance to diamine, a thiol-specific oxidizing agent. A similar regulatory system involved in zinc uptake and resistance to H_2O_2 was described in *L. lactis*. The two FNR-like (fumarate/nitrate reduction regulator) proteins, FlpA and FlpB, control the expression of a zinc-uptake system that increases cellular zinc, and they enhance the resistance to H_2O_2 (Gostick et al. 1999).

On the other hand, excess zinc can inhibit protein function by blocking pivotal thiols or by competing with other metal ions for binding to the active sites of proteins. Zinc at high concentrations can also bind to negatively charged domains of proteins that are crucial for function. It was shown, for example, that zinc inhibits cytochrome *c* oxidase, presumably by binding to the negatively charged proton entry site of the enzyme (Aagaard and Brzezinski 2001). Clearly, zinc levels in the cell must be tightly regulated.

The first zinc-resistance protein was identified in the extremely metal-resistant bacterium *Ralstonia metallidurans*, followed by the CnrA protein from the same bacterium (see Nies (2003) for review). They are members of the RND protein family, which was first described as a related group of bacterial transport proteins involved in heavy metal resistance (*R. metallidurans*), nodulation (*Mesorhizobium loti*), and cell division (*E. coli*) (Saier et al. 1994). This family has grown into a huge superfamily that includes seven protein families that can be found in all major kingdoms of life. In *R. metallidurans*, three genes are organized into the *czcCBA* operon. CzcCBA mediates resistance to Co^{2+}, Zn^{2+}, and Cd^{2+}, driven by the proton motive force. Similar systems are also involved in nickel and manganese efflux (Claverys 2001). These metal transporters belong to cluster nine of the family of ABC transporters, or ATP-binding cassette permeases. ABC transporters typically consist of a cytosolic metal-binding protein, a membrane permease, and an ATPase and can serve in the uptake as well as in the secretion of metal ions. ABC transporters can be found in the genomes of all bacterial species, but the function has been characterized only in a few cases.

Zinc homeostasis has so far received little attention in LAB, but transport systems for zinc uptake as well as for zinc efflux have been described for other Gram-positive bacteria (Hantke 2005). In *Streptococcus pneumoniae*, it has been proposed that the *adcCBA* operon encodes an ATP-binding cassette transporter for zinc uptake, and the *psa* one for manganese uptake (Dintilhac et al. 1997). A similar ABC-type manganese uptake system that is important for virulence has been described in *S. gordonii* (Dintilhac et al. 1997; Hantke 2005; Hazlett et al. 2003; Jakubovics et al. 2000; Janulczyk et al. 1999), and an ABC transporter of *Streptococcus pyogenes* has been shown to bind copper, iron, and zinc, but no transport studies were performed (Janulczyk et al. 1999). In *L. lactis* IL1403, ZitSQP is an ABC transporter putatively involved in high-affinity Zn^{2+} uptake (Bolotin et al. 2001). The ABC transporter–encoding genes, *zitSQP*, are organized into the putative *zitRSQP* operon, also encoding the *zitR* repressor. Sequence similarities of the putative *zitR* metalloregulator suggest that *zit* expression could be regulated by zinc present in the environment, as already shown for other zinc transport operons in Gram-positive bacteria (Dintilhac et al. 1997; Gaballa and Helmann 1998; Hantke 2005). Several zinc-uptake systems in bacteria have been shown to be under the control of a similar zinc-sensing Fur homolog, the zinc-uptake repressor Zur (Dalet et al. 1999; Gaballa and Helmann 2002; Lindsay and Foster 2001; Patzer and Hantke 2000).

An expression system P_{Zn}*zitR*, based on the regulatory signals (P_{Zn} promoter and *zitR* putative zinc repressor gene) of the *L. lactis* IL1403 *zit* operon, has been developed and shown to be highly inducible upon divalent cation starvation and strongly repressed in the presence of excess Zn^{2+}, thereby reinforcing the hypothesis of the involvement of the *zit* operon in Zn^{2+} high-affinity uptake and regulation in *L. lactis* IL1403 (Llull and Poquet 2004).

Zinc efflux in Gram-positive bacteria has received even less attention than zinc uptake. In *Streptococcus thermophilus* 4134, the chromosomal *cadC* and *cadA* genes were shown to constitute a cadmium/zinc resistance system (Schirawski et al. 2002). They are organized in an operon, and their transcription is cadmium-dependent in vivo. The predicted gene product of *cadA* is a CPx-type cadmium efflux ATPase and that of *cadC* an ArsR-type regulatory protein. The two *cad* genes could also confer cadmium and zinc resistance to *L. lactis*. Clearly, the understanding of zinc homeostasis in LAB is still in its infancy, and considerable more work is required to understand the uptake, regulation, and secretion of zinc in these bacteria.

9.4.3 Response to Manganese

Cytoplasmic manganese can help to protect bacteria against oxidative stress, and the induction of manganese uptake by H_2O_2 has been observed in many bacteria (Horsburgh et al. 2002b). *L. plantarum* can accumulate manganese to over 30 mM (Archibald and Duong 1984); uptake is accomplished by *mntA*, encoding an Mn^{2+}- and Cd^{2+}-transporting P-type ATPase (Hao et al. 1999). In *S. gordonii*, the scaCBA operon encodes an ABC-type manganese permease that is induced by low-ambient manganese via the ScaR repressor (Jakubovics et al. 2000). ABC-type manganese permeases as described in *S. gordonii* are also widespread in LAB (Claverys 2001). In addition to the ABC-type manganese permease MntABC, *Staphylococcus aureus* also possesses an Nramp-type manganese-uptake system, MntH (Horsburgh et al. 2002a). These systems are regulated by the manganese-dependent MntR repressor and the PerR oxidative stress regulator in a concerted fashion. Similar transporters and regulators have also been described in other bacteria (see Horsburgh et al. (2002b) for a review), but studies in LAB have remained scarce.

Manganese homeostasis plays a key role in many organisms, chiefly to defy oxidative stress and/or during infection of a human host (Jakubovics et al. 2000). In many Gram-positive bacteria, the major SOD that protects against oxidative stress is a manganese-containing enzyme (McEwan 2009). In addition to its role as a cofactor of SOD, manganese is able to directly protect against oxidative stress. In vitro, it has been shown that complexes of Mn(II) with bicarbonate can rapidly dismutate H_2O_2 (Stadtman et al. 1990). The most recently proposed scheme for this reaction involves cycling between reactions (9.8) and (9.9):

$$Mn^{2+} + H_2O_2 \rightarrow Mn^{3+} + 2\,OH^- \qquad 9.8$$

$$Mn^{3+} + H_2O_2 \rightarrow Mn^{2+}O_2 + 2H^+ \qquad 9.9$$

The active species of manganese is an $Mn^{2+}(HCO_3^-)_2$ complex in which HCO_3^- acts as an acceptor for protons (Tikhonov et al. 2006). HCO_3^- also lowers the redox potential of the Mn(II)–Mn(III) couple, which makes the reaction with H_2O_2 more favorable. Nevertheless, the rate constants of these reactions are still far lower than those of enzymic manganese-dependent SODs and Mn catalases, and it was recently shown that imported Mn does not significantly scavenge H_2O_2 in *E. coli* (Anjem et al. 2009). Rather, the beneficial effects of manganese appear to lie in its ability to metallate mononuclear enzymes in lieu of iron. When the iron is not deeply buried in iron-loaded enzymes, it can engage in the Fenton reaction and cause oxidative stress. The substitution of such iron by manganese under oxidative stress conditions could thus prevent protein damage. In line with this concept, *E. coli* mutants that could not import manganese were found to suffer high rates of protein oxidation (Anjem et al. 2009). Clearly, the protection of bacteria against oxidative stress by manganese remains an interesting area of investigation for the future.

9.4.4 Response to Nickel

Nickel is an essential trace nutrient for some bacteria, required at nanomolar concentrations. To date, nine nickel-containing enzymes are known: urease, NiFe-hydrogenase, carbon monoxide dehydrogenase, acetyl-CoA decarboxylase/synthase, methyl coenzyme M reductase, certain SODs, some glyoxylases, aci-reductone dioxygenase, and methylenediurease (Mulrooney and Hausinger 2003; Ragsdale 2009). None of these enzymes appears to play a role in LAB. Consequently, nickel is probably not an essential trace nutrient of these bacteria. Nickel in excess can induce oxidative stress in cells by cycling through the three redox states Ni^{+}, Ni^{2+}, and Ni^{3+} (Costa et al. 2002). In organisms requiring nickel, it is taken up by dedicated nickel-uptake transporters, such as the NikABCDE import pump in *E. coli* (De Pina et al. 1999), or by high-affinity nickel/cobalt permeases (Eitinger et al. 2005). Members of these HoxN-type permeases have been identified in Gram-negative and Gram-positive bacteria (Eitinger and Mandrand-Berthelot 2000). Different Ni(II)-responsive metalloregulators that maintain nickel homeostasis in Gram-positive bacteria have been characterized. In *M. tuberculosis,* the transcription factor NmtR of the Ars/SmtB family inhibits the expression of the gene for NmtA, an ATP-dependent transporter responsible for the efflux of nickel and cobalt. NmtR tightly binds to the promoter region of the *nmtR* and *nmtA* genes and releases the DNA when nickel, or to a certain extent cobalt, is abundant (Cavet et al. 2002). A second nickel and cobalt sensor in *M. tuberculosis*, KmtA, represses the expression of a putative cation diffusion facilitator (CDF) metal exporter. NmtR and KmtA differ in their nickel-/cobalt-sensing affinity. It appears that first KmtR detects basal levels of cytosolic nickel or cobalt, which are then exported following the expression of the CDF transporter. Only when a higher threshold of these metals accumulates does NmtR sense them and allow expression of the P-type ATPase (Campbell et al. 2007). In *Streptomyces coelicolor,* the nickel-responsive regulator Nur, belonging to the Fur family, was characterized. This regulator represses the transcription of Fe-SOD and simultaneously induces the transcription of Ni-SOD under nickel stress (Ahn et al. 2006). It is not known how LAB respond to nickel stress. Natural environments are generally low in nickel and there may not have been a need for these organisms to evolve nickel detoxification systems. Indeed, the genome analysis of sequenced LAB does not reveal any genes that are obviously connected to nickel.

9.4.5 Response to Cobalt

Cobalt is a transition metal with the two naturally occurring oxidation states Co^{2+} and Co^{3+} and is primarily found in the corrin ring of coenzyme B12. To date, several noncorrin-cobalt-containing enzymes have been isolated and characterized (Kobayashi and Shimizu 1999). Cobalt undergoes redox chemistry and can thus participate in Fenton-type reactions, a fact making it potentially toxic at higher concentrations (Valko et al. 2005). Cobalt homeostasis is closely related to the homeostasis of nickel and other divalent ions. Both cobalt and nickel are taken up by the cell via secondary metal transporters with different ion preferences, ranging from strict selectivity for nickel through unbiased transport of both ions to a strong preference for cobalt (Eitinger et al. 2005; Komeda et al. 1997). In *S. aureus*, the zinc-/cobalt-responsive transcriptional repressor CzrA, which belongs to the ArsR/SmtB family, regulates the expression of the *czr* operon encoding a cobalt/zinc pump (Pennella et al. 2003). Transcriptional repressors with high-sequence identities to CzrA can be found in the genomes of many LAB, but no characterization of their function has been performed to date.

9.4.6 Response to Chromium

The widespread industrial use of the heavy metal chromium has caused it to be considered a serious environmental pollutant. It is mostly found in its trivalent or hexavalent forms in nature. Cr^{6+} is highly toxic to all forms of life, whereas Cr^{3+} is an essential micronutrient for many higher organisms (De Flora et al. 1990; Megharaj et al. 2003). However, for microorganisms and plants, chromium is nonessential. Chromate (CrO_4^{2-}) crosses biological membranes by means of the sulfate-uptake pathway (Ramirez-Diaz et al. 2008). Inside the cell, Cr^{6+} is reduced to Cr^{3+}, a process in which free radicals may be formed (Liu and Shi 2001). Bacterial chromium-resistance systems related to plasmid genes usually encode membrane transporters that catalyze the efflux of chromate ions from the cytoplasm. The best-studied example is the *Pseudomonas aeruginosa* ChrA protein, which functions as a chemiosmotic pump that extrudes chromate from the cytoplasm using the proton motive force (Alvarez et al. 1999). A broad phylogenic analysis for *chrA* transporter genes revealed homologous genes in bacteria, archaea, and fungi (Diaz-Perez et al. 2007). Several bacilli possess homologous gene sequences, but none of the known chromium-defense genes is present in the sequenced LAB genomes.

9.4.7 Response to Cadmium

Cadmium is a heavy metal with an oxidation state of +2. It is chemically similar to zinc and occurs naturally with zinc and lead in sulfide ores. Cadmium is not generally believed to have a biological function; however, one enzyme (cadmium-carbonic anhydrase) incorporating cadmium under low-zinc conditions has been found in the marine diatom *Thalassiosira weissflogii* (Lane and Morel 2000). In spite of not being a Fenton metal, cadmium is capable of inducing oxidative stress in cell culture models and in experimental animals (Joseph et al. 2001; Nigam et al. 1999) and may exhibit its toxicity in microorganisms in a similar way. Cadmium is accumulated by cells via uptake systems responsible for essential cations. In Gram-positive bacteria, such as *B. subtilis*, *S. aureus*, or *L. plantarum*, Cd^{2+} competes for transport with Mn^{2+} (Archibald and Duong 1984; Burke and Pfister 1986; Tynecka et al. 1981). To prevent toxic effects by cadmium, active efflux mechanisms have evolved in prokaryotes. The best-characterized cadmium efflux system is that in the Gram-positive bacterium *S. aureus*, which consists of two plasmid-encoded genes, *cadA* and *cadC*. CadA, a CPx-type ATPase, catalyzes the efflux of Cd^{2+} (and probably also Zn^{2+}) and CadC is a transcriptional repressor (Nucifora et al. 1989). CadC binds specifically to the *cad* operator DNA and is released by the addition of Cd^{2+}, Pb^{2+}, and Bi^{3+} (Endo and Silver 1995). Genetic analyses in *L. lactis* and *Oenococcus oeni* have shown the occurrence of similar plasmid-encoded cadmium-resistance systems in LAB (Bon et al. 2009; Liu et al. 1997). Additionally, cadmium may be pumped out of the cell by multidrug transporters. The ATP-binding cassette- (ABC) type multidrug transporters LmrA (*L. lactis*) and OmrA (*O. oeni*) could confer cadmium resistance to an *E. coli* mutant strain, which was hypersensitive to this heavy metal (Achard-Joris et al. 2005; Bourdineaud et al. 2004; Van Veen et al. 1996).

9.5 The Phosphate–Metal Connection

Most bacteria store phosphate in phosphate polymers of up to hundreds of residues called *polyphosphates*. It has been shown in a number of cases that polyphosphates are degraded under metal stress, such as by growth in the presence of lead or cadmium (Keasling 1997). Presumably, phosphate derived from the degradation of polyphosphate is exported as complexes with toxic metal ions, thereby detoxifying the cytoplasm. For example, an *E. coli* mutant defective in both polyphosphate kinase and polyphosphatase exhibited greatly increased cadmium sensitivity (Keasling and Hupf 1996). The extrusion of neutral metal phosphate complexes of the form $MeHPO_4$ has, in fact, been directly demonstrated in *Acinetobacter johnsonii* and has been shown to generate electron-motif force (Van Veen et al. 1994b). Species of *Sulfolobus* have also been shown to accomplish high copper tolerance by the induction of polyphosphatase and secretion of copper phosphate (Remonsellez et al. 2006). The extrusion of metal–phosphate complexes takes place via the same Pit systems that also work in phosphate uptake (see ahead). Pit systems have been shown to catalyze the translocation of phosphate complexed to Mg^{2+}, Ca^{2+}, Co^{2+}, Cu^{2+}, or Mn^{2+} (Van Veen et al. 1994a). Conceivably, complexes of phosphate with other divalent metal ions may also be translocated. LAB are generally not considered to produce polyphosphate, but there are findings suggesting that at least some of them do (Benthin et al. 1994).

The phosphate transport systems that can participate in $MeHPO_4$ extrusion can also serve in phosphate uptake under phosphate-limiting conditions. In *E. coli*, the low-affinity PitA and PitB phosphate transport systems were shown to catalyze the uptake of neutral metal–phosphate complexes (interestingly, PitA is nonfunctional in *E. coli* K-12 lab strains due to a mutation; (Harris et al. 2001)). The mutation of *E. coli pitA* conferred increased zinc resistance (Beard et al. 2000), and growth in the presence of zinc reduced the intracellular magnesium concentration and increased intracellular zinc, presumably due to competition between the two ions (Jackson et al. 2008). High-affinity phosphate uptake by *L. lactis* and related organisms is accomplished by an ATP-driven ABC-type transporter encoded by the *pstFEDCBA* operon. Mutations in *pstFEDCBA* were found to increase the resistance to copper and zinc by lowering the intracellular reactivity of these metals, which in turn also reduced the sensitivity of the cells to oxygen (Cesselin et al. 2009). This suggests that the *pst* system can (or must) also transport metal–phosphate complexes.

The observation of Pit- and Pst-catalyzed metal–phosphate cotransport is a surprising aspect of these transporters, which was not taken into consideration in most studies of either phosphate or metal transport. This masquerade may have disguised the true function of many transporters. The magnesium transporter CorA, which is ubiquitous in Gram-negative bacteria, may in fact be a metal–phosphate transporter, and the magnesium transporter MgtE, which also occurs in Gram-positive bacteria, may similarly be a metal–phosphate transporter. Clearly, much more work is required for a detailed understanding of bacterial metal transport.

Acknowledgments Some of the work described in this chapter has been supported by Grant 3100A0_122551 from the Swiss National Foundation, a grant from the Swiss State Secretary for Education & Research, and a grant from the International Copper Association.

References

Aagaard A, Brzezinski P (2001) Zinc ions inhibit oxidation of cytochrome *c* oxidase by oxygen. FEBS Lett 494:157–160

Achard-Joris M, van den Berg van Saparoea HB, Driessen AJ, Bourdineaud JP (2005) Heterologously expressed bacterial and human multidrug resistance proteins confer cadmium resistance to *Escherichia coli*. Biochemistry 44:5916–5922

Ahn BE, Cha J, Lee EJ, Han AR, Thompson CJ, Roe JH (2006) Nur, a nickel-responsive regulator of the Fur family, regulates superoxide dismutases and nickel transport in *Streptomyces coelicolor*. Mol Microbiol 59:1848–1858

Alvarez AH, Moreno-Sanchez R, Cervantes C (1999) Chromate efflux by means of the ChrA chromate resistance protein from *Pseudomonas aeruginosa*. J Bacteriol 181:7398–7400

Andrews SC, Robinson AK, Rodriguez-Quinones F (2003) Bacterial iron homeostasis. FEMS Microbiol Rev 27:215–237

Anjem A, Varghese S, Imlay JA (2009) Manganese import is a key element of the OxyR response to hydrogen peroxide in *Escherichia coli*. Mol Microbiol 72:844–858

Archibald FS, Duong MN (1984) Manganese acquisition by *Lactobacillus plantarum*. J Bacteriol 158:1–8
Arciero DM, Pierce BS, Hendrich MP, Hooper AB (2002) Nitrosocyanin, a red cupredoxin-like protein from *Nitrosomonas europaea*. Biochemistry 41:1703–1709
Arguello JM, Gonzalez-Guerrero M (2008) Cu^{+} -ATPases brake system. Structure 16:833–834
Arnesano F, Banci L, Bertini I, Ciofi-Baffoni S, Molteni E, Huffman DL, O'Halloran TV (2002) Metallochaperones and metal-transporting ATPases: a comparative analysis of sequences and structures. Genome Res 12:255–271
Axelsen KB, Palmgren MG (1998) Evolution of substrate specificities in the P-type ATPase superfamily. J Mol Evol 46:84–101
Baichoo N, Wang T, Ye R, Helmann JD (2002) Global analysis of the *Bacillus subtilis* Fur regulon and the iron starvation stimulon. Mol Microbiol 45:1613–1629
Balasubramanian R, Rosenzweig AC (2008) Copper methanobactin: a molecule whose time has come. Curr Opin Chem Biol 12:245–249
Barkay T, Miller SM, Summers AO (2003) Bacterial mercury resistance from atoms to ecosystems. FEMS Microbiol Rev 27:355–384
Barré O, Mourlane F, Solioz M (2007) Copper induction of lactate oxidase of *Lactococcus lactis*: a novel metal stress response. J Bacteriol 189:5947–5954
Battistoni A (2003) Role of prokaryotic Cu, Zn superoxide dismutase in pathogenesis. Biochem Soc Trans 31:1326–1329
Beard SJ, Hashim R, Wu G, Binet MR, Hughes MN, Poole RK (2000) Evidence for the transport of zinc(II) ions via the pit inorganic phosphate transport system in *Escherichia coli*. FEMS Microbiol Lett 184:231–235
Benthin S, Nielsen J, Villadsen J (1994) Galactose expulsion during lactose metabolism in *Lactococcus lactis* subsp. *cremoris* FD1 due to dephosphorylation of intracellular galactose 6-phosphate. Appl Environ Microbiol 60:1254–1259
Bird AJ (2008) Metallosensors, the ups and downs of gene regulation. Adv Microb Physiol 53:231–267
Bolotin A, Wincker P, Mauger S, Jaillon O, Malarme K, Weissenbach J, Ehrlich SD, Sorokin A (2001) The complete genome sequence of the lactic acid bacterium *Lactococcus lactis* ssp. *lactis* IL1403. Genome Res 11:731–753
Bon E, Delaherche A, Bilhere E, De Daruvar A, Lonvaud-Funel A, Le MC (2009) *Oenococcus oeni* genome plasticity is associated with fitness. Appl Environ Microbiol 75:2079–2090
Bourdineaud JP, Nehme B, Tesse S, Lonvaud-Funel A (2004) A bacterial gene homologous to ABC transporters protect *Oenococcus oeni* from ethanol and other stress factors in wine. Int J Food Microbiol 92:1–14
Brazeau BJ, Johnson BJ, Wilmot CM (2004) Copper-containing amine oxidases. Biogenesis and catalysis; a structural perspective. Arch Biochem Biophys 428:22–31
Brooijmans RJW, Poolman B, Schuurman-Wolters GK, De Vos WM, Hugenholtz J (2007) Generation of a membrane potential by *Lactococcus lactis* through aerobic electron transport. J Bacteriol 189:5203–5209
Bsat N, Helmann JD (1999) Interaction of *Bacillus subtilis* Fur (ferric uptake repressor) with the *dhb* operator *in vitro* and *in vivo*. J Bacteriol 181:4299–4307
Burke BE, Pfister RM (1986) Cadmium transport by a Cd^{2+}-sensitive and a Cd^{2+}-resistant strain of *Bacillus subtilis*. Can J Microbiol 32:539–542
Bury NR, Grosell M, Grover AK, Wood CM (1999) ATP-dependent silver transport across the basolateral membrane of rainbow trout gills. Toxicol Appl Pharmacol 159:1–8
Campbell DR, Chapman KE, Waldron KJ, Tottey S, Kendall S, Cavallaro G, Andreini C, Hinds J, Stoker NG, Robinson NJ, Cavet JS (2007) Mycobacterial cells have dual nickel-cobalt sensors: sequence relationships and metal sites of metal-responsive repressors are not congruent. J Biol Chem 282:32298–32310
Cantini F, Banci L, Solioz M (2009) The copper-responsive repressor CopR of *Lactococcus lactis* is a "winged helix" protein. Biochem J 417:493–499
Cavet JS, Meng W, Pennella MA, Appelhoff RJ, Giedroc DP, Robinson NJ (2002) A nickel-cobalt-sensing ArsR-SmtB family repressor. Contributions of cytosol and effector binding sites to metal selectivity. J Biol Chem 277:38441–38448

Cavet JS, Borrelly GP, Robinson NJ (2003) Zn, Cu and Co in cyanobacteria: selective control of metal availability. FEMS Microbiol Rev 27:165–181

Cesselin B, Ali D, Gratadoux JJ, Gaudu P, Duwat P, Gruss A, El KM (2009) Inactivation of the *Lactococcus lactis* high affinity phosphate transporter confers oxygen and thiol resistance and alters metal homeostasis. Microbiology 155:2274–2281

Cha JS, Cooksey DA (1991) Copper resistance in *Pseudomonas syringae* mediated by periplasmic and outer membrane proteins. Proc Natl Acad Sci USA 88:8915–8919

Chan SI, Chen KH, Yu SS, Chen CL, Kuo SS (2004) Toward delineating the structure and function of the particulate methane monooxygenase from methanotrophic bacteria. Biochemistry 43:4421–4430

Changela A, Chen K, Xue Y, Holschen J, Outten CE, O'Halloran TV, Mondragon A (2003) Molecular basis of metal-ion selectivity and zeptomolar sensitivity by CueR. Science 301:1383–1387

Chillappagari S, Miethke M, Trip H, Kuipers OP, Marahiel MA (2009) Copper acquisition is mediated by YcnJ and regulated by YcnK and CsoR in *Bacillus subtilis*. J Bacteriol 191:2362–2370

Claverys JP (2001) A new family of high-affinity ABC manganese and zinc permeases. Res Microbiol 152:231–243

Cobine P, Wickramasinghe WA, Harrison MD, Weber T, Solioz M, Dameron CT (1999) The *Enterococcus hirae* copper chaperone CopZ delivers copper(I) to the CopY repressor. FEBS Lett 445:27–30

Cobine P, Jones CE, Wickramasinghe WA, Solioz M, Dameron CT (2002a) Interaction of copper binding proteins from *Enterococcus hirae*. In: Massaro EJ (Ed.), Handbook of copper pharmacology and toxicology. Humana Press, Totowa, pp. 177–186

Cobine PA, George GN, Jones CE, Wickramasinghe WA, Solioz M, Dameron CT (2002b) Copper transfer from the Cu(I) chaperone, CopZ, to the repressor, Zn(II)CopY: Metal coordination environments and protein interactions. Biochemistry 41:5822–5829

Cobine PA, Jones CE, Dameron CT (2002c) Role for zinc(II) in the copper(I) regulated protein CopY. J Inorg Biochem 88:192–196

Coleman JE (1998) Zinc enzymes. Curr Opin Chem Biol 2:222–234

Costa M, Salnikow K, Sutherland JE, Broday L, Peng W, Zhang Q, Kluz T (2002) The role of oxidative stress in nickel and chromate genotoxicity. Mol Cell Biochem 234–235:265–275

Crichton RR, Pierre J-L (2001) Old iron, young copper: from Mars to Venus. Biometals 14:99–112

Dalet K, Gouin E, Cenatiempo Y, Cossart P, Hechard Y (1999) Characterisation of a new operon encoding a Zur-like protein and an associated ABC zinc permease in *Listeria monocytogenes*. FEMS Microbiol Lett 174:111–116

Davis AV, O'Halloran TV (2008) A place for thioether chemistry in cellular copper ion recognition and trafficking. Nat Chem Biol 4:148–151

De Flora S, Bagnasco M, Serra D, Zanacchi P (1990) Genotoxicity of chromium compounds. A review. Mutat Res 238:99–172

De Pina K, Desjardin V, Mandrand-Berthelot MA, Giordano G, Wu LF (1999) Isolation and characterization of the *nikR* gene encoding a nickel-responsive regulator in *Escherichia coli*. J Bacteriol 181:670–674

Diaz-Perez C, Cervantes C, Campos-Garcia J, Julian-Sanchez A, Riveros-Rosas H (2007) Phylogenetic analysis of the chromate ion transporter (CHR) superfamily. FEBS J 274: 6215–6227

Dintilhac A, Alloing G, Granadel C, Claverys JP (1997) Competence and virulence of *Streptococcus pneumoniae*: Adc and PsaA mutants exhibit a requirement for Zn and Mn resulting from inactivation of putative ABC metal permeases. Mol Microbiol 25:727–739

Dunn KL, Farrant JL, Langford PR, Kroll JS (2003) Bacterial [Cu,Zn]-cofactored superoxide dismutase protects opsonized, encapsulated *Neisseria meningitidis* from phagocytosis by human monocytes/macrophages. Infect Immun 71:1604–1607

Eitinger T, Mandrand-Berthelot MA (2000) Nickel transport systems in microorganisms. Arch Microbiol 173:1–9

Eitinger T, Suhr J, Moore L, Smith JA (2005) Secondary transporters for nickel and cobalt ions: theme and variations. Biometals 18:399–405

Ellis MJ, Grossmann JG, Eady RR, Hasnain SS (2007) Genomic analysis reveals widespread occurrence of new classes of copper nitrite reductases. J Biol Inorg Chem 12:1119–1127

Endo G, Silver S (1995) CadC, the transcriptional regulatory protein of the cadmium resistance system of *Staphylococcus aureus* plasmid pI258. J Bacteriol 177:4437–4441

Gaballa A, Helmann JD (1998) Identification of a zinc-specific metalloregulatory protein, Zur, controlling zinc transport operons in *Bacillus subtilis*. J Bacteriol 180:5815–5821

Gaballa A, Helmann JD (2002) A peroxide-induced zinc uptake system plays an important role in protection against oxidative stress in *Bacillus subtilis*. Mol Microbiol 45:997–1005

Galvez A, Abriouel H, Lopez RL, Ben ON (2007) Bacteriocin-based strategies for food biopreservation. Int J Food Microbiol 120:51–70

Garcia-Castellanos R, Mallorqui-Fernandez G, Marrero A, Potempa J, Coll M, Gomis-Ruth FX (2004) On the transcriptional regulation of methicillin resistance: MecI repressor in complex with its operator. J Biol Chem 279:17888–17896

Gonzalez-Guerrero M, Arguello JM (2008) Mechanism of Cu^+-transporting ATPases: soluble Cu^+ chaperones directly transfer Cu^+ to transmembrane transport sites. Proc Natl Acad Sci USA 105:5992–5997

Gostick DO, Griffin HG, Shearman CA, Scott C, Green J, Gasson MJ, Guest JR (1999) Two operons that encode FNR-like proteins in *Lactococcus lactis*. Mol Microbiol 31:1523–1535

Gupta A, Matsui K, Lo JF, Silver S (1999) Molecular basis for resistance to silver cations in *Salmonella*. Nat Med 5:183–188

Hanson SR, Donley SA, Linder MC (2001) Transport of silver in virgin and lactating rats and relation to copper. J Trace Elem Med Biol 15:243–253

Hantke K (2005) Bacterial zinc uptake and regulators. Curr Opin Microbiol 8:196–202

Hao Z, Reiske HR, Wilson DB (1999) Characterization of cadmium uptake in *Lactobacillus plantarum* and isolation of cadmium and manganese uptake mutants. Appl Environ Microbiol 65:4741–4745

Harris RM, Webb DC, Howitt SM, Cox GB (2001) Characterization of PitA and PitB from *Escherichia coli*. J Bacteriol 183:5008–5014

Hasman H, Kempf I, Chidaine B, Cariolet R, Ersboll AK, Houe H, Bruun Hansen HC, Aarestrup FM (2006) Copper resistance in *Enterococcus faecium*, mediated by the *tcrB* gene, is selected by supplementation of pig feed with copper sulfate. Appl Environ Microbiol 72:5784–5789

Hazlett KR, Rusnak F, Kehres DG, Bearden SW, La Vake CJ, La Vake ME, Maguire ME, Perry RD, Radolf JD (2003) The *Treponema pallidum tro* operon encodes a multiple metal transporter, a zinc-dependent transcriptional repressor, and a semi-autonomously expressed phosphoglycerate mutase. J Biol Chem 278:20687–20694

Himeno T, Imanaka T, Aiba S (1986) Nucleotide sequence of the penicillinase repressor gene penI of *Bacillus licheniformis* and regulation of *penP* and *penI* by the repressor. J Bacteriol 168:1128–1132

Horsburgh MJ, Wharton SJ, Cox AG, Ingham E, Peacock S, Foster SJ (2002a) MntR modulates expression of the PerR regulon and superoxide resistance in *Staphylococcus aureus* through control of manganese uptake. Mol Microbiol 44:1269–1286

Horsburgh MJ, Wharton SJ, Karavolos M, Foster SJ (2002b) Manganese: elemental defence for a life with oxygen. Trends Microbiol 10:496–501

Huffman DL, O'Halloran TV (2001) Function, structure, and mechanism of intracellular copper trafficking proteins. Annu Rev Biochem 70:677–701

Hullo MF, Moszer I, Danchin A, Martin-Verstraete I (2001) CotA of *Bacillus subtilis* is a copper-dependent laccase. J Bacteriol 183:5426–5430

Imbert M, Blondeau R (1998) On the iron requirement of lactobacilli grown in chemically defined medium. Curr Microbiol 37:64–66

Jackson RJ, Binet MR, Lee LJ, Ma R, Graham AI, McLeod CW, Poole RK (2008) Expression of the PitA phosphate/metal transporter of *Escherichia coli* is responsive to zinc and inorganic phosphate levels. FEMS Microbiol Lett 289:219–224

Jakubovics NS, Smith AW, Jenkinson HF (2000) Expression of the virulence-related Sca (Mn^{2+}) permease in *Streptococcus gordonii* is regulated by a diphtheria toxin metallorepressor-like protein ScaR. Mol Microbiol 38:140–153

Janulczyk R, Pallon J, Bjorck L (1999) Identification and characterization of a *Streptococcus pyogenes* ABC transporter with multiple specificity for metal cations. Mol Microbiol 34:596–606

Joseph P, Muchnok TK, Klishis ML, Roberts JR, Antonini JM, Whong WZ, Ong T (2001) Cadmium-induced cell transformation and tumorigenesis are associated with transcriptional activation of c-fos, c-jun, and c-myc proto-oncogenes: role of cellular calcium and reactive oxygen species. Toxicol Sci 61:295–303

Kaim W, Rall J (1996) Copper – a "modern" bioelement. Angew Chem Int Ed Engl 35:43–60

Kanamaru K, Kashiwagi S, Mizuno T (1994) A copper-transporting P-type ATPase found in the thylakoid membrane of the cyanobacterium *Synechococcus* species PCC7942. Mol Microbiol 13:369–377

Kaneko T, Takahashi M, Suzuki H (1990) Acetoin fermentation by citrate-positive *Lactococcus lactis* subsp. *lactis* 3022 grown aerobically in the presence of hemin or Cu. Appl Environ Microbiol 56:2644–2649

Karlin KD (1993) Metalloenzymes, structural motifs, and inorganic models. Science 261: 701–708

Kasting JF, Siefert JL (2002) Life and the evolution of Earth's atmosphere. Science 296: 1066–1068

Keasling JD (1997) Regulation of intracellular toxic metals and other cations by hydrolysis of polyphosphate. Ann NY Acad Sci 829:242–249

Keasling JD, Hupf GA (1996) Genetic manipulation of polyphosphate metabolism affects cadmium tolerance in *Escherichia coli*. Appl Microbiol Biotechnol 62:743–746

Kehres DG, Zaharik ML, Finlay BB, Maguire ME (2000) The NRAMP proteins of *Salmonella typhimurium* and *Escherichia coli* are selective manganese transporters involved in the response to reactive oxygen. Mol Microbiol 36:1085–1100

Kiermeier F, Kyrein HJ (1971) Einfluß des Kupfers auf den Acetoin-Diacetyl- und Pyruvatgehalt von Käse [in German]. Z Lebensmittelunters -Forschung A 147:128–133

Kihlken MA, Leech AP, Le Brun NE (2002) Copper-mediated dimerization of CopZ, a predicted copper chaperone from *Bacillus subtilis*. Biochem J 368:729–739

Kihlken MA, Singleton C, Le Brun NE (2008) Distinct characteristics of Ag^{+} and Cd^{2+} binding to CopZ from *Bacillus subtilis*. J Biol Inorg Chem 13:1011–1023

Kim BE, Nevitt T, Thiele DJ (2008) Mechanisms for copper acquisition, distribution and regulation. Nat Chem Biol 4:176–185

Kim HJ, Graham DW, DiSpirito AA, Alterman MA, Galeva N, Larive CK, Asunskis D, Sherwood PM (2004) Methanobactin, a copper-acquisition compound from methane-oxidizing bacteria. Science 305:1612–1615

Kobayashi M, Shimizu S (1999) Cobalt proteins. Eur J Biochem 261:1–9

Komeda H, Kobayashi M, Shimizu S (1997) A novel transporter involved in cobalt uptake. Proc Natl Acad Sci USA 94:36–41

Kuper J, Llamas A, Hecht HJ, Mendel RR, Schwarz G (2004) Structure of the molybdopterin-bound Cnx1G domain links molybdenum and copper metabolism. Nature 430:803–806

Lane TW, Morel FM (2000) A biological function for cadmium in marine diatoms. Proc Natl Acad Sci USA 97:4627–4631

Lindsay JA, Foster SJ (2001) *zur*: a Zn^{2+}-responsive regulatory element of *Staphylococcus aureus*. Microbiology 147:1259–1266

Liu CQ, Khunajakr N, Chia LG, Deng YM, Charoenchai P, Dunn NW (1997) Genetic analysis of regions involved in replication and cadmium resistance of the plasmid pND302 from *Lactococcus lactis*. Plasmid 38:79–90

Liu KJ, Shi X (2001) *In vivo* reduction of chromium (VI) and its related free radical generation. Mol Cell Biochem 222:41–47

Liu T, Ramesh A, Ma Z, Ward SK, Zhang L, George GN, Talaat AM, Sacchettini JC, Giedroc DP (2007) CsoR is a novel *Mycobacterium tuberculosis* copper-sensing transcriptional regulator. Nat Chem Biol 3:60–68

Llull D, Poquet I (2004) New expression system tightly controlled by zinc availability in *Lactococcus lactis*. Appl Environ Microbiol 70:5398–5406

Lopez-Serrano D, Solano F, Sanchez-Amat A (2004) Identification of an operon involved in tyrosinase activity and melanin synthesis in *Marinomonas mediterranea*. Gene 342:179–187

Lu ZH, Solioz M (2001) Copper-induced proteolysis of the CopZ copper chaperone of *Enterococcus hirae*. J Biol Chem 276:47822–47827

Lutsenko S, Kaplan JH (1995) Organization of P-type ATPases: significance of structural diversity. Biochemistry 34:15607–15613

Ma Z, Cowart D, Scott R, Giedroc DP (2009) Molecular insights into the metal selectivity of the Cu(I)-sensing repressor CsoR from *Bacillus subtilis*. Biochemistry 48:3325–3334

Macomber L, Imlay JA (2009) The iron-sulfur clusters of dehydratases are primary intracellular targets of copper toxicity. Proc Natl Acad Sci USA 106:8344–8349

Macomber L, Rensing C, Imlay JA (2007) Intracellular copper does not catalyze the formation of oxidative DNA damage in *Escherichia coli*. J Bacteriol 189:1616–1626

Magnani D, Barré O, Gerber SD, Solioz M (2008) Characterization of the CopR regulon of *Lactococcus lactis* IL1403. J Bacteriol 190:536–545

Makui H, Roig E, Cole ST, Helmann JD, Gros P, Cellier MF (2000) Identification of the *Escherichia coli* K-12 Nramp orthologue (MntH) as a selective divalent metal ion transporter. Mol Microbiol 35:1065–1078

Marty-Teysset C, de la Torre F, Garel J (2000) Increased production of hydrogen peroxide by *Lactobacillus delbrueckii* subsp. *bulgaricus* upon aeration: involvement of an NADH oxidase in oxidative stress. Appl Environ Microbiol 66:262–267

McEwan AG (2009) New insights into the protective effect of manganese against oxidative stress. Mol Microbiol 72:812–814

Megharaj M, Avudainayagam S, Naidu R (2003) Toxicity of hexavalent chromium and its reduction by bacteria isolated from soil contaminated with tannery waste. Curr Microbiol 47:51–54

Mitrakul K, Loo CY, Hughes CV, Ganeshkumar N (2004) Role of a *Streptococcus gordonii* copper-transport operon, *copYAZ*, in biofilm detachment. Oral Microbiol Immunol 19:395–402

Miyoshi A, Rochat T, Gratadoux JJ, Le Loir Y, Oliveira SC, Langella P, Azevedo V (2003) Oxidative stress in *Lactococcus lactis*. Genet Mol Res 2:348–359

Moore CM, Helmann JD (2005) Metal ion homeostasis in *Bacillus subtilis*. Curr Opin Microbiol 8:188–195

Mulrooney SB, Hausinger RP (2003) Nickel uptake and utilization by microorganisms. FEMS Microbiol Rev 27:239–261

Multhaup G, Strausak D, Bissig K-D, Solioz M (2001) Interaction of the CopZ copper chaperone with the CopA copper ATPase of *Enterococcus hirae* assessed by surface plasmon resonance. Biochem Biophys Res Commun 288:172–177

Narindrasorasak S, Zhang X, Roberts EA, Sarkar B (2004) Comparative analysis of metal binding characteristics of copper chaperone proteins, Atx1 and ATOX1. Bioinorg Chem Appl 2:105–123

Neilands JB (1995) Siderophores: structure and function of microbial iron transport compounds. J Biol Chem 270:26723–26726

Nies DH (2003) Efflux-mediated heavy metal resistance in prokaryotes. FEMS Microbiol Rev 27:313–339

Nigam D, Shukla GS, Agarwal AK (1999) Glutathione depletion and oxidative damage in mitochondria following exposure to cadmium in rat liver and kidney. Toxicol Lett 106:151–157

Nucifora G, Chu L, Misra TK, Silver S (1989) Cadmium resistance from *Staphylococcus aureus* plasmid pI258 *cadA* gene results from a cadmium-efflux ATPase. Proc Natl Acad Sci USA 86:3544–3548

Odermatt A, Solioz M (1995) Two *trans*-acting metalloregulatory proteins controlling expression of the copper-ATPases of *Enterococcus hirae*. J Biol Chem 270:4349–4354

Odermatt A, Suter H, Krapf R, Solioz M (1992) An ATPase operon involved in copper resistance by *Enterococcus hirae*. Ann NY Acad Sci 671:484–486

Odermatt A, Krapf R, Solioz M (1994) Induction of the putative copper ATPases, CopA and CopB, of *Enterococcus hirae* by Ag^+ and Cu^{2+}, and Ag^+ extrusion by CopB. Biochem Biophys Res Commun 202:44–48

Outten CE, O'Halloran TV (2001) Femtomolar sensitivity of metalloregulatory proteins controlling zinc homeostasis. Science 292:2488–2492
Outten FW, Outten CE, Hale J, O'Halloran TV (2000) Transcriptional activation of an *Escherichia coli* copper efflux regulon by the chromosomal MerR homologue, CueR. J Biol Chem 275:31024–31029
Outten FW, Huffman DL, Hale JA, O'Halloran TV (2001) The independent *cue* and *cus* systems confer copper tolerance during aerobic and anaerobic growth in *Escherichia coli*. J Biol Chem 276:30670–30677
Patzer SI, Hantke K (2000) The zinc-responsive regulator Zur and its control of the znu gene cluster encoding the ZnuABC zinc uptake system in *Escherichia coli*. J Biol Chem 275:24321–24332
Pedersen PL, Carafoli E (1987) Ion motive ATPases. I. Ubiquity, properties, and significance to cell function. Trends Biochem Sci 12:146–150
Penaud S, Fernandez A, Boudebbouze S, Ehrlich SD, Maguin E, van de Guchte M (2006) Induction of heavy-metal-transporting CPX-type ATPases during acid adaptation in *Lactobacillus bulgaricus*. Appl Environ Microbiol 72:7445–7454
Pennella MA, Shokes JE, Cosper NJ, Scott RA, Giedroc DP (2003) Structural elements of metal selectivity in metal sensor proteins. Proc Natl Acad Sci USA 100:3713–3718
Perez JM, Calderon IL, Arenas FA, Fuentes DE, Pradenas GA, Fuentes EL, Sandoval JM, Castro ME, Elias AO, Vasquez CC (2007) Bacterial toxicity of potassium tellurite: unveiling an ancient enigma. PLoS ONE 2:e211
Portmann R, Magnani D, Stoyanov JV, Schmechel A, Multhaup G, Solioz M (2004) Interaction kinetics of the copper-responsive CopY repressor with the *cop* promoter of *Enterococcus hirae*. J Biol Inorg Chem 9:396–402
Portmann R, Poulsen KR, Wimmer R, Solioz M (2006) CopY-like copper inducible repressors are putative "winged helix" proteins. Biometals 19:61–70
Powell SR (2000) The antioxidant properties of zinc. J Nutr 130:1447S–1454S
Pufahl RA, Singer CP, Peariso KL, Lin S, Schmidt PJ, Fahrni CJ, Culotta VC, Penner-Hahn JE, O'Halloran TV (1997) Metal ion chaperone function of the soluble Cu(I) receptor Atx1. Science 278:853–856
Radford DS, Kihlken MA, Borrelly GP, Harwood CR, Le Brun NE, Cavet JS (2003) CopZ from *Bacillus subtilis* interacts *in vivo* with a copper exporting CPx-type ATPase CopA. FEMS Microbiol Lett 220:105–112
Ragsdale SW (2009) Nickel-based enzyme systems. J Biol Chem 284:18571–18575
Ramirez-Diaz MI, Diaz-Perez C, Vargas E, Riveros-Rosas H, Campos-Garcia J, Cervantes C (2008) Mechanisms of bacterial resistance to chromium compounds. Biometals 21:321–332
Rapisarda VA, Chehin RN, De Las Rivas J, Rodriguez-Montelongo L, Farias RN, Massa EM (2002) Evidence for Cu(I)-thiolate ligation and prediction of a putative copper-binding site in the *Escherichia coli* OADH dehydrogenase-2. Arch Biochem Biophys 405:87–94
Reeve WG, Tiwari RP, Kale NB, Dilworth MJ, Glenn AR (2002) ActP controls copper homeostasis in *Rhizobium leguminosarum* bv. *viciae* and *Sinorhizobium meliloti* preventing low pH-induced copper toxicity. Mol Microbiol 43:981–991
Remonsellez F, Orell A, Jerez CA (2006) Copper tolerance of the thermoacidophilic archaeon *Sulfolobus metallicus*: possible role of polyphosphate metabolism. Microbiology 152:59–66
Richer E, Courville P, Bergevin I, Cellier MF (2003) Horizontal gene transfer of "prototype" Nramp in bacteria. J Mol Evol 57:363–376
Ridge PG, Zhang Y, Gladyshev VN (2008) Comparative genomic analyses of copper transporters and cuproproteomes reveal evolutionary dynamics of copper utilization and its link to oxygen. PLoS One 3:e1378
Rochat T, Gratadoux JJ, Gruss A, Corthier G, Maguin E, Langella P, van de Guchte M (2006) Production of a heterologous nonheme catalase by *Lactobacillus casei*: an efficient tool for removal of H_2O_2 and protection of *Lactobacillus bulgaricus* from oxidative stress in milk. Appl Environ Microbiol 72:5143–5149
Rodriguez-Montelongo L, Volentini SI, Farias RN, Massa EM, Rapisarda VA (2006) The Cu(II)-reductase NADH dehydrogenase-2 of *Escherichia coli* improves the bacterial growth in

extreme copper concentrations and increases the resistance to the damage caused by copper and hydroperoxide. Arch Biochem Biophys 451:1–7
Rosenzweig AC (2001) Copper delivery by metallochaperone proteins. Acc Chem Res 34:119–128
Saier MH Jr, Tam R, Reizer A, Reizer J (1994) Two novel families of bacterial membrane proteins concerned with nodulation, cell division and transport. Mol Microbiol 11:841–847
Schirawski J, Hagens W, Fitzgerald GF, Van Sinderen D (2002) Molecular characterization of cadmium resistance in *Streptococcus thermophilus* strain 4134: an example of lateral gene transfer. Appl Environ Microbiol 68:5508–5516
Schwarz G, Mendel RR (2006) Molybdenum cofactor biosynthesis and molybdenum enzymes. Annu Rev Plant Biol 57:623–647
Scott C, Rawsthorne H, Upadhyay M, Shearman CA, Gasson MJ, Guest JR, Green J (2000) Zinc uptake, oxidative stress and the FNR-like proteins of *Lactococcus lactis*. FEMS Microbiol Lett 192:85–89
Sharma VK, Hackbarth CJ, Dickinson TM, Archer GL (1998) Interaction of native and mutant MecI repressors with sequences that regulate *mecA*, the gene encoding penicillin binding protein 2a in methicillin-resistant staphylococci. J Bacteriol 180:2160–2166
Singleton C, Le Brun NE (2009) The N-terminal soluble domains of *Bacillus subtilis* CopA exhibit a high affinity and capacity for Cu(I) ions. Dalton Trans 28:688–696
Singleton C, Banci L, Ciofi-Baffoni S, Tenori L, Kihlken MA, Boetzel R, Le Brun NE (2008) Structure and Cu(I)-binding properties of the N-terminal soluble domains of *Bacillus subtilis* CopA. Biochem J 411:571–579
Solioz M, Odermatt A (1995) Copper and silver transport by CopB-ATPase in membrane vesicles of *Enterococcus hirae*. J Biol Chem 270:9217–9221
Solioz M, Odermatt A, Krapf R (1994) Copper pumping ATPases: common concepts in bacteria and man. FEBS Lett 346:44–47
Solioz M, Stoyanov JV (2003) Copper homeostasis in *Enterococcus hirae*. FEMS Microbiol Rev 27:183–195
Solioz M, Vulpe C (1996) CPx-type ATPases: a class of P-type ATPases that pump heavy metals. Trends Biochem Sci 21:237–241
Sondi I, Salopek-Sondi B (2004) Silver nanoparticles as antimicrobial agent: a case study on *E. coli* as a model for Gram-negative bacteria. J Colloid Interface Sci 275:177–182
Stadtman ER, Berlett BS, Chock PB (1990) Manganese-dependent disproportionation of hydrogen peroxide in bicarbonate buffer. Proc Natl Acad Sci USA 87:384–388
Steffen C, Eberhad P, Bosset JM, Rüegg M (2009) Swiss-type varieties. In: Fox PF (Ed.) Cheese: chemistry, physics and microbiology. Aspen Publishers, Gaithersburg, pp. 83–110
Stoyanov JV, Magnani D, Solioz M (2003) Measurement of cytoplasmic copper, silver, and gold with a *lux* biosensor shows copper and silver, but not gold, efflux by the CopA ATPase of *Escherichia coli*. FEBS Lett 546:391–394
Strausak D, Solioz M (1997) CopY is a copper-inducible repressor of the *Enterococcus hirae* copper ATPases. J Biol Chem 272:8932–8936
Sun HW, Plapp BV (1992) Progressive sequence alignment and molecular evolution of the Zn-containing alcohol dehydrogenase family. J Mol Evol 34:522–535
Tikhonov KG, Zastrizhnaya OM, Kozlov YN, Klimov VV (2006) Composition and catalase-like activity of Mn(II)-bicarbonate complexes. Biochemistry (Mosc) 71:1270–1277
Tottey S, Rich PR, Rondet SA, Robinson NJ (2001) Two Menkes-type atpases supply copper for photosynthesis in Synechocystis PCC 6803. J Biol Chem 276:19999–20004
Tottey S, Harvie DR, Robinson NJ (2005) Understanding how cells allocate metals using metal sensors and metallochaperones. Acc Chem Res 38:775–783
Toyoshima C, Mizutani T (2004) Crystal structure of the calcium pump with a bound ATP analogue. Nature 430:529–535
Toyoshima C, Nomura H, Sugita Y (2003) Crystal structures of Ca^{2+}-ATPase in various physiological states. Ann NY Acad Sci 986:1–8
Tsai TY, Lee YH (1998) Roles of copper ligands in the activation and secretion of *Streptomyces* tyrosinase. J Biol Chem 273:19243–19250

Turner MS, Tan YP, Giffard PM (2007) Inactivation of an iron transporter in *Lactococcus lactis* results in resistance to tellurite and oxidative stress. Appl Environ Microbiol 73:6144–6149

Tynecka Z, Gos Z, Zajac J (1981) Reduced cadmium transport determined by a resistance plasmid in *Staphylococcus aureus*. J Bacteriol 147:305–312

Valko M, Morris H, Cronin MT (2005) Metals, toxicity and oxidative stress. Curr Med Chem 12:1161–1208

Vallee BL, Falchuk KH (1993) The biochemical basis of zinc physiology. Physiol Rev 73: 79–118

Van Melckebeke H, Vreuls C, Gans P, Filee P, Llabres G, Joris B, Simorre JP (2003) Solution structural study of BlaI: implications for the repression of genes involved in β-lactam antibiotic resistance. J Mol Biol 333:711–720

Van Veen HW, Abee T, Kortstee GJ, Konings WN, Zehnder AJ (1994a) Translocation of metal phosphate via the phosphate inorganic transport system of *Escherichia coli*. Biochemistry 33:1766–1770

Van Veen HW, Abee T, Kortstee GJ, Pereira H, Konings WN, Zehnder AJ (1994b) Generation of a proton motive force by the excretion of metal-phosphate in the polyphosphate-accumulating *Acinetobacter johnsonii* strain 210A. J Biol Chem 269:29509–29514

Van Veen HW, Venema K, Bolhuis H, Oussenko I, Kok J, Poolman B, Driessen AJ, Konings WN (1996) Multidrug resistance mediated by a bacterial homolog of the human multidrug transporter MDR1. Proc Natl Acad Sci USA 93:10668–10672

Vats N, Lee SF (2001) Characterization of a copper-transport operon, *copYAZ*, from *Streptococcus mutans*. Microbiology 147:653–662

Weinberg ED (1997) The Lactobacillus anomaly: total iron abstinence. Perspect Biol Med 40:578–583

Wimmer R, Herrmann T, Solioz M, Wüthrich K (1999) NMR structure and metal interactions of the CopZ copper chaperone. J Biol Chem 274:22597–22603

Wittman V, Wong HC (1988) Regulation of the penicillinase genes of *Bacillus licheniformis*: interaction of the *pen* repressor with its operators. J Bacteriol 170:3206–3212

Wunderli-Ye H, Solioz M (2001) Purification and functional analysis of the copper ATPase CopA of *Enterococcus hirae*. Biochem Biophys Res Commun 280:713–719

Zhang Y, Gladyshev VN (2008) Molybdoproteomes and evolution of molybdenum utilization. J Mol Biol 379:881–899

Zhang Y, Turanov AA, Hatfield DL, Gladyshev VN (2008) *In silico* identification of genes involved in selenium metabolism: evidence for a third selenium utilization trait. BMC Genomics 9:251

Zhang Y, Rodionov DA, Gelfand MS, Gladyshev VN (2009) Comparative genomic analyses of nickel, cobalt and vitamin B12 utilization. BMC Genomics 10:78–103

Part III
Stress Responses of Lactic Acid Bacteria in the Context of Species or Genera

Chapter 10
Stress Responses of *Lactococcus lactis*

Juan Zhang, Chongde Wu, Feng Xue, Guocheng Du, and Jian Chen

10.1 Introduction

Lactococcus lactis is a Gram-positive bacterium commonly used as a starter culture for the manufacture of cheese, fermented milks, and wine products. Based on its applications in the food industry worldwide and its identification as GRAS ("Generally Recognized as Safe"), this mesophilic and microaerophilic bacterium plays a crucial role in food biotransformations and is considered to be the model organism of lactic acid bacteria (LAB). Moreover, the use of this microorganism may extend to the area of health benefits, like probiotics, or may even be a potential delivery system of therapeutic molecules to the gastrointestinal tract. Because of the great prospects, increasing attention has been given to optimizing the performance of *L. lactis* not only at the start of production, but also in the process of eating.

However, as a natural microflora from plant materials and dairy products, *L. lactis* always encounters certain stress conditions such as oxidation, heating, acid, high osmolarity (i.e., dehydration), and nutrient starvation. All of these stresses greatly affect its viability, and fermentative capability. Meanwhile, serious cold and

G. Du (✉)
School of Biotechnology, Jiangnan University, Wuxi, Jiangsu 214122, China

The Key Laboratory of Industrial Biotechnology, Ministry of Education,
Wuxi, Jiangsu 214122, China
e-mail: gcdu@jiangnan.edu.cn

J. Chen (✉)
School of Biotechnology, Jiangnan University, Wuxi, Jiangsu 214122, China

State Key Laboratory of Food Science and Technology,
Jiangnan University, Wuxi, Jiangsu 214122, China
e-mail: jchen@jiangnan.edu.cn

E. Tsakalidou and K. Papadimitriou (eds.), *Stress Responses of Lactic Acid Bacteria*,
Food Microbiology and Food Safety, DOI 10.1007/978-0-387-92771-8_10,

heat challenges resulting from freeze-dried and spray-dried preparations for transport and storage also threaten the viability and stability of *L. lactis* cells. Moreover, as a probiotic, *L. lactis* must be able to adapt to harsh conditions, such as hydrochloric acid stress in human gastric juice and bile stress in the small bowel. To optimize its application in industrial fermentations and its performance as a probiotic, it is of great importance to explore the stress-sensing systems and defenses of *L. lactis* against stress, which allow it to withstand harsh conditions.

Recently, certain physiological changes, labeled *stress responses*, were found in *L. lactis* cells under stressful environments. Usually, adaptation to mild changes may allow cells to increase tolerance when subsequently exposed to extreme levels of the same stress or to other environmental conditions that may normally be lethal. This phenomenon has drawn increasing attention in recent years, and some of the regulatory mechanisms responding to an environmental stress condition are related to those found in other bacteria, while some are unique in *L. lactis*. The unraveling of the underlying regulatory systems would possibly give insight into the development of stress resistance. In this part of the chapter, we present an overview of the mechanisms of stress responses in *L. lactis*. Meanwhile, a number of strategies can be derived for the manipulation of stress-regulatory elements for more effective application of *L. lactis*.

10.2 Stress Responses and the Metabolic Pathways in *L. lactis*

10.2.1 Mainstream Metabolic Pathways

L. lactis plays an important role in the manufacture of fermented foods with ideal flavor (e.g., diacetyl) and texture (e.g., exopolysaccharides). The fact that LAB usually lack functional electron chains, grow under low-oxygen tension, and rely mostly on fermentation processes to provide energy considerably limits their metabolic versatility. Ordinarily, these organisms metabolize glucose homofermentatively to lactate. The relative simplicity of the *L. lactis* metabolism that converts sugars via the glycolytic (homofermentative) pathway to pyruvate makes it an attractive target for the development of effective cell factories (Neves et al. 2005). However, under various stress conditions, such as acid, cold, and bile salt, a shift from homolactic to mixed acid fermentation with productions of formate, acetate, ethanol, and CO_2 is observed (Fig. 10.1).

10.2.2 Metabolic Activities and Cellular Energy State

Generally, stresses such as acid, temperature, and starvation inhibit the metabolic activities of *L. lactis* in the glycolytic pathway, thus diminishing the catabolic flux of glycolysis and decreasing the rate of biochemical energy synthesis (Even et al. 2002).

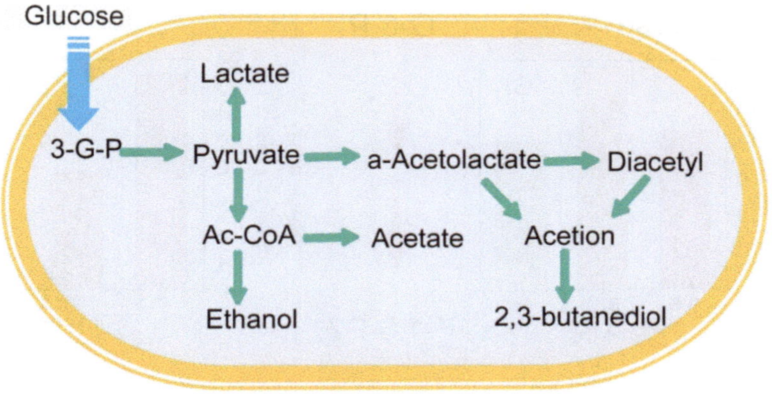

Fig. 10.1 Mixed-fermentation pathway of glucose metabolism in *Lactococcus lactis*. *3-G-P* 3-glyceraldehyde phosphate; *Ac-CoA* acetyl-coenzyme A

Meanwhile, as *L. lactis* cells lack the cytochrome system and are unable to produce ATP by oxidative phosphorylation, they rely on glycolysis and substrate-level phosphorylation to generate compounds that serve as energy donors for solute transport and growth. Unfortunately, stress conditions decrease the ability of *L. lactis* cells to synthesize ATP, generate proton motive force (PMF), and maintain their viability over time.

Acid-stress response may be particularly important in LAB, whose growth and transition to the stationary phase are accompanied by the production of lactic acid, which, on the one hand, inhibits the competition of contaminating microorganisms, but on the other hand results in acidification of the media, arrest of cell multiplication, and possible cell death. During acid treatment, cytoplasmic acidification decreases enzyme activities and, consequently, diminishes the catabolic flux through glycolysis (Fig. 10.2) (Even et al. 2002).

Moreover, acidification leads to changes in the energy state of *L. lactis* cells. The decreased catabolic flux through glycolysis results in a decrease of the ATP synthesis, causing a clear reduction of the biomass (Piard and Desmazeaud 1991). In addition, the H^+-ATPase pumps protons out of the cells at the expense of ATP and also diminishes the available ATP for biomass synthesis (Siegumfeldt et al. 2000). Therefore, a strong decline in the efficiency of biomass synthesis relative to the energy supply (Y_{ATP}) is observed (Mercade et al. 2000).

The same is to be encountered under starvation stress. In optimal conditions, *L. lactis* is a homofermentative LAB that produces lactic acid and ATP. In response to carbohydrate starvation, lactococci become VBNC (viable but nonculturable), a state during which cells continue to transport and metabolize nutrients rather than die and lyze, but do not form colonies on solid agar (Stuart et al. 1999). In the first hours of starvation, glycolytic activities decline rapidly. The loss of glycolytic activity results in the loss of glyceraldehyde 3-phosphate dehydrogenase, phosphoglycerate mutase, and pyruvate kinase activities (Ganesan et al. 2007). Furthermore, after long periods of starvation, decreased glycolytic activity slows down the

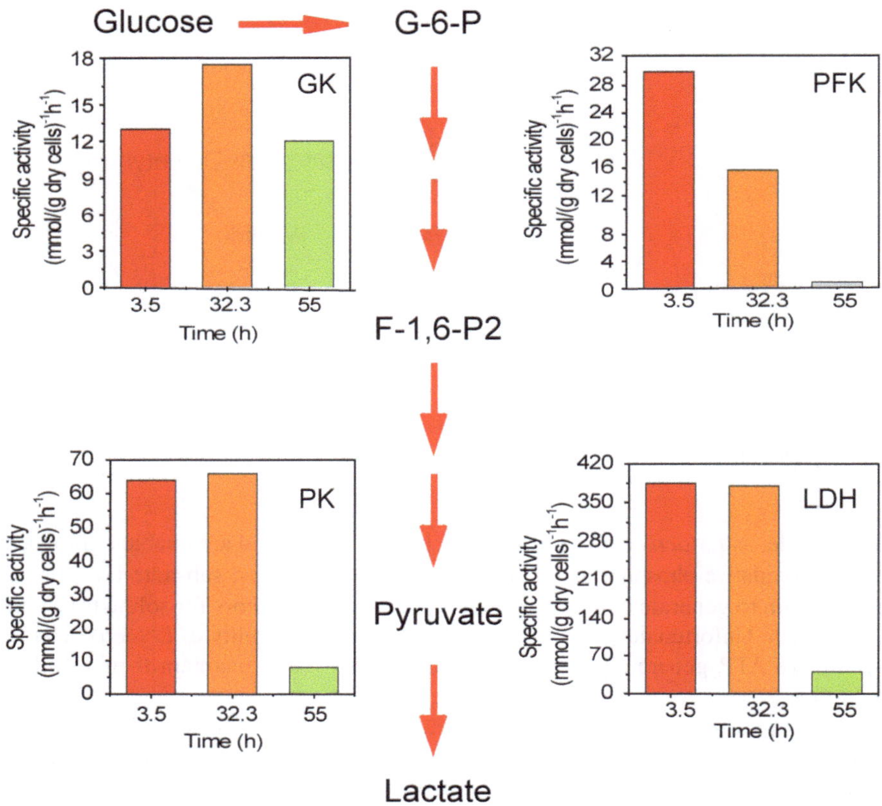

Fig. 10.2 Enzyme-specific activities involved in glycolysis and lactate pathway in cell extracts of *L. lactis* MG 1363 on glucose under uncontrolled pH conditions. *GK* glucokinase; *PFK* phosphofructokinase; *PK* pyruvate kinase; *LDH* lactate dehydrogenase

ATP generation that supplies the PMF, and as a result, the *L. lactis* cells fail to accumulate nutrients that are essential for growth, even with a renewed carbon source (Kunji et al. 1993; Ganesan et al. 2006). In detail, the majority of amino acids are transported by the PMF-driven transport that links amino acid uptake to the PMF (Fig. 10.3). As the driving force for the H^+ translocation and the PMF is usually supplied by the free energy of ATP hydrolysis, the decreased ATP supply leads to a limited utilization of nutrients (Kunji et al. 1993).

As a response to temperature stress, the metabolic activity of *L. lactis* changes and eventually leads to a modification of the growth state (Arnau et al. 1996; Wouters et al. 2000). For example, the *L. lactis* cells stopped growing when the culture was shifted from its optimal growth temperature of 30–42°C, and the growth resumed after a downshift to 30°C, with the growth rate reaching the preshock level after 1 h at 30°C (Whitaker and Batt 1991). Unfortunately, cells were hardly capable of recovering from exposure to 50°C for 30 min (Whitaker and Batt 1991).

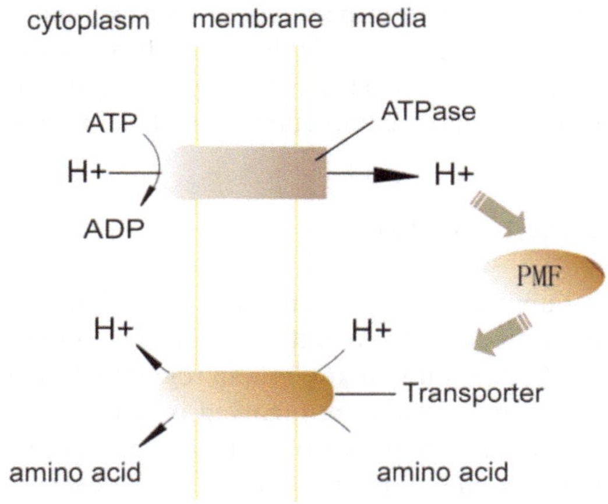

Fig. 10.3 Proton motive force-driven symport

It seems that growth at high temperatures is deleterious to a cell, while growth at low temperatures merely slows down biological processes, and these modulations may partly be related to the metabolic activity of *L. lactis* cells (Panoff et al. 1994).

10.3 Intracellular Stress Response Signals in *L. lactis*

10.3.1 Cytoplasmic pH Homeostasis

The growth of *L. lactis* is accompanied by the generation of acidic end products of sugar fermentation, which results in acidification of the medium. The pronounced organic acid production creates an environment that is unfavorable for many other organisms. It is well established, however, that acids can passively diffuse through the cell membrane and, after entry into the cytoplasm, rapidly dissociate into protons and charged derivatives to which the cell membrane is impermeable (Presser et al. 1997). The intracellular accumulation of protons may lead to a decrease in the internal pH (pH_{in}). A decrease in internal pH is a major signal in the induction of acid tolerance response (ATR), and the magnitude of the ATR is dependent on the degree of acidification of the cell cytoplasm (pH_{in}) (O'Sullivan and Condon 1997). The ability to regulate pH_{in} is a fundamental requirement to withstand acidic environments (Kashket 1987). So far, the importance of pH homeostasis and the effects of pH_{in} on metabolic activities in microorganisms have become increasingly recognized. The pH_{in} affected the uptake of nutrients, such as K^+, phosphate, and amino acids. Also, pH_{in} is an important component of the PMF and therefore has a profound effect on the bioenergetic state of the cell (Hutkins and Nannen 1993). The factors contributing to pH homeostasis in lactococci are (1) ATPases, (2) the arginine deiminase pathway, (3) decarboxylation reactions and electrogenic transport, and

(4) the citrate transport system (Kashket and Barker 1977; Marquis et al. 1987; Garcia-Quintans et al. 1998; O'Sullivan and Condon 1999; van de Guchte et al. 2002). All four mechanisms may reduce the acidification of the internal compartment and thus are important in maintaining cell viability.

10.3.1.1 ATPases

A number of mechanisms have been shown to confer stress resistance. There is good evidence from both batch and chemostat culture studies that the membrane H^+-ATPase plays a key role in regulating the pH_{in} of LAB and may be the most important mechanism involved in pH_{in} regulation (Kobayashi et al. 1986). H^+-ATPase extrudes protons out of the cell via ATP hydrolysis. This reaction requires energy (in the form of ATP) because the expulsion of protons from a relatively alkaline environment (i.e., the cytoplasm) into an acidic environment (i.e., the medium) requires the movement of protons against a concentration gradient (Belli and Marquis 1991). Acidification of the cytoplasm increases both the amount and the activity of ATPase (Nannen and Hutkins 1991). Nannen and Hutkins reported that the greatest level of H^+-ATPase activity in lactococci occurred in cells harvested from media at an extracellular pH (pH_{out}) of 4.9–5.9 (corresponding to a pH_{in} of approximately 6.0–6.5) (Nannen and Hutkins 1991). Below a pH value of about 5.0, however, the H^+-ATPase activity decreased sharply, which correlated with a rapid decrease in the pH_{in} (Sanchez et al. 2006). In this case, an acid-sensitive *L. lactis* mutant is unable to maintain a neutral pH_{in} in an acidic environment (Amachi et al. 1998), and its acid sensitivity may have derived from a mutation of the ATPase structural gene, which results in reduced enzymatic activity (Amachi et al. 1998). Meanwhile, in *L. lactis*, cation transport ATPases such as K^+-ATPase can contribute to pH homeostasis. The K^+/H^+ exchange converts the transmembrane potential generated by the K^+-ATPase into a transmembrane pH gradient (ΔpH). This ion exchange allows the establishment of a ΔpH and can participate in pH homeostasis (Kashket and Barker 1977).

10.3.1.2 Arginine Deiminase Pathway

Another mechanism for pH homeostasis is the arginine deiminase pathway (ADI). This pathway allows *L. lactis* to neutralize its environment by NH_3 production (Marquis et al. 1987), and the generated ATP enables the extrusion of cytoplasmic protons by F_0F_1-ATPase (Fig. 10.4). The ADI pathway consists of three cytoplasmic enzymes, arginine deiminase, ornithine carbamoyltransferase, and carbamate kinase, which catalyze the conversion of arginine into ornithine, ammonia, and carbon dioxide with the formation of 1 mol of ATP per mol of consumed arginine (Marquis et al. 1987). Arginine and ornithine are exchanged by a membrane-located antiporter without an energy requirement, and the activity of the pathway is induced

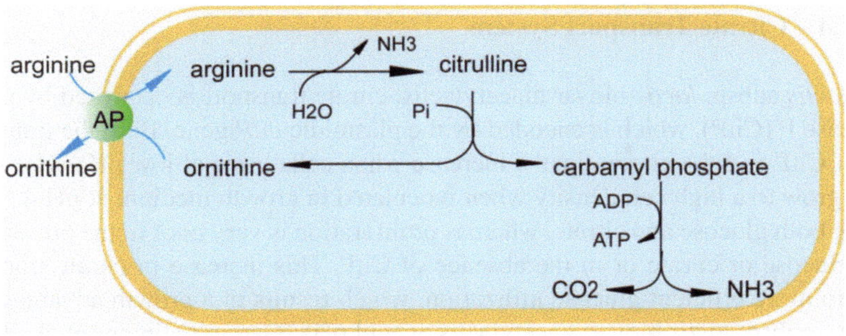

Fig. 10.4 Arginine deiminase pathway

three- to fivefold in the presence of arginine (Poolman et al. 1987). As of now, the ADI pathway seems to be a unique phenotypic property of *L. lactis* subsp. *lactis* (Deibel and Seeley 1974), as it is rarely observed in the subsp. *cremoris*.

10.3.1.3 Decarboxylation Reactions and Electrogenic Transport

Several systems based on decarboxylation and an electrogenic transporter found in LAB may contribute to pH homeostasis. In these reactions, a carboxylic acidic compound (e.g., an amino acid) is transported into the cell to be decarboxylated. A proton is consumed in the reaction and the product is exported from the cell via a transporter. The effect of this reaction is to increase the alkalinity of the cytoplasm. Moreover, the coupling of the decarboxylation to an electrogenic transporter (antiport or uniport) allows ATP to be generated via the PMF (van de Guchte et al. 2002).

In *L. lactis*, a putative glutamate decarboxylase-γ-aminobutyrate (GABA) antiporter system has been shown to confer acid resistance in the presence of chloride and glutamate (Sanders et al. 1998a, b). The system is encoded by an operon consisting of two genes, *gadC* and *gadB*, which specify a putative glutamate-γ-aminobutyrate antiporter and a glutamate decarboxylase, respectively. The combined action of these two proteins may confer acid resistance by removing a proton from the cytoplasm and exporting γ-aminobutyrate, which is more basic than the imported glutamate (Sanders et al. 1998a, b). Alternatively, it may also result in the formation of a PMF that could generate ATP in the presence of glutamate (Higuchi et al. 1997). In addition, the expression of *L. lactis gadCB* in acidified cultures is higher than in buffered medium and may also depend on glutamate (Sanders et al. 1998a, b).

Malolactic fermentation (MLF) is the conversion of the dicarboxylic malic acid to the monocarboxylic lactic acid. It has been identified in *L. lactis* that the generation of metabolic energy during MLF arises from electrogenic malate/lactate antiport and electrogenic malate uptake (in combination with the outward diffusion of lactic acid), together with proton consumption as a result of the decarboxylation of L-malate. The net energy gain would be equivalent to one proton translocated from the inside to the outside per L-malate metabolized (Poolman et al. 1991).

10.3.1.4 Citrate Transport System

In *L. lactis* subsp. *lactis* biovar diacetylactis, citrate transport is catalyzed by citrate permease P (CitP), which is encoded by the plasmidic *citP* gene. Both the transcription of CitP and the citrate uptake increase when cells grow at low pHs. The strain could grow to a high cell density when inoculated in growth medium at pH 4.5 containing both glucose and citrate, whereas proliferation is very poor in the presence of only glucose or citrate or in the absence of CitP. This increase in citrate transport leads to more efficient glucose utilization, which results in a growth advantage for *L. lactis* subsp. *lactis* biovar diacetylactis at acid pHs (Garcia-Quintans et al. 1998).

10.3.2 Accumulation of Specific Intermediates as a Response to Stressful Conditions

10.3.2.1 Production of Oxygen Metabolites Under Oxidative Stress

Oxygen, as another stressor encountered during storage, processing, and drying, may induce the appearance of certain intermediates (Miyoshi et al. 2003). In fact, oxygen itself is unable to cause any damage to the cell, while during the cellular processes, O_2 is partially reduced to water and leads to the formation of reactive O_2 metabolites, which are the superoxide anion radical (O_2^-), the hydroxyl radical ($OH^\cdot$), and hydrogen peroxide (H_2O_2) (Table 10.1). In this case, the intermediates presented above are responsible for cellular oxygen toxicity (Storz and Imlayt 1999). To overcome this challenge, *L. lactis* employs an NADH oxidase-NADH peroxidase system, through which intracellular O_2 oxidizes NADH into NAD^+ by NADH oxidase, and the generated H_2O_2 is reduced to H_2O by NADH peroxidase (Table 10.1).

Meanwhile, under aerobic conditions, two NADH molecules, generated from the oxidation of glyceraldehyde-3-phosphate, are reoxidized to favor the reduction of pyruvate to lactic acid by the action of lactate dehydrogenase (LDH) (Yamada and Carlsson 1975). The increased expression of NADH oxidase and NADH peroxidase competing with LDH for NADH molecules leads to the reduced production of lactic acid and shifted the glycolytic flux from homolactic acid to mixed-acid fermentation (Thomas et al. 1979; Murphy and Condon 1984). Consequently, acetate,

Table 10.1 Redox reaction and related enzymes in *L. lactis*

Enzymatic reactions	Catalytic enzymes
$NADH + H^+ + O_2 \rightarrow NAD^+ + H_2O_2$	NADH: H_2O_2 oxidase
$2NADH + 2H^+ + O_2 \rightarrow 2NAD^+ + 2H_2O$	NADH: H_2O_2 oxidase
Pyruvate + phosphate + $O_2 \rightarrow$ acetylphosphate + $CO_2 + H_2O_2$	Pyruvate oxidase
α-Glycerophosphate + $O_2 \rightarrow$ dihydroxyacetone phosphate + H_2O_2	α-Glycerophosphate oxidase
$2\ O_2{-} + 2H^+ \rightarrow H_2O_2 + O_2$	Superoxide dismutase
$NADH + H^+ + H_2O_2 \rightarrow NAD^+ + 2H_2O$	NADH peroxidase

ethanol, acetoin, diacetyl, and CO_2 (mixed-fermentation) are generated by the participation of pyruvate dehydrogenase (PDH), pyruvate-formate lyase, and α-acetolactate synthase (Miyoshi et al. 2003).

Cells containing glutathione (GSH) have an alternative mechanism, namely, a glutathione-glutathione peroxidase-glutathione reductase system to protect themselves against damage from H_2O_2 treatment. In this system H_2O_2 is reduced by GSH peroxidase, which uses GSH as a hydrogen donor, giving rise to H_2O, and then the oxidized GSH (GSSG) formed is reduced by glutathione reductase (GR) and NADPH. However, some *L. lactis* subsp. *lactis* strains that cannot accumulate GSH still demonstrate a strong resistance to H_2O_2, suggesting that there are other mechanisms for H_2O_2 resistance, such as thioredoxin and thioredoxin reductase (Li et al. 2003).

10.3.2.2 Accumulation of Specific Compatible Solutes as a Response to Osmotic Stress

External osmolarity is one of the most common types of stress factors for *L. lactis* (Morbach and Kramer 2002). Cells try to keep their volume and/or turgor pressure constant. Thus, either a decrease (hypoosmotic stress) or increase (hyperosmotic stress) in the solute concentration (increase or decrease in water activity) in the environment will be deleterious to cellular metabolism and survival (Morbach and Kramer 2002) (Fig. 10.5).

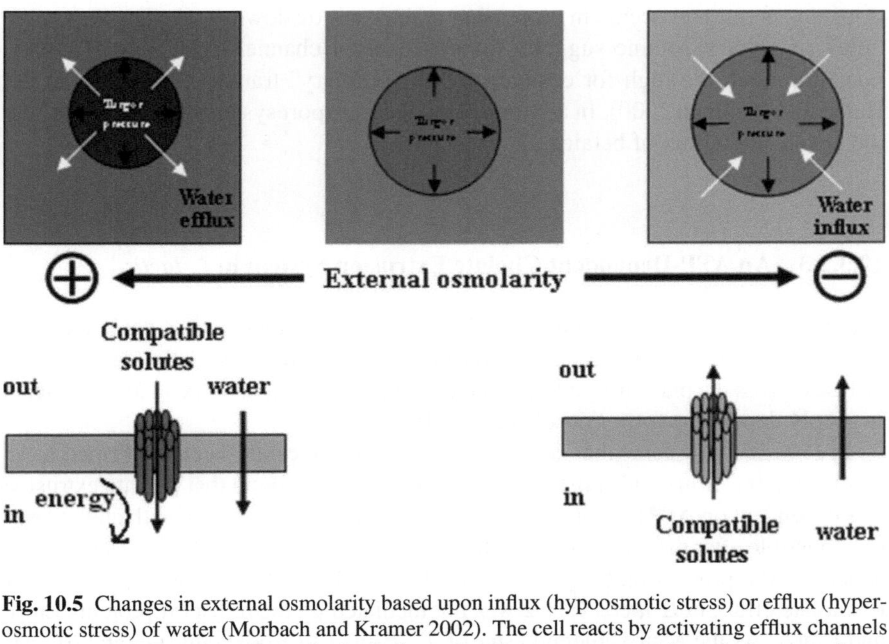

Fig. 10.5 Changes in external osmolarity based upon influx (hypoosmotic stress) or efflux (hyperosmotic stress) of water (Morbach and Kramer 2002). The cell reacts by activating efflux channels in the former case and uptake carrier in the latter. In all cases, an attempt is made for water fluxes to be redirected

Without the capability to synthesize compatible solutes, *L. lactis* cells take up some key compatible solutes such as glycine betaine, carnitine, and proline from the environment to maintain turgor pressure (Hutkins et al. 1987; Fougere and Le Rudulier 1990; Molenaar et al. 1993). In *L. lactis*, a high-affinity uptake system for glycine betaine (betaine) (Km = 1.5 μM) is constitutively expressed (Molenaar et al. 1993). This is an osmoregulated ABC transporter (OpuA) that is made up of an ATP-binding/hydrolyzing subunit (OpuAA) and a protein (OpuABC) that contains both the translocator and the substrate-binding domain. The activity of this system increases not only with the extent of the osmotic upshock but also with the growth temperature and the supplementation of Tween-80 (Guillot et al. 2000). In addition, it varies strongly with the medium pH. Moreover, a low-affinity proline-uptake system ($K_m > 5$ mM) is also active but only in a chemically defined medium (CDM) of high osmolarity and not in rich media or in the absence of KCl (Molenaar et al. 1993). Proline transport is inhibited by betaine (K_i of between 50 and 100 μM), and the exchange of proline for betaine suggests that the proline transport system may also transport betaine (Molenaar et al. 1993). Additionally, the proline uptake in *L. lactis* is enzyme-mediated and independent of passive diffusion. The energy and pH dependency and the impact of ionophores on the activity of both transport systems show that these systems are not PMF-driven (Molenaar et al. 1993).

When the uptake of glycine betaine and proline is activated by an osmotic upshock, no activation of glutamate uptake takes place. However, the osmotic downshock causes a rapid efflux of some compatible solutes, among which are proline, glycine betaine, and glutamate (van de Guchte et al. 2002). Glycine betaine efflux is instantaneous and occurs in proportion to the osmotic downshock. The very rapid efflux of glycine betaine suggests the existence of channel-like activities, as the rates are much too high for catalysis by an "ordinary" transport system (van der Heide and Poolman 2000). In addition, separate transport systems are postulated for the uptake and efflux of betaine.

10.3.2.3 An ATP-Dependent Cholate Extrusion System in *L. lactis*

Bile acids (also regarded as bile salts), the major constituents of bile, are derivatives of cholic acid (CA). CA is a weak acid (pKa, 6.4), which, in its undissociated form, diffuses across the plasma membrane and becomes trapped as cholate in the cytoplasm in a ΔpH-dependent form (Yokota et al. 2000).

In *L. lactis,* a system (that is ATP-dependent) has recently been confirmed to be involved in the efflux and resistance to CA. Yokota et al. found that cholate extrusion was dependent on ATP hydrolysis in *L. lactis* C41-2, and the loading with cholate was only possible after the inactivation of the ATP-dependent cholate efflux system due to a lack of ATP production by glycolysis (Yokota et al. 2000). In addition, the activity of cholate efflux in *L. lactis* C41-2 was not mediated by the transport proteins LmrP and LmrA. As the acetoxymethyl ester of 2', 7'-bis-(2-carboxyethyl)–5(and 6) -carboxyfluorescein (BCECF) and cholate inhibit each other's transport, it suggests that they are effluxed by a common ATP-dependent transport system (Yokota et al. 2000).

10.3.2.4 Multidrug-Resistance Systems in *L. lactis*

Besides the system to fight against environmental stresses, the *L. lactis* cells have developed versatile mechanisms to resist antibiotics and other cytotoxic drugs. A significant group of resistance mechanisms were found in the specific drug-resistance systems (SDR) and the multidrug-resistance systems (MDR) (Konings et al. 1997). These systems prohibit the entrance of specific (SDR) and various (MDR) toxic compounds, in which the activity of most bacterial MDR transporters depends on ATP hydrolysis and PMF (Poolman and Konings 1993).

Concretely speaking, three MDR systems have been found in *L. lactis*. One system provides resistance against anionic compounds, while the other two are responsible for cation excretion (Konings et al. 1997). An ATP-dependent multidrug extrusion system that is specific for organic anions was discovered due to its ability to extrude the fluorescent pH indicator BCECF, which transports carboxyfluorescein derivatives and glutathione conjugates such as dinitrophenyl-glutathione (Van Veen et al. 1996). Another two cationic multidrug extrusion systems in *L. lactis* are the PMF-dependent drug efflux system encoded by the *lmrP* gene and the ATP-dependent drug extrusion system encoded by the *lmrA* gene, separately (Fig. 10.6).

10.4 Cross-Protection in *L. lactis*

As the emerging studies confirmed the interaction of multiple stresses, related physiological responses caused by the cross-protection of *L. lactis* cells seem to be extremely meaningful. It is well known that subjection to a mild stress increases cells resistant to a lethal challenge with the same stress condition. Moreover, preadaptation to one

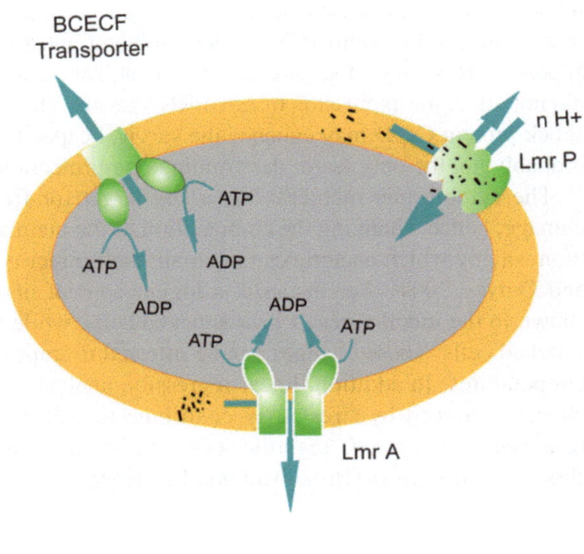

Fig. 10.6 Three characterized multidrug-resistance systems in *L. lactis*

stress condition can also help cells fight against other harmful conditions (Leyer and Johnson 1993; O'Driscoll et al. 1996; Lou and Yousef 1997; Desmond et al. 2001). Cross-protection was widely demonstrated in the 1980s, but the first report about such a phenomenon in *L. lactis* was described in 1991 (Boutibonnes et al. 1991). In this study, the response of *L. lactis* ML3 to heat shock was found to be similar with what happened under the condition with 4.0% (v/v) ethanol. Four years later (Panoff et al. 1995), the authors further found that puromycin or chemicals such as cadmium chloride, mercury chloride, sodium azide, and β-mercaptoethanol can also help *L. lactis* to set up a line of defense against heat shock.

10.4.1 Cross-Protection During Cold Stress in L. lactis

The physiology of the cold-shock response in *L. lactis* subsp. *lactis* IL1403 at a subzero temperature and the cold-induced adaptation to heat shock were studied (Panoff et al. 1994, 1995). Results showed that preadaptation of cells at 8°C leads to the development of cold resistance. Interestingly, the same condition can also increase their thermotolerance to a 52°C challenge, while preincubation at 42°C did not improve cells' ability to survive in freezing–thawing stress. These results indicated that cold- and heat-shock responses may be physiologically linked by a complex relationship. Further study demonstrates that such a phenomenon is subject to the adaptation temperature in slight changes: Exposure to 8°C for 48 h leads to 95% survival, and 4°C gives no adaptation to the freezing–thawing challenge, while exposing cells to 16°C gives intermediate results. Apparently, an optimal adaptation temperature (8°C for *L. lactis*) to a subzero challenge corresponds to a sufficiently low temperature in order to permit a metabolic response to the stimulus (Panoff et al. 1995). The proteins induced in low-temperature adaptation and cryoprotection were then studied in *L. lactis* MG1363 (Wouters et al. 1999). Exposure to 4, 10, and 20°C for several hours increased the cells' freeze survival. At the same time, the expression of cold-induced proteins (CSPs) increased and, most significantly, the 7-kDa CSPs appeared. However, other stresses (i.e., heat, salt, acid, and stationary-phase stress) did not affect the expression of *csp* mRNA levels. This indicates that the 7-kDa cold-shock protein CspD may enhance the survival capacity after freezing, although it is probably not the only factor determining cryoprotection and cross-protection.

The cytoplasmic membrane is a key target for freeze- or desiccation-induced damage; hence, changing the composition of the membrane's fatty acids is the common way by which bacteria can maintain their integrity (Guerzoni et al. 2001; Fozo and Quivey 2004). For instance, a higher content of 19:0 cyclopropane acid was shown in the membranes of heat-shocked cells, while the value was lower in cold-shocked cells. These changes can be inferred to improve membrane fluidity at low temperatures. In addition, linear regression analysis revealed that membrane lipid changes induced by stress may contribute to enhanced freeze and lyophilization resistances in *L. lactis* regardless of whether the change is contributed by cold-shock or other stress (Broadbent and Lin 1999).

10.4.2 Cross-Protection During Acid Stress in *L. lactis*

O'Sullivan et al. have shown that the exposure of exponentially growing *Lactococcus lactis* subsp. *cremoris* 712 cells to a mild acid pH induced the synthesis of proteins that confer protection not only against lethal acid concentrations but also against other lethal environmental stresses such as heat, ethanol, sodium chloride, and hydrogen peroxide (O'Sullivan and Condon 1997). In contrast to the spectrum specificity of tolerance induced by mild acid pH, the adaption of the other environmental stresses tested, except heat, did not induce a tolerance to acid. Actually, the phenomenon is the so-called ATR, which has been reported widely in many microorganisms. In this process, intracellular pH is considered to be an important but not unique factor for establishing the line of defense. Hartke et al. found that in *L. lactis* subsp. *lactis*, the acid adaptation caused protein synthesis, and at least 33 proteins were upregulated in the acid-adapted cells (Hartke et al. 1997). Further study confirmed that many of these acid-induced polypeptides, including the chaperone DnaK and chaperonine GroEL, were also defined as heat-inducible (42°C) proteins. In particular, proteins such as DnaK and GroEL were synthesized under moderate conditions, while others such as GroES were induced by a serious pH shift below 3.8. Two-dimensional gel electrophoresis was used to investigate the protein expression of *L. lactis* subsp. *cremoris* MG1363 in low pH (Champomier-Verges et al. 2002). Results showed that certain oxidative stress proteins, such as superoxide dismutase, autoinducer synthesis protein, LuxS, and alkylhydroperoxidase, were significantly induced during acid treatment, while the induced heat-shock proteins can be characterized as two different groups: As members of the CtsR regulon, ClpE and ClpP were induced between pH 5.5–4.5, while the HrcA-regulated chaperones, GroEL, GroES, DnaK, and GrpE, were induced only at pH 4.5. These findings, once again, confirmed the potential links among heat stress, oxygen stress, and ATR in *L. lactis*. Surprisingly, the addition of chloramphenicol did not inhibit the tolerance response to lactic acid in *L. lactis*, indicating that de novo protein synthesis is unnecessary in an acidic environment, which is consistent with the behavior of Gram-negative microorganisms when the pH was adjusted by HCl (Hartke et al. 1996).

10.4.3 Cross-Protection During Starvation Stress in *L. lactis*

The stationary phase is always regarded as the best stage to explore cross-protection and global stress regulation (Sanders et al. 1999). Under starvation conditions, the size and fatty acid composition of cells changed. Although the protein synthesis generally decreased in this process, certain proteins essential for starvation resistance were induced, some of which have also been related to heat, oxidative, or osmotic stress (Hartke et al. 1994). In *E. coli*, the product of the *katF* gene has been identified to be the major regulator of the general starvation response, as it encodes a specific sigma factor named σ^S (Lange and Hengge-Aronis 1991; McCann et al. 1991). In *L. lactis* cells such as *L. lactis* subsp. *lactis* IL1403, cross-protection was also observed in starved cells, which, after carbohydrate limitation, enhanced their resistance to heat, ethanol, acid, osmotic, and oxidative stress (Hartke et al. 1994). This kind of global resistance seems to be more significant during the transition from growth to the stationary phase. Interestingly, chloramphenicol or rifampicin treatment, which inhibits de novo protein synthesis, does not abolish the development of tolerant cells, but, on the contrary, seems to provoke this response in *L. lactis* subsp. *lactis* (Hartke et al. 1994).

10.4.4 Other Cross-Protections Occurring Under Stress Conditions in L. lactis

The relationship between UV irradiation–induced tolerance to other environmental stresses and changes in protein synthesis was examined in *L. lactis* subsp *lactis* IL1403 (Hartke et al. 1995). The results showed that preirradiation with UV (254 nm) increased cell resistance to lethal challenges of acid, ethanol, H_2O_2, and heat. At least 14 polypeptides were found to be induced by UV irradiation, and some of them were also induced by acid, ethanol, H_2O_2, and heat treatments (Hartke et al. 1995). However, the cross-protection pattern in *L. lactis* was quite different from that of *E. coli*, since UV irradiation did not induce the RecA protein in a drastic fashion, and the major heat-shock proteins in *E. coli*, GroEL and DnaK, were not overexpressed (Krueger and Walker 1984; Auffray et al. 1991). Interestingly, a polypeptide with a molecular weight of 50 and a pI of 5.5 was found as the sole protein induced by all of the five treatments, and it may have a critical role in the general stress response of *L. lactis* subsp. *lactis* (Hartke et al. 1995).

Two-dimensional electrophoresis was performed to study the overlap of induced proteins between heat-shock– and salt-stress responses (Kilstrup et al. 1997). In *L. lactis* subsp. *cremoris* MG1363, after a temporal induction, proteins upregulated by twofold during heat stress could be divided into two groups. One set of proteins was rapidly synthesized in the first 10 min, but their synthesis rate declined after 15 min (except for GroEL and GroES, which maintained a high rate of synthesis consistently). The other set of proteins exhibited a slowly increasing synthesis rate after the onset of the stress treatment. Interestingly, all of the fast-induced proteins during heat processing also appeared under salt stress, and it was confirmed that they were regulated by CIRCE (controlling inverted repeat of chaperone expression) elements in the promoter region of heat-shock genes (Yuan and Wong 1995; Schulz and Schumann 1996; Kilstrup et al. 1997). In addition, DnaK, GroEL, and GroES showed similar temporal patterns of induction during salt stress with heat stress, although at a lower induction level.

In *L. lactis* cells, the correlated responses under oxidative and thermal stresses were focused on the study of the *recA* gene, which is essential for homologous recombination and DNA repair and refers to oxidation and heat-shock regulation (Duwat et al. 1995). In fact, the *recA*-deficient mutant lost heat-shock proteins

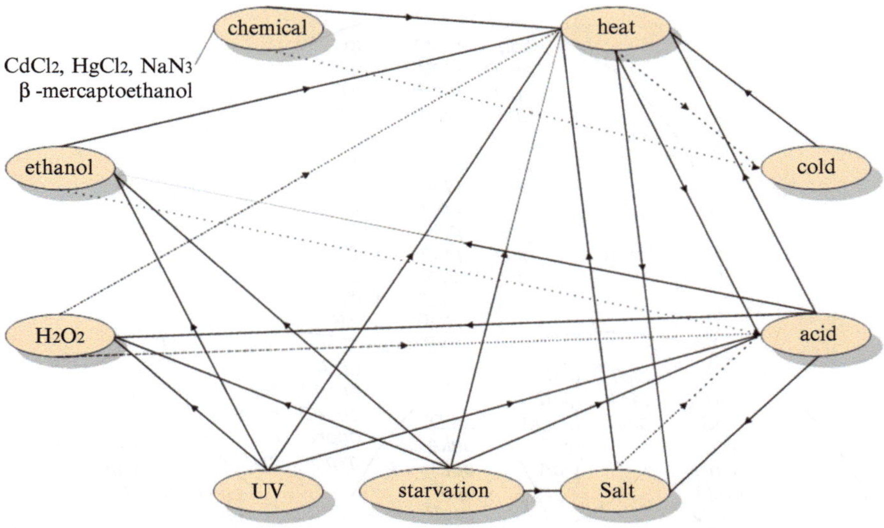

Fig. 10.7 Cross-protection network of *L. lactis*. *Solid lines* indicate cross-tolerance, while *dotted lines* indicate combinations of stresses where cross-protection was not observed. The direction arrows reveal the sequence from adaptation stress to the lethal condition

DnaK, GroEL, and GrpE and therefore grew poorly at increased temperatures, with a diminished heat-shock response. Levels of the HflB analog FtsH were elevated in the lactococcal *recA* strain. Because HflB downregulated heat-shock genes' transcription in *E. coli* (Herman et al. 1995), it is reasonable to consider that *recA* might regulate the heat-shock response in *L. lactis* by controlling the levels of HflB. Figure 10.7 displays the relationship of cross-protection under multiple stresses.

DNA microarrays were used to systematically evaluate the impact of environmental stresses on gene expression in *L. lactis* subsp. *lactis* IL1403 (Xie et al. 2004). A total of 375 genes during heat, acid, and osmotic processing were profiled, and their expression patterns were drawn (Fig. 10.8). Based on this data, further insight into the stress-resistance mechanisms of *L. lactis* cells may be gained. Furthermore, this technology will provide more information about the cross-protection responses of *L. lactis* cells.

10.5 Concluding Remarks

A greater understanding of the self-protection mechanisms of environmental stress responses may inspire us to improve the application of *L. lactis* as a starter, a probiotic, as well as a genetic host. As *L. lactis* is a complex system, numerous information resources implicated in its stress responses need to be collected and investigated further. Fortunately, these studies can also benefit from the knowledge already

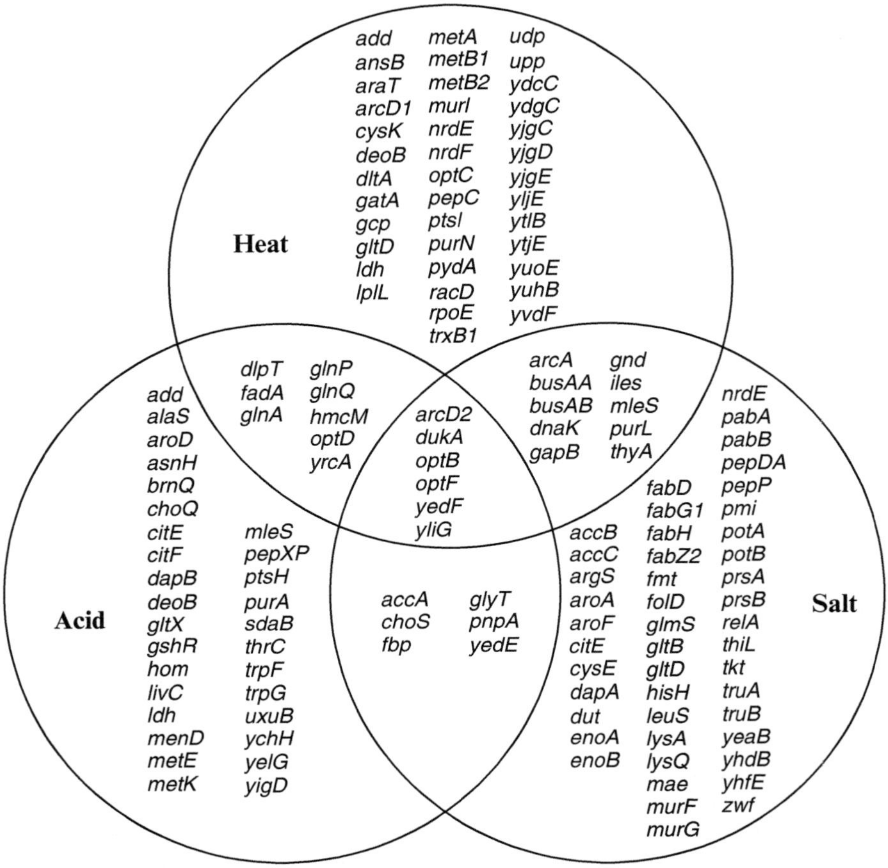

Fig. 10.8 Venn diagram of shared stress responses of *L. lactis* subsp. *lactis* IL1403 (Xie et al. 2004). Genes are grouped according to their response to each stress treatment and are presented in each circle. Common genes among different stress responses are represented in the intersection area among the circles

acquired in other bacteria, for example, the analysis of several stress proteins whose role has been demonstrated in other microorganisms. So far, we know that the stress responses of *L. lactis* are related to metabolic activities and the energy state of cells. Moreover, certain intracellular regulators such as the cytoplasmic pH, the ADI pathway, several decarboxylation reactions, and the electrogenic transport also favor the stress tolerance of *L. lactis* cells. With the rise of omic studies, enriched knowledge about genomics, proteomics, transcriptomics, and bioinformatics will reveal the responses of cross-protection that take place under a series of stresses and will enlighten us to draw the map of stress responses in *L. lactis* globally. All of these findings will undoubtedly increase the tolerance to stress and shed new light on the rational use of *L. lactis* in the food industry.

Acknowledgments Preparation of this chapter was supported in part by a grant from the Major Program of National Natural Science Foundation of China (No. 20836003) and the National Natural Science Foundation of China (No. 30900013). We also thank Chongde Wu and Feng Xue for their numerous contributions to this work.

References

Amachi S, Ishikawa K, Toyoda S, Kagawa Y, Yokota A, Tomita F (1998) Characterization of a mutant of *Lactococcus lactis* with reduced membrane-bound ATPase activity under acidic conditions. Bioscience Biotech Biochem 62:1574–1580

Arnau J, Sorensen KI, Appel KF, Vogensen FK, Hammer K (1996) Analysis of heat shock gene expression in *Lactococcus lactis* MG1363. Microbiology 142:1685–1691

Auffray Y, Thammavongs B, Boutibonnes P (1991) Identification of a RecA-like protein in *Lactococcus lactis*. Biochimie 73:231–233

Belli WA, Marquis RE (1991) Adaptation of *Streptococcus mutans* and *Enterococcus hirae* to acid stress in continuous culture. Appl Environ Microbiol 57:1134–1138

Boutibonnes P, Gillot B, Auffray Y, Thammavongs B (1991) Heat shock induces thermotolerance and inhibition of lysis in a lysogenic strain of *Lactococcus lactis*. Int J Food Microbiol 14:1–9

Broadbent JR, Lin C (1999) Effect of heat shock or cold shock treatment on the resistance of *Lactococcus lactis* to freezing and lyophilization. Cryobiology 39:88–102

Champomier-Verges MC, Maguin E, Mistou MY, Anglade P, Chich JF (2002) Lactic acid bacteria and proteomics: current knowledge and perspectives. J Chrom B 771:329–342

Deibel RH, Seeley HW Jr (1974) *Streptococcaceae*. Bergey's manual of determinative bacteriology, 8th ed. Williams & Wilkins, Baltimore, pp. 490–509

Desmond C, Stanton C, Fitzgerald GF, Collins K, Ross RP (2001) Environmental adaptation of probiotic lactobacilli towards improvement of performance during spray drying. Int Dairy J 11:801–808

Duwat P, Ehrlich SD, Gruss A (1995) The *recA* gene of *Lactococcus lactis*: characterization and involvement in oxidative and thermal stress. Mol Microbiol 17:1121–1131

Even S, Lindley ND, Loubiere P, Cocaign-Bousquet M (2002) Dynamic response of catabolic pathways to autoacidification in *Lactococcus lactis*: transcript profiling and stability in relation to metabolic and energetic constraints. Mol Microbiol 45:1143–1152

Fougere F, Le Rudulier D (1990) Uptake of glycine betaine and its analogues by bacteroids of *Rhizobium meliloti*. Microbiology 136:157–163

Fozo EM, Quivey RG Jr (2004) Shifts in the membrane fatty acid profile of *Streptococcus mutans* enhance survival in acidic environments. ApplEnviron Microbiol 70:929–936

Ganesan B, Dobrowolski P, Weimer BC (2006) Identification of the leucine-to-2-methylbutyric acid catabolic pathway of *Lactococcus lactis*. Appl Environ Microbiol 72:4264–4273

Ganesan B, Stuart MR, Weimer BC (2007) Carbohydrate starvation causes a metabolically active but nonculturable state in *Lactococcus lactis*. Appl Environ Microbiol 73:2498–2512

Garcia-Quintans N, Magni C, de Mendoza D, Lopez P (1998) The citrate transport system of *Lactococcus lactis* subsp. *lactis* biovar diacetylactis is induced by acid stress. Appl Environ Microbiol 64:850–857

Guerzoni ME, Lanciotti R, Cocconcelli PS (2001) Alteration in cellular fatty acid composition as a response to salt, acid, oxidative and thermal stresses in *Lactobacillus helveticus*. Microbiology 147:2255–2264

Guillot A, Obis D, Mistou MY (2000) Fatty acid membrane composition and activation of glycine-betaine transport in *Lactococcus lactis* subjected to osmotic stress. Int J Food Microbiol 55:47–51

Hartke A, Bouché S, Gansel X, Boutibonnes P, Auffray Y (1994) Starvation-induced stress resistance in *Lactococcus lactis* subsp. *lactis* IL1403. Appl Environ Microbiol 60(9):3474–3478

Hartke A, Bouché S, Laplace JM, Benachour A, Boutibonnes P, Auffray Y (1995) UV-inducible proteins and UV-induced cross-protection against acid, ethanol, H_2O_2 or heat treatments in *Lactococcus lactis* subsp. *lactis*. Arch Microbiol 163:329–336

Hartke A, Bouché S, Giard JC, Benachour A, Boutibonnes P, Auffray Y (1996) The lactic acid stress response of *Lactococcus lactis* subsp. *lactis*. Curr Microbiol 33:194–199

Hartke A, Frère J, Boutibonnes P, Auffray Y (1997) Differential induction of the chaperonin GroEL and the co-chaperonin GroES by heat, acid, and UV-irradiation in *Lactococcus lactis* subsp. *lactis*. Curr Microbiol 34:23–26

Herman C, Thevenet D, d'Ari R, Bouloc P (1995) Degradation of sigma 32, the heat shock regulator in *Escherichia coli*, is governed by HflB. Proc Natl Acad Sci 92:3516–3520

Higuchi T, Hayashi H, Abe K (1997) Exchange of glutamate and gamma-aminobutyrate in a *Lactobacillus* strain. J Bacteriol 179:3362–3364

Hutkins RW, Ellefson WL, Kashket ER (1987) Betaine transport imparts osmotolerance on a strain of *Lactobacillus acidophilus*. Appl Environ Microbiol 53:2275–2281

Hutkins RW, Nannen NL (1993) pH homeostasis in lactic acid bacteria. J Dairy Sci 76:2354–2365

Kashket ER (1987) Bioenergetics of lactic acid bacteria: cytoplasmic pH and osmotolerance. FEMS Microbiol Lett 46:233–244

Kashket ER, Barker SL (1977) Effects of potassium ions on the electrical and pH gradients across the membrane of *Streptococcus lactis* cells. J Bacteriol 130:1017–1023

Kilstrup M, Jacobsen S, Hammer K, Vogensen FK (1997) Induction of heat shock proteins DnaK, GroEL, and GroES by salt stress in *Lactococcus lactis*. Appl Environ Microbiol 63:1826–1837

Kobayashi H, Suzuki T, Unemoto T (1986) *Streptococcal* cytoplasmic pH is regulated by changes in amount and activity of a proton-translocating ATPase. J Biol Chem 261:627–630

Konings WN, Lolkema JS, Bolhuis H, Van Veen HW, Poolman B, Driessen AJM (1997) The role of transport processes in survival of lactic acid bacteria, Energy transduction and multidrug resistance. Antonie van Leeuwenhoek 71:117–128

Krueger JH, Walker GC (1984) *groEL* and *dnaK* genes of *Escherichia coli* are induced by UV irradiation and nalidixic acid in an htpR+–dependent fashion. Proc Natl Acad Sci 81:1499–1503

Kunji ERS, Ubbink T, Matin A, Poolman B, Konings WN (1993) Physiological responses of *Lactococcus lactis* ML3 to alternating conditions of growth and starvation. Arch Microbiol 159:372–379

Lange R, Hengge-Aronis R (1991) Identification of a central regulator of stationary-phase gene expression in *Escherichia coli*. Mol Microbiol 5:49–59

Leyer GJ, Johnson EA (1993) Acid adaptation induces cross-protection against environmental stresses in *Salmonella typhimurium*. Appl Environ Microbiol 59:1842–1847

Li Y, Hugenholtz J, Abee T, Molenaar D (2003) Glutathione protects *Lactococcus lactis* against oxidative stress. Appl Environ Microbiol 69:5739–5745

Lou Y, Yousef AE (1997) Adaptation to sublethal environmental stresses protects *Listeria monocytogenes* against lethal preservation factors. Appl Environ Microbiol 63:1252–1255

Marquis RE, Bender GR, Murray DR, Wong A (1987) Arginine deiminase system and bacterial adaptation to acid environments. Appl Environ Microbiol 53:198–200

McCann MP, Kidwell JP, Matin A (1991) The putative sigma factor KatF has a central role in development of starvation-mediated general resistance in *Escherichia coli*. J Bacteriol 173:4188–4194

Mercade M, Lindley ND, Loubiere P (2000) Metabolism of *Lactococcus lactis* subsp *cremoris* MG 1363 in acid stress conditions. Int J Food Microbiol 55:161–165

Miyoshi A, Rochat T, Gratadoux JJ, Le Loir Y, Oliveira SC, Langella P, Azevedo V (2003) Oxidative stress in *Lactococcus lactis*. Genet Mol Res 2:348–359

Molenaar D, Hagting A, Alkema H, Driessen AJ, Konings WN (1993) Characteristics and osmoregulatory roles of uptake systems for proline and glycine betaine in *Lactococcus lactis*. J Bacteriol 175:5438–5444

Morbach S, Kramer R (2002) Body shaping under water stress: osmosensing and osmoregulation of solute transport in bacteria. ChemBioChem 3:384–397

Murphy MG, Condon S (1984) Correlation of oxygen utilization and hydrogen peroxide accumulation with oxygen induced enzymes in *Lactobacillus plantarum* cultures. Arch Microbiol 138:44–48

Nannen NL, Hutkins RW (1991) Proton-translocating adenosine triphosphatase activity in lactic acid bacteria. J Dairy Sci 74:747–751

Neves AR, Pool WA, Kok J, Kuipers OP, Santos H (2005) Overview on sugar metabolism and its control in *Lactococcus lactis*-the input from *in vivo* NMR. FEMS Microbiol Rev 29:531–554

O'Driscoll B, Gahan CG, Hill C (1996) Adaptive acid tolerance response in *Listeria monocytogenes*: isolation of an acid-tolerant mutant which demonstrates increased virulence. Appl Environ Microbiol 62:1693–1698

O'Sullivan E, Condon S (1997) Intracellular pH is a major factor in the induction of tolerance to acid and other stresses in *Lactococcus lactis*. Appl Environ Microbiol 63:4210–4215

O'Sullivan E, Condon S (1999) Relationship between acid tolerance, cytoplasmic pH, and ATP and H+–ATPase levels in chemostat cultures of *Lactococcus lactis*. Appl Environ Microbiol 65:2287–2293

Panoff JM, Legrand S, Thammavongs B, Boutibonnes P (1994) The cold shock response in *Lactococcus lactis* subsp. *lactis*. Curr Microbiol 29:213–216

Panoff JM, Thammavongs B, Laplace JM, Hartke A, Boutibonnes P, Auffray Y (1995) Cryotolerance and cold adaptation in *Lactococcus lactis* subsp. *lactis* IL1403. Cryobiology 32:516–520

Piard JC, Desmazeaud M (1991) Inhibiting factors produced by lactic acid bacteria. 1. Oxygen metabolites and catabolism end-products. Lait 71:525–541

Poolman B, Driessen AJ, Konings WN (1987) Regulation of arginine-ornithine exchange and the arginine deiminase pathway in *Streptococcus lactis*. J Bacteriol 169:5597–5604

Poolman B, Konings WN (1993) Secondary solute transport in bacteria. Biochim Biophys Acta 1183:5–39

Poolman B, Molenaar D, Smid EJ, Ubbink T, Abee T, Renault PP, Konings WN (1991) Malolactic fermentation: electrogenic malate uptake and malate/lactate antiport generate metabolic energy. J Bacteriol 173:6030–6037

Presser KA, Ratkowsky DA, Ross T (1997) Modelling the growth rate of *Escherichia coli* as a function of pH and lactic acid concentration. Appl Environ Microbiol 63:2355–2360

Sanchez B, Reyes-Gavilan CGD, Margolles A (2006) The F_1F_0-ATPase of Bifidobacterium animalis is involved in bile tolerance. Environ Microbiol 8:1825–1833

Sanders JW, Leenhouts K, Burghoorn J, Brands JR, Venema G, Kok J (1998a) A chloride-inducible acid resistance mechanism in *Lactococcus lactis* and its regulation. Mol Microbiol 27:299–310

Sanders JW, Venema G, Kok J, Leenhouts K (1998b) Identification of a sodium chloride-regulated promoter in *Lactococcus lactis* by single-copy chromosomal fusion with a reporter gene. Mol Gen Genet 257:681–685

Sanders JW, Venema G, Kok J (1999) Environmental stress responses in *Lactococcus lactis*. FEMS Microbiol Rev 23:483–501

Schulz A, Schumann W (1996) hrcA, the first gene of the *Bacillus subtilis* dnaK operon encodes a negative regulator of class I heat shock genes. J Bacteriol 178:1088–1093

Siegumfeldt H, Rechinger KB, Jakobsen M (2000) Dynamic changes of intracellular pH in individual lactic acid bacterium cells in response to a rapid drop in extracellular pH. Appl Environ Microbiol 66:2330–2335

Storz G, Imlayt JA (1999) Oxidative stress. Curr Opin Microbiol 2:188–194

Stuart MR, Chou LS, Weimer BC (1999) Influence of carbohydrate starvation and arginine on culturability and amino acid utilization of *Lactococcus lactis* subsp. *lactis*. Appl Environ Microbiol 65:665–673

Thomas TD, Ellwood DC, Longyear VMC (1979) Change from homo-to heterolactic fermentation by *Streptococcus lactis* resulting from glucose limitation in anaerobic chemostat cultures. J Bacteriol 138:109–117

van de Guchte M, Serror P, Chervaux C, Smokvina T, Ehrlich SD, Maguin E (2002) Stress responses in lactic acid bacteria. Antonie van Leeuwenhoek 82:187–216

van der Heide T, Poolman B (2000) Glycine betaine transport in *Lactococcus lactis* is osmotically regulated at the level of expression and translocation activity. J Bacterio 182(1):203–206

Van Veen HW, Bolhuis H, Putman M, Konings WN (1996) Multidrug resistance in prokaryotes: Molecular mechanisms of drug efflux. Handbook Biological Phys 2:165–185

Whitaker RD, Batt CA (1991) Characterization of the heat shock response in *Lactococcus lactis* subsp. *lactis*. Appl Environ Microbiol 57:1408–1412

Wouters JA, Jeynov B, Rombouts FM, de Vos WM, Kuipers OP, Abee T (1999) Analysis of the role of 7 kDa cold-shock proteins of *Lactococcus lactis* MG1363 in cryoprotection. Microbiology 145:3185–3194

Wouters JA, Kamphuis HH, Hugenholtz J, Kuipers OP, de Vos WM, Abee T (2000) Changes in glycolytic activity of *Lactococcus lactis* induced by low temperature. Appl Environ Microbiol 66:3686–3691

Xie Y, Chou SL, Cutler A, Weimer B (2004) DNA macroarray profiling of *Lactococcus lactis* subsp. *lactis* IL1403 gene expression during environmental stresses. Utah Agricultural Experiment Station contribution no. 7539. Appl Environ Microbiol 70:6738–6747

Yamada T, Carlsson J (1975) Regulation of lactate dehydrogenase and change of fermentation products in *Streptococci*. J Bacteriol 124:55–61

Yokota A, Veenstra M, Kurdi P, van Veen HW, Konings WN (2000) Cholate resistance in *Lactococcus lactis* is mediated by an ATP-dependent multispecific organic anion transporter. J Bacteriol 182:5196–5201

Yuan G, Wong SL (1995) Isolation and characterization of *Bacillus subtilis groE* regulatory mutants: evidence for *orf39* in the *dnaK* operon as a repressor gene in regulating the expression of both *groE* and *dnaK*. J Bacteriol 177:6462–6468

Chapter 11
Stress Responses of Lactobacilli

Maria De Angelis and Marco Gobbetti

11.1 Introduction

The genus *Lactobacillus* belongs to the filum *Firmicutes*, class *Bacilli*, order II *Lactobacillales,* and family *Lactobacillaceae*. *Lactobacillus* is the largest genus within the group of lactic acid bacteria. As of this writing, in July 2010, it contains 166 species, which are mainly used in the manufacture of fermented dairy, sourdough, meat, and vegetable foods or used as probiotics. The general utility of the *Lactobacillus* species is related to their generally recognized as safe (GRAS) status and will be dependent on the availability of cost-effective methods for the production and delivery of viable cultures (Walker et al. 1999).

The industrial applications all imply that lactobacilli are exposed to various environmental stress conditions, such as extremes in temperature, pH, osmotic pressure, oxygen, and starvation, which may affect the physiological status and properties of the cells. Knowledge about the stress response of *Lactobacillus* may permit (1) tools to be developed for screening tolerant or sensitive strains, (2) an enhanced use in food processes and for medical purposes through the optimization of growth, acidification, proteolysis, bacteriophage resistance, bacteriocin synthesis, and probiotic effects, (3) an enhanced growth and/or survival by appropriate preservation methods or by the use of genetic engineering to build new food-grade starters, and (4) the fitness and level of adaptation of a culture to be evaluated. This chapter describes the physiological and molecular mechanisms of environmental stress responses of lactobacilli.

M. Gobbetti (✉)
Department of Biologia e Chimica Agro-Forestale ed Ambientale, University of Bari, Via G. Amendola 165/a, 70125 Bari, Italy
e-mail: gobbetti@agr.uniba.it

E. Tsakalidou and K. Papadimitriou (eds.), *Stress Responses of Lactic Acid Bacteria*, Food Microbiology and Food Safety, DOI 10.1007/978-0-387-92771-8_11,

11.2 Overall Adaptation

The ability of the *Lactobacillus* species to adapt to different environments is variable. Several species are isolated from several habitats, but some are restricted to specific niches (Table 11.1). This is the case of *Lactobacilllus sanfranciscensis,* which is only found in sourdoughs (Gobbetti 1998). On the contrary, *Lactobacillus casei* and *Lactobacillus plantarum* are more versatile species that are isolated from fermented foods and are natural inhabitants of the human gastrointestinal tract (Cai et al. 2009; Sturme et al. 2007; Di Cagno et al. 2009). Comparative genome analysis of the *Lactobacillus* species indicated that the combination of gene gain and gene loss occurred during adaptation (Cai et al. 2009; Goh and Klaenhammer 2009). The adaptation to dairy niches was associated with the trend toward a metabolic simplification. It resulted in the loss of genes involved in the metabolism of carbohydrates and the biosynthesis of amino acids and cofactors as well as in the gain of genes for peptide transport and hydrolysis (van de Guchte et al. 2006; Callanan et al. 2008). Species (e.g., *Lactobacillus gasseri* and *Lactobacillus johnsonii*) usually found in the gastrointestinal tract showed genetic features that contribute to gastric survival and promote interactions with intestinal mucosa. *L. casei* and *L. plantarum* have the largest chromosome size (2.9 and 3.3 Mb, respectively) (Table 11.1) and contain a large number of regulatory and transport functions (Kleerebezem et al. 2003; Cai et al. 2009). Lifestyle adaptation regions were identified in the genomes of *L. casei* and *L. plantarum*. This suggested their recent acquisition via horizontal gene transfer and the loss of dispensable ancestral functions, probably as the response to natural environmental conditions (Goh and Klaenhammer 2009; Cai et al. 2009). *L. plantarum* encodes a high number of regulatory genes (262 genes, or 8.5% of total proteins), which is comparable to those of some pathogenic microorganisms (Kleerebezem et al. 2003). Three of these genes encode for the primary sigma factor RpoD and for the alternative sigma factors RpoN and SigH. The regulation of transcription in lactobacilli is usually mediated by two-component regulatory systems (TCSs). TCSs consist of a membrane-located histidine protein kinase (HPK), which monitors one or more environmental factors, and a cytoplasmic response regulator (RR), which modulates the expression of specific genes (Fig. 11.1). Fifteen and 13 TCSs were identified in the genome of *L. casei* and *L. plantarum*, respectively (Goh and Klaenhammer 2009; Kleerebezem et al. 2003). TCSs monitor and respond to environmental changes, such as temperature, acids, bile, osmolarity, and nutrient availability (Altermann et al. 2005; Azcarate-Peril et al. 2005; Pfeiler et al. 2007), and to specific secreted signaling molecules involved in quorum sensing (QS) (Sturme et al. 2002, 2005). Nevertheless, transcriptional regulatory systems may differ among lactobacilli. *Lactobacillus delbrueckii* subsp. *bulgaricus* has only about one fifth of the putative transcriptional regulatory genes of *L. plantarum*. This dairy species has two putative extra-cytoplasm function (ECF)-type sigma factors, a putative sigma factor of unknown type and an antisigma factor.

Table 11.1 Main source of isolation/habitat and genome sequences of the most frequently isolated species belonging to the genus *Lactobacillus*

Source of isolation/habitat	Species	Strain with identified genome	Genome size (Mb)	GenBank accession number
Human gastrointestinal tract	*L. gasseri*	ATCC33323	1.89	NC_008530
	L. johnsonii	NCC533	1.99	NC_005362
	L. plantarum	WCFS1	3.31	NC_004567
	L. reuteri	DSM20016	1.99	NC_009513;
		JMC1112	2.04	NC_010609
	L. salivarius	UCC118	1.83	NC_007929
	L. casei, L. crispatus, L. rhamnosus, L. ruminis	NA		
Pig gastrointestinal tract	*L. amylovorus, L. plantarum; L. reuteri, L. salivarius*	NA		
Bovine gastrointestinal tract	*L. johnsonii, L. reuteri, L. ruminis*	NA		
Vegetables	*L. plantarum, L. brevis, L. fermentum*	NA		
Compost	*L. curvatus, L. coryniformis*	NA		
Dirty water	*L. fermentum, L. reuteri, L. brevis*	NA		
Silage	*L. plantarum, L. casei, L. brevis, L. buchneri*	NA		
Sauerkraut	*L. sakei, L. curvatus, L. plantarum, L. brevis*	NA		
Table olives	*L. plantarum, L. brevis*	NA		
Leavened baked goods	*L. brevis*	ATCC367	2.34	NC_008497
	L. sanfranciscensis, L. pontis, L. plantarum	NA		
Cheeses	*L. casei*	ATCC334	2.89	NC_008526
	L. helveticus	DPC4571	2.08	NC_010080
	L. delbrueckii subsp. *bulgaricus*	ATCC11842	1.86	NC_008054
		ATCC BAA-365	1.85	NC_008529

(continued)

Table 11.1 (continued)

Source of isolation/habitat	Species	Strain with identified genome	Genome size (Mb)	GenBank accession number
	L. delbrueckii subsp. *lactis*, *L. plantarum*	NA		
Fermented milk	*L. delbrueckii* subsp. *bulgaricus*, *L. acidophilus*, *L. kefir*, *L. kefiranofaciens*	NA		
Sausages	*L. sakei* subsp. *sakei*	23 K	1.88	NC_007576
	L. plantarum, *L. curvatus*	NA		
Wine	*L. plantarum*, *L. brevis*	NA		

NA, not available; L, *Lactobacillus*

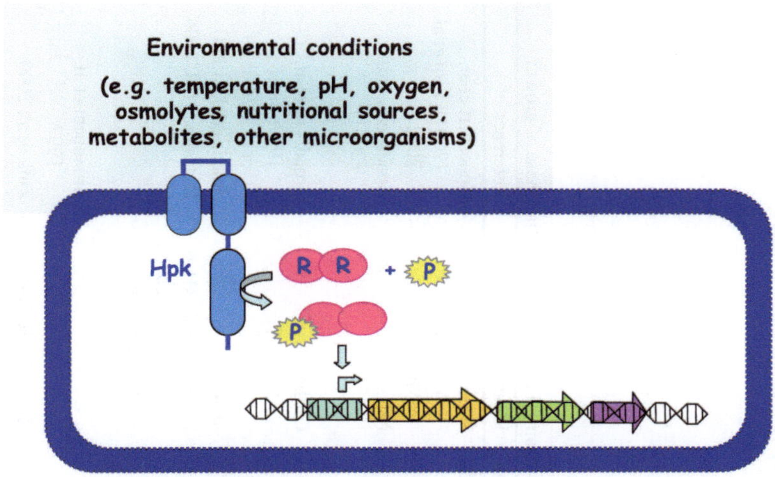

Fig. 11.1 Schematic representation of the general signal transduction mechanism based on the two-component regulatory systems (TCSs). Upon a change in the external conditions, a membrane-located HPK activates the cytoplasmic response regulator (RR) via transfer of a phosphoryl group (P). Activated RR modulates the expression of specific genes

11.3 Heat Stress Response

The mechanisms of heat resistance in lactobacilli are based upon the induction of a specific set of proteins (e.g., heat-shock proteins, HSPs; membrane serine proteinase of the HtrA/DegP family; Clp multimeric complex; and small heat-shock proteins, sHSPs). The induction of these proteins is found after exposure to sublethal heat stress (specific response) or after exposure to other types of environmental stress (generic response).

The level of heat resistance and the type of induced proteins vary within the *Lactobacillus* species (Table 11.2). Exponential-phase cells of *L. delbrueckii* subsp. *bulgaricus* became heat-sensitive at a temperature challenge of 10 min at 65°C, and the cell recovery was 0.015–0.0001% (Gouesbert et al. 2001). When exponential-phase cells were submitted to 50°C for 30 min before the lethal temperature challenge, the viability increased from 10- to 1,000-fold depending on the strain. The induction of heat tolerance after exposure to a moderate heat shock was also found for *Lactobacillus helveticus* LH212 and CNRZ32 (Broadbent et al. 1997), *Lactobacillus acidophilus* NCFM, *L. casei* LC301, *Lactobacillus paracasei* NFBC338, *L. plantarum* DPC2739 (De Angelis et al. 2004), and *Lactobacillus collinoides* (Laplace et al. 1999) (Table 11.2). A small heat shock, such as 10°C above the normal growth temperature, is usually followed by the transient induction of HSPs in *L. helveticus, L. casei, L. acidophilus, L. johnsonii* (Broadbent et al. 1997; Zink et al. 2000; Smeds et al. 1998; Di Cagno et al. 2006), *L. plantarum* (De Angelis et al. 2004), *L. collinoides* (Laplace et al. 1999), *Lactobacillus sakei* (Schmidt et al.

Table 11.2 Effect of heat stress on survival, heat tolerance, and total protein synthesis of *Lactobacillus* strains during the exponential phase of growth

Strain (*Lactobacillus*)	Stress condition	Adaptation condition	Tolerance factor[a]	Induced protein	Identified gene/protein	Reference
L. delbrueckii subsp. *bulgaricus* 546	65°C, 10 min	50°C, 30 min	1,000	20 (I.F.=2–5.5)	GroEL, DnaK, PtsI, Ywsc (ClpL, ClpX, MrEB, GAP)	Gouesbert et al. (2001); Lim et al. 2002
L. delbrueckii subsp. *bulgaricus* ATCC11842	65°C, 10 min	50°C, 30 min	10	nd	Nd	Gouesbert et al. (2001)
L. helveticus LH212	63°C, 20 min	52°C, 20 min	11	18	GroES, GroEL	Broadbent et al. 1997; Smeds et al. (1998)
L. helveticus CNRZ32	nd	52°C, 10–60 min	nd	nd	*htrA*	Smeds et al. (1998)
L. helveticus PR4	nd	Gradient of temperature 55–20°C	1,000	48	DnaK, GroEL, sHsp, glycolysis-related and regulatory proteins	Di Cagno et al. 2006
L. acidophilus NCFM	63°C, 20 min	50°C, 20 min	27	24		Broadbent et al. 1997
L. acidophilus		45–55°C, 60 min			HrcA, GrpE, DnaK, DnaJ	Zink et al. 2000
L. acidophilus LA1-1	60°C, 30 min	53°C, 30 min	166	nd	Nd	Kim et al. 2001

L. sakei LTH681	nd	42°C, 5–15 min	10	nd	HrcA, GrpE, DnaK, DnaJ, ClpE	Schmidt et al. (1999); Hertel and Hammes (1999); Stentz et al. (2000)
L. johnsonii VPI11088	nd	55°C, 30–60 min	20	nd	GroES, GroEL,	Walker et al. (1999)
L. johnsonii	nd	45–55°C, 60 min	10	nd	HrcA, GrpE, DnaK, DnaJ	Zink et al. (2000)
L. casei LC301	54°C, 20 min	42°C, 20 min	5	15	Nd	Broadbent et al. (1997)
L. paracasei NFBC338	60°C, 10 min	52°C, 15 min	300–700	nd	Nd	Desmond et al. (2001)

Adapted from (De Angelis and Gobbetti 2004)
[a]Tolerance factor is calculated as the ratio: survival of adapted cells (%)/survival of control cells (%)

1999), *L. paracasei* (Desmond et al. 2001), *L. delbrueckii* subsp. *bulgaricus* (Gouesbert et al. 2001), and *L. gasseri* (Suokko et al. 2008). Two-dimensional gel electrophoresis (2-DE) analyses revealed variable numbers of significantly induced proteins: 18 and 48 in *L. helveticus* LH212 and PR4, respectively (Broadbent et al. 1997), 24 in *L. acidophilus* NCFM, 15 in *L. casei* LC301, and 36 in *L. collinoides* (Laplace et al. 1999) (Table 11.2). When mid-exponential- or stationary-phase cells of *L. plantarum* DPC2739 were adapted to 42°C for 60 min, the heat resistance at 72°C for 1.5 min increased by about 3 and 2 log cycles, respectively. The 2-DE of proteins expressed by control and heat-adapted cells revealed changes in the levels of expression of 31 and 18 proteins in mid-exponential- and stationary-phase cells, respectively (De Angelis et al. 2004).

The two major groups of HSPs consisted of 70- (DnaK) and 60- (GroEL) kDa families that function, with accessory proteins, as chaperone machines. The components of the DnaK chaperone machine typically consist of DnaK, DnaJ, and GrpE, while that of GroEL is composed of GroEL and GroES (Table 11.2). In situ reconstitution experiments showed that DnaK/DnaJ and GroES/GroEL chaperones can interact successively to bind, refold, and release a chemically denaturated protein, giving evidence that the two machines execute sequential stages of the same chaperone process. The organization of the *groE* operon is highly conserved. It was sequenced from *L. johnsonii* (Walker et al. 1999), *L. acidophilus* (Zink et al. 2000), and *L. helveticus* (Broadbent et al. 1998). The maximum *groESL* transcription activity of *L. johnsonii* VPI11088 was found after a shift at 55°C for 15–30 min. The organization of the *dnaK* operon changed during evolution (Segal and Ron 1996). The organization *hrcA-grpE-dnaK-dnaJ* is the common structure found in *L. sakei* LTH681 (Schmidt et al. 1999), *L. johnsonii,* and *L. acidophilus* (Zink et al. 2000). The *L. plantarum dnaK* and *groESL* operons are characterized by the presence of the *cis* acting sequence controlled inverted repeat of chaperone expression (CIRCE) in the promoter region, suggesting a negative regulation by the HrcA/CIRCE system, which is the common type of control among the class I heat-shock operons of Gram-positive bacteria. An additional system of regulation is based on the positive control exerted by the CcpA protein, which interacts with catabolite responsive element (*cre*) sequences of the regulatory region of the *dnaK* and *groESL* operons (Castaldo et al. 2006). The analysis of the transcription start site revealed that the *dnaK* operon of *L. sakei* is preceded by an α^{A}-type promoter, and this site varied according to the stress conditions (Schmidt et al. 1999). The analysis of the mRNA of *L. sakei*, *L. johnsonii,* and *L. acidophilus* revealed that the *dnaK* operon is regulated at the transcriptional level and that transcription is induced by heat as well as salt and ethanol stresses (Zink et al. 2000; Schmidt et al. 1999). It was shown that the HrcA protein of *L. sakei* serves as a negative regulator repressing the expression of the *dnaK* operon under nonheat shock through the interaction with the *cis*-element CIRCE (Hecker et al. 1996).

The gene *htrA*, coding for a putative membrane serine proteinase of the HtrA/DegP family, could also be involved in heat-stress response (Bass et al. 1996; Pallen and Wren 1997). The expression of *htrA* was induced eightfold at the transcriptional level in *L. helveticus* CNRZ32 as the response to exposure to 4% NaCl

(Smeds et al. 1998) (Table 11.2). The exposure of growing cells to ethanol or heat stress (37–52°C) resulted in a lower but still considerable induction. Interestingly, the heat-shock response of *L. helveticus* cells exposed to a less severe heat shock (37–48°C) exhibited an induction pattern that was quite distinct. In particular, at 48°C the amount of *htrA* transcripts rapidly declined between 10 and 20 min after heat shock and remained at very low levels thereafter. The reporter gene *gusA* was integrated in the *Lactobacillus* chromosome downstream of the *htrA* promoter by a double-crossover event that also interrupted the wild-type gene. The expression of *gusA* in stress conditions was similar to that of *htrA* itself. The presence of an intact *htrA* gene facilitated growth under heat stress (Smeds et al. 1998).

The ClpP multimeric complex is a serine protease that degrades peptides less than seven amino acids in length. The genome sequence of probiotic *L. gasseri* ATCC33323 indicated the presence of ClpC, ClpE, ClpL, and ClpX from the Clp ATPase family of stress proteins (Table 11.2). After 30 min of heat shock, the expression of ClpC, ClpE, and ClpL proteins increased (Suokko et al. 2008). The negative ClpL mutant strain exhibited drastically decreased survival at a lethal temperature and was unable to induce thermotolerance. It has been shown that *clpP* and *clpE* are preceded by CIRCE boxes in *L. delbrueckii* subsp. *bulgaricus*, *L. acidophilus*, and *L. johnsonii* (van de Guchte et al. 2006). This suggests that the HrcA-dependent regulation of *clp* gene expression is conserved in these lactobacilli (Suokko et al. 2008). The expression of *clpE* increased in *L. sakei* (Stentz et al. 2000), *Lactobacillus brevis*, *Lactobacillus reuteri*, and *Lactobacillus rhamnosus* (Nousiainen et al. 2002) under heat stress. As shown by 2-DE, the expression of the ClpX, ClpQ, and ClpL proteins of *L. delbrueckii* subsp. *bulgaricus* increased after heat shock (Lim et al. 2002). Except for *clpX*, all the promoter regions of the *clp* genes exhibited a sequence homologous to the CtsR box (Derré et al. 1999). The CtsR operators were also found upstream of several *clp* genes of *L. sakei*. FtsH proteins having dual chaperone-protease activities (Ito and Akiyama 2005) were involved in the protection of *L. plantarum* cells against stress, mainly heat shock. The FtsH proteins are controlled by the class III gene repressor CtsR (Fiocco et al. 2009).

The sHSPs were also found within the lactobacilli genomes. The genomes of *L. acidophilus* (Altermann et al. 2005), *L. delbrueckii* subsp. *bulgaricus* (van de Guchte et al. 2006), and *L. johnsonii* (Pridmore et al. 2004) encode one sHSP, while the genome of *L. plantarum* encodes three sHSPs (Kleerebezem et al. 2003; Spano et al. 2005) (Table 11.2). This protein family has the following features: (1) a molecular mass between 12–30 kDa; (2) a conserved domain of 80–100 amino acids, located in the C-terminal region and called the alpha-crystallin domain; (3) the formation of large oligomeric complexes ranging from 150–800 kDa; and (4) adenosine triphosphate-independent chaperone activity (Narberhaus 2002). Although some sHSPs are constitutive, other sHSPs of *L. plantarum* are induced in response to various stresses, including heat shock (Fiocco et al. 2007; De Angelis et al. 2004).

Overall, the heat stress response has an effect on the technological performance of lactobacilli due to the induction of some metabolic pathways (Table 11.3).

Table 11.3 Examples of physiological and technological effects of stress exposure in lactobacilli

Stress	*Lactobacillus*	Food system	Biochemical effect	Technological effect	Reference
Heat stress: exposure to a transient temperature from 55 to 40°C	*L. helveticus*	Cheese whey for Grana cheese manufacturing	Increased protease and peptidase activities	Contribution to cheese ripening	Di Cagno et al. (2006)
Heat stress: 49°C for 30 min	*L. gasseri*	Probiotics, cheese manufacturing	Increased neutral endopeptidase (PepO)	Contribution to cheese ripening	Suokko et al. (2008)
Cold stress	*L. delbrueckii* subsp. *bulgaricus*	Cheese manufacturing	Removal of water from the cell	Improved cryotolerance	Castro et al. (1997); Pannoff et al. (2000)
Acid stress	*L. plantarum*, *L. paracasei*	Sourdough, cheese	Synthesis of GABA	Manufacture of functional fermented foods	Siragusa et al. (2007); Rizzello et al. (2008)
Acid stress	*L. sanfranciscensis*	Sourdough	Enhanced arginine deiminase pathway and synthesis of 2-acetyl-pirroline	Improvement of crust aroma	De Angelis et al. (2002); Cotter and Hill (2003)
Acid stress	*Lactobacillus*	Sourdough, cheese	Synthesis of exopolysaccharides	Improvement of texture	Tieking and Gaenzle (2005)
Osmotic stress	*Lactobacillus*	Sourdough (DY <160)	Increased synthesis of GABA	Manufacture of functional fermented foods	Rizzello et al. (2008)
Osmotic stress	*L. sanfranciscensis*	Sourdough	Overproduction of isovaleric acid, acetic acid, g-decalactones and alcohols	Improvement of flavor	Guerzoni et al. (2007); Vernocchi et al. (2008)
Oxidative stress	*L. sakei*, *L. sanfranciscensis*	Sourdough	Synthesis of H_2O_2, synthesis of (E,E)-2,4-decadienal (E)-2-nonenol	Improvement of flavor	Vermeulen et al. (2007)
Oxidative stress	*L. sanfranciscensis*	Sourdough	Increased level of thiols by GshR activity	Improvement of texture	Jänsch et al. (2007)

High hydrostatic pressure (HHP)	*L. sanfranciscensis*	Sourdough	Enhanced survival as the result of translational repair	Increased heat tolerance	Scheyhing et al. (2004); Pavlovic et al. (2008)
HHP	*L. helveticus*	Cheese whey	Activation of aminopeptidases	Contribution to cheese ripening	Miyakawa et al. (1994)
HHP	*L. delbrueckii* subsp. *bulgaricus*	Cheeses	Activation of peptidases (PepN, PepX, and PepA)	Contribution to cheese ripening	Katsaros et al. (2009)
HHP	*L. sanfranciscensis*				
Quorum sensing (QS)	*L. sanfranciscensis*	Sourdough	Modification of enzyme activities (e.g., peptidases and synthesis of organic volatile compounds)	Improvement of flavor	Di Cagno et al. (2007)
QS	*L. plantarum*	Sourdough	Modification of enzyme activities (e.g., synthesis of organic volatile compounds)	Improvement of flavor	Di Cagno et al. (2009, 2010)
QS	*L. rhamnosus* and *L. acidophilus*	Functional foods and human gastrointestinal tract	Synthesis of autoinducer-2	Improved survival under acidic conditions	Moslehi-Jenabian et al. (2009)

Adapted from Serrazanetti et al. (2009)

L. helveticus, traditionally used for making Parmigiano Reggiano and Grana Padano cheeses, is subjected to 55–45°C during growth in cheese whey under the decreasing gradient of temperature. The level of expression of a large number of HSPs (e.g., DnaK and GroEL) increased during the first 3 h of incubation from 55–40°C (Di Cagno et al. 2006). Additionally, proteinase and peptidase activities increased and glycolytic enzymes such as glyceraldheyde-3-phosphate dehydrogenase, enolase, and pyruvate kinase were overexpressed during exposure to the temperature gradient, thus affecting the performance during fermentation and cheese ripening. Probiotic *L. gasseri* subjected to heat stress (e.g., 49°C for 30 min) showed an increase in the endopeptidase PepO activity, which contributes to secondary proteolysis during cheese ripening (Suokko et al. 2008) (Table 11.3).

11.4 Cold Stress Response

To overcome the deleterious effects of cold stress and to ensure that cell activity will be resumed or maintained at low temperature, bacteria have to develop a transient adaptive cold-shock response (Horn et al. 2007). The cold stress response involves adaptation to suboptimal temperature. During adaptation, cells may modify (1) the proportion of shorter and/or unsaturated fatty acids in membrane lipids, (2) the level of compatible solutes, (3) the induction of proteins involved in heat stress response (HSPs and sHSPs), and (4) the induction of a specific set of cold-induced proteins (CIPs).

When cells cultivated at 30°C were cold-adapted at 15°C for 2 h before freezing, the cell recovery increased approximately tenfold for *L. sanfranciscensis* CB1, 25-fold for *L. plantarum* DB200 and *L. brevis* H12, and 100-fold for *L. plantarum* 20B (De Angelis and Gobbetti 2004). The acquisition of the cryotolerance during adaptation at suboptimal temperatures was also documented for *L. acidophilus* (Lorca and Font de Valdez 1999; Bâati et al. 2000) and *L. delbrueckii* subsp. *bulgaricus* (Pannoff et al. 2000). Overall, as the growth temperature decreases, the proportion of shorter and/or unsaturated fatty acids in membrane lipids increases. This allows an optimal degree of fluidity in the cytoplasmic membrane, which modulates the activity of intrinsic proteins that perform functions such as ion pumping and nutrient uptake (Russel and Fukanage 1990).

Freezing of bacterial cells in the presence of suitable cryoprotectants results in a lower loss of viability, since it counteracts the removal of water from the cell during freeze-drying (Castro et al. 1997) (Table 11.3). Compatible solutes such as glycine betaine, proline, and carnitine play a crucial role in osmoprotection and cold adaptation. The tolerance of *L. delbrueckii* subsp. *bulgaricus* CIP 101027^T to freezing at −20°C and thawing at 37°C was induced by pretreatment with various solutes. Me_2SO had a higher cryoprotective effect than glycerol, while lactose, sucrose, and trehalose had better cryoadaptive than cryoprotective properties (Pannoff et al. 2000). Adonitol, which cannot be metabolized by most of the lactobacilli, showed a protective effect during freeze-drying. This may be due to the steric conformations of its hydroxyl groups, which replace the water molecules in the protein structures.

Exposure to other environmental stresses also enhanced survival during freezing. The survival to freezing of *L. johnsonii* and *L. acidophilus* was increased by exposing the cells to 55°C for 30–45 min or to 1.25% NaCl, respectively (Walker et al. 1999). Indeed, the response to cold stress involved proteins such as HSPs (GroES, GroEL) in *L. paracasei* (Corcoran et al. 2006) and sHSPs (sHSP 18.5, sHSP 18.55, and sHSP 19.3) in *L. plantarum* (Fiocco et al. 2007). As shown for other lactobacilli, the protective effect of NaCl was increased when combined with glycine betaine (Zink et al. 2000). Acid- and bile-resistant variants of *L. acidophilus* were more stable to freezing than the parental strain (Chou and Weimer 1999).

A rapid induction of CIPs was found in lactobacilli upon cold shock. CIPs are probably synthesized in order to maintain (1) membrane fluidity by increasing the proportion of shorter and/or unsaturated fatty acids in the lipids, (2) DNA supercoiling, and (3) the efficiency of transcription and translation (Phadtare et al. 2000). Upon cold adaptation, 2-DE analyses revealed the overexpression of 14 proteins in *L. sanfranciscensis* CB1, 18 in *L. plantarum* 20B, and 13 in *L. brevis* H12 (De Angelis and Gobbetti 2004). Phosphofructokinase (PFK), glycerol-3P-dehydrogenase (GlpA), hydroperoxide resistance (Ohr), peptide methionine sulfoxide reductase (MsrA), and universal stress protein (Usp) were the CIPs induced in *L. sakei* (Marceau et al. 2002). Low-molecular-mass (ca. 7 kDa) CIPs are often distinguished from the others since they putatively belong to the cold-shock protein (Csp) family. Csps, consisting of three to nine members, were characterized in *Escherichia coli* ($CspA^E$) and *Bacillus subtilis* ($CspB^B$) (Wouters et al. 2000; Yamanaka et al. 1998). $CspA^E$ and $CspB^B$ were capable of binding to single-stranded DNA and RNA, suggesting a function of transcriptional activators and of antiterminators (Bae et al. 2000). *Csp* genes (*cspL*, *cspP*, and *cspC*) were cloned and sequenced in *L. plantarum* (Mayo et al. 1997; Derzelle et al. 2000). All three genes encode small 66-amino-acid proteins with 73–88% homology among them. The universal primers CSPU5 and CSPU3 were used to amplify DNA sequences of *L. plantarum* 20B and DB200, *L. brevis* H12, and *L. sanfranciscensis* CB1 (De Angelis and Gobbetti 2004). *Csp* genes were also detected in *L. helveticus*, *L. acidophilus*, and *L. casei* (Kim and Dunn 1997; Francis and Stewart 1997). Comparative northern blot analysis of *L. plantarum* revealed that the relative abundance of *cspC*, *cspL*, and *cspP* transcripts varied after a temperature downshift, as well as during growth under optimal conditions. Upon cold shock of exponentially growing cells, *cspL* underwent a significant and transient induction, whereas the amount of *cspC* and *cspP* mRNAs remained unchanged or increased slightly. It becomes clear that Csps are not only involved in cold adaptation but also play an important role during growth at optimal temperature. The levels of *cspC*, *cspL*, and *cspP* transcripts in *L. plantarum* markedly fluctuated during growth at 28°C, indicating that specific regulation mechanisms operate toward the three genes to direct their expression at different times of the life cycle (Derzelle et al. 2000).

11.5 Acid-Stress Response

Acid-stress response in lactobacilli is a process that mobilizes a large spectrum of different cellular functions. Several mechanisms regulate the homeostasis of the intracellular pH (pH_i) and the proton-translocating in lactobacilli (Fig. 11.2). They include the F_0F_1-ATPase proton pumps, amino acid decarboxylation/catabolism, the expression of general stress proteins (GSPs) and chaperones that repair or degrade damaged DNA and proteins, the synthesis of alkaline compounds, and the modification of cell membrane composition (Cotter and Hill 2003).

F_0F_1-ATPase is the most important for fermentative bacteria (Hutkins and Nannen 1993) (Fig. 11.2). This enzyme was characterized from *L. casei* and *L. plantarum*, and its activity was optimal at values of pH 5.0–5.5 (Bender and Marquis 1987; Hong et al. 1999; Nannen and Hutkins 1991) lower than those (pH 7.0–7.5) found for *Streptococcus thermophilus* and *Lactococcus lactis* subsp. *lactis* (Nannen and Hutkins 1991). The overall proton permeability of the plasma membrane also contributes to the regulation of the pH_i. The minimal membrane permeability of *L. casei* and *L. plantarum* was recorded at pH 4.0 (Bender and Marquis 1987; Hong et al. 1999), while that in the acid-sensitive organism *Acinomyces viscosus* was found at pH 6.0 (Bender and Marquis 1987). This difference could explain the rapid decrease of pH_i in lactobacilli since they may not actively regulate the pH_i until the extracellular pH reaches very low values (Siegumfeldt et al. 2000).

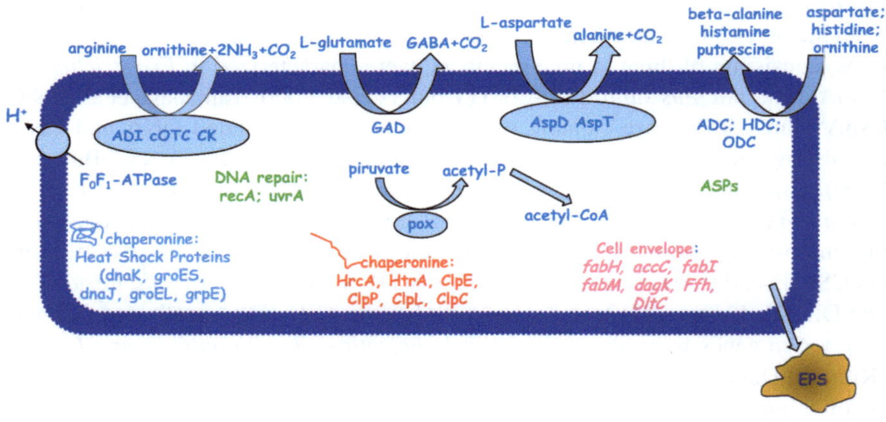

Fig. 11.2 Schematic representation of the general acidic stress response in *Lactobacillus* spp. F_0F_1-ATPase, proton pump; *ADI* arginine deiminase; *cOTC* cOTC; *CK* carbamate kinase; *GAD* glutamate decarboxylase; *AspD* L-aspartate-β-decarboxylase; *AspT* aspartate:alanine antiporter; *ADC* arginine decarboxylase; *HDC* histidine decarboxylase; *ODC* ornithine decarboxylase; *ASPs* acid shock proteins; GroES GroEL, GrpE, DnaK, and DnaJ, heat shock proteins; HrcA, repressor protein; HtrA, periplasmic protease/chaperone; ClpE, ClpP, ClpC, and ClpL, chaperonine; *Pox* pyruvate oxidase; Ffh, chaperone; FabH, β-ketoacyl-acyl carrier protein synthase III; FabI, enoyl-acyl carrier protein (ACP) reductase; FabM, poly(A)-binding protein (PABP); AccC, biotin carboxyl carrier protein; *DagK* diacylglycerol kinase; *DltC* D-alanyl carrier protein; RecA, DNA-dependent ATPase; and UvrA, subunit of the damage-specific UvrABC endonuclease

Another effective mechanism for resistance to acid stress is the glutamate decarboxylase (GAD) [EC 4.1.1.15] activity (Fig. 11.2). The expression of GAD genes is assumed to control the acidification of the cytosolic environment by decarboxylating an acid substrate (glutamate) into a neutral compound (γ-aminobutirric acid, GABA) via the incorporation of H^+. GABA would then be exported into the extracellular environment, thereby contributing to alkalinization [for a review, see Cotter and Hill (2003)]. GABA-producing strains of *L. paracasei, L. delbrueckii* subsp. *bulgaricus,* and *L. plantarum* survived and synthesized GABA under simulated gastrointestinal conditions (Siragusa et al. 2007). When used for cheese making or sourdough fermentation, the above GABA-producing strains enhanced the functional value of fermented foods (Table 11.3). The arginine deiminase (ADI) pathway is another acid stress-response mechanism (Fig. 11.2). The ADI pathway is composed of three enzymes: ADI (EC 3.5.3.6), catabolic ornithine transcarbamoylase (cOTC, EC 2.1.3.3), and carbamate kinase (CK, EC 2.7.2.2). A fourth membrane transport protein, which catalyzes an electroneutral exchange between arginine and ornithine, is also fundamental for this pathway. The increase of the acid resistance in lactobacilli may be due to the restoration of the optimal pH_i through arginine catabolism and NH_3 production (Sanders et al. 1999; Marquis et al. 1987). In addition, the extra energy (ATP) produced via the ADI pathway enables the extrusion of cytoplasmic protons by the F_0F_1-ATPase (Sanders et al. 1999) and may help cells to survive longer after depletion of the primary energy source (Stuart et al. 1999; Arena et al. 1999). ADI, cOTC, and CK were purified to homogeneity in *L. sanfranciscensis* CB1 (De Angelis et al. 2002). The presence of a carbon source such as glucose and/or maltose, low arginine supply (6 mM), low oxygen concentration, and cell adaptation to arginine are all conditions leading to the expression of the ADI pathway in *L. sanfranciscensis* during sourdough fermentation. This enhanced cell growth and survival under acid stress conditions and favored the synthesis of 2-acetyl-pirroline from ornithine, which improved the flavor of the bread crust (Table 11.3). *L. sakei* has developed regulatory mechanisms, including the ADI pathway, to gain competitive advantages in its natural meat environment (Champomier Vergès et al. 1999). Genetic analysis of *L. sakei* revealed the presence of five genes (*arcABCTD*) encoding for the four components of the ADI pathway and a putative transaminase gene (*arcT*) (Zúñiga et al. 1998). Aspartate-alanine antiporter and histidine decarboxylation, combined with electrogenic histidine-histamine antiport, stimulated the formation of ATP and the generation of proton motive force in *Lactobacillus* subsp. M3 (Abe et al. 1996) and *Lactobacillus buchneri* (Molenaar et al. 1993) (Fig. 11.2). The genome of *L. acidophilus* encodes several amino acid decarboxylases, a cation transport ATPase, and the chaperone Ffh that are likely involved in the regulation of pH_i (Altermann et al. 2005). Four genes that encode homologs involved in amino acid decarboxylation (amino acid antiporter and permease, ornithine decarboxylase, and transcriptional regulator) were associated with the acid tolerance of *L. acidophilus* (Azcarate-Peril et al. 2004).

The survival under acidic conditions is positively affected by an adaptation to low pH, a mechanism known as *acid-tolerance response* (*ATR*) (Foster and Hall 1991). Adapted cells (30 min at pH 4.75) of *L. delbrueckii* subsp. *bulgaricus* were approximately 250 times more tolerant to lethal acid stress (30 min at pH 3.6) than nonadapted ones. The incubation of *L. collinoides* at pH 3.5 for 30 min resulted in 0.015% survival. Cells adapted at pH 5.0 for 90 min increased the tolerance to the acid challenge 1,600-fold compared to nonadapted cells. Acid- and bile-resistant variants of *L. acidophilus* capable of growing at pH 3.5 with 0.3% bile were isolated by using natural selection techniques after the sequential exposure to hydrochloric acid (Chou and Weimer 1999). Two constitutively acid-tolerant mutants of *L. sanfranciscensis* CB1 were isolated following prolonged exposure (24 h) to the challenge pH 3.2. ATR seemed to be linked to a modification of the protein synthesis by the induction of acid-shock proteins (ASPs). The inhibition of ASP neosynthesis by chloramphenicol reversed ATR in *L. sanfranciscensis* (De Angelis et al. 2001). The blocking of protein synthesis reduced the ATR of *L. acidophilus* during adaptation at pH 5.0, but had no effect on ATR at pH 4.2, suggesting that both the inducible preexisting system and the mechanism dependent on the de novo protein synthesis coexist, which together afford the maximal protection against extreme acid stress (Lorca et al. 1998). Studies on the acid-adaptive responses of lactobacilli by 2-DE revealed the induction of a large number of ASPs: Twenty-one in *L. collinoides* (Laplace et al. 1999), 15 in *L. sanfranciscensis* (De Angelis et al. 2001), 30 in *L. delbrueckii* subsp. *bulgaricus* (Lim et al. 2000), and 40 in *L. reuteri* (Lee et al. 2008). ASPs induced in *L. reuteri* were distributed into six major classes: (1) transport and binding proteins; (2) transcription–translation; (3) nucleotide metabolism and amino acid biosynthesis; (4) carbon energy metabolism; (5) pH homeostasis and stress; and (6) unassigned. Some HSPs (e.g., GroES, GroEL, DnaK, and DnaJ) were oversynthesised in *L. delbrueckii* subsp. *bulgaricus* during acid adaptation, while the expression of GroES, DnaK, and DnaJ remained unaltered during the acid adaptation of *L. sanfranciscensis* with the induction of GrpE. 2-DE analyses of the protein expression in nonadapted, acid-adapted, and acid-tolerant mutant cells of *L. sanfranciscensis* showed changes in the levels of 63 proteins. The acid-adapted and acid-tolerant mutant showed similarities: eight proteins that decreased, as well as 8 or 11 proteins (depending on the mutant) that increased their level of expression compared to nonadapted cells. A comparison between the two mutants revealed that of the total of 27 oversynthesized proteins, 21 had the same level of expression (De Angelis et al. 2001). This suggests that most of the overexpressed proteins were specifically involved in the ATR. The same was found for heat-tolerant variants of *L. delbrueckii* subsp. *bulgaricus* (Gouesbert et al. 2001). The mutant strains of *L. sanfranciscensis* showed other different phenotypic features compared to the parental strain: (1) increased aminopeptidase activity under acid conditions; (2) better growth at 10°C and in the presence of 5% NaCl; and (3) an increased acidification rate during sourdough fermentation under acidic conditions (De Angelis et al. 2001) (Table 11.3). Proteins involved in the pyruvate and acetyl-CoA metabolism are also affected during acid adaptation (Fig. 11.2). Under acid-stress conditions, the transcription of *pox*, encoding pyruvate oxidase, which is responsible for the conversion of pyruvate to acetyl-P, increased, while the transcription of *ldhA*, encoding D-lactate dehydrogenase, decreased (Pieterse et al. 2005). Regarding damage to DNA due to acid stress, bacteria have developed different ways to defend their genome. One of the most important repair systems is the nucleotide excision repair (NER). NER is carried out by Uvr endonucleases and repairs DNA by the removal of a 12–13-base-long oligonucleotide containing the lesion (Fig. 11.2). UvrA, UvrB, and UvrC proteins recognize and cleave damaged DNA in a multistep reaction and are able to repair minor single-base modifications, bulky nucleotide adducts, noncovalent modifications, and intra- and interstrand cross-links (Cappa et al. 2005).

The acid-stress response of *L. delbrueckii* subsp. *bulgaricus* caused the induction of genes (e.g., *fabH*, *accC*, *fabI*) involved in the biosynthesis of fatty acids of the cell membrane (Streit et al. 2007, 2008) (Fig. 11.2). Lactic acid specifically induced the expression of several cell surface proteins of unidentified functions that might be responsible for altering the *ffh* of *L. acidophilus* (Azcarate-Peril et al. 2004). Acid stress induced the synthesis of exopolysaccharide (EPS) in sourdough lactobacilli (Gaenzle and Schwab 2009). Overall, EPS increased the survival of microbial cells to acid stress and markedly contributes to the texture, rheology, mouth feel, taste perception, and stability of several fermented foods (Ruas-Madiedo et al. 2000; Tieking and Gaenzle 2005) (Table 11.3).

11.6 Osmotic Stress Response

To retain water within the cell and thus to maintain turgor pressure, lactobacilli have systems for accumulating specific solutes that do not interfere with the cell physiology (Jordan et al. 2008; van de Guchte et al. 2002).

In silico genome analysis indicated that *L. plantarum* has at least three systems for the accumulation and biosynthesis of quaternary ammonium compounds such as glycine betaine/carnitine/choline (Kleerebezem et al. 2003). In the presence of these solutes, the inhibition of growth occurred at much higher stress conditions (Glaasker et al. 1996, 1998). Although quaternary ammonium compounds are preferred by *L. plantarum*, proline and glutamate are also used to increase the internal osmolarity. Gene inactivation studies showed that a single hyperosmotic stress-activated transport system protects *L. plantarum* against low-water activity. The quaternary ammonium compound transport (QacT) system accepts various quaternary ammonium compounds with high affinity and proline with low affinity (Glaasker et al. 1998). Activation by hyperosmotic stress mainly takes place at the enzyme level and rarely at the level of gene expression. Since the rate of accumulation of glycine betaine by *L. plantarum* decreased as its internal concentration increased, and since this apparent inhibition of the uptake by *trans*-substrate is diminished upon osmotic upshock, the activation mechanism most likely involves the regulation of the transporter through conformational changes at an internal binding site (Glaasker et al. 1998). The genome of *L. sakei* encodes three putative ABC transporters and a sodium symporter that have predicted specificities for glycine betaine and carnitine, which contribute to survival under meat curing (Chaillou et al. 2005).

Compatible solutes are rapidly released from the cells upon a hypoosmotic shock. When *L. plantarum* is subjected to an osmotic downshock, a rapid efflux of glycine betaine, proline, and some glutamate occurs, whereas the pools of the other amino acids remain unaffected (Glaasker et al. 1996). Although the molecular nature of these efflux activities is still not precisely known, the systems exhibit properties that mimic mechanosensitive channels (Martinac et al. 1990; Sukharev et al. 1997). Some features that discriminate the systems from ordinary carrier proteins, driven either by ATP or by electrochemical ion gradients, are the following: (1) The efflux is extremely fast and affected by an osmotic downshock as well as by amphipathic compounds that insert into the membrane; (2) the efflux is independent of the metabolic energy; (3) the efflux is unaffected by substrates at the *trans* site of the membrane; and (4) in many cases, the efflux is inhibited by gadolinium ions (Gd^{3+}), an unspecific channel blocker. The efflux of glycine betaine and proline by *L. plantarum* is characterized by two kinetic components, the fast one with a $t_{0.5}$ <1 s, and the slow one with a $t_{0.5}$ of 4–5 min (Glaasker et al. 1998). The component with the slow kinetic is affected by the metabolic state of the cell and may represent a specific efflux system. The kinetic of the fast efflux component is too rapid to be analyzed accurately, but an estimation of the turnover of the putative channel in *L. plantarum* indicates that it releases glycine betaine at a rate of more than 10^5–10^6 s^{-1}, which is orders of magnitude faster than the turnover number of known carrier proteins (10^{-1} to at most 10^3 s^{-1}).

Hyperosmotic conditions activated responses to other stresses. The addition of NaCl in the presence of glycine betaine significantly improved the survival of *L. acidophilus*, *L. johnsonii,* and other lactobacilli to freeze-drying (Zink et al. 2000). The addition of NaCl increased the concentration of *dnaK*-specific mRNA in *L. acidophilus* and *L. johnsonii* (Zink et al. 2000). Hyperosmotic conditions induced the transcription of the *dnaK* and *htrA* operons in *L. sakei* LTH681 (Schmidt et al. 1999) and *L. helveticus* CNRZ32 (Smeds et al. 1998), respectively. The synthesis of GABA by *Lactobacillus* strains increased under osmotic stress during sourdough fermentation (Rizzello et al. 2008) (Table 11.3). In addition, some volatile organic compounds increased under osmotic stress (Guerzoni et al. 2007; Vernocchi et al. 2008) (Table 11.3).

11.7 Oxidative Stress Response

Lactobacilli use enzymes (e.g., NADH oxidase, NADH peroxidase, and superoxide dismutase, Sod) (Table 11.4) or nonenzymatic compounds (e.g., Mn^{2+}, ascorbate, tocopherols, and glutathione) to reduce oxygen radicals originating from the oxidative stress (Archibald and Fridovich 1981; Warriner and Morris 1995). The ability to cope with oxidative stress varied among different groups of lactobacilli depending on the levels and types of antioxidative mechanisms.

L. delbrueckii subsp. *bulgaricus* could reduce O_2 into H_2O_2 with a NADH oxidase to eliminate or lower the levels of O_2 (Table 11.4). The same was found for *L. plantarum* ATCC8014 through the oxidation of NADH or pyruvate. Three moles of NADH were consumed per mole of O_2, indicating that O_2 was reduced to H_2O

Table 11.4 Reactions involving oxygen or its toxic derivatives, which are catalyzed by enzymes of *Lactobacillus* strains

Enzyme	Reaction	Host (*Lactobacillus*)	Reference
NADH:H_2O_2 oxidase	$NADH + H^+ + O_2 \rightarrow NAD^+ + H_2O_2$	*L. plantarum*, *L. sanfranciscensis*, *L. delbrueckii* subsp. *bulgaricus*	Stolz et al. (1995); Götz et al. (1980); Marty-Teysset et al. (2000)
NADH:H_2O oxidase	$2\ NADH + 2H^+ + O_2 \rightarrow 2NAD^+ + H_2O$	*L. plantarum*, *L. casei*	Condon (1987)
NADH: peroxidase	$NADH + H^+ + H_2O_2 \rightarrow NAD^+ + 2H_2O$	*L. plantarum*, *L. sanfranciscensis*	Stolz et al. (1995); Götz et al. (1980)
Superoxide dismutase	$O_2^{\cdot-} + 2H^+ \rightarrow H_2O_2$	*L. sanfranciscensis* CB1, *L. sakei*	De Angelis and Gobbetti (1999); Amanatidou et al. (2001)
Heme-dependent catalase	$2H_2O_2 \rightarrow 2H_2O + O_2$	*L. sakei*	Knauf et al. (1992)
Thioredoxin reductase	Thioredoxin + $NADP^+ \rightarrow$ thioredoxin disulfite + NADH + H^+	*L. johnsonii*, *L. delbrueckii* subsp. *bulgaricus*	Elli et al. (2002); van de Guchte and Serron (unpublished)
Pyruvate oxidase	Pyruvate + phosphate + $O_2 \rightarrow$ acetylphosphate $+ CO_2 + H_2O_2$	*L. delbrueckii* subsp. *bulgaricus*, *L. casei*, *L. plantarum*	Götz et al. (1980)
L-Cysteine uptake system	L-Cysteine → reducing LMW sulfhydryl compound thiocysteine	*L. fermentum*	Turner et al. (1999)
Glutathione reductase	Glutathione disulfite + $NADP^+ \rightarrow$ glutathione + NADPH + H^+	*L. sanfranciscensis*	Jänsch et al. (2007)

Adapted from De Angelis and Gobbetti (2004)

via H_2O_2, involving an NADH oxidase and an NADH peroxidase. Pyruvate oxidation with O_2 leads to the formation of H_2O_2 (Götz et al. 1980). The detoxification of O_2 led to an overproduction of H_2O_2 that caused oxidative stress and triggered an early entry of the cells into the stationary phase (Marty-Teysset et al. 2000). Dialyzable, EDTA-sensitive $O_2^{\cdot-}$-scavenging activity, presumably due to Mn^{2+}, was found in *L. plantarum*, *L. casei*, *L. fermentun*, and *Lactobacillus ruminis* (Archibald and Fridovich 1981). The genome of *L. plantarum* encodes at least five putative manganese transporters, which supports previous findings that Mn^{2+} ions serve either as scavengers for oxygen radicals or as cofactors for manganese-dependent catalase (Serrano et al. 2007; Groot et al. 2005; Rochat et al. 2006). An approximately 12.5-KDa manganese-dependent SodA was identified from *L. sanfranciscensis* CB1 (De Angelis and Gobbetti 1999) (Table 11.4). The enzyme was insensitive to H_2O_2

treatment, was not induced by the presence of paraquat under aerobic conditions, and was relatively stable at pH 4.0. As determined by the percentage inhibition of NBT reduction, the oxygen supply, with or without Mn^{2+}, was related to the highest Sod activity in the late-exponential-phase cells. Agitation, aeration, and Mn^{2+} availability reduced the length of the lag phase and increased the growth rate and cell yield of *L. sanfranciscensis*. As *L. sanfranciscensis* does not exhibit catalase activity, NADH peroxidase is responsible for H_2O_2 breakdown (Stolz et al. 1995) (Table 11.4). In the absence of fermentable substrate, H_2O_2 accumulates because, unlike catalase, NADH peroxidases require continuous metabolism to produce the necessary NADH (Stolz et al. 1995). Indeed, air-grown *L. sanfranciscensis* supplied with Mn^{2+} showed a considerable accumulation of H_2O_2, which coincided with complete maltose depletion (De Angelis and Gobbetti 1999). The capacity to grow at different O_2 levels differed in *L. sakei* strains isolated from different foods. A high H_2O_2 splitting capacity and low specific rates of H_2O_2 production were shown in the O_2-insensitive *L. sakei* NCFB2813 (*L. sakei*^{ins}), which could grow at 90% O_2 (Amanatidou et al. 2001). *L. sakei* DSM6333 (*L. sakei*^{sens}), unlike *L. sakei*^{ins}, showed a low growth rate in the presence of 90% O_2 and a rapid loss in viability shortly after entry into the stationary phase. The steady-state cytosolic superoxide radical concentration in *L. sakei*^{sens} was tenfold higher than in the oxygen-insensitive strain, indicating the significance of the $O_2^{\cdot-}$-scavenging system in protecting against elevated O_2 (Table 11.4). The Sod activity was 10- to 20-fold higher in *L. sakei*^{ins} than in *L. sakei*^{sens}. Protein damage in *L. sakei*^{sens} was marked by the increased protein carbonyl content and reduced activities of the (Fe-S)-cluster-containing enzymes fumarase and fumarase reductase (Amanatidou et al. 2001). The gene *katA*, encoding the true catalase of *L. sakei* LTH677, was characterized and cloned into *L. casei* LK1 and *Lactobacillus curvatus* LTH1432 (Hertel et al. 1998). The regulation of *katA* expression was found to be the same in *L. sakei* LTH677 and in the recombinant strains. The addition of H_2O_2 to anaerobic cultures, as well as a switch to aerobic conditions, resulted in a strong increase in the KatA activity. The recombinant strain *L. curvatus* LTH4002 did not accumulate H_2O_2 under glucose-limited aerobic conditions and remained viable in the stationary phase. Under inductive conditions, the *katA*-specific mRNA and the apoenzyme were synthesized de novo. By fusing the *katA* promoter to the β-glucuronidase reporter gene, it was shown that the regulatory sequence for *katA* expression is part of the promoter (Hertel et al. 1998). The synthesis of *gusA*-specific mRNA, as well as of the GusA protein, correlated with the expression of *katA*. A small putative regulatory sequence of at least 25 bp, located upstream of the −35 site, was identified. The specific properties of the *katA* promoter activity may also be useful in food fermentation. The gene *trxB* encoding the thioredoxin reductase was identified from *L. delbrueckii* subsp. *bulgaricus* (M. van de Guchte and P. Serron, unpublished) and *L. johnsonii* (Elli et al. 2002). This gene, functioning as an accessory enzyme for the ribonuclease reductase, is responsible for an essential step in DNA synthesis: the reduction of ribonucleotides to deoxyribonucleotides. It was postulated that *trxB* plays a role in maintaining the reduced state of cytoplasmic proteins (Table 11.4). The *trxB* gene of *L. johnsonii* was expressed in a thioredoxin reductase-deficient strain of

E. coli AD494. The increased oxygen tolerance of this recombinant strains strongly supports the possible involvement of *trxB* in the protection of *Lactobacillus* cells against the accumulation of detrimental cytoplasmic disulfide bonds (Elli et al. 2002). The overexpression of thioredoxin also enhanced the resistance of *L. plantarum* to oxidative stress (Serrano et al. 2007). The overexpression of thioredoxin triggered the transcription of 16 genes that encode proteins involved in purine metabolism, protein biosynthesis, stress response, and manganese transport. The enzyme glutathione reductase (GshR) catalyzes the NADPH-dependent reduction of glutathione disulfide in *L. sanfranciscensis* (Jänsch et al. 2007) (Table 11.4). Glutathione has an important function as a redox-buffering compound in bacterial cells. Glutathione is the major nonprotein thiol compound in living cells and is involved in the resistance to osmotic stress, toxic electrophiles, and oxidative stress. Glutathione also acts as an electron donor for both the scavenging of reactive oxygen (e.g., from respiration) and metabolic reactions such as the reduction of hydroperoxides and lipid peroxides. GshR plays an essential role in cell defense against oxygen stress by maintaining a high intracellular GSH/GSSG status (Jänsch et al. 2007). In *L. sanfranciscensis,* GshR activity increased under oxidative stress conditions that favored an improved texture of baked goods. Under oxidative stress conditions, some lactobacilli increased the synthesis of volatile organic compounds, thus enhancing the flavor of baked goods (Vermeulen et al. 2007) (Table 11.3).

11.8 High-Hydrostatic-Pressure Stress Response

The primary effect of high-pressure hydrostatic treatment (HHP) involves the conformation of the ribosome, notably the 30S subunit, and therefore inhibits the binding of the aa-tRNA to ribosome, resulting in a decrease in the translational capacity.

The protein patterns of atmospherically growing *L. sanfranciscensis* were compared to that of pressure-treated cells (Drews et al. 2002). Variations in response to pressure upshift show that cells tried to counteract the decrease in the translational capacity by (1) the regulation of translational factors, (2) the regulation of genes changing translational accuracy, and (3) inducing stress proteins. The treatment of 80 MPa for 1 h showed that the response to HHP overlaps with other stress responses (Hörmann et al. 2006). Peptidases/proteases contribute to the degradation of peptides/proteins that cannot be folded by molecular chaperones. The protease ClpL of *L. sanfranciscensis* was upregulated by HHP treatment. The transcriptomic study of *L. sanfranciscensis* under HHP (45 MPa for 30 min) revealed the induction of 42 genes. The majority of these genes encode translation factors (EF-G, EF-TU), ribosomal proteins (S2, L6, L11), and genes changing translational accuracy or molecular chaperones (GroEL, ClpL) (Pavlovic et al. 2005). Several ribosomal proteins (RplJ, RpsF) were induced in *L. sakei* subjected to HHP stress. Another protein, usually induced by cold shock (NusG), was also accumulated (Jofré et al. 2007). Sublethal HHP stress also favored an increased baro- and heat tolerance (Kilimann et al. 2006a, b; Vogel et al. 2005) and caused the modification of the end products from maltose fermentation by *L. sanfranciscensis* (Korakli et al. 2002; Pavlovic et al. 2008; Hörmann et al. 2006). An increase in the activity of various peptidases was found for cheese-related lactobacilli when treated to a sublethal dose of HHP (Katsaros et al. 2009) (Table 11.3).

11.9 Starvation Response

Bacteria, including lactobacilli, adapt to nutritional limitation (e.g., carbohydrate, phosphate, and nitrogen starvation) by assuming a physiological state that is mainly characterized by the downregulation of nucleic acid and protein synthesis and the simultaneous upregulation of protein degradation and amino acid synthesis (Chatterji and Ojha 2001). The exhaustion of an essential nutrient and/or accumulation of a fermentation end product (e.g., lactic acid) limit the exponential phase of growth of a microorganism, which then enters into the stationary phase. Entry into the stationary phase is also caused by heat, cold, osmotic, and oxidative stresses. The adaptive response during growth usually involves the induction of specific groups of genes or regulons to cope with a specific stress condition, while the stress response during the stationary phase is mediated by numerous regulons that cope with numerous stress conditions. Maintaining an active metabolic state is a common aspect for lactobacilli survival during the stationary phase of growth.

The catabolism of amino acids plays an important role in the survival of *L. sakei* (Champomier Vergès et al. 1999) during the stationary phase. Two-DE analyses showed that starvation induced the synthesis of 16 proteins in *L. acidophilus*. Seven of these proteins were specifically induced by the stationary phase itself, while the synthesis of the others was attributed to the low pH (Lorca and Font de Valdez 2001). In *L. plantarum*, most of the stress-response proteins are induced after the onset of the stationary phase (Cohen et al. 2006). Numerous glycolytic enzymes were differentially regulated under lactose starvation in *L. casei* (Hussain et al. 2009). The differential expression of these glycolytic enzymes suggests a potential survival strategy under harsh growth conditions. GSPs of *E. coli* and *B. subtilis* are induced under various conditions and especially under glucose starvation. The alternative sigma factor σ^B controls the expression of the majority of GSPs, and the activation of this regulon is induced by distinct pathways that depend on the stress conditions. At least 125 genes are regulated by σ^B in *B. subtilis,* and some of the GSPs have a role in survival during the stationary phase (Guerzoni et al. 2001). No σ^B homologs were found in lactobacilli, and the regulation of starvation-induced proteins in these bacteria is not yet understood. From a technological point of view, it is well known that bacteria develop a general stress resistance after entering into the stationary phase and become more resistant to various types of stresses (Lorca and Font de Valdez 2001; van de Guchte et al. 2002). The resistance to heat, acid, and bile stresses increased in lactose-starved cells of *L. delbrueckii* subsp. *bulgaricus* cells (Chervaux et al. unpublished data; Gouesbert et al. 2001).

11.10 Quorum Sensing

Bacteria synthesize, release, detect, and respond to small signaling hormone-like molecules called *autoinducers*. When a critical threshold concentration, the *quorum*, of the signal molecule is achieved, bacteria detect its presence and initiate a signaling cascade, resulting in changes of the target gene expression (for a review, see Gobbetti et al. (2007)). One class of bacterial QS signaling molecules corresponds to autoinducer 2 (AI-2) synthesized through the activity of the LuxS enzyme. In addition, most of the Gram-positive bacteria (including lactobacilli) use autoinducing peptides (AIP, or peptide pheromones), which act as species-specific communication signals. The gene for AIP often flanks the QS-TCS gene cassette.

The synthesis of AI-2 from probiotic lactobacilli (e.g., *L. rhamnosus* and *L. acidophilus*) was induced under acidic shock, and the *luxS* gene appears to have a clear role in acidic stress response (Moslehi-Jenabian et al. 2009). The growth of sourdough *L. sanfranciscensis* CB1 in mono-culture was compared with that in co-cultures with *L. plantarum* DC400, *L. brevis* CR13, or *Lactobacillus rossiae* A7 (Di Cagno et al. 2007). Compared to mono-culture, *L. sanfranciscensis* CB1 overexpressed 48, 42, and 14 proteins when cocultured with strains DC400, CR13, and A7. Induced polypeptides, only in part common to all cocultures, were identified as stress proteins, energy metabolism-related enzymes, proline dehydrogenase, GTP-binding protein, S-adenosyl-methyltransferase, and Hpr phosphocarrier protein. By using primers designed from consensus amino acid sequences of phylogenetically related bacteria, two QS-involved genes, *luxS* and *metF*, were shown to be expressed in *L. sanfranciscensis* CB1. After the stationary phase of growth was reached, the expression of the *luxS* gene was only found in the coculture of *L. sanfranciscensis* CB1 with *L. brevis* CR13. Phenotypically, the rate of dead cells, fermentation end products, and proteolytic activities reflected the type of associations. Similar results were found by using another *L. sanfranciscensis* strain (Di Cagno et al. 2009). The growth and survival of *L. plantarum* DC400 were not affected when cocultivated with *L. sanfranciscensis* DPPMA174 or *L. rossiae* A7. Nevertheless, 2-DE analysis showed that the level of protein expression of *L. plantarum* DC400 increased under coculture conditions. Although several proteins were commonly induced in both cocultures, the highest induction was found in coculture with *L. rossiae* A7. Overexpressed proteins, related to QS and stress-response mechanisms, were identified: DnaK, GroEL, 30S ribosomal protein S1 and S6, ATP synthase subunit beta, MetK, phosphopyruvate hydratase, phosphoglycerate kinase, elongation factor Tu, putative manganese-dependent inorganic pyrophosphatase, D-lactate dehydrogenase, triosephosphate isomerase, fructose-bisphosphate aldolase, and nucleoside-diphosphate kinase. As shown by real-time PCR, the expression of the *luxS* gene of *L. plantarum* DC400 was also affected during cocultivation. According to the overexpression of *metK* and *luxS* during cocultivation, the synthesis of AI-2-like substances was also influenced by the type of cocultures (Di Cagno et al. 2009). Five peptide-based QS-TCS were predicted in the genome of *L. plantarum* WCFS1 (Kleerebezem et al. 2003). Four of these systems contained an upstream gene encoding a putative AIP (Sturme et al. 2007). One of these was identical to the *plnABCD* regulatory system of *L. plantarum* C11 and regulated the synthesis of plantaricin (e.g., plnA) in *L. plantarum* WCFS1. The synthesis of plnA by *L. plantarum* DC400 was affected by cocultivation with other lactobacilli. The highest synthesis of plnA was found when strain DC400 was cocultured with *L. sanfranciscensis* DPPMA174. The addition of plnA to the culture medium caused a decrease in cell growth and survival as well as the overexpression of several stress proteins (e.g., GroES, DnaK) in *L. sanfranciscensis* DPPMA174 (Di Cagno et al. 2010).

11.11 Concluding Remarks

A greater understanding of the molecular mechanisms of environmental stress responses may provide targets for improving the use of lactobacilli as starters and probiotics. The efficient induction of HSPs, sHSPs, or CIPs could increase heat tolerance during food processing or survival during freeze-dried starter preparation. The introduction of HSPs with plasmids into lactobacilli is especially interesting, as it may offer a convenient way to introduce HSP genes into different hosts, leading to a possible increase in specific or multiple stress tolerances in recipient strains that are of industrial importance. The addition of osmolytes to the growth medium could favor the uptake and/or could activate accumulation mechanisms prior to subjecting the cells to drying conditions. The selection and use of strains that express the ADI pathway may provide an alternative source of energy, give protection against acid stress, and influence the sensory properties of foods. The ATR is becoming fundamental in most fermented foods to ensure optimal growth and acidification rates and to maintain cell viability in probiotic strains. Defense against ROS may improve the competitiveness of lactobacilli within complex microbial communities. The understanding of extracellular signaling and the mechanisms of QS under stressful conditions, such as cocultivation with competing bacteria, may give new insights into the mechanisms of bacterial adaptation. On the other hand, stress-induced proteins are clear molecular markers for the fitness of starters, indicating that a culture is fully adapted to resist an upcoming stress condition.

References

Abe K, Hayashi H, Maloney PC (1996) Exchange of aspartate and alanine. J Biol Chem 271:3079–3084

Altermann E, Russell WM, Azcarate-Peril MA, Barrangou R, Buck BL, McAuliffe O, Souther N, Dobson A, Duong T, Callanan M, Lick S, Hamrick A, Cano R, Klaenhammer TR (2005) Complete genome sequence of the probiotic lactic acid bacterium *Lactobacillus acidophilus* NCFM. Proc Natl Acad Sci USA 102:3906–3912

Amanatidou A, Bennik MHJ, Gorris LGM, Smid EJ (2001) Superoxide dismutase plays an important role in the survival of *Lactobacillus sake* upon exposure to elevated oxygen. Arch Microbiol 176:79–88

Archibald FS, Fridovich I (1981) Manganese, superoxide dismutase, and oxygen tolerance in some lactic acid bacteria. J Bacteriol 146:928–936

Arena ME, Saguir FM, Manca de Nadra MC (1999) Arginine dihydrolase pathway in *Lactobacillus plantarum* from orange. Int J Food Microbiol 47:203–209

Azcarate-Peril MA, Altermann E, Hoover-Fitzula RL, Cano RJ, Klaenhammer TR (2004) Identification and inactivation of genetic loci involved with *Lactobacillus acidophilus* acid tolerance. Appl Environ Microbiol 70:5315–5322

Azcarate-Peril MA, McAuliffe O, Altermann E, Lick S, Russell WM, Klaenhammer TR (2005) Microarray analysis of a two-component regulatory system involved in acid resistance and proteolytic activity in *Lactobacillus acidophilus*. Appl Environ Microbiol 71:5794–5804

Bâati L, Fabre-Gea C, Auriol D, Blanc PJ (2000) Development of bioactive food packaging materials using immobilised bacteriocins Lacticin 3147 and Nisaplin®. Int J Food Microbiol 59:241–247

Bae W, Xia B, Inouye M, Severinov K (2000) *Escherichia coli* CspA-family RNA chaperones are transcription antiterminators. Proc Natl Acad Sci USA 97:7784–7789

Bass S, Gu Q, Christen A (1996) Multicopy suppressors of prc mutant *Escherichia coli* include two HtrA (DegP) protease homologs (HhoAB), DksA, and a truncated R1pA. J Bacteriol 178:1154–1161

Bender GR, Marquis, RE (1987) Membrane ATPases and acid tolerance of *Actinomyces viscosus* and *Lactobacillus casei*. Appl Environ Microbiol 53:2124–2128

Broadbent JR, Oberg JC, Wang H, Wie L (1997) Attributes of the heat shock response in three species of dairy *Lactobacillus*. Syst Appl Microbiol 20:12–19

Broadbent JR, Oberg CJ, Wie L (1998) Characterization of the *Lactobacillus helveticus* groESL operon. Res Microbiol 149:247–253

Cai H, R Thompson, Budinich MF, Broadbent JR, Steele JL (2009) Genome sequence and comparative genome analysis of *Lactobacillus casei*: insights into their niche-associated evolution. Gen Biol Evol 9:239–257

Callanan M, Kaleta P, O'Callaghan J, O'Sullivan O, Jordan K, McAuliffe O, Sangrador-Vegas A, Slattery L, Fitzgerald, Beresford T, Ross RP (2008) Genome sequence of *Lactobacillus helveticus*, an organism distinguished by selective gene loss and insertion sequence element expansion. J Bacteriol 190:727–735

Cappa F, Cattivelli D, Cocconcelli PS (2005) The *uvrA* gene is involved in oxidative and acid stress responses in *Lactobacillus helveticus* CNBL1156. Res Microbiol 156:1039–1047

Castaldo C, Siciliano RA, Muscariello L, Marasco R, Sacco M (2006) CcpA affects expression of the groESL and dnaK operons in *Lactobacillus plantarum*. Microb Cell Fact 5:35

Castro HP, Teixeira PM, Kirby R (1997) Evidence of membrane damage in *Lactobacillus bulgaricus* following freeze drying. J Appl Microbiol 82:87–94

Chaillou S, Champomier-Vergès MC, Cornet M, Crutz-Le Coq AM, Dudez AM, Martin V, Beaufils S, Darbon-Rongere E, Bossy R, Loux V, Zagorec M (2005) The complete genome sequence of the meat-borne lactic acid bacterium *Lactobacillus sakei* 23 K. Nat Biotechnol 23:1527–1533

Champomier Vergès MC, Zú iga M, Morel-Deville F, Pèrez-Martinez G, Zagorec M, Ehrlich SD (1999) Relationships between arginine degradation, pH and survival in *Lactobacillus sakei*. FEMS Microbiol Lett 180:297–304

Chatterji D, Ojha A K (2001) Revisiting the stringent response, ppGpp and starvation signaling. Curr Opinion Microbiol 4(2):160–165

Chou LS, Weimer B (1999) Isolation and characterization of acid- and bile-tolerant isolates from strains of *Lactobacillus acidophilus*. J Dairy Sci 82:23–31

Cohen DPA, Renes J, Bouwman FG, Zoetendall EG, Mariman E, de Vos WM, Vaughan EE (2006) Proteomic analysis of log to stationary growth phase *Lactobacillus plantarum* cells and a 2-DE database. Proteomics 6:6485–6493

Condon S (1987) Responses of lactic acid bacteria to oxygen. FEMS Microbiol Rev 46:269–280

Corcoran BM, Ross RP, Fitzgerald GF, Dockery P, Stanton C (2006) Enhanced survival of GroESL-overproducing *Lactobacillus paracasei* NFBC 338 under stressful conditions induced by drying. Appl Environ Microbiol 72:5104–5107

Cotter PD, Hill C (2003) Surviving the acid test: responses of Gram-positive bacteria to low pH. Microbiol Mol Biol Rev 67:429–453

De Angelis M, Gobbetti M (1999) *Lactobacillus sanfranciscensis* CB1: manganese, oxygen, superoxide dismutase and metabolism. Appl Microbiol Biotechnol 51:358–363

De Angelis M, Gobbetti M (2004) Environmental stress responses in *Lactobacillus*: a review. Proteomics 4:106–122

De Angelis M, Bini L, Pallini V, Cocconcelli PS, Gobbetti M (2001) The acid-stress response in *Lactobacillus sanfranciscensis* CB1. Microbiology 147:1863–1873

De Angelis M, Mariotti L, Rossi J, Servili M, Fox P, Rollan G, Gobbetti M (2002) Arginine catabolism by sourdough lactic acid bacteria: purification and characterization of the arginine deiminase pathway enzymes from *Lactobacillus sanfranciscensis* CB1. Appl Environ Microbiol 68:6193–6201

De Angelis M, Di Cagno R, Huet C, Crecchio C, Fox PF, Gobbetti M (2004) Heat shock response in *Lactobacillus plantarum*. Appl Environ Microbiol 70:1336–1346

Derré I, Rapoport G, Msadek T (1999) CtsR, a novel regulator of stress and heat shock response, controls clp and molecular chaperone gene expression in Gram-positive bacteria. Mol Microbiol 31:117–131

Derzelle S, Hallet B, Francis KP, Ferain T, Delcour J, Hols P (2000) Changes in cspL, cspP, and cspC mRNA abundance as a function of cold shock and growth phase in *Lactobacillus plantarum*. J Bacteriol 182:5105–5113

Desmond C, Stanton C, Fitzgerald GF, Collins K, Ross RP (2001) Environmental adaptation of probiotic lactobacilli towards improvement of performance during spray drying. Int Dairy J 11:801–808

Di Cagno R, De Angelis M, Limitone A, Fox PF, Gobbetti M (2006) Response of *Lactobacillus helveticus* PR4 to heat stress during propagation in cheese whey with a gradient of decreasing temperatures. Appl Environ Microbiol 72:4503–4514

Di Cagno R, De Angelis M, Limitone A, Minervini F, Simonetti C, Buchin S, Gobbetti M (2007) Cell-cell communication in sourdough lactic acid bacteria: a proteomic study in *Lactobacillus sanfranciscensis* CB1. Proteomics 7:2430–2446

Di Cagno R, De Angelis M, Coda R, Minervini F, Gobbetti M (2009) Molecular adaptation of sourdough *Lactobacillus plantarum* DC400 under co-cultivation with other lactobacilli. Microbiol Res 160:358 –366

Di Cagno R, De Angelis M, Calasso M, Vicentini O, Vernocchi P, Ndagijimana M, De Vincenzi M, Dessì M, Guerzoni ME, Gobbetti M (2010) Quorum sensing in sourdough *Lactobacillus plantarum* DC400: induction of plantaricin A (PlnA) under co-cultivation with other lactic acid bacteria and effect of PlnA on bacterial and Caco-2 cells. Proteomics, in press

Drews O, Weiss W, Reil G, Parlar H, Wait R, Görg A (2002) High pressure effects step-wise altered protein expression in *Lactobacillus sanfranciscensis*. Proteomics 2:765–774

Elli M, Morelli L, Zink R (2002) In: Book of Abstracts, "Seventh Symposium on Lactic Acid Bacteria—genetics, metabolism and applications," FEMS Netherlands, p. G9

Fiocco D, Capozzi V, Goffin P, Hols P, Spano G (2007) Improved adaptation to heat, cold, and solvent tolerance in *Lactobacillus plantarum*. Appl Microbiol Biotechnol 77:909–915

Fiocco D, Collins M, Muscariello L, Hols P, Kleerebezem M, Msadek T, Spano G (2009) The *Lactobacillus plantarum* ftsH gene is a novel member of the CtsR stress response regulon. J Bacteriol 191:1688–1694

Foster JW, Hall HK (1991) Inducible pH homeostasis and the acid tolerance response of *Salmonella typhimurium*. J Bacteriol 173:5129–5135

Francis KP, Stewart GSAB (1997) Detection and speciation of bacteria through PCR using universal major cold-shock protein primer oligomers. J Ind Microbiol Biotechnol 19:286–293

Gaenzle MG, Schwab C (2009) Ecology of exopolysaccharide formation by lactic acid bacteria: sucrose utilisation, stress tolerance, and biofilm formation. In: Ulrich M (Ed.), Bacterial polysaccharides—current innovation and trends. Horizon Press

Glaasker E, Konings WN, Poolman B (1996) Osmotic regulation of intracellular solute pools in *Lactobacillus plantarum*. J Bacteriol 178:575–582

Glaasker E, Tjan FSB, Tergesteeg PF, Konings WN, Poolman B (1998) Physiological response of *Lactobacillus plantarum* to salt and nonelectrolyte stress. J Bacteriol 180:4718–4723

Gobbetti M (1998) The sourdough microflora: interactions of lactic acid bacteria and yeasts. Trends Food Sci Technol 9:267–274

Gobbetti M, De Angelis M, Di Cagno R, Minervini F, Limitone A (2007) Cell cell communication in food related bacteria. Int J Food Microbiol 120:34–45

Goh Y, Klaenhammer TR (2009) Genomic features of *Lactobacillus* species. Front Biosci 14:1362–1386

Götz F, Sedewitz B, Elstner EF (1980) Oxygen utilization by *Lactobacillus plantarum*. Arch Microbiol 125:209–214

Gouesbert G, Jan G, Boyaval P (2001) *Lactobacillus delbrueckii* ssp. *bulgaricus* thermotolerance. Lait 81:301–309

Groot MN, Klaassens E, de Vos WM, Delcour J, Hols P, Kleerebezem M (2005) Genome-based *in silico* detection of putative manganese transport systems in *Lactobacillus plantarum* and their genetic analysis. Microbiology 151:1229–1238

Guerzoni ME, Lanciotti R, Cocconcelli PS (2001) Alteration in cellular fatty acid composition as a response to salt, acid, oxidative and thermal stresses in *Lactobacillus helveticus*. Microbiology 147:2255–2264

Guerzoni ME, Vernocchi P, Ndagijimana M, Giannotti A, Lanciotti R (2007) Generation of aroma compounds in sourdough: effects of stress exposure and lactobacilli–yeasts interactions. Food Microbiol 24:139–148

Hecker M, Schumann W, Völker U (1996) Heat-shock and general stress response in *Bacillus subtilis*. Mol Microbiol 19:417–428

Hertel C, Schmidt G, Fisher M, Oellers K, Hammes WP (1998) Oxygen-dependent regulation of the expression of the catalase gene *katA* of *Lactobacillus sakei* LTH677. Appl Environ Microbiol 64:1359–1365

Hong SI, Kim YJ, Pyun YR (1999) Acid tolerance of *Lactobacillus plantarum* from Kimchi. Food Sci Technol Lebensm Wiss Technol 32:142–148

Hörmann S, Scheyhing C, Behr J, Pavlovic M, Ehrmann M, Vogel RF (2006) Comparative proteome approach to characterize the high-pressure stress response of *Lactobacillus sanfranciscensis* DSM 20451 T. Proteomics 6:1878–1885

Horn G, Hofweber R, Kremer W, Kalbitzer HR (2007) Structure and function of bacterial cold shock proteins. Cell Mol Life Sci 64:1457–1470

Hussain MA, Knight MI, Britz ML (2009) Proteomic analysis of lactose-starved *Lactobacillus casei*. J Appl Microbiol 106(3):764–773

Hutkins RW, Nannen NL (1993) pH homeostasis in lactic acid bacteria. J Dairy Sci 76:2354–2365

Ito K, Akiyama Y (2005) Cellular functions, mechanism of action, and regulation of FtsH protease. Annu Rev Microbiol 59:211–231

Jänsch A, Korakli M, Vogel RF, Gänzle MG (2007) Glutathione reductase from *Lactobacillus sanfranciscensis* DSM20451T: contribution to oxygen tolerance and thiol exchange reactions in wheat sourdoughs. Appl Environ Microbiol 73:4469–4476

Jofré A, Champomier-Vergès SM, Anglade P, Baraige F, Martìn B, Garriga M, Zagorec M, Aymerich T (2007) Protein synthesis in lactic acid and pathogenic bacteria during recovery from a high pressure treatment. Res Microbiol 158:512–520

Jordan S, Hutchings MI, Mascher T (2008) Cell envelope stress response in Gram-positive bacteria. FEMS Microbiol Rev 32:107–146

Katsaros GI, Giannoglou MN, Taoukis PS (2009) Kinetic study of the combined effect of high hydrostatic pressure and temperature on the activity of *Lactobacillus delbrueckii* ssp. *bulgaricus* aminopeptidases. J Food Sci 74:E219–E225

Kilimann KV, Doster W, Vogel RF, Hartmann C, Gänzle MG (2006a) Protection by sucrose against heat-induced lethal and sublethal injury of *Lactococcus lactis*: an FT-IR study. Biochim Biophys Acta 1764:1188–1197

Kilimann KV, Hartmann C, Delgado A, Vogel RF, Gänzle MG (2006b) Combined high pressure and temperature induced lethal and sublethal injury of *Lactococcus lactis*—application of multivariate statistical analysis. Int J Food Microbiol 109:25–33

Kim SW, Dunn NW (1997) Identification of a cold shock gene in lactic acid bacteria and the effect of cold shock on cryotolerance. Curr Microbiol 35:59–63

Kim WS, Perl L, Park JH, Tandianus JE, Dunn NW (2001) Assessment of stress response of the probiotic *Lactobacillus acidophilus*. Curr Microbiol 43:346–350

Kleerebezem M, Boekhorst J, van Kranenburg R, Molenaar D, Kuipers OP, Leer R, Tarchini R, Peters SA, Sandbrink HM, Fiers MW, Stiekema W, Lankhorst RM, Bron PA, Hoffer SM, Groot MN, Kerkhoven R, de Vries M, Ursing B, de Vos WM, Siezen RJ (2003) Complete genome sequence of *Lactobacillus plantarum* WCFS1. Proc Natl Acad Sci USA 100:1990–1995

Knauf HJ, Vogel RF, Hammes WP (1992) Cloning, sequence, and phenotypic expression of katA, which encodes the catalase of *Lactobacillus sake* LTH677. Appl Environ Microbiol 58:832–839

Korakli M, Gänzle MG, Knorr R, Frank M, Rossmann A, Vogel RF (2002) Metabolism of *Lactobacillus sanfranciscensis* under high-pressure: investigations using stable carbon isotopes. In: Hayashi R (Ed.), Trends in high pressure bioscience and biotechnology. Elsevier, Amsterdam, pp. 287–294

Laplace JM, Sauvageot N, Harke A, Auffray Y (1999) Characterization of *Lactobacillus collinoides* response to heat, acid and ethanol treatments. Appl Microbiol Biotechnol 51:659–663

Lee K, Lee HG, Pi KB, Choi YJ (2008) The effect of low pH on protein expression by the probiotic bacterium *Lactobacillus reuteri*. Proteomics 8:1624–1630

Lim EM, Ehrlich SD, Maguin E (2000) Identification of stress-inducible proteins in *Lactobacillus delbrueckii* subsp. *bulgaricus*. Electrophoresis 21:2557–2561

Lim EM, Smokvina T, Chervaux C, Ehrlich SD, Maguin E (2002) In: Book of Abstracts, "Seventh Symposium on Lactic Acid Bacteria—genetics, metabolism and applications," FEMS Netherlands, p. G85

Lorca GL, Raya RR, Taranto MP, Font de Valdez G (1998) Adaptative acid tolerance response in *Lactobacillus acidophilus*. Biotechnol Lett 20:239–241

Lorca GL, Font de Valdez GF (1999) Effect of suboptimal growth temperature and growth phase on resistance of *Lactobacillus acidophilus* to environmental stress. Cryobiology 39:144–149

Lorca GL, Font de Valdez G (2001) Acid tolerance mediated by membrane ATPases in *Lactobacillus acidophilus*. Curr Microbiol 42:21–25

Marceau A, Zagorec M, Champonier-Vergès MC (2002) In: Book of Abstracts, "Seventh Symposium on Lactic Acid Bacteria—genetics, metabolism and applications," FEMS Netherlands, p. G62

Marquis RE, Mèra T, Bender GR, Murray DR, Pesce de Ruiz Holgado A (1987) Arginine deiminase system and bacterial adaptation to acid environments. Appl Environ Microbiol 53:198–200

Martinac B, Adler J, Kung C (1990) Mechanosensitive ion channels of *E. coli* activated by amphipaths. Nature 348:261–263

Marty-Teysset C, De La Torre F, Garel JR (2000) Increased production of hydrogen peroxide by *Lactobacillus delbrueckii* subsp. *bulgaricus* upon aeration: involvement of an NADH oxidase in oxidative stress. Appl Environ Microbiol 66:262–267

Mayo B, Derzelle S, Fernandez L, Leonard C, Ferain T, Hols P, Suarez JE, Delcour J (1997) Cloning and characterization of cspL and cspP, two cold-inducible genes from *Lactobacillus plantarum*. J Bacteriol 179:3039–3042

Miyakawa H, Anjitsu K, Ishibashi N, Shimamura S (1994) Effects of pressure on enzyme activities of *Lactobacillus helveticus* LHE-511. Biosci Biotech Biochem 58:606–607

Molenaar D, Bosscher JS, Ten Brink B, Driessen AJM, Konings WN (1993) Generation of a proton motive force by histidine decarboxylation and electrogenic histidine/histamine antiport in *Lactobacillus buchneri*. J Bacteriol 175:2864–2870

Moslehi-Jenabian S, Gori K, Jespersen L (2009) AI-2 signalling is induced by acidic shock in probiotic strains of *Lactobacillus* spp. Int J Food Microbiol 135:295–302

Nannen NL, Hutkins RW (1991) Proton-translocating adenosine triphosphatase activity in lactic acid bacterial. J Dairy Sci 74:747–751

Narberhaus F (2002) α-Crystallin-type heat shock proteins: socializing minichaperones in the context of a multichaperone network. Microbiol Mol Biol Rev 66:64–93

Nousiainen LL, Savijoki K, Palva A, Varmanen P (2002) In: Book of Abstracts, "Seventh Symposium on Lactic Acid Bacteria—genetics, metabolism and applications," FEMS, Netherlands, p. G44

Pallen MJ, Wren BW (1997) The HtrA family of serine proteases. Mol Microbiol 26:209–221

Pannoff JM, Thammavongs B, Gueguen M (2000) Cryoprotectants lead to phenotypic adaptation to freeze–thaw stress in *Lactobacillus delbrueckii* ssp. *bulgaricus* CIP 101027 T. Cryobiology 40:264–269

Pavlovic M, Hörmann S, Vogel RF, Ehrmann MA (2005) Transcriptional response reveals translation machinery as target for high pressure in *Lactobacillus sanfranciscensis*. Arch Microbiol 184:11–17

Pavlovic M, Hörmann S, Vogel RF, Ehrmann MA (2008) Characterisation of a piezotolerant mutant of *Lactobacillus sanfranciscensis*. Z Naturforsch 63b:791–797

Pfeiler EA, Azcarate-Peril MA, Klaenhammer TR (2007) Characterization of a novel bile-inducible operon encoding a two-component regulatory system in *Lactobacillus acidophilus*. J Bacteriol 189:4624–4634

Phadtare S, Yamanata K, Inouye M (2000) In: Stortz G, Hengge-Aronis R (Eds.), Bacterial stress response, ASM Press, Washinngton, DC, pp. 33–45

Pieterse B, Leer RJ, Schuren FHJ, van der Werf M (2005) Unravelling the multiple effects of lactic acid stress on *Lactobacillus plantarum* by transcription profiling. Microbiology 151:3881–3894

Pridmore RD, Berger B, Desiere F, Vilanova D, Barretto C, Pittet AC, Zwahlen MC, Rouvet M, Altermann E, Barrangou R, Mollet B, Mercenier A, Klaenhammer T, Arigoni F, Schell MA (2004) The genome sequence of the probiotic intestinal bacterium *Lactobacillus johnsonii* NCC 533. Proc Natl Acad Sci USA 101:2512–2517

Rizzello CG, Cassone A, Di Cagno R, Gobbetti M (2008) Synthesis of angiotensin I-converting enzyme (ACE)-inhibitory peptides and g-aminobutyric acid (GABA) during sourdough fermentation by selected lactic acid bacteria. J Agric Food Chem 56:6936–6943

Rochat T, Gratadoux JJ, Gruss A, Corthier G, Maguin E, Langella P, van de Guchte M (2006) Production of a heterologous nonheme catalase by *Lactobacillus casei*: an efficient tool for removal of H_2O_2 and protection of *Lactobacillus bulgaricus* from oxidative stress in milk. Appl Environ Microbiol 72:5143–5149

Ruas-Madiedo P, Hugenholtz J, Zoon P (2000) An overview of the functionality of exopolysaccharides produced by lactic acid bacteria. Int Dairy J 12:163–171

Russel NJ, Fukanage M (1990) A comparison of thermal adaptation of membrane lipids in psychrophilic and thermo-philic bacteria. FEMS Microbiol Rev 75:171–182

Sanders JW, Vemena G, Kok J (1999) Environmental stress responses in *Lactococcus lactis*. FEMS Microbiol Rev 23:483–501

Scheyhing CH, Hörmann S, Ehrmann MA, Vogel RF (2004) Barotolerance is inducible by preincubation under hydrostatic pressure, cold-, osmotic- and acid-stress conditions in *Lactobacillus sanfranciscensis* DSM20451^T. Lett Appl Microbiol 39:284–289

Schmidt G, Hertel C, Hammes WP (1999) Characterisation of the dnaK operon of *Lactobacillus sakei* LTH681. System Appl Microbiol 22:321–328

Segal G, Ron EZ (1996) Regulation and organization of the *groE* and *dnaK* operons in *Eubacteria*. FEMS Microbiol Lett 138:1–10

Serrano LM, Molenaar D, Wels M, Teusink B, Bron PA, de Vos WM, Smid EJ (2007) Thioredoxin reductase is a key factor in the oxidative stress response of *Lactobacillus plantarum* WCFS1. Microb Cell Fact 6:29

Serrazanetti D, Guerzoni ME, Corsetti A, Vogel R (2009) Metabolic impact and potential exploitation of the stress reactions in lactobacilli. Food Microbiol 26:700–711

Siegumfeldt H, Rechinger KB, Jakobsen M (2000) Dynamic changes of intracellular pH in individual lactic acid bacterium cells in response to a rapid drop in extracellular pH. Appl Environ Microbiol 66:2330–2335

Siragusa S, De Angelis M, Di Cagno R, Rizzello CG, Coda R, Gobbetti M (2007) Production of γ-aminobutyric acid (GABA) by lactic acid bacteria isolated from Italian cheese varieties. Appl Environ Microbiol 73(22):7283–7290

Smeds A, Varmanen P, Palva A (1998) Molecular characterization of a stress-inducible gene from *Lactobacillus helveticus*. J Bacteriol 180:6148–6153

Spano G, Beneduce L, Perrotta C, Massa S (2005) Cloning and characterization of the *hsp* 18.55 gene, a new member of the small heat shock genes family isolated from wine *Lactobacillus plantarum*. Res Microbiol 156:219–224

Stentz R, Loizel C, Mallert C, Zagorec M (2000) Development of genetic tools for *Lactobacillus sakei*: disruption of the β-galactosidase gene and Use of *lacZ* as a reporter gene to study regulation of the putative copper ATPase, AtkB. Appl Environ Microbiol 66:4272–4278

Stolz P, Böcker G, Hammes WP, Vogel RF (1995) Utilization of electron acceptors by lactobacilli isolated from sourdough. II. *Lactobacillus pontis*, *L. reuteri*, *L. amylovorus*, *L. fermentum*. Z Lebensm Unters Forsch 201:402–410

Streit F, Corrieu G, Béal C (2007) Acidification improves cryotolerance of *Lactobacillus delbrueckii* subsp. *bulgaricus* CFL1. J Biotechnol 128:659–667

Streit F, Delettre J, Corrieu G, Béal C (2008) Acid adaptation of *Lactobacillus delbrueckii* subsp. *bulgaricus* induces physiological responses at membrane and cytosolic levels that improves cryotolerance. J Appl Microbiol 105:1071–1080

Stuart MR, Chou LS, Weimer BC (1999) Influence of carbohydrate starvation and arginine on culturability and amino acid utilization of *Lactococcus lactis* subsp. *lactis*. Appl Environ Microbiol 65:665–673

Sturme MHJ, Kleerebezem M, Nakayama J, Akkermans ADL, Vaughan EE, de Vos WM (2002) Cell to cell communication by autoinducing peptides in Gram-positive bacteria. Antonie van Leeuwenhoek 81:233–243

Sturme MH, Nakayama J, Molenaar D, Murakami Y, Kunugi R, Fujii T, Vaughan EE, Kleerebezem M, de Vos WM (2005) An agr-like twocomponent regulatory system in *Lactobacillus plantarum* is involved in production of a novel cyclic peptide and regulation of adherence. J Bacteriol 187:5224–5235

Sturme MH, Francke JC, Siezen RJ, de Vos WM, Kleerebezem M (2007) Making sense of quorum sensing in lactobacilli: a special focus on *Lactobacillus plantarum* WCFS1. Microbiology 153:3939–3947

Sukharev SI, Blount P, Martinac B, Kung C (1997) Mechanosensitive channels of *Escherichia coli*: the *MscL* gene, protein, and activities. Annu Rev Physiol 59:633–657

Suokko A, Poutanen M, Savijoki K, Kalkkinen N, Varmanen P (2008) ClpL is essential for induction of thermotolerance and is potentially part of the HrcA regulon in *Lactobacillus gasseri*. Proteomics 8:1029–1041

Tieking M, Gaenzle MG (2005) Exopolysaccharides from cereal-associated lactobacilli. Trends Food Sci Technol 16:79–84

Turner MS, Woodberry T, Hafner LM, Giffard PM (1999) The *bspA* locus of *Lactobacillus fermentum* BR11 encodes an L-cystine uptake system. J Bacteriol 181:2192–2198

van de Guchte M, Serror P, Chervaux C, Smokvina T, Ehrlich SD, Maguin E (2002) Stress responses in lactic acid bacteria. Antonie van Leeuwenhoek 82:187–216

van de Guchte M, Penaud S, Grimaldi C, Barbe V, Bryson K, Nicolas P, Robert C, Oztas S, Mangenot S, Couloux A, LouxV, Dervyn R, Bossy R, Bolotin A, Batto JM, Walunas T, Gibrat JF, Bessieres P, Weissenbach J, Ehrlich SD, Maguin E (2006) The complete genome sequence of *Lactobacillus bulgaricus* reveals extensive and ongoing reductive evolution. Proc Natl Acad Sci USA 103:9274–9279

Vermeulen N, Czerny M, Gänzle MG, Schieberle P, Vogel RF (2007) Reduction of (E)-2-nonenal and (E,E)-2,4-decadienal during sourdough fermentation. J Cereal Sci 45:78–87

Vernocchi P, Ndagijimana M, Serrazanetti D, Gianotti A, Vallicelli M, Guerzoni ME (2008) Influence of starch addition and dough microstructure on fermentation aroma production by yeasts and lactobacilli. Food Chem 108:1217–1225

Vogel RF, Pavlovic M, Hörmann S, Ehrmann MA (2005) High pressure-sensitive gene expression in *Lactobacillus sanfranciscensis*. Braz J Med Biol Res 38:1247–1252

Walker DC, Girgis HS, Klaenhammer TR (1999) The *groESL* chaperone operon of *Lactobacillus johnsonii*. Appl Environ Microbiol 65:3033–3041

Warriner KRS, Morris JG (1995) The effects of aeration on the bioreductive abilities of some heterofermentative lactic acid bacteria. Lett Appl Microbiol 20:323–327

Wouters JA, Rombouts M, Kuipers OP, de Vos WM, Abee T (2000) The role of cold-shock proteins in low-temperature adaptation of food-related bacteria. Syst Appl Microbiol 23:165–173

Yamanaka K, Fang L, Inouye M (1998) The CspA family in *Escherichia coli*: multiple gene duplication for stress adaptation. Mol Microbiol 27:247–255

Zink R, Walker C, Schmidt G, Elli M, Pridmore D, Reniero R (2000) Impact of multiple stress factors on the survival of dairy lactobacilli. Sci Alim 20:119–126

Zúñiga M, Champomier-Verges M, Zagorec M, Pèrez-Martinez G (1998) Structural and functional analysis of the gene cluster encoding the enzymes of the arginine deiminase pathway of *Lactobacillus sakei*. J Bacteriol 180:4154–4159

Chapter 12
Stress Responses of Streptococci

José A. Lemos, Effie Tsakalidou, and Konstantinos Papadimitriou

12.1 Introduction

Streptococci were among the first bacterial entities to be recognized by microbiologists because of their involvement in a large number of human and animal diseases (Billroth 1874). Nevertheless, it was realized early on that, the so-called lactic acid streptococci of dairy origin displayed some different characteristics from the pathogenic ones (Sherman and Albus 1918). The actual taxonomy of the *Streptococcus* genus has been a complicated field of research, with a number of historical breakthroughs leading to the current classification. In 1884, Rosenbach was the first to use the name *Streptococcus* in the generic sense (Rosenbach 1884). Initial attempts for the classification of streptococci in the first part of the twentieth century relied mainly upon phenotypic traits. The use of hemolytic reactions given by streptococci when cultured on blood agar, as introduced by Shottmuller in 1903, and Lancefield's serological testing based on the distribution of carbohydrate antigens described in 1933 were probably the most influential approaches for discriminating streptococci at that time (Shottmuller 1903; Lancefield 1933). Using the aforementioned criteria, Sherman proposed in 1937 the first widely accepted classification scheme, separating the genus into four divisions, designated pyogenic, viridans, lactic, and enterococci (Sherman 1937). Despite the early success of Sherman's taxonomy, it became

J.A. Lemos (✉)
Center for Oral Biology and Department of Microbiology and Immunology,
University of Rochester Medical Center, 601 Elmwood Avenue,
Box 611, Rochester, NY 14642, USA
e-mail: jose_lemos@urmc.rochester.edu

K. Papadimitriou (✉)
Laboratory of Dairy Research, Department of Food Science and Technology,
Agricultural University of Athens,
Iera Odos 75, 118 55 Athens, Greece
e-mail: kpapadimitriou@aua.gr

E. Tsakalidou and K. Papadimitriou (eds.), *Stress Responses of Lactic Acid Bacteria*,
Food Microbiology and Food Safety, DOI 10.1007/978-0-387-92771-8_12,

obvious that its application was hampered by intrinsic vulnerabilities because in many cases physiological heterogeneous strains had to be grouped together, leading to species definitions that were often qualified by a number of exceptions (Hardie and Whiley 1997). For the next decades, a number of attempts allowed some improvement of the situation by using more complex biochemical and immunological tests but still failed to provide the natural association among streptococci. This did not happen until the mid-1980s, when Schleifer and coworkers managed to unravel the first "true" genetic relationship of streptococci that derived from advanced chemotaxonomic traits and DNA-based techniques, like DNA-DNA and DNA-rRNA hybridization, as well as 16S rRNA cataloging. They clearly demonstrated that the *Streptococcus* genus should be split into three genetically distinct genera: *Streptococcus senso stricto*, *Enterococcus*, and *Lactococcus* (Schleifer and Kilpper-Balz 1984; Schleifer et al. 1985). Currently, *Lactococcus* spp. are important for the food industry and are generally considered nonpathogenic (Casalta and Montel 2008), whereas *Enterococcus* spp. may play a role in food fermentation but also contain pathogenic species (Ogier and Serror 2008). Today, the classification system for the *Streptococcus* genus is governed by molecular taxonomy, and the phylogenetic tree of all species that takes into account molecular genotypical information is available (Facklam 2002; Kohler 2007). In the present chapter, which concerns the stress physiology of streptococci, we will present streptococcal species according to their pathogenicity, ecological niche, and food compatibility. We came to this decision because these factors probably have been the most decisive in driving the relevant stress physiology research of each group up to now.

In detail, we will initially focus our review on the current understanding of the stress responses of the most important pathogens. This would include the dental pathogen *S. mutans*, the beta-hemolytic Lancefield's group A streptococci (GAS) and group B streptococci (GBS) (i.e., *S. pyogenes* and *S. agalactiae*, respectively), as well as the nonbeta-hemolytic *S. pneumoniae*. During their commensal state, it is decisive for these bacteria to be able to adapt to the host's environment. In fact, the host's environment can be truly adverse because the physicochemical conditions of each organ can markedly differ, requiring a constant adaptation during the passive or invasive translocation of these microorganisms (Smoot et al. 2001; Brown et al. 2004). Moreover, the host organism is in a continuous "search and destroy" mode for presumed pathogens. The first line of the host's defense is the innate immune system. Phagocytic cells like neutrophils and macrophages have evolved a sophisticated antimicrobial arsenal to fight invading microbes. During phagocytosis, these cells unleash a complex biochemical warfare against the engulfed bacterium involving the abrupt production of reactive oxygen species (ROS, also known as respiratory burst), a rapid pH drop due to the activity of the vacuolar ATPase, and the subsequent release of several bactericidal proteins and peptides targeting the cell envelope (e.g., lactoferrin, lipocalin, lysozyme, α-defensin-1, LL-37, and cathepsin G) (Voyich et al. 2004a, b; Urban et al. 2006). Pathogens have either to evade phagocytosis or to avert destruction after phagocytosis. Streptococcal pathogens can use both strategies to circumvent the attack of phagocytes. It is obvious though that the latter strategy relies on their ability to turn on survival and stress responses (Urban et al. 2006). The mechanisms by which streptococci are able to endure the bactericidal conditions imposed by the innate immunity components are of outmost importance for understanding disease due to these pathogens.

We will begin with the description of the stress physiology of *S. mutans*, a dental pathogen that serves as a model organism for oral streptococci. Today, *S. mutans* is probably the best-studied *Streptococcus* species in terms of stress physiology, and the knowledge of this bacterium that has accumulated over the past years exceeds that of most streptococcal species. Next, we will describe the mechanisms used by GAS, GBS, and *S. pneumoniae*, the three most frequent streptococcal species associated with severe infections. Finally, we will concentrate on the food-compatible streptococci. To date, only two streptococcal species, namely, *S. thermophilus* and *S. macedonicus*, have been clearly associated with the food environment as their primary ecological niche (Bolotin et al. 2004; De Vuyst and Tsakalidou 2008). All streptococci are exposed to the same types of stressful conditions; however, the actual physicochemical parameters of these stresses can markedly differ for food-related streptococci. Food processes have been designed to produce a sanitary – from a microbiological perspective – product and thus involve adverse conditions that lead to bacteriostatic or bactericidal effects. Such extreme conditions in most cases far exceed the natural capacity of the host's defenses and, as a result, are rarely met by pathogenic and/or commensal streptococci of humans and animals. Because food-related streptococci actively participate in the preservation as well as the texture and flavor formation of fermented foods, the failure to withstand and perform during industrial processes can seriously influence the overall quality of the final product (van de Guchte et al. 2002).

While extensive, the investigation of the stress physiology of streptococci has been uneven, despite their significance. There are specific groups or species within the genus, that have been under rigorous research whereas others are still lagging behind. Throughout the chapter we will attempt to highlight all areas in the field that have been scrutinized as well as areas that deserve further investigation.

12.2 Streptococcus mutans

Bacteria colonizing the teeth form dental plaque, a multispecies biofilm that is normally beneficial to the host because it prevents colonization by exogenous species. Usually, oral biofilms exist in equilibrium with the host, with no detrimental impact on the tissues of the mouth (Marsh 2003). In some cases, though, environmental perturbations can change the composition and metabolic activities of biofilm communities, leading to an increase in the proportions of pathogenic species. In the case of dental caries, prolonged periods of plaque acidification allow for the emergence of an aciduric flora and demineralization of the tooth enamel. Dental caries remain among the most common infectious diseases afflicting humans and result in an enormous health and economic burden worldwide (Centers for Disease Control and Prevention 2001).

Two species of mutans streptococci, *S. mutans* and *Streptococcus sobrinus*, are considered to be the primary causative agents of human dental caries. Decades of research reveal that mutans streptococci virulence resides in three core attributes: their abilities to form biofilms on the tooth surface, to produce large quantities of organic acids (acidogenicity) from a wide range of carbohydrates, and to tolerate environmental stresses, particularly low pH (aciduricity) (Lemos and Burne 2008). In addition to dental caries, mutans streptococci are often implicated in subacute bacterial endocarditis, a life-threatening inflammation of heart valves. Although both *S. mutans* and *S. sobrinus* are believed to be highly cariogenic, the majority of genetic studies have been carried out with *S. mutans*, primarily because of the strong association of this species with caries in developed nations, but also because of the relative ease of genetic manipulation of *S. mutans* compared to *S. sobrinus*. In particular, studies concerning *S. sobrinus* biology have been hampered by the lack of a reference genome and by the extremely refractory nature of this organism to genetic manipulations. Notably, despite the close relatedness of these two species, the characterization of the acid-tolerance responses of *S. sobrinus* revealed that there are important differences in the mechanisms of acid adaptation between *S. mutans* and *S. sobrinus* (Nascimento et al. 2004; Martinez et al. 2010).

Different than most diseases, through which a virulence factor such as a toxin or protease causes tissue damage, dental caries is a disease associated with bacterial metabolism, which creates stresses in the form of acids, ROS, and other highly reactive radicals. In particular, the intermittent ingestion of food by the host results in dramatic changes in nutrient availability and pH. Although it has been proposed that oral biofilms experience a "feast or famine" lifestyle (Carlsson 1983), organisms residing in the oral cavity are not exposed to severely oligotrophic environments. For this reason, studies of the stress responses of *S. mutans* have served as an excellent model to reveal critical differences in the ways that obligately host-associated bacteria cope with environmental stresses when compared with bacteria that have both free-living and host-associated lifestyles.

The availability of the complete genome sequence of a serotype *c* strain of *S. mutans* (Ajdic et al. 2002) allowed oral microbiologists to apply functional genomic, transcriptomic, and proteomic approaches to dissect the mechanisms of stress tolerance utilized by this dental pathogen. In fact, since the release of the first *S. mutans* genome sequence in 2002, researchers have made remarkable advances toward understanding how *S. mutans* integrates the stress regulon to coordinate responses to environmental fluctuations and pathogenesis. *S. mutans* can be considered the model organism among lactic acid bacteria (LAB) in terms of acid-stress physiology studies. It should also be noted that the initial research performed with *S. mutans* and other oral streptococci and their particular lifestyle of biofilms led to the appreciation of this physiological state as a means to withstand multiple environmental stresses in other LAB species.

Proteomic studies have been instrumental in unveiling proteins and pathways that participate in acid tolerance and acid adaptation (Wilkins et al. 2002; Welin et al. 2003; Len et al. 2004a, b; Rathsam et al. 2005). By coupling proteomic data with measurements of the end products of carbon utilization, Len et al. (2004a) were able to propose that *S. mutans* tolerates growth at low pH by expending energy to extrude H^+, by modulating the production of acid end products to reduce H^+ production, and possibly by using branched-chain amino acid (BCAA) biosynthesis to reduce acid production and buffer the intracellular pH. A comparison of the proteome of mature biofilms and planktonic cells of *S. mutans* grown at neutral pH revealed that multiple proteins associated with carbon uptake and cell division were downregulated, whereas competence-associated proteins were upregulated (Rathsam et al. 2005). This finding has particular significance in terms of plaque ecology because the coordinated production of bacteriocins from *S. mutans* (mutacins) and the development of competence have been documented in high-cell density environments, suggesting that the organism could use competence-induced cell lysis to acquire DNA from neighboring species (Kreth et al. 2005–2007). Recent studies with *S. mutans* have implicated the presence of DNA released from competence-induced cell lysis in the extracellular matrix with proper biofilm maturation (Petersen et al. 2005). It remains to be determined whether *S. mutans* biofilms acquire DNA from the external environment as a nutrient source, to increase genetic diversity, or both.

S. mutans possesses several acid-adaptive strategies, some of which are distributed among all *Streptococcus* species and some that are unique (Lemos et al. 2005). In contrast to enteric bacteria, which usually maintain a fairly constant internal pH (pH_i) at 7.6–7.8 through the use of antiporters, pH_i in streptococci fluctuates in response to extracellular pH, with the organisms working to maintain a ΔpH of 0.5–1 unit above the external environment. The membrane-bound F_1F_0-ATPase is the primary mechanism of proton extrusion to maintain pH homeostasis of all streptococci. In oral streptococci, acid tolerance correlates well with the pH optimum of the F-ATPase enzyme (Sturr and Marquis 1992). For example, in the highly acid-tolerant *S. mutans*, the optimal pH for the F-ATPase enzyme is 6.0, whereas in the less acid-tolerant *Streptococcus salivarius* and *Streptococcus sanguinis*, the pH optima of the ATPase enzymes are 7.0 and 7.5, respectively. In addition to functioning as a proton pump, it was recently demonstrated that the F-ATPase of *S. mutans* and other LAB could also function as an ATP synthase in starved cells grown at low pH (Sheng and Marquis 2006). Thus, the F-ATPase may play a dual role in acid tolerance by extruding protons out of the cells and, under certain conditions, by generating ATP for growth and maintenance.

Although *S. mutans* is not considered capable of generating significant quantities of alkali because it lacks urease and the arginine deiminase system (ADS) pathways (Ajdic et al. 2002), a functional agmatine deiminase system (AgDS), analogous to the ADS, was found in the genome of *S. mutans* UA159 (Griswold et al. 2006). The AgDS converts agmatine, a decarboxylated derivative of arginine that is found in human dental plaque, to putrescine, ammonia, and CO_2. Whereas the ADS and urease pathways catalyze substantial environmental alkalinization and appear to be associated with caries resistance, the AgDS of *S. mutans* is expressed at relatively low levels and is unlikely to elicit a significant alkalinization of the environment. However, the production of ammonia from agmatine is believed to contribute to the competitive fitness of *S. mutans* at low pH by increasing the cytoplasmic pH and by providing ATP, which could be used for growth or to extrude protons (Griswold et al. 2006).

Another contributor to the acid tolerance of *S. mutans* is malolactic fermentation (MLF), which catalyzes the conversion of dicarboxylic L-malate, a major acid in fruits such as apple, to the monocarboxylic lactic acid and CO_2. It was demonstrated that although malate did not serve as a catabolite for the growth of *S. mutans*, it did protect the organism against acid killing by increasing the pH of the cytoplasm via production of CO_2 (Sheng and Marquis 2007).

The importance of cell membrane integrity and composition in relation to changes that affect proton permeability and F-ATPase activity in *S. mutans* has been well documented. Quivey and coworkers showed that *S. mutans* strains grown at pH 5.0 had increased levels of monounsaturated fatty acids and longer chain fatty acids than cells grown at pH 7.0 (Quivey et al. 2000). The inactivation of the gene responsible for the biosynthesis of monounsaturated fatty acids, *fabM*, resulted in a strain that was extremely sensitive to low pH and unable to maintain ΔpH (Fozo and Quivey 2004b). Rats infected with the *fabM* mutant exhibited substantially reduced caries, as compared to the parent strain (Fozo et al. 2007). One of the main mechanisms by which changes in membrane composition are thought to affect acid tolerance is through decreased proton permeability, although an equally important potential mechanism is that the altered lipid environment strongly influences the activity of the F-ATPase and transport proteins in a manner that influence acid resistance.

The importance of membrane protein biogenesis to stress tolerance has been further demonstrated in a study with mutated strains lacking the signal recognition particle-translocation (SRP) pathway or the membrane-localized chaperone YidC, both involved in the translocation and assembly of membrane proteins. Once considered essential for the viability of all organisms, the SRP pathway was found dispensable in *S. mutans* (Hasona et al. 2005), although mutants lacking proteins of the SRP pathway or YidC were severely impaired in growth under a variety of stress conditions (Hasona et al. 2005). The authors observed that YidC and a functional SRP pathway is necessary for the optimal insertion of membrane proteins, including the F-ATPase, providing a partial explanation for the diminished stress tolerance of strains lacking YidC or components of the SRP pathway (Hasona et al. 2005, 2007). Similar observations have been made with strains defective in D-alanyl-lipoteichoic acid (LTA) synthesis and diacylglycerol kinase (DagK) that showed an acid-sensitive phenotype (Yamashita et al. 1993; Boyd et al. 2000). In both cases, increased proton permeability due to changes in membrane architecture and composition was directly implicated with the failure to tolerate low pH.

Many of the stresses encountered by oral bacteria induce DNA damage; in particular, acid and oxidative stresses increase the formation of abasic sites in DNA. Hahn and coworkers identified an AP endonuclease in *S. mutans* that showed higher levels of activity in cells grown at low pH compared with cells grown at pH 7.0 (Hahn et al. 1999), providing the first evidence linking DNA repair enzymes with acid adaptation in oral streptococci. In another study, differential display PCR identified an *S. mutans* gene, with similarity to the UV repair excinuclease UvrA of *Bacillus subtilis*, which showed an increased expression in response to low pH (Hanna et al. 2001). In *Escherichia coli*, UvrA participates in the nucleotide excision repair pathway, which is responsible for excising larger DNA lesions caused by acid and other DNA-damaging agents. An *S. mutans uvrA* mutant was extremely sensitive to UV irradiation and had enhanced sensitivity to acid killing when compared to the parent strain (Hanna et al. 2001). Finally, Faustoferri et al. characterized the Smx exonuclease in *S. mutans* and showed that an *smx* mutant strain was highly sensitive to DNA damage caused by the production of hydroxyl radicals via the Fenton reaction (Faustoferri et al. 2005).

One consequence of the exposure to environmental stresses is the accumulation of abnormal proteins due to increased errors in transcription and translation. Moreover, aging cells present in mature biofilms are prone to mistranslation and aggregation. In this context, molecular chaperones and proteases, which modulate the stability of proteins and prevent the accumulation of misfolded proteins, are central to physiologic homeostasis. The GroEL and DnaK chaperones take part in several cellular processes, including protein folding, renaturation, and the presentation of proteins for degradation. In *S. mutans*, the DnaK and GroEL chaperones appear to be indispensable (Lemos et al. 2001), and the essential nature of these chaperones was confirmed by the downregulation of *groEL* and *dnaK* expression (Lemos et al. 2007b). Lowering DnaK levels resulted in an impaired capacity to form biofilms in the presence of glucose and rendered the strain more sensitive to low pH, elevated temperature, and H_2O_2 stress (Lemos et al. 2007b). The acid sensitivity of the strain with low levels of DnaK was attributed, at least in part, to the DnaK chaperone participating in the biogenesis or stabilization of the F-ATPase complex (Lemos et al. 2007b). The downregulation of GroEL also resulted in high-temperature sensitivity and an impaired capacity to form biofilms, but did not affect growth at low pH or in the presence of H_2O_2 (Lemos et al. 2007b). Wen et al. showed that the ribosome-associated peptidyl-prolyl isomerase RopA (trigger factor) is important for the adherence and formation of biofilms and for the tolerance to low pH and H_2O_2 (Wen et al. 2005). Inactivation of the surface-associated HtrA protease or the cytoplasmic ClpP peptidase generated multiple stress-sensitive phenotypes in *S. mutans* and was also linked to altered biofilm formation and reduced genetic competence (Chattoraj et al. 2010; Lemos and Burne 2002; Ahn et al. 2005; Biswas and Biswas 2005; Deng et al. 2007b). Recently, the inactivation of all members of the Clp system indicated that, in addition to the ClpP peptidase, the inactivation of the genes encoding for the ClpL and ClpX ATPases also affected the expression of virulence attributes in *S. mutans* (Kajfasz et al. 2009). Compared to its parent strain, the Δ*clpL* strain displayed an enhanced capacity to form biofilms in the presence of sucrose, had reduced long-term viability, and was more sensitive to acid killing (Kajfasz et al. 2009). The Δ*clpP* and Δ*clpX* strains displayed several phenotypes in common, including an enhanced capacity to form biofilms in sucrose, a reduced autolysis, and a reduced ability to grow under stress. Unexpectedly, the Δ*clpP* and Δ*clpX* mutants were more resistant to acid killing and demonstrated an enhanced viability in long-term survival assays (Kajfasz et al. 2009).

The Spx (suppressor for *clpP* and *clpX*) global regulator is conserved among low-G+C Gram-positive bacteria whose expression and activity are controlled at the transcriptional level and by ClpP-dependent proteolysis (Zuber 2004). Spx was first identified in the soil organism *B. subtilis*, when it was demonstrated that the accumulation of Spx was responsible for the pleiotropic phenotypes associated with *clpP* and *clpX* mutations (Zuber 2004). The *B. subtilis* Spx was shown to participate in the oxidative stress response by interacting with the C-terminal domain of the RNA polymerase α-subunit (α-CTD). The inactivation of either one of the genes coding for the Spx regulator in *S. mutans* (designated SpxA and SpxB) bypassed phenotypes of the *clpP* and *clpX* mutants, suggesting that the underlying mechanisms by which *S. mutans* ClpXP protease affects virulence traits are associated with the accumulation of these two Spx orthologs (Kajfasz et al. 2009). The subsequent characterization of strains lacking the *spx* genes revealed that Spx global regulation participates in the regulation of processes associated with *S. mutans* pathogenesis (Kajfasz et al. 2010). More specifically, the *spxA* mutant displayed an impaired ability to grow under acidic and oxidative stress conditions and had diminished long-term viability at low pH. Although the *spxB* mutant strain did not show any inherent stress-sensitive phenotype, the phenotypes observed in the *spxA* mutant were more pronounced in the Δ*spxA* Δ*spxB* double mutant (Kajfasz et al. 2010). In addition, in vivo studies strongly supported the association of Spx with the virulence of *S. mutans* (Kajfasz et al. 2010).

Despite the need to endure periods of nutrient limitation, an abrupt exposure to an excess amount of carbohydrate in the diet can result in the rapid accumulation of toxic glycolytic intermediates, acidification of the environment, and osmotic stress. To survive nutrient starvation, to cope with the detrimental effects of glycolytic intermediates, and to maintain proper NAD/$NADH^+$ balances, *S. mutans* developed a sophisticated regulatory network that combines transcriptional regulation with allosteric modulation of enzymatic activities to coordinate an optimal flow of carbohydrates.

Carbohydrate source and availability is a key factor affecting the pathogenic potential of oral biofilms. The sugar phosphotransferase system (PTS) is the major carbohydrate transport system in oral streptococci, especially under carbohydrate-limiting conditions. In addition to participating in sugar uptake, PTS components influence many other cellular processes. Mutations in the ManL PTS permease influence biofilm development, the regulation of acid tolerance, and the global control of gene expression, in particular, carbon catabolite repression (Abranches et al. 2003, 2006). Two global regulators of central metabolism genes, CcpA and CodY, have been shown to impact acid tolerance and the expression of other virulence traits of *S. mutans* (Abranches et al. 2008; Lemos et al. 2008). In addition to serving as a global regulator in response to carbohydrate availability, a CcpA-deficient strain was substantially more acid-resistant than its parent (Abranches et al. 2008). The enhanced acid tolerance of the CcpA mutant was associated with increases in the expression of the PTS that result in higher rates of ATP generation through glycolysis. Microarrays revealed that CodY is indeed a global regulator of gene expression in *S. mutans*, and phenotypic studies revealed that the *codY* mutant had reduced capacities to form biofilms and was more sensitive to growth at low pH (Lemos et al. 2008).

The nutritional alarmone (p)ppGpp also appears to play an important role in orchestrating an appropriate response to multiple environmental and physiologic inputs that *S. mutans* encounters in the oral cavity. In *S. mutans*, the bifunctional RelA enzyme was shown to play a major role in the regulation of phenotypic traits that are required for establishment, persistence, and survival (Lemos et al. 2004;

Nascimento et al. 2008), further supporting an overlap between circuits that govern nutrient starvation, general stress tolerance, and biofilm formation. Until recently, RelA was considered the sole enzyme responsible for the synthesis and degradation of (p)ppGpp in Gram-positive bacteria. However, Lemos and coworkers recently identified two novel enzymes, designated RelP and RelQ, with (p)ppGpp-synthase activities in *S. mutans* that could be found in a number of related Gram-positive bacteria (Lemos et al. 2007a). A *relA/relP/relQ* triple mutant was auxotrophic for the BCAAs leucine and valine, but not isoleucine, a phenotype that was directly related to the CodY-dependent repression of genes involved in the synthesis of BCAAs (Lemos et al. 2008). Interestingly, RelP is cotranscribed with, and apparently regulated by, the RelRS two-component signal transduction system (TCS) (Lemos et al. 2007a), suggesting that *S. mutans* may use environmental signals to optimize cell growth and survival in a manner that allows the organism to balance growth during dietary intake by the host with the capacity to rapidly mount an adaptive response during fasting periods. Consistent with the role of (p)ppGpp in bacteria, homologs of RelRS in *S. pyogenes*, designated SptRS, were shown to be critical for this bacterium to survive in saliva (Shelburne et al. 2005).

Metal ions, including iron and manganese, have also been implicated in stress tolerance and the regulation of virulence expression by *S. mutans*. In particular, the SloR metalloregulator was shown to modulate *S. mutans* biofilm formation, genetic competence, and oxidative stress tolerance in response to manganese availability (Rolerson et al. 2006). Work from the Spatafora lab has linked SloR repression of the transcriptional regulator *gcrR* with acid-stress tolerance (Dunning et al. 2008). More specifically, a *gcrR* mutant was more sensitive to low pH, and this phenotype was linked to the capacity of the mutant to maintain ΔpH homeostasis.

S. mutans lives almost exclusively in densely populated biofilms that form on the tooth surface. As is typical of bacteria with highly specialized niches, a very small number of alternative sigma factors were found in the *S. mutans* genome (Ajdic et al. 2002). Thus, regulatory systems that sense intrinsic and extrinsic signals to coordinate appropriate gene expression patterns appear to play a very large role in the organism's stress tolerance.

Quorum-sensing (QS) systems coordinate gene expression in a cell density–dependent manner, and QS genes have been shown to participate in the regulation of several physiological activities, including biofilm differentiation and stress management. There are at least two QS systems in *S. mutans* and related oral streptococci, and both of these are tied intimately in the tolerance of acid and other stresses and in biofilm formation. The first is the competence regulon, which relies on the CSP (competence stimulating peptide) signaling molecule that is detected by the *comDE* TCS. The inactivation of *comD* and *comE* conferred an acid-sensitive phenotype to *S. mutans* and resulted in the formation of biofilms with reduced biomass (Li et al. 2001, 2002a). The second QS in *S. mutans* is the widely distributed *luxS* system, which produces autoinducer-2 (AI-2), a furanosyl borate diester that regulates a large panel of genes in a variety of microorganisms. A strain of *S. mutans* lacking LuxS was shown to have altered biofilm architecture and to have increased sensitivity to acid killing (Wen and Burne 2004).

Sequence analysis identified 14 TCSs in *S. mutans* UA159 (Ajdic et al. 2002; Biswas et al. 2008). The global transcriptional analysis of acid-inducible genes revealed that genes encoding multiple TCSs, including CiaHR, LevSR, ScnKR, LiaSR, and ComDE, were upregulated during acid adaptation (Gong et al. 2009). Over the past few years, several studies have evaluated the role of TCS in *S. mutans* (Li et al. 2001, 2002a, b; Qi et al. 2004; Chen et al. 2008; Deng et al. 2007a; Levesque et al. 2007; Biswas et al. 2008; Senadheera et al. 2009), revealing that TCSs regulate virulence gene expression, the induction of competence, biofilm development, bacteriocin production, and stress tolerance. In particular, two studies from independent laboratories concerning the systematic inactivation of genes encoding sensor kinases of the TCSs evaluated their role in stress tolerance by *S. mutans* (Levesque et al. 2007; Biswas et al. 2008). In the study by Levesque et al., *scnK* and *levS* mutants displayed significantly slower growth at pH 5.5, whereas the *ciaH* mutant grew better than the parental strain in the presence of NaCl or H_2O_2 (Levesque et al. 2007). Biswas and coworkers found that the inactivation of three sensor kinases, *liaS*, *ciaH*, and *vicK*, affected the stress tolerance of strain UA159 (Biswas et al. 2008). The *liaS* and *ciaH* mutants showed an increased sensitivity to puromycin and reduced growth when incubated in aerobic conditions or on agar medium supplemented with H_2O_2 (Biswas et al. 2008). The *ciaH* mutant also showed a significant reduction of growth at pH 5.0 and displayed an increased sensitivity to DNA damage caused by mitomycin C (Biswas et al. 2008). Notably, previous reports have also shown that the inactivation of *ciaH* resulted in an acid-sensitive phenotype in strains UA159 and UA140 (Qi et al. 2004; Ahn et al. 2006). The *S. mutans* VicRK TCS system was shown to respond to, and protect against, oxidative stress in one particular study (Deng et al. 2007a), but not in the studies discussed above (Levesque et al. 2007; Biswas et al. 2008). A role in oxidative stress response was also assigned to ScnRK, as *scnRK* mutants were more sensitive to H_2O_2 and more susceptible to phagocytic killing in nonactivated macrophages (Chen et al. 2008). Finally, studies from the Cvitkovitch lab have shown that the inactivation of *liaS* or *comDE* conferred an acid-sensitive phenotype to strains NG8 and BM71, respectively (Li et al. 2001, 2002a), although *comD* and *comE* do not appear to affect acid tolerance in strain UA159 (Ahn et al. 2006). Collectively, these data support the idea that there may be substantial heterogeneity among strains in the role of specific TCSs, not only in the genes they regulate, but also in the external stimuli to which they respond. Nevertheless, the CiaRH TCS has been consistently found to play a role in the stress responses by *S. mutans*, and has also been implicated in competence development, bacteriocin production, and biofilm formation (Qi et al. 2004; Ahn et al. 2006; Levesque et al. 2007). More recently, it was demonstrated that the *ciaRH* operon of *S. mutans* consists of three genes, with the first gene, named *ciaX*, encoding a small, double-glycine signaling peptide that allows CiaRH to modulate its own expression in response to calcium (He et al. 2008). The inactivation of *ciaX*, or point mutations in its calcium-binding domain, resulted in a diminished biofilm formation that was rescued by added calcium. Human saliva is saturated in calcium, and calcium is the principal cation in tooth enamel, so calcium signaling may be an important regulator, through CiaRH, of stress responses in *S. mutans*.

Although oxidative stress is an important factor in plaque ecology (Marquis 1995), few detailed studies of the molecular mechanisms regulating oxidative stress responses in *S. mutans* are available. Yet these organisms are constantly exposed to oxidative agents from either intracellular or extracellular origin. During respiration, ROS are generated inside the cell from single-electron reductions of oxygen, and the cells must cope with oxygen metabolites in order to survive. External sources of ROS include host defenses, such as the lactoperoxidase system, as well as peroxide-containing oral hygiene products and Fenton chemistry. The production of H_2O_2 by some oral bacteria, such as *S. sanguinis*, *Streptococcus gordonii*, and *Streptococcus oligofermentans*, is also believed to add to the oxidative stress burden of oral biofilm bacteria (Ryan and Kleinberg 1995; Tong et al. 2007).

Streptococci do not possess cytochromes and therefore do not carry out oxidative phosphorylation. Instead, the bulk of the respiration activity in these organisms is due to NADH oxidases (Nox) (Marquis 1995). Oral streptococci also lack catalase, a major protective enzyme against oxidative insults, but harbor superoxide dismutase (SOD), NADH peroxidase, glutathione reductase (Gor), and alkyl hydroperoxide reductase (Marquis 1995). The growth mode of bacteria, that is, planktonic or as biofilms, largely influences the respiratory rates and the levels of activity of protective enzymes like SOD and Nox. In a study comparing the levels of NADH-oxidase and SOD in planktonic and biofilm populations of several oral streptococci species, Nguyen et al. demonstrated that both enzymes were repressed in the biofilm state (Nguyen et al. 2002). It was suggested that oxidative stress appears to be minimized in streptococcal biofilms because of reduced oxygen metabolism by the bacteria. Reducing oxygen metabolism and consequently the production of ROS could be an important strategy to live in a slow-growth mode and crowded environment such as a biofilm.

A small number of studies have assessed the roles of the oxidative stress genes in *S. mutans*. Although the *S. mutans* H_2O_2-forming Nox and alkyl hydroperoxide reductase (AhpCF) enzymes were shown to have a protective role against oxidative stress in *E. coli*, the inactivation of one or both enzymes did not affect peroxide tolerance or the ability of *S. mutans* to grow aerobically (Higuchi et al. 1999). Similarly, a strain lacking the SOD enzyme (SodA) was able to grow aerobically, albeit more slowly than the parent strain (Nakayama 1992). However, the inactivation of the iron-biding Dpr significantly impaired oxygen tolerance in *S. mutans* (Yamamoto et al. 2000). In particular, double Δ*dpr* Δ*sod* and triple Δ*ahpC* Δ*nox* Δ*dpr* mutants were completely unable to grow in the presence of air (Yamamoto et al. 2000), indicating that Dpr plays a protective role against oxidative stress. Based on its iron-binding ability, the most logical interpretation is that Dpr indirectly contributes to oxygen tolerance through the titration of free intracellular iron ions, avoiding the formation of toxic hydroxyl radicals via the Fenton reaction.

Few other gene products have been implicated in oxidative stress tolerance. Recently, random insertion mutagenesis was used to identify genes involved in superoxide tolerance (Zhang and Biswas 2009). One of the identified genes was shown to encode a 3′-phosphoadenosine-5′-phosphate phosphatase (pAp), but the molecular mechanisms by which pAp confers superoxide tolerance remains unclear.

In another study, the cell envelope-associated eukaryotic serine/threonine protein kinase (STPK) was shown to play a protective role against H_2O_2 (Zhu and Kreth 2010). Of note, STPK was also shown to be involved in biofilm formation, competence, and acid tolerance (Hussain et al. 2006). The surface-associated BrpA was also found to play a role in H_2O_2 tolerance as well as biofilm development, autolysis, and acid tolerance (Wen et al. 2006). A comparison of the transcriptomes of the *brpA* mutant and the parent revealed significant alterations in the expression of genes involved in cell wall biogenesis, stress tolerance, and adherence (Wen et al. 2006). It was proposed that increases in autolysis, as a result of BrpA inactivation, could contribute to the increased susceptibility to acid and oxidative stresses in the Δ*brpA* strain (Wen et al. 2006).

It is noteworthy that many of the gene products addressed above are required both for an appropriate response by organisms to environmental stress and to form biofilms, suggesting that the stress regulon of *S. mutans* may be responsible for controlling a broader set of biological functions when compared to organisms with more complex genomes. The gradients that develop during biofilm maturation allow for the accumulation of end products, the concentration of diffusible molecules, the generation of heterogeneity in the growth domain within the populations, and reduced access to nutrients. Thus, an appropriately regulated response, both temporally and spatially, by the microorganisms to the environmental stresses they encounter within a maturing biofilm may have a profound influence on biofilm structure.

12.3 *Streptococcus pyogenes* (Group A *Streptococcus*)

Along with *S. pneumoniae*, *S. pyogenes* (also referred to as GAS) is responsible for the great majority of deaths associated with this genus. It has been estimated that more than 500,000 deaths due to GAS infections occur each year worldwide (Carapetis et al. 2005). GAS is a strictly human pathogen that causes a broad spectrum of diseases, ranging from relatively mild suppurative throat and skin infections to often-fatal invasive infections such as necrotizing fasciitis (flesh-eating disease), toxic shock syndrome, and septicemia. Moreover, a large number of deaths are due to the development of poststreptococcal infections sequelae, which include rheumatic fever, acute glomerulonephritis, and reactive arthritis (Cunningham 2000).

The ability of GAS to survive and replicate in diverse anatomic sites such as the skin, throat, and blood implies that this organism can efficiently adapt to different microenvironments, including different temperatures, pH, oxygen tension (pO_2), as well as fluctuations in essential nutrients. For example, temperatures in the human body can range from the high 20s (°C) at the extremes of superficial skin to 40°C during fever episodes (Smoot et al. 2001). During invasive infection, GAS must cope with the host's innate immune response, which includes the oxidative burst mediated by macrophages and polymorphonuclear neutrophils. These professional phagocytic cells are capable of producing a number of highly bactericidal ROS, including superoxide anions ($O_2^{\bullet-}$), H_2O_2, and hydroxyl radicals ($OH^{\bullet}$). Also, during invasive infections, the pH may fall to 6.0 or less following the formation of necrotic lesions or abscesses (Simmen and Blaser 1993). While the pH of the skin is normally acidic, the release of perspiration results in increased salt concentrations that increase osmolarity and pH. Finally, glucose concentrations in the host can range from 3 to 6 mM in blood to as low as 0.02–0.4 mM in saliva (Shelburne et al. 2008b). In light of their remarkable ability to adapt to a variety of human tissues, numerous studies have specifically demonstrated that GAS sense and respond to environmental changes by regulating the expression of important virulence factors (Chaussee et al. 1997; Graham et al. 2005; Virtaneva et al. 2005; Loughman and Caparon 2006a).

Classic virulence factors are considered those encoding products that directly cause damage to the host and are unique to the pathogen. However, it is also clear that factors that allow an organism to survive the host microenvironments are also crucial for pathogenesis. In the specific case of GAS, considerable effort has been put into understanding the global responses of this pathogen to the host environment. Microarray analysis has been used to identify genes that contribute to survival in human polymorphonuclear leukocytes, saliva, and blood, to different temperatures, during mouse soft tissue infection, and during experimental pharyngitis in macaques (Smoot et al. 2001; Graham et al. 2002, 2005, 2006; Virtaneva et al. 2003, 2005; Shelburne et al. 2005). These studies have highlighted the importance of several TCSs, in particular, CovRS (*co*ntrol of *v*irulence), in coordinating the expression of genes encoding virulence factors, as well as catabolic and biosynthetic enzymes (Graham et al. 2002, 2005; Virtaneva et al. 2005). As mentioned above, another TCS, named *sptRS*, for *s*aliva *p*ersis*t*ence, was shown to contribute to GAS survival in saliva by regulating metabolic pathways and virulence factor production (Shelburne et al. 2005).

In addition to TCS regulation, trancriptome studies have revealed that complex carbohydrate metabolism has an important role in the pathogenesis of invasive GAS diseases (Shelburne et al. 2008a, b). In particular, the catabolite control protein (CcpA) and the aldolase LacD.1 were shown to control transcript levels of several carbohydrate utilization genes as well as key virulence factors, such as the cysteine protease SpeB (Loughman and Caparon 2006b; Shelburne et al. 2008a). Moreover, transcriptome analysis has also revealed that the CovRS and SptRS TCSs control the expression of carbohydrate metabolism, coming full circle with the concept that GAS uses its nutritional cues to couple complex carbohydrate metabolism with virulence factor production (Shelburne et al. 2008b).

Because GAS is a polyauxotrophic organism, global nutrient sense regulators such as (p)ppGpp and CodY have been linked to GAS survival and virulence gene expression under nutritional stress conditions. These regulators are likely to play an important role during conditions that resemble in vitro stationary-phase growth, such as during initial colonization of the throat or skin. The nutritional alarmone (p)ppGpp accumulates during amino acid starvation, triggering the stringent response, namely, the downregulation of genes for macromolecular biosynthesis and the upregulation of genes involved in stress tolerance, and amino acid biosynthesis and transport (Potrykus and Cashel 2008). In Gram-positive bacteria, RelA is a bifunctional enzyme with potent (p)ppGpp-synthetic activity and -degradative activities, whereas one or two small (p)ppGpp synthetases appear to be responsible for maintaining low basal levels of (p)ppGpp (Lemos et al. 2007a). CodY is a highly conserved global regulator of Gram-positive bacteria that helps cells to adapt to poor nutritional conditions (Sonenshein 2005). Studies using *relA* and *codY* deletional mutants revealed that (p)ppGpp and CodY regulate the expression of key metabolic enzymes, transporters, and virulence factors (Malke et al. 2006; Malke and Ferretti 2007).

In addition to global transcriptional regulators and regulators of central metabolic pathways, two intimately related processes, metal regulation and oxidative stress responses, have also been implicated in GAS pathogenesis. Metal ions are essential micronutrients to bacterial homeostasis. For example, iron is an essential cofactor of bacterial metabolism that regulates the virulence gene expression of many pathogenic bacteria. In the human body, very little free iron is present, as most of the iron in the body is found intracellularly or bound to glycoproteins, such as transferrins and lactoferrins. In GAS, at least three ABC transporters (FtsABCD, HtsABC, and MtsABC) have been associated with iron uptake (Ferretti et al. 2001). However, an excess of iron is toxic as a result of hydroxyl radicals that are produced by the Fenton reaction. Thus, the processes associated with iron acquisition must be tightly regulated. The GAS genome encodes two metalloregulators, PerR and MtsR, which regulate the expression of proteins involved in metal ion uptake. MtsR was shown to regulate the expression of *mtsABC* and *htsABC*, suggesting that this regulator may play an important role in GAS virulence (Hanks et al. 2006). It has been reported that the inactivation of *mtsR* reduces GAS virulence in a zebrafish infection model (Bates et al. 2005), but not in a mouse model of subcutaneous infection (Hanks et al. 2006). Interestingly, genomewide molecular analysis revealed that a naturally occurring single-nucleotide insertion in the *mtsR* gene that created a premature stop codon was epidemiologically associated with a significantly decreased number of human necrotizing fasciitis cases (Beres et al. 2006). Subsequent work revealed that the inactivation of *mtsR* resulted in the dysregulation of the extracellular cysteine protease SpeB (Olsen et al. 2010). PerR is a transcriptional repressor that has been linked to peroxide stress resistance and metalloregulation (Brenot et al. 2005). Although a *perR* mutant was hyperresistant to peroxide, animal studies have shown that the inactivation of *perR* results in attenuated GAS virulence in skin, soft tissue, and oropharynx infections (Ricci et al. 2002; Brenot et al. 2007; Gryllos et al. 2008). In the case of oropharynx infections, the attenuated virulence was associated with a reduced resistance to phagocytic killing (Gryllos et al. 2008). Surprisingly, the PerR regulon appears to include genes involved in sugar uptake and utilization, suggesting a link among central metabolic processes, oxidative stress, and virulence (Gryllos et al. 2008).

Given the nature of GAS infections, which are highly inflammatory, resulting in the production of ROS, the oxidative stress responses of GAS have been investigated in some detail. One product that lies directly in the intersection between iron and oxidative stress is Dpr (MrgA) (Brenot et al. 2005; Tsou et al. 2008). In vitro studies have shown that PerR regulates *dpr* expression and that Dpr provides protection against multiple stresses by preventing the Fenton reaction (Tsou et al. 2008). Although Dpr functions as a general stress protein protecting the cells against multiple stresses, a *dpr* mutant was fully virulent in two different animal models (Brenot et al. 2005). Few other genes involved in oxidative stress management have been studied in GAS. Nox is a flavoprotein that catalyzes the four-electron reduction of O_2 to H_2O. The inactivation of *nox* resulted in a reduced aerotolerance and an increased sensitivity to the superoxide-generating agent paraquat (Gibson et al. 2000). Functional analysis identified two other peroxidases in the GAS genome, alkyl hydroperoxide reductase (AhpC) and glutathione peroxidase (GpoA) (King et al. 2000). The inactivation of one or both peroxidases resulted in an increased sensitivity to superoxide-generating agents, but neither enzyme was essential for growth under aerobic conditions (King et al. 2000). Notably, GpoA was essential for GAS pathogenesis in several murine models that mimic streptococcal suppurative diseases (Brenot et al. 2004) but, on the other hand, was dispensable for virulence in the zebrafish model of streptococcal myositis, an infection characterized by the absence of the oxidative stress that is triggered by inflammatory responses (Brenot et al. 2004).

QS systems have long been considered important for modulating bacterial gene expression in response to environmental changes that affect population density. By inactivating *luxS*/AI-2 in strains belonging to two different serotypes, serotypes M1 and M19, Siller et al. demonstrated that the loss of LuxS leads to increased acid tolerance, increased cell invasion, and intracellular survival, suggesting that low levels of AI-2 provide an advantage for bacterial survival under conditions that are relevant during infection (Siller et al. 2008). The increased internalization of the *luxS* mutants in the study above is in agreement with a previous study that demonstrated that a *luxS* mutant from an invasive serotype strain (serotype 3) was internalized by Hep-2 cells more efficiently than its parent strain (Marouni and Sela 2003). Although the expression of virulence-associated factors was affected in the *luxS* mutants (Marouni and Sela 2003; Siller et al. 2008), the exact role of the LuxS/AI-2 system in *S. pyogenes* pathogenesis remains to be elucidated.

While we have seen some reasonable progress toward understanding the mechanisms controlling the stress responses of GAS, several questions remain to be answered. Given the relevance of stress survival and adaptation to GAS virulence expression, there is a major need to continue to identify and characterize the stress-response pathways of this dangerous pathogen.

12.4 *Streptococcus agalactiae* (Group B *Streptococcus*)

S. agalactiae (also referred to as GBS) is a commensal organism colonizing the gastrointestinal or genital tract of 10–30% of healthy human adults, depending on the gender, ethinicity, and geographical region (Shet and Ferrieri 2004; Phares et al. 2008). However, it can also cause life-threatening invasive infections in susceptible hosts, such as newborn infants, pregnant women, and nonpregnant adults with underlying chronic illnesses (Shet and Ferrieri 2004; Phares et al. 2008). In particular, GBS can be vertically transmitted from a colonized mother to her newborn by the aspiration of contaminated amniotic fluid during delivery and is the leading cause of invasive neonatal infections such as pneumonia, meningitis, and septicemia. Despite antibiotic therapy, mortality rates in neonatal meningitis are extremely high, and 25–50% of surviving infants can be left with permanent neurological sequelae (Edwards et al. 1985). In addition to human infections, GBS is also a leading cause of bovine mastitis (Keefe 1997).

The pathophysiology of GBS infections indicates that this bacterium is capable of adapting to large environmental fluctuations within the human host, including changes in pH, osmolarity, oxygen tension, and temperature. For example, while the pH of the amniotic fluid or the fetal lung is close to neutral, GBS can survive the acidic pH environment of the vagina or intracellular endocytic compartments. The levels of pO_2 in tissues infected by GBS can also vary dramatically, ranging from atmospheric O_2 levels to near anaerobiosis (Johri et al. 2003). During invasive infection, GBS must cope with the host's innate immune response, which includes the oxidative burst mediated by macrophages and polymorphonuclear neutrophils.

In comparison to other medically relevant *Streptococcus*, studies on the stress responses of GBS have been rather limited. Still, a few of the mechanisms used by GBS to cope with oxygen and ROS have been uncovered. Liu and coworkers demonstrated that a characteristic GBS pigment is used as a physical barrier at the bacterial surface to protect against ROS (Liu et al. 2004). Another study revealed that ROS detoxification is highly dependent on the Mn-dependent SOD (SodA), as an *sodA* mutant strain was highly susceptible to oxidative stresses in vitro, showed increased susceptibility to macrophage killing, and survived poorly in the blood of infected mice (Poyart et al. 2001). The H_2O-forming Nox-2 is another oxidative stress system that was shown to participate in GBS virulence (Yamamoto et al. 2006). Interestingly, while the inactivation of *nox2* resulted in an aerobic growth arrest, this defect was not due to the accumulation of toxic ROS, but due to an underlying defect in fatty acid biosynthesis. It was speculated that NAD^+ depletion in the *nox2* mutant resulted in reduced acetyl-CoA production, which in turn affected fatty acid biosynthesis. Another interesting observation was that GBS, while primarily a fermentative organism, undergoes respiration metabolism given that exogenous sources of quinone and heme are available (Yamamoto et al. 2005). This finding appears to be particularly important, as respiration metabolism was shown to facilitate GBS dissemination and virulence by promoting survival in blood (Yamamoto et al. 2005).

Apart from the studies described above, the publication of the complete genome sequence of GBS strains offered some insights into the stress-adaptation mechanisms of GBS (Glaser et al. 2002; Tettelin et al. 2002, 2005). In addition to the SodA and Nox-2 enzymes previously studied (Poyart et al. 2001; Yamamoto et al. 2006), the genome of GBS encodes several other putative enzymes involved in ROS detoxification and repair, including a thiol peroxidase (Tpx), a thioredoxin/thioredoxin reductase (TrxAB), an alkylhydroperoxidase (AhpCF), and a glutathionine reductase (Gor). However, the role of these enzymes in protection against oxidative stress and virulence remains to be investigated.

Molecular chaperones and proteases are another important group of proteins involved in stress protection. In addition to the ubiquitous GroEL and DnaK chaperones, the GBS core genome encodes subunits of the Clp proteolytic system, which include the ClpP peptidase and four ATP-binding subunits (ClpC, ClpE, ClpL, and ClpX) (Glaser et al. 2002; Tettelin et al. 2005). A mutant lacking the ClpP peptidase was more sensitive to multiple stresses, a phenotype that was linked to increases in total protein oxidation (Nair et al. 2003).

Alternative sigma factors, transcriptional regulators, and TCSs participate in the coordinated regulation of gene expression in response to environmental changes playing a critical role in stress adaptation. The GBS genome encodes a high proportion of regulators (approximately 5% of the genome encode regulatory genes), suggesting that GBS is highly capable of adapting to various environments. In particular, GBS encodes an alternative ECF-type sigma factor, which is present in *Streptococcus equi* but absent in other medically important streptococci (for example, *S. pneumoniae*, *S. mutans*, and GAS), and encodes a higher number of TCSs (17–21 depending on the sequenced strain) in comparison to related species: 14 in *S. pneumoniae*, 14 in *S. mutans*, and 13 in GAS. Only a few TCSs in GBS have been studied to date, and only two have been assigned a role in stress adaptation. The CsrRS (also called CovRS), a major regulator of virulence in GAS (see Sect. 12.3 for details), was shown to also regulate virulence expression in GBS. The inactivation of *csrRS* affected the expression of several virulence factors, such as reduced CAMP activity and increased hemolytic activity, and resulted in a significant reduction of virulence in rodents (Lamy et al. 2004; Jiang et al. 2005). The phenotypes of the *csrRS* mutants were linked to a large number of genes controlled by CsrRS, including several secreted or cell surface–associated proteins (Lamy et al. 2004). Interestingly, a number of genes involved in pH adaptation are controlled by CsrRS, suggesting that CsrRS participates in the adaptation of GBS to different pH environments (Santi et al. 2009). It has been hypothesized that the translocation of GBS from the acidic milieu of the vagina to the neutral pH of neonatal tissues signals the conversion from a colonizing to an invasive phenotype in a CsrRS-dependent manner (Santi et al. 2009). In addition to CsrRS, the CiaRH TCS was also implicated in adaptation to the host environment. Specifically, inactivation of the *ciaR* response regulator resulted in a significant decrease in intracellular survival within neutrophils, macrophages, and brain microvascular endothelial cells, which was linked to an increased susceptibility to antimicrobial peptides, lysozyme, and ROS (Quach et al. 2009).

The completion of the GBS genome sequence also facilitated the application of expression microarray analysis to investigate the responses of GBS to different environments. A handful of studies have analyzed the responses of GBS to different temperature, growth phase, as well as growth in the human amniotic fluid, or blood (Mereghetti et al. 2008a, b, 2009; Sitkiewicz et al. 2009; Sitkiewicz and Musser 2009). Transition from 30 to 40°C resulted in the upregulation of genes encoding virulence factors such as hemolysin and extracellular secreted proteins and of genes involved in purine metabolism and iron acquisition (Mereghetti et al. 2008b). Unexpectedly, the expression of genes involved in adaptation and stress responses was not induced by growth in the amniotic fluid, suggesting that the organism can readily adapt to this environment (Sitkiewicz et al. 2009). On the other hand, incubation in whole human blood resulted in complex and dynamic changes in the expression of transcriptional regulators, stress-response genes, and genes involved in carbohydrate metabolism (Mereghetti et al. 2008a).

12.5 *Streptococcus pneumoniae*

The human pathogen *S. pneumoniae*, also known as pneumococcus, is the causative agent of acute respiratory infections, otitis media, sepsis, and meningitis. Pneumococcus is the leading cause of bacterial pneumonia, meningitis, and sepsis in children, which are estimated to result in over one million child deaths worldwide (WHO 2007). In addition, *S. pneumoniae* is also the leading agent of severe community-acquired pneumonia and meningitis in the elderly and in immunocompromised individuals (WHO 2007). While multivalent capsular vaccines and antibiotic therapies are useful to control *S. pneumoniae* infections, multiple antibiotic-resistant strains and the evasion of current vaccines by serotype selection continue to increase worldwide.

Normal disease progression begins with colonization of the nasopharyngeal niche followed by invasion of normally sterile sites such as the lung, meninges, and bloodstream. During the course of infection, *S. pneumoniae* encounters numerous stress conditions, including changes in temperature (from 30 to 34°C at the upper respiratory tract to 37–40°C at deeper tissues), pH fluctuations, exposure to ROS generated by host phagocytes, nutrient deprivation, and, in some cases, antibiotic stress. In addition, *S. pneumoniae* normally resides in the oxygen-rich upper respiratory tract and therefore must be able to cope with toxic levels of oxygen present in this environment.

Compared to GAS and GBS, the stress-response mechanisms of *S. pneumoniae* have been studied in a little more detail. The surface-associated serine protease HtrA was shown to play a role in the ability of pneumococcus to grow at high temperatures, resist oxidative stress, and undergo genetic transformation (Ibrahim et al. 2004b). Also, the inactivation of *htrA* resulted in a major attenuation of virulence in mouse infection models, indicating that HtrA is an important virulence factor in pneumococcus (Ibrahim et al. 2004b). The Clp proteolytic/molecular chaperone system has been the subject of extensive studies in pneumococcus. Similar to GBS (Glaser et al. 2002), the *S. pneumoniae* core genome encodes four Clp ATPases (ClpC, ClpE, ClpL, and ClpX) and a single ClpP peptidase (Tettelin et al. 2001). Among the four Clp ATPases, ClpL is mainly found in Gram-positive organisms and does not have the recognition tripeptide responsible for the interaction with ClpP. Thus, ClpL is expected to function mainly as a molecular chaperone performing essential housekeeping functions, including protein reactivation activities, folding, and translocation. Although growth of the *clpL* mutant was not affected at 30 or 37°C, growth of Δ*clpL* was significantly impaired at 43°C, suggesting a role for ClpL during heat stress (Kwon et al. 2003). The Δ*clpL* strain was able to adhere to and invade epithelial cells much more efficiently than the wild-type strain but was more susceptible to macrophage killing and showed a similar level of virulence when compared to the parent strain in three different murine models of infection (Kwon et al. 2003; Tu le et al. 2007). Among the remaining four *clp* genes encoding for ClpP, and the ClpC, ClpE, and ClpX ATPases, the *clpX* gene was found to be essential, whereas *clpC*, *clpE*, and *clpP* were dispensable (Robertson et al. 2003). The essentiality of *clpX* but not of *clpP* was unexpected given that ClpX functions as a regulatory subunit for the ClpXP protease. Transcriptional analysis of *clpC*, *clpE*, and *clpP* revealed that the expression of these genes was induced in response to high temperatures, a stress response that was mediated by the activity of the CtsR repressor (Charpentier et al. 2000; Chastanet et al. 2001). The ClpC ATPase was found to participate in several important aspects of pneumococcal physiology, including growth under heat stress, cell division, autolysis, adherence, and competence development (Charpentier et al. 2000). However, these findings are a subject of controversy, as a second report proposed that ClpC was not involved in growth at high temperatures, autolysis, and competence (Chastanet et al. 2001). This controversy has been partially settled by a third report that indicated that the contribution of ClpC to thermotolerance and autolysis may be strain-dependent (Ibrahim et al. 2005). In addition to being required for growth under heat stress, ClpE was shown to contribute to pneumococcal virulence as assessed in a mouse intraperitoneal infection model (Zhang et al. 2009). Finally, the ClpP peptidase appears to play a major role in the expression of virulence of *S. pneumoniae*. Several studies have shown that ClpP-mediated proteolysis participates in many processes of pneumococcal physiology, including stress responses, competence development, and virulence (Kwon et al. 2003). Subsequent work linked the attenuated virulence of *clpP* mutants to an increased sensitivity to oxidative stress in macrophages, in particular to reactive nitric oxide (NO) intermediates (Park et al. 2010). The virulence of Δ*clpP* strains was found highly attenuated in several murine infection models (colonization of the nasopharynx, lung survival, intraperitoneal infection, and sepsis) (Robertson et al. 2002; Kwon et al. 2003, 2004). Notably, immunization of mice with ClpP elicited a protective immune response against invasive pneumococcal infections, making ClpP an attractive antigen candidate for the development of a broad-range pneumococcal vaccine (Kwon et al. 2004; Cao et al. 2007, 2009).

In *S. pneumoniae*, two Spx-like proteins, designated SpxA1 and SpxA2 (homologous to the *S. mutans* SpxA and SpxB global regulators, respectively), were identified and the gene encoding SpxA1 was partially characterized (Turlan et al. 2009). Single mutations in either of the *S. pneumoniae spx* genes could be stably isolated, but the simultaneous inactivation of *spxA1* and *spxA2* was lethal. Although the association between the accumulation of Spx and the underlying mechanisms by which ClpP proteolysis affects virulence traits in *S. pneumoniae* was not explored, it was demonstrated that transcriptional repression by SpxA1 has a negative effect on the development of the X-state (competence), a proposed general stress-response mechanism in *S. pneumoniae* (Claverys et al. 2006) also linked to cannibalism or fratricide (Claverys et al. 2007). It was hypothesized that SpxA1-dependent repression could act to sense environmental or metabolic stresses and prevent the initiation of the X-state development cascade in the absence of stress (Turlan et al. 2009). However, the role of Spx in the regulation of the *S. pneumoniae* oxidative stress responses and the possible involvement of SpxA1 and SpxA2 with the phenotypes observed in the *clpP* mutant strain remain to be explored.

The role of the global nutrient sense regulators (p)ppGpp and CodY (see Sects. 12.2 and 12.3 for details) have also been explored in *S. pneumoniae*. The inactivation of the bifunctional (p)ppGpp synthetase/hydrolase RelA homolog, *rsh* or rel_{Spn}, did not affect growth in a complex medium, but the Δrel_{Spn} did not grow in a chemically defined medium unless supplemented with the metals copper and manganese (Kazmierczak et al. 2009). Most importantly, it was demonstrated that Rel_{Spn} is a major virulence factor in a murine pneumonia/sepsis model of infection (Kazmierczak et al. 2009). The inactivation of the nutritional repressor *codY* significantly affected colonization of the nasopharynx, a finding that was supported by the diminished ability of the *codY* mutant to adhere to nasopharyngeal cells in vitro (Hendriksen et al. 2008).

In addition to Spx and the nutrient-sensing regulators (p)ppGpp and CodY, other regulatory systems that coordinate *S. pneumoniae* gene regulation in response to metabolic and environmental stresses have been investigated. Among those are signal transduction pathways such as TCSs and the eukaryotic-type STPKs. In streptococci, a single STPK-like gene (*stkP*) is genetically linked to a putative protein phosphatase (*phpP*) (Saskova et al. 2007). The autophosphorylated StpK is a substrate of PhpP, suggesting the existence of a functional coupling between StpK and PhpP (Novakova et al. 2005). The inactivation of *stpK* in *S. pneumoniae* resulted in sensitivity to multiple stress factors, including acidic, osmotic, and oxidative stresses (Saskova et al. 2007). Transcriptome analysis revealed that StpK positively controls the expression of genes involved in cell wall metabolism, DNA repair, iron uptake, and oxidative stress response (Saskova et al. 2007).

The *S. pneumoniae* genome encodes 13 TCSs and an additional orphan unpaired response regulator (Throup et al. 2000). The systematic inactivation of the 14 response regulators identified that VicRK was essential, whereas seven of the *S. pneumoniae* TCS gene pairs and the orphan response regulator appear to be important for growth and pathogenicity in the mouse respiratory tract infection model (Throup et al. 2000). At least two TCSs, ComDE and CiaHR, were implicated in competence development (X-state) (Cheng et al. 1997; Dagkessamanskaia et al. 2004), and a third, LiaSR, was shown to coordinate protection against fratricide-induced self-lysis (Eldholm et al. 2010). The CiaHR TCS has been the subject of extensive research and, in addition to a role in competence, it has also been implicated in β-lactam resistance, stationary-phase autolysis, and virulence (Throup et al. 2000; Dagkessamanskaia et al. 2004; Mascher et al. 2006). The CiaHR-dependent virulence has been attributed, at least in part, to the regulation of the surface-associated HtrA protease by CiaHR (Ibrahim et al. 2004a). As described above, HtrA has been implicated in virulence by influencing the ability of pneumococcus to colonize the nasopharynx of infant rats (Ibrahim et al. 2004b). Notably, the reduced virulence of the *ciaR* mutant could be restored by increasing the expression of HtrA (Ibrahim et al. 2004a). Finally, CiaHR and two additional uncharacterized TCSs (TCS03 and

TCS11) have been implicated in the regulation of the *S. pneumoniae* response to vancomycin stress (Haas et al. 2005).

Because the upper respiratory tract is an oxygen-rich environment, the mechanisms used by *S. pneumoniae* to cope with oxygen and its reactive derivatives have been explored in some detail. Transcriptome analysis of *S. pneumoniae* cells grown in aerobiosis or anaerobiosis revealed that the expression of approximately 5% of the genome is altered by the presence of oxygen (Bortoni et al. 2009). Among the genes upregulated under aerobiosis was the gene encoding for the transcriptional repressor Rgg. The inactivation of *rgg* resulted in an increased sensitivity to oxygen and paraquat, but not to H_2O_2. In addition, the virulence of the *rgg* mutant strain was attenuated in mice (Bortoni et al. 2009). The *S. pneumoniae* genome encodes several genes that participate in oxidative stress responses, including two types of SOD (MnSOD and FeSOD), an alkylhydroperoxidase, and a Nox. A pneumococcus strain lacking the Nox protein displayed normal growth under anaerobic conditions, but was unable to grow under vigorous aeration, and showed attenuated virulence in an animal model (Yu et al. 2001). SODs catalyze the conversion of superoxide molecules to H_2O_2 and O_2. The inactivation of MnSOD affected growth under aerobic conditions and resulted in attenuated virulence in mice (Yesilkaya et al. 2000). The alkyl hydroperoxidase AhpCF is an enzyme that functions by breaking down toxic peroxide compounds to alcohol and H_2O. A strain lacking the *S. pneumoniae ahpCF* genes was attenuated in competitive infections with the parent strain (Paterson et al. 2006).

A few other gene products have been linked to the ability of *S. pneumoniae* to survive oxidative stress conditions. The inactivation of *psaA* and *psaD*, members of an ABC Mn^{2+} transport complex and a thiol peroxidase, respectively, resulted in strains highly sensitive to oxidative stress, with significant reduction in the expression of the SOD and Nox enzymes, and attenuated virulence (Tseng et al. 2002; McAllister et al. 2004). The SpxB pyruvate oxidase enzyme, responsible for the synthesis of acetyl-phosphate and the production of endogenous H_2O_2, was shown to be required for survival during exposure to exogenous H_2O_2. During superoxide stress, ATP levels decreased more rapidly in *spxB* mutants than in wild-type cells, suggesting that the increased sensitivity of *spxB* mutants to H_2O_2 was due to the rapid depletion of ATP (Pericone et al. 2003).

12.6 *Streptococcus thermophilus*

S. thermophilus is one of the most economically important starter cultures for the dairy industry since it is widely used in the production of yogurt and many cheeses, such as hard-cooked cheeses (Emmental, Parmigiano, Grana types, etc.), Mozzarella, Cheddar, and Feta (Delorme 2008). The complete genome sequence of three *S. thermophilus* strains is available to date (Bolotin et al. 2004; Makarova et al. 2006), allowing access to the molecular evolution of the species. It has been established that *S. thermophilus* emerged recently throughout its adaptation for growth in milk by loss-of-function events at the gene level (Bolotin et al. 2004; Hols et al. 2005).

This recessive evolutionary process seems to have led to the inactivation or extinction of most streptococcal virulence-related genes, supporting the safe status of *S. thermophilus* (Bolotin et al. 2004). In parallel, genomic islands acquired by horizontal gene transfer carry important industrial traits (e.g., polysaccharides biosynthesis, bacteriocin production, etc.) (Hols et al. 2005).

There has been great interest in the bacterium's general metabolism and metabolic traits related with organoleptic characteristics of dairy produce. On the contrary, studies concerning the stress physiology of *S. thermophilus* are rather infrequent and the picture that emerges from the literature for this significant aspect of the bacterium's physiology is only fragmented. It is worth noting that despite the fact that *S. thermophilus* was one of the first LAB species to have fully sequenced genomes, research on its stress responses has unfortunately hitherto been only slightly affected.

Among the different stresses to which *S. thermophilus* is exposed, oxidative stress is the most documented. Strains of the bacterium are facultative anaerobes, with most of them showing a preference to anaerobic conditions (Pebay et al. 1995). During dairy processes, a variable degree of aeration leads to the unavoidable incorporation of O_2 in *S. thermophilus* metabolism. Like other streptococci, *S. thermophilus* is unable to respire by using O_2 as an electron acceptor and is devoid of catalase and pseudocatalase activities (van de Guchte et al. 2002). Under these circumstances, the cellular accumulation of O_2 leads to the formation of ROS, such as $O_2^{\bullet-}$, $OH^{\bullet}$, and H_2O_2, through which oxidative stress exerts its deleterious effect by targeting macromolecules and interfering with metabolic pathways (van de Guchte et al. 2002; Miyoshi et al. 2003).

Initial research concerned with the exposure of *S. thermophilus* to O_2 concentrated on surveying activities of key enzymes also identified in other LAB and *Enterobacteriaceae*, involved in either the generation or detoxification of ROS. It was shown that *S. thermophilus* cell-free extracts exhibit Nox, NADH peroxidase, and SOD activities along with a generalized antioxidative potential through metal ion chelating, scavenging of ROS, and a reducing capability (Smart and Thomas 1987; Teraguchi et al. 1987; Lin and Yen 1999). Due to the absence of other major oxidases like glucose, lactose, lactate, or pyruvate oxidases, Nox is the primary oxygen-metabolizing enzyme in *S. thermophilus* (Smart and Thomas 1987). In the case of strain TS2, evidence supported a tightly coupled Nox/NADH peroxidase activity, indicating that Nox caused a two-electron reduction of O_2 to H_2O_2 that served as a substrate for NADH peroxidase, leading to detoxification (Smart and Thomas 1987). In contrast, strains STH450 and ATCC19258 exhibited very low NADH peroxidase activity, supporting a four-electron reduction of O_2 directly to H_2O through Nox (Teraguchi et al. 1987). Similarly to *Lactococcus lactis*, the aeration of *S. thermophilus* cultures shifts homolactic fermentation to mixed fermentation with end products like acetate, α-acetolactate, acetoin, diacetyl, and CO_2 (Smart and Thomas 1987; Teraguchi et al. 1987). The exposure of strain TS2 to O_2 caused an increase in Nox/NADH peroxidase activity and a reduction of lactate dehydrogenase activity (Smart and Thomas 1987). These findings clearly indicate that lactate dehydrogenase under aerobic conditions has to compete for NADH with Nox/

NADH peroxidase, leading to decreased lactate production. At the same time, pyruvate corresponding to the oxidized NADH is committed to form all the aforementioned alternative end products through the action of pyruvate dehydrogenase and α-acetolactate synthetase (Teraguchi et al. 1987). These early observations provided an insight into the aerotolerant nature of several *S. thermophilus* strains.

Soon after, advances in the genetics of LAB allowed the identification of the first gene involved in the oxidative stress defense of *S. thermophilus*. The *gor* gene was characterized, showing a significant identity in the predicted amino acid sequence with its *E. coli* counterpart (Pebay et al. 1995). In fact, it was the first report for the existence of a *gor* gene in Gram-positive bacteria, as it had been described previously only in eukaryotic cells and Gram-negative bacteria (Pebay et al. 1995). Gor catalyzes the regeneration of glutathione from glutathione disulfide at the expense of NADPH. Glutathione is involved in the maintenance of a reducing environment by reacting with toxic oxygen species like H_2O_2 (van de Guchte et al. 2002). In strain CNRZ368, Gor activity was increased under aerated conditions, reaching its maximum value during the stationary phase, implicating the enzyme in the oxidative stress response (Pebay et al. 1995). In parallel, the SOD system of *S. thermophilus* was studied in greater detail. The biochemical analysis of strain AO54 showed that it possesses a single MnSOD necessary for the dismutation of $O_2^{\bullet-}$ into H_2O_2 and H_2O (Chang and Hassan 1997). Interestingly, MnSOD activity was insensitive to aeration or to the presence of paraquat, while its expression was growth phase–dependent (Chang and Hassan 1997). The complete gene sequence of *sodA* of strain AO54 was later determined (Andrus et al. 2003). The gene was disrupted and the constructed *sodA* mutant was hypersensitive to aerobic growth (Andrus et al. 2003). A partial restoration of the phenotype could be achieved by the addition of $MnCl_2$ in the medium under microaerophilic, but not under aerobic, conditions. This indicates that Mn^{2+} may partially compensate for the deficiency of SodA in protecting *S. thermophilus* against O_2 toxicity by scavenging $O_2^{\bullet-}$, as in the case of LAB that inherently lack SODs (Archibald and Fridovich 1981).

It was not until 2001 that reports demonstrating a true adaptive response of *S. thermophilus* under oxidative stress conditions appeared in the literature (Thibessard et al. 2001a, b). Strain CNRZ368 exhibited a biphasic H_2O_2 dose-response curve (Thibessard et al. 2001a). The viability of cells reduced sharply between 0 and 0.5 mM H_2O_2, to increase again at concentrations above 0.5–1 mM, while above 1 mM, a further loss of viability was observed. These findings established a two-mode killing mechanism by H_2O_2 for *S. thermophilus*, similarly to what had been previously reported for *E. coli* (Imlay and Linn 1987, 1988; Imlay et al. 1988). The first mode (i.e., at low concentrations) has been attributed to DNA damage caused by $OH^{\bullet}$ produced by the Fenton reaction. At the concentration range at which survival is increased (i.e., 0.5–1 mM), it has been proposed that cells may actively attempt to reduce $OH^{\bullet}$ toxicity by diminishing available NAD(P)H and by utilizing O_2 itself to scavenge active free radicals into $O_2^{\bullet-}$ that can then be detoxified by SOD activity. The second mode is thought to start taking place when the system is saturated by H_2O_2. Its cellular targets are yet unidentified even though there is evidence to support that some type of DNA damage is also involved (Thibessard et al. 2001a).

A dose-response curve similar to H_2O_2 was described for strain CNRZ368 after exposure to menadione (Thibessard et al. 2001b). Menadione is a $O_2^{\bullet-}$ generator and its biphasic killing effect on *S. thermophilus* indicates the induction of a protective response (Thibessard et al. 2001b). However, the mechanisms behind the two-mode killing in this case remain unknown. The strain was also able to adapt to H_2O_2 since the preexposure of exponentially growing cells to sublethal conditions greatly enhanced their survival after a lethal challenge (Thibessard et al. 2001a). Furthermore, the resistance of strain CNRZ368 was growth-phase–dependent, with stationary-phase cells being the most tolerant to the presence of H_2O_2. This behavior is consistent with all previous reports showing that relevant to O_2 metabolism, enzymes could exhibit differential activities under aerobic conditions and/or during growth, proving the existence of inducible defense mechanisms of *S. thermophilus* against oxidative stress (Pebay et al. 1995; Chang and Hassan 1997).

Another important observation that has been reported for *S. thermophilus* CNRZ368 is a conditional colony instability that depends on O_2 availability. Under aerobic growth, four different colony phenotypes were observed (Thibessard et al. 2001b). The predominant phenotype was a diffused colony accounting for 99% of the population, while the three others (ring, edged, and opaque) constituted almost 1%. Interestingly enough, it was established that minor colonial variants contained an increased number of menadione-resistant mutants with a rate of 26–38%, in contrast to diffused clones exhibiting only 3.5% (Thibessard et al. 2001b). In addition, in the case of ring and opaque colonies, approximately 8 and 6% were menadione-sensitive mutants, respectively, with none detected for the diffused phenotype (Thibessard et al. 2001b). Originally, no conclusion was drawn from these findings, but it is tempting to speculate that they offer a strong indication for the existence of cell subpopulations that are to some degree hypermutable and produce stress-induced mutants, as has been clearly demonstrated in the case of *E. coli* and other organisms (Galhardo et al. 2007). The generation of such mutants may contribute to adapted evolution, enhancing the fitness of the organism under unfavorable conditions (Gonzalez et al. 2008). However, this hypothesis needs to be verified.

Some of the molecular events underlying *S. thermophilus* response to ROS were uncovered by four consecutive studies involving the isolation of sensitive or resistant *S. thermophilus* CNRZ368 mutants to oxidative stress through IS*S*1 mutagenesis (Thibessard et al. 2002, 2004; Fernandez et al. 2004a, b). Ten menadione-sensitive mutants were originally isolated (Thibessard et al. 2004). The mutants were further tested for their resistance to high temperatures in an attempt to examine their potential specificity to oxidative stress. Five of the mutants, impaired in the genes *mreD*, *rodA*, *pbp2b* (all three implicated in peptidoglycan biosynthesis and cell segregation), *cpsX* (putatively involved in exopolysaccharide translocation), and *iscU* (participating in [Fe-S] cluster assembly), were found also to be sensitive to heat stress. Consequently, the action of these genes seems to be generalized to other stressful conditions as well. The remaining five mutants exhibited the same resistance to high temperature as the wild type, and thus it was concluded that their phenotype might be specific to oxidative stress defense (Thibessard et al. 2004). In detail, two of the mutants were impaired in tRNA modification, since disrupted genes encoded for a tRNA guanine transglycosylase and an RNA methyltransferase. The modification of tRNA precursors can seriously affect the efficiency and fidelity of translation. As a result, it was suggested that tRNA maturation or stabilization under oxidative stress conditions may be necessary for the proper expression of genes involved in *S. thermophilus* defense mechanisms (Thibessard et al. 2004). The third mutant was deficient in a putative Fe^{3+} ABC transporter (FatD). However, the mutant was found equally sensitive to the wild type toward streptonigrin (an iron-activated antibiotic), showing that FatD was not involved in iron transport in *S. thermophilus*. It was proposed that FatD could have a possible role in Mn^{2+} transport, affecting the MnSOD activity and as a consequence triggering sensitivity to oxidative stress in the *fatD* mutant (Thibessard et al. 2004). The fourth mutant was disrupted in the *sufD* locus also involved in [Fe-S] cluster assembly and repair, as in the case of *iscU*. Genetic analysis of the Δ*sufD* and $iscU_{97}$ nonpolar mutants proved the requirement of these proteins in an iron-depleted medium, indicating their involvement in intracellular iron balance. Since [Fe-S] clusters are O_2-sensitive prosthetic groups incorporated into proteins, it is likely that *sufD* and *iscU* mutations are implicated in oxidative stress deficiency through the impairment of proper [Fe-S] cluster assembly (Thibessard et al. 2004). Interestingly, the fifth mutant that was also specifically sensitive to $O_2^{\bullet-}$ stress was found disrupted in a gene encoding for a conserved protein whose function was unknown at the time of publication (Thibessard et al. 2004). We updated the analysis; we found that this protein possesses a DisA domain, which has been suggested to be involved in monitoring the presence of stalled replication forks or recombination intermediates (Witte et al. 2008). The sensitivity of this strain to oxidative stress may, indicate the necessity for proper identification of the damaged sites of DNA.

Furthermore, eight menadione-resistant mutants were also characterized (Fernandez et al. 2004a, b). Disrupted loci included *deoB*, *gst*, *rggC*, and *osrD*, coding for a phosphopentomutase, a proton-glutamate/aspartate symport protein, an Rgg-like transcriptional regulator, and a putative DyP-like peroxidase, respectively. The rest of the mutants were deficient in proteins of unknown function. The phosphopentomutase is involved in the purine salvage pathway and the uptake of exogenous-free bases or nucleosides. The deletion of *deoD*, which is potentially cotranscribed with *deoB*, caused increased thermotolerance in *S. thermophilus* probably through the regulation of the (p)ppGpp stringent response alarmone. A putative increase in the ppGpp levels in a *deoB* mutant, as described for the *deoD* mutant, could result in a constitutive preadapted state of cells, leading to enhanced resistance to oxidative stress (Fernandez et al. 2004b). The Gst protein, which is probably coexpressed with an alpha-L-glutamate ligase as part of the same operon, imports glutamate residues needed for the activity of the latter protein. Knocking out Gst could result in decreased levels of glutamate, which, if acting as a stress sensor, could trigger a stringent response (Fernandez et al. 2004b). In the case of the Rgg-like transcriptional regulator, production of the full-length protein requires ribosome or RNA polymerase slippage to an alternative frame, since its gene contains a frameshift mutation (Fernandez et al. 2004a, b). Nevertheless, it has been suggested that RggC may act as a repressor of oxidative stress gene expression in

S. thermophilus (Fernandez et al. 2004a). Finally, the DyP-like peroxidase, which has an atypical heme-binding region, could also be coexpressed with its downstream protein, a lipoprotein involved in the transport of iron that is necessary for peroxidase activity. The inactivation of this operon could cause a decrease in iron concentration, suppressing the generation of ROS through the Fenton reaction and thus increasing survival (Fernandez et al. 2004b).

A recent comparative genomics analysis of *S. thermophilus* derived conclusions that are in agreement with most of the information presented up to now and shed more light onto the putative protection mechanisms against oxidative stress (Hols et al. 2005). The bacterium possesses both glutathione and thioredoxin systems, which are responsible for the low intracellular redox potential and the maintenance of proteins in a reduced state. It was verified that *S. thermophilus* carries a single H_2O-forming Nox (i.e., catalyzing a four-electron reduction of O_2 directly to H_2O) and an MnSOD. All these enzymatic activities and the involvement to oxidative stress defense had previously been reported for *S. thermophilus* with the exception of the thioredoxin system. Whole-genome sequence investigation revealed additional interesting information. In detail, apart from the *gor* gene, one ortholog of the *S. agalactiae gshA* gene was identified in the genomes of *S. thermophilus* LMG18311 and CNRZ1066. The *gshA* of the *S. agalactiae* gene has been shown to encode a bifunctional γ-glutamylcysteine synthetase-glutathione synthetase responsible for glutathione synthesis, while in most bacteria the synthesis of this tripeptide is accomplished by the action of two separate enzymes (Janowiak and Griffith 2005). Thus, like GBS and GAS, *S. thermophilus* may be able to synthesize glutathione in contrast to *S. mutans*, which has to assimilate it from the medium (Sherrill and Fahey 1998; Hols et al. 2005). Furthermore, thioredoxins and glutaredoxins are small proteins with redox-active cysteine residues that play an established role in the defense against oxidative stress, either by reducing protein disulfide bonds produced by various oxidants or by scavenging ROS (Carmel-Harel and Storz 2000). Both types of proteins are present in the genome of *S. thermophilus* (Hols et al. 2005). In addition, as mentioned previously, the SodA activity of *S. thermophilus* is not altered by O_2, in contrast to some other streptococci, and many well-known peroxidases are absent of its genome sequence (Hols et al. 2005). These last observations support a shortcoming in the oxidative stress-protection mechanisms of *S. thermophilus* in relation to pathogenic streptococci and have led to the hypothesis that this deficiency can be the reason for the bacterium's lack of pathogenicity (Hols et al. 2005).

The differential analysis of the *S. thermophilus* ATCC19258 proteome exposed to H_2O_2 has been performed (Arena et al. 2006). Induction was observed for Nox, thioredoxin reductase, MnSOD, SufB, SufC, Gor, and Dpr. In addition, chaperones like GroEL, DnaK, hypothetical protein STU163, general stress protein 24, ClpL, and ClpP were also induced, indicating a necessity for protein repair and/or degradation due to oxidative stress. Proteins found repressed were mainly involved in energy metabolism and protein synthesis. Among them was the phosphate ABC transporter, which may contribute to the level of stress induction by controlling phosphate pools (Rallu et al. 2000). Even though from the existing literature the responses of *S. thermophilus* to counteract the toxicity of oxidative stress have been elucidated to an adequate degree, there is certainly room for advancement since several questions remain to be answered.

S. thermophilus is a moderate thermophilic bacterium. Most strains can grow at temperatures ranging from around 20°C to slightly above 50°C. However, during industrial processes, *S. thermophilus* has to match various types of temperature shifts, and thus the adaptive responses of this bacterium to both heat and cold stresses have been also investigated. It was established early on that *S. thermophilus* is able to exhibit heat adaptation. The exposure of strain Sfi39 to 5–10°C above the optimum temperature for growth (42°C) was sufficient to trigger thermotolerance (Auffray et al. 1995). The proteomic analysis of *S. thermophilus* NCDO573 during transfer from 42 to 52°C resulted in a significant number of downregulated proteins, 22 polypeptides that were either upregulated or newly synthesized after the heat shock, and a number of proteins whose expression levels remained unaffected (Auffray et al. 1995). The availability of genomic sequences of *S. thermophilus* strains along with advances in proteomics technology allowed a more precise characterization of heat adaptation for *S. thermophilus* ATCC19258 (Arena et al. 2006). In accordance with the previous study, a significant number of proteins were downregulated, and the subsequent identification demonstrated their involvement in energy metabolism and amino acid and protein synthesis, indicating a major effect to the overall bacterial metabolism during the triggering of thermotolerance. Among these proteins was DeoD, namely, the purine nucleoside phosphorylase A. This was in agreement with previously reported data demonstrating that the *deoD* mutant of *S. thermophilus* Sfi39 was significantly more thermotolerant than the wild type even though it exhibited a reduced growth rate at 20°C (Varcamonti et al. 2003). As mentioned earlier, the *deoD* mutant had increased the basal concentration of (p)ppGpp, which was correlated with the observed tolerance. The number of upregulated polypeptide species varied with the time of heat shock, ranging from 25 to 37 spots after 20 and 180 min of treatment, respectively (Arena et al. 2006). These included members of the well-characterized chaperone machines like GroEL/GroES, DnaK/DnaJ/GrpE, ClpL, and the ClpP peptidase. Notably, a *clpL* mutant of *S. thermophilus* Sfi39 has also been reported to be deficient in heat tolerance (Varcamonti et al. 2006). In addition, the translation elongation factors G and Ts participating in polypeptide chain elongation in protein biosynthesis were found upregulated, and chaperone-like properties have been suggested for both proteins (Caldas et al. 2000; Krab et al. 2001). The overall data obtained in this study confirm that like other streptococci and LAB, among the targets during thermal adaptation of *S. thermophilus* is the maintenance of the proteins' proper folding, in order to ensure their functionality and avoid aggregation, coupled with the proteolytic degradation of proteins that are beyond repair (Arena et al. 2006).

To date, no global stress-response regulator has been identified in LAB. A σ^B ortholog of *B. subtilis* is absent in lactococci, streptococci, and many other LAB species (van de Guchte et al. 2002). *S. thermophilus* harbors copies of both the HrcA and CtsR repressors that regulate the expression of chaperones and Clp proteases and ATPases. In fact, the HrcA protein of *S. thermophilus* ST11 was the first to be characterized among LAB (Martirani et al. 2001). The protein is active as a homodimer of 74.5 kDa and has a sequence-specific DNA-binding activity against controlling inverted repeats of chaperone expression (CIRCE) that are present in the promoter region of *dnaK* and *groE* operons known as class I stress-response genes. On the other hand, CtsR has been shown to bind to the so-called CtsR-box (a directly repeated heptad) and controls the expression of several *clp* genes known as class III stress-response genes (Derré et al. 1999). A recent study was performed to investigate the effect of knocking out these two major repressors along with a putative *covR* ortholog annotated as the *rr01* gene in the stress responses of *S. thermophilus* Sfi39, including heat adaptation (Zotta et al. 2009). The latter repressor was studied since the TCS CovRS regulates a number of stress-responsive genes in GAS (Dalton and Scott 2004). In all cases, nonadapted mid-log-phase cells, stationary-phase cells, and mid-log-phase cells after acid or heat adaptation were exposed to different lethal challenges (Zotta et al. 2009). Mid-log-phase and stationary-phase Δ*ctsR* and Δ*hrcA* cells were significantly more resistant to heat stress in comparison to the nonadapted wild-type control cells. After acid or heat adaptation, Δ*hrcA* cells retained their increased survival rates in comparison to the respective adapted wild-type cells, while Δ*ctsR* cells retained their increased survival only after thermal adaptation. The Δ*rr01* mutant exhibited the most sensitive phenotype concerning thermal stress. The growth characteristics and the overall stress patterns of any of the mutants formed a rather complicated picture since the inactivation of the stress response regulator did not necessarily result in increased stress tolerance. A preliminary analysis of 2D gels verified in some instances the derepression of known protein targets of these repressors (Zotta et al. 2009). Even though findings presented in this report are very interesting, a direct evaluation of their actual meaning is seriously hindered by the fact that mutations were generated in a manner that cannot rule out polar effects.

In *L. lactis*, RecA has been suggested to regulate thermal adaptation, since an *recA* mutant was sensitive to thermal stress and DNA damage, while displaying diminished levels of DnaK, GroEL, and GrpE chaperones and an increased amount of FtsH (a membrane-bound ATP-dependent metalloprotease) (Duwat et al. 1995; Savijoki et al. 2003). The *recA* mutant of *S. thermophilus* Sfi39 was deficient in heat-shock and nutrient starvation responses and exhibited alterations of cell morphology (Giliberti et al. 2002). However, the mutant was capable of the normal induction of GroEL and ClpL proteins under both stresses, indicating that RecA does not have a regulatory effect in *S. thermophilus*. Interestingly, using an *recA*::*gusA* translational fusion in *S. thermophilus* Sfi39, it was demonstrated that RecA activity was induced only after DNA damage and remained at basal levels after heat shock or nutrient starvation (Giliberti et al. 2006).

The other side of the coin for LAB concerning stress due to temperature is the temperature downshifts. In many cases, such as the duration of cheese ripening or during storage of dairy products, *S. thermophilus* encounters temperatures far below the optimal (Nicodeme et al. 2004). Moreover, as a starter culture, the bacterium has to be stored under freezing conditions (Wouters et al. 1999). Similarly to other LAB, *S. thermophilus* exhibits a rapid cold-stress adaptive response when exposed for a rather brief period (1–4 h) to approximately 20°C below the optimal for growth temperature (42°C), as revealed by enhanced survival after freezing (Wouters et al. 1999). Nevertheless, it can be inferred from experiments performed with strain CNRZ302 that if the treatment is performed at temperatures below the minimal for growth temperature (i.e., at 10°C), cold adaptation is significantly hampered (Wouters et al. 1999). This could explain previous observations where strain TS2 failed to induce a cold-shock response (Kim and Dunn 1997). The proteomic analysis of strain CNRZ302 verified the existence of several cold-induced proteins of 7 kDa, which is typical for the cold-shock proteins (CSPs) (Wouters et al. 1999). Their expression was found to be maximal at 20°C, while their induction was diminished at 10°C. The exposure of cells to chloramphenicol prior to cold adaptation completely blocked the adaptive response to freezing and resulted in no increase in CSPs. The partial sequence of *cspA* was determined, showing that it was highly identical (up to 95%) to the partial *csp* sequences of *S. thermophilus* ST1-1 (Kim et al. 1998), GBS and GAS (Francis and Stewart 1997), and to several of the *csp* sequences of *L. lactis* (Wouters et al. 1998). Northern blot analysis proved that this gene is strongly expressed at low temperatures (Wouters et al. 1999). This finding, along with the inhibitory effect of chloramphenicol to cold adaptation, supports that the overproduction of CSPs is regulated at the transcriptional level. Proteomic analysis in strain PB18 after the exposure of cells to 20°C identified two cold-stress responsive proteins, a 7.5 and a 21.5 kDa (Perrin et al. 1999). N-terminal sequencing of these two proteins showed that the first one belonged to the family of CSPs, while the second was later characterized as an iron-binding protein of the Dps family (Perrin et al. 1999; Nicodeme et al. 2004). It was suggested that the latter protein could be expressed during cold shock in *S. thermophilus* in order to sequester ferrous iron ions, since at low temperatures, the time for the toxic Fe^{2+} oxidation into the nontoxic Fe^{3+} is increased and could allow the generation of ROS. The recent proteomic analysis of strain ATCC19258 under cold shock also verified the induction of several isoforms of CspA and CspB and the induction of Dpr in accordance to previous observations and added some more information such as the involvement of GroEL, ClpL, translation elongation factor Tu, and 6-phosphofructokinase in the response (Arena et al. 2006). Thereby, it can be concluded that *S. thermophilus* uses CSPs like other bacteria according to their putative function, that is, as RNA chaperones or alternative translation initiation factors, in order to protect nucleic acids or to participate to the synthesis of cold-induced proteins that are necessary for cryoprotection (De Angelis and Gobbetti 2004). The induction of GroEL and ClpL indicates that chaperones are also important at low temperatures, as they are in heat stress (Ferrer et al. 2004). A *clpL* mutant of *S. thermophilus* Sfi39 was deficient not only in heat-stress resistance, as mentioned earlier, but also in cold tolerance, showing that this protein is necessary for both responses (Varcamonti et al. 2006). Nevertheless, the cross-protection of heat stress toward cold stress observed in the wild type of this strain was not abolished in the *clpL* mutant, indicating the existence of an interaction between heat and cold adaptive responses that is ClpL-independent. Interestingly, cold shock results in the enhanced thermotolerance of *L. lactis* subsp. *lactis* IL1403 (Panoff et al. 1995). All these findings suggest that heat- and cold-shock responses may not be entirely independent.

Apart from the stresses described so far for *S. thermophilus*, the physiology of the bacterium in response to other stresses has very rarely been examined. It is astonishing that there are extremely few reports concerning the acid and starvation adaptation of this bacterium, with no existing reports concerning its osmotic stress response. Acid adaptation was examined at the proteome level for *S. thermophilus* ATCC19258 (Arena et al. 2006). It is remarkable that no downregulated proteins were identified after exposure to pH 5.0, which indicates no major effect to the overall cells' status. Among the induced proteins was the acid-shock protein, belonging to the small heat-shock proteins (sHSPs), also characterized to be induced in *S. thermophilus* strain PB18 due to acid stress (Gonzalez-Marquez et al. 1997). Most of the acid responsive proteins like Dpr, GroEL, GroES and the translation elongation factors G, Tu, and Ts have been shown to participate in *S. thermophilus* adaptation to other stress responses as well (Arena et al. 2006). It can thus be suggested that the acid adaptation of *S. thermophilus* also requires general stress-adaptive mechanisms such as increased chaperone activity and a tight quality control of protein expression. Only a few of the upregulated proteins were considered specific for acid stress, including the subunits of the F_0F_1-ATPase that are necessary to extrude protons and ensure pH homeostasis, the lactate dehydrogenase that may cause reduction of intracellular lactic acid, and the urease that can lead to increased ammonia production in an attempt to alkalize the environment (Arena et al. 2006).

Starvation is also a very important stress for *S. thermophilus*. Even though an actual deprivation of nutrients is unlikely to take place when the bacterium is used as a starter culture during milk fermentation, it has been suggested that other stresses may provoke a starvation effect (van de Guchte et al. 2002). For example, some transporters may be downregulated due to acidification of the growth medium and thereby cause a diminished availability of essential nutrients. Another interesting feature of *S. thermophilus* is the bacterium's preference for lactose instead of glucose as its primary carbon source as a result of its adaptation to the milk environment (van den Bogaard et al. 2000). The proteomic analysis of lactose-starved cells of strain ATCC19258 revealed a global reprogramming of the cells' metabolism toward energy saving and an attempt to utilize secondary energy sources (Arena et al. 2006). The majority of the 95 proteins found downregulated were involved in the glycolysis, translation, transport, and biosynthesis of polypeptides, nucleosides, and exopolysaccharides. Once more, DeoD was identified among this group of proteins, implicating the levels of (p)ppGpp in the starvation response (Varcamonti et al. 2003). Several proteins also involved in other stress conditions were found upregulated, including GroEL, ClpL, ClpP, Dpr, and the general stress protein 24 (Arena et al. 2006). Most importantly, enzymes involved in lactose and galactose catabolism, like β-galactosidase and proteins of the Leloir pathway, were induced. These findings are in accordance with previous reports concerning the overall catabolite repression mechanism of *S. thermophilus* (Gunnewijk and Poolman 2000; van den Bogaard et al. 2000; Vaillancourt et al. 2002). A poor phosphorylation of Hpr was observed due to the reduced glycolytic activity of lactose-starved cells (Arena et al. 2006). The amount of Hpr(Ser-P) was significantly decreased, influencing the expression of catabolite control genes and operons, since the formation of a complex between this protein and CcpA is necessary for the binding to catabolite responsive elements (CRE sites) (Gunnewijk and Poolman 2000; van den Bogaard et al. 2000). As expected, this reduction of the CcpA/Hpr(Ser-P) complex resulted in the derepression of sugar-degrading proteins (e.g., β-galactosidase) accompanied by a decline in glycolytic enzymes (e.g., lactate dehydrogenase) (Arena et al. 2006). The activation of the Leloir pathway suggests an increased galactose metabolism in *S. thermophilus*–starved cells (Vaillancourt et al. 2002). The bacterium also induced various amino acid ABC transporters and several secondary carbon substrate-metabolizing proteins, such as NADP-dependent glyceraldehyde-3-phophate dehydrogenase, NADP-specific glutamate dehydrogenase, and L-2-hydroxyisocaproate dehydrogenase, among others (Arena et al. 2006). These findings support the induction of gluconeogenesis and the redirection to alternative energy sources (Bernhardt et al. 2003; De Angelis and Gobbetti 2004).

It is obvious that while some aspects of the *S. thermophilus* stress physiology have attracted much attention, other parts need to be further investigated. The benefits of a detailed insight into LAB's stress behavior are demonstrated by some of the conclusions drawn by the authors of the most recent and the most thorough analysis of *S. thermophilus* stress responses at the proteome level (Arena et al. 2006). The authors hypothesize that several proteins found to be induced (e.g., various chaperones, elongation factors, ClpL, ClpP, etc.) or repressed (e.g., DeoD) in a number of stressful conditions could be used as markers for the selection of *S. thermophilus* strains with a generalized robust stress-resistant phenotype. Although this hypothesis had been actually proposed as a goal of studying LAB's stress physiology in the literature much earlier (van de Guchte et al. 2002), it has not really been investigated yet. We believe that if this hypothesis is verified, it would be a very valuable tool for the food industry.

12.7 *Streptococcus macedonicus*

S. macedonicus was first isolated from traditional Greek Kasseri cheese (Tsakalidou et al. 1998). It was originally assigned to the bovis group based on 16S rRNA sequence, but it was recently suggested to be reclassified as *Streptococcus gallolyticus* subsp. *macedonicus* in the cluster formerly known as *Streptococcus bovis* biotype II (Schlegel et al. 2003). Since its first characterization, it has been identified in the fermenting flora of all cheese varieties (soft, semi-hard, and hard) in both commercial and artisanal products, mostly in South European countries, where the per capita cheese consumption is the highest in the world (Maragkoudakis et al. 2009). Strains of *S. macedonicus* have been shown to possess multifunctional properties similar to those of *S. thermophilus* that are important for the food industry (De Vuyst and Tsakalidou 2008). In a recent study concerning the in vitro and in vivo safety evaluation of the bacteriocin producer *S. macedonicus* ACA-DC198, it was found to be nonhemolytic and a significant number of *S. pyogenes* pathogenicity genes were demonstrated to be absent from its genome (Maragkoudakis et al. 2009). Furthermore, it was concluded that the strain did not appear to induce any pathogenic effect in mice. In contrast to *S. thermophilus*, much less is known about the overall physiology and metabolism of *S. macedonicus*. This is mainly due to the fact that the species was only recently described. However, nowadays there is a clear shift in the interest of a number of research groups toward the exploration and exploitation of this microorganism, because its widespread presence in fermented dairy products is now documented, and several strains with significant technological properties have been identified (De Vuyst and Tsakalidou 2008). Most importantly, there is noticeable indirect evidence that supports the hypothesis that *S. macedonicus* may be the second nonpathogenic food-compatible *Streptococcus* along with *S. thermophilus* (Zoumpopoulou et al. 2008; Maragkoudakis et al. 2009).

Initially, our laboratory concentrated on the acid-stress physiology of *S. macedonicus*. For this, in situ flow cytometric assays for the assessment of the physiological status of the cells were developed (Papadimitriou et al. 2006). This culture-independent approach has distinct advantages over typical growth techniques. Culture-based assessment of the viability of bacterial cells imposes a two-value logic; that is, cells are characterized solely as viable or dead if they retain or lose their ability to propagate on a certain medium under constant conditions, respectively (Davey and Kell 1996). Nevertheless, it is to date well documented that the exposure of bacteria to stressful environments results in large heterogeneities of the cells' actual status that escape this two-value logic (Booth 2002; Brehm-Stecher and Johnson 2004). Indeed, we were able to demonstrate that after acidic lethal challenge, *S. macedonicus* cells could be categorized as viable, injured, or dead according to their metabolic activity and membrane integrity (Papadimitriou et al. 2006). These features were assessed by employing the viability fluorochrome carboxyfluorescein diacetate (cFDA) and the membrane-impermeable fluorescent dye propidium iodide (PI), respectively. True single-cell analysis was achieved by the development of a moderate sonication procedure that led to the disaggregation of typical streptococcal cell chains into separate cells. The in situ attributed cell viability status was further validated by single-cell sorting experiments and subsequent culturability tests.

Four different acid-tolerant phenotypes of *S. macedonicus* were determined (Papadimitriou et al. 2007). Cells were able to exhibit logarithmic-phase acid-tolerance response (L-ATR) after transient exposure to sublethal pH (pH 6.0 or 5.5 for 30 min), autoacidification of the growth medium, or continuous growth at suboptimal acidic conditions (acid habituation). L-ATR phenotypes after transient exposure to nonlethal acidic pH and acid habituation resulted in equal protection against lethal challenge (pH 3.5, 15 min), while autoacidification led to a lesser protective effect. The highest protection was obtained with stationary-phase cells (12 h); however, as cells persisted in the stationary phase (i.e., after 24 or 48 h), they showed an inherent gradual increase in dead cells that was accompanied by a clear reduction of the ATR phenotype magnitude until the resistance to the acidic lethal challenge was completely lost (after 48 h). This indicated that even though early stationary-phase cells displayed the most robust phenotype, residence at this phase undoubtedly imposes a stress on its own and leads to the degeneration of the cell population.

A similar detrimental effect of the stationary-phase environment to the cells has also been reported for *L. lactis* (Niven and Mulholland 1998) and *S. mutans* (Renye et al. 2004). The epigenetic nature of the four acid-tolerant phenotypes was verified since the regrowth of adapted cells under nonadapting conditions resulted in abolishment of the acquired resistance (Papadimitriou et al. 2007).

The kinetics of L-ATR induction during transient exposure to nonlethal acidic pH (pH 5.5, for 0–30 min) was also analyzed (Papadimitriou et al. 2007). Acid tolerance was evident within the first 10 min of acid adaptation and was fully induced at 20 min. The simultaneous monitoring of the three distinct subpopulations (i.e., viable, injured, and dead cells) after lethal challenge revealed a complicated interplay among them, allowing the behavior of a clonal population during such a response to be appreciated. Our experiments demonstrated that L-ATR is acquired at the single-cell level equivalently to what has been previously suggested for *Oenococcus oeni* during its adaptation to ethanol stress (Graca da Silveira et al. 2002). *S. macedonicus* cells also exhibited a significant difference among them in their ability to induce the response (Papadimitriou et al. 2007). Indeed, during transient exposure to nonlethal acidic pH, we could detect fully adapted and partially adapted cells and cells that totally failed to adapt. Furthermore, there were significant cell-to-cell deviations in the rate of the induction. Such intrapopulation heterogeneity has previously received a number of explanations, such as stochastic variations, aging, differences in cell cycle progression, or metabolic status, and so forth (Davey and Kell 1996; Brehm-Stecher and Johnson 2004). We believe that this is a phenomenon that has been underestimated in the field of LAB stress physiology, and it deserves more scrutiny since it may have important technological implications.

We then attempted to identify biochemical events that underpin L-ATR (Papadimitriou et al. 2007). A number of inhibitors of key metabolic processes were tested for their effect in the induction of L-ATR after transient exposure to sublethal acidic pH. All inhibitors, including chloramphenicol, actinomycin D, cerulenin, and penicillin, caused a partial inhibition of L-ATR, with the exception of vancomycin, which totally abrogated the response. These findings show that L-ATR requires de novo protein synthesis and changes in the chemical composition of the cell membrane as well as the cell wall. Similar observations have been reported for *L. lactis* (Rallu et al. 1996), *Lactobacillus sanfranciscensis* (De Angelis et al. 2001), *Listeria monocytogenes* (O'Driscoll et al. 1996), and *S. mutans* (Boyd et al. 2000). The fact that in most cases the manifestation of the response was reduced only to some extent in the presence of the inhibitors supports the idea that none of these changes is truly essential (Papadimitriou et al. 2007).

It is well established that during acid adaptation, cellular alterations that take place aid to decrease the excessive proton influx from the extracellular environment in a passive or active manner (Cotter and Hill 2003). Changes in cell envelope chemical composition have been previously characterized as a passive protection mechanism. The commercially available fluorescent derivative of vancomycin (Bodipy FL vancomycin) was used for the assessment of possible alterations in the cell wall formation profile due to acid adaptation (Papadimitriou et al. 2007).

Our analysis revealed that acid-habituated cells displayed longer cell chains, with an abnormal pattern of active cell wall biosynthesis sites that correspond to recent and forthcoming septa. The increase in the length of the cell chains of acid-habituated cells could be attributed to impairment of the cell segregation mechanisms, similar to those described for *S. mutans* (Chatfield et al. 2005), *S. pneumoniae* (Lopez et al. 2000), and *S. thermophilus* (Borges et al. 2005), due to low pH. In addition, the disturbance of septa formation may indicate severe perturbations of cell cycle progression as a result of continuous growth at low pH, as it has been reported that *S. mutans* grown at pH 5.2 exhibited enhanced levels of FtsZ and FtsA proteins, which belong to the cell division machinery and are located at the septum (Wilkins et al. 2002). Furthermore, in the case of streptococci and *L. lactis*, the main mechanism that actively counteracts the attack of protons under low pH conditions by increasing their efflux from the cytosol is mediated through the activity of the F_0F_1-ATPase, which also generates membrane potential in these bacteria since they lack a respiratory chain (Bender et al. 1986; O'Sullivan and Condon 1999). An assessment of the membrane potential of nonadapted and acid-adapted cells after acidic lethal challenge was performed with the membrane potential probe $DiBAC_4(3)$ and flow cytometry (Papadimitriou et al. 2007). It was established that nonadapted cells were totally depolarized, while the majority of adapted cells retained normal potential, showing that L-ATR protects the membrane potential of cells at low pH. With the use of *N,N′*-dicyclohexylcarbodiimide (DCCD), a specific inhibitor of F_0F_1-ATPase, it was found that this enzymatic activity was necessary for the induction of L-ATR. Nevertheless, once L-ATR was fully launched, F_0F_1-ATPase activity was only partially required. Of note, the same effect has been described for *L. monocytogenes* (Cotter et al. 2000). These observations suggest that F_0F_1-ATPase activity is important during the induction of L-ATR in order to maintain proper membrane potential and ΔpH. On the other hand, in the case of already adapted cells, F_0F_1-ATPase activity may be dispensable. As described earlier, the L-ATR of oral streptococci also involves the upregulation of F_0F_1-ATPase, accompanied by the downregulation of the glucose-specific PTS (PEP-PTS), with the exception of *S. sobrinus*, which upregulates only the glucose-specific PEP-PTS while retaining stable F_0F_1-ATPase levels (Quivey et al. 2001; Nascimento et al. 2004). All L-ATR *S. macedonicus* phenotypes exhibited elevated activity of both F_0F_1-ATPase and glucose-specific PEP-PTS (Papadimitriou et al. 2007). The enhanced activity of F_0F_1-ATPase under acidic conditions is tightly coupled with the extensive consumption of ATP generated through glycolysis, which is necessary for the translocation of the protons and ultimately leads to energy depletion and cell growth arrest (van de Guchte et al. 2002; Cotter and Hill 2003). Thus, the upregulation of both F-ATPase and PTS activities in *S. macedonicus* seems to be an important feature that distinguishes this bacterium from oral streptococci. However, the exact physiological meaning of this differentiation needs further investigation.

Two distinct experimental procedures have been the most common in assessing the ability of bacteria to induce ATR. The bacteria are either transiently exposed to or continuously grown in suboptimal acidic pH. The actual physicochemical parameters of these experiments vary greatly since ATR characteristics are species- or even subspecies-dependent (Cotter and Hill 2003). This causes a problematic situation, making the comparison of the behavior among different strains under acid-stress conditions very difficult. It is obvious that a unique experimental framework for the study of ATR induction may be unfeasible. All these practical limitations obscure our overall understanding of the ATR, since it remains unknown whether the response of a strain at a specific growth phase is manifested through an inimitable mechanism or, if not, to what extent the mechanism is influenced by the treatment used to induce it. There have been some contradicting reports about the specifics of ATR in certain microorganisms, such as *L. lactis* (Hartke et al. 1996; Rallu et al. 1996) and *S. mutans* (Jayaraman et al. 1997; Wilkins et al. 2002), that have been attributed to the different procedures undertaken to stimulate acid adaptation (Wilkins et al. 2002; Frees et al. 2003). Additionally, short-term vs. long-term acidic shock influenced gene expression in *S. pneumoniae* differently (Martin-Galiano et al. 2005), while the type of proteins induced in *L. lactis* during transient exposure to low pH was affected by the pH value itself (Frees et al. 2003). With these disperse observations in mind, we undertook the direct comparison among the four epigenetic acid-tolerant phenotypes of *S. macedonicus* based on their transcriptional profiles and the whole-cell chemical composition as revealed by RNA arbitrarily primed PCR (RAP-PCR) and Fourier-transform infrared (FT-IR) spectroscopy, respectively (Papadimitriou et al. 2008). As the genome sequence of *S. macedonicus* remains unknown, RAP-PCR, a differential display-like method, offered the possibility of assessing changes in the bacterium's gene expression. The clustering analysis of RAP-PCR fingerprints revealed significant differences among the phenotypes. In detail, the transcriptome profile of stationary-phase cells was the most divergent and was clearly segregated from all the profiles of mid-log-phase cells. Auto-acidified cells exhibited a more similar RNA pattern compared to control nonadapted cells, while transiently adapted and acid-habituated cells were more divergent, though having very few, but distinct, differences between them. Cloning and sequencing of the differentially expressed cDNA bands determined that the genes coding for the mannose-specific IID component, the 3-oxoacyl-acyl carrier protein, the 1,2-diacylglycerol 3-glucosyltransferase, the large subunit of carbamoyl-phosphate synthase, and a hypothetical protein of unknown function were found to be induced at least under some of the acid-adapting conditions. Subsequent RT-PCR experiments specific for the above-mentioned genes validated the expression patterns observed in the RAP-PCR fingerprints, where the induction rates were, in most cases, significantly different among the acid-tolerant phenotypes. These findings supported our notion that the mechanisms underlying ATR are influenced not only by the growth phase but also by the treatment employed to induce the response.

In most streptococci, glucose transport is mediated through the mannose-specific PEP-PTS (Cochu et al. 2003). The upregulation of this transcript is consistent with our previous observation that all L-ATR phenotypes exhibited enhanced enzymatic activity of this system and may reflect the necessity for increased glucose uptake under low-pH conditions in order to compensate for the depletion of ATP pools caused by the concurrent increased activity of F_0F_1-ATPase (Papadimitriou et al. 2007, 2008).

The involvement of the 3-oxoacyl-acyl carrier protein (encoded by *fabF*) is also consistent with the fact that treatment of *S. macedonicus* cells during acid adaptation with cerulenin caused a significant decrease in the magnitude of the acquired tolerance, since cerulenin is a specific inhibitor of this enzyme (Papadimitriou et al. 2007). In *S. mutans*, cerulenin had the same effect as in the case of *S. macedonicus*, and it was demonstrated that this inhibitor actually blocked the biosynthesis of monounsaturated fatty acids, a necessary alteration of the cell membrane's chemical composition during acid adaptation, probably to reduce cells' permeability to protons (Fozo and Quivey 2004a). As mentioned earlier, FabM, a *trans*-2, *cis*-3-decenoyl-ACP isomerase that catalyzes the isomerization of *trans* unsaturated bonds to their *cis* isomers, was found to be the key player for the biosynthesis of monounsaturated fatty acids in *S. mutans*, and the respective mutant was very sensitive to acid stress (Fozo and Quivey 2004b). However, it has been reported for *S. pneumoniae* that it is the competition between FabK (an enoyl reductase) and FabF, both acting downstream of FabM, that determines the ratio of saturated to monounsaturated fatty acids' biosynthesis (Lu and Rock 2006). Our findings indicate that apart from FabM, FabF may also play an important role in acid adaptation (Papadimitriou et al. 2008).

In principle, cell wall teichoic acids could have a significant role in ATR, since the polyanionic nature of this matrix is responsible for the homeostasis of anions and assists in their trafficking (Neuhaus and Baddiley 2003). This is suggested by the induction of the 1,2-diacylglycerol 3-glucosyltransferase, which catalyzes the anchoring of LTA on the bacterial cell membrane during the acid adaptation of *S. macedonicus* (Papadimitriou et al. 2008). In addition, a *dltC* mutant of *S. mutans*, with a deficiency in the biosynthesis of LTAs, exhibited acid sensitivity, disturbed ATR manifestation, and increased proton permeability (Boyd et al. 2000).

Finally, the induction of the large subunit of carbamoyl-phosphate synthase (encoded by *carB*) indicates an increased production of carbamoyl-phosphate, an intermediate for the biosynthesis of pyrimidines and arginine (Papadimitriou et al. 2008). In agreement with our observation, acid-adapted *S. pneumoniae* exhibited increased *carB* expression along with other genes involved in the biosynthesis of pyrimidine nucleotides (Martin-Galiano et al. 2005). This necessity for enhanced nucleotide synthesis may reflect the requirement to support DNA repair, since it has been demonstrated that DNA is damaged at low pH (Hanna et al. 2001). Nevertheless, the biosynthesis of arginine and the subsequent release of NH_3 after its degradation through the ADS as an attempt to alkalize extracellular pH, a common feature of oral streptococci, cannot be excluded (Quivey et al. 2001).

Many studies concerning the ATR induction in a variety of bacteria, performed with transcriptomic and proteomic approaches, show significant changes in the expression patterns of genes and proteins involved in the central or intermediary metabolism and in the biosynthesis and/or degradation of different macromolecules and smaller molecules during acid adaptation (Cotter and Hill 2003). Thus, it can be speculated that the physiological state of acid-adapted cells coincides with changes in the global chemical composition of the cells. In the past, only some of these predicted changes have been verified directly (Fozo and Quivey 2004a). FT-IR spectroscopy is a physicochemical method that can produce the IR fingerprint of intact cells. The technique is not destructive, and it is well established that certain regions of the IR spectra are influenced by the actual chemical composition of certain cellular constituents (Naumann et al. 1991). Principal component analysis of the second derivative–transformed FT-IR spectra segregated *S. macedonicus* acid-tolerant phenotypes individually in all spectral regions that are characteristic for a wide variety of substances, including polysaccharides of the cell wall, fatty acids of the cell membrane, proteins, nucleic acids, and all other compounds of the cell that may absorb in these spectral regions (Papadimitriou et al. 2008). Overall, our data demonstrated the plasticity in the ATR of *S. macedonicus*, which may be necessary in order to maximize its adaptability by adjusting its response to the distinctiveness of the imposed stress condition. In our opinion, this may be an inherent ability not only of *S. macedonicus* but also of other LAB as well.

S. macedonicus ACA-DC198 produces the food-grade lantibiotic macedocin, which is active against *Clostridium tyrobutyricum* among other food spoilage or pathogenic bacteria; thus, it can be used to prevent clostridial spoilage of semi-hard cheeses (Georgalaki et al. 2000, 2002; Anastasiou et al. 2009). Since in situ bacteriocin-producing cultures have considerable advantages over using purified bacteriocin preparations (De Vuyst and Tsakalidou 2008), the effect of stressful conditions during cheese manufacturing on the growth and bacteriocin production of *S. macedonicus* was investigated. The optimal temperature for bacteriocin production was found between 20 and 25°C, which is significantly lower than that for growth (42°C), while the optimum pH was similar for both parameters (pH 6.0–6.4) (Van den Berghe et al. 2006). A simulation of Kasseri cheese (a Pasta filata cheese variety) manufacturing revealed that *S. macedonicus* was able to survive heat injury caused by heating at 55°C or above, a curd-heating step necessary for this type of cheese (Poirazi et al. 2007). Growth was abolished when the physicochemical parameters mimicked those of the stabilized cheese (25°C, pH 5.0, 3% NaCl) (Poirazi et al. 2007). Two types of predictive approaches based on either the gamma concept or artificial neural networks (ANNs) were used to model growth and bacteriocin production kinetics at different combinations of temperatures (25, 40, and 55°C), pH (constant pH 5.0 or 6.0), and added salt (0, 2, and 4% NaCl) (Poirazi et al. 2007). The obtained functions showed that the physicochemical parameters of the environment to which *S. macedonicus* is exposed during cheese making can indeed seriously influence its performance, as estimated by the biokinetic parameters of the maximum specific growth rate, the cell count decrease rate, and the specific bacteriocin production.

Interestingly, while the bacterium is able to grow in milk with NaCl concentrations up to 4%, macedocin's production stops abruptly at 1.5% (Papadimitriou et al., unpublished results). A genetic screen for the isolation of osmotic stress-resistant mutants (*osm*) using IS*S1* transposition was performed. When tested, all *osm* mutants were able to produce macedocin to a different degree in milk with NaCl concentrations significantly higher than 1.5%, with some of them retaining bacteriocin production up to 4% (Papadimitriou et al., unpublished results). Our findings indicate a correlation between acquired osmotic stress resistance through mutation and bacteriocin production, which seems to be rather complicated since no direct relationship could be established between the magnitude of resistance of the *osm* mutants to hyperosmosis and the extent of preservation of the bacteriocin production. Even though this study is in progress, it is a useful example of how the stress physiology of LAB can be manipulated in order to enhance the applicability of industrially important secondary traits, like bacteriocin production. The *osm* mutants were generated in a food-grade manner and thus may be valuable for applications in cheese manufacturing, an ecosystem characterized by high osmolality.

12.8 Concluding Remarks

In this chapter, we attempted to provide a comprehensive review of the literature concerning the stress responses of streptococci, a very diverse bacterial genus capable of colonizing a variety of ecological niches ranging from innocuous and friendly species such as *S. thermophilus* to life-threatening pathogenic species like *S. pyogenes* and *S. pneumoniae*. Despite differences in habitat and pathogenic potential, the data highlighted in this chapter demonstrate that bacteria in this group have developed a common set of stress-resistance pathways to cope with the wide diversity of environmental assaults encountered during each organism's life cycle. Clearly, molecular chaperones, selected proteases, and enzymes that confer protection against oxidative damage are part of a core general stress response shared by all species in the group. It is also clear that, due to the paucity of stress-dedicated sigma factors, streptococci appear to rely on response regulators to orchestrate an appropriate response to environmental challenges. In particular, two-component signal transduction systems comprised by a membrane-bound histidine kinase and a corresponding response regulator that mediates the cellular response appear to play a major role in stress tolerance and might be considered desirable targets for the development of novel therapeutic treatments. Furthermore, the identification and dissection of resistance traits unique to a particular species may facilitate the development of mechanisms to specifically control a particular pathogen, or to enhance the usefulness of food-related streptococci. In this regard, the data reviewed here suggest that the dental pathogen *S. mutans* and the dairy starter *S. thermophilus* appear to have an edge over other species. In fact, a novel therapeutic approach based on the QS interference of pathogenic bacteria or targeted antimicrobial therapy has been successfully used to specifically eliminate *S. mutans* from dental biofilms (Li et al. 2010).

While substantial progress in understanding the mechanisms used by streptococcal species to cope with stresses has been made over the years, it is quite clear that certain groups and species are still understudied. In particular, the limited number of investigations on the stress responses of *S. pyogenes* and *S. agalactiae*, two of the most medically important streptococcal species, is rather surprising. Given that the core stress regulon is often found intertwined with virulence, the continued dissection of the mechanisms controlling stress tolerance could facilitate the development of novel therapeutic agents against streptococcal diseases. In addition, a further characterization of the intricate interactions between each species and their environment is still warranted. Notably, the advent of breakthrough technologies, such as next-generation sequencing, allows researchers to obtain high-throughput sequencing information at an unprecedent speed and, most importantly, to evaluate the behavior of microbial communities in their natural environment (metagenomics) at a reasonable cost. With very few exceptions, the microbial interactions within their communities, and between these communities and the host or environment, are still poorly understood. As we begin to experience the scientific revolution of high-throughput genomic analysis, the knowledge gathered from the efforts of many hundreds of scientists discussed in this chapter is here to provide the frameshift for the years ahead.

References

Abranches J, Chen YY, Burne RA (2003) Characterization of *Streptococcus mutans* strains deficient in EIIAB Man of the sugar phosphotransferase system. Appl Environ Microbiol 69:4760–4769

Abranches J, Candella MM, Wen ZT, Baker HV, Burne RA (2006) Different roles of EIIABMan and EIIGlc in regulation of energy metabolism, biofilm development, and competence in *Streptococcus mutans*. J Bacteriol 188:3748–3756

Abranches J, Nascimento MM, Zeng, L, Browngardt, CM, Wen, ZT, Rivera MF, Burne RA (2008) CcpA regulates central metabolism and virulence gene expression in *Streptococcus mutans*. J Bacteriol 190:2340–2349

Ahn SJ, Lemos JA, Burne RA (2005) Role of HtrA in growth and competence of *Streptococcus mutans* UA159. J Bacteriol 187:3028–3038

Ahn SJ, Wen ZT, Burne RA (2006) Multilevel control of competence development and stress tolerance in *Streptococcus mutans* UA159. Infect Immun 74:1631–1642

Ajdic D, McSham WM, McLaughlin RE, Savic G, Chang, J, Carson, MB, Primeaux C, Tian R, Kenton, S, Jia H, Lin S, Qian Y, Li S, Zhu H, Najar F, Lai H, White J, Roe BA, Ferretti JJ (2002) Genome sequence of *Streptococcus mutans* UA159, a cariogenic dental pathogen. Proc Natl Acad Sci USA 99:14434–14439

Anastasiou R, Aktypis A, Georgalaki M, Papadelli M, De Vuyst L, Tsakalidou E (2009) Inhibition of *Clostridium tyrobutyricum* by *Streptococcus macedonicus* ACA–DC 198 under conditions mimicking Kasseri cheese production and ripening. Int Dairy J 19:330–335

Andrus JM, Bowen SW, Klaenhammer TR, Hassan HM (2003) Molecular characterization and functional analysis of the manganese–containing superoxide dismutase gene (*sodA*) from *Streptococcus thermophilus* AO54. Arch Biochem Biophys 420:103–113

Archibald FS, Fridovich I (1981) Manganese, superoxide dismutase, and oxygen tolerance in some lactic acid bacteria. J Bacteriol 146:928–936

Arena S, D'Ambrosio C, Renzone G, Rullo R, Ledda L, Vitale F, Maglione G, Varcamonti M, Ferrara L, Scaloni A (2006) A study of *Streptococcus thermophilus* proteome by integrated analytical procedures and differential expression investigations. Proteomics 6:181–192

Auffray Y, Lecesne E, Hartke A, Boutibonnes P (1995) Basic features of the *Streptococcus thermophilus* heat shock response. Curr Microbiol 30:87–91

Bates CS, Toukoki C, Neely MN, Eichenbaum Z (2005) Characterization of MtsR, a new metal regulator in group A *Streptococcus*, involved in iron acquisition and virulence. Infect Immun 73:5743–5753

Bender GR, Sutton SV, Marquis RE (1986) Acid tolerance, proton permeabilities, and membrane ATPases of oral streptococci. Infect Immun 53:331–338

Beres SB, Ritcher EW, Nagiec MJ, Sumby P, Porcella SF, DeLeo FR, Musser JM (2006) Molecular genetic anatomy of inter– and intraserotype variation in the human bacterial pathogen group A *Streptococcus*. Proc Natl Acad Sci USA 103:7059–7064

Bernhardt J, Weibezahn J, Scharf C, Hecker M (2003) *Bacillus subtilis* during feast and famine: visualization of the overall regulation of protein synthesis during glucose starvation by proteome analysis. Genome Res 13:224–237

Billroth T (1874) Untersuchungen uber die Vegetationsformen von Coccobacteria Septica [in German]. Berlin

Biswas I, Drake L, Erkina D, Biswas S (2008) Involvement of sensor kinases in the stress tolerance response of *Streptococcus mutans*. J Bacteriol 190:68–77

Biswas S, Biswas I (2005) Role of HtrA in surface protein expression and biofilm formation by *Streptococcus mutans*. Infect Immun 73:6923–6934

Bolotin A, Quinquis B, Renault P, Sorokin A, Ehrlich SD, Kulakauskas S, Lapidus A, Goltsman E, Mazur M, Pusch GD, Fonstein M, Overbeek R, Kyprides N, Purnelle B, Prozzi D, Ngui K, Masuy D, Hancy F, Burteau S, Boutry M, Delcour J, Goffeau A, Hols P (2004) Complete sequence and comparative genome analysis of the dairy bacterium *Streptococcus thermophilus*. Nat Biotechnol 22:1554–1558

Booth IR (2002) Stress and the single cell: intrapopulation diversity is a mechanism to ensure survival upon exposure to stress. Int J Food Microbiol 78:19–30

Borges F, Layec S, Thissebard A, Fernandez A, Gintz B, Hols P, Decaris B, Leblond-Bourget N (2005) *cse*, A chimeric and variable gene, encodes an extracellular protein involved in cellular segregation in *Streptococcus thermophilus*. J Bacteriol 187:2737–2746

Bortoni ME, Terra VS, Hinds J, Andrew PW, Yesilkaya H (2009) The pneumococcal response to oxidative stress includes a role for Rgg. Microbiology 155:4123–4134

Boyd DA, Cvitkovitch DG, Bleiweis AS, Kiriukhin MY, Debabov DV, Neuhaus FC, Hamilton IR (2000) Defects in D-alanyl-lipoteichoic acid synthesis in *Streptococcus mutans* results in acid sensitivity. J Bacteriol 182:6055–6065

Brehm-Stecher BF, Johnson EA (2004) Single-cell microbiology: tools, technologies, and applications. Microbiol Mol Biol Rev 68:538–559

Brenot A, King KY, Janowiak B, Griffith O, Caparon MG (2004) Contribution of glutathione peroxidase to the virulence of *Streptococcus pyogenes*. Infect Immun 72:408–413

Brenot A, King KY, Caparon MG (2005) The PerR regulon in peroxide resistance and virulence of *Streptococcus pyogenes*. Mol Microbiol 55:221–234

Brenot A, Weston BF, Caparon MG (2007) A PerR-regulated metal transporter (PmtA) is an interface between oxidative stress and metal homeostasis in *Streptococcus pyogenes*. Mol Microbiol 63:1185–1196

Brown JS, Gilliland SM, Basavanna S, Holden DW (2004) *phgABC*, a three-gene operon required for growth of *Streptococcus pneumoniae* in hyperosmotic medium and *in vivo*. Infect Immun 72:4579–4588

Caldas T, Laalami S, Richarme G (2000) Chaperone properties of bacterial elongation factor EF-G and initiation factor IF2. J Biol Chem 275:855–860

Cao J, Chen D, Xu W, Chen T, Xu S, Luo J, Zhao Q, Liu B, Wang D, Zhang X, Shan Y, Yin Y (2007) Enhanced protection against pneumococcal infection elicited by immunization with the combination of PspA, PspC, and ClpP. Vaccine 25:4996–5005

Cao J, Li D, Gong Y, Yin N, Chen T, Wong CK, Xu W, Luo J, Zhang X, Lam CW, Yin Y (2009) Caseinolytic protease: a protein vaccine which could elicit serotype-independent protection against invasive pneumococcal infection. Clin Exp Immunol 156:52–60

Carapetis JR, Steer AC, Mulholland EK, Weber M (2005) The global burden of group A streptococcal diseases. Lancet Infect Dis 5:685–694

Carlsson J (1983) Regulation of sugar metabolism in relation to feast-and-famine existence of plaque. In: Guggenheim B (Ed.), Cariology today. Karger, Basel

Carmel-Harel O, Storz G (2000) Roles of the glutathione- and thioredoxin-dependent reduction systems in the *Escherichia coli* and *Saccharomyces cerevisiae* responses to oxidative stress. Annu Rev Microbiol 54:439–461

Casalta E, Montel MC (2008) Safety assessment of dairy microorganisms: the *Lactococcus* genus. Int J Food Microbiol 126:271–273

Chang SK, Hassan HM (1997) Characterization of superoxide dismutase in *Streptococcus thermophilus*. Appl Environ Microbiol 63:3732–3735

Charpentier E, Novak R, Tuomanen E (2000) Regulation of growth inhibition at high temperature, autolysis, transformation and adherence in *Streptococcus pneumoniae* by *clpC*. Mol Microbiol 37:717–726

Chastanet A, Prudhomme M, Claverys JP, Msadek T (2001) Regulation of *Streptococcus pneumoniae clp* genes and their role in competence development and stress survival. J Bacteriol 183:7295–7307

Chatfield CH, Koo H, Quivey RG Jr (2005) The putative autolysin regulator LytR in *Streptococcus mutans* plays a role in cell division and is growth-phase regulated. Microbiology 151:625–631

Chattoraj P, Banerjee A, Biswas S, Biswas I (2010) ClpP of *Streptococcus mutans* differentially regulates expression of genomic islands, mutacin production, and antibiotic tolerance. J Bacteriol 192:1312–1323

Chaussee MS, Phillips ER, Ferretti JJ (1997) Temporal production of streptococcal erythrogenic toxin B (streptococcal cysteine proteinase) in response to nutrient depletion. Infect Immun 65:1956–1959

Chen PM, Chen HC, Ho CT, Jung CJ, Lien HT, Chen JY, Chia JS (2008) The two-component system ScnRK of *Streptococcus mutans* affects hydrogen peroxide resistance and murine macrophage killing. Microbes Infect 10:293–301

Cheng Q, Campbell EA, Naughton AM, Johnson S, Masure HR (1997) The *com* locus controls genetic transformation in *Streptococcus pneumoniae*. Mol Microbiol 23:683–692

Claverys JP, Prudhomme M, Martin B (2006) Induction of competence regulons as a general response to stress in gram-positive bacteria. Annu Rev Microbiol 60:451–475

Claverys JP, Martin B, Havarstein LS (2007) Competence-induced fratricide in streptococci. Mol Microbiol 64:1423–1433

Cochu A, Vadeboncoeur C, Moineau S, Frenette M (2003) Genetic and biochemical characterization of the phosphoenolpyruvate:glucose/mannose phosphotransferase system of *Streptococcus thermophilus*. Appl Environ Microbiol 69:5423–5432

Cotter PD, Hill C (2003) Surviving the acid test: responses of Gram-positive bacteria to low pH. Microbiol Mol Biol Rev 67:429–453

Cotter PD, Gahan CG, Hill C (2000) Analysis of the role of the *Listeria monocytogenes* F0F1 -ATPase operon in the acid tolerance response. Int J Food Microbiol 60:137–146

Cunningham MW (2000) Pathogenesis of group A streptococcal infections. Clin Microbiol Rev 13:470–511

Dagkessamanskaia A, Moscoso M, Henard V, Guiral S, Overweg K, Reuter M, Martin B, Wells J, Claverys JP (2004) Interconnection of competence, stress and CiaR regulons in *Streptococcus pneumoniae*: competence triggers stationary phase autolysis of *ciaR* mutant cells. Mol Microbiol 51:1071–1086

Dalton TL, Scott JR (2004) CovS inactivates CovR and is required for growth under conditions of general stress in *Streptococcus pyogenes*. J Bacteriol 186:3928–3937

Davey HM, Kell DB (1996) Flow cytometry and cell sorting of heterogeneous microbial populations: the importance of single-cell analyses. Microbiol Rev 60:641–696

De Angelis M, Gobbetti M (2004) Environmental stress responses in *Lactobacillus*: a review. Proteomics 4:106–122

De Angelis M, Bini L, Pallini V, Cocconcelli PS, Gobbetti M (2001) The acid-stress response in *Lactobacillus sanfranciscensis* CB1. Microbiology 147:1863–1873

De Vuyst L, Tsakalidou E (2008) *Streptococcus macedonicus*, a multi-functional and promising species for dairy fermentations. Int Dairy J 18:476–485

Delorme C (2008) Safety assessment of dairy microorganisms: *Streptococcus thermophilus*. Int J Food Microbiol 126:274–277
Deng DM, Liu MJ, ten Cate JM, Crielaard W (2007a) The VicRK system of *Streptococcus mutans* responds to oxidative stress. J Dent Res 86:606–610
Deng DM, ten Cate JM, Crielaard W (2007b) The adaptive response of *Streptococcus mutans* towards oral care products: involvement of the ClpP serine protease. Eur J Oral Sci 115:363–370
Derré I, Rapoport G, Msadek T (1999) CtsR, a novel regulator of stress and heat shock response, controls *clp* and molecular chaperone gene expression in Gram-positive bacteria. Mol Microbiol 31:117–131
Dunning DW, McCall LW, Powell WF, Arscott WT, McConocha EM, McClurg CJ, Goodman SD, Spatafora GA (2008) SloR modulation of the *Streptococcus mutans* acid tolerance response involves the GcrR response regulator as an essential intermediary. Microbiology 154:1132–1143
Duwat P, Ehrlich SD, Gruss A (1995) The *recA* gene of *Lactococcus lactis*: characterization and involvement in oxidative and thermal stress. Mol Microbiol 17:1121–1131
Edwards MS, Rench MA, Haffar AA, Murphy MA, Desmond MM, Baker CJ (1985) Long-term sequelae of group B streptococcal meningitis in infants. J Pediatr 106:717–722
Eldholm V, Gutt B, Johnsborg O, Bruckner R, Maurer P, Hakenbeck R, Mascher T, Havarstein LS (2010) The pneumococcal cell envelope stress-sensing system LiaFSR is activated by murein hydrolases and lipid II-interacting antibiotics. J Bacteriol 192:1761–1773
Facklam R (2002) What happened to the streptococci: overview of taxonomic and nomenclature changes. Clin Microbiol Rev 15:613–630
Faustoferri RC, Hahn K, Weiss K, Quivey RG Jr (2005) Smx nuclease is the major, low-pH-inducible apurinic/apyrimidinic endonuclease in *Streptococcus mutans*. J Bacteriol 187:2705–2714
Fernandez A, Borges F, Thibessard A, Gintz B, Decaris B, Leblond-Bourget N (2004a) Characterization of *Streptococcus thermophilus* CNRZ368 oxidative stress-resistant mutants: involvement of a potential Rgg-like transcriptional regulator. Lait 84:77–85
Fernandez A, Thibessard A, Borges F, Gintz B, Decaris B, Leblond-Bourget N (2004b) Characterization of oxidative stress-resistant mutants of *Streptococcus thermophilus* CNRZ368. Arch Microbiol 182:364–372
Ferrer M, Lunsdorf H, Chernikova TN, Yakimov M, Timmis KN, Golyshin PN (2004) Functional consequences of single:double ring transitions in chaperonins: life in the cold. Mol Microbiol 53:167–182
Ferretti JJ, McShan WM, Ajdic D, Savic DJ, Savic G, Lyon, K, Primeaux C, Sezate S, Suvorov, AN, Kenton S, Lai HS, Lin SP, Qian Y, Jia HG, Najar FZ, Ren Q, Zhu H, Song L, White, J, Yuan X, Clifton SW, Roe, BA, McLaughlin R (2001) Complete genome sequence of an M1 strain of *Streptococcus pyogenes*. Proc Natl Acad Sci USA 98:4658–4663
Fozo EM, Quivey RG Jr (2004a) Shifts in the membrane fatty acid profile of *Streptococcus mutans* enhance survival in acidic environments. Appl Environ Microbiol 70:929–936
Fozo EM, Quivey RG Jr (2004b) The *fabM* gene product of *Streptococcus mutans* is responsible for the synthesis of monounsaturated fatty acids and is necessary for survival at low pH. J Bacteriol 186:4152–4158
Fozo EM, Scott-Anne K, Koo H, Quivey RG Jr (2007) Role of unsaturated fatty acid biosynthesis in virulence of *Streptococcus mutans*. Infect Immun 75:1537–1539
Francis KP, Stewart GS (1997) Detection and speciation of bacteria through PCR using universal major cold-shock protein primer oligomers. J Ind Microbiol Biotechnol 19:286–293
Frees D, Vogensen FK, Ingmer H (2003) Identification of proteins induced at low pH in *Lactococcus lactis*. Int J Food Microbiol 87:293–300
Galhardo RS, Hastings PJ, Rosenberg SM (2007) Mutation as a stress response and the regulation of evolvability. Crit Rev Biochem Mol Biol 42:399–435

Georgalaki MD, Sarantinopoulos P, Ferreira ES, De Vuyst L, Kalantzopoulos G, Tsakalidou E (2000) Biochemical properties of *Streptococcus macedonicus* strains isolated from Greek Kasseri cheese. J Appl Microbiol 88:817–825

Georgalaki MD, Van Den Berghe E, Kritikos D, Devreese B, Van Beeumen J, Kalnatzpoulos G, De Vuyst L, Tsakalidou E (2002) Macedocin, a food-grade lantibiotic produced by *Streptococcus macedonicus* ACA-DC 198. Appl Environ Microbiol 68:5891–5903

Gibson CM, Mallett TC, Claiborne A, Caparon MG (2000) Contribution of NADH oxidase to aerobic metabolism of *Streptococcus pyogenes*. J Bacteriol 182:448–455

Giliberti G, Naclerio G, Martirani L, Ricca E, De Felice M (2002) Alteration of cell morphology and viability in a *recA* mutant of *Streptococcus thermophilus* upon induction of heat shock and nutrient starvation. Gene 295:1–6

Giliberti G, Baccigalupi L, Cordone A, Ricca E, De Felice M (2006) Transcriptional analysis of the *recA* gene of *Streptococcus thermophilus*. Microb Cell Fact 5:29

Glaser P, Rusniok C, Buchrieser C, Chevalier F, Frangeul L, Msadek T, Zouine M, Couve E, Lalioui L, Poyart C, Trieu-Cuot P, Kunst F (2002) Genome sequence of *Streptococcus agalactiae*, a pathogen causing invasive neonatal disease. Mol Microbiol 45:1499–1513

Gong Y, Tian XL, Sutherlan T, Sisson G, Mai J, Ling J, Li YH (2009) Global transcriptional analysis of acid-inducible genes in *Streptococcus mutans*: multiple two-component systems involved in acid adaptation. Microbiology 155:3322–3332

Gonzalez C, Hadany L, Ponder RG, Price M, Hastings PJ, Rosenberg SM (2008) Mutability and importance of a hypermutable cell subpopulation that produces stress-induced mutants in *Escherichia coli*. PLoS Genet 4:e1000208

Gonzalez-Marquez H, Perrin C, Bracquart P, Guimont C, Linden G (1997) A 16 kDa protein family overexpressed by *Streptococcus thermophilus* PB18 in acid environments. Microbiology 143 (Pt 5):1587–1594

Graca da Silveira M, Vitoria San Romao M, Loureiro-Dias MC, Rombouts FM, Abee T (2002) Flow cytometric assessment of membrane integrity of ethanol-stressed *Oenococcus oeni* cells. Appl Environ Microbiol 68:6087–6093

Graham MR, Smoot LM, Migliaccio CA, Virtaneva K, Sturdevant DE, Porcella SF, Federle MJ, Adams GJ, Scott JR, Musser JM (2002) Virulence control in group A *Streptococcus* by a two-component gene regulatory system: global expression profiling and *in vivo* infection modeling. Proc Natl Acad Sci USA 99:13855–13860

Graham MR, Virtaneva K, Porcella SF, Barry WT, Gowen BB, Johnson CR, Wright FA, Musser JM (2005) Group A *Streptococcus* transcriptome dynamics during growth in human blood reveals bacterial adaptive and survival strategies. Am J Pathol 166:455–465

Graham MR, Virtaneva K, Porcella SF, Gardner DJ, Long RD, Welty DM, Barry WT, Johnson CA, Parkins LD, Wright FA, Musser JM (2006) Analysis of the transcriptome of group A *Streptococcus* in mouse soft tissue infection. Am J Pathol 169:927–942

Griswold AR, Jameson-Lee M, Burne RA (2006) Regulation and physiologic significance of the agmatine deiminase system of *Streptococcus mutans* UA159. J Bacteriol 188:834–841

Gryllos I, Grifantini R, Colaprico A, Cary ME, Hakansson A, Carey DW, Suarez-Chavez M, Kalish LA, Mitchell PD, White GL, Wessels MR (2008) PerR confers phagocytic killing resistance and allows pharyngeal colonization by group A *Streptococcus*. PLoS Pathog 4:e1000145

Gunnewijk MG, Poolman B (2000) Phosphorylation state of HPr determines the level of expression and the extent of phosphorylation of the lactose transport protein of *Streptococcus thermophilus*. J Biol Chem 275:34073–34079

Haas W, Kaushal D, Sublett J, Obert C, Tuomanen EI (2005) Vancomycin stress response in a sensitive and a tolerant strain of *Streptococcus pneumoniae*. J Bacteriol 187:8205–8210

Hahn K, Faustoferri RC, Quivey RG Jr (1999) Induction of an AP endonuclease activity in *Streptococcus mutans* during growth at low pH. Mol Microbiol 31:1489–1498

Hanks TS, Liu M, McClure MJ, Fukumura M, Duffy A, Lei B (2006) Differential regulation of iron- and manganese-specific MtsABC and heme-specific HtsABC transporters by the metalloregulator MtsR of group A *Streptococcus*. Infect Immun 74:5132–5139

Hanna MN, Ferguson RJ, Li YH, Cvitkovitch DG (2001) *uvrA* is an acid-inducible gene involved in the adaptive response to low pH in *Streptococcus mutans*. J Bacteriol 183:5964–5973

Hardie JM, Whiley RA (1997) Classification and overview of the genera *Streptococcus* and *Enterococcus*. Soc Appl Bacteriol Symp Ser 26:1S–11S

Hartke A, Bouche S, Giard JC, Benachour A, Boutibonnes P, Auffray Y (1996) The lactic acid stress response of *Lactococcus lactis* subsp. *lactis*. Curr Microbiol 33:194–199

Hasona A, Crowley PJ, Levesque CM, Mair RW, Cvitkovitch DG, Bleiweis AS, Brady LJ (2005) Streptococcal viability and diminished stress tolerance in mutants lacking the signal recognition particle pathway or YidC2. Proc Natl Acad Sci USA 102:17466–17471

Hasona A, Zuobi-Hasona K, Crowley PJ, Abranches J, Ruelf MA, Bleiweis AS, Brady LJ (2007) Membrane composition changes and physiological adaptation by *Streptococcus mutans* signal recognition particle pathway mutants. J Bacteriol 189:1219–1230

He X, Wu C, Yarbrough D, Sim L, Niu G, Merritt J, Shi W, Qi F (2008) The *cia* operon of *Streptococcus mutans* encodes a unique component required for calcium-mediated autoregulation. Mol Microbiol 70:112–126

Hendriksen WT, Bootsma HJ, Estevao S, Hoogenboezem T, de Jong A, de Groot R, Kuipers OP, Hermans PW (2008) CodY of *Streptococcus pneumoniae*: link between nutritional gene regulation and colonization. J Bacteriol 190:590–601

Higuchi M, Yamamoto Y, Poole LB, Shimada M, Sato Y, Takahashi N, Kamio Y (1999) Functions of two types of NADH oxidases in energy metabolism and oxidative stress of *Streptococcus mutans*. J Bacteriol 181:5940–5947

Hols P, Hancy F, Fontaine L, Grossiord B, Prozzi D, Leblond-Bourget N, Decaris B, Bolotin A, Delorme C, Dusko Ehrlich S, Guedon E, Monnet V, Renault P, Kleerebezem M (2005) New insights in the molecular biology and physiology of *Streptococcus thermophilus* revealed by comparative genomics. FEMS Microbiol Rev 29:435–463

Hussain H, Branny P, Allan E (2006) A eukaryotic-type serine/threonine protein kinase is required for biofilm formation, genetic competence, and acid resistance in *Streptococcus mutans*. J Bacteriol 188:1628–1632

Ibrahim YM, Kerr AR, McCluskey J, Mitchell TJ (2004a) Control of virulence by the two-component system CiaR/H is mediated via HtrA, a major virulence factor of *Streptococcus pneumoniae*. J Bacteriol 186:5258–5266

Ibrahim YM, Kerr AR, McCluskey J, Mitchell TJ (2004b) Role of HtrA in the virulence and competence of *Streptococcus pneumoniae*. Infect Immun 72:3584–3591

Ibrahim YM, Kerr AR, Silva NA, Mitchell TJ (2005) Contribution of the ATP-dependent protease ClpCP to the autolysis and virulence of *Streptococcus pneumoniae*. Infect Immun 73:730–740

Imlay JA, Linn S (1987) Mutagenesis and stress responses induced in *Escherichia coli* by hydrogen peroxide. J Bacteriol 169:2967–2976

Imlay JA, Linn S (1988) DNA damage and oxygen radical toxicity. Science 240:1302–1309

Imlay JA, Chin SM, Linn S (1988) Toxic DNA damage by hydrogen peroxide through the Fenton reaction *in vivo* and *in vitro*. Science 240:640–642

Janowiak BE, Griffith OW (2005) Glutathione synthesis in *Streptococcus agalactiae*. One protein accounts for gamma-glutamylcysteine synthetase and glutathione synthetase activities. J Biol Chem 280:11829–11839

Jayaraman GC, Penders JE, Burne RA (1997) Transcriptional analysis of the *Streptococcus mutans hrcA, grpE and dnaK* genes and regulation of expression in response to heat shock and environmental acidification. Mol Microbiol 25:329–341

Jiang SM, Cieslewicz MJ, Kasper DL, Wessels MR (2005) Regulation of virulence by a two-component system in group B *Streptococcus*. J Bacteriol 187:1105–1113

Johri AK, Padilla J, Malin G, Paoletti LC (2003) Oxygen regulates invasiveness and virulence of group B *Streptococcus*. Infect Immun 71:6707–6711

Kajfasz JK, Martinez AR, Rivera-Ramos I, Abranches J, Koo H, Quivey RG, Lemos JA (2009) Role of Clp proteins in expression of virulence properties of *Streptococcus mutans*. J Bacteriol 191:2060–2068

Kajfasz JK, Rivera-Ramos I, Abranches J, Martinez AR, Rosalen PL, Derr AM, Quivey RG, Lemos JA (2010) Two Spx proteins modulate stress tolerance, survival, and virulence in *Streptococcus mutans*. J Bacteriol 192:2546–2556

Kazmierczak KM, Wayne KJ, Rechtsteiner A, Winkler ME (2009) Roles of *rel* in stringent response, global regulation and virulence of serotype 2 *Streptococcus pneumoniae* D39. Mol Microbiol 72:590–611

Keefe GP (1997) *Streptococcus agalactiae* mastitis: a review. Can Vet J 38:429–437

Kim WS, Dunn NW (1997) Identification of a cold shock gene in lactic acid bacteria and the effect of cold shock on cryotolerance. Curr Microbiol 35:59–63

Kim WS, Khunajakr N, Ren J, Dunn NW (1998) Conservation of the major cold shock protein in lactic acid bacteria. Curr Microbiol 37:333–336

King KY, Horenstein JA, Caparon MG (2000) Aerotolerance and peroxide resistance in peroxidase and PerR mutants of *Streptococcus pyogenes*. J Bacteriol 182:5290–5299

Kohler W (2007) The present state of species within the genera *Streptococcus* and *Enterococcus*. Int J Med Microbiol 297:133–150

Krab IM, te Biesebeke R, Bernardi A, Parmeggiani A (2001) Elongation factor Ts can act as a steric chaperone by increasing the solubility of nucleotide binding-impaired elongation factor-Tu. Biochemistry 40:8531–8535

Kreth J, Merritt J, Shi W, Qi F (2005) Co-ordinated bacteriocin production and competence development: a possible mechanism for taking up DNA from neighbouring species. Mol Microbiol 57:392–404

Kreth J, Merritt J, Zhu L, Shi W, Qi F (2006) Cell density- and ComE-dependent expression of a group of mutacin and mutacin-like genes in *Streptococcus mutans*. FEMS Microbiol Lett 265:11–17

Kreth J, Hung DC, Merritt J, Perry J, Zhu L, Goodman SD, Cvitkovitch DG, Shi W, Qi F (2007) The response regulator ComE in *Streptococcus mutans* functions both as a transcription activator of mutacin production and repressor of CSP biosynthesis. Microbiology 153:1799–1807

Kwon HY, Kim SW, Choi MH, Ogunniyi AD, Paton JC, Park SH, Pyo SN, Rhee DK (2003) Effect of heat shock and mutations in ClpL and ClpP on virulence gene expression in *Streptococcus pneumoniae*. Infect Immun 71:3757–3765

Kwon HY, Ogunniyi AD, Choi MH, Pyo SN, Rhee DK, Paton JC (2004) The ClpP protease of *Streptococcus pneumoniae* modulates virulence gene expression and protects against fatal pneumococcal challenge. Infect Immun 72:5646–5653

Lamy MC, Zouine M, Fert J, Vergassola M, Couve E, Pellegrini E, Glaser P, Kunst F, Msadek T, Trieu-Cuot P, Poyart C (2004) CovS/CovR of group B *Streptococcus*: a two-component global regulatory system involved in virulence. Mol Microbiol 54:1250–1268

Lancefield RC (1933) A serological differentiation of human and other groups of streptococci. J Exp Med 59:571–595

Lemos JA, Burne RA (2002) Regulation and Physiological Significance of ClpC and ClpP in *Streptococcus mutans*. J Bacteriol 184:6357–6366

Lemos JA, Burne RA (2008) A model of efficiency: stress tolerance by *Streptococcus mutans*. Microbiology 154:3247–3255

Lemos JA, Chen YY, Burne RA (2001) Genetic and physiologic analysis of the *groE* operon and role of the HrcA repressor in stress gene regulation and acid tolerance in *Streptococcus mutans*. J Bacteriol 183:6074–6084

Lemos JA, Brown TA Jr, Burne RA (2004) Effects of RelA on key virulence properties of planktonic and biofilm populations of *Streptococcus mutans*. Infect Immun 72:1431–1440

Lemos JA, Abranches J, Burne RA (2005) Responses of cariogenic streptococci to environmental stresses. Curr Issues Mol Biol 7:95–107

Lemos JA, Lin VK, Nascimento MM, Abranches J, Burne RA (2007a) Three gene products govern (p)ppGpp production by *Streptococcus mutans*. Mol Microbiol 65:1568–1581

Lemos JA, Luzardo Y, Burne RA (2007b) Physiologic effects of forced down-regulation of *dnaK* and *groEL* expression in *Streptococcus mutans*. J Bacteriol 189:1582–1588

Lemos JA, Nascimento MM, Lin VK, Abranches J, Burne RA (2008) Global regulation by (p) ppGpp and CodY in *Streptococcus mutans*. J Bacteriol 190:5291–5299
Len AC, Harty DW, Jacques NA (2004a) Proteome analysis of *Streptococcus mutans* metabolic phenotype during acid tolerance. Microbiology 150:1353–1366
Len AC, Harty DW, Jacques NA (2004b) Stress-responsive proteins are upregulated in *Streptococcus mutans* during acid tolerance. Microbiology 150:1339–1351
Levesque CM, Mair RW, Perry JA, Lau PC, Li YH, Cvitkovitch DG (2007) Systemic inactivation and phenotypic characterization of two-component systems in expression of *Streptococcus mutans* virulence properties. Lett Appl Microbiol 45:398–404
Li LN, Guo LH, Lux R, Eckert R, Yarbrough D, He J, Anderson M, Shi WY (2010) Targeted antimicrobial therapy against *Streptococcus mutans* establishes protective non-cariogenic oral biofilms and reduces subsequent infection. Int J Oral Sci 2:66–73
Li YH, Hanna MN, Svensater G, Ellen RP, Cvitkovitch DG (2001) Cell density modulates acid adaptation in *Streptococcus mutans*: implications for survival in biofilms. J Bacteriol 183:6875–6884
Li YH, Lau PC, Tang N, Svensater G, Ellen RP, Cvitkovitch DG (2002a) Novel two-component regulatory system involved in biofilm formation and acid resistance in *Streptococcus mutans*. J Bacteriol 184:6333–6342
Li YH, Tang N, Aspiras MB, Lau PC, Lee JH, Ellen RP, Cvitkovitch DG (2002b) A quorum-sensing signaling system essential for genetic competence in *Streptococcus mutans* is involved in biofilm formation. J Bacteriol 184:2699–2708
Lin MY, Yen CL (1999) Antioxidative ability of lactic acid bacteria. J Agric Food Chem 47:1460–1466
Liu GY, Doran KS, Lawrence T, Turkson N, Puliti M, Tissi L, Nizet V (2004) Sword and shield: linked group B streptococcal beta-hemolysin/cytolysin and carotenoid pigment function to subvert host phagocyte defense. Proc Natl Acad Sci USA 101:14491–14496
Lopez R, Gonzalez MP, Garcia E, Garcia JL, Garcia P (2000) Biological roles of two new murein hydrolases of *Streptococcus pneumoniae* representing examples of module shuffling. Res Microbiol 151:437–443
Loughman JA, Caparon M (2006a) Regulation of SpeB in *Streptococcus pyogenes* by pH and NaCl: a model for *in vivo* gene expression. J Bacteriol 188:399–408
Loughman JA, Caparon MG (2006b) A novel adaptation of aldolase regulates virulence in *Streptococcus pyogenes*. EMBO J 25:5414–5422
Lu YJ, Rock CO (2006) Transcriptional regulation of fatty acid biosynthesis in *Streptococcus pneumoniae*. Mol Microbiol 59:551–566
Makarova K, Slesarev A, Wolf Y, Sorokin A, Mirkin B, Koonin E, Pavlov A, Pavlova N, Karamychev V, Polouchine N, Shakhova V, Grigoriev I, Lou Y, Rohksar D, Lucas S, Huang K, Goodstein DM, Hawkins T, Plengvidhya V, Welker D, Hughes J, Goh Y, Benson A, Baldwin K, Lee JH, Diaz-Muniz I, Dosti B, Smeianov V, Wetcher W, Barabote R, Lorca G, Altermann E, Barrangou R, Ganesah B, Xie Y, Rawsthorne H, Tamir D, Parker C, Breidt F, Broadbent J, Hutkins R, O'Sullivan D, Steele J, Unlu G, Saier M, Klaenhammer T, Richardson P, Kozyavkin S, Weimer B, Mills D (2006) Comparative genomics of the lactic acid bacteria. Proc Natl Acad Sci USA 103:15611–15616
Malke H, Ferretti JJ (2007) CodY-affected transcriptional gene expression of *Streptococcus pyogenes* during growth in human blood. J Med Microbiol 56:707–714
Malke H, Steiner K, McShan WM, Ferretti JJ (2006) Linking the nutritional status of *Streptococcus pyogenes* to alteration of transcriptional gene expression: the action of CodY and RelA. Int J Med Microbiol 296:259–275
Maragkoudakis PA, Papadelli M, Gerogalaki M, Panayotopoulos EG, Martinez-Gonzalez B, Mentis AF, Petraki K, Sgouras DN, Tsakalidou E (2009) *In vitro* and *in vivo* safety evaluation of the bacteriocin producer *Streptococcus macedonicus* ACA-DC 198. Int J Food Microbiol 133:141–147
Marouni MJ, Sela S (2003) The *luxS* gene of *Streptococcus pyogenes* regulates expression of genes that affect internalization by epithelial cells. Infect Immun 71:5633–5639

Marquis RE (1995) Oxygen metabolism, oxidative stress and acid-base physiology of dental plaque biofilms. J Ind Microbiol 15:198–207

Marsh PD (2003) Are dental diseases examples of ecological catastrophes? Microbiology 149:279–294

Martin-Galiano AJ, Overweg K, Ferrandiz MJ, Reuter M, Wells JM, de la Campa AG (2005) Transcriptional analysis of the acid tolerance response in *Streptococcus pneumoniae*. Microbiology 151:3935–3946

Martinez AR, Abranches A, Kajfasz JK, Lemos JA (2010) Characterization of the *Streptococcus sobrinus* acid-stress response by interspecies microarrays and proteomics. Molec Oral Microbiol 25:331–342

Martirani L, Raniello R, Naclerio G, Ricca E, De Felice M (2001) Identification of the DNA-binding protein, HrcA, of *Streptococcus thermophilus*. FEMS Microbiol Lett 198:177–182

Mascher T, Heintz M, Zahner D, Merai M, Hakenbeck R (2006) The CiaRH system of *Streptococcus pneumoniae* prevents lysis during stress induced by treatment with cell wall inhibitors and by mutations in *pbp2x* involved in beta-lactam resistance. J Bacteriol 188:1959–1968

McAllister LJ, Tseng HJ, Ogunniyi AD, Jennings MP, McEwan AG, Paton JC (2004) Molecular analysis of the psa permease complex of *Streptococcus pneumoniae*. Mol Microbiol 53:889–901

Mereghetti L, Sitkiewicz I, Green NM, Musser JM (2008a) Extensive adaptive changes occur in the transcriptome of *Streptococcus agalactiae* (group B *Streptococcus*) in response to incubation with human blood. PLoS One 3:e3143

Mereghetti L, Sitkiewicz I, Green NM, Musser JM (2008b) Remodeling of the *Streptococcus agalactiae* transcriptome in response to growth temperature. PLoS One 3:e2785

Mereghetti L, Sitkiewicz I, Green NM, Musser JM (2009) Identification of an unusual pattern of global gene expression in group B *Streptococcus* grown in human blood. PLoS One 4:e7145

Miyoshi A, Rochat T, Gratadoux JJ, Le Loir Y, Oliveira SC, Langella P, Azevedo V (2003) Oxidative stress in *Lactococcus lactis*. Genet Mol Res 2:348–359

Nair S, Poyart C, Beretti JL, Veiga-Fernandes H, Berche P, Trieu-Cuot P (2003) Role of the *Streptococcus agalactiae* ClpP serine protease in heat-induced stress defence and growth arrest. Microbiology 149:407–417

Nakayama K (1992) Nucleotide sequence of *Streptococcus mutans* superoxide dismutase gene and isolation of insertion mutants. J Bacteriol 174:4928–4934

Nascimento MM, Lemos JA, Abranches J, Goncalves RB, Burne RA (2004) Adaptive acid tolerance response of *Streptococcus sobrinus*. J Bacteriol 186:6383–6390

Nascimento MM, Lemos JA, Abranches J, Lin VK, Burne RA (2008) Role of RelA of *Streptococcus mutans* in global control of gene expression. J Bacteriol 190:28–36

Naumann D, Helm D, Labischinski H (1991) Microbiological characterizations by FT-IR spectroscopy. Nature 351:81–82

Neuhaus FC, Baddiley J (2003) A continuum of anionic charge: structures and functions of D-alanyl-teichoic acids in Gram-positive bacteria. Microbiol Mol Biol Rev 67:686–723

Nguyen PT, Abranches J, Phan TN, Marquis RE (2002) Repressed respiration of oral streptococci grown in biofilms. Curr Microbiol 44:262–266

Nicodeme M, Perrin C, Hols P, Bracquart P, Gaillard JL (2004) Identification of an iron-binding protein of the Dps family expressed by *Streptococcus thermophilus*. Curr Microbiol 48:51–56

Niven GW, Mulholland F (1998) Cell membrane integrity and lysis in *Lactococcus lactis*: the detection of a population of permeable cells in post-logarithmic phase cultures. J Appl Microbiol 84:90–96

Novakova L, Saskova L, Pallova P, Janececk J, Novotna J, Ulrych A, Echenique J, Trombe MC, Branny P (2005) Characterization of a eukaryotic type serine/threonine protein kinase and protein phosphatase of *Streptococcus pneumoniae* and identification of kinase substrates. FEBS J 272:1243–1254

O'Driscoll B, Gahan CG, Hill C (1996) Adaptive acid tolerance response in *Listeria monocytogenes*: isolation of an acid-tolerant mutant which demonstrates increased virulence. Appl Environ Microbiol 62:1693–1698

Ogier JC, Serror P (2008) Safety assessment of dairy microorganisms: the *Enterococcus* genus. Int J Food Microbiol 126:291–301

Olsen RJ, Sitkiewicz I, Ayeras AA, Gonulal VE, Cantu C, Beres SB, Green NM, Lei B, Humbird T, Greaver J, Chang E, Ragasa WP, Montgomery CA, Cartwright J, McGeer A, Low DE, Whitney AR, Cagle PT, Blasdel TL, DeLeo FR, Musser JM (2010) Decreased necrotizing fasciitis capacity caused by a single nucleotide mutation that alters a multiple gene virulence axis. Proc Natl Acad Sci USA 107:888–893

O'Sullivan E, Condon S (1999) Relationship between acid tolerance, cytoplasmic pH, and ATP and H+-ATPase levels in chemostat cultures of *Lactococcus lactis*. Appl Environ Microbiol 65:2287–2293

Panoff J-M, Thammavongs B, Laplace J-M, Hartke A, Boutibonnes P, Auffray Y (1995) Cryotolerance and cold adaptation in *Lactococcus lactis* subsp. *lactis* IL1403. Cryobiology 32:516–520

Papadimitriou K, Pratsinis H, Nebe-von-Caron G, Kletsas D, Tsakalidou E (2006) Rapid assessment of the physiological status of *Streptococcus macedonicus* by flow cytometry and fluorescence probes. Int J Food Microbiol 111:197–205

Papadimitriou K, Pratsinis H, Nebe-von-Caron G, Kletsas D, Tsakalidou E (2007) Acid tolerance of *Streptococcus macedonicus* as assessed by flow cytometry and single-cell sorting. Appl Environ Microbiol 73:465–476

Papadimitriou K, Boutou E, Zoumpopoulos G, Tarantilis PA, Polissiou M, Vorgias CE, Tsakalidou E (2008) RNA arbitrarily primed PCR and Fourier transform infrared spectroscopy reveal plasticity in the acid tolerance response of *Streptococcus macedonicus*. Appl Environ Microbiol 74:6068–6076

Park CY, Kim EH, Choi SW, Tran TD, Kim IH, Kim SN, Pyo S, Rhee DK (2010) Virulence attenuation of *Streptococcus pneumoniae clpP* mutant by sensitivity to oxidative stress in macrophages via an NO-mediated pathway. J Microbiol 48:229–235

Paterson GK, Blue CE, Mitchell TJ (2006) An operon in *Streptococcus pneumoniae* containing a putative alkylhydroperoxidase D homologue contributes to virulence and the response to oxidative stress. Microb Pathog 40:152–160

Pebay M, Holl AC, Simonet JM, Decaris B (1995) Characterization of the gor gene of the lactic acid bacterium *Streptococcus thermophilus* CNRZ368. Res Microbiol 146:371–383

Pericone CD, Park S, Imlay JA, Weiser JN (2003) Factors contributing to hydrogen peroxide resistance in *Streptococcus pneumoniae* include pyruvate oxidase (SpxB) and avoidance of the toxic effects of the Fenton reaction. J Bacteriol 185:6815–6825

Perrin C, Guimont C, Bracquart P, Gaillard JL (1999) Expression of a new cold shock protein of 21.5 kDa and of the major cold shock protein by *Streptococcus thermophilus* after cold shock. Curr Microbiol 39:342–347

Petersen FC, Tao L, Scheie AA (2005) DNA binding-uptake system: a link between cell-to-cell communication and biofilm formation. J Bacteriol 187:4392–4400

Phares CR, Lynfield R, Farley MM, Mohle-Boetani J, Harrison LH, Petit S, Craig AS, Schaffner W, Zansky SM, Gershman K, Stefonek KR, Albanese BA, Zell ER, Schuchat A, Schrag SJ (2008) Epidemiology of invasive group B streptococcal disease in the United States, 1999–2005. JAMA 299:2056–2065

Poirazi P, Leroy F, Georgalaki MD, Aktypis A, De Vuyst L, Tsakalidou E (2007) Use of artificial neural networks and a gamma-concept-based approach to model growth of and bacteriocin production by *Streptococcus macedonicus* ACA-DC 198 under simulated conditions of Kasseri cheese production. Appl Environ Microbiol 73:768–776

Potrykus K, Cashel M (2008) (p)ppGpp: still magical? Annu Rev Microbiol 62:35–51

Poyart C, Pellegrini E, Gaillot O, Boumaila C, Baptista M, Trieu-Cuot P (2001) Contribution of Mn-cofactored superoxide dismutase (SodA) to the virulence of *Streptococcus agalactiae*. Infect Immun 69:5098–5106

Qi F, Merritt J, Lux R, Shi W (2004) Inactivation of the *ciaH* gene in *Streptococcus mutans* diminishes mutacin production and competence development, alters sucrose-dependent biofilm formation, and reduces stress tolerance. Infect Immun 72:4895–4899

Quach D, van Sorge NM, Kristian SA, Bryan JD, Shelver DW, Doran KS (2009) The CiaR response regulator in group B *Streptococcus* promotes intracellular survival and resistance to innate immune defenses. J Bacteriol 191:2023–2032

Quivey RG Jr, Faustoferri R, Monahan K, Marquis R (2000) Shifts in membrane fatty acid profiles associated with acid adaptation of *Streptococcus mutans*. FEMS Microbiol Lett 189:89–92

Quivey RG, Kuhnert WL, Hahn K (2001) Genetics of acid adaptation in oral streptococci. Crit Rev Oral Biol Med 12:301–314

Rallu F, Gruss A, Maguin E (1996) *Lactococcus lactis* and stress. Antonie van Leeuwenhoek 70:243–251

Rallu F, Gruss A, Ehrlich SD, Maguin E (2000) Acid- and multistress-resistant mutants of *Lactococcus lactis*: identification of intracellular stress signals. Mol Microbiol 35:517–528

Rathsam C, Eaton RE, Simpson CL, Browne GV, Berg T, Harty DW, Jacques NA (2005) Up-regulation of competence- but not stress-responsive proteins accompanies an altered metabolic phenotype in *Streptococcus mutans* biofilms. Microbiology 151:1823–1837

Renye JA Jr, Piggot PJ, Daneo-Moore L, Buttaro BA (2004) Persistence of *Streptococcus mutans* in stationary-phase batch cultures and biofilms. Appl Environ Microbiol 70:6181–6187

Ricci S, Janulczyk R, Bjorck L (2002) The regulator PerR is involved in oxidative stress response and iron homeostasis and is necessary for full virulence of *Streptococcus pyogenes*. Infect Immun 70:4968–4976

Robertson GT, Ng WL, Foley J, Gilmour R, Winkler ME (2002) Global transcriptional analysis of *clpP* mutations of type 2 *Streptococcus pneumoniae* and their effects on physiology and virulence. J Bacteriol 184:3508–3520

Robertson GT, Ng WL, Gilmour R, Winkler ME (2003) Essentiality of *clpX*, but not *clpP*, *clpL*, *clpC*, or *clpE*, in *Streptococcus pneumoniae* R6. J Bacteriol 185:2961–2966

Rolerson E, Swick A, Newlon L, Palmer C, Pan Y, Keeshan B, Spatafora G (2006) The SloR/Dlg metalloregulator modulates *Streptococcus mutans* virulence gene expression. J Bacteriol 188:5033–5044

Rosenbach FJ (1884) Mikro-organismen bei den Wund-Infections-Krankheiten des Menschen [in German]. Weisbaden

Ryan CS, Kleinberg I (1995) Bacteria in human mouths involved in the production and utilization of hydrogen peroxide. Arch Oral Biol 40:753–763

Santi I, Grifantini R, Jiang SM, Brettoni C, Grandi G, Wessels MR, Soriani M (2009) CsrRS regulates group B *Streptococcus* virulence gene expression in response to environmental pH: a new perspective on vaccine development. J Bacteriol 191:5387–5397

Saskova L, Novakova L, Basler M, Branny P (2007) Eukaryotic-type serine/threonine protein kinase StkP is a global regulator of gene expression in *Streptococcus pneumoniae*. J Bacteriol 189:4168–4179

Savijoki K, Ingmer H, Frees D, Vogensen FK, Palva A, Varmanen P (2003) Heat and DNA damage induction of the LexA-like regulator HdiR from *Lactococcus lactis* is mediated by RecA and ClpP. Mol Microbiol 50:609–621

Schlegel L, Grimont F, Ageron E, Grimont PA, Bouvet A (2003) Reappraisal of the taxonomy of the *Streptococcus bovis/Streptococcus equinus* complex and related species: description of *Streptococcus gallolyticus* subsp. *gallolyticus* subsp. nov., *S. gallolyticus* subsp. *macedonicus* subsp. nov. and *S. gallolyticus* subsp. *pasteurianus* subsp. nov. Int J Syst Evol Microbiol 53:631–645

Schleifer KH, Kilpper-Balz R (1984) Transfer of *Streptococcus faecalis* and *Streptococcus faecium* to the genus *Enterococcus* nom. rev. as *Enterococcus faecalis* comb. nov. and *Enterococcus faecium* comb. nov. Int J Syst Bacteriol 34:31–34

Schleifer KH, Kraus J, Dvorak C (1985) Transfer of *Streptococcus lactis* and related *Streptococci* to the genus lactococcus gen. nov. Syst Appl Microbiol 6:183–195

Senadheera D, Krastel K, Mair R, Persadmehr A, Abranches J, Burne RA, Cvitkovitch DG (2009) Inactivation of VicK affects acid production and acid survival of *Streptococcus mutans*. J Bacteriol 191:6415–6424

Shelburne SA, III, Sumby P, Sitkiewicz I, Granville C, DeLeo FR, Musser JM (2005) Central role of a bacterial two-component gene regulatory system of previously unknown function in pathogen persistence in human saliva. Proc Natl Acad Sci USA 102:16037–16042

Shelburne SA, III, Keith D, Horstmann N, Sumby P, Davenport MT, Graviss EA, Brennan RG, Musser JM (2008a) A direct link between carbohydrate utilization and virulence in the major human pathogen group A *Streptococcus*. Proc Natl Acad Sci USA 105:1698–1703

Shelburne SA, Davenport MT, Keith DB, Musser JM (2008b) The role of complex carbohydrate catabolism in the pathogenesis of invasive streptococci. Trends Microbiol 16:318–325

Sheng J, Marquis RE (2006) Enhanced acid resistance of oral streptococci at lethal pH values associated with acid-tolerant catabolism and with ATP synthase activity. FEMS Microbiol Lett 262:93–98

Sheng J, Marquis RE (2007) Malolactic fermentation by *Streptococcus mutans*. FEMS Microbiol Lett 272:196–201

Sherman JM (1937) The streptococci. Bacteriol Rev 1:3–97

Sherman JM, Albus WR (1918) Some characters which differentiate the lactic-acid *Streptococcus* from streptococci of the *pyogenes* type occurring in milk. J Bacteriol 3:153–174

Sherrill C, Fahey RC (1998) Import and metabolism of glutathione by *Streptococcus mutans*. J Bacteriol 180:1454–1459

Shet A, Ferrieri P (2004) Neonatal & maternal group B streptococcal infections: a comprehensive review. Ind J Med Res 120:141–150

Shottmuller H (1903) Die Artunterscheidung der fur den menschen Pathogen Streptokokken durch Blutagar [in German]. Munch Med Wochenschr 50:849–853

Siller M, Janapatla RP, Pirzada ZA, Hassler C, Zinkl D, Charpentier E (2008) Functional analysis of the group A streptococcal *luxS*/AI-2 system in metabolism, adaptation to stress and interaction with host cells. BMC Microbiol 8:188

Simmen HP, Blaser J (1993) Analysis of pH and pO2 in abscesses, peritoneal fluid, and drainage fluid in the presence or absence of bacterial infection during and after abdominal surgery. Am J Surg 166:24–27

Sitkiewicz I, Musser JM (2009) Analysis of growth-phase regulated genes in *Streptococcus agalactiae* by global transcript profiling. BMC Microbiol 9:32

Sitkiewicz I, Green NM, Guo N, Bongiovanni AM, Witkin SS, Musser JM (2009) Transcriptome adaptation of group B *Streptococcus* to growth in human amniotic fluid. PLoS One 4:e6114

Smart JB, Thomas TD (1987) Effect of oxygen on lactose metabolism in lactic streptococci. Appl Environ Microbiol 53:533–541

Smoot LM, Smoot JC, Graham MR, Somerville GA, Sturdevant DE, Migliaccio CA, Sylva GL, Musser JM (2001) Global differential gene expression in response to growth temperature alteration in group A *Streptococcus*. Proc Natl Acad Sci USA 98:10416–10421

Sonenshein AL (2005) CodY, a global regulator of stationary phase and virulence in Gram-positive bacteria. Curr Opin Microbiol 8:203–207

Sturr MG, Marquis RE (1992) Comparative acid tolerances and inhibitor sensitivities of isolated F-ATPases of oral lactic acid bacteria. Appl Environ Microbiol 58:2287–2291

Centers for Disease Control and Prevention (2001) Promoting oral health: interventions for preventing dental caries, oral and pharyngeal cancers, and sports-related craniofacial injuries. MMWR Recomm Rep 50(RR21):1–13

Teraguchi S, Ono J, Kiyosawa I, Okonogi S (1987) Oxygen uptake activity and aerobic metabolism of *Streptococcus thermophilus* STH450. J Dairy Sci 70:514–523

Tettelin H, Nelson KE, Paulsen IT, Elsen JA, Read TD, Peterson S, Heidelberg J, DeBoy RT, Haft DH, Dodson RJ, Durkin AS, Gwinn M, Kolonay JF, Nelson WC, Peterson JD, Umayam LA, White O, Salzberg SL, Lewis MR, Radune D, Holtzapple E, Khouri H, Wolf AM, Utterback TR, Hansen CL, McDonald LA, Feldblyum TV, Angiuoli S, Dickinson T, Hickey EK, Holt IE, Loftus BJ, Yang F, Smith HO, Venter JC, Dougherty BA, Morrison DA, Holingshead SK, Fraser CM (2001) Complete genome sequence of a virulent isolate of *Streptococcus pneumoniae*. Science 293:498–506

Tettelin H., Masignani V, Cieslewicz MJ, Eisen JA, Peterson S, Wessels MR, Paulsen I, Nelson KE, Margarit I, Read TD, Madoff LC, Wolf AM, Beanan MJ, Brinkac LM, Daugherty SC, DeBoy RT, Durkin AS, Kolonay JF, Madupu R, Lewis MR, Radune D, Fedorova NB, Scanlan D, Khouri H, Mulligan S, Carty HA, Cline RT, Van Aken SE, Gill J, Scarselli M, Mora M, Iacobini ET, Brettoni C, Galli G, Mariani M, Vegni F, Maione D, Rinaudo D, Rappuoli R, Telford JL, Kasper DL, Grandi G, Fraser CM (2002) Complete genome sequence and comparative genomic analysis of an emerging human pathogen, serotype V *Streptococcus agalactiae*. Proc Natl Acad Sci USA 99:12391–12396

Tettelin H, Msignani V, Cieslewicz MJ, Donati C, Medini D, Ward NL, Angiuoli SV, Crabtree J, Jones AL, Durkin AS, Deboy RT, Davidsen TM, Mora M, Scarselli M, Margarit Ros I, Peterson JD, Hauser CR, Sundaram JP, Nelson WC, Madupu R, Brinkac LM, Dodson RJ, Rosovitz MJ, Sullivan SA, Daugherty SC, Haft DH, Selengut J, Gwinn ML, Zhou L, Zafar N, Khouri H, Radune D, Dimitrov G, Watkins K, O'Connor KJ, Smith S, Utterback TR, White O, Rubens CE, Grandi G, Madoff LC, Kasper DL, Telford JL, Wessels MR, Rappuoli R, Fraser CM (2005) Genome analysis of multiple pathogenic isolates of *Streptococcus agalactiae*: implications for the microbial "pan-genome." Proc Natl Acad Sci USA 102:13950–13955

Thibessard A, Fernandez A, Gintz B, Leblond-Bourget N, Decaris B (2001a) Hydrogen peroxide effects on *Streptococcus thermophilus* CNRZ368 cell viability. Res Microbiol 152:593–596

Thibessard A, Leblond-Bourget N, Fernandez A, Gintz B, Decaris B (2001b) Response of *Streptococcus thermophilus* CNRZ368 and its colonial variants to oxidative stress: evidence for an inducible defence system. Lait 81:311–316

Thibessard A, Fernandez A, Gintz B, Leblond-Bourget N, Decaris B (2002) Effects of *rodA* and *pbp2b* disruption on cell morphology and oxidative stress response of *Streptococcus thermophilus* CNRZ368. J Bacteriol 184:2821–2826

Thibessard A, Borges F, Fernandez A, Gintz B, Decaris B, Leblond-Bourget N (2004) Identification of *Streptococcus thermophilus* CNRZ368 genes involved in defense against superoxide stress. Appl Environ Microbiol 70:2220–2229

Throup JP, Koretke KK, Bryant AP, Ingraham KA, Chalker AF, Ge Y, Marra A, Wallis NG, Brown JR, Holmes DJ, Rosenberg M, Burnham MK (2000) A genomic analysis of two-component signal transduction in *Streptococcus pneumoniae*. Mol Microbiol 35:566–576

Tong H, Chen W, Merritt J, Qi F, Shi W, Dong X (2007) *Streptococcus oligofermentans* inhibits *Streptococcus mutans* through conversion of lactic acid into inhibitory H2O2: a possible counteroffensive strategy for interspecies competition. Mol Microbiol 63:872–880

Tsakalidou E, Zoidou E, Pot B, Wassill L, Ludwig W, Devriese LA, Kalantzopoulos G, Schleifer KH, Kersters K (1998) Identification of streptococci from Greek Kasseri cheese and description of *Streptococcus macedonicus* sp. nov. Int J Syst Bacteriol 48 (Pt 2):519–527

Tseng HJ, McEwan AG, Paton JC, Jennings MP (2002) Virulence of *Streptococcus pneumoniae*: PsaA mutants are hypersensitive to oxidative stress. Infect Immun 70:1635–1639

Tsou CC, Chiang-Ni C, Lin YS, Chuang WJ, Lin MT, Liu CC, Wu JJ (2008) An iron-binding protein, Dpr, decreases hydrogen peroxide stress and protects *Streptococcus pyogenes* against multiple stresses. Infect Immun 76:4038–4045

Tu le N, Jeong HY, Kwon HY, Ogunniyi AD, Paton JC, Pyo SN, Rhee DK (2007) Modulation of adherence, invasion, and tumor necrosis factor alpha secretion during the early stages of infection by *Streptococcus pneumoniae* ClpL. Infect Immun 75:2996–3005

Turlan C, Prudhomme M, Fichant G, Martin B, Gutierrez C (2009) SpxA1, a novel transcriptional regulator involved in X-state (competence) development in *Streptococcus pneumoniae*. Mol Microbiol 73:492–506

Urban CF, Lourido S, Zychlinsky A (2006) How do microbes evade neutrophil killing? Cell Microbiol 8:1687–1696

Vaillancourt K, Moineau S, Frenette M, Lessard C, Vadeboncoeur C (2002) Galactose and lactose genes from the galactose-positive bacterium *Streptococcus salivarius* and the phylogenetically related galactose-negative bacterium *Streptococcus thermophilus*: organization, sequence, transcription, and activity of the *gal* gene products. J Bacteriol 184:785–793

van de Guchte M, Serror P, Chervaux C, Smokvina T, Ehrlich SD, Maguin E (2002) Stress responses in lactic acid bacteria. Antonie van Leeuwenhoek 82:187–216

Van den Berghe E, Skourtas G, Tsakalidou E, De Vuyst L (2006) *Streptococcus macedonicus* ACA-DC 198 produces the lantibiotic, macedocin, at temperature and pH conditions that prevail during cheese manufacture. Int J Food Microbiol 107:138–147

van den Bogaard PT, Kleerebezem M, Kuipers OP, de Vos WM (2000) Control of lactose transport, beta-galactosidase activity, and glycolysis by CcpA in *Streptococcus thermophilus*: evidence for carbon catabolite repression by a non-phosphoenolpyruvate-dependent phosphotransferase system sugar. J Bacteriol 182:5982–5989

Varcamonti M, Graziano MR, Pezzopane R, Naclerio G, Arsenijevic S, De Felice M (2003) Impaired temperature stress response of a *Streptococcus thermophilus deoD* mutant. Appl Environ Microbiol 69:1287–1289

Varcamonti M, Arsenijevic S, Martirani L, Fusco D, Naclerio G, De Felice M (2006) Expression of the heat shock gene *clpL* of *Streptococcus thermophilus* is induced by both heat and cold shock. Microb Cell Fact 5:6

Virtaneva K, Graham MR, Porcella SF, Hoe NP, Su H, Graviss EA, Gardner TJ, Allison JE, Lemon WJ, Bailey JR, Parnell MJ, Musser JM (2003) Group A *Streptococcus* gene expression in humans and cynomolgus macaques with acute pharyngitis. Infect Immun 71:2199–2207

Virtaneva K, Porcella SF, Graham MR, Ireland RM, Johnson CA, Ricklefs SM, Babar I, Parkins LD, Romero RA, Corn GJ, Gardner DJ, Bailey JR, Parnell MJ, Musser JM (2005) Longitudinal analysis of the group A *Streptococcus* transcriptome in experimental pharyngitis in cynomolgus macaques. Proc Natl Acad Sci USA 102:9014–9019

Voyich JM, Braughton KR, Sturdevant DE, Vuong C, Kobayashi SD, Porcella SD, Otto M, Musser JM, DeLeo FR (2004a) Engagement of the pathogen survival response used by group A *Streptococcus* to avert destruction by innate host defense. J Immunol 173:1194–1201

Voyich JM, Musser JM, DeLeo FR (2004b) *Streptococcus pyogenes* and human neutrophils: a paradigm for evasion of innate host defense by bacterial pathogens. Microbes Infect 6:1117–1123

Welin J, Wilkins JC, Beighton D, Wrzesinski K, Fey SJ, Mose-Larsen P, Hamilton IR, Svensater G (2003) Effect of acid shock on protein expression by biofilm cells of *Streptococcus mutans*. FEMS Microbiol Lett 227:287–293

Wen ZT, Burne RA (2004) LuxS-mediated signaling in *Streptococcus mutans* is involved in regulation of acid and oxidative stress tolerance and biofilm formation. J Bacteriol 186:2682–2691

Wen ZT, Suntharaligham P, Cvitkovitch DG, Burne RA (2005) Trigger factor in *Streptococcus mutans* is involved in stress tolerance, competence development, and biofilm formation. Infect Immun 73:219–225

Wen ZT, Baker HV, Burne RA (2006) Influence of BrpA on critical virulence attributes of *Streptococcus mutans*. J Bacteriol 188:2983–2992

WHO (2007) Pneumococcal conjugate vaccine for childhood immunization – World Health Organization position paper. Wkly Epidemiol Rec 82:93–104

Wilkins JC, Homer KA, Beighton D (2002) Analysis of *Streptococcus mutans* proteins modulated by culture under acidic conditions. Appl Environ Microbiol 68:2382–2390

Witte G, Hartung S, Buttner K, Hopfner KP (2008) Structural biochemistry of a bacterial checkpoint protein reveals diadenylate cyclase activity regulated by DNA recombination intermediates. Mol Cell 30:167–178

Wouters JA, Sanders JW, Kok J, de Vos WM, Kuipers OP, Abee T (1998) Clustered organization and transcriptional analysis of a family of five csp genes of *Lactococcus lactis* MG1363. Microbiology 144 (Pt 10):2885–2893

Wouters JA, Rombouts FM, de Vos WM, Kuipers OP, Abee T (1999) Cold shock proteins and low-temperature response of *Streptococcus thermophilus* CNRZ302. Appl Environ Microbiol 65:4436–4442

Yamamoto Y, Higuchi M, Poole LB, Kamio Y (2000) Role of the *dpr* product in oxygen tolerance in *Streptococcus mutans*. J Bacteriol 182:3740–3747

Yamamoto Y, Poyart C, Trieu-Cuot P, Lamberet G, Gruss A, Gaudu P (2005) Respiration metabolism of group B *Streptococcus* is activated by environmental haem and quinone and contributes to virulence. Mol Microbiol 56:525–534

Yamamoto Y, Pargade V, Lamberet G, Gaudu P, Thomas F, Texereau J, Gruss A, Trieu-Cuot P, Poyart C (2006) The group B *Streptococcus* NADH oxidase Nox-2 is involved in fatty acid biosynthesis during aerobic growth and contributes to virulence. Mol Microbiol 62:772–785

Yamashita Y, Takehara T, Kuramitsu HK (1993) Molecular characterization of a *Streptococcus mutans* mutant altered in environmental stress responses. J Bacteriol 175:6220–6228

Yesilkaya H, Kadioglu A, Gingles N, Alexander JE, Mitchell TJ, Andrew PW (2000) Role of manganese-containing superoxide dismutase in oxidative stress and virulence of *Streptococcus pneumoniae*. Infect Immun 68:2819–2826

Yu J, Bryant AP, Marra A, Lonetto MA, Ingraham KA, Chalker AF, Holmes DJ, Holden D, Rosenberg M, McDevitt D (2001) Characterization of the *Streptococcus pneumoniae* NADH oxidase that is required for infection. Microbiology 147:431–438

Zhang J, Biswas I (2009) 3′-Phosphoadenosine-5′-phosphate phosphatase activity is required for superoxide stress tolerance in *Streptococcus mutans*. J Bacteriol 191:4330–4340

Zhang Q, Xu SX, Wang H, Xu WC, Zhang XM, Wu KF, Liu L, Yin YB (2009) Contribution of ClpE to virulence of *Streptococcus pneumoniae*. Can J Microbiol 55:1187–1194

Zhu L, Kreth J (2010) Role of *Streptococcus mutans* eukaryotic-type serine/threonine protein kinase in interspecies interactions with *Streptococcus sanguinis*. Arch Oral Biol 55:385–390

Zotta T, Asterinou K, Rossano R, Ricciardi A, Varcamonti M, Parente E (2009) Effect of inactivation of stress response regulators on the growth and survival of *Streptococcus thermophilus* Sfi39. Int J Food Microbiol 129:211–220

Zoumpopoulou G, Foligne B, Christodoulou K, Grangette C, Pot B, Tsakalidou E (2008) *Lactobacillus fermentum* ACA-DC 179 displays probiotic potential *in vitro* and protects against trinitrobenzene sulfonic acid (TNBS)-induced colitis and *Salmonella* infection in murine models. Int J Food Microbiol 121:18–26

Zuber P (2004) Spx-RNA polymerase interaction and global transcriptional control during oxidative stress. J Bacteriol 186:1911–1918

Chapter 13
Stress Responses of Enterococci

Yanick Auffray, Abdellah Benachour, Aurélie Budin-Verneuil, Jean-Christophe Giard, Axel Hartke, Vianney Pichereau, Alain Rincé, Nicolas Sauvageot, and Nicolas Verneuil

13.1 Introduction

Enterococci are Gram-positive bacteria that fit within the general definition of lactic acid bacteria (Stiles and Holzapfel 1997). The genus *Enterococcus* is composed of species associated with animals and plants, but only those from humans and domestic animals have been studied in detail (Flahaut et al. 1997a). Enterococci are most often considered to be components of the intestinal flora of humans and animals and are generally searched for in wastewater and food products, where their detection may indicate the risk of the presence of enteropathogenic organisms. Once they leave the oro-fecal lifestyle, their survival is linked to their exceptional aptitude to resist or grow in hostile environments that are usually detrimental to the development of most mesophilic microorganisms. However, a certain ambiguity exists concerning their relationship with human beings.

Indeed, some *Enterococcus* species or strains constitute a component of the microflora of fermented foods such as cheeses, sausages, green olives, and others that contributes to ripening and to the development of flavor (Foulquié Moreno et al. 2006). They also produce bacteriocins against pathogens such as *Listeria* and *Clostridium* (Cintas et al. 1998). This beneficial role led to the introduction of enterococci strains in starter cultures. Moreover, some strains of enterococci are used as probiotics to improve the microbial balance of the intestine and to treat gastroenteritis in humans and animals (Lewenstein et al. 1979; O'Sullivan et al. 1992).

On the other hand, over the last two decades, enterococci, formerly viewed as organisms of minimal clinical impact, have emerged as important hospital-acquired pathogens in immunocompromised patients (Low et al. 1994; Morrison et al. 1997).

Y. Auffray (✉)
Laboratoire de Microbiologie de l'Environnement, USC INRA2017/EA UCBN956, Université de Caen, 14032 Caen Cedex, France
e-mail: yanick.auffray@unicaen.fr

E. Tsakalidou and K. Papadimitriou (eds.), *Stress Responses of Lactic Acid Bacteria*, Food Microbiology and Food Safety, DOI 10.1007/978-0-387-92771-8_13,

According to epidemiological data, enterococci rank among the top four pathogens responsible for nosocomial infections (Chenoweth and Schaberg 1990). Enterococcal infections include urinary tract infections, hepatobiliary sepsis, endocarditis, surgical wound infections, bacteremia, and neonatal sepsis (Poh et al. 2006; de Fátima Silva Lopes et al. 2005). These data also indicate that *Enterococcus faecalis* is the most common species isolated from human illnesses, whereas *Enterococcus faecium* is linked with a higher threat of antibiotic resistance (Huycke et al. 1998; Giraffa 2002). In fact, enterococci show intrinsic (chromosomal) resistance to cephalosporins, lincosamides, and many beta-lactams in combination with acquired resistances to all classes of antibiotics through the exchange of resistance genes carried on conjugative transposons, pheromone-responsive plasmids, and a broad host range plasmids. Among these acquired resistances, vancomycin-resistant enterococci (VRE) constitute the most serious concern that has emerged in human clinical infections (Gilmore 2002).

Currently, antibiotic resistance alone cannot explain the virulence of enterococci. To cause infection, enterococci have to express virulence factors. Enterococci possess subtle virulence traits, which are not easily identified, including various adhesins, resistance to and modulation of the host's defense mechanisms, secretion of cytolysin, and production of plasmid-encoded pheromones. These determinants are harbored mainly by *E. faecalis* and to a lesser degree by *E. faecium* (Jett et al. 1994).

How enterococci sense environmental changes in the soil, water, host organisms, and so forth and respond to the various stimuli with adaptative behavior, which virulence factors are essential for disease and how these factors are expressed and regulated, and what the connections are between stress responses and pathogenesis are the most urgent questions to be answered in order to better understand the physiology of these organisms and address the increasing threat they pose to human health.

13.2 Physiological Responses to Environmental Fluctuations

A bacterial cell, in general, has very limited abilities to choose and modify its environment actively. In consequence, bacteria have to cope with fluctuations in their environment, and those microorganisms with very variable lifestyles, such as enterococci, are frequently confronted with these variations. Enterococci are naturally part of the digestive tract flora in many organisms, including humans. In the different parts of the digestive tract, they are exposed to pH variations, bile salts, fluctuations of osmolarity, high cellular densities, and nutrient limitations. After release by the host, they are confronted with a drastic decrease in temperature and with oligotrophic conditions. Enterococci can also be isolated from food, such as milk and meat and their byproducts. Stresses potentially experienced in the food industry are extreme temperatures and high osmolarity. Enterococci are also well equipped to thrive in hospitals, where they can be exposed to high temperatures, detergents, antibiotics, and oxidants.

Due to these changes between different ecological niches, these bacteria have evolved sophisticated mechanisms that allow rapid and efficient adaptation to environmental changes in order to face these challenges. Enterococci are well known for their high intrinsic resistances; these characteristics, known as the Sherman criteria of classification (Sherman 1937, 1938), have been used to distinguish them from other Gram-positive cocci. They are able to grow at 10 and 45°C and under "hostile" conditions, such as in broth adjusted to pH 9.6 or containing 6.5% NaCl or 40% of bile, or in milk containing 0.1% of methylene blue. Enterococci belong to the most thermotolerant of the nonsporulating bacteria since they survive exposure for 30 min at 60°C. In the case of *E. faecalis*, extraordinary resistances to heavy metals (Laplace et al. 1996), high osmolarity (Flahaut et al. 1996a), extreme pH values (Flahaut et al. 1996c, 1997a, b; Brändle et al. 2008), oxidants (Flahaut et al. 1998; Laplace et al. 1997), and detergents (Flahaut et al. 1996b) have also been documented.

As other bacteria, enterococci can also develop an adaptive stress response when exposed to moderate stress conditions before coping with a challenge dose of the same (homologous adaptation) or another (heterologous adaptation) agent. For example, the exposure of exponentially growing cultures of *E. faecalis* to moderate heat shock (50°C), cold shock (8–16°C), acid (pH 4.8) or basic pH (pH 10.5), H_2O_2 (2.4 mM), cadmium chloride (50 μg/mL), or detergent (0.01% sodium dodecyl sulfate [SDS] or 0.08% bile salts) develop increased resistance to exposure to a normally lethal homologous treatment (Boutibonnes et al. 1993; Flahaut et al. 1996a–c, 1997a, b; Laplace et al. 1996; Thammavongs et al. 1996). Except for acid pH, the pretreatment of the cultures under the above-mentioned sublethal conditions also led to multiple resistances toward heterologous stresses. Sometimes very high tolerance factors have been evidenced. The most spectacular examples are given in Table 13.1. As can be seen, the survival of adapted cells can increase by several orders of magnitude in comparison to control cultures.

Table 13.1 Development of tolerance of adapted cultures of *E. faecalis*

Adaptation condition	Challenge condition	Tolerance factor[a]	References
50°C	62°C	6,500	Flahaut et al. (1996c)
pH 4.8	pH 3.2	13,000	Flahaut et al. (1996c)
2.4 mM H_2O_2	45 mM H_2O_2	225	Flahaut et al. (1998)
Bile salts 0.08%	Bile salts 0.3%	1,700	Flahaut et al. (1996b, c)
50°C	Ethanol 22%	12,000	Boutibonnes et al. (1993)
Bile salts 0.08%	SDS 0.017%	2,600	Flahaut et al. (1996b)
SDS 0.01%	Bile salts 0.3%	1,000	Flahaut et al. (1996b)
NaCl	Bile salts	500 (6,800)	Flahaut et al. (1996a) and Pichereau et al. (1999)
NaCl	SDS	100	Flahaut et al. (1996a)
50°C	Bile salts	370	Flahaut et al. (1996c)
pH 10.9	Bile salts	1,500	Flahaut et al. (1997a, b)

[a] The tolerance factor is defined as the ratio of the survival of adapted cultures to the survival of control cultures

Moreover, the entrance of *E. faecalis* into the stationary phase due to exhaustion of glucose from the growth medium (Giard et al. 1996) or when cells are incubated under oligotrophic conditions (Hartke et al. 1998) led to the development of multiple, nonspecific prospective stress resistances in anticipation of future stresses. For example, cultures harvested under either condition were more resistant than their growing counterparts against exposition to heat and acid pH. Furthermore, glucose-starved cells developed resistance to H_2O_2 and ethanol (Giard et al. 1996), and those harvested from the oligotrophic microcosm became more resistant to NaOCl and UV irradiation (Hartke et al. 1998). For some stresses, maximal resistances are observed early in the stationary phase, whereas in other cases, cells have to spend more time in a nongrowing state in order to resist the challenge efficiently.

These physiological studies on the responses to stresses potentially encountered by enterococci during their lifestyles were hitherto only conducted with the opportunistic pathogen *E. faecalis*. The combined results demonstrate that this enterococcal species is a very hardy, stress-resistant bacterium that shows a tenacity for life, which could be important traits for successful infection. In order to strengthen this last hypothesis, it would be interesting to compare the results with those obtained with other enterococcal species, including *E. faecium*, the second enterococcal species of clinical importance.

13.3 Identification of Effectors of the Stress Response

13.3.1 Proteomic Studies

In addition to candidate genes approaches, global experimental strategies were successfully used to identify the proteins involved in enterococcal stress adaptation. In particular, proteomic methods were widely used to identify the proteins deregulated following stress exposure. The main technique to separate and examine a whole set of proteins encoded by a given genome is the two-dimensional electrophoresis (2-DE) (O'Farrell 1975). Such a proteomic approach is performed qualitatively and quantitatively even if only 10% of the theoretical proteins can be revealed by 2-DE of a whole-cell protein extract. The major interest in these comparative studies is to create a proteome phenotyping, allowing disclosure of the modifications of the protein metabolism in response to a specific stimulus. Showing and comparing the expression of stress-induced proteins is crucial to understanding how bacteria are able to cope with environmental changes. Moreover, global approaches such as 2-DE are performed to identify targets of regulatory proteins, which are known to be involved in the stress response.

Only a few studies on the induction of stress proteins have been performed by 2-DE on *Enterococcus*. The first stress proteome of *E. faecalis* was performed by Giard et al. (2001). 2-DE studies were performed on *E. faecalis* in sublethal stress conditions, such as NaCl, ethanol, bile salts, H_2O_2, or pH. A comparison of the different patterns revealed a total of 167 protein spots deregulated (mainly upregulated) during different stresses, approximately 50 of which were identified (Flahaut et al. 1996a–c; Giard et al. 2001, 2004). For example, 37 proteins were found overexpressed under alkaline conditions (Flahaut et al. 1997a, b), 20 after an ethanol treatment (Rincé et al. 2003), 34 and 45 during exposure to SDS and bile salts, respectively (Flahaut et al. 1996b), and 96 during salt stress (Flahaut et al. 1996a). These proteins, listed by Giard et al. (2004), were classified according to their expression patterns. Some of them were specific to an individual stress, but a few showed an increase of synthesis under multiple-stress situations. Such polypeptides are of particular interest because it was legitimate to believe that they play a significant role in the stress response. In *E. faecalis*, only seven polypeptides could be described as "general stress proteins" (Gsp): the six proteins Gsp62, 63, 64, 65, 66, and 67 (Rincé et al. 2000; Le Breton et al. 2002b), and Gls24 (Giard et al. 2001).

The two spots, Gsp66 and 67, corresponded to two major heat-shock proteins, DnaK and GroEL, respectively. In addition, the analysis of mutant strains allowed Gsp64 to be characterized as responsible for survival against organic hydroperoxide [*tert*-butyl hydroperoxide (tBOOH)], ethanol, and basic pH, and Gsp65, encoding Ohr, to be classified as involved in tBOOH and ethanol resistances (Rincé et al. 2001; Le Breton et al. 2002b). Unfortunately, no phenotype could be observed with the *gsp62* mutant. Moreover, the proteomic analysis of this strain did not show modification of the protein metabolism. Although Gsp62 is a general stress protein, its implication in the protection or repair of the damage caused by lethal treatments remains to be elucidated (Rincé et al. 2002).

In addition to proteome phenotyping of the stress response, different physiological states, such as glucose starvation, oligotrophic conditions, and viable but nonculturable state (VBNC), were also performed by 2-DE on *E. faecalis*. For example, 50 proteins were overexpressed after oligotrophic conditions (Hartke et al. 1998), three proteins were upregulated and five downregulated in a VBNC state (Heim et al. 2002), and, finally, 42 proteins were overexpressed by glucose starvation (Giard et al. 1997). Among these, Gls24 was first classified as a glucose starvation protein, but it was later revealed to be essential for multiple stress adaptation and virulence of *E. faecalis* (Giard et al. 2001; Teng et al. 2005). Phenotypic and proteomic analysis of a *gls24* mutant strain showed that the Gls24 protein was involved in the formation of chains in the exponential growth phase, in the acquisition of bile salt resistance in the stationary phase, and in the expression of genes involved in pyruvate metabolism (Giard et al. 2000). It should be noted that clinical isolates were observed to be significantly enriched for *gls24* family genes and that *gls24* was induced in response to biological media such as serum and urine (Shepard and Gilmore 2002).

Proteome phenotyping of stress regulators were also performed. For example, the protein expression analysis of a mutant of CcpA (Catabolite control protein), the transcriptional regulator of the catabolite repression, revealed that six proteins were underexpressed as compared to the wild-type strain, and 16 were overexpressed (Leboeuf et al. 2000). Interestingly, 13 out of 16 proteins were also induced by glucose starvation, including the arginine deiminase (Gls14) and Gls24 (Giard et al. 1997).

The inactivation of the two-component system (TCS) Ehk10-Err10, a regulator involved in the acid, salt, bile salt, and heat-stress responses, appeared to regulate the expression of the two chaperones DnaK and GroEL (Le Breton et al. 2003). Finally, in an *hypR* mutant (hydrogen peroxide regulator), an important transcriptional regulator involved in the oxidative stress response and virulence, four proteins were repressed under H_2O_2 challenge (Verneuil et al. 2004b). Three of them were involved in metabolic pathways (Eno, TpiA, DhaK), and Ef1744 was a stress protein. The BS10 (bile salts) transcriptional regulator is involved in the bile salt-stress response in *E. faecalis*, and the proteome phenotype of the mutated strain appears different from the wild-type one, showing seven and eight proteins down- and upregulated, respectively (Le Breton et al. 2002a).

13.3.2 Random and Targeted Strategies

An inescapable and obligatory step in the study of genes involved in the bacterial stress response is to inactivate them in order to determine their implication in the physiological response of the cell. However, genetic analysis in enterococci has often been restricted by limitations of the available genetic tools. Since the start of molecular studies in enterococci, few genetic tools have been used to introduce chromosomal mutations. Until now, two kinds of tools have been applied in *Enterococcus* to disrupt target genes and create mutants; most of them were originally developed for *Lactococcus lactis*. Briefly, one corresponds to the use of non-replicating plasmids for targeted or random mutagenesis, and the second is the utilization of transposons for the construction of mutant libraries. In all strategies, the insertion of the mutagenic element can be screened by antibiotic selection. Among the large scale of antibiotics available and used to construct mutants in bacteria, only a few are currently used in enterococci. Antibiotic-resistance markers commonly found in different vectors are often the same, namely, erythromycin, kanamycine, tetracycline, and chloramphenicol.

Among the proteins identified to be involved in the stress response and characterized in *Enterococcus* (Rincé et al. 2000), the first inactivations were carried out by homologous recombination with the insertional vector pUCB300, which corresponds to a plasmid harboring an erythromycin marker. This plasmid allowed the inactivation of genes encoding the glucose starvation protein of *E. faecalis* Gls24 (Giard et al. 2000), but also Gsp62 (Rincé et al. 2002; Capiaux et al. 2000), the catabolite repression protein CcpA (Leboeuf et al. 2000), and the oxidative stress response regulators (RRs) PerR and HypR (Verneuil et al. 2004b, 2005). Several other plasmids with similar properties have been used to achieve targeted mutagenesis, pTEX4577 and p3tet, which contain a kanamycin and tetracycline resistance gene, respectively (Qin et al. 1998; Hancock and Perego 2004b). These plasmids, still in use, were complemented by other, more sophisticated plasmids. In 2004, Arnaud et al. constructed the pMAD vector using a thermosensitive replication associated with a colorimetric blue-white discrimination of bacteria that have lost the plasmid (Arnaud et al. 2004). This vector, developed for low-G+C-content Gram-positive bacteria, has been used in *Enterococcus*. It allowed the inactivation of proteins involved in oxidative stress (peroxidase and hydroperoxide reductase) or in lysozyme resistance (La Carbona et al. 2007; Hébert et al. 2007). But most genes encoding stress proteins have been mutated with the strategy described by Law et al. (1995) for *L. lactis*, which combines the use of two pWV01 derivatives: the cloning vector pORI19, and the thermosensitive helper plasmid pG+host3. This technique allows the stable integrant mutant formation in *E. faecalis*, with high efficiencies, and the fast analysis of targeted genes. Some examples include a *gsp64* mutant (Le Breton et al. 2002b), a *gsp65* encoding an organic hydroperoxide resistance protein (Ohr) (Rincé et al. 2001), a superoxide dismutase (SOD) (Verneuil et al. 2006), and *e*xtra*c*ytoplasmic *f*unction (ECF) sigma and antisigma factor mutants (Benachour et al. 2005).

Parallel to this, the study of genes involved in the stress response of enterococci led to the development of global approaches, such as the inactivation of two-component signal transduction systems (Hancock and Perego 2004b; Le Breton et al. 2003) or the construction of a library of 9,600 random insertion mutants (Le Breton et al. 2002a). These genome-wide searches allowed the highlighting of unidentified genetic determinants that promote the resistance of *Enterococcus* to different stresses such as SagA (Le Breton et al. 2002a). More recently, random mutant libraries by transposition have been attempted. At least three transposons have previously been used for comprehensive random mutagenesis (Tn916, Tn917, and a mariner-based system) (Scott et al. 1994; Garsin et al. 2004; Kristich et al. 2008). However, some of these systems show significantly biased insertion-site preferences that prevent random coverage of the whole genome. The problems have been circumvented by a mutagenesis system based on the mariner transposable element, which allowed the selection of approximately 15,000 insertions around the entire chromosome (Kristich et al. 2008).

Moreover, due to low transformation efficiencies and recombination in enterococci, recombinants are difficult to obtain, and any such recombinants suffer from significant drawbacks inherent in this type of mutation (for example, mutants that carry antibiotic-resistance markers, mutations that are polar, and reversion to the wild-type genotype by plasmid excision). To facilitate efficient genetic manipulation, new tools were developed in *E. faecalis*, like a conjugative delivery system for the high-frequency introduction of cloned DNA into strains based on the enterococcal pheromone-responsive conjugative plasmid pCF10 (Kristich et al. 2007).

13.3.3 Research and Analysis of Stress Effectors by Transcriptomics Approaches

Few works have used global approaches (microarray or RT-qPCR) in order to find stress effectors by comparing standard growth and stress conditions or mutant and wild-type strains in *E. faecalis*. To our knowledge, the first work performed with

DNA chips to study the modification of gene expressions due to the presence of erythromycin was that of Aakra et al. (2005). More recently, microarrays were used to compare transcriptional profiles of the wild-type strain vs. regulator deleted mutants and then characterize the regulon (Bourgogne et al. 2006). The only global transcriptional analysis in *E. faecalis* confronted with a stress is the response to bovine bile (BB) and SDS (Solheim et al. 2007). This study shows that 308, 209, and 254 genes were identified as differentially expressed in the presence of BB, SDS, or BB + SDS, respectively. Since detergents affect the bacterial membrane, it is not surprising to find several genes that encode proteins with membrane-associated functions. However, none of the putative bile salt hydrolase showed differential expression during bile exposure. The major conclusion of this study is that different mechanisms are involved in detergent resistance in *E. faecalis* (Solheim et al. 2007).

RT-qPCR analysis of the transcription of genes suspected or known to be involved in the oxidative stress response has been carried out in *E. faecalis* wild-type strain and in mutants affected in the transcriptional regulators HypR, PerR, and Ers (Verneuil et al. 2004b, 2005; Giard et al. 2006). Results revealed that the transcriptional regulation of oxidative stress effectors is mainly due to the activity of HypR in *E. faecalis* (Verneuil et al. 2004b, 2005; La Carbona et al. 2007).

As mentioned earlier, several general stress proteins of *E. faecalis* (Gls24, Gsp62, Gsp64, and Gsp65) have been identified following proteomic studies (Giard et al. 2000, 2001). The transcriptional analysis by Northern blot of the corresponding genes showed that they were induced under several stress conditions, including temperature, acid pH, presence of detergents, ethanol, NaCl, H_2O_2, and tBOOH. Interestingly, the induction rates observed for the *gsp62*, *64*, and *65* genes were similar for each condition (Rincé et al. 2000). In addition to the transcriptional induction under environmental stress conditions, the onset of glucose starvation also induced the transcription of *gsp62*.

13.4 Transcriptional Regulation Involved in the Stress Response

13.4.1 Two-Component Systems

TCSs are mechanisms involved in perception and signal transduction in order to respond to environmental changes. They are composed of one histidine kinase (HK), which acts as a sensor, and one cytoplasmic RR. In response to a specific extracellular signal, the HK is phosphorylated, and then the phosphoryl group is transferred to the corresponding RR. Phosphorylation of the latter allows its activation to modulate the cellular response to the stimulus (Stock et al. 2000). The activated RR often acts as a transcriptional regulator of targeted genes, while in some other cases, it acts by modulating enzymatic activity. Based on similarities in the structure and sequence of the HK and RR proteins, the TCSs were grouped into six classes (Others-B, NarL, OmpR, NtrB, Others-A, and CheY) (Mizuno 1997; Fabret et al. 1999).

In *E. faecalis*, 17 TCSs (named Ehk-Err for enterococcal HK and enterococcal RR, and consisting of an HK and the corresponding RR), plus one orphan RR (i.e., without coupled HK), have been identified following the analysis of the sequenced genome of the V583 strain (Hancock and Perego 2002). Their involvement in stress response and virulence was tested using different *E. faecalis* strains.

Teng et al. (2002) constructed seven TCS mutants in the *E. faecalis* OG1RF strain and showed, using peritoneal injection in a mouse model, that the Err10-Ehk10 (EtaR) and Ehk13-Err13 (EtbR) systems are involved in virulence. These TCSs were then studied in the context of the stress response. Compared to the wild-type strain, the *err10* mutant (*etaR*) showed a reduced survival at pH 3.4 and an increased resistance at 55°C, unlike the *ehk13* mutant (*etbS*), which is more sensitive to high temperature.

In order to study TCSs from the OmpR class in *E. faecalis* JH2-2, their expression was evaluated under several stress conditions. Mutants affected in the RR genes were constructed and used to test their stress behavior and intracellular survival in mouse macrophages (Le Breton et al. 2003; Muller et al. 2008). The intracellular survival of the mutants *err04*, *err05*, and *err06* was shown to be reduced compared to that of the wild-type JH2-2 strain. In addition, Err04-Ehk04 is induced in phosphate deprivation conditions and plays a role in the expression of *pstF*, a gene encoding a phosphate-binding protein; the corresponding mutant is more sensitive to heat. The *err05* mutant (*croR*) presented a growth defect, abnormal cell morphology, and alteration of the expression of genes encoding antigenic determinant expressed during *E. faecalis* human endocarditis infections (Muller et al. 2006; Le Breton et al. 2007). Err06-Ehk06 was shown to be involved in the oxidative stress response. Indeed, the *err06-ehk06* operon is induced by H_2O_2 and the *err06* mutant is more sensitive to H_2O_2. The Err07-Ehk07 system appeared essential to the development of the bacteria; its expression is induced by low and high temperatures. The *err10* mutant, in which survival was less affected inside macrophages compared to that of the wild-type strain, is also more resistant to heat and bile salts (two stresses that induced its expression) while being more susceptible to acid and NaCl stresses. Finally, the expression of the Ehk13-Err13 and Ehk14-Err14 TCSs is induced by heat.

Hancock and Perego (2004b) performed a systematic inactivation of genes encoding the RR in the V583 strain. The growth of each mutant under standard or environmental stress conditions and their sensitivity to antibiotics have been studied. The *err04* mutant presented a growth defect, while the *err06* mutant showed a sensitivity to high temperatures (46°C) and SDS.

It was also shown that the TCS Err15-Ehk15 (FsrCA) is activated by mechanisms of quorum sensing (Nakayama et al. 2001) and plays a role in the biofilm formation (Hancock and Perego 2004a), virulence (Qin et al. 2000; Mylonakis et al. 2002; Sifri et al. 2002), and transcriptional regulation of the *gelE* and *sprE* genes encoding extracellular proteases (Qin et al. 2001).

In addition to being directly involved in stress and/or virulence, some TCSs contribute to pathogenicity by altering the sensitivity or resistance to some antibiotics. Indeed, Ehk01-Err01 and Ehk03-Err03 participate in the resistance to bacitracin (Hancock and Perego 2004b), and Err05-Ehk05 (CroRS) is involved in the resistance to cell wall active antimicrobials such as β-lactams (Comenge et al. 2003). Moreover, Err11-Ehk11 regulates genes involved in the resistance to vancomycin (Arthur et al. 1992; Evers and Courvalin 1996). Glycopeptides (including vancomycin) act by blocking cell wall formation. Several types of acquired glycopeptide resistance have been described. They allow the synthesis of modified peptidoglycan precursors with reduced affinity for the antibiotic and, like Err11-Ehk11 for the VanB-type mechanism, are encoded by genes associated with a TCS, which ensures their regulation. In enterococci, such mechanisms were described mainly in *E. faecium*, *E. faecalis, and Enterococcus casseliflavus*; they were also reported in other species (e.g., *Enterococcus raffinosus*, *Enterococcus avium*, *Enterococcus hirae*) (Werner et al. 2008).

Finally, more recently, Del Papa and Perego (2008) have shown that the *E. faecalis* Err17-Ehk17 TCS controls the utilization of ethanolamine. Err17-Ehk17 is activated by this carbon source founded in abundance in the human intestine, the site of bacterial invasion.

13.4.2 Alternative Sigma Factors

Bacterial sigma factors are a class of proteins constituting essential dissociable subunits of RNA polymerase to direct the initiation of transcription from specific promoter sequences (Haldenwang 1995). In addition to the "housekeeping" sigma factor (i.e., RpoD or σ^{70} of *Escherichia coli*), most bacteria have multiple alternative sigma factors that they use to coordinately regulate the expression of genes whose products are involved in diverse functions, such as stress responses, iron uptake, virulence, morphological development, and chemotaxis. In bacteria, a common mode of adaptation to any environmental changes including those encountered during host-cell interactions is through the switch in sigma factors utilized by the RNA polymerase core enzyme. In this context, bacteria utilize alternative sigma factors, among which is one distinct σ^{70} subfamily known to regulate ECFs (Lonetto et al. 1994). Many ECF sigma factors were effectively shown to be involved in the stress responses and virulence of different bacteria (Bashyam and Hasnain 2004; Kazmierczak et al. 2005).

Like those of *Streptococcus pyogenes*, *Streptococcus pneumoniae*, and *L. lactis*, the enterococci genomes available in public databases lack a "general stress RR" homologous to σ^B of *Bacillus subtilis* (Yother et al. 2002). This raises the question of how these bacteria regulate gene expression during stress responses. It is also assumed that large sets of alternative sigma factors are found in bacteria with varied lifestyles (Mittenhuber 2002). This is in agreement with the in silico analysis of the *E. faecalis* available genome, which revealed the presence of four potential ECF sigma factors [EF3180 (SigV), EF3020 (SigX), EF0049 (SigH), and EF2290], while, paradoxically, the related *E. faecium* DO strain possesses only one (EfaeDraft_1628), homologous to SigH (http://www.microbesonline.org/). On the other hand, the representative model of lactic acid bacteria *L. lactis* also has only one ECF sigma factor (Mittenhuber 2002). The structural analysis of these enterococcal sigma factors detected in silico showed that only the *sigV* and *sigX* genes have features and evidence for cluster organization consistent with ECF sigma factors; that is, their corresponding antisigma factor is included in the same operonic structure.

Among these ECF sigma factors, the *E. faecalis* SigV was shown to be involved in the response to heat shock, acid and ethanol treatments. It contributes to long-term survival in rich medium or under oligotrophic conditions, and it is also involved in the lysozyme resistance of *E. faecalis* (Benachour et al. 2005). Preliminary results revealed that SigV is also involved in the virulence of the *E. faecalis* JH2-2 strain (Benachour, personal communication). Attempts with the *E. faecalis* OG1RF *sigH2* mutant strain failed to detect any defect following growth kinetics in several rich and semidefined media nor under a number of other assays, including sensitivity to a panel of antibiotics, biofilm formation in a microtiter-plate assay, adherence to collagen-coated surfaces, and ability to serve as a donor or recipient for conjugative plasmid transfer (Kristich et al. 2007). On the other hand, while not belonging to the alternative sigma factors, the *E. faecalis* σ^{54} (RpoN) was also characterized and shown to be involved in susceptibility to the bacteriocin mesentericin Y105 (Héchard et al. 2001).

The functions of the ECF sigma factors remain, therefore, largely undetermined in these organisms. Thus, their study will have great impact on the public health field, since it is expected to open up new insights and strategies to help overcome enterococcal stress responses, infection, and virulence.

13.4.3 Other Regulatory Mechanisms

In *B. subtilis*, the Gram-positive bacterial model, modifications at the transcriptional levels in response to oxidative stress are managed by the metalloregulatory protein PerR (peroxide regulator) member of the Fur (ferric uptake regulator) family (Bsat et al. 1998). Similarly to *Bacillus*, *E. faecalis* phenotype analysis showed that the *perR* mutant was significantly more resistant to a hydrogen peroxide challenge (Verneuil et al. 2005). However, only slight differences in the transcriptional activity of oxidative stress genes in the *perR* mutant vs. wild-type strain were observed (Verneuil et al. 2005). It was then concluded that the molecular basis of the oxidative stress response in *E. faecalis* is distinct from the *B. subtilis* model.

OxyR was characterized in *E. coli* as the main transcriptional activator in response to H_2O_2 treatment (Storz and Imlay 1999). It has been shown that the purified OxyR protein from *E. coli* can interact with the promoter region of *npr* (encoding the NADH peroxidase) of *E. faecalis* and that a specific antibody against OxyR recognized a 34-kDa protein of *E. faecalis* (Ross and Claiborne 1997). In *E. faecalis*, the locus *ef2958* encodes a regulator of the LysR family (like OxyR) and, in spite of the absence of the conserved cysteine residues and of the difference in the regulated genes, it was the most closely related gene to *oxyR* of *E. coli* (Verneuil et al. 2004a). Moreover, a mutation in the *ef2958* gene, renamed *hypR* (for *hy*drogen *p*eroxide *r*egulator), sensitized *E. faecalis* to an H_2O_2 treatment (Verneuil et al. 2004b).

Transcriptional analysis revealed that *hypR* itself, *ahpCF* (encoding the alkyl hydroperoxide reductase), *tpx* (encoding a thiol peroxidase), *ef3270* (encoding the glutathione reductase), *sodA* (encoding the SOD), and *katA* (encoding the catalase) gene expressions were controlled by HypR only under H_2O_2 stress conditions. Until now, only the expressions of *ahpCF*, *tpx*, and *hypR* have been shown to be directly regulated by HypR (Verneuil et al. 2004b; La Carbona et al. 2007). These results give evidence for a new oxidative stress regulon in *E. faecalis*.

Ef2063, a member of the AraC family, is the *E. faecalis* protein with the highest degree of sequence identity (35%) to the SoxS regulator from *E. coli* (Verneuil et al. 2004a). The *soxRS* regulon has been characterized as a defense against oxidative stress in response to $O_2^{\bullet-}$-generating compounds and contains genes such as SOD (*sodA*) (Storz and Imlay 1999). The *ef2063* gene has been mutated in *E. faecalis*, but the resulting strain did not exhibit increased sensitivity toward several oxidants (Verneuil et al. 2004a).

In *B. subtilis*, CtsR has been characterized as a regulator of stress and heat-shock response (Derré et al. 1999). It acts as a repressor of *clpP*, *clpC*, and *clpE* that encodes ATP-dependent proteases important for stress survival, including growth at high temperatures. In this study, the authors showed that a homolog of CtsR, with a very similar predicted helix-turn-helix DNA-binding motif, is present in the *E. faecalis* genome. In addition, the *E. faecalis* sequence upstream *clpC*, *clpP*, and *clpE* harbors the CtsR recognition sequence (Derré et al. 1999). Taken together, these data strongly suggest that a CtsR-like regulator exists in *E. faecalis* and that it could play a role in the stress response.

The crystal structure of *E. faecalis* SlyA-like transcriptional factor has been carried out. SlyA from *E. faecalis* belongs to the MarR/SlyA family and shows significant sequence similarity to members of both the SlyA and MarR subfamilies (Wu et al. 2003). Since the *slyA* gene is involved in virulence and environmental adaptation in *Salmonella typhimurium* and in *Yersinia* (Spory et al. 2002; Nagel et al. 2001), it is tempting to speculate that the SlyA-like transcriptional factor of *E. faecalis* could also play a role in the pathogenicity and/or stress response.

In recent years, small noncoding RNAs (sncRNAs) have been identified as key regulators of several cellular processes, including stress response in multicellular eukaryotes and in prokaryotes. One recent study allowed the prediction and annotation of putative sRNA-encoding genes in 932 bacterial genomes, including *E. faecalis*. In this microorganism, this high-throughput computational search identified 17 sRNAs (14 on the chromosome and three on plasmids) (Livny et al. 2008). This is the first step toward the identification of new regulatory networks that would be undoubtedly involved in the stress response of *E. faecalis*.

13.5 Concluding Remarks

The stress response of enterococci has been intensively studied in the last two decades. This led to the discovery of impressive physiological capacities of these microorganisms that allow them to adapt to and survive under harsh conditions.

These properties, at least in part, may be the basis, and of prime importance, for these bacteria to persist in hospital settings despite their being permanently exposed to aggressive treatments that aim to kill them. The adaptive stress responses are accompanied by the induction of the synthesis of complex sets of polypeptides. Several regulators have been identified to control the expression of stress-responsive genes. The challenges will now be to assign a function to those stress proteins with yet unknown roles, to prove their direct implication in the corresponding stress responses, and to build regulatory networks for these systems. The molecular tools available for these bacteria need to be completed. Urgently needed is a sensitive reporter-based system for the study of gene expression. Of prime importance also is the discovery of a general transducing bacteriophage allowing the rapid transfer of mutations between strains, with the aim of analyzing the diversity of stress responses among different isolates. Recently, an interesting relationship between the oxidative stress response and the killing ability of bactericidal antibiotics was discovered in *E. faecalis* (Bizzini et al. 2009). Understanding the events induced by the antibiotic leading to an oxidative burst, as well as the identification and characterization of the oxidant-sensitive targets, may provide new treatments that should allow the bactericidal effects of antibiotics to be increased. This important finding, which may lead to a direct application for human health, as well as the growing evidence of a close relationship between stress and bacterial virulence should encourage the continued study of the stress responses of enterococci.

References

Aakra A, Vebo H, Snipen L, Hirt H, Aastveit A, Kapur V, Dunny G, Murray BE, Nes IF (2005) Transcriptional response of *Enterococcus faecalis* V583 to erythromycin. Antimicrob Agents Chemother 49:2246–2259

Arnaud M, Chastanet A, Débarbouillé M (2004) New vector for efficient allelic replacement in naturally nontransformable, low-GC-content, Gram-positive bacteria. Appl Environ Microbiol 70:6887–6891

Arthur MC, Molinas C, Courvalin P (1992) The VanS-VanR two-component regulatory system controls synthesis of depsipeptide peptidoglycan precursors in *Enterococcus faecium* BM4147. J Bacteriol 174:2582–2591

Bashyam MD, Hasnain SE (2004) The extracytoplasmic function sigma factors: role in bacterial pathogenesis. Infect Genet Evol 4:301–308

Benachour A, Muller C, Dabrowski-Coton M, Le Breton Y, Giard JC, Rincé A, Auffray Y, Hartke A (2005) The *Enterococcus faecalis* SigV protein is an extracytoplasmic function sigma factor contributing to survival following heat, acid, and ethanol treatments. J Bacteriol 187:1022–1035

Bizzini A, Zhao C, Auffray Y, Hartke A (2009) The *Enterococcus faecalis* superoxide dismutase is essential for its tolerance to vancomycin and penicillin. J Antimicrob Chemother 64:1196–1202

Bourgogne A, Hilsenbeck SG, Dunny GM, Murray BE (2006) Comparison of OG1RF and an isogenic *fsrB* deletion mutant by transcriptional analysis: the Fsr system of *Enterococcus faecalis* is more than the activator of gelatinase and serine protease. J Bacteriol 188:2875–2884

Boutibonnes P, Giard JC, Hartke A, Thammavongs B, Auffray Y (1993) Characterization of the heat shock response of *Enterococcus faecalis*. Antonie von Leeuwenhoek 64:47–55

Brändle N, Zehnder M, Weiger R, Waltimo T (2008) Impact of growth conditions on susceptibility of five microbial species to alkaline stress. J Endod 34:579–582

Bsat N, Herbig A, Casillas-Martinez L, Setlow P, Helmann JD (1998) *Bacillus subtilis* contains multiple Fur homologs: identification of the iron uptake (Fur) and peroxide regulon (PerR) repressors. Mol Microbiol 29:189–198

Capiaux H, Giard JC, Lemarinier S, Auffray Y (2000) Characterization and analysis of a new gene involved in glucose starvation response in *Enterococcus faecalis*. Int J Food Microbiol 55:99–102

Chenoweth C, Schaberg D (1990) The epidemiology of enterococci. Eur J Clin Microbiol Infect Dis 9:80–89

Cintas LM, Casaus P, Holo H, Hernandez PE, Nes IF, Håvarstein LS (1998) Enterocins L50A and L50B, two novel bacteriocins from *Enterococcus faecium* L50, are related to staphylococcal hemolysins. J Bacteriol 180:1988–1994

Comenge YR, Quintiliani R Jr, Li L, Dubost L, Brouard JP, Hugonnet JE, Arthur M (2003) The CroRS two-component regulatory system is required for intrinsic beta-lactam resistance in *Enterococcus faecalis*. J Bacteriol 185:7184–7192

de Fátima Silva Lopes M, Ribeiro T, Abrantes M, Figueiredo Marques JJ, Tenreiro R, Crespo MT (2005) Antimicrobial resistance profiles of dairy and clinical isolates and type strains of enterococci. Int J Food Microbiol 103:191–198

Del Papa MF, Perego M (2008) Ethanolamine activates a sensor histidine kinase regulating its utilization in *Enterococcus faecalis*. J Bacteriol 190:7147–7156

Derré I, Rapoport G, Msadek T (1999) CtsR, a novel regulator of stress and heat shock response, controls *clp* and molecular chaperone gene expression in Gram-positive bacteria. Mol Microbiol 31:117–131

Evers S, Courvalin P (1996) Regulation of VanB-type vancomycin resistance gene expression by the VanS(B)-VanR(B) two-component regulatory system in *Enterococcus faecalis* V583. J Bacteriol 178:1302–1309

Fabret C, Feher VA, Hoch JA (1999) Two-component signal transduction in *Bacillus subtilis*: how one organism sees its world. J Bacteriol 181:1975–1983

Flahaut S, Benachour A, Giard JC, Boutibonnes P, Auffray Y (1996a) Defense against lethal treatments and *de novo* protein synthesis induced by NaCl in *Enterococcus faecalis* ATCC 19433. Arch Microbiol 165:317–324

Flahaut S, Frère J, Boutibonnes P, Auffray Y (1996b) Comparison of the bile salts and sodium dodecyl sulfate stress responses in *Enterococcus faecalis*. Appl Environ Microbiol 62: 2416–2420

Flahaut S, Hartke A, Giard JC, Benachour A, Boutibonnes P, Auffray Y (1996c) Relationships between stress responses toward bile salts, acid and heat treatments in *Enterococcus faecalis*. FEMS Microbiol Lett 138:49–54

Flahaut S, Boutibonnes P, Auffray Y (1997a) Les enterocoques dans l'environnement proche de l'homme [in French]. Can J Microbiol 43:699–708

Flahaut S, Hartke A, Giard JC, Auffray Y (1997b) Alkaline stress response in *Enterococcus faecalis*: adaptation, cross-protection and changes in protein synthesis. Appl Environ Microbiol 63:812–814

Flahaut S, Laplace JM, Frère J, Auffray Y (1998) The oxidative stress response in *Enterococcus faecalis*: relationship between H_2O_2 tolerance and H_2O_2 stress proteins. Lett Appl Microbiol 26:259–264

Foulquié Moreno MR, Sarantinopoulos P, Tsakalidou E, De Vuyst L (2006) The role and application of enterococci in food and health. Int J Food Microbiol 106:1–24

Garsin DA, Urbach J, Huguet-Tapia JC, Peters JE, Ausubel FM (2004) Construction of an *Enterococcus faecalis* Tn917-mediated-gene-disruption library offers insight into Tn917 insertion patterns. J Bacteriol 186:7280–7289

Giard JC, Hartke A, Flahaut S, Benachour A, Boutibonnes P, Auffray Y (1996) Starvation induced multi-resistance in *Enterococcus faecalis* JH2-2. Curr Microbiol 32:264–271

Giard JC, Hartke A, Flahaut S, Boutibonnes P, Auffray Y (1997) Glucose starvation response in *Enterococcus faecalis* JH2-2: survival and protein analysis. Res Microbiol 148:27–35

Giard JC, Rincé A, Capiaux H, Auffray Y, Hartke A (2000) Inactivation of the stress- and starvation-inducible *gls*24 operon has a pleiotropic effect on cell morphology, stress sensitivity and gene expression in *Enterococcus faecalis*. J Bacteriol 182:4512–4520

Giard JC, Laplace JM, Rincé A, Pichereau V, Benachour A, Leboeuf C, Flahaut S, Auffray Y, Hartke A (2001) The stress proteome of *Enterococcus faecalis*. Electrophoresis 14:2947–2954

Giard JC, Auffray Y, Benachour A, Hartke A, Laplace JM, Rincé A, Verneuil N, Pichereau V (2004) Proteomics analysis: a powerful tool to identify proteome phenotype and proteome signature in *Enterococcus faecalis*. Curr Proteomics 1:273–282

Giard JC, Riboulet E, Verneuil N, Sanguinetti M, Auffray Y, Hartke A (2006) Characterization of Ers, a PrfA-like regulator of *Enterococcus faecalis*. FEMS Immunol Med Microbiol 46:410–418

Gilmore MS (2002) The enterococci: pathogenesis, molecular biology and antimicrobial resistance. American Society for Microbiology, Washington, DC

Giraffa G (2002) Enterococci from foods. FEMS Microbiol Rev 26:163–171

Haldenwang WG (1995) The sigma factors of *Bacillus subtilis*. Microbiol Rev 59:1–30

Hancock L, Perego M (2002) Two-component signal transduction in *Enterococcus faecalis*. J Bacteriol 184:5819–5825

Hancock L, Perego M (2004a) The *Enterococcus faecalis* Fsr two-component system controls biofilm development through production of gelatinase. J Bacteriol 186:5629–5639

Hancock L, Perego M (2004b) Systematic inactivation and phenotypic characterization of two-component signal transduction systems of *Enterococcus faecalis* V583. J Bacteriol 186:7951–7958

Hartke A, Giard JC, Laplace JM, Auffray Y (1998) Survival of *Enterococcus faecalis* in an oligotrophic microcosm: changes in morphology, development of general stress resistance and analysis of protein synthesis. Appl Environ Microbiol 64:4238–4245

Hébert L, Courtin P, Torelli R, Sanguinetti M, Chapot-Chartier MP, Auffray Y, Benachour A (2007) *Enterococcus faecalis* constitutes an unusual bacterial model in lysozyme resistance. Infect Immun75:5390–5398

Héchard Y, Pelletier C, Cenatiempo Y, Frère J (2001) Analysis of σ^{54}-dependent genes in *Enterococcus faecalis*: a mannose PTS permease (EII^{Man}) is involved in sensitivity to a bacteriocin, mesentericin Y105. Microbiology 147:1575–1580

Heim S, Lleo MM, Bonato B, Guzman CA, Canepari P (2002) The viable but nonculturable state and starvation are different stress responses of *Enterococcus faecalis*, as determined by proteome analysis. J Bacteriol 184:6739–6745

Huycke MM, Sahm DF, Gilmore MS (1998) Multiple-drug resistant enterococci: the nature of the problem and an agenda for the future. Emerg Infect Dis 4:239–249

Jett BD, Huycke MM, Gilmore MS (1994) Virulence of enterococci. Clin Microbiol Rev 7:462–478

Kazmierczak MJ, Wiedmann M, Boor KJ (2005) Alternative sigma factors and their roles in bacterial virulence. Microbiol Mol Biol Rev 69:527–543

Kristich CJ, Chandler JR, Dunny GM (2007) Development of a host-genotype-independent counterselectable marker and a high-frequency conjugative delivery system and their use in genetic analysis of *Enterococcus faecalis*. Plasmid 57:131–144

Kristich CJ, Nguyen VT, Le T, Barnes AMT, Grindle S, Dunny GM (2008) Development and use of an efficient system for random mariner transposon mutagenesis to identify novel genetic determinants of biofilm formation in the core *Enterococcus faecalis* genome. Appl Environ Microbiol 74:3377–3386

La Carbona S, Sauvageot N, Giard JC, Benachour A, Posteraro B, Auffray Y, Sanguinetti M, Hartke A (2007) Comparative study of the physiological roles of three peroxidases (NADH peroxidase, Alkyl hydroperoxide reductase and Thiol peroxidase) in oxidative stress response, survival inside macrophages and virulence of *Enterococcus faecalis*. Mol Microbiol 66:1148–1163

Laplace JM, Boutibonnes P, Auffray Y (1996) Unusual resistance and acquired tolerance to cadmium chloride in *Enterococcus faecalis*. J Basic Microbiol 36:311–317

Laplace JM, Thuault M, Hartke A, Boutibonnes P, Auffray Y (1997) Sodium hypochlorite stress in *Enterococcus faecalis*: influence of antecedent growth conditions and induced proteins. Curr Microbiol 34:284–289

Law J, Buist G, Haandrikman A, Kok J, Venema G, Leenhouts K (1995) A system to generate chromosomal mutations in *Lactococcus lactis* which allows fast analysis of targeted genes. J Bacteriol 177: 7011–7018

Le Breton Y, Mazé A, Hartke A, Lemarinier S, Auffray Y, Rincé A (2002a) Isolation and characterization of bile salts-sensitive mutants of *Enterococcus faecalis*. Curr Microbiol 45:434–439

Le Breton Y, Pichereau V, Flahaut S, Auffray Y, Rincé A (2002b) Identification of new genes related to osmotic adaptation in *Enterococcus faecalis*. Sci Alim 22:87–96

Le Breton Y, Boël G, Benachour A, Prévost H, Auffray Y, Rincé A (2003) Molecular characterization of *Enterococcus faecalis* two-component signal transduction pathways related to environmental stresses. Environ Microbiol 5:329–337

Le Breton Y, Muller C, Auffray Y, Rincé A (2007) New insights about the *Enterococcus faecalis* CroRS two-component system using a differential-display RAP-PCR approach. Appl Environ Microbiol 73:3738–3741

Leboeuf C, Leblanc L, Auffray Y, Hartke A (2000) Characterization of the *ccpA* gene of *Enterococcus faecalis*: identification of starvation-inducible proteins regulated by *ccpA*. J Bacteriol 182:5799–5806

Lewenstein A, Frigerio G, Moroni M (1979) Biological properties of SF68, a new approach for the treatment of diarrheal diseases. Curr Ther Res 26:967–981

Livny J, Teonadi H, Livny M, Waldor MK (2008). High-throughput, kingdom-wide prediction and annotation of bacterial non-coding RNAs. PLoS One. 3: e3197

Lonetto MA, Brown KL, Rudd KE, Buttner MJ (1994) Analysis of the *Streptomyces coelicolor sigE* gene reveals the existence of a subfamily of eubacterial RNA polymerase σ factors involved in the regulation of extracytoplasmic functions. Proc Natl Acad Sci USA 91:7573–7577

Low DE, Willey BM, Betschel S, Kreiswirth B (1994) Enterococcis: pathogens of the 90s. Eur J Surg Suppl 573:19–24

Mittenhuber G (2002) An inventory of genes encoding RNA polymerase sigma factors in 31 completely sequenced eubacterial genomes. J Mol Microbiol Biotechnol 4:77–91

Mizuno T (1997) Compilation of all gene encoding two-component phosphotransfer signal transducers in the genome of *Escherichia coli*. DNA Res 4:161–168

Morrison D, Woodford N, Cookson B (1997) Enterococci as emerging pathogens of humans. Soc Appl Bacteriol Symp Suppl 83:89S–99S

Muller C, Le Breton Y, Morin T, Benachour A, Auffray Y, Rincé A (2006) The response regulator CroR modulates expression of the secreted stress-induced SalB protein in *Enterococcus faecalis*. J Bacteriol 188:2636–2645

Muller C, Sanguinetti M, Riboulet E, Hébert L, Posteraro B, Fadda G, Auffray Y, Rincé A (2008) Characterization of two signal transduction systems involved in intracellular macrophage survival and environmental stress response in *Enterococcus faecalis*. J Mol Microbiol Biotechnol 14:59–66

Mylonakis E, Engelbert M, Qin X, Sifri CD, Murray BE, Ausubel FM, Gilmore MS, Calderwood SB (2002). The *Enterococcus faecalis fsrB* gene, a key component of the *fsr* quorum-sensing system, is associated with virulence in the rabbit endophthalmitis model. Infect Immun 70:4678–4681

Nagel, G, Lahrz A, Dersch P (2001) Environmental control of invasin expression in *Yersinia pseudotuberculosis* is mediated by regulation of RovA, a transcriptional activator of the SlyA/Hor family. Mol Microbiol 41:1249–1269

Nakayama J, Cao Y, Horii T, Sakuda S, Akkermans DL, de Vos W, Nagasawa H (2001) Gelatinase biosynthesis-activating pheromone: a peptide lactone that mediates a quorum-sensing in *Enterococcus faecalis*. Mol Microbiol 41:145–154

O'Farrell PH (1975) High resolution two-dimensional electrophoresis of proteins. J Biol Chem 250:4007–4021

O'Sullivan MG, Thornton GM, O'Sullivan GC, Collins JK (1992) Probiotic bacteria: myth or reality? Trends Food Sci Technol 3:309–314

Pichereau V, Bourot S, Flahaut S, Blanco C, Auffray Y, Bernard T (1999) The osmoprotectant glycine betaine inhibits salt-induced cross-tolerance towards lethal treatment in *Enterococcus faecalis*. Microbiology 145:427–435

Poh CH, Oh HM, Tan AL (2006) Epidemiology and clinical outcome of enterococcal bacteraemia in an acute care hospital. J Infect 52:383–386

Qin X, Singh KV, Xu Y, Weinstock GM, Murray BE (1998) Effect of disruption of a gene encoding an autolysin of *Enterococcus faecalis* OG1RF. Antimicrob Agents Chemother 42:2883–2888

Qin X, Singh KV, Weinstock GM, Murray BE (2000) Effects of *Enterococcus faecalis fsr* genes on production of gelatinase and a serine protease and virulence. Infect Immun 68:2579–2586

Qin X, Singh KV, Weinstock GM, Murray BE (2001) Characterization of Fsr, a regulator controlling expression of gelatinase and serine protease in *Enterococcus faecalis* OG1RF. J Bacteriol 183:3372–3382

Rincé A, Flahaut S, Auffray Y (2000) Identification of general stress genes in *Enterococcus faecalis*. Int J Food Microbiol 55:87–91

Rincé A, Giard JC, Pichereau V, Flahaut S, Auffray Y (2001) Identification and characterization of gsp65, an organic hydroperoxide resistance (ohr) gene encoding a general stress protein in *Enterococcus faecalis*. J Bacteriol 183:1482–1488

Rincé A, Uguen M, Le Breton Y, Giard J, Flahaut S, Dufour A, Auffray Y (2002) The *Enterococcus faecalis* gene encoding the novel general stress protein Gsp62. Microbiology 148:703–711

Rincé A, Le Breton Y, Verneuil N, Giard JC, Hartke A, Auffray Y (2003) Physiological and molecular aspects of *Enterococcus faecalis* bile salts response. Int J Food Microbiol 88:207–213

Ross RP, Claiborne A (1997) Evidence for regulation of the NADH peroxidase gene (*npr*) from *Enterococcus faecalis* by OxyR. FEMS Microbiol Lett 151:177–183

Scott JR, Bringel F, Marra D, Van Alstine G, Rudy CK (1994) Conjugative transposition of Tn916: preferred targets and evidence for conjugative transfer of a single strand and for a double-stranded circular intermediate. Mol Microbiol 11:1099–1108

Shepard BD, Gilmore MS (2002) Differential expression of virulence-related genes in *Enterococcus faecalis* in response to biological cues in serum and urine. Infect Immun 70:4344–4352

Sherman JM (1937) The enterococci. Bacteriol Rev 1:3–97

Sherman JM (1938) The enterococci and related streptococci. J Bacteriol 35:81–93

Sifri CD, Mylonakis E, Singh KV, Qin X, Garsin DA, Murray BE, Ausubel FM, Calderwood SB (2002) Virulence effect of *Enterococcus faecalis* protease genes and quorum-sensing locus *fsr* in *Caenorhabditis elegans* and mice. Infect Immun 70:5647–5650

Solheim M, Aakra A, Vebo H, Snipen L, Nes I (2007) Transcriptional response of *Enterococcus faecalis* V583 to bovine bile and sodium dodecyl sulphate. J Bacteriol 73:5767–5774

Spory A, Bosserhoff A, von Rhein C, Goebel W, Ludwig A (2002) Differential regulation of multiple proteins of *Escherichia coli* and *Salmonella enterica* serovar *Typhimurium* by the transcriptional regulator SlyA. J Bacteriol 184: 3549–3559

Stiles ME, Holzapfel WH (1997) Lactic acid bacteria of foods and their current taxonomy. Int J Food Microbiol 36:1–29

Stock AM, Robinson VL, Goudreau PN (2000) Two-component signal transduction. Annu Rev Biochem 69:183–215

Storz G, Imlay JA (1999) Oxidative stress. Curr Opin Microbiol 2:188–194

Teng F, Wang L, Singh KV, Murray BE, Weinstock GM (2002) Involvement of PhoP-PhoS homologs in *Enterococcus faecalis* virulence. Infect Immun 70:1991–1996

Teng F, Nannini EC, Murray BE (2005) Importance of *gls24* in virulence and stress response of *Enterococcus faecalis* and use of the Gls24 protein as a possible immunotherapy target. J Infect Dis 191:472–480

Thammavongs B, Corroler D, Panoff JM, Auffray Y, Boutibonnes P (1996) Physiological response of *Enterococcus faecalis* JH2-2 to cold shock: growth at low temperatures and freezing/thawing challenge. Lett Appl Microbiol 23:398–402

Verneuil N, Le Breton Y, Hartke A, Auffray Y, Giard JC (2004a) Identification of a new oxidative stress transcriptional regulator in *Enterococcus faecalis*. Lait 84:69–76

Verneuil N, Sanguinetti M, Le Breton Y, Posteraro B, Fadda G, Auffray Y, Hartke A, Giard JC (2004b) Effects of the *Enterococcus faecalis hypR* gene encoding a new transcriptional regulator on oxidative stress response and intracellular survival within macrophages. Infect Immun 72:4424–4431

Verneuil N, Rincé A, Sanguinetti M, Posteraro B, Fadda G, Auffray Y, Hartke A, Giard JC (2005) Contribution of a PerR-like regulator to the oxidative-stress response and virulence of *Enterococcus faecalis*. Microbiology 151:3997–4004

Verneuil N, Mazé A, Sanguinetti M, Laplace J, Benachour A, Auffray Y, Giard J, Hartke A (2006) Implication of (Mn)superoxide dismutase of *Enterococcus faecalis* in oxidative stress responses and survival inside macrophages. Microbiology 152:2579–2589

Werner G, Coque TM, Hammerum AM, Hope R, Hryniewicz W, Johnson A, Klare I, Kristinsson KG, Leclercq R, Lester CH, Lillie M, Novais C, Olsson-Liljequist B, Peixe LV, Sadowy E, Simonsen GS, Top J, Vuopio-Varkila J, Willems RJ, Witte W, Woodford N (2008) Emergence and spread of vancomycin resistance among enterococci in Europe. Eur Surveill 13:1–11

Wu RY, Zhang RG, Zagnitko O, Dementieva I, Maltzev N, Watson JD, Laskowski R, Gornicki P, Joachimiak A (2003) Crystal structure of *Enterococcus faecalis* SlyA-like transcriptional factor. J Biol Chem 278: 20240–20244

Yother J, Trieu-Cuot P, Klaenhammer TR, de Vos WM (2002) Genetics of streptococci, lactococci, and enterococci: review of the Sixth International Conference. J Bacteriol 84:6085–6092

Chapter 14
Stress Responses of Bifidobacteria

Marco Ventura, Abelardo Margolles, Francesca Turroni, Aldert Zomer, Clara G. de los Reyes-Gavilán, and Douwe van Sinderen

14.1 Introduction

The genus *Bifidobacterium* belongs to the *Bifidobacteriaceae* family, the *Bifidobacteriales* order, and the *Actinobacteria* phylum, which represents one of the largest taxonomic units among the *Eubacteria* (Stackebrandt et al. 1997), and which are Gram-positive microorganisms with a high guanine and cytosine DNA content (ranging between 51% for some *Corynebacterium* species to more than 70% for members of the genera *Streptomyces* and *Frankia*) (Ventura et al. 2007b). Bifidobacteria are common inhabitants of the gastrointestinal tract (GIT) of mammals and insects, and the genus currently contains more than 30 species and subspecies, which have mostly been isolated from intestine-associated ecological niches, including the intestines of humans, animals, and insects; the oral cavity; human blood; sewage; and food (Table 14.1) (Ventura et al. 2007b). Bifidobacteria found in the last three niches are probably contaminants originating from an intestinal source. Moreover, according to their phylogenetic relationships based on 16S rRNA sequence similarity, all bifidobacterial species can be subdivided into six different groups: *Bifidobacterium boum, Bifidobacterium asteroides*, *Bifidobacterium adolescentis*, *Bifidobacterium pullorum*, *Bifidobacterium longum*, and *Bifidobacterium pseudolongum* (Miyake et al. 1998). Interestingly, multigene-based phylogenetic analysis suggests that *B. asteroides*, which was originally isolated from the hindgut of a honeybee, is most closely related to the common ancestor of all members of the *Bifidobacterium* genus (Ventura et al. 2007a). Several ecological studies have demonstrated that these microorganisms are among the dominant bacterial groups in the mammalian gut (Turroni et al. 2009a, 2009b). Together with

M. Ventura (✉)
Laboratory of Probiogenomics, Department of Genetics, Biology of Microorganisms, Anthropology, Evolution, University of Parma, Parco Area delle Scienze 11A, 43124 Parma, Italy
e-mail: marco.ventura@unipr.it

E. Tsakalidou and K. Papadimitriou (eds.), *Stress Responses of Lactic Acid Bacteria*, Food Microbiology and Food Safety, DOI 10.1007/978-0-387-92771-8_14,

Table 14.1 Bifidobacterial species so far identified in the genus *Bifidobacterium* grouped according to their ecological distribution

Bifidobacterium breve	Human gut	*Bifidobacterium pseudo-longum* subsp. *pseudolongum*	Animal gut
Bifidobacterium bifidum		*Bifidobacterium thermacido-philum* subsp. *porcinum*	
Bifidobacterium pseudocatenulatum		*Bifidobacterium psychraerophilum*	
Bifidobacterium longum subsp. *infantis*		*Bifidobacterium thermophilum*	
Bifidobacterium longum subsp. *longum*		*Bifidobacterium longum* biotype suis	
Bifidobacterium adolescentis		*Bifidobacterium choerinum*	
Bifidobacterium catenulatum		*Bifidobacterium animalis* subsp. *animalis*	
Bifidobacterium angulatum		*Bifidobacterium cuniculi*	
Bifidobacterium gallicum		*Bifidobacterium magnum*	
Bifidobacterium dentium	Oral cavity	*Bifidobacterium saeculare*	
Bifidobacterium tsurumiense		*Bifidobacterium ruminantium*	
Bifidobacterium scardovii	Human blood	*Bifidobacterium merycicum*	
Bifidobacterium coryneforme	Insect	*Bifidobacterium pseudo-longum* subsp. *globosum*	
Bifidobacterium indicum		*Bifidobacterium boum*	
Bifidobacterium asteroides		*Bifidobacterium gallinarum*	
Bifidobacterium animalis subsp. *lactis*	Food	*Bifidobacterium pullorum*	
		Bifidobacterium thermacido-philum subsp. *thermacidophilum*	Sewage
		Bifidobacterium minimum	

genome data, these studies highlight the impact of these bacteria on host physiology. An ecological study of the bifidobacterial population present in animal feces revealed that they are present in the gut of a wide variety of animals (e.g., birds, ungulates, lagomorphs, and rodents) (Lamendella et al. 2008). Interestingly, some species are present in different animal species and thus show a cosmopolitan life-style, whereas other bifidobacterial species display a highly specialized ecological adaptation to the GIT of a particular animal (e.g., *Bifidobacterium cuniculi* for rabbits, *Bifidobacterium angulatum* for cows, and *Bifidobacterium gallinarum* for chickens). In the case of the human GIT, several reports have analyzed the contribution of bifidobacteria to the gut microbiota using traditional culturing techniques on fecal or colonoscopic samples (Harmsen et al. 2000; Fanaro et al. 2003; Turroni et al. 2009a). The above-mentioned host-specific ecological adaptation was also noticed in humans, but the distribution seems to be related to host age and seems to apply to certain species only. For example, some bifidobacterial species are exclusively found in the gut microbiota of a human adult (i.e., *B. pseudolongum* and *Bifidobacterium bifidum*), whereas others are more widely distributed irrespective of host age (i.e., *B. longum*, *Bifidobacterium breve*, *Bifidobacterium pseudocatenulatum*, and *B. adolescentis*) (Turroni et al. 2009a).

Certain bifidobacterial strains have been commercialized and are commonly added as health-promoting bacteria or probiotics to foods and beverages. However, despite the ubiquitous use of these probiotics, it is surprising that the molecular mechanisms responsible for the presumed health-promoting activities are only poorly understood. With the advent of the (functional) probiogenomic era (Ventura et al. 2008), the genetic determinants responsible for probiotic action will be targeted for mechanistic exploration. Many bifidobacterial genomes are currently freely accessible in public databases (at the time of this writing, the NCBI database contained seven complete *Bifidobacterium* genomes and 12 draft assembly *Bifidobacterium* genomes), while many other bifidobacterial sequencing projects are ongoing. All currently published bifidobacterial genome information is illustrative of extensive genetic adaptations to the intestinal ecological niche (Schell et al. 2002; Ventura et al. 2007b, 2008; Lee et al. 2008; Sela et al. 2008). This is supported by the discovery that these genomes contain a large arsenal of genes involved in the breakdown of carbohydrates (e.g., fructo-oligosaccharides, gluco-oligosaccharides, xylo-oligosaccharides, lactulose, and raffinose) (Schell et al. 2002; Ventura et al. 2007a; Sela et al. 2008), which are not hydrolyzed by host-encoded enzymes and for this reason reach the distal part of the human GIT, where they are fermented by enteric bacteria, including bifidobacteria, to produce a variety of short-chain fatty acids. In such a mutualistic relationship, the host gains carbon and energy through assimilation of these short-chain fatty acids, while the enteric bacteria are provided with a rich supply of glycans (Ventura et al. 2007a).

14.2 Brief Overview of the Molecular Basis of Stress Response in Bacteria

Proteins represent one of the main macromolecules of living cells. Their functionality is dependent on an accurate transformation from a linear amino acid sequence to a properly folded structure. Misfolding and aggregation are risks to which a protein is exposed, even during the first stages of synthesis. Moreover, even if some cellular proteins may fold independently following the completion of the primary structure, the support of molecular chaperones during elongation and release from the ribosome is essential as a prerequisite for proper protein folding. When protein damage is considered to be beyond repair, the action of ATP-dependent proteases promote protein breakdown into amino acid residues. This proteolytic action is also important to prevent the detrimental cellular accumulation of damaged or misfolded proteins, which otherwise may harm bacterial metabolic processes (Gottesman 1996; Gottesman et al. 1997; Teter and Klionsky 1999). Prokaryotic cells maintain a delicately balanced protein quality-control process, which tries to avoid the proteolytic degradation of properly folded proteins or proteins that are misfolded/damaged but can still be repaired/refolded.

The ability to check protein quality when misfolded/damaged proteins are encountered is fundamental in deciding between repair/refolding or degradation (Keiler et al. 1996; Wickner et al. 1999).

The repertoire of molecular chaperones and proteases so far identified in bacteria include nine major families: HSP100, ClpP, HSP90, HSP70, HSP60, DnaJ, GrpE, HSP33, and small HSPs (Bukau and Horwich 1998; Narberhaus 2002). With the advent of the genomic era, it became clear that bacterial genomes contain varying numbers of each of the above-mentioned molecular chaperone families. In fact, in the case of bifidobacteria, genome analysis revealed that this group of bacteria, compared to many other *Actinobacteria*, display a relatively small chaperone set (Ventura et al. 2006b), which consists of one or more members of the HSP100, ClpP, HSP70, HSP60, DnaJ, GrpE, and small HSPs chaperone/protease families, most of which will be discussed in the following sections.

14.3 The Main Molecular Actors of Stress Response in Bifidobacteria

In *Eubacteria,* one of the main players that contends with the detrimental effects of environmental stress is represented by the HSP100 protein family, including ATP-hydrolyzing proteins (Grimaud et al. 1998). This family includes two distinct groups, which are further subdivided according to specific signature sequence motifs or the lengths of the interdomain region separating the nucleotide-binding domain (Schirmer et al. 1996). The first group is comprised of proteins (with a mass of approximately 83 kDa) with two ATP-binding domains, represented by ClpA, ClpB, ClpC, ClpD, and ClpE, whereas the second group comprises a group of proteins (with an approximate mass of 21–22 kDa) with a single ATP-binding domain (ClpM, ClpN, ClpX, and ClpY) (Schirmer et al. 1996; Derré et al. 1999a, 1999b). The model organism in bifidobacteria for studying stress response is the *B. breve* UCC2003 strain, whose genome sequence has been determined and which has been subjected to extensive research efforts regarding the principal players involved in stress response.

Two members of the HSP100 family, ClpB and ClpC, have been characterized in *B. breve* UCC2003 (Ventura et al. 2005a, 2005b, 2006a), whose genome contains nonadjacent *clp*B and *clp*C genes, both transcribed as monocistronic units (Ventura et al. 2005a, 2005b). The inducibility of the *B. breve* UCC2003 *clp*B and *clp*C genes following the exposure to stressful conditions was investigated by slot-blot hybridization (Ventura et al. 2005a, 2005b) and microarray analysis (Zomer et al. 2009), revealing that both are highly transcribed upon the exposure of cultures to severe heat treatment (e.g., 50°C), conditions that considerably increase the level of expression (between 18- and 90-fold in the case of *clp*B) (Ventura et al. 2005b).

Another crucial protein player of stress response is the ClpP protein, which participates with other members of the HSP100 family protein in the formation of the Clp holoenzyme and that performs a proteolytic role by degrading misfolded proteins (Hlavacek and Vachova 2002). The intensively studied Clp complex of *Escherichia coli* consists of two functionally distinct subunits: ClpA functions as the ATP-binding regulatory subunit, providing substrate specificity, whereas the smaller ClpP subunit provides the proteolytic activity (Maurizi et al. 1990). The *E. coli* ClpP protein on its own only displays peptidase activity and requires the presence of ClpA to hydrolyze polypeptides longer than six amino acids. The Clp protease complex includes two central heptameric rings of ClpP, flanked by two hexameric rings of ClpA, which in *Actinobacteria* is assumed to be replaced by ClpC (Ventura et al. 2006b).

Genome analyses revealed that in most *Firmicutes,* just a single copy of the *clp*P gene is present on the chromosome (Fedhila et al. 2002). In contrast, in *actinobacterial* genomes, more than five *clp*P paralogous genes have been identified (Viala and Mazodier 2002). It has been suggested that the presence of multiple copies of *clp*P genes is correlated to enhanced protection against certain stressful conditions or facing different environmental conditions (Viala et al. 2000; Viala and Mazodier 2002). In the case of *Streptomyces lividans*, the five identified *clp*P-like genes are arranged into two operons, one corresponding to *clp*P1 and *clp*P2, the second corresponding to *clp*P3 and *clp*P4, and finally an additional monocistronic transcription unit corresponding to *clp*P5 (Engels et al. 2004).

In contrast, in bifidobacterial genomes, a single *clp*P operon containing two *clp*P paralogs, designated *clp*P1 and *clp*P2, is present (Ventura et al. 2005c; Lee et al. 2008; Sela et al. 2008; Barrangou et al. 2009). Transcriptional analyses by Northern blot hybridization of the *B. breve* UCC2003 *clp*P operon showed that the *clp*P1 and *clp*P2 genes are transcribed in response to heat shock as a bicistronic unit from a single promoter.

Two other important molecular players of bacterial stress response are part of the HSP70 and HSP60 families, which include the DnaK and GroEL proteins, respectively. Such molecular chaperones act together with cochaperone molecules (DnaJ and GrpE in the case of DnaK, and GroES in the case of GroEL) (Bukau and Horwich 1998). Through cycles of protein binding and release, a process that is energized by ATP, the DnaK and GroEL chaperone complexes facilitate protein (re) folding to the native state (Hartl 1996). With respect to the GroELS complex, it has been described that in the presence of ATP, GroES constitutes an equimolar complex with GroEL, which binds the protein substrate (Langer et al. 1992). The release of the correctly folded protein is contingent upon ATP hydrolysis, and multiple binding and release events may be necessary for a protein to reach its native conformation (Weissman et al. 1994). Three-dimensional modeling of the GroEL protein of *B. breve* UCC2003 demonstrated that this protein possesses the classical bacterial GroEL architecture, which consists of GroEL subunits assembled into a cylindrical complex consisting of two stacked homoheptameric rings with a large cylindrical chamber that accommodates substrate proteins (De Dea Lindner et al. 2007). In contrast, in various bacteria, the DnaK protein has been described to function as a monomer (Teter and Klionsky 1999), and the 3D structure prediction of the *B. breve* UCC2003 DnaK protein demonstrated that it very much resembles that of other characterized DnaK proteins (De Dea Lindner et al. 2007).

As described in previous reports (Houry et al. 1999; Mogk et al. 1999), these 3D structures are reflections of the different substrate characteristics of the GroEL and

DnaK complexes. In fact, DnaK binds to short hydrophobic segments of nonnative proteins, whereas the GroEL ring encloses the entire substrate protein. In the case of GroEL, this effectively limits the size of the protein that can be refolded to 55 kDa (Ewalt et al. 1997), although in the presence of ATP, the GroELS chaperonin machinery can enclose proteins up to 70 kDa (Xu et al. 1997).

In *Eubacteria*, the genes encoding the molecular chaperones GroES and GroEL are frequently arranged in an operon, where *gro*ES is the first gene of such an operon. In contrast, the unconnected genetic organization of the *gro*EL and *gro*ES loci in all sequenced bifidobacterial genomes is without any precedent in the bacterial world (Ventura et al. 2004a, 2004b; Lee et al. 2008; Sela et al. 2008; Barrangou et al. 2009). The reason, if any, for this physical separation is presently unknown. In bifidobacteria, the gene encoding the DnaK chaperone is located within the *dna*K operon, which also encompasses the *grp*E, *dna*J1, and *hsp*R genes in a genetic constellation that is highly conserved within *actinobacterial* genomes (Ventura et al. 2005a, 2006b).

Transcriptional analyses of the *dna*K locus in *B. breve* UCC2003 using Northern blot assays revealed a large polycistronic mRNA encompassing the *dna*K-*grp*E-*dna*J-*hsp*R genes (Ventura et al. 2005a). However, this primary transcript seems to be very unstable and subject to rapid processing. It has been suggested that such a high level of instability is caused by the presence of many stem-loop structures located within the coding regions as well as in the intergenic region between the *dna*J and *grp*E genes, which may act as processing sites. A similar situation has previously been described for *Bacillus subtilis*, where four hairpin-like structures were implicated in the processing of *dna*KJ-encompassing mRNA (Homuth et al. 1997).

Other molecular chaperones that are widespread in prokayotes as well as in eukaryotes include the small heat-shock proteins (sHSPs). The role of these proteins in bacteria is to avoid irreversible protein denaturation and the aggregation of heat-damaged proteins (Narberhaus 1999). The sHSP family is characterized by a small molecular mass, ranging between 14 and 27 kDa; in contrast to the large-molecular-weight HSPs, the sHSPs are highly divergent in primary sequences except for the presence of a conserved α-crystallin domain. Furthermore, in contrast to the large-molecular-weight HSPs, which usually exist as well-defined oligomers, the sHSPs from different organisms exist as oligomers that vary in size and shape (Hartl and Hayer-Hartl 2002). According to the present model, sHSPs act as ATP-independent chaperones by binding to denaturated proteins accumulated under stressful conditions, and by maintaining such proteins in a folding-competent state (Lee et al. 1997). Interestingly, the sHSPs are widely distributed throughout all kingdoms of life, from bacteria to animals, although not all organisms encode sHSPs, for example, *Mycoplasma genitalium* (Fraser et al. 1995). In contrast, *Rhizobia* genomes contain a large set of genes specifying sHSPs (Muchowski et al. 1999). With respect to bifidobacteria, the inspection of the *B. breve* UCC2003 genome as well as that of *B. longum* subsp. *longum* NCC2705 revealed the presence of a putative HSP20-encoding gene, whose transcription is induced upon heat stress (Ventura et al. 2007c). However, a genetic survey of all the bifidobacterial species described so far revealed that this *hsp*20 gene is not uniformly conserved in the *Bifidobacterium* genus (Ventura et al. 2007c).

14.4 Analyses of Temporal Expression of Heat-Shock Genes in Bifidobacteria

Transcriptomic and/or proteomic analyses constitute appropriate means to investigate the changes in the expression patterns of stress-related genes or proteins upon the submission of bacterial cells to stressfull conditions. Currently, in the case of bifidobacteria, these investigations are restricted to heat, acid, and bile salt stresses (Sanchez et al. 2005; Savijoki et al. 2005; Rezzonico et al. 2007; Zomer et al. 2009). With respect to heat-shock genes, transcription analyses of the major chaperones and protease-encoding genes have also been performed using classical molecular tools such as Northern blot hybridization and primer extension assays (Ventura et al. 2004a, b, 2005a, 2006a). These analyses have targeted various *hsp* genes from *B. breve* UCC2003, such as *gro*EL, *gro*ES, *dna*K, *grp*E, *dna*J1, *dna*J2, *clp*B, *clp*C, and *clp*P1P2. Based on the temporal pattern of induction, these genes can be classified into two main groups. The *gro*EL, *gro*ES, *clp*C, and *clp*P1P2 genes are rapidly induced to high levels (up to 10-fold) 150 min after moderate heat-shock regimes (a temperature upshift of 5°C, $\Delta T5$) but do not appear to be induced upon exposure to severe heat treatment (a temperature upshift of 13°C, $\Delta T13$). In contrast, genes in the other induction class, *dna*K, *grp*E, *dna*J1, and *clp*B, were shown to be strongly induced upon severe heat-shock ($\Delta T13$) regimes, although they do not exhibit increased transcription following moderate heat-shock challenges. These results are consistent with results from a proteomic study, in which the protein synthesis in *B. longum* 3A cells was monitored following heat shock by 2D gel electrophoresis (Savijoki et al. 2005). The global analysis of the *B. breve* UCC2003 stress transcriptome using microarray technology revealed a less strict subdivision of molecular chaperones into the aforementioned two groups based on their temporal and temperature-dependent pattern of induction (Zomer et al. 2009). Using the latter approach, it was shown that the *gro*EL, *gro*ES, *dna*J2, *clp*C, and *clp*P1/P2 genes were induced upon heat shock at temperatures ranging between 44 and 47°C (with an induction level of up to a circa fourfold increase), while the *dna*K, *grp*E, *dna*J1, *clp*B, and *hsp*20 genes were demonstrated to exhibit the strongest transcriptional induction at 50°C (up to approximately 400-fold), although many of these genes already showed some induction following a temperature upshift to 42°C.

14.5 Identification of Regulators of the Heat-Shock Response

Although genes whose encoded proteins are overexpressed upon the exposure of bacterial cells to heat shock are highly conserved in bacterial genomes, the control mechanisms of their expression may vary considerably. In the main representatives of the *Actinobacteria* phylum, *Streptomyces* and *Mycobacterium*, a complete regulatory system of heat-shock response has not been described so far.

Very often, however, the heat-shock response in bacteria is regulated by the interplay of various control systems, such as selective promoter recognition, transcriptional repression, and activation. In addition, alternative sigma factors redirect the RNA polymerase to recognize specific heat-shock promoters. However, despite the fact that the genomes of bifidobacteria (e.g., *B. longum* subsp. *longum* NCC2705 and *B. breve* UCC2003) contain *rpo*E (σ^E) homologs, no data are currently available regarding their possible role in coping with different stress challenges.

Regulatory mechanisms that are based on transcriptional repression involve molecules that are synthesized under noninducing conditions and that bind to conserved repeats positioned upstream of the promoter or operons, thereby preventing their transcription. Heat treatment removes such molecules and their associated repressive action, consequently allowing the transcription of particular heat-shock genes. The analysis of bifidobacterial genomes identified two transcriptional repressors, which are commonly found in *Actinobacteria*. These include the heat regulation at CIRCE (HrcA) protein, which binds to a controlling inverted repeat of chaperones expression (CIRCE) that is positioned upstream of the promoter of certain heat-shock genes and operons, and the heat-shock protein repressor (HspR), which binds to an HspR-associated inverted repeat (HAIR) motif (Ventura et al. 2005a, 2005b; Zomer et al. 2009). Analyses of currently available bifidobacterial genome sequences for HAIR motifs have shown the presence of such DNA motifs in the putative promoter region of many genes, such as *clp*B, *dna*K, *clg*R, *nfo,* and *hrd*B, which together may constitute the HspR regulon of bifidobacteria (Ventura et al. 2005b; Zomer et al. 2009). Although the presence of a so-called HAIR motif was detected in the promoters of *clp*B, *dna*K, *clg*R, *nfo,* and *hrd*B, binding of His-tagged HspR was shown only for the promoters of the *dna*K operon, the *clg*R operon, and the *clp*B gene. No binding was observed for the promoters of *nfo* and *hrd*B. Closer inspection of the HAIR motif revealed the presence of a larger inverted repeat structure. A mutation of this extended HAIR motif resulted in reduced binding (Zomer et al. 2009), implicating its importance in recognition of the HAIR motif by HspR. The CIRCE motif was found in the promoters of *gro*ES, *gro*EL, and *hrc*A, and gel-shift retardation assays showed that His-tagged HrcA indeed forms complexes with these DNA regions (Zomer et al. 2009).

The regulation of heat-shock genes by transcriptional activation relies on molecules that positively control gene expression. The only transcriptional activator of stress response described thus far in *Actinobacteria* is the Clp gene regulator (ClgR) protein. This gene regulator has been molecularly characterized in *B. breve* UCC2003 (Ventura et al. 2005c). In this bifidobacterial strain, the purified His-tagged ClgR was shown to bind to the promoter regions of the *clp*C, *clp*P1P2, and *hrc*A genes (Ventura et al. 2005a, 2005b; Zomer et al. 2009). Nevertheless, it was observed that the *clp*P1 and *clp*C promoter regions were only bound by purified ClgR protein in the concomitant presence of a crude lysate of heat-stressed *B. breve* UCC2003 cells (Ventura et al. 2005a, 2005c). The binding activity was lost upon proteolytic treatment of the crude lysate, thus indicating the existence of a proteinaceous cofactor involved in ClgR binding. DNase I footprinting experiments showed that the *clp*P1 promoter region protected by ClgR binding contains an imperfect palindromic motif (CGCT-N4-GCCNA), which may represent the operator site for the ClgR protein. This motif is similar to the operator site of the ClgR homolog in other *Actinobacteria*, namely,

Streptomyces and *Corynebacterium* (Bellier and Mazodier 2004). Aligments of the ClgR motifs found in the promoter regions of *clp*C, *clp*P, and *hrc*A, combined with the EMSA experiments on mutated ClgR-binding motifs, suggest that the full ClgR-binding site is TNCGCTNNNGGCGNAA, larger than the motif found in *S. lividans*, GTTCGC-N[5]-GCG, but with similarities to the motif WNNWCGCYNANR GCGWWS proposed for *Corynebacterium* species (Zomer et al. 2009). In all these species it was shown that the ClgR regulon includes heat-shock genes, such as *clp*C, *clp*P, and *lon* (Engels et al. 2005). However, in the case of bifidobacteria, the ClgR seems to possess properties rendering this activator unique in the bacterial regulatory mechanisms so far investigated because it requires a proteinaceous cofactor(s) as a prerequisite to act as a binding protein. In pull-down assays using whole-cell extracts from heat-stressed *B. breve* UCC2003 cultures, a protein of 56 kDa was shown to copurify with His-tagged ClgR (hClgR) (Ventura et al. 2005c). Moreover, the hClgR-56-kDa protein co-eluate mixture was able to bind to the *clp*P1 promoter region without the assistance of any other cofactor. Preliminary results identified the 56-kDa protein to be the *B. breve* UCC2003 GroEL chaperone. Interestingly, the bifidobacterial ClgR possesses an N-terminus that is 72–92 amino acids longer than homologous proteins in other members of the *Actinobacteria* phylum (*Streptomyces* or *Corynebacterium* spp.). In the absence of a cofactor, the full-length His-tagged form of ClgR is unable to form complexes with its target sequences, while the removal of this N-terminal extension relieves the requirement for this cofactor (Zomer et al. 2009). These data, along with the fact that ClgR homologs in other *Actinobacteria* do not require any cofactor molecules to bind to the *clp*P1 promoter region, suggest that this N-terminal extension found only in bifidobacterial ClgR interacts with the cofactor molecule. It is possible that the N-terminal domain interferes with the ability of ClgR to reach its active state in the absence of a chaperone, such as GroEL. Taken together, these results represent a novel type of positive cofactor-mediated regulation of gene expression in high-G+C Gram-positive bacteria.

14.6 The Osmotic Stress Response

The current information concerning the molecular basis of the osmotic stress response in bifidobacteria is rather limited. Nevertheless, the need to expand our knowledge on the stress response of commercially relevant bifidobacteria is crucial in order to select strains that resist stressful conditions encountered during food manufacture. In fact, rather than heat or acid stress, gut bacteria are often subjected to osmotic stresses, which are caused by frequent changes of diet composition. Consequently, it is reasonable to assume that members of the intestinal microbiota like bifidobacteria may have evolved an effective response mechanism to enable them to cope with osmotic stress.

Transcriptomics experiments based on Northern blot hybridizations showed that in the model bifidobacterial strain, *B. breve* UCC2003, the transcription of molecular chaperone-encoding genes, such as *dna*K, *dna*J1, *grp*E, *clp*B, *dna*J2, and *hsp*20, is highly upregulated following the exposure of cells to a high amount of salt (0.5 M and 0.7 M NaCl) (Ventura et al. 2005a, 2005b, 2005c). Furthermore, with the exception of the *dna*J2 gene, the same genes were shown to be induced upon severe heat treatment, which suggests the existence of an overlapping regulatory network between the osmotic stress- and heat-induced genes, implicating HspR as the regulator in this process.

14.7 Acid pH Stress

Acidic environments have an elevated proton concentration, causing undissociated acids to passively diffuse across membranes following an electrochemical gradient. These acids, once internalized, dissociate in the neutral cytoplasm, thereby acidifying the cytoplasm and promoting metabolic failure and eventually death. Therefore, molecular mechanisms allowing the maintenance of cytoplasmic pH at physiological values are crucial for the survival of microorganisms in acidic environments. Bifidobacteria produce acid metabolites as a consequence of their carbohydrate catabolism, by the fermentation of sugars and the formation of lactic and acetic acids as the main end products (Vernazza et al. 2006). In fermented dairy products, bifidobacteria support the acid conditions during the shelf-life of the product and the postacidification impairs their viability (Jayamanne and Adams 2006). After being ingested, bifidobacteria face severe acidic conditions when they arrive in the human stomach, where food stays for around 90 min at pH values close to 2.0. Probiotic bacteria must retain high enough viability levels through the gastrointestinal transit in order to exert their beneficial effects on the human host. Thus, a tolerance to low pH is crucial for the success of a strain as a probiotic.

The inducible mechanism in bacteria triggered by the exposure to mild acidic conditions, which leads to an adaptation to lower, and otherwise lethal, pH values, is known as *acid-tolerance response* (*ATR*) (Davis et al. 1996; Samelis et al. 2003). The survival of bifidobacteria in acidic environments can be enhanced by the previous exposure to sublethal pH conditions through ATR induction (Maus and Ingham 2003), and stable low-pH resistant mutants can be obtained in this way (Collado et al. 2006; Collado and Sanz 2006). This somehow results in the acquisition of cross-resistance to other stresses (Chung et al. 1999; Collado and Sanz 2007). In general, the resistance of bifidobacteria to a low pH is poor, with the exception of the species *Bifidobacterium animalis* (Masco et al. 2007; Ritter et al. 2009), which could partially explain the good survival of this species through the human GIT (Alander et al. 2001; Ritter et al. 2009). The acquisition of acid resistance can induce other pleiotropic changes in *Bifidobacterium* (Collado and Sanz 2006). Remarkably, acid adaptation may cause increased resistance to some antibiotics in several strains (Collado and Sanz 2007; Kheadr et al. 2007).

ATR is a complex process involving changes in the transcription of several genes and in the synthesis of many proteins (Davis et al. 1996; Len et al. 2004; Bore et al. 2007). In bifidobacteria, molecular studies of the ATR are very scarce (Table 14.2), but currently it is known that the process is strongly dependent on the strain and the species (Saarela et al. 2004; Takahashi et al. 2007). Two of the general mechanisms involved in low-pH tolerance are the active extrusion of protons and the

Table 14.2 Proteins or genes involved in the stress response in *B. animalis* and *B. longum*

Putative function (name)		Activity	Reference
Bile response			
B. animalis			
Priming glycosyltransferase (*gtf01207)*	+	EPS production	Ruas-Madiedo et al. (2009)
H(+)-ATPase, beta subunit (AtpD)	+ *	Regulation of intracellular pH	Sánchez et al. (2006)
Xylulose-5-phosphate/fructose-6-phosphate phosphoketolase	+	Shift to increased glucose consumption	Sánchez et al. (2007b)
Glyceraldehyde-3-phosphate dehydrogenase	+/– *	Shift to increased glucose consumption	Sánchez et al. (2007b)
Methionine synthase	– *	ND	Sánchez et al. (2007b)
O-Acetylhomoserine sulfhydrolase	– *	ND	Sánchez et al. (2007b)
Long-chain-fatty-acid-CoA ligase	– *	Shift in fatty acid composition	Sánchez et al. (2007b); Ruiz et al. (2009a)
Heat-shock protein ClpB	+ *	ND	Sánchez et al. (2007b)
Trypsin-like serine protease (HtrA homolog)	+	ND	Sánchez et al. (2007b)
Thioredoxin-dependent thiol peroxidase	+	ND	Sánchez et al. (2007b)
Cochaperone GrpE	+	ND	Sánchez et al. (2007b)
Chaperone protein GroES	+	ND	Sánchez et al. (2007b)
Chaperone protein GroEL	+	ND	Sánchez et al. (2007b)
Chaperone protein DnaK	+	ND	Sánchez et al. (2007b)
DNA protection during starvation protein	+ *	ND	Sánchez et al. (2007b)
Cell division initiation protein	*	ND	Sánchez et al. (2007b)
Chromosome-segregating ATPase	*	ND	Sánchez et al. (2007b)
Bile salt hydrolase	*	Increased BSH activity	Sánchez et al. (2007b); Noriega et al. (2005)

(continued)

Table 14.2 (continued)

Putative function (name)		Activity	Reference
B. longum			
Sodium-dependent bile acid transporter (Ctr)		Bile salt efflux	Price et al. (2006)
Efflux transporter protein (BL0920)	+	Bile salt efflux	Gueimonde et al. (2009)
Protease (HtrA)	+	ND	Savijoki et al. (2005)
Molecular chaperone (DnaK)	+	ND	Savijoki et al. (2005); Sánchez et al. (2005)
Chaperonin (GroEL)	+	ND	Savijoki et al. (2005); Sánchez et al. (2005)
Fructose-6-P phosphoketolase (Xfp)	+	Increased glycolytic flux	Sánchez et al. (2005)
Glyceraldehyde-3-phosphate dehydrogenase C (Gap)	+	Increased glycolytic flux	Sánchez et al. (2005)
Probable branched-chain amino acid aminotransferase (IlvE)	+	ND	Sánchez et al. (2005)
Enolase (Eno)	+	ND	Ruiz et al. (2009a, b)
Cell division protein FtsI (PbpA)	+	ND	Ruiz et al. (2009a, b)
Acid-tolerance response			
B. animalis			
H(+)-ATPase, beta subunit (AtpD, *atpD*)	+	Regulation of intracellular pH	Ventura et al. (2004a, b); Sánchez et al. (2006)
H(+)-ATPase, alpha subunit (*atpA*)	+	Regulation of intracellular pH	Ventura et al. (2004a, b)
B. longum			
H(+)-ATPase, alpha subunit (AtpA)	+	Regulation of intracellular pH	Sánchez et al. (2007a)
H(+)-ATPase, beta subunit (AtpD)	+	Regulation of intracellular pH	Sánchez et al. (2007a)
Methionine synthase (MetE)	+ *	ND	Sánchez et al. (2007a)
Cystathionine gamma-synthase (MetB)	+ *	ND	Sánchez et al. (2007a)
O-Acetylhomoserine (thiol)-lyase (CysD)	+ *	ND	Sánchez et al. (2007a)
Bile salt hydrolase (BSH)	– *	Decreased BSH activity	Sánchez et al. (2007a)
Probable BCAA aminotransferase (IlvE)	+	Increased valine synthesis	Sánchez et al. (2007a)

(continued)

Table 14.2 (continued)

Putative function (name)		Activity	Reference
Dihydroxy acid dehydratase (IlvD)	+	Increased valine synthesis	Sánchez et al. (2007a)
Ketol-acid reductoisomerase (IlvC2)	+	Increased valine synthesis	Sánchez et al. (2007a)
Glutamine synthetase 1 (GlnA1)	+	Increased ammonia synthesis	Sánchez et al. (2007a)
General stress chaperone GroES (GroES)	+/–	ND	Sánchez et al. (2007a)
Chaperone protein (DnaJ)	–	ND	Sánchez et al. (2007a)

+ induced; – repressed, * involved in stress adaptation; *ND* not determined

alkalinization of the cytoplasm. In anaerobic bacteria, it is known that the proton excess occurring in acidic conditions is counteracted by the action of the F_0F_1-ATPase enzyme (Cotter and Hill 2003; Matsumoto et al. 2004). This enzyme catalyzes the hydrolysis of ATP, enabling the translocation of protons across the cytoplasmic membrane against the cellular electrochemical gradient. This helps to maintain the cytoplasmic pH at physiological values under harsh acidic conditions. Ammonia formation is another common mechanism of the tolerance to low pH widespread among bacteria; as a basic compound, ammonia can capture one proton, yielding ammonium and thus helping to buffer the cytoplasm under acidic conditions (Cotter and Hill 2003).

In bifidobacteria, the increased activity of the F_0F_1-ATPase in the ATR response (Miwa et al. 1997) has been supported with some recent molecular data (Sanchez et al. 2007a). Proteomic data currently available on the resistance of *B. longum* to acidic pH have been obtained through the isolation of an acid-resistant mutant from the acid-sensitive strain *B. longum* NCIMB8809 (Sanchez et al. 2005). A comparison of electrophoretic patterns indicated that the F_0F_1-ATPase is upregulated under acidic conditions. In the genus *Bifidobacterium*, the eight subunits of the enzyme are encoded in an operon and arranged into two groups, one for the membrane-integral complex (F_0) and the other for the cytoplasmic complex (F_1). Proteomic analyses showed that *B. longum* synthesizes higher amounts of the two main cytoplasmic subunits of the F_0F_1-ATPase (Sanchez et al. 2007a). In the same way, the content of one of the cytoplasmic subunits of the enzyme was higher in *B. animalis* when grown in acidic conditions (Sanchez et al. 2006). Genetic studies carried out by Ventura and coworkers indicated that the F_0F_1-ATPase-encoding operon of *B. animalis* subsp. *lactis* is transcribed as two mRNAs, one containing the information for translation of all the subunits, while the other exclusively encompasses the cytoplasmic subunits (Ventura et al. 2004a, b). Both mRNAs are detected at higher levels following exposure to acidic pH, suggesting a specific transcriptional regulation of the F_0F_1-ATPase. Additional proteomic and physiological data of *B. longum* showed a shift in the relative amounts of some proteins of the glycolytic pathway whose biological activity may contribute to fuel the bifid shunt (Sanchez et al. 2007a).

Theoretically, this reorganization of the glycolytic pathway would increase the synthesis of ATP, needed for F_0F_1-ATPase activity.

In addition to the clear involvement of F_0F_1-ATPase of *Bifidobacterium* in the low-pH response, the concentration of enzymes responsible for the biosynthesis of branched-chain amino acids, as well as glutamine synthetase, was increased in *B. longum* under acidic conditions (Sanchez et al. 2007a). These proteomic data support the hypothesis of Len and coworkers (Len et al. 2004) in which branched-chain amino acid biosynthesis is coupled to the conversion of glutamate into oxoglutarate, glutamate being obtained by the deamination of glutamine. Ammonia released in this last reaction would act as a buffer in acidic conditions by capturing one proton (van de Guchte et al. 2002). Physiological experiments corroborated this assumption, since higher concentrations of the branched-chain amino acid valine and ammonium were found in *B. longum* cells grown under acidic conditions (Sanchez et al. 2007a).

Other proteins and enzymes, whose role in the resistance to acidic pH is less clear than those commented on above, showed variations in their relative production by *B. longum* (Sanchez et al. 2007a). Among them, a mention should be given to the downregulation of bile salt hydrolase production as a response to acidic pH in the sensitive *B. longum* strain as well as in the adapted strain 8809dpH. Some enzymes involved in the biosynthesis of certain intermediates of the sulfur-containing amino acids cysteine and methionine were overexpressed in the acid-resistant strain. Finally, the acid-resistant strain showed an increase in the amount of the general stress-response chaperone GroES under acidic conditions, as observed in other bacteria (van de Guchte et al. 2002; Cotter and Hill 2003).

14.8 Bile Response

Bile is a heterogeneous mixture of electrolytes, bile salts, phospholipids, cholesterol, bilirubin, and proteins, although other endogenous (vitamins, steroids) and exogenous (antibiotics and drugs) compounds can also be detected. Bile is secreted into the duodenum during digestion to facilitate the emulsification and absorption of liposoluble nutrients. Due to the presence of sodium, potassium, and chloride, its main inorganic ions, bile is isotonic with plasma. Of the organic part, the major constituents are bile salts (Kristiansen et al. 2004; Begley et al. 2005), which are compounds with strong antimicrobial activity. After the delivery of bile into the small intestine, bile salts are absorbed through active transport in the distal ileum and returned to the liver through the portal vein. This process is known as the enterohepatic circulation of bile salts. However, during their passage through the small intestine, a significant fraction of bile salts escape active transport and pass to the colon, where they exert a strong selective pressure on the local microbiota. Their amphipathic nature endows them with a detergent action, which induces membrane damage and severely impairs membrane functionality. Furthermore, bile salts cause oxidative stress, DNA damage, and protein misfolding and trigger a series of molecular mechanisms, such as chaperone overproduction, variations of sugar and amino acid metabolism, or the production of cell-protection layers, directed to fight the harmful action of these compounds. Thus, the ability of bifidobacteria to adapt to and tolerate bile salts is critical for their colonization/persistence in their host.

Bile acids represent about 50% of the bile's organic components (Begley et al. 2002). Those bile acids that are synthesized from cholesterol in the hepatocytes are termed *primary* bile acids. They are C-24 cyclopentanephenanthrene sterols, and the two main primary bile acids are cholic acid and chenodeoxycholic acid (Fuchs 2003; Ridlon et al. 2006). Before secretion, all bile acids are transformed by liver amidases via conjugation to the amino acids glycine or taurine; the ratio of glyco- to tauro-conjugates in human bile is usually around 3:1 (Sjovall 1959; Hardison 1978). This conjugation decreases the pKa of the acid, and, at physiological pH, conjugated bile acids are almost fully ionized (in contrast to the unconjugated ones) and are called bile salts. Furthermore, it is remarkable to point out the significant influence of the gut microbiota on bile salt modification. During intestinal transit, primary bile acids are modified by bacterial enzymes, and the resulting bile acids are termed *secondary* bile acids. These modifications are mainly carried out by fermentative bacteria in the large intestine, the most prevalent of which are deconjugation, oxidation of the hydroxyl groups at C-7, and 7α-dehydroxylation (Begley et al. 2005; Ridlon et al. 2006; Hofmann and Hagey 2008).

The bile resistance of bifidobacteria has been the focus of numerous scientific studies. Bile tolerance is an in vitro selection criterion for probiotic bacteria, and it is generally considered necessary to evaluate their ability to resist the effects of bile acids (Tharmaraj and Shah 2003; Vinderola et al. 2009). In relation to the targets of bile action, the bacterial cell surface is the first barrier of defense against its deleterious activity. Bile salts modulate the expression of membrane proteins in bifidobacteria (Sanchez et al. 2006) and they profoundly impact their fatty acid composition, phospholipid composition, and cell membrane functionality (Kociubinski et al. 2002; Ruiz et al. 2009a). The mechanism of growth inhibition by bile salts is likely due to the dissipation of the proton motive force (Kurdi et al. 2006). Electron-microscope analyses have demonstrated the influence of bile on the cell surface of bifidobacteria (Ruas-Madiedo et al. 2009; Ruiz et al. 2009b). The kind of bile salt interacting with the bacterial membrane is also directly correlated to its cytotoxicity, hydrophobic bile salts being more toxic since their affinity for the membrane is higher. Furthermore, conjugated bile acids are strong acids, are ionized at physiological pH, and are unable to cross the membrane through passive diffusion (unless a transporter internalizes them). In contrast, free bile acids are weak acids and can pass across the lipid bilayer without the need for a specific transporter to enter the cell (Kurdi et al. 2003). As a consequence, cholate and deoxycholate are much more inhibitory for cells than the corresponding taurine or glycine conjugated forms. Also, bifidobacteria may display other mechanisms to counteract bile toxicity, such as active extrusion of bile salts (Price et al. 2006; Gueimonde et al. 2009). Another physiological surface response to bile in bifidobacteria is the production of exocellular polymers. Bile has been shown to induce the production of exopolysaccharide in *B. animalis* subsp. *lactis*, perhaps in order to coat its surface with a protective layer against the membrane-damaging action of bile salts (Ruas-Madiedo et al. 2009).

On the other hand, oxidative damage is a well-known deleterious effect of bile salts in cells (Bernstein et al. 1999). Many of the genes that are consistently induced in bacteria by bile salts are responsive to oxidative stress (Leverrier et al. 2003; Bron et al. 2006; Sanchez et al. 2007a, 2007b). The action of bile on cellular macromolecules has been described previously and includes promoting aberrant secondary structure formation in RNA, inducing DNA damage and thus activating enzymes involved in DNA repair, and modifiying protein conformation and turnover (Bernstein et al. 1999; Begley et al. 2005; Sanchez et al. 2005). Finally, the intracellular dissociation of bile salts lowers the intracellular pH, and it is likely that mechanisms to neutralize (or extrude) the excess of protons are required to maintain a viable cell status (Len et al. 2004; Sanchez et al. 2006, 2007b).

Independently of the natural (intrinsic) resistance levels to bile, bacteria can develop a stable resistance phenotype as a consequence of environmental conditions (by adaptive mutations), which can be transmitted to the next bacterial generation (Sanchez et al. 2007b). Furthermore, an adaptation to bile can confer protection against other environmental stress factors, namely, acid or heat (Saarela et al. 2004; Sanchez et al. 2006). For instance, when a *B. animalis* subsp. *lactis* strain, the species most commonly used in functional dairy products, was adapted to bile, this procedure also resulted in cross-resistance to other stresses, such as an increased resistance to low pH (Sanchez et al. 2006, 2007b).

The availability of bifidobacterial genome information has allowed the scientific community to undertake molecular and functional genomic studies that were unthinkable a decade ago. The genome sequences of *B. longum* (subsp. *longum* and subsp. *infantis*), *B. animalis* subsp. *lactis*, and *B. adolescentis* are currently available in the NCBI public database. Furthermore, many companies and research institutes are executing their own genome projects, while several human-gut microbiome projects are currently in progress (Gill et al. 2006; Ley et al. 2008). This will undoubtedly make an enormous amount of genetic information available for gut microbiologists. In this regard, genomic, transcriptomic, and proteomic research can be extremely useful to unravel the molecular mechanisms leading to bile response and adaptation, and this is providing functional information on the molecular players of bile response (Table 14.2). However, although significant advances have been achieved very recently (O'Connell-Motherway et al. 2008, 2009; Yasui et al. 2009), the lack of efficient transformation systems and effective molecular tools has limited functional studies in *Bifidobacterium*. Despite these bottlenecks, there are a few studies that have shed some light on the mechanisms of bifidobacterial response to bile.

In *B. longum,* two bile efflux pumps have been characterized. The overexpression of the two transporters, Ctr and BL0920, has been shown to confer bile resistance and bile salt extrusion activity (Price et al. 2006; Gueimonde et al. 2009). The *ctr* gene of *B. longum* codes for a putative sodium-dependent bile acid transporter. When transformed in an *E. coli* efflux-negative strain, this protein was responsible for an increase in the resistance to cholic acid and other antimicrobial compounds (i.e., chloramphenicol and erythromycin) and caused the efflux of radioactive cholate (Price et al. 2006). In addition, the product of the gene BL0920, a putative efflux pump, was overexpressed more than 30-fold when *B. longum* cells were grown in the presence of subinhibitory concentrations of ox bile. The expression of BL0920 in *E. coli* was shown to confer resistance to bile and also caused a higher efflux of fluorescently labeled ursodeoxycholic acid, indicating that active efflux is the mechanism underlying the bile-resistance phenotype. Interestingly, the BL0920 homolog in *B. breve*, Bbr_0838, is highly induced in the presence of bile as well, suggesting a similar function (Gueimonde et al. 2009).

Several studies have demonstrated the influence of bile on the central metabolic pathways of Gram-positive microorganisms, that is, carbohydrate catabolism. In this way the bacteria are able to reroute the metabolites to different branches, with the aim of modifying the production of energy and the redox equivalents of the cell (Begley et al. 2005; Sanchez et al. 2007b). Related to this, the proteomic analysis of *B. longum* and *B. animalis* subsp. *lactis* revealed different ways as to how these two related bacteria respond to bile salt stress (Sanchez et al. 2005, 2007b). While *B. longum* showed an overexpression of most of the enzymes of the glycolytic pathway, *B. animalis* subsp. *lactis* displayed an increment of the synthesis of enzymes involved in the formation of fructose-6-phosphate from more complex sugars, with Xfp and Gap as the sole enzymes of the bifid shunt that varied their expression levels. These data point to an increase in glucose degradation to lactate in *B. longum*, but not in *B. animalis*, in the presence of bile. In fact, glucose consumption by both species in the presence of bile salts was completely different, higher in *B. longum* and lower in *B. animalis*. Thus, it seems that, under bile challenge, *B. animalis* subsp. *lactis* increases the acetic acid formation through the bifid shunt that, theoretically, could allow bifidobacteria to synthesize more ATP by substrate phosphorylation. It has been shown that the intracellular ATP concentration decreases as the bile salt concentration in the growth medium increases, suggesting that the response to bile salts is ATP-dependent. Also, fluctuations in the redox status of cells, which are dependent on the metabolic reactions occurring at a given time, were also shown to be promoted by bile (Sánchez et al. 2007b).

Bifidobacteria are also able to trigger mechanisms to cope with the oxidative damage promoted by bile. The concentration of a DPS protein and that of a thioredoxin-dependent thiol peroxidase increased in cytoplasmic extracts of *B. animalis* grown in the presence of bile (Sanchez et al. 2007b). These two enzymes have previously been shown to be involved in the SOS response (Comtois et al. 2003; Nair and Finkel 2004). Furthermore, in *B. animalis,* bile promotes the reduced expression of two enzymes, methionine synthase and O-acetylhomoserine sulfhydrolase, involved in the last two steps of the biosynthetic pathway of methionine, whose sulfur group is susceptible to oxidation. Remarkably, in a bile-adapted strain of *B. animalis,* these two enzymes were highly overproduced (Sanchez et al. 2007b), suggesting that methionine metabolism is crucial for the regulation of oxidative stress, and its variations could reflect modifications in redox signaling in the bacterial cytoplasm (Hondorp and Matthews 2004; Sanchez et al. 2007b).

There is also a chaperone/protease-mediated bile response in *Bifidobacterium*, although it seems that *B. animalis* deploys a wider variety of these molecules than

B. longum to cope with this specific stress. Six putative chaperones and proteases were found to be overproduced in the presence of bile in *B. animalis*: ClpB, a trypsin-like serine protease ortholog of HtrA, GrpE, GroES, GroEL, and DnaK (Sanchez et al. 2007b). HtrA, GroEL, and DnaK were also found to be present in higher amounts in *B. longum* grown in the presence of bile (Sanchez et al. 2005; Savijoki et al. 2005). Thus, and despite the lack of molecular studies in other *Bifidobacterium* species, these two chaperones, together with HtrA, seem to constitute a common chaperone/protease bile-mediated response in *Bifidobacterium*.

As previously mentioned, the first target of bile action is the bacterial cell surface. In bifidobacteria, it has been shown that bile can induce the expression of genes involved in EPS production and modify the concentration of proteins involved in fatty acid synthesis, cell wall-binding peptides, cell division proteins, membrane-embedded ATPases, and cell wall "moonlighting" proteins important for the colonization of the gut environment (Sanchez et al. 2006, 2007b; Ruas-Madiedo et al. 2009; Ruiz et al. 2009a). Bile was shown to induce the expression of *gtf01207* in *B. animalis*, which codes for a putative priming glycosyltransferase involved in EPS synthesis (Ruas-Madiedo et al. 2009). Also, changes in the levels of *B. longum* surface proteins with important biological functions were detected in the presence of bile, such as the enolase. Surface-displayed enolase is able to bind human plasminogen, suggesting a role in gut colonization (Candela et al. 2007; Knaust et al. 2007). The production of long-chain-fatty-acid-CoA ligase was found to be reduced in *B. animalis* in the presence of bile, while this enzyme was produced at a constitutively low level in a *B. animalis* bile-resistant mutant as compared with the wild type. This result is in agreement with studies of *B. animalis* subsp. *lactis* BB-12, in which certain genes involved in fatty acid biosynthesis were downregulated in the presence of bile salts (Garrigues et al. 2005).

The molecular study of a bile-resistant mutant of *B. animalis* subsp. *lactis*, and the correlation between its different protein levels and the bile-resistance phenotype, allowed a closer look at the mechanisms of bile adaptation. Two proteins directly related to the cell division process (cell division initiation protein and chromosome segregation ATPase) were detected at higher amounts in the bile salt–adapted strain, which correlates with one of the more notorious consequences of bile salt adaptation in bifidobacteria: a decrease in cell size (Margolles et al. 2003; Sanchez et al. 2007b). Remarkably, the cell division protein FtsI was also induced by bile in *B. longum* (Ruiz et al. 2009a). Moreover, the synthesis of a bile salt hydrolase (BSH) and its activity increase in *B. animalis* after bile adaptation (Noriega et al. 2005; Sanchez et al. 2007b). However, no induction was found for BSH in both *B. longum* and *B. animalis* as a response to bile salts. Thus, the expression of this enzyme seems to be constitutive in bifidobacteria, but it could play a role in the bile salt adaptation phenomenon. Compared to the wild type, bile-sensitive strain, the *B. animalis* bile-adapted strain is able to tolerate bile by increasing the intracellular ATP reserve, and by inducing proton pumping by the F-ATPase, therefore tightly regulating the internal pH (Sanchez et al. 2006).

Finally, a brief mention should be made of the increased concentration of two enzymes involved in the biosynthesis of branched-chain amino acids, a branched-chain amino acid aminotransferase and a ketol-acid reducto isomerase, overproduced in *B. longum* and *B. animalis*, respectively, in response to bile (Sanchez et al. 2005, 2007b). The overexpression of enzymes involved in the biosynthesis of branched-chain amino acids has been linked to a similar phenomenon observed during the response to acid stress (Len et al. 2004; Sanchez et al. 2007a). Since some bile salts cause acidification of the cytoplasm, this may present a helper mechanism to counteract the acid toxicity of bile salts.

14.9 Concluding Remarks

Bifidobacteria have been extensively used as viable additives for the production of functional food preparations and probiotic foods. However, both during food manufacture and storage, as well as during passage through the human gut, probiotic bifidobacteria are exposed to several environmental insults (e.g., heat, bile salt, acidic and osmotic stresses). So, in order to isolate and select suitable probiotic strains, an accurate knowledge of the molecular players facing both industrial as well as environmental stresses is essential. Bifidobacterial genome-based approaches have advanced our knowledge considerably in this regard, through the identification and characterization of an ever-growing list of stress-related genes. Functional genomics, namely, transcriptomics as well as proteomics, have enabled global analyses of stress-related genes in bifidobacteria and have shown that this group of bacteria depends on a smaller set of chaperone-encoding HSPs compared to other characterized members of the *Actinobacteria* group (Ventura et al. 2006b). This can perhaps be explained by the fact that bifidobacteria live in a more or less isothermal niche and do not require an elaborate molecular heat-stress system responding to temperature fluctuations. Bifidobacteria are nevertheless expected to possess a more sophisticated system to protect themselves against the more frequently occurring bile salt and osmotic stresses. Although gene-specific regulators have been documented in bifidobacteria, the bacterial stress response is displayed as a general reaction to the growth inhibition elicited by environmental stress or physiological imbalances (Ventura et al. 2006a). In *Streptomyces* and *Corynebacterium*, shock-related genes generally display individual expression patterns (Puglia et al. 1995), presumably reflecting gene-specific regulatory systems (Servant and Mazodier 1995). However, these genes are regulated by at least three coordinate systems, which are dependent on the nutritional sources supporting growth, the nature or starvation response that eventually limits growth, or the specific environmental stress. In contrast, in bifidobacteria (e.g., *B. breve* UCC2003), some degree of overlap between genes induced by heat and osmotic stress has been observed, consistent with their common regulatory mechanisms (Ventura et al. 2005c).

The knowledge on the genetics of the bifidobacterial stress response could constitute a basis for comparisons with well-characterized model organisms (e.g., *B. subtilis* or *E. coli*). Such comparisons may show the specificity of the bifidobacterial response to environmental challenges, which may have evolved to fit the specific constraints of a given ecological niche (e.g., human intestine). The limited comparisons made so far have revealed a similar content of chaperone-encoding genes but significant differences in regulatory circuits. Our understanding of the regulation of bifidobacterial stress-related genes, however, is still at a very primitive level. A major challenge for future research will be to explore how environmental stresses are being sensed to identify the principal signals and to investigate how the cell perceives such signals. In addition, it will be extremely interesting to shed more light on the interactiveness of the regulators of the stress response in bacteria and how the various regulons involved are cross-talking.

References

Alander M, Mättö J, Kneifel W, Johansson M, Kögler B, Crittenden R, Mattila-Sandholm T, Saarela M (2001) Effect of galacto-oligosaccharide supplementation on human faecal microflora and on survival and persistence of *Bifidobacterium lactis* Bb-12 in the gastrointestinal tract. Int Dairy J 11:817–825

Barrangou R, Briczinski EP, Traeger LL, Loquasto JR, Richards M, Horvath P, Coute-Monvoisin AC, Leyer G, Rendulic S, Steele JL, Broadbent JR, Oberg T, Dudley EG, Schuster S, Romero DA, Roberts RF (2009) Comparison of the complete genome sequences of *Bifidobacterium animalis* subsp. *lactis* DSM 10140 and Bl-04. J Bacteriol 191:4144–4151

Begley M, Gahan CG, Hill C (2002) Bile stress response in *Listeria monocytogenes* LO28: adaptation, cross-protection, and identification of genetic loci involved in bile resistance. Appl Environ Microbiol 68:6005–6012

Begley M, Gahan CG, Hill C (2005) The interaction between bacteria and bile. FEMS Microbiol Rev 29:625–651

Bellier A, Mazodier P (2004) ClgR, a novel regulator of *clp* and *lon* expression in *Streptomyces*. J Bacteriol 186:3238–3248

Bernstein C, Bernstein H, Payne CM, Beard SE, Schneider J (1999) Bile salt activation of stress response promoters in *Escherichia coli*. Curr Microbiol 39:68–72

Bore E, Langsrud S, Langsrud O, Rode TM, Holck A (2007) Acid-shock responses in *Staphylococcus aureus* investigated by global gene expression analysis. Microbiology 153:2289–2303

Bron PA, Molenaar D, de Vos WM, Kleerebezem M (2006) DNA micro-array-based identification of bile-responsive genes in *Lactobacillus plantarum*. J Appl Microbiol 100:728–738

Bukau B, Horwich AL (1998) The Hsp70 and Hsp60 chaperone machines. Cell 92:351–366

Candela M, Bergmann S, Vici M, Vitali B, Turroni S, Eikmanns BJ, Hammerschmidt S, Brigidi P (2007) Binding of human plasminogen to *Bifidobacterium*. J Bacteriol 189:5929–5936

Chung HS, Kim YB, Chun SL, Ji GE (1999) Screening and selection of acid and bile resistant bifidobacteria. Int J Food Microbiol 47:25–32

Collado MC, Sanz Y (2006) Method for direct selection of potentially probiotic *Bifidobacterium* strains from human feces based on their acid-adaptation ability. J Microbiol Methods 66:560–563

Collado MC, Sanz Y (2007) Induction of acid resistance in *Bifidobacterium*: a mechanism for improving desirable traits of potentially probiotic strains. J Appl Microbiol 103:1147–1157

Collado MC, Gueimonde M, Sanz Y, Salminen S (2006) Adhesion properties and competitive pathogen exclusion ability of bifidobacteria with acquired acid resistance. J Food Prot 69:1675–1679

Comtois SL, Gidley MD, Kelly DJ (2003) Role of the thioredoxin system and the thiol-peroxidases Tpx and Bcp in mediating resistance to oxidative and nitrosative stress in *Helicobacter pylori*. Microbiology 149:121–129

Cotter PD, Hill C (2003) Surviving the acid test: responses of Gram-positive bacteria to low pH. Microbiol Mol Biol Rev 67:429–453

Davis MJ, Coote PJ, O'Byrne CP (1996) Acid tolerance in *Listeria monocytogenes*: the adaptive acid tolerance response (ATR) and growth-phase-dependent acid resistance. Microbiology 142(Pt 10):2975–2982

De Dea Lindner J, Canchaya C, Zhang Z, Neviani E, Fitzgerald GF, van Sinderen D, Ventura M (2007) Exploiting *Bifidobacterium* genomes: the molecular basis of stress response. Int J Food Microbiol 120:13–24

Derré I, Rapoport G, Devine K, Rose M, Msadek T (1999a) ClpE, a novel type of HSP100 ATPase, is part of the CtsR heat shock regulon of *Bacillus subtilis*. Mol Microbiol 32:581–593

Derré I, Rapoport G, Msadek T (1999b) CtsR, a novel regulator of stress and heat shock response, controls *clp* and molecular chaperone gene expression in Gram-positive bacteria. Mol Microbiol 31:117–131

Engels S, Schweitzer JE, Ludwig C, Bott M, Schaffer S (2004) *clpC* and *clpP1P2* gene expression in *Corynebacterium glutamicum* is controlled by a regulatory network involving the transcriptional regulators ClgR and HspR as well as the ECF sigma factor sigmaH. Mol Microbiol 52:285–302

Engels S, Ludwig C, Schweitzer JE, Mack C, Bott M, Schaffer S (2005) The transcriptional activator ClgR controls transcription of genes involved in proteolysis and DNA repair in *Corynebacterium glutamicum*. Mol Microbiol 57:576–591

Ewalt KL, Hendrick JP, Houry WA, Hartl FU (1997) *In vivo* observation of polypeptide flux through the bacterial chaperonin system. Cell 90:491–500

Fanaro S, Vigi V, Chierici R, Boehm G (2003) Fecal flora measurements of breastfed infants using an integrated transport and culturing system. Acta Paediatr 92:634–635

Fedhila S, Msadek T, Nel P, Lereclus D (2002) Distinct *clpP* genes control specific adaptive responses in *Bacillus thuringiensis*. J Bacteriol 184:5554–5562

Fraser CM, Gocayne JD, White O, Adams MD, Clayton RA, Fleischmann RD, Bult CJ, Kerlavage AR, Sutton G, Kelley JM, Fritchman RD, Weidman JF, Small KV, Sandusky M, Fuhrmann J, Nguyen D, Utterback TR, Saudek DM, Phillips CA, Merrick JM, Tomb JF, Dougherty BA, Bott KF, Hu PC, Lucier TS, Peterson SN, Smith HO, Hutchison CA 3rd, Venter JC (1995) The minimal gene complement of *Mycoplasma genitalium*. Science 270:397–403

Fuchs M (2003) Bile acid regulation of hepatic physiology: III. Regulation of bile acid synthesis: past progress and future challenges. Am J Physiol Gastrointest Liver Physiol 284:G551–557

Garrigues C, Stuer-Lauridsen B, Johansen E (2005) Characterisation of *Bifidobacterium animalis* subsp. *lactis* BB-12 and other probiotic bacteria using genomics, transcriptomics and proteomics. Aust J Dairy Technol 60:84–92

Gill SR, Pop M, Deboy RT, Eckburg PB, Turnbaugh PJ, Samuel BS, Gordon JI, Relman DA, Fraser-Liggett CM, Nelson KE (2006) Metagenomic analysis of the human distal gut microbiome. Science 312:1355–1359

Gottesman S (1996) Proteases and their targets in *Escherichia coli*. Annu Rev Genet 30:465–506

Gottesman S, Wickner S, Maurizi MR (1997) Protein quality control: triage by chaperones and proteases. Genes Dev 11:815–823

Grimaud R, Kessel M, Beuron F, Steven AC, Maurizi MR (1998) Enzymatic and structural similarities between the *Escherichia coli* ATP-dependent proteases, ClpXP and ClpAP. J Biol Chem 273:12476–12481

Gueimonde M, Garrigues C, van Sinderen D, de los Reyes-Gavilan CG, Margolles A (2009) Bile-inducible efflux transporter from *Bifidobacterium longum* NCC2705, conferring bile resistance. Appl Environ Microbiol 75:3153–3160

Hardison WG (1978) Hepatic taurine concentration and dietary taurine as regulators of bile acid conjugation with taurine. Gastroenterology 75:71–75

Harmsen HJ, Wildeboer-Veloo AC, Raangs GC, Wagendorp AA, Klijn N, Bindels JG, Welling GW (2000) Analysis of intestinal flora development in breast-fed and formula-fed infants by using molecular identification and detection methods. J Pediatr Gastroenterol Nutr 30:61–67

Hartl FU (1996) Molecular chaperones in cellular protein folding. Nature 381:571–579

Hartl FU, Hayer-Hartl M (2002) Molecular chaperones in the cytosol: from nascent chain to folded protein. Science 295:1852–1858

Hlavacek O, Vachova L (2002) ATP-dependent proteinases in bacteria. Folia Microbiol (Praha) 47:203–212

Hofmann AF, Hagey LR (2008) Bile acids: chemistry, pathochemistry, biology, pathobiology, and therapeutics. Cell Mol Life Sci 65:2461–2483

Homuth G, Masuda S, Mogk A, Kobayashi Y, Schumann W (1997) The *dnaK* operon of *Bacillus subtilis* is heptacistronic. J Bacteriol 179:1153–1164

Hondorp ER, Matthews RG (2004) Oxidative stress inactivates cobalamin-independent methionine synthase (MetE) in *Escherichia coli*. PLoS Biol 2:e336

Houry WA, Frishman D, Eckerskorn C, Lottspeich F, Hartl FU (1999) Identification of *in vivo* substrates of the chaperonin GroEL. Nature 402:147–154

Jayamanne VS, Adams MR (2006) Determination of survival, identity and stress resistance of probiotic bifidobacteria in bio-yoghurts. Lett Appl Microbiol 42:189–194

Keiler KC, Waller PR, Sauer RT (1996) Role of a peptide tagging system in degradation of proteins synthesized from damaged messenger RNA. Science 271:990–993

Kheadr E, Dabour N, Le Lay C, Lacroix C, Fliss I (2007) Antibiotic susceptibility profile of bifidobacteria as affected by oxgall, acid, and hydrogen peroxide stress. Antimicrob Agents Chemother 51:169–174

Knaust A, Weber MV, Hammerschmidt S, Bergmann S, Frosch M, Kurzai O (2007) Cytosolic proteins contribute to surface plasminogen recruitment of *Neisseria meningitidis*. J Bacteriol 189:3246–3255

Kociubinski G, Zavaglia AG, Perez PF, Disalvo EA, De Antoni GL (2002) Effect of bile components on the surface properties of bifidobacteria. J Dairy Res 69:293–302

Kristiansen TZ, Bunkenborg J, Gronborg M, Molina H, Thuluvath PJ, Argani P, Goggins MG, Maitra A, Pandey A (2004) A proteomic analysis of human bile. Mol Cell Proteomics 3:715–728

Kurdi P, Tanaka H, Van Veen HW, Asano K, Tomita F, Yokota A (2003) Cholic acid accumulation and its diminution by short-chain fatty acids in bifidobacteria. Microbiology 149:2031–2037

Kurdi P, Kawanishi K, Mizutani K, Yokota A (2006) Mechanism of growth inhibition by free bile acids in lactobacilli and bifidobacteria. J Bacteriol 188:1979–1986

Lamendella R, Santo Domingo JW, Kelty C, Oerther DB (2008) Bifidobacteria in feces and environmental waters. Appl Environ Microbiol 74:575–584

Langer T, Pfeifer G, Martin J, Baumeister W, Hartl FU (1992) Chaperonin-mediated protein folding: GroES binds to one end of the GroEL cylinder, which accommodates the protein substrate within its central cavity. EMBO J 11:4757–4765

Lee BH, Hibino T, Jo J, Viale AM, Takabe T (1997) Isolation and characterization of a *dnaK* genomic locus in a halotolerant cyanobacterium *Aphanothece halophytica*. Plant Mol Biol 35:763–775

Lee JH, Karamychev VN, Kozyavkin SA, Mills D, Pavlov AR, Pavlova NV, Polouchine NN, Richardson PM, Shakhova VV, Slesarev AI, Weimer B, O'Sullivan DJ (2008) Comparative genomic analysis of the gut bacterium *Bifidobacterium longum* reveals loci susceptible to deletion during pure culture growth. BMC Genom 9:247

Len AC, Harty DW, Jacques NA (2004) Proteome analysis of *Streptococcus mutans* metabolic phenotype during acid tolerance. Microbiology 150:1353–1366

Leverrier P, Dimova D, Pichereau V, Auffray Y, Boyaval P, Jan G (2003) Susceptibility and adaptive response to bile salts in *Propionibacterium freudenreichii*: physiological and proteomic analysis. Appl Environ Microbiol 69:3809–3818

Ley RE, Hamady M, Lozupone C, Turnbaugh PJ, Ramey RR, Bircher JS, Schlegel ML, Tucker TA, Schrenzel MD, Knight R, Gordon JI (2008) Evolution of mammals and their gut microbes. Science 320:1647–1651

Margolles A, García L, Sánchez B, Gueimonde M, de los Reyes-Gavilán CG (2003) Characterisation of a *Bifidobacterium* strain with acquired resistance to cholate – a preliminary study. Int J Food Microbiol 82:191–198

Masco L, Crockaert C, Van Hoorde K, Swings J, Huys G (2007) *In vitro* assessment of the gastrointestinal transit tolerance of taxonomic reference strains from human origin and probiotic product isolates of *Bifidobacterium*. J Dairy Sci 90:3572–3578

Matsumoto M, Ohishi H, Benno Y (2004) H+-ATPase activity in *Bifidobacterium* with special reference to acid tolerance. Int J Food Microbiol 93:109–113

Maurizi MR, Clark WP, Kim SH, Gottesman S (1990) ClpP represents a unique family of serine proteases. J Biol Chem 265:12546–12552

Maus JE, Ingham SC (2003) Employment of stressful conditions during culture production to enhance subsequent cold- and acid-tolerance of bifidobacteria. J Appl Microbiol 95:146–154

Miwa T, Esaki H, Umemori J, Hino T (1997) Activity of H(+)-ATPase in ruminal bacteria with special reference to acid tolerance. Appl Environ Microbiol 63:2155–2158

Miyake T, Watanabe K, Watanabe T, Oyaizu H (1998) Phylogenetic analysis of the genus *Bifidobacterium* and related genera based on 16S rDNA sequences. Microbiol Immunol 42:661–667

Mogk A, Tomoyasu T, Goloubinoff P, Rüdiger S, Röder D, Langen H, Bukau B (1999) Identification of thermolabile *Escherichia coli* proteins: prevention and reversion of aggregation by DnaK and ClpB. EMBO J 18:6934–6949

Muchowski PJ, Hays LG, Yates JR 3rd, Clark JI (1999) ATP and the core "alpha-crystallin" domain of the small heat-shock protein alphaB-crystallin. J Biol Chem 274:30190–30195

Nair S, Finkel SE (2004) Dps protects cells against multiple stresses during stationary phase. J Bacteriol 186:4192–4198

Narberhaus F (1999) Negative regulation of bacterial heat shock genes. Mol Microbiol 31:1–8

Narberhaus F (2002) Alpha-crystallin-type heat shock proteins: socializing minichaperones in the context of a multichaperone network. Microbiol Mol Biol Rev 66:64–93; table of contents

Noriega L, de los Reyes-Gavilan CG, Margolles A (2005) Acquisition of bile salt resistance promotes antibiotic susceptibility changes in *Bifidobacterium*. J Food Prot 68:1916–1919

O'Connell-Motherway M, Fitzgerald GF, Neirynck S, Ryan S, Steidler L, van Sinderen D (2008) Characterization of ApuB, an extracellular type II amylopullulanase from *Bifidobacterium breve* UCC2003. Appl Environ Microbiol 74:6271–6279

O'Connell-Motherway M, O'Driscoll J, Fitzgerald GF, van Sinderen D (2009) Overcoming the restriction barrier to plasmid transformation and targeted mutagenesis in *Bifidobacterium breve* UCC2003. Microbiol Biotechnol 2:321–332

Price CE, Reid SJ, Driessen AJ, Abratt VR (2006) The *Bifidobacterium longum* NCIMB 702259T *ctr* gene codes for a novel cholate transporter. Appl Environ Microbiol 72:923–926

Puglia AM, Vohradsky J, Thompson CJ (1995) Developmental control of the heat-shock stress regulon in *Streptomyces coelicolor*. Mol Microbiol 17:737–746

Rezzonico E, Lariani S, Barretto C, Cuanoud G, Giliberti G, Delley M, Arigoni F, Pessi G (2007) Global transcriptome analysis of the heat shock response of *Bifidobacterium longum*. FEMS Microbiol Lett 271:136–145

Ridlon JM, Kang DJ, Hylemon PB (2006) Bile salt biotransformations by human intestinal bacteria. J Lipid Res 47:241–259

Ritter P, Kohler C, von Ah U (2009) Evaluation of the passage of *Lactobacillus gasseri* K7 and bifidobacteria from the stomach to intestines using a single reactor model. BMC Microbiol 9:87

Ruas-Madiedo P, Gueimonde M, Arigoni F, de los Reyes-Gavilan CG, Margolles A (2009) Bile affects the synthesis of exopolysaccharides by *Bifidobacterium animalis*. Appl Environ Microbiol 75:1204–1207

Ruiz L, Coute Y, Sanchez B, de los Reyes-Gavilan CG, Sanchez JC, Margolles A (2009a) The cell-envelope proteome of *Bifidobacterium longum* in an *in vitro* bile environment. Microbiology 155:957–967

Ruiz L, Sanchez B, de Los Reyes-Gavilan CG, Gueimonde M, Margolles A (2009b) Coculture of *Bifidobacterium longum* and *Bifidobacterium breve* alters their protein expression profiles and enzymatic activities. Int J Food Microbiol 133:148–153

Saarela M, Rantala M, Hallamaa K, Nohynek L, Virkajarvi I, Matto J (2004) Stationary-phase acid and heat treatments for improvement of the viability of probiotic lactobacilli and bifidobacteria. J Appl Microbiol 96:1205–1214

Samelis J, Ikeda JS, Sofos JN (2003) Evaluation of the pH-dependent, stationary-phase acid tolerance in *Listeria monocytogenes* and *Salmonella Typhimurium* DT104 induced by culturing in media with 1% glucose: a comparative study with *Escherichia coli* O157:H7. J Appl Microbiol 95:563–575

Sanchez B, de los Reyes-Gavilan CG, Margolles A (2006) The F1F0-ATPase of *Bifidobacterium animalis* is involved in bile tolerance. Environ Microbiol 8:1825–1833

Sanchez B, Champomier-Verges MC, Anglade P, Baraige F, de Los Reyes-Gavilan CG, Margolles A, Zagorec M (2005) Proteomic analysis of global changes in protein expression during bile salt exposure of *Bifidobacterium longum* NCIMB 8809. J Bacteriol 187:5799–5808

Sanchez B, Champomier-Verges MC, Collado Mdel C, Anglade P, Baraige F, Sanz Y, de los Reyes-Gavilan CG, Margolles A, Zagorec M (2007a) Low-pH adaptation and the acid tolerance response of *Bifidobacterium longum* biotype *longum*. Appl Environ Microbiol 73:6450–6459

Sanchez B, Champomier-Verges MC, Stuer-Lauridsen B, Ruas-Madiedo P, Anglade P, Baraige F, de los Reyes-Gavilan CG, Johansen E, Zagorec M, Margolles A (2007b) Adaptation and response of *Bifidobacterium animalis* subsp. *lactis* to bile: a proteomic and physiological approach. Appl Environ Microbiol 73:6757–6767

Savijoki K, Suokko A, Palva A, Valmu L, Kalkkinen N, Varmanen P (2005) Effect of heat-shock and bile salts on protein synthesis of *Bifidobacterium longum* revealed by [35S]methionine labelling and two-dimensional gel electrophoresis. FEMS Microbiol Lett 248:207–215

Schell MA, Karmirantzou M, Snel B, Vilanova D, Berger B, Pessi G, Zwahlen MC, Desiere F, Bork P, Delley M, Pridmore RD, Arigoni F (2002) The genome sequence of *Bifidobacterium longum* reflects its adaptation to the human gastrointestinal tract. Proc Natl Acad Sci USA 99:14422–14427

Schirmer EC, Glover JR, Singer MA, Lindquist S (1996) HSP100/Clp proteins: a common mechanism explains diverse functions. Trends Biochem Sci 21:289–296

Sela DA, Chapman J, Adeuya A, Kim JH, Chen F, Whitehead TR, Lapidus A, Rokhsar DS, Lebrilla CB, German JB, Price NP, Richardson PM, Mills DA (2008) The genome sequence of *Bifidobacterium longum* subsp. *infantis* reveals adaptations for milk utilization within the infant microbiome. Proc Natl Acad Sci USA 105:18964–18969

Servant P, Mazodier P (1995) Characterization of *Streptomyces* albus 18-kilodalton heat shock-responsive protein. J Bacteriol 177:2998–3003

Sjovall J (1959) Dietary glycine and taurine on bile acid conjugation in man; bile acids and steroids 75. Proc Soc Exp Biol Med 100:676–678

Stackebrandt E, Sproer C, Rainey FA, Burghardt J, Pauker O, Hippe H (1997) Phylogenetic analysis of the genus *Desulfotomaculum*: evidence for the misclassification of *Desulfotomaculum guttoideum* and description of *Desulfotomaculum orientis* as *Desulfosporosinus orientis* gen. *nov*., comb. *nov*. Int J Syst Bacteriol 47:1134–1139

Takahashi N, Xiao JZ, Miyaji K, Iwatsuki K (2007) H+-ATPase in the acid tolerance of *Bifidobacterium longum*. Milk Sci Int 62:151–153

Teter SA, Klionsky DJ (1999) How to get a folded protein across a membrane. Trends Cell Biol 9:428–431

Tharmaraj N, Shah NP (2003) Selective enumeration of *Lactobacillus delbrueckii* ssp. *bulgaricus*, *Streptococcus thermophilus*, *Lactobacillus acidophilus*, bifidobacteria, *Lactobacillus casei*, *Lactobacillus rhamnosus*, and propionibacteria. J Dairy Sci 86:2288–2296

Turroni F, Foroni E, Pizzetti P, Giubellini V, Ribbera A, Merusi P, Cagnasso P, Bizzarri B, de'Angelis GL, Shanahan F, van Sinderen D, Ventura M (2009a) Exploring the diversity of the bifidobacterial population in the human intestinal tract. Appl Environ Microbiol 75:1534–1545

Turroni F, Marchesi JR, Foroni E, Gueimonde M, Shanahan F, Margolles A, van Sinderen D, Ventura M (2009b) Microbiomic analysis of the bifidobacterial population in the human distal gut. ISME J 3:745–751

van de Guchte M, Serror P, Chervaux C, Smokvina T, Ehrlich SD, Maguin E (2002) Stress responses in lactic acid bacteria. Antonie van Leeuwenhoek 82:187–216

Ventura M, Canchaya C, van Sinderen D, Fitzgerald GF, Zink R (2004a) *Bifidobacterium lactis* DSM 10140: identification of the *atp* (*atpBEFHAGDC*) operon and analysis of its genetic structure, characteristics, and phylogeny. Appl Environ Microbiol 70:3110–3121

Ventura M, Canchaya C, Zink R, Fitzgerald GF, van Sinderen D (2004b) Characterization of the *groEL* and *groES* loci in *Bifidobacterium breve* UCC 2003: genetic, transcriptional, and phylogenetic analyses. Appl Environ Microbiol 70:6197–6209

Ventura M, Fitzgerald GF, van Sinderen D (2005a) Genetic and transcriptional organization of the *clpC* locus in *Bifidobacterium breve* UCC 2003. Appl Environ Microbiol 71:6282–6291

Ventura M, Kenny JG, Zhang Z, Fitzgerald GF, van Sinderen D (2005b) The *clpB* gene of *Bifidobacterium breve* UCC 2003: transcriptional analysis and first insights into stress induction. Microbiology 151:2861–2872

Ventura M, Zhang Z, Cronin M, Canchaya C, Kenny JG, Fitzgerald GF, van Sinderen D (2005c) The ClgR protein regulates transcription of the *clpP* operon in *Bifidobacterium breve* UCC 2003. J Bacteriol 187:8411–8426

Ventura M, Canchaya C, Del Casale A, Dellaglio F, Neviani E, Fitzgerald GF, van Sinderen D (2006a) Analysis of bifidobacterial evolution using a multilocus approach. Int J Syst Evol Microbiol 56:2783–2792

Ventura M, Canchaya C, Zhang Z, Bernini V, Fitzgerald GF, van Sinderen D (2006b) How high G+C Gram-positive bacteria and in particular bifidobacteria cope with heat stress: protein players and regulators. FEMS Microbiol Rev 30:734–759

Ventura M, Canchaya C, Fitzgerald GF, Gupta RS, van Sinderen D (2007a) Genomics as a means to understand bacterial phylogeny and ecological adaptation: the case of bifidobacteria. Antonie van Leeuwenhoek 91:351–372

Ventura M, Canchaya C, Tauch A, Chandra G, Fitzgerald GF, Chater KF, van Sinderen D (2007b) Genomics of Actinobacteria: tracing the evolutionary history of an ancient phylum. Microbiol Mol Biol Rev 71:495–548

Ventura M, Canchaya C, Zhang Z, Fitzgerald GF, van Sinderen D (2007c) Molecular characterization of *hsp20*, encoding a small heat shock protein of *Bifidobacterium breve* UCC2003. Appl Environ Microbiol 73:4695–4703

Ventura M, O'Flaherty S, Claesson MJ, Turroni F, Klaenhammer TR, van Sinderen D, O'Toole PW (2008) Genome-scale analyses of health-promoting bacteria: probiogenomics. Nat Rev Microbiol 7:61–71

Vernazza CL, Gibson GR, Rastall RA (2006) Carbohydrate preference, acid tolerance and bile tolerance in five strains of *Bifidobacterium*. J Appl Microbiol 100:846–853

Viala J, Mazodier P (2002) ClpP-dependent degradation of PopR allows tightly regulated expression of the *clpP3 clpP4* operon in *Streptomyces lividans*. Mol Microbiol 44:633–643

Viala J, Rapoport G, Mazodier P (2000) The *clpP* multigenic family in *Streptomyces lividans*: conditional expression of the *clpP3 clpP4* operon is controlled by PopR, a novel transcriptional activator. Mol Microbiol 38:602–612

Vinderola G, Prosello W, Molinari F, Ghiberto D, Reinheimer J (2009) Growth of *Lactobacillus paracasei* A13 in Argentinian probiotic cheese and its impact on the characteristics of the product. Int J Food Microbiol 135:171–174

Weissman JS, Kashi Y, Fenton WA, Horwich AL (1994) GroEL-mediated protein folding proceeds by multiple rounds of binding and release of nonnative forms. Cell 78:693–702

Wickner S, Maurizi MR, Gottesman S (1999) Posttranslational quality control: folding, refolding, and degrading proteins. Science 286:1888–1893

Xu Z, Horwich AL, Sigler PB (1997) The crystal structure of the asymmetric GroEL-GroES-(ADP)7 chaperonin complex. Nature 388:741–750

Yasui K, Kano Y, Tanaka K, Watanabe K, Shimizu-Kadota M, Yoshikawa H, Suzuki T (2009) Improvement of bacterial transformation efficiency using plasmid artificial modification. Nucl Acids Res 37:e3

Zomer A, Fernandez M, Kearney B, Fitzgerald GF, Ventura M, van Sinderen D (2009) An interactive regulatory network controls stress response in *Bifidobacterium breve* UCC2003. J Bacteriol 191:7039–7049

Chapter 15
Stress Responses of *Oenococcus oeni*

Jean Guzzo

15.1 Introduction

Oenococcus oeni is a lactic acid bacterium (LAB) with acidophilic properties that enable the cells to grow in a hostile medium such as wine. It is a facultative anaerobe and is characterized by coccoid-shaped cells in pairs or chains. *O. oeni* exhibits a heterofermentative metabolism and consequently transforms sugar into lactic acid, ethanol, and carbon dioxide. *O. oeni* is found naturally in the winery or in the must and performs malolactic fermentation (MLF) (Bartowsky 2005). *O. oeni* is the new name for the LAB *Leuconostoc oenos* (Dicks et al. 1995). Molecular data confirmed that *Oenococcus* was different from the genus *Leuconostoc*. *O. oeni* is a bacterium with a high tolerance to ethanol (10% v/v or higher) and acidic pH (pH 3.5 or lower).

This bacterium is often responsible for MLF, an important step in the vinification process. MLF is a secondary fermentation that occurs in wine once alcoholic fermentation by yeast has been completed. It is the bacterial conversion of L-malate into L-lactate and carbon dioxide, leading to a natural decrease in acidity, together with an enhanced stability and quality of wine (Kunkee 1991).

In fact, wine is a very harsh medium. It contains ethanol, has a pH of 3–3.5, and has a high concentration of sulfite (a powerful antimicrobial agent). Wine is, therefore, a poor medium with growth-limiting conditions for bacteria. In addition, temperatures in wineries are low (Guzzo and Desroche 2009). Improving the control of MLF is of technical and economical importance for the winemaker. MLF is often a step that is not completely controlled because wine is such a harsh medium with multiple stresses. Such stress conditions are often a cause of delayed MLF.

J. Guzzo (✉)
Equipe de Recherche en Vigne et Vin, Institut Jules Guyot, Université de Bourgogne, BP 27877–21078, Dijon, France
e-mail: jguzzo@u-bourgogne.fr

E. Tsakalidou and K. Papadimitriou (eds.), *Stress Responses of Lactic Acid Bacteria*, Food Microbiology and Food Safety, DOI 10.1007/978-0-387-92771-8_15,

Winemakers can use industrial cultures of *O. oeni* strains selected for their technological performance to resolve this problem. In general, these strains have been optimized for MLF by an acclimatization process that improves survival and prevents a high mortality of lyophilized cells after direct inoculation into wine. An understanding of the physiology of the stress responses of this bacterium is clearly required to optimize the preparation of *O. oeni* starter cultures. The recent sequencing of the genome of two *O. oeni* strains is also an important step toward understanding the potential and diversity of this bacterium.

O. oeni seems to be a good model for the study of mechanisms involved in stress resistance at the molecular and physiological levels. It is an interesting bacterium to study, but there are difficulties, particularly related to the low growth rate, even in a rich medium, compared with other LAB. Moreover, researchers have not been able to genetically modify the cells even using the current standard protocol of electroporation (Assad-Garcia et al. 2008). Nethertheless, the stress responses of *O. oeni* are well documented, especially stress gene regulation and the function of heat-shock proteins (HSPs). Moreover, several works focus on the mechanisms involved in the maintenance of membrane integrity and the activity of associated enzymes. Knowledge of the genome sequence provided new insight that will be discussed at the end of this chapter.

15.2 Impact of Adaptation on Survival

Cells submitted to sublethal stress can acquire a transient tolerance. This physiological state enables cells to resist lethal stress conditions. This acquired tolerance was demonstrated by applying a heat shock of 42°C, which enabled the direct inoculation of synthetic wine and improved survival (Guzzo et al. 1994). Similarly, preadaptation involving a period of growth at pH 3.5 enhanced survival in synthetic wine containing ethanol (11% v/v), at a temperature of 18°C (instead of 30°C) and limited nutrients (Beltramo et al. 2006). These results provide evidence of cross-protection between stresses. Another example is the increase in resistance to sulfites of *O. oeni* cells following preadaptation to pH 3.5 (Guzzo et al. 1998).

15.3 Regulation of Stress Gene Expression

Many stress genes have been characterized in *O. oeni*. Several belong to the large family of heat-shock genes. The principal stress genes, characterized mainly by cloning, sequencing, and transcriptional analysis, are listed in Table 15.1. The *hsp18* gene encoding the small HSP (sHSP) called Lo18 and the cluster harboring the transcriptional regulator *ctsR* and the *clpC* gene, a member of the Clp family, were extensively studied, as described ahead. Other genes such as *trxA* encoding thioredoxine and *clpX* were also studied in detail, but the regulation mechanisms involved remain to be elucidated.

Table 15.1 Stress genes characterized in *O. oeni*

Stress genes	Gene products	References
hsp18	Molecular chaperone	Jobin et al. (1997)
trxA	Thioredoxine	Jobin et al. (1999a)
clpX	Regulatory ATPase subunit of ClpP–molecular chaperone	Jobin et al. (1999b)
arcA	Arginine deiminase	Tonon et al. (2001)
atp	H^+-ATPase	Fortier et al. (2003)
ftsH	Membrane protease	Bourdineaud et al. (2003)
clpL	Regulatory ATPase subunit of ClpP–molecular chaperone	Beltramo et al. (2004a, b)
clpP	Protease	Beltramo et al. (2004a, b)
omrA	MDR	Bourdineaud et al. (2004)
ctsR	Repressor	Grandvalet et al. (2005)
clpC	Regulatory ATPase subunit of ClpP–molecular chaperone	Grandvalet et al. (2005)
cfa	Cyclic fatty acid synthase	Grandvalet et al. (2008)

15.3.1 Stress Gene Expression as a Function of Growth

A comparison of the stress gene expression profiles for *O. oeni* cultured in synthetic wine or rich medium has shown strong similarities in the genes studied. In both culture conditions, *clpL* and *hsp18* appear to be strongly expressed at the start of the stationary growth phase. Such genes are considered to be specific physiological state markers of cells in the stationary phase. Moreover, the strong expression of *clpX* at the start of the growth phase suggests the involvement of the ClpX protein in the early growth. Another gene, *trxA,* is expressed throughout the growth period, and there is no variation related to the physiological state of the cell (data not shown). *clpP* is preferentially induced in the exponential phase. To conclude, the expression of stress genes is a function of growth. Moreover, the kinetics of gene expression during growth appear to be similar for *O. oeni* cells grown in synthetic wine or in standard laboratory medium, as shown in Fig. 15.1 (Beltramo et al. 2006; Cavin et al. 1989).

15.3.2 The CtsR Regulon

The repressor CtsR plays a central role in the regulation of stress gene expression in many LAB. The first heat-shock gene characterized in *O.oeni* was *hsp18* (Jobin et al. 1997). A 2.3-kb DNA fragment carrying the *hsp18* gene was shown to encode an sHSP. Upstream from the coding sequence was a promoter region similar to consensus promoter sequences of Gram- positive bacteria and *Escherichia coli.* The study conducted in *Bacillus subtilis* (Derré et al. 1999) demonstrated the

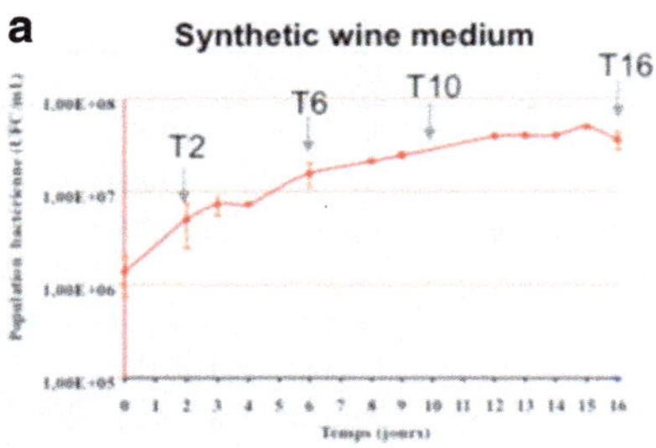

RT-PCR

	T2	T6	T10	T16
clpL1	2,2	1,0	2,1	11,2
clpP	1,3	2,9	1,0	2,8
hsp18	2,5	1,0	1,5	3,5
clpX	3,2	1,4	1,0	1,6

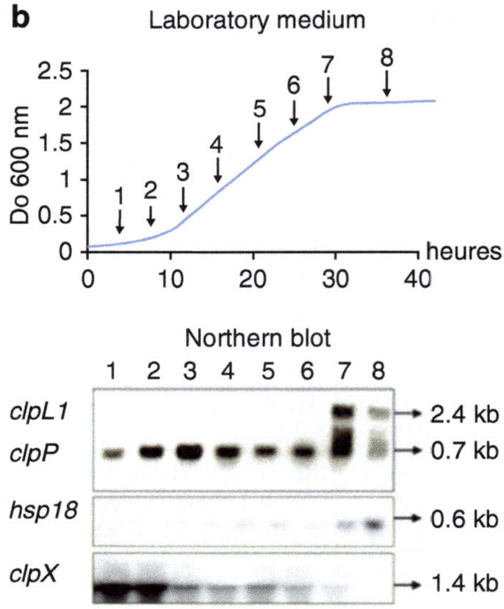

Fig. 15.1 Kinetics of stress gene expression during growth in (**a**) synthetic wine and (**b**) laboratory medium. Gene expression was analyzed by RT-qPCR (relative expression levels calculated using the comparative critical threshold (ΔΔCt) method were presented) or Northern blot (Beltramo et al. 2004b, 2006; Jobin et al. 1999b)

– **Transcriptional regulation mechanism**

```
                    -35                       -10        +1
TTTTTTATTTGACCTTTCTTGACTTTCTGACCAATACTGACTATTATAATAACTGTAAGCG

                      RBS          >>>hsp18>>>

GTATGTAATATCACTTAGTTAATAGGAGGTATTTGATATGGCAAATGAATTAATGGATAGA

                                     M  A  N  E  L  M  D  R
```

Fig. 15.2 Organization of the *hsp18* gene and nucleotide sequence of the promoter region (Grandvalet et al. 2005). The putative –10 and –35 operator sequences are in *bold*. The ribosome-binding site (RBS) is *underlined* and is in *bold*. *Arrows* indicate the likely CtsR operator sites

presence of two putative operator sequences for CtsR in the *hsp18* promoter region by sequence comparison analysis (Fig. 15.2). This operator was a heptanucleotide direct repeat well conserved in Gram-positive bacteria and involved in the regulation of *clp* and molecular chaperone genes (Derré et al. 1999).

Despite the presence of consensus sequences upstream from the *hsp18* coding sequence, the CtsR-dependent regulation of *hsp18* has not yet been demonstrated in *O. oeni*. However, using *B. subtilis* as a heterologous host, Grandvalet et al. (2005) demonstrated that the expression of several *O. oeni* genes can be deregulated in a *B. subtilis ctsR* mutant and downregulated by the CtsR from *B. subtilis* in the wild type. They proposed CtsR as the master regulator of stress gene expression in *O. oeni*. Indeed, these authors argue that CtsR must play a central role due to the absence in the genome of genes encoding alternative sigma factors and the other known repressor HrcA and its cognate CIRCE (controlling inverted repeat of chaperone expression) operator sequence. Other genes such as *dnaK* and *groESL* and genes belonging to the *clp* family appeared to be regulated via CtsR in *O. oeni* (Beltramo et al. 2004b; Grandvalet et al. 2005). A unique feature of the *O. oeni* response is the fact that the *dnaK* and *groESL* genes are regulated by CtsR but not by HrcA.

The *clpP-clpL* locus has been extensively studied by Beltramo et al. (2004b). They demonstrated that *clpP* and *clpL* could be expressed as two distinct transcripts or as a large cotranscript. The *clpP* gene appears to belong to the CtsR regulon, but not the *clpL* gene. It has been suggested that the regulation of *clpL* expression is at the post-transcriptional level involving an mRNA stability process, as discussed later.

The expression of *ctsR-clpC* also appeared to be dependent on CtsR. A CtsR-binding site in the promoter region was identified (Grandvalet et al. 2005). The regulation of CtsR stability as a function of stress conditions has been proposed in *B. subtilis.* CtsR is degraded under stress by the ClpP-ClpC complex with the help of MscA and MscB, two modulator proteins with phosphorylation-dependent activity (Kirstein et al. 2007). Although the regulation of many stress genes in *O. oeni* is CtsR-dependent, there are no data available on the mechanism involved in the process of deregulation under stress conditions. The organization of *ctsR* together with *clpC* in an operon structure suggests that the product of *clpC* might induce stress gene expression by acting with other protein partners that still have to be identified.

O. oeni is subjected to multiple stresses in wine, in particular, the presence of ethanol and acidic pH. The signal-perception mechanisms of cell damage that induces stress gene expression appear complex and diverse. Recently, the induction of the *hsp18* gene in *O. oeni* cells incubated with benzyl alcohol suggested that changes in membrane fluidity may trigger signal transduction, inducing stress gene expression (Coucheney et al. 2005b). *ctsR* gene expression is also induced in the presence of benzyl alcohol (Stéphanie Weidmann, personal communication). The intermediate factors involved in this signal transduction still have to be investigated. These findings may improve our understanding of stress tolerance in *O. oeni*. A rapid induction of stress genes is required for survival. This involves an efficient system focused on the central regulator CtsR. Perception and transduction are essential steps in the adaptation to adverse environmental conditions. This applies to *O. oeni* following direct inoculation into wine and immediate confrontation with stress conditions.

15.3.3 Other Putative Regulation Mechanisms

Among the stress genes characterized in *O. oeni*, *trxA*, which encodes thioredoxin, appeared to be induced by heat and oxidative stress (Jobin et al. 1999a). Thioredoxin is known to be involved in various metabolic processes, such as nucleic acid biosynthesis and cell wall synthesis. Analysis of the DNA sequence upstream from the coding part of the *trxA* gene in *O. oeni* has not revealed the signature of any known regulation system. Consequently, there is no current information on the mechanism of expression induction for the *trxA* gene in *O. oeni*. The *clpX* gene, which encodes a regulatory ATPase subunit of ClpP, has a housekeeping promoter like *trxA* (Jobin et al. 1999b). No CtsR-binding site or CIRCE sequence was noted. A 5′UTR sequence of about 400 nucleotides may be involved in posttrancriptional regulation. A similar but shorter 5′UTR sequence is also present in the upstream region of the *clpL* gene. Northern blot and RT-qPCR suggested that instability of the *clpL* messenger was a function of growth conditions. Indeed, RT-qPCR using the 16S rRNA transcript as an internal control showed that the half-life of the *clpL* transcript was about 1 min in the exponential growth phase, 3 min under heat shock, and at least 10 min in the stationary phase. Thus, instability of the *clpL* mRNA appears to act as a form of posttranscriptional regulation (Beltramo et al. 2004b).

15.4 Heat-Shock Proteins

HSPs are known to act as molecular chaperones (Bukau et al. 2006). Such proteins are highly conserved in all free-living organisms. They have a variety of functions in cell biology (Watson 1990). They are involved in nascent polypeptide folding, transit across membranes, refolding of misfolded or aggregated proteins, targeting for proteolysis, and regulation of protein conformational changes (signaling). Other HSPs exhibit protease activity, for example, FtsH protein. The *ftsH* gene from *O. oeni* was cloned and characterized (Bourdineaud et al. 2003). The production of this protein in an *E. coli ftsH* mutant made *E. coli* resistant to wine toxicity, and the authors concluded that this protease may constitute a well-conserved stress-protective device. The 18-kDa sHSP Lo18 has been extensively studied at the physiological and functional levels in *O. oeni*. sHSPs are ubiquitous proteins and are similar to the α-crystallin protein of the vertebrate eye lens. The main properties of sHSPs are the following: a monomer ranging in size from 12 to 43 kDa; a conserved domain of about 100 amino acid residues called the α-crystallin domain; a large oligomeric structure with conserved structural organization; and multiple cellular localization (Nakamoto and Vigh 2007).

15.4.1 Induction Process

Lo18 is strongly induced by heat (42°C), ethanolic (12% v/v) or acidic (pH 3.0) shocks, and during the stationary growth phase in *O. oeni* (Guzzo et al. 1997).

Labeling cells with S^{35} methionine and separation of total proteins by bidimensional electrophoresis followed by autoradiography revealed the induction process. Using antiserum against Lo18, a rapid synthesis of Lo18 under heat shock was observed (Jobin et al. 1997). Moreover, induction also occurred after the addition of sulfite to the medium (Guzzo et al. 1998).

Lo18 appears to be induced by various stresses. Thus, this sHSP was considered to be a good stress-response marker for *O. oeni*. As suggested by Coucheney et al. (2005a), a high level of Lo18 may correspond to the start of an adaptation process enabling *O. oeni* to survive better following direct inoculation into wine.

15.4.2 Cellular Localization and In Vitro Activity

One of the most notable features of sHSPs is their diverse cellular localization, which enables various roles to be fulfilled. Jobin et al. (1997) showed that Lo18 is peripherally associated with the membrane of *O. oeni*. However, Lo18 is also detected in the cytoplasmic fraction. In fact, the membrane association of Lo18 depends on an increase in temperature. Repartition between the two fractions appeared to be the same up to 44°C, while after a heat shock at 46°C, more Lo18 was found in the membrane fraction (Delmas et al. 2001) (Fig. 15.3).

Immunocytochemistry using specific purified antibodies against Lo18 enabled the protein to be located in situ and the pattern of distribution in *O. oeni* cells after heat shock to be established. After heat shock at 42°C, most of the labeling was found in the cytoplasm and some of the gold particles were located at the cell periphery (Fig. 15.4).

Delmas et al. (2001) demonstrated that Lo18 prevents the in vitro thermal aggregation of citrate synthase at 45°C and lactate dehydrogenase (LDH) at 60°C in the absence of ATP. However, Lo18 appeared unable to reactivate LDH, suggesting that other components may be needed to refold LDH into an active conformation.

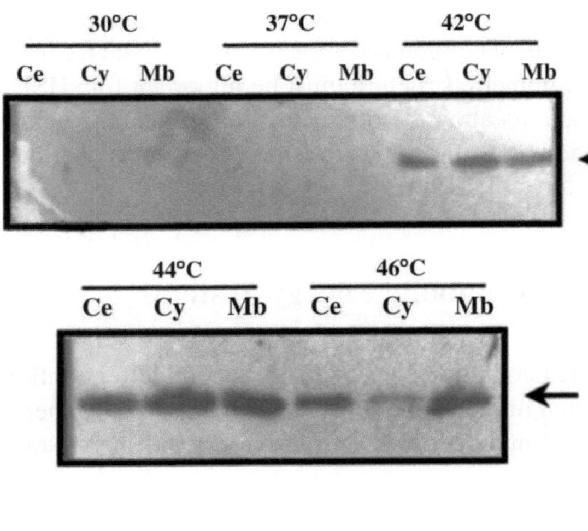

Fig. 15.3 Immunoblot analysis of Lo18 in *O. oeni*-fractionated cells, after heat shock following a rise in temperature. *Ce* cellular extract; *Cy* cytoplasmic fraction; *Mb* membrane fraction (Delmas et al. 2001)

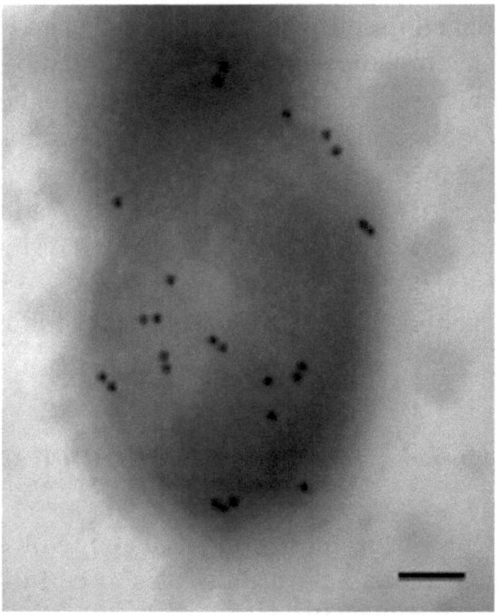

Fig. 15.4 Immunolocalization of Lo18 in *O. oeni* cells using polyclonal antibodies against Lo18 followed by secondary antibodies conjugated to 15-nm gold particles. Cells were heat-shocked at 42°C (Coucheney et al. 2005b)

Thus, Lo18 located in the cytoplasm may, in conjunction with an ATP-dependent chaperone system, assist denaturated polypeptides (Lee and Vierling 2000). The putative role of Lo18 at the membrane level will be described later on, in Sect. 15.6.2.

15.4.3 Multimeric Structure of the Small Heat-Shock Protein Lo18

A common characteristic of sHSPs is their ability to form oligomeric structures (Giese and Vierling 2004). Several systems of gene expression and protein purification were used to produce and characterize in vitro Lo18 at the biochemical level (Delmas et al. 2001; Coucheney et al. 2005b). Results obtained from size exclusion chromatography and chemical cross-linking suggested that Lo18 forms multimeric structures. A dimer, a trimer, and higher oligomeric forms were observed. There is evidence that oligomerization is a prerequisite for in vitro chaperone activity; hence, the oligomeric structure must be important for sHSPs in vivo. In the case of Lo18, the dual location in both the cytoplasm and membrane fractions may suggest various activities in cells under stress conditions and may involve different oligomeric structures of sHSPs.

15.5 Metabolic Energy Systems

Growth conditions in wine after alcoholic fermention are stringent, with a lack of residual fermentative sugar. *O. oeni* thus has to generate metabolic energy in various ways. A number of solute transporters and energy transduction systems are involved in regulating the change in internal and external pH to enable the survival of *O. oeni* cells. The enzymatic activities that are most characterized to date are the membrane-bound F_1F_0-ATPase (H^+-ATPase), which functions as a proton exclusion pump, and the malolactic system, with the transport of malate and its decarboxylation (Salema et al. 1996).

15.5.1 ATPase Activity

Low pH appears to limit bacterial growth in wine. H^+-ATPase is known for its key role in acid tolerance by bacteria. Its induction has been associated with increased resistance to low pH (Kobayashi et al. 1986). Fortier et al. (2003) studied profiles of *O. oeni* H^+-ATPase activity in various growth conditions. They observed that *O. oeni* cells growing at low pH (pH 3.5) had 1.6 times the H^+-ATPase activity of cells grown at pH 5.3.

The pH of the growth medium was shown to be stable in the presence of malic acid, but a decrease in pH from 5.3 to 3.9 occured during growth in the absence of malic acid, with H^+-ATPase activity increasing by a factor of 1.5. Partial cloning of the genes encoding the H^+-ATPase β- and ε-subunits suggested the existence of a typical F_1F_0-ATPase genetic organization in *O. oeni*. Slot blots were used to detect *atp* mRNA. Cells subjected to acid shock were shown to contain a higher level of *atp* mRNA than cells grown at pH 5.3. These results indicate that H^+-ATPase is induced at low pH in *O. oeni* and that regulation occurs at the transcription level; they are in agreement with those obtained with cells grown in a synthetic wine medium (Beltramo et al. 2006). RT-qPCR was used to quantify gene expression, and the relative expression level of the *atpB* gene increased about 3 times. These studies support the role of the product of this gene in the regulation of cytoplasmic pH and acid tolerance of *O. oeni*. The characterization of a spontaneous neomycin-resistant mutant deficient in H^+-ATPase activity and sensitive to acidic environments confirmed the major role played by this membrane-associated protein in *O. oeni* acid tolerance (Tourdot-Maréchal et al. 1999).

Surprisingly, the H^+-ATPase-deficient mutants were also devoid of malolactic activity. Transcriptional analysis of the malolactic operon (*mleA-mleP*) and the putative regulator *mleR* revealed that the genes at this locus were not transcribed in the H^+-ATPase-deficient mutant. This link between the H^+-ATPase and malolactic systems was not present in other LAB such as *Lactococcus lactis* and *Leuconostoc mesenteroides* and may be specific to bacteria with acidophilic behavior (Galland et al. 2003). Carreté et al. (2002) showed that certain compounds, such as SO_2, fatty acids, and copper, may reduce ATPase activity with a decrease in growth rate and delayed MLF.

15.5.2 The Malolactic System

MLF is the decarboxylation of L-malate, giving L-lactate and CO_2. This reaction requires the malolactic enzyme encoded by the *mleA* gene and a malate permease encoded by the *mleP* gene (Labarre et al. 1996). The free energy generated by MLF is conserved by a chemiosmotic mechanism (Salema et al. 1996). Two events, (1) the transport of a monoanionic form of L-malate resulting in a net translocation of one charge, and (2) the consumption of one proton during decarboxylation, generate a membrane potential ($\Delta\psi$) and a pH gradient (ΔpH), respectively. The proton motive force (PMF) derived from both processes is used to control the internal pH and to optimize the uptake of nutrients. *O. oeni* is unable to grow on malate as the sole substrate, but malate metabolism stimulates the growth of *O. oeni* in acidic conditions and participates in the adaptation of the bacteria to conditions in wine. Moreover, L-malate conversion to the monoacid form (L-lactate) results in an increase in wine pH.

15.5.3 Other Key Metabolic Processes

Citrate metabolism may affect the acid tolerance of *O. oeni* (Ramos et al. 1995). According to Augagneur et al. (2007), citrate was not directly responsible for the inhibition observed in cultures at low pH, but probably its end products. Other authors have studied the transcriptional response of the citrate pathway genes of *O. oeni* in the presence of ethanol and at low pH (Olguin et al. 2009). The expression of these genes was mainly affected by ethanol. The differences in gene expression under stress were related to a variation in end-product content such as acetic acid and diacetyl. These variations have to be taken into account in the development of wine organoleptic characteristics.

The decarboxylation of organic acids or amino acids with the consumption of a proton may assist in stabilizing the internal pH in *O. oeni*. Moreover, the arginine deiminase pathway is known to enhance acid tolerance by converting arginine into an alkaline product and by increasing external pH (Tonon et al. 2001).

The composition of the nitrogen source in wine may cause deprivation conditions that impair the growth of *O. oeni*. This bacterium exhibits multiple auxotrophies; therefore, the uptake of peptides and amino acids from the medium is crucial (Remize et al. 2006). Recent work showed protease activity in the supernatant of *O. oeni* strains (Remize et al. 2005) and demonstrated the effect of yeast metabolism on the qualitative aspects of nitrogen source composition (Alexandre et al. 2004; Guilloux-Bénatier et al. 2006). However, nitrogen metabolism in *O. oeni* in relation to the diversity and complexity of the composition of wine still has to be studied. The way in which low-molecular-weight peptides are assimilated as a nitrogen source by *O. oeni* remains unclear. Nevertheless, this appears to be essential to understanding MLF (Ritt et al. 2008). Another interaction between yeast and bacteria concerns the production of peptides by yeasts that inhibit the growth of *O. oeni* and consequently MLF. This inhibitory effect seems to be strain-dependent (Osborne and Edwards 2007).

O. oeni in wine may be confronted with another kind of stress due to phenolic compounds. These molecules contribute to the sensory characteristics of wine but may inhibit the growth of microorganisms, especially *O. oeni* (Figueiredo et al. 2008).

15.6 Membrane Integrity

Stress conditions in wine may damage the membrane integrity, decrease cell viability, and delay the start of MLF. The maintenance of a semipermeable barrier is required to preserve membrane function, which is essential to control ionic permeability and to regulate solute exchange between the cell and the external medium. The fatty acid composition of the *O. oeni* membrane is modified in response to stress conditions. One of the principal effects of this change is the regulation of membrane fluidity, which maintains membrane function. Garbay and Lonvaud-Funel (1996) showed that the ratio of phospholipids to proteins in the membrane decreased in *O. oeni* cells submitted to stress conditions. The modification of this ratio may limit the area for passive diffusion of ions and preserve the semipermeability of the membrane. Membrane fluidity regulation in bacteria in response to stress seems to be more complex and involves the universal chaperone GroEL (Török et al. 1997) and sHSPs (Török et al. 2001). The role of the sHSP Lo18, which is encoded by the *hsp18* gene, in the regulation of membrane fluidity has been studied in *O. oeni*.

15.6.1 Modification of the Fatty Acid Composition

Changes in membrane fatty acid composition occur during the transition from the exponential phase to the stationary phase. The principal modification is the disappearance of *cis*-vaccenic acid in favor of lactobacillic acid (Garbay et al. 1995). This cyclization of unsaturated fatty acids stabilizes the membrane and enables cells to physiologically adapt to stress conditions.

Many studies focused on the deleterious effect of ethanol on membrane integrity. This physicochemical stress is of primary importance in the survival of *O. oeni* in wine and consequently has been extensively studied. Using a proteomic approach, Da Silveira et al. (2004) demonstrated an active ethanol adaptation response of *O. oeni* at the cytoplasmic and membrane protein levels. Interestingly, they noticed an increase in the level of proteins known to be involved in cell wall biosynthesis. This suggests a new line of investigation involving the regulation of ethanol adaptation via the cell wall.

Dols-Lafargue et al. (2008) showed that *O. oeni* strains harboring a functional *gtf* gene involved in polysaccharide synthesis are more resistant to several of the stresses occurring in wine (alcohol, pH, and SO_2).

Ethanol affects the physical state and biological functions of the cell membrane. It interacts with membranes at the lipid–water interface, weakening the hydrophobic barrier to the free exchange of polar molecules. The fluidity of the cytoplasmic membrane in *O. oeni* cells increased instantaneously after ethanol shock (10–14% v/v). This fluidizing effect was transitory (Chu-Ky et al. 2005), but low cell viability was observed.

Other work on cells grown in the presence of 8% (v/v) ethanol highlighted the modification of the membrane fatty acid profile. The authors observed an increase in membrane cyclopropane fatty acid (CFA) that may counteract the effect of ethanol on membrane fluidity (Teixeira et al. 2002; Grandvalet et al. 2008). A decrease in the oleic acid content was observed. CFA may reduce the effects of ethanol on the membrane fluidity because cyclopropane rings restrict the overall mobility and disorder of the acyl chains more than the *cis* double bonds.

Changes in membrane fatty acid composition are less documented in the case of acidic stress. However, recent work (Grandvalet et al. 2008) demonstrated similar, but smaller, changes in the membrane fatty acid profile for *O. oeni* cells grown in medium with a pH of 3.5 in comparison to ethanol-grown cells. An increase in CFA and a decrease in oleic acid content were also noticed. Chu-Ky et al. (2005) showed that acid shock (pH 3.0) exerted a rigidifying effect on the membrane without affecting cell viability.

Membrane CFA also increases in cells entering the stationary phase, and we can therefore conclude that these adjustments in the fatty acid composition appear necessary for *O. oeni* adaptation. Although there is support for the role of CFA in the stress tolerance of *O. oeni,* there is possible variation among the strains regarding fatty acid distribution in response to stress (Da Silveira et al 2003; Drici-Cachon et al. 1996).

The biosynthesis of CFAs from unsaturated fatty acids is catalyzed by the enzyme CFA synthase. The induction of *cfa* gene expression was demonstrated in *O. oeni* cells grown in the presence of ethanol or at low pH and during entry in the stationary growth phase. The increased levels of *cfa* mRNA transcripts correlated with increased amounts of CFA in the membrane. These results suggest that the expression of this gene may be regulated at the transcriptional level (Grandvalet et al. 2008).

In wine, a cumulative effect of different stresses may modify the effect of a single stress on membrane integrity and cell viability (Fig. 15.5). A study on the effects of combined stresses showed that cold (14°C) and ethanol (14% v/v) shocks

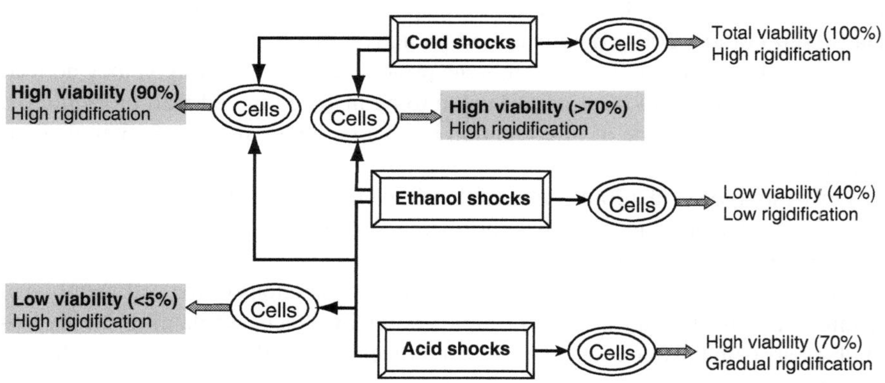

Fig. 15.5 Diagram of the effects of single and combined cold, ethanol, and acid shocks on the physical state of the membrane and the cell viability of *O. oeni* (Chu-Ky et al. 2005)

applied together resulted in strong membrane rigidification and better survival than a single ethanol shock. However, combined acid (pH 3.5) and ethanol (10% v/v) shocks induced rigidification but also high cell mortality (Chu-Ky et al. 2005).

15.6.2 *Membrane Stabilization by Small Heat-Shock Proteins*

Following the fractionation of *O. oeni* heat-shocked cells, a significant quantity of the sHSP Lo18 was detected in the membrane fraction (Jobin et al. 1997). This was confirmed after heat shocks at various temperatures (42, 44, and 46°C) in the presence of ethanol and a fluidizing agent, benzyl alcohol. These stress conditions all had a fluidizing effect on the membrane. A role in the regulation of membrane fluidity was therefore envisaged, as demonstrated for other HSPs (Török et al. 1997). This association of sHSPs with the membrane may normalize the fluidity and prevent the formation of membrane damaging non-bilayer lipid structures (Nakamoto and Vigh 2007).

The *hsp18* gene was introduced into *E. coli* using the pET28a expression vector for the overproduction and purification of the protein (Coucheney et al. 2005b). The purified Lo18 was incubated with liposomes formed from lipids extracted from *O. oeni.* Using fluorescence anisotropy of diphenylhexatriene (DPH) and generalized polarization of Laurdan, it was demonstrated that Lo18 interacts with liposomes and increases the molecular order of the lipid bilayer when the temperature reaches 33.8°C. Thus, it has been suggested that Lo18 contributes to the maintenance of membrane integrity under stress conditions. Further investigations are being carried out to identify the structural domains of Lo18 involved in the stabilizing effect on proteins and membranes and the relationship with the oligomerization process.

15.7 Concluding Remarks

The molecular and physiological aspects of the stress responses of *O. oeni* in relation to its capacity to grow in a hostile environment like wine have been investigated. The regulation of stress genes and, in some cases, the function of their products have been successfully studied, but only for a limited number of genes. Moreover, these genes had previously been studied in other bacteria. Nonetheless, this approach enabled the development of several interesting tools, presently useful to select and characterize *O. oeni* strains isolated from wine (Coucheney et al. 2005a). The maintenance of the functional components associated with the extracytoplasmic compartments of cells also appears to contribute to the process of adaptation to wine. Data on the capacity of the strain to regulate membrane fluidity has provided information on the adaptive response. Another method based on the active extrusion of the fluorescent probe carboxy fluorescein has also been used to assess malolactic activity in ethanol-stressed *O. oeni* cells (Da Silveira and Abee 2009).

However, cell wall biosynthesis and exopolysaccharide production require investigation at the molecular and physiological levels (Ibarburu et al. 2007; Da Silveira et al. 2004).

Despite the complexity of the cellular mechanisms involved and the phenotypic variation in wine resistance in *O. oeni* isolates, the results obtained for acquired stress tolerance in *O. oeni* have helped improve the industrial production of a malolactic starter culture for direct inoculation into wine.

Unfortunately, the lack of genetic tools has hindered the analysis of *O. oeni*. However, the development of plasmid delivery (Beltramo et al. 2004a) and efficient electroporation methods (Assad-Garcia et al. 2008) should remove these barriers in the near future.

Another characteristic of the *Oenococcus* genus is its genome plasticity. Multilocus sequence typing clearly revealed a high level of allelic diversity in *O. oeni* (De las Rivas et al. 2004). Two genome sequences for *O. oeni* are currently available, *O. oeni* strain PSU-1 (Mills et al. 2005) and *O. oeni* strain ATCC BAA-1163 (J. Guzzo, A. Lonvaud et al., unpubl. data). On the basis of recent genomic analysis, Marcobal et al. (2008) showed that *O. oeni* lacks the ubiquitous mismatch-repair system genes *mutS* and *mutL* (Matic et al. 1995), which normally contribute to lowering the rate of spontaneous mutations and recombinations. The loss of such genes suggests that the genus *Oenococcus* is hypermutable. The authors proposed that this explains the observed level of allelic polymorphism in known *O. oeni* isolates and probably contributes to the unique adaptation of this genus to acidic and alcoholic environments.

Genomic subtractive hybridization between two isolates with differing enological potentials was used to elucidate the genetic basis of intraspecific diversity and identify novel genes involved in the adaptation to wine (Bon et al. 2009). This approach demonstrated that events such as recombination, insertion/deletion, and the acquisition of novel functions by horizontal transfer may contribute to the genomic diversity observed in *O. oeni*. These results on genome plasticity suggest that *O. oeni* has various strategies to adapt to restrictive environments. The comparative genomic is a high-performance tool enabling particular sequences to be related to enological potential. Nevertheless, the presence of a specific gene has to be related not only to efficient expression under stress conditions but also to the activity of the gene product in a technological situation, and this will be a big challenge.

References

Alexandre H, Costello PJ, Remize F, Guzzo J, Guilloux-Benatier M (2004) *Saccharomyces cerevisiae-Oenococcus oeni* interactions in wine: current knowledge and perspectives. Int J Food Microbiol 93:141–154

Assad-Garcia, JS, Guzzo J, Grandvalet C (2008) An improved protocol for electroporation of *oenococcus oeni* ATCC BAA-1163 using ethanol as immediate membrane fluidizing agent. Lett Appl Microbiol 47:333–338

Augagneur Y, Ritt JF, Linares DM, Remize F, Tourdot-Maréchal R, Garmyn D, Guzzo J (2007) Dual effect of organic acids as a function of external pH in *Oenococcus oeni*. Arch Microbiol 188:147–157

Bartowsky EJ (2005) *Oenococcus oeni* and malolactic fermentation – moving into the molecular arena. Austral J Grape Wine Res 11:174–187

Beltramo C, Oraby M, Bourel G, Garmyn D, Guzzo J (2004a) A new vector, pGID052, for genetic transfer in *Oenococcus oeni*. FEMS Microbiol Lett 236:53–60

Beltramo C, Grandvalet C, Guzzo J (2004b) Evidence for multiple levels of regulation of *Oenococcus oeni clpP-clpL* locus expression in response to stress. J Bacteriol 186:2200–2205

Beltramo C, Desroche N, Tourdot-Maréchal R, Grandvalet C, Guzzo J (2006) Real-time PCR for characterizing the stress response of *Oenococcus oeni* in a wine-like medium. Res Microbiol 157:267–274

Bon E, Delaherche A, Bilhère E, De Daruvar, A, Lonvaud-Funel A, Le Marrec C (2009) *Oenococcus oeni* genome plasticity is associated with fitness. Appl Environ Microbiol 75:2079–2090

Bourdineaud JP, Nehmé B, Tesse S, Lonvaud-Funel A (2003) The *ftsH* gene of the wine bacterium *Oenococcus oeni* is involved in protection against environmental stress. Appl Environ Microbiol 69:2512–2520

Bourdineaud JP, Nehmé B, Tesse S, Lonvaud-Funel A (2004) A bacterial gene homologous to ABC transporters protect *Oenococcus oeni* ethanol and other stress factors in wine. Int J Food Microbiol 92:1–14

Bukau B, Weissman J, Horwich A (2006) Molecular chaperones and protein quality control. Cell 125:443–451

Carreté R, Vidal MT, Bordons A, Constanti M (2002) Inhibitory effect of sulfur dioxide and other stress compounds in wine on the ATPase activity of *Oenococcus oeni*. FEMS Microbiol Lett 211:155–159

Cavin JF, Prevost H, Lin J, Schmitt P, Divies C (1989) Medium for screening *Leuconostoc oenos* strains defective in malolactic fermentation. Appl Environ Microbiol 62:1274–1282

Chu-Ky S, Tourdot-Marechal R, Marechal PA, Guzzo J (2005) Combined cold, acid, ethanol shocks in *Oenococcus oeni*: effects on membrane fluidity and cell viability. Biochim Biophys Acta 1717:118–124

Coucheney F, Desroche N, Bou M, Tourdot-Maréchal R, Dulau L, Guzzo J (2005a) A new approach for selection of *Oenococcus oeni* strains in order to produce malolactic starter. Int J Food Microbiol 105:463–470

Coucheney F, Gal L, Beney L, Lherminier J, Gervais P, Guzzo J (2005b) A small HSP, Lo18 interacts with the cell membrane and modulates lipid physical state under heat shock conditions in a lactic acid bacterium. Biochim Biophys Acta 1720:92–98

Da Silveira MG, Abee T (2009) Activity of ethanol-stressed *Oenococcus oeni* cells: a flow cytometric approach. J Appl Microbiol 106:1690–1696

Da Silveira MG, Golovina EA, Hoekstra FA, Rombouts FM, Abee T (2003) Membrane fluidity adjustments in ethanol-stressed *Oenococcus oeni* cells. Appl Environ Microbiol 69:5826–5832

Da Silveira MG, Buamgärtner M, Rombouts FM, Abee T (2004) Effect of adaptation to ethanol on cytoplasmic and membrane protein profiles of *Oenococcus oeni*. Appl Env Microbiol 70: 2748–2755

De las Rivas B, Marcobal A, Munoz R (2004) Allelic diversity and population structure in *Oenococcus oeni* as determined from sequence analysis of housekeeping genes. Appl Environ Microbiol 70:7210–7219

Delmas F, Diviès C, Guzzo J (2001) *In vitro* and *in vivo* studies of the membrane associated small heat shock protein Lo18 from the lactic acid bacterium *Oenococcus oeni*. J Mol Microbiol Biotechnol 3:601–610

Derré I, Rapoport G, Msadek T (1999) CtsR, a novel regulator of stress and heat shock response, controls *clp* and molecular chaperone gene expression in Gram-positive bacteria. Mol Microbiol 31:117–131

Dicks LM, Dellaglio F, Collins MD (1995) Proposal to reclassify *Leuconostoc oenos* as *Oenococcus oeni*. Int J Syst Bacteriol 45:395–397

Dols-Lafargue M, Lee HY, Le Marrec C, Heyraud A, Chambat G, Lonvaud-Funel A (2008) Characterization of *gtf*, a glucosyltransferase gene in the genome of *Pediococcus parvulus* and *Oenococcus oeni*, two bacterial species commonly found in wine. Appl Environ Microbiol 74:4079–4090

Drici-Cachon Z, Cavin JF, Divies C (1996) Effect of pH and age of culture on cellular fatty acid composition of *Leuconostoc oenos*. Lett Appl Microbiol 22:331–334

Figueiredo AR, Campos F, de Freitas V, Hogg T, Couto JA (2008) Effect of phenolic aldehydes and flavonoids on growth and inactivation of *Oenococcus oeni* and *Lactobacillus hilgardii*. Food Microbiol 25:105–112

Fortier LC, Tourdot-Maréchal R, Lee BH, Diviès C, Guzzo J (2003) Induction of *Oenococcus oeni* H^+-ATPase activity and mRNA transcription under acidic conditions of growth. FEMS Microbiol Lett 222:165–169

Galland D, Tourdot-Maréchal R, Abraham M, Chu Ky S, Guzzo J (2003) Absence of malolactic activity is a characteristic of H^+-ATPase deficient mutants from the lactic acid bacterium *Oenococcus oeni*. Appl Environ Microbiol 69:1973–1979

Garbay S, Lonvaud-Funel A (1996) Response of *Leuconostoc oenos* to environmental changes. J Appl Bacteriol 81:619–625

Garbay S, Rozès N, Lonvaud-Funel A (1995) Fatty acid composition of *Leuconostoc oenos*, incidence of growth conditions and relationship with malolactic efficiency. Food Microbiol 12:387–395

Giese KC, Vierling E (2004) Mutants in a small heat shock protein that affect the oligomeric state. J Biol Chem 279:32674–32683

Grandvalet C, Coucheney F, Beltramo F, Guzzo J (2005) CtsR is the master regulator of stress response gene expression in *Oenococcus oeni*. J Bacteriol 187:5614–5623

Grandvalet C, Assad-Garcia JS, Chu Ky S, Tollot M, Guzzo J, Gresti J, Tourdot-Maréchal R (2008) Changes in membrane lipid composition in ethanol and acid adapted *Oenococcus oeni* cells: characterization of the CFA gene by heterologous recombination. Microbiology 154:2611–2619

Guilloux-Bénatier M, Remize F, Gal L, Guzzo J, Alexandre H (2006) Effects of yeast proteolytic activity on *Oenococcus oeni* and malolactic fermentation. FEMS Microbiol Lett 263:183–188

Guzzo J, Desroche N (2009) Physical and chemical stress factors in lactic acid bacteria. In: König H, Unden G, Fröhlich J (Eds.), Biology of microorganisms on grapes, in must and in wine. Springer-Verlag, Berlin, pp. 295–308

Guzzo J, Cavin JF, Divies C (1994) Induction of stress proteins in *Leuconostoc oenos* to perform direct inoculation of wine. Biotechnol Lett 16:1189–1194

Guzzo J, Delmas F, Pierre F, Jobin MP, Samyn B, Van Beeumen J, Cavin JF, Diviès C (1997) A small heat shock protein from *Leuconostoc oenos* induced by multiple stresses and during stationary growth phase. Lett Appl Microbiol 24:393–396

Guzzo J, Jobin MP, Diviès C (1998) Increase of sulfite tolerance in *Oenococcus oeni* by means of acidic adaptation. FEMS Microbiol Lett 160:43–47

Ibarburu I, Soria-Diaz ME, Rodriguez-Carvajal MA, Velasco SE, Tejero-Mateo P, Gil-Serrano AM, Irastorza A, Duenas MT (2007) Growth and exopolysaccharide (EPS) production by *Oenococcus oeni* I4 and structural characterization of their EPSs. J Appl Microbiol 103:477–486

Jobin MP, Delmas F, Garmyn D, Diviès C, Guzzo J (1997) Molecular characterization of the gene encoding a 18-kDa small heat shock protein associated with the membrane of *Leuconostos oenos*. Appl Env Microbiol 63:609–614

Jobin MP, Garmyn D, Diviès C, Guzzo J (1999a) Expression of the *Oenococcus oeni trxA* gene is induced by hydrogen peroxide and heat shock. Microbiology 145:1245–1251

Jobin MP, Garmyn D, Diviès C, Guzzo J (1999b) The *Oenococcus oeni clpX* homologue is a heat shock gene preferentially expressed in exponential growth phase. J Bacteriol 181:6634–6641

Kirstein J, Dougan DA, Gerth U, Kecker M, Turgay K (2007) The tyrosine kinase McsB is a regulated adaptor protein for ClpCP. EMBO J 26:2061–2070

Kobayashi H, Suzuki T, Unemoto T (1986) Streptococcal cytoplasmic pH is regulated by changes in amount and activity of a proton-translocating ATPase. J Biol Chem 261:627–630

Kunkee RE (1991) Some roles of malic acid in the malolactic fermentation in wine making. FEMS Microbiol Lett 88:55–71

Labarre C, Guzzo J, Cavin JF, Diviès C (1996) Cloning and characterization of the genes encoding the malolactic enzyme and the malate permease of *Leuconostoc oenos*. Appl Environ Microbiol 62:1274–1282

Lee GJ, Vierling E (2000) A small heat shock protein cooperates with heat shock protein 70 systems to reactivate a heat-denaturated protein. Plant Physiol 122:189–198

Marcobal AM, Sela DA, Wolf YI, Makarova KS, Mills DA (2008) Role of hypermutability in the evolution of the genus *Oenococcus*. J Bacteriol 190:564–570

Matic I, Rayssiguier C, Radman M (1995) Interspecies gene exchange in bacteria: the role of SOS and mismatch repair systems in evolution of species. Cell 80:507–515

Mills DA, Rawsthorne H, Parker C, Tamir D, Makarova K (2005) Genomic analysis of *Oenococcus oeni* PSU-1 and its relevance to winemaking. FEMS Microbiol Lett 29:465–475

Nakamoto H, Vigh L (2007) The small heat shock proteins and their clients. Cell Mol Life Sci 64:294–306

Olguin N, Bordons A, Reguant C (2009) Influence of ethanol and pH on the gene expression of the citrate pathway in *Oenococcus oeni*. Food Microbiol 26:197–203

Osborne JP, Edwards CG (2007) Inhibition of malolactic fermentation by a peptide produced by *Saccharomyces cerevisiae* during alcoholic fermentation. Int J Food Microbiol 118:27–34

Ramos A, Lolkema JS, Konings, WN, Santos H (1995) Enzyme basis for pH regulation of citrate and pyruvate metabolism by *Leuconostoc oenos*. Appl Environ Microbiol 61:1303–1310

Remize F, Augagneur Y, Guilloux-Bénatier M, Guzzo J (2005) Effect of nitrogen limitation and nature of the feed upon *Oenococcus oeni* metabolism and extracellular protein production. J Appl Microbiol 98:652–661

Remize F, Gaudin A, Kong Y, Guzzo J, Alexandre H, Krieger S, Guilloux-Bénatier M (2006) *Oenococcus oeni* preference for peptides: qualitative and quantitative analysis of nitrogen assimilation. Arch Microbiol 185:459–469

Ritt JF, Guilloux-Benatier M, Guzzo J, Alexandre H, Remize F (2008) Oligopeptides assimilation and transport by *Oenococcus oeni*. J Appl Microbiol 104:573–580

Salema M, Lolkema JS, San Romao MV, Loureiro Dias MC (1996) The proton motive force generated in *Leuconostoc oenos* by malate fermentation. J Bacteriol 178:3127–3132

Teixeira H, Gonçalves MG, Rozès N, Ramos A, San Romao MV (2002) Lactobacillic acid accumulation in the plasma membrane of *Oenococcus oeni*: a response to ethanol stress. Microbiol Ecol 43:146–153

Tonon T, Bourdineaud JP, Lonvaud-Funel A (2001) The *arcABC* gene cluster encoding the arginine deiminase pathway of *Oenococcus oeni*, and arginine induction of a CRP-like gene. Res Microbiol 152:653–661

Török Z, Horvath I, Goloubinoff P, Kovacs E, Glatz A, Balogh G, Vigh L (1997) Evidence for a lipochaperonin: association of active protein-folding GroESL oligomers with lipids can stabilize membranes under heat shock conditions. Proc Natl Acad Sci USA 94:2192–2197

Török Z, Goloubinoff P, Horvath I, Tsvetkova NM, Glatz A, Balogh G, Varvasovszki V, Los DA, Vierling E, Crowe JH, Vigh L (2001) *Synechocystis* HSP17 is an amphitropic protein that stabilizes heat-stresses membranes and binds denatured proteins for subsequent chaperone-mediated refolding. Proc Natl Acad Sci USA 98:3098–3103

Tourdot-Maréchal R, Fortier LC, Guzzo J, Lee B, Diviès C (1999) Acid sensitivity of neomycin resistant mutants of *Oenococcus oeni*: a relationship between reduction of ATPase activity and lack of malolactic activity. FEMS Microbiol Lett 178:319–326

Watson K (1990) Microbial stress proteins. Adv Microbial Physiol 31:183–223

Part IV
Current Applications and Future Aspects

Chapter 16
Engineering Robust Lactic Acid Bacteria

Peter A. Bron, Hermien van Bokhorst-van de Veen, Michiel Wels, and Michiel Kleerebezem

16.1 Introduction

The natural habitat of lactic acid bacteria (LAB) varies from plants to animals and humans, including the oral, genital, and gastrointestinal (GI) tracts. LAB have long been thought to be strictly fermentative and to convert sugar to lactic acid as one of the main end-point metabolites. However, more recently it has been shown that the addition of heme to growth media enables aerobic respiration in lactococcal cultures, supporting increased biomass yields without acidification and enhanced stationary-phase survival (Duwat et al. 2001). Analogously, a recent survey confirmed that heme and/or menaquinone could also stimulate respiration in a subset of *Lactobacillus* species (Brooijmans et al. 2009). Nevertheless, LAB-containing fermented food and beverages, including fruits, vegetables, cereal grains, meat, and milk (Hufner et al. 2007; Bachmann et al. 2008), have been used for centuries because the lactic acid produced acts as a preservative due to the pH-lowering effect. Moreover, these bacteria greatly contribute to the flavor and texture of the fermentation end products (Smit et al. 2005). More recently, specific strains of *Lactobacillus* have been associated with health-promoting effects in consumers, including a suppressive effect of *Lactobacillus johnsonii* (Gotteland and Cruchet 2003) and *Lactobacillus acidophilus* (Cocconier et al. 1998) on *Helicobacter pylori* infection, as well as alleviation of lactose intolerance (Martini et al. 1991) and inflammatory bowel disease (O'Mahony et al. 2005). Although the exact numbers depend on the strain and type

PAB and HvB-vdV contributed equally to this chapter.

M. Kleerebezem (✉)
TI Food and Nutrition, Nieuwe Kanaal 9A, 6709 PA Wageningen, The Netherlands

NIZO Food Research, P.O. Box 20, 6710 BA Ede, The Netherlands

Laboratory for Microbiology, Wageningen University, Dreijeplein 10, 6703 HB Wageningen, The Netherlands
e-mail: Michiel.Kleerebezem@nizo.nl

E. Tsakalidou and K. Papadimitriou (eds.), *Stress Responses of Lactic Acid Bacteria*, Food Microbiology and Food Safety, DOI 10.1007/978-0-387-92771-8_16,

of application, it is recommended that probiotic products contain at least 10^7 microorganisms per gram or milliliter (Corcoran et al. 2008). Moreover, by definition, probiotics must reach their target site alive to confer a health benefit to the host (FAO/WHO 2002). Hence, an important prerequisite for the industrial application of these starter and probiotic cultures is their persistence toward the stresses encountered in the industrial pipeline, ranging from temperature, osmotic, and oxidative and/or solvent stress during industrial fermentation to industrial processing stresses such as freeze-drying (Fig. 16.1). For probiotics, the plethora of stresses encountered is even greater because they require survival during the shelf-life of the fermentation end product. Subsequently, more stresses are met during the residence in and the travel through the different parts of the host's GI tract, such as the gastric acidity in the stomach, bile salt and digestive enzyme challenges in the duodenum, a relatively high osmolarity in the colon, as well as stress conditions associated with oxygen gradients that are steep at the mucosal surface, whereas the colonic lumen is virtually anoxic (Kleerebezem et al. 2010).

During the last decade, genome sequencing of LAB and the application of functional genomics have drastically enhanced our insight into this group of industrially important bacteria, specifically their overall molecular makeup, metabolic capacities, evolutionary relatedness, and molecular adaptation to environmental conditions, including those associated with industrial applications and/or their residence in the mammalian GI tract. The fact that starter cultures and probiotics require either metabolic activity to contribute to the taste and texture of the fermentation end products or vitality to exert their in situ beneficial effect on the consumer, respectively, justifies the increasing interest in the molecular mechanisms

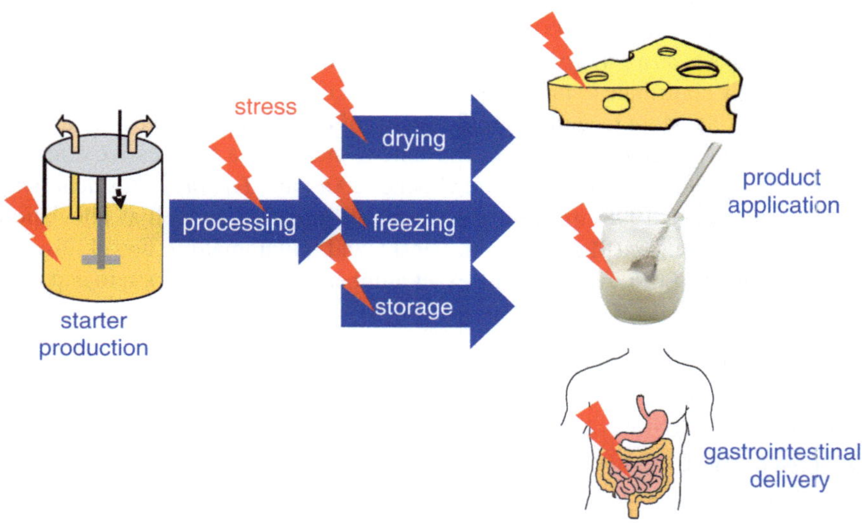

Fig. 16.1 Schematic representation of the various stresses encountered by starter and probiotic cultures in the industrial pipeline

behind the observed stress responses in these bacteria (Fig. 16.1). This chapter describes the state-of-the-art tools available in the postgenomic era to identify specific LAB stress responses. Furthermore, examples will be presented in which the acquired knowledge could be applied rationally to design more robust strains.

16.2 The Wealth of Genomics

16.2.1 Lactic Acid Bacteria Genomics

Following the first publication of the genome sequence of an autonomously growing microbe, *Haemophilus influenzae*, in 1995 (Fleischmann et al. 1995), the field of genomics initially concentrated strongly on the determination of genome sequences of pathogenic bacteria and several model organisms that were traditionally used for molecular research. The genome sequencing of microbes of biotechnological importance, including LAB, lagged behind, but has caught up in the last 5 years. A landmark study in the field of LAB genomics was the release of the complete genome sequence of *Lactococcus lactis* subsp. *lactis* strain IL1403 (Bolotin et al. 2001). This 2.4-Mbp genome was annotated to encode 2,310 proteins, and its comparison to the microbial genomes available at that time confirmed the relatedness of *Lactococcus* to the streptococcal genus and revealed genes predicted to be involved in fermentative and respiratory pathways. Shortly after this first LAB genome, the genome sequences of two lactobacilli, *Lactobacillus plantarum* (Kleerebezem et al. 2003) and *L. johnsonii* (Pridmore et al. 2004), were determined. Their initial comparison already highlighted the relatively high diversity encountered within the genus *Lactobacillus* (Boekhorst et al. 2004), whereas the determination of the genome sequence of *L. acidophilus* (Altermann et al. 2005) underlined the higher degree of similarity within subgroups of the *Lactobacillus* genus, such as the "acidophilus complex." Since these initial LAB genome sequence releases, the database today contains more than 25 genome sequences of industrially important LAB, whereas at least 100 genome-sequencing projects are reported to be ongoing (Siezen and Wilson 2010). Genomic research is thereby contributing enormously to our knowledge of the genetics of this group of organisms. Nevertheless, we should not ignore that many industrially relevant traits of LAB are encoded on mobile genetic elements, such as plasmids and/or transposons, rather than on the chromosome (Rauch and De Vos 1992; Siezen et al. 2005; van Kranenburg et al. 2005), underlining that the sequence determination of these genetic elements should be included to complement the function-blueprint prediction of an organism.

The wealth of genomic sequence information of LAB has stimulated a variety of in silico analyses to compare the available LAB genomes. A landmark study in this field is associated with the release of nine novel LAB genome sequences of various species and their comprehensive comparison, which includes the definition of so-called LaCOGs that represent a refinement of the existing categories of orthologous genes (COGs) dedicated to the LAB genomes. LaCOG distribution analyses were subsequently used to reconstruct an initial view of the evolutionary relationship among LAB (Makarova et al. 2006). The same LaCOG and COG analyses were also employed to identify a set of LAB genes that are associated with various stress responses in these bacteria (Table 16.1). This overview shows that in most LAB HrcA is involved in the control of the heat-shock protein expression (all except *Oenococcus oeni*; see Table 16.1). Notably, for *O. oeni* it has been proposed that the heat-shock and general stress response may be controlled via a complex regulatory network encompassing various regulatory proteins (Guzzo et al. 2000). In agreement with HrcA conservation, the canonical heat-shock proteins that are commonly under HrcA control and perform chaperonin-like functions (GroELS, DnaJK, GrpE) are universally conserved, while the majority of LAB species also encode the additional chaperones HtpX and HSP20 (IbpA). The involvement of CtsR in the regulation of class III stress proteins, including the Clp proteases and related functions, is predicted for all LAB genomes except *Leuconostoc mesenteroides* and the lactobacilli belonging to the "acidophilus complex." Nevertheless, the corresponding Clp proteases (COG associated gene names: ClpA, X, Q, and P) appear to be universally present in these LAB genomes, although ClpYQ's (also designated HslUV) presence varies. The conservation of the oxidative stress components involved in thioredoxin metabolism (TrxA and B) supports an important role for this module in protection against oxidative stress, which has recently been experimentally confirmed in *L. plantarum* (Serrano et al. 2007). All LAB genomes appear to encode a virtually complete machinery associated with DNA-damage stress responses, including the pathways for homologous recombination and double-strand break repair (RecABDFJNOR, RuvAB, and Ssb) and its homology-independent facilitator complex (GyrAB, TopA), as well as the global genome-repair pathway involved in base and nucleotide excision repair (Mfd, UvrABCD, and Xth), although the endonuclease IV (Nfo), which plays a role in base excision repair, appears to be absent in many species. Notably, the canonical DNA-mismatch repair function encoded by *mutS* and *mutL* appears absent from *O. oeni*, while this species as well as *Streptococcus thermophilus* also lack an *recQ*-like gene. The lack of *recQ* in *S. thermophilus* has been proposed to partially explain the genome decay observed in this species, which is characterized by a high frequency of pseudogenes and function loss and is likely due to its extensive adaptation to the benign and nutrient-rich environment encountered during growth in milk (Bolotin et al. 2004; Hols et al. 2005). These examples illustrate how genome sequencing and comparative genomics may accelerate our understanding of the conserved and differential mechanisms underlying LAB stress tolerance and its control. In the next section, this strategy will be further refined, to illustrate how the genome diversity among strains of a species might be exploited to identify chromosomally encoded genes that are involved in functional properties of interest, including stress-tolerance phenotypes.

Table 16.1 Overview of stress-associated genes and COGs (sorted by their degree of conservation among LAB species) predicted in 12 LAB genomes, focusing on variable COG categories

S. thermophilus LMD-9	*L. lactis* subsp. *lactis* IL1403	*L. lactis* subsp. *cremoris* SK-11	*Lb. brevis* ATCC367	*Lb. plantarum* WCFS1	*P. pentosaceus* ATCC25745	*Le. mesenteroides*	*O. oeni* PSU-1	*Lb. johnsonii* NC533	*Lb. gasseri* ATCC33323	*L. bulgaricus* ATCC BAA-365	*Lb. casei* ATCC334	Number of organisms	Number of proteins	COG Functional group	COG gene	Corresponding COG	COG annotation
2	2	5	1	3	2	1	1	1	1	2	3	12	24	K	*cspC*	COG1278	Cold shock protein
1	1	1	2	2	2	1	1	1	1	1	2	12	16	KL	*dinG*	COG1199	Rad3-related DNA helicase
1	1	1	2	3	1	1	2	2	3	1	2	12	20	L	–	COG2256	ATPase of helicase subunit of the holliday junction resolvase
1	1	2	1	1	1	1	2	1	1	2	1	12	15	L	*comFA*	COG4098	Superfamily II DNA/RNA helicase required for DNA uptake
4	2	3	3	2	2	4	1	1	2	2	2	12	28	L	*dinP*	COG0389	Nucleotidyltransferase/DNA polymerase involved in DNA repair
2	2	2	2	2	2	2	2	2	2	2	2	12	24	L	*gyrA, B*	#	DNA gyrase
1	2	1	1	1	1	1	1	2	2	2	1	12	16	L	*mutS2*	COG1193	DNA structure-specific ATPase, suppression of recombination
1	1	1	1	1	1	1	1	1	1	1	1	12	12	L	*recA, B, D, F, J, N, O, R*	#	Homologous recombination, DNA repair[a]
1	1	1	1	1	1	1	1	1	1	1	1	12	12	L	*ruvA, B*	#	Holliday junction resolvasome
1	1	1	2	2	1	1	1	1	1	1	1	12	14	L	*sbcC, D*	#	ATPase involved in DNA repair
2	4	3	1	1	3	3	1	1	1	1	2	12	23	L	*ssb*	COG0629	Single-stranded DNA-binding protein
1	1	1	1	1	1	2	3	1	1	2	1	12	16	L	*topA*	COG0550	Topoisomerase IA
1	1	1	3	3	1	2	2	1	1	1	3	12	20	L	*uvrA*	COG0178	Excinuclease ATPase subunit

(continued)

Table 16.1 (continued)

S. thermophilus LMD-9	*L. lactis* subsp. *lactis* IL1403	*L. lactis* subsp. *cremoris* SK-11	*Lb. brevis* ATCC367	*Lb. plantarum* WCFS1	*P. pentosaceus* ATCC25745	*Le. mesenteroides*	*O. oeni* PSU-1	*Lb. johnsonii* NC533	*Lb. gasseri* ATCC33323	*L. bulgaricus* ATCC BAA-365	*Lb. casei* ATCC334	Number of organisms	Number of proteins	COG Functional group	COG gene	Corresponding COG	COG annotation
1	1	1	1	1	1	1	1	1	1	1	1	12	12	L	*uvrB, C, D*	#	Excision repair complex
1	2	4	2	2	4	2	2	3	3	2	2	12	29	L	*xerC*	COG0582	Integrase
1	1	1	1	1	1	1	1	1	1	1	1	12	12	L	*xseA, B*	#	Exonuclease VII
1	1	1	1	1	1	1	1	2	2	1	1	12	14	L	*xthA*	COG0708	Exonuclease III
1	1	1	1	1	1	1	1	1	1	1	1	12	12	LK	*mfd*	COG1197	Transcription-repair coupling factor (superfamily II helicase)
6	8	9	4	8	2	3	3	4	5	8	6	12	66	LR	*mutT*	COG0494	NUDIX family hydrolase
3	3	3	3	4	4	5	3	5	4	5	5	12	47	O	*clpA*	COG0542	ATP-binding subunit of Clp protease and DnaK/DnaJ chaperones
1	1	1	1	1	1	1	1	1	1	1	2	12	13	O	*clpX*	COG1219	ATP-dependent protease Clp, ATPase subunit
1	1	1	2	2	1	2	2	1	1	1	2	12	17	O	*degQ*	COG0265	Serine protease (heat shock protein)
1	1	1	1	1	1	1	1	1	1	1	1	12	12	O	*groEL,S, dnaJ, K, grpE*	#	Chaperonin functions
2	2	2	2	2	2	3	2	2	2	2	2	12	25	O	*trxB*	COG0492	Thioredoxin reductase
4	3	3	3	4	3	5	3	2	2	2	4	12	38	OC	*trxA*	COG0526	Thiol-disulfide isomerase or thioredoxin
1	2	1	1	1	2	2	1	1	2	1	1	12	16	OU	*clpP*	COG0740	Protease subunit of ATP-dependent Clp protease
1	1	1	1	1	1	1		1	1	1	1	11	11	K	*hrcA*	COG1420	Transcriptional regulator of heat shock gene

1	1	1	1	1	1	1		1	1	1	1	11	11	L	*mutS, L*	#	DNA mismatch repair enzyme, ATPase
1	1	1	1	1	2	1		1	1	1	1	11	12	L	*radC*	COG2003	DNA repair proteins
	1	1	2	2	2	1		1	1	1	2	10	14	L	*recQ*	COG0514	Superfamily II DNA helicase
1			1	3	1	1	1	1	1	1	2	10	13	O	*ibpA*	COG0071	Molecular chaperone (HSP20 family)
1	1	1	1	1				1	1	1	1	9	9	L	*nth*	COG0177	Predicted EndoIII-related endonuclease
1	1	1	1	1	1		1				1	8	8	K	*ctsR*	COG4463	Transcriptional repressor of class III stress genes
2	3	2	2					2	1	2	1	8	15	L	*xerD*	COG4974	Site-specific recombinase XerD
1			1	1	1	1		2	2	2		8	11	O	*htpX*	COG0501	Zn-dependent protease with chaperone function
1	1	1	1	1		1					1	7	7	L	*mutY*	COG1194	A/G-specific DNA glycosylase
			1	1	1			1	1	1	1	7	7	O	*hslU, V*	#	ATP-dependent protease HslVU (ClpYQ)
1	1	1			1				3	4		6	11	KL	*hepA*	COG0553	Superfamily II DNA/RNA helicase, SNF2 family
1		1	1		1			1	1	4		6	10	KL	*SSL2*	COG1061	DNA or RNA helicase of superfamily II
		2	2		2		2			1	3	6	12	L	*pinR*	COG1961	Site-specific recombinase, DNA invertase Pin homolog
			1	1	1						1	4	4	L	*nfo*	COG0648	Endonuclease IV
			1	1				1	2			4	5	L	*relB*	COG3077	DNA-damage-inducible protein J
	1			1		1						3	3	L	*recT*	COG3723	Recombinational DNA repair protein, RecT
	1	2										2	3	L	*rus*	COG4570	Holliday junction resolvase

Variations in the presence vs. absence per genome as well as variations in the predicted number of gene copies per genome are provided. Adapted from Makarova et al. (2006)

Functions that were not assigned a COG gene name are indicated with "–", and functions that lack a corresponding COG assignment are indicated with "#"

[a]Two copies of *recA* and *recD* are predicted for *Lb. casei*

16.2.2 Species Diversity Mining to Elucidate Genotype–Phenotype Correlations

Although many LAB species are currently represented by a genome sequence of an exemplary isolate, it is clear that many phenotypic differences exist among strains of a certain species. This phenotypic variation among strains has a major impact on their performance in fermentation applications and has been an important source of product diversification and innovation in the past decades. As an example, the application of different strains of *L. lactis* in cheese production can dramatically impact the flavor (and texture) characteristics of the end product [for a review, see Smit et al. (2005)], which has stimulated the development of high-throughput, miniaturized cheese manufacturing procedures that enable the product-related functionality screening of individual strains to accelerate product diversification (Bachmann et al. 2009a). This phenotypic variation among strains is at least partially due to their diversity in gene content. Several approaches are available to determine the genomic diversity among strains. The comparative genome hybridization (CGH) approach employs one-directional comparison of gene-content profiles per strain using genomewide microarrays that are designed on the basis of the genome of a single strain. This approach enables the construction of high-resolution genomewide presence–absence patterns for each of the strains that is analyzed. Many array platforms that are currently used for transcriptome analyses contain several probes per annotated gene and are generally suitable for CGH. However, an even higher resolution can be achieved by using so-called tiling arrays that contain probes that cover the entire genome sequence through minimal tiling probe design. CGH has been applied to determine the genomic diversity of several LAB, including *L. plantarum* (Molenaar et al. 2005; Siezen et al. 2010) and *Lactobacillus sakei* (McLeod et al. 2008). The gene-specific diversity database obtained in this way can readily be applied to identify the gene(s) responsible for specific phenotypic traits that are variable among the strains analyzed by genotype–phenotype matching (GPM). This approach is exemplified by the diversity-based identification of the mannose-specific adhesin (Msa) of *L. plantarum*, which is proposed to be involved in its probiotic functionality related to reducing the severity of infection of enterotoxic *Escherichia coli* in humans by competitive exclusion. Subsequently, the role of Msa in mannose-specific adherence proposed by GPM could be confirmed by *msa* mutation analysis (Pretzer et al. 2005). Intriguingly, transcriptome analyses of pig intestinal mucosa revealed that mucosal interaction with the *msa* mutant of *L. plantarum* fails to elicit the expression of the host bactericidal PAP (pancreatitis-associated protein), in amounts comparable to those observed for the wild-type strain, suggesting an *msa*-dependent interaction with the host innate immune system (Gross et al. 2008). Moreover, the sequence of the *msa* gene in different *L. plantarum* strains appeared to encode a protein with strain-specific domain composition, which can be associated with strain-specific quantitative mannose adherence capacities (Gross et al. 2010). This work underlines the discovery power of the GPM approach for the elucidation of genetic determinants underlying specific phenotypes. Analogously, GPM approaches using the same CGH datasets have elucidated several genes involved in the immunomodulatory capacity of *L. plantarum* (Meijerink et al. 2010; van Hemert et al. 2010).

Analogous to the approach above, strain-specific genomic diversity may be correlated to strain-specific survival rates under defined stress conditions, to elucidate candidate genes involved in stress tolerance. To illustrate the potential of this approach, the relative survival rate of more than 40 *L. plantarum* strains in an in vitro model mimicking the stress conditions encountered during human GI tract passage (acid challenge, followed by pH neutralization combined with pancreatic enzyme and bile acid exposure) revealed strain-specific stress-tolerance phenotypes that span a more than 10^6-fold difference in survival rates between the best- and poorest-surviving strains. However, the GPM of this highly variable phenotype to the CGH profiles of the corresponding strains is not straightforward and requires advanced correlation analysis algorithms like the so-called random forest (RF) module. The RF algorithm offers a classification method that is particularly well suited for data sets that have a large number of individual features, for example, comparative genome hybridization features for more than 3,000 genes, and a relatively small sample size, such as survival and gene-diversity features for "only" 40 strains (Breiman 2001). RF combines the power of decision trees (dividing the data into different classes on the basis of the feature information) and bagging (randomly picking multiple subsets from the data to use for classification). In so doing, RF constructs a multitude of decision trees consisting of different subsets of the data, which are combined into a "forest". The overall outcome of RF is based on the combination of all the individual decision trees into majority-vote decisions. Importance values can be assigned to all genes (features) depending on their position in the decision trees, where genes with the highest contribution to the prediction (decision) end up in the top of all trees and are assigned the highest importance. Such genes are the most likely candidate genes involved in the phenotype studied. An advantage of RF is the allowance for the inclusion of gene hierarchy, which allows the detection of genes that are only important determinants upon the presence or absence of another gene, thereby permitting the detection of underlying dependencies of genes in relation to the phenotype and enabling more subtle GPM.

The application of RF to identify genetic determinants of *L. plantarum* that are involved in high-intestinal-tract stress survival revealed a list of candidate genes that are currently evaluated by mutation analysis to confirm their contribution to this relevant phenotype (data not shown). An extrapolation of this approach to other forms of stress can be applied when an appropriate survival assay is available (e.g., oxidative stress, heat, UV-irradiation tolerance). These studies can lead to the definition of gene sets that contribute to LAB robustness, which can be applied to select or construct strains or mutants with improved survival capacity under harsh conditions.

In addition to the CGH approach illustrated above, strain diversity can nowadays also be addressed by the determination of multiple genome sequences of individual isolates of a particular species. The emergence of the highly effective next-generation sequencing technologies (Gresham and Kruglyak 2008; Shendure and Ji 2008) especially facilitates this approach, which is illustrated by the appearance of multiple genome sequences of specific LAB species in the public domain, including the *Lactobacillus* species *L. plantarum*, *Lactobacillus casei*, *Lactobacillus delbrueckii*, *Lactobacillus reuteri*, and *Lactobacillus rhamnosus*. This trend is bound to accelerate gene-function assignment, including the identification of genes involved in relevant phenotypes. A clear example of such novel gene-function assignment potential is provided by the recent completion of the genome sequence of the best-documented probiotic strain, *L. rhamnosus* GG (Kankainen et al. 2009), and its comparison to the closely related strain LC705. The two *L. rhamnosus* genomes (both approximately 3.0 Mbp) are very similar and syntenous, but also contain strain-specific genomic islands. One of the GG-specific genome islands encodes a pilin-like surface structure that is important in adherence to intestinal mucus and is proposed to aid the persistence of *L. rhamnosus* GG in vivo in the intestine (Kankainen et al. 2009). Analogously, a comparative genome sequence analysis of two or more LAB strains of the same species that display a high difference in survival capacity under specific stress conditions might enable the identification of the genetic determinants underlying this phenotypic difference.

However, despite the successes of GPM approaches just described, it is also clear that many phenotypes do not depend on the presence or absence of specific genes, but are predominantly determined by the difference in the level of expression of conserved genes. A clear illustration of this is provided by the very high diversity in gene expression-regulation phenotypes observed in individual strains of *L. lactis*, which was based on the comparative analysis of the activity levels of five enzymes in two different growth media. The enzymes analyzed are considered relevant for their flavor-forming capacities during cheese making, illustrating the potential impact of this regulatory diversity on eventual product properties (Bachmann et al. 2009b). Moreover, it is likely that the majority of stress-tolerance genes are conserved among strains of a particular species and that strain-specific survival capacities depend on their relative levels of expression rather than their presence or absence. Therefore, to unravel the contribution of conserved genes in stress-tolerance phenotypes, comparative genomics should be performed at the functional (e.g., transcriptome, proteome) level.

16.3 Functional Genomics Approaches to Unravel Lactic Acid Bacteria Stress Responses

16.3.1 In Vitro *Approaches to Identify Robustness Genes in Lactic Acid Bacteria*

The intrinsic underrepresentation of conserved stress factors identified utilizing different LAB strains and GPM approaches just described can be complemented by comparing transcriptome profiles derived from an individual strain grown under normal and stress conditions. To this end, DNA microarray technology has been exploited widely to identify several of the (conserved) genetic factors regulated during stress imposed on LAB during industrial fermentation (e.g., lactate production), processing (e.g., hydrostatic pressure), and storage (e.g., high-osmolarity/low-water activity) or after consumption by the human host (low pH encountered in the stomach and pancreatic enzyme and bile-associated stress in the duodenum) (Fig. 16.1). For example, a transcriptome-profiling approach revealed the effect of lactic acid stress in *L. plantarum* strain WCFS1 (Pieterse et al. 2005). Strikingly, three cell surface complex (*csc*) operons (Siezen et al. 2006) were found to be among the highly induced gene clusters in response to lactic acid stress, suggesting that the corresponding proteins are abundantly present on the cell surface. Indeed, cells preexposed to lactic acid displayed striking morphological changes, including a rough morphology, compared to the smooth appearance of unstressed control cells. The observed morphological changes might be associated with the observed lactic acid tolerance (Pieterse et al. 2005). Unfortunately, a subsequent dedicated mutagenesis approach in which these *csc* gene clusters were deleted could not confirm the involvement of these cell surface proteins in lactate stress tolerance, as the mutants displayed tolerance levels comparable to the wild type (Pieterse 2006), which may be due to genetic compensation, as it is suggested by the high degree of redundancy of the *csc* clusters in the *L. plantarum* genome (Siezen et al. 2006). Similarly, transcriptome analysis in *Lactobacillus sanfranciscensis* revealed the high-pressure-regulated gene expression of genes of several (conserved) functional classes, including protein and fatty acid biosynthesis, energy metabolism, as well as transport and cell envelope proteins (Vogel et al. 2005).

The transcriptome profiling of *L. reuteri* ATCC55730 after exposure to acid revealed the induction of several genes with potential functions in membrane fluidity regulation or peptidoglycan biosynthesis and organization, including a putative phosphatidyl glycerophosphatase and a putative esterase gene, belonging to the family of penicillin-binding proteins (Wall et al. 2007). A mutant lacking the latter gene displayed a gastric juice and bile sensitivity phenotype (Wall et al. 2007), establishing a definite role for the penicillin-binding protein of this LAB in its robustness under GI conditions. Similarly, DNA microarray experiments using bile-exposed *L. acidophilus* NCFM (Pfeiler et al. 2007) or *L. plantarum* WCFS1 (Bron et al. 2006) revealed the induction of several genes potentially involved in cell envelope and surface protein biosynthesis. These data corroborate earlier observations made when the bile response in *L. plantarum* WCFS1 was investigated utilizing a genetic screen (Bron et al. 2004b). Furthermore, *L. acidophilus* gene-disruption mutants in a cell division protein (*cdpA*) and surface layer protein A (*slpA*) displayed an increased bile resistance, while their osmotolerance was negatively affected (Altermann et al. 2004; Klaenhammer et al. 2005), further highlighting the importance of subtle modifications in cell envelope composition on the robustness of LAB to persist in different stress conditions relevant for industrial processing and GI tract survival. The DNA microarray studies in *L. plantarum* WCFS1 also revealed the induction of the *dlt* operon during bile stress, suggesting the importance of D-Ala decoration of wall- and/or lipo-teichoic acid (WTA and LTA) for the cell envelope integrity and robustness of this LAB (Bron et al. 2006). Notably, an *L. rhamnosus dltD* mutant displayed a reduced survival capacity in simulated gastric juice (Perea Velez et al. 2007), whereas a similar approach in *L. reuteri* revealed a pronounced effect on in vitro growth at low pH (Walter et al. 2007).

Although the DNA microarray analyses described above performed for *L. acidophilus* (Pfeiler et al. 2007) and *L. reuteri* (Whitehead et al. 2008) also indicated that genes involved in their respective EPS production are regulated upon bile exposure, no phenotypic analysis of dedicated mutants have been reported to date (Lebeer et al. 2008). To this end, a detailed physiological characterization of the recently constructed *L. rhamnosus* GG (Lebeer et al. 2009) and *L. plantarum* WCFS1 (Bron PA, Remus D, van Swam II, Hols P, and Kleerebezem M, unpubl. data) mutants unable to produce (specific forms of) EPS could establish a more definitive role for EPS in the robustness of these LAB.

Overall, these studies investigating alterations in transcriptome profiles under industrially relevant stress conditions have led to substantial insight into the candidate factors important for bacterial robustness when encountering these stresses. Importantly, several of these studies have been followed by dedicated mutagenesis approaches and the subsequent reassessment of stress robustness, establishing a definite role for the robustness factors in the industrial and GI performance of LAB (Fig. 16.2). However, most data have been obtained in simplified laboratory systems that fail to accurately assess the physiochemical complexity encountered during industrial fermentation and processing (Walter et al. 2003; Hufner et al. 2007) or the multitude of stresses and bacterial competition of the intestinal environment (de Vos et al. 2004; Kleerebezem and Vaughan 2009). These issues have been addressed by more recent in vivo approaches that are discussed ahead.

16.3.2 (Resolvase-Based) **In Vivo** *Expression Technology*

In vivo expression technology (IVET) and its resolvase-based variant (R-IVET) are powerful methodologies that allow the genomewide identification of in vivo induced (*ivi*) promoters and their corresponding genes utilizing a promoter trapping system (Rainey and Preston 2000; Rediers et al. 2005). By applying IVET to *L. sakei* 23 K, 15 genetic loci could be identified that display increased expression levels during raw-sausage fermentation. These in carne induced genes included several genes that are likely to contribute to known stress-related functions, as well as a gene involved in the acquisition of ammonia from amino acids and several genes encoding unknown functions. Subsequently, mutants in the *ivi* genes encoding an L-asparaginase, a hypothetical metallo-β-lactamase, and a hypothetical membrane protein displayed a hampered in carne performance, establishing a definite role for these proteins during raw-meat fermentation (Hufner et al. 2007). Similarly, Bachmann et al. developed an optimized R-IVET system that enables the double-positive selection of responding clones by the implementation of a MelA and a luciferase-based promoter probe system into the R-IVET vector (Bachmann et al. 2008). Following the initial validation of this system by the identification of genes specifically induced in minimal media as compared to rich laboratory media (Bachmann et al. 2008), this system was applied to identify genes that are induced specifically during cheese manufacturing (Bachmann et al. 2010). Subsequent

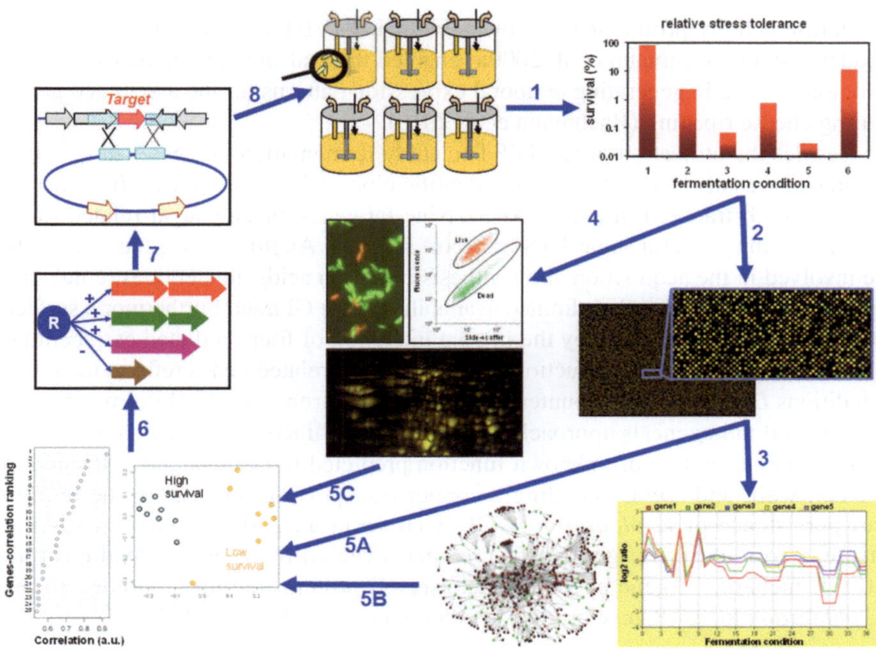

Fig. 16.2 Schematic representation of the strategic approach to identify and validate stress tolerance–associated genes and functions, including a pivotal role for "omics"-based technologies. Variations in fermentation conditions, including several mild forms of relevant stress conditions, can be directly coupled to assays aiming to determine the stress tolerance (relative survival under stress conditions) associated with these growth conditions (*arrow 1*). Transcriptome profiles of the bacteria obtained from the different fermentations (*arrow 2*) can be determined and can either directly or indirectly, via gene expression pattern comparison and regulatory network reconstructions (*arrow 3*) be correlated to stress-tolerance characteristics using correlation algorithms like random forest (*arrows 5A* and *5B*, respectively). Alternatively, viable and killed subpopulations (*arrow 4*) generated in the stress-tolerance assays can be differentially labeled by fluorescence-viability probes (central figure, *top left* image) and can subsequently be separated by flow-cytometry coupled sorting (central figure, *top right* image). The molecular makeup can be differentially analyzed using quantitative proteomic technologies such as Cy-DIGE (central figure, *lower* image), which is directly coupled to stress tolerance and leads to the correlation of specific genes and proteins to stress tolerance using correlation analysis algorithms like random forest (*arrow* 5C). These approaches lead to the identification of candidate genes, proteins, or regulons (and their regulators) that are potentially of importance for the observed differential stress tolerance generated by the varying growth conditions (*arrow 6*). The mutagenesis (*arrow 7*) of these candidate stress factors (either knockout or overexpression strategies) and their subsequent evaluation using the same fermentation-coupled stress-tolerance setup (*arrow 8*) enable validation of the postulated correlation. Overall, this strategy identifies and validates genetic determinants of stress tolerance, which can be modulated by growth conditions and lead to the rational design of fermentation schemes to maximize the robustness of the cultures produced. In addition, the genetic determinants of tolerance can be targeted by dedicated (as illustrated by *arrow 7*) or classical engineering approaches to change their level of expression by mutagenesis in order to enhance strain robustness

luciferase activity profile analysis of individual R-IVET clones in a micro-cheese model system (Bachmann et al. 2009a) enabled the real-time in situ assessment of promoter strength, generating temporal expression patterns for the associated genes during cheese ripening (Bachmann et al. 2010).

Besides the utilization of (R)-IVET to study fermentation processes in situ, this technique was also exploited for the identification of 72 *L. plantarum ivi* genes in the mouse GI tract (Bron et al. 2004a). Nine *ivi* genes encode sugar-related functions, including several sugar PTS transport systems. Another nine genes appear to be involved in the acquisition and synthesis of amino acids, nucleotides, cofactors, and vitamins, indicating their limited availability in the GI tract. Furthermore, surface adaptations were suggested by the in vivo induction of four predicted extracellular proteins, while the in situ induction of several stress-related genes reflects the harsh conditions *L. plantarum* encounters in the GI tract (Bron et al. 2004a). Importantly, a dedicated mutagenesis approach (Fig. 16.2) underlined the critical contribution of Lp_2940, a protein of unknown function predicted to be covalently attached to the cell wall, and *copA*, encoding a copper-transporting ATPase, to the murine gut persistence of *L. plantarum* WCFS1 (Bron et al. 2007). Notably, a similar strategy to identify *ivi* genes in *L. reuteri* revealed three genes (Walter et al. 2003), including a gene encoding a conserved protein-sharing homology to the *L. plantarum* lp_2718 gene product (Walter et al. 2003; Bron et al. 2004a).

16.3.3 In Situ *Transcriptome Profiling*

In situ transcriptome profiling during fermentation, application, or intestinal residence is a complimentary approach to reveal gene expression patterns elicited by the physicochemical stress conditions encountered during these complex processes. Subsequently, this information can be harnessed for the production and/or selection of more robust strains or cultures on the basis of stress-response mimicking expression patterns of these candidate tolerance or robustness factors. Recently, this strategy was used to unravel the adaptive behavior of *S. thermophilus* during the late stages of milk fermentation, revealing a strong regulation in sugar metabolism pathways. Moreover, the induction of nitrogen metabolism was eminent, particularly in the transport and biosynthetic pathways for sulfur-containing amino acids (Herve-Jimenez et al. 2008). Although insightful, this study ignores the fact that *S. thermophilus* is naturally associated with *L. delbrueckii* subsp. *bulgaricus* during industrial yogurt production. Hence, a follow-up study from the same research group investigated *S. thermophilus* gene and protein expression profiles in milk while being cocultured with *L. delbrueckii*. Specific coculture regulation could be established for 77 genes, including several genes encoding functions in nitrogen metabolism, reiterating the importance of a prompt response toward the limited availability of (specific) amino acids during milk fermentation. Moreover, the expression of nearly all genes predicted to be involved in iron transport were downregulated, whereas that of iron-chelating *dpr* and that of the *fur* regulator were induced, suggesting a reduction in the intracellular iron concentration, likely in reponse to H_2O_2 production by *L. delbrueckii* (Herve-Jimenez et al. 2009).

Due to recent technical advances, the isolation of high-quality bacterial RNA derived from intestinal samples nowadays is a routine laboratory procedure (Zoetendal et al. 2006), allowing in situ transcriptome approaches to monitor changes in bacterial gene expression in the GI tract. For example, transcriptomes of *L. plantarum* were obtained from samples derived from the ceca of mono-associated mice that were fed differential diets [either Western-style (high-fat, low-fiber) or standard chow (low-fat, high–fiber)] (Marco et al. 2009), and in intestinal biopsies removed from patients diagnosed with colon cancer who volunteered to participate in a probiotic trial prior to surgery (De Vries 2006). Comparative analyses of these human-derived transcriptomes and the mouse cecum–derived and more than 100 in vitro transcriptomes revealed a significant convergence of the *L. plantarum* response to human and mouse intestinal conditions. Altered carbohydrate acquisition and cell surface composition were among the most pronounced altered functional classes. For example, the capsular polysaccharide biosynthesis operon *cps3* and the cell surface protein clusters *cscI* and *cscVIII* were consistently induced in all in vivo samples (Macro et al. 2010). These overlapping responses for *L. plantarum* in different GI compartments and using different mammalian model systems support a diet-, host-, and microbiota-independent core response in *L. plantarum*. Hence, the cognate extracellular molecules of this LAB are key performance factors involved in (probiotic) functionalities, likely to include robustness, in the GI tract (Kleerebezem et al. 2010). In another study, in situ transcriptomes of *L. johnsonii* residing in different compartments of the mouse GI tract were obtained (Denou et al. 2007). Colon-specific gene expression was not detected, whereas the induction of specific sugar PTS transport systems was demonstrated in the jejunum, the stomach, and the cecum. Moreover, the stomach-specific genes include several multidrug transport systems, a cation-efflux protein, as well as a copper-transporting ATPase, closely resembling the alterations in gene expression found in the *L. plantarum* R-IVET approach described previously (Bron et al. 2004a).

16.3.4 Assessment of Multiple Stress Responses and Regulatory Network Reconstruction

Although these in situ studies have shed light on the molecules involved in the stress responses in LAB, these analyses have generally focused on one particular stress during one aspect of the LAB application pipeline. However, a few studies have reported the LAB stress response toward multiple industrially relevant stresses; for example, alterations in *L. lactis* gene expression after exposure to heat, acid, and osmotic stress were assessed utilizing DNA macroarrays, focusing on 375 metabolic genes. Although the majority of stress-regulated genes were specific for an individual stress condition, a number of stress responses were common for the different stresses, including the repression of several transporters and the induction of two nucleotide kinases (Xie et al. 2004). Another elegant study described the elucidation of the response of *Bifidobacterium breve* toward heat, osmotic, and solvent stress (Zomer et al. 2009). Data obtained from transcriptome analysis, DNA–protein interaction data, and GusA reporter fusion studies were combined with an in silico analysis, allowing the construction of a model for an interacting regulatory network for stress responses in this probiotic bacterium. This model revealed HspR controls the SOS response and the ClgR operon, which in turn regulates and is regulated by HrcA. Unfortunately, to our knowledge, such extensive network reconstruction based on multiple stress-induced transcriptome profiles has not been reported for LAB to date. As exemplified by the bifidobacterial study described earlier here, such a multiple-variable "stressomics" approach is highly valuable for comprehensive stress-response analyses, as it not only identifies the genes directly involved in robustness and/or stress survival, but can also reveal the regulatory networks and complete regulons involved.

In conclusion, the tools that the postgenomics era offers have led to an explosion in the knowledge on LAB stress factors and their corresponding regulators. We are now at the exciting stage where this wealth of information can be exploited to improve the robustness of industrially relevant strains.

16.4 Improving Robustness

Several strategies have been exploited to achieve more robust bacterial strains, ranging from pregenomics approaches such as modifications in the growth medium and cross-protection strategies, to postgenomics strategies, including the construction of mutant strains over- or underproducing relevant stress factors and regulators, as well as the advanced exploitation of fermentation genomics platforms for the prediction and improvement of robustness effector molecules. These strategies are addressed next.

16.4.1 Additives

A relatively straightforward way to improve survival is achieved by the addition of specific compounds to the medium. For example, the addition of 19.4 mM of glucose resulted in up to a 6-log enhanced survival of *L. rhamnosus* GG following 90 min of exposure to simulated gastric juice (pH 2.0) (Corcoran et al. 2005). Notably, only *L. rhamnosus* cells preexposed to metabolizable sugars, such as glucose or fructose, displayed improved survival characteristics, suggesting that a highly energized state is essential for the robustness of these bacterial cells (Corcoran et al. 2005). In another study, a Tween-80 addition to the growth medium of several *L. rhamnosus, Lactobacillus paracasei,* and *Lactobacillus salivarius* strains resulted in up to a 3-log increased survival during exposure to gastric juice. The subsequent analysis of the fatty acid composition of *L. rhamnosus* GG revealed a 55-fold higher oleic acid content as well as a significantly higher overall unsaturated/saturated fatty acids ratio in the membranes of cells grown in the presence of Tween-80. The authors suggest that these changes in the membrane composition and fluidity are the most likely explanation for the observed enhanced survival characteristics (Corcoran et al. 2007). Yet another study revealed that the addition of the compatible solute betaine improved the survival of *L. plantarum*, *Lactobacillus halotolerans*, and *Enterococcus faecium* during drying by 14, 1.5, and 1.7 fold, respectively (Kets et al. 1996). Similarly, gum acacia improved the survival of *L. paracasei* 20 fold during spray-drying and subsequent storage for 4 weeks at 4°C (Desmond et al. 2002).

16.4.2 Preadaptation

Although subtle strain-specific differences in LAB stress regulons have been described (Spano and Massa 2006), LAB have developed several conserved stress responses, especially the widely distributed regulatory networks centering around the CtsR and HrcA regulators [see above and earlier chapters; for reviews, see van de Guchte et al. (2002), De Angelis and Gobbetti (2004), Corcoran et al. (2008)]. For instance, the *L. plantarum* HrcA is predicted to regulate the expression of chaperonins (GroEL, GroES, GrpE, and DnaK), and CtsR is predicted to regulate Clp proteases and ATPases (ClpE, ClpP, and ClpB) (Wels et al. 2006; Wels 2008). Not surprisingly, many reports describe the induction of protein expression levels in the CtsR and HrcA regulons upon bacterial exposure to different stresses. Examples include the increased expression of GroE and ClpP in *L. lactis* after exposure to heat, acid, or UV irradiation (Hartke et al. 1997; Frees and Ingmer 1999), the induction of *L. mesenteroides* DnaK and GroEL upon cold shock (Salotra et al. 1995), and GroES, GroEL, and DnaK induction upon an osmotic upshift in *L. lactis* (Kilstrup et al. 1997).

The fact that these relatively mild stress conditions can induce an adaptive response suggests that the preexposition of industrially relevant strains to a sublethal stress condition will improve their robustness toward a correlated but more severe stress encountered in the industrial pipeline (Corcoran et al. 2008). Indeed, preexposure to a sublethal acid stress was reported to improve the survival to more severe acidic conditions for *L. sanfranciscensis* (De Angelis et al. 2001), *L. lactis* (Hartke et al. 1996), and *L. acidophilus* (Lorca et al. 2002). Moreover, when *L. acidophilus* encountered low concentrations of bile, its subsequent survival in an environment containing a relatively high bile concentration was enhanced (Kim et al. 2001).

16.4.3 Cross-Protection

The observed overlap in different stress regulons described above, likely to be caused by comparable DNA, protein, and cell envelope damage or denaturation induced by different stresses, implies that stress tolerance can also be induced by the prior exposure to a different stress than the actual stress encountered in the industrial pipeline (Fig. 16.1). This phenomenon is known as *cross-protection* and can be exploited to improve robustness while circumventing the addition of undesired compounds, such as (bile) salts, during the fermentation process. For example, the preexposure of *Propionibacterium freudenreichii* to heat induced bile tolerance (Leverrier et al. 2003). Furthermore, *L. plantarum* preexposed to sublethal heat treatment displayed enhanced growth at 6% NaCl and at pH 5.0 (De Angelis et al. 2004). Other examples include an acid pretreatment in *L. lactis,* which results in improved heat, ethanol, H_2O_2, and NaCl robustness, as well as the generation of an acid-tolerant phenotype by heat adaptation (O'Sullivan and Condon 1997). Moreover, *L. paracasei* displayed an increased robustness toward spray-drying after mild osmotic or sublethal heat stress, and similar pretreatments in *L. rhamnosus* significantly improved subsequent storage at 30°C for 14 weeks (Prasad et al. 2003). In conclusion, these approaches have led to the significant improvement of several important starter and probiotic strains. However, the exact genes and corresponding molecular mechanisms are not always fully understood, hampering the full exploitation of these strategies.

16.4.4 Overexpression of Stress Regulon Members by Genetic Modification

Nowadays, a vast genetic toolbox for the manipulation of LAB gene expression levels is available, allowing complementary approaches to alter expression levels of genes encoding stress regulon members or regulators, prior to stress exposure. The overproduction of GroESL in *L. paracasei* and *L. lactis* resulted in an improved salt tolerance (Desmond et al. 2004), and manganese-dependent catalase overexpression in *L. casei* resulted in a strain that displayed better survival characteristics upon H_2O_2 exposure (Rochat et al. 2005a). Similarly, the heterologous expression of superoxide dismutase in *L. acidophilus* (Bruno-Barcena et al. 2004, 2005), as well as heme catalase in *L. plantarum* (Noonpakdee et al. 2004) or *L. lactis* (Rochat et al. 2005b), resulted in oxidative stress robust strains. Yet another example describes *L. plantarum* variants engineered to overexpress the heat-shock proteins HSP 18.55 or HSP 19.3, leading to improved heat resistance and enhanced survival in the presence of ethanol (Fiocco et al. 2007).

An alternative approach, termed *patho-biotechnology* by Sleator et al., exploits the heterologous expression of the sophisticated compatible solute accumulation system derived from the foodborne pathogen *Listeria monocytogenes* in industrially relevant strains (Sleator and Hill 2006). For example, the introduction of the *betL* gene, encoding the betaine-uptake system of *L. monocytogenes*, into *L. salivarius* resulted in a strain that displayed a significant increased capability of betaine accumulation and a corresponding increased salt tolerance (Sheehan et al. 2006), as well as an improved barotolerance (Smiddy et al. 2004).

16.4.5 Targeted Mutagenesis of Stress Regulators

Besides the approaches discussed above, which engineer strains toward the overexpression of one genetic locus important for stress tolerance, several studies describe the manipulation of complete stress regulons by targeting the canonical stress regulator CtsR. Following this strategy, a *ctsR* deletion mutant in *S. thermophilus* (Zotta et al. 2009) and *L. sakei* (Hufner et al. 2007) exhibited improved heat stress during exponential growth and more efficient growth during sausage fermentation, respectively. However, the *S. thermophilus* gene-deletion mutant displayed an osmotic- and oxidative-sensitive phenotype (Fiocco et al. 2010; Zotta et al. 2009). Furthermore, a *ctsR* mutant in *L. plantarum* was ethanol- and heat-sensitive as compared to the wild type, despite the fact that several of the genes in the CtsR regulon (*hsp1*, *clpB*, *clpC*, *clpE*, and *clpP*) were demonstrated to be derepressed (Fiocco et al. 2010; Zotta et al. 2009). The observed highly variable consequences of *ctsR* deletion assessing different stresses within one species suggest subtle stress-dependent differences in the induced regulon. Moreover, the highly variable phenotypic effects observed for *ctsR* mutants in different LAB species underline the interspecies differences in the *ctsR* regulon, even when the same stress is applied, severely complicating the generic application of these mutants for the design of more robust strains. Moreover, the debate on the application of genetically modified organisms (GMOs) in the food industry is momentarily undecided, and legal issues and general public opinion are hampering the industrial application of many of the more robust strains described here. Notably, the biological containment strategies of GMOs have been exploited in the pharma industry (Steidler et al. 2003; Braat et al. 2006), and similar approaches could circumvent the release of GMOs into the environment in food applications.

16.4.6 Fermentation-Enhanced Probiotic Function

Our laboratory has recently developed a functional-genomics fermentation platform for the identification and optimization of expression of specific probiotic functionality parameters, including robustness effector molecules (Van Bokhorst-van de Veen et al., unpublished data). In the first set of 29 fermentations, *L. plantarum* was grown in chemically defined medium according to a combinatorial fermentation scheme that included variations in NaCl concentration, pH, and oxygen and amino acid availability. The scheme was designed such that single and combined effects of these conditions can be assessed. These conditions include mild stress conditions known to affect growth rate, GI persistence, and/or cell wall properties. Samples harvested from these fermentations were assessed for their respective transcriptome, proteome, and glycome profiles. In parallel, these samples were analyzed for GI stress-tolerance phenotypes. The data obtained were stored in a random forest–based correlation database that was used to identify molecular features enhancing survival capacity (Fig. 16.2). Transcriptome trait matching revealed candidate effector molecules important for GI survival, including four transport and three hypothetical proteins, as well as a regulator with homology to members of the AraC-regulatory family. Current efforts focus on the construction of deletion and overexpression mutants, depending on the direction of the correlation of gene expression to survival. Subsequently, the constructed mutants will be assessed for their anticipated altered survival capacities, which could establish a definite role for these molecules in the GI stress response of *L. plantarum*. The generic approaches presented here for *L. plantarum* are applicable to other probiotic bacteria, in which the correlation of survival characteristics to the "omics" data sets might lead to the identification of the molecular properties and their encoding genes that are involved in (probiotic) functionality. The obtained molecular knowledge can be exploited to predict the efficacy of starter and probiotic cultures by assessing expression levels of the genes encoding the bacterial effector molecules. Moreover, subtle alterations in fermentation processes can be implemented to maximize the delivery efficacy and health impact of (subpopulations of) these strains. Another interesting possibility that such a functional genomics platform offers employs subpopulation analysis approaches, for example, combining life–death staining with fluorescence-activated cell-sorting (FACS) technology and subsequent proteomics to assess the differences in the proteinaceous makeup of the surviving cells as compared to the killed subpopulation (Fig. 16.2). Although significant insight has been gained from similar approaches in bacteria such as *B. subtilis* (Veening et al. 2008a, b), single-cell and subpopulation biology is a virtually unexplored research area in LAB that is expected to gain momentum in the near future.

16.5 Concluding Remarks

Overall, this chapter underlines the value of genomics to increase our understanding of the molecular biology of LAB in general and illustrates how a variety of post-genomic approaches can accelerate the identification of genes involved in stress response and tolerance, in particular. Such a knowledge base is essential for the design of rational approaches for the selection, production, and/or engineering of LAB starter and probiotic cultures with increased robustness. In particular, the correlation between a multitude of (mild) stress conditions during LAB growth, their consequences for survival or tolerance under particular stress conditions, and the postgenomic analysis of their molecular makeup will provide a highly effective platform for the identification of stress-tolerance effector molecules that are key determinants for LAB robustness (Fig. 16.2). Such determinants enable the design of stress-tolerance biomarkers based on gene expression patterns and allow robustness development to be monitored during industrial fermentation. Moreover, such approaches also directly point to specific fermentation conditions that can be employed to enhance the culture's robustness during subsequent processing and/or application. Overall, the postgenomics era offers great promises for improving our capacity to produce, select, or engineer LAB that are more resistant to the harsh conditions encountered during industrial processes, or in the GI tract of humans.

References

Altermann E, Buck LB, Cano R, Klaenhammer TR (2004) Identification and phenotypic characterization of the cell-division protein CdpA. Gene 342:189–197

Altermann E, Russell WM, Azcarate-Peril MA, Barrangou R, Buck BL, McAuliffe O, Souther N, Dobson A, Duong T, Callanan M, Lick S, Hamrick A, Cano R, Klaenhammer TR (2005) Complete genome sequence of the probiotic lactic acid bacterium *Lactobacillus acidophilus* NCFM. Proc Natl Acad Sci USA 102:3906–3912

Bachmann H, Kleerebezem M, van Hylckama Vlieg JE (2008) High-throughput identification and validation of *in situ*-expressed genes of *Lactococcus lactis*. Appl Environ Microbiol 74:4727–4736

Bachmann H, Kruijswijk Z, Molenaar D, Kleerebezem M, van Hylckama Vlieg JE (2009a) A high-throughput cheese manufacturing model for effective cheese starter culture screening. J Dairy Sci 92:5868–5882

Bachmann H, Starrenburg MJC, Dijkstra A, Molenaar D, Kleerebezem M, Rademaker JLW, van Hylckama Vlieg JET (2009b) Regulatory phenotyping reveals important diversity within the species *Lactococcus lactis*. Appl Environ Microbiol 75:5687–5694

Bachmann H, de Wilt L, Kleerebezem M, van Hylckama Vlieg JE (2010) Timeresolved genetic responses of *Lactococcus lactis* to a dairy environment. Environ Microbiol 12:1260–1270

Boekhorst J, Siezen RJ, Zwahlen MC, Vilanova D, Pridmore RD, Mercenier A, Kleerebezem M, de Vos WM, Brüssow H, Desiere F (2004) The complete genomes of *Lactobacillus plantarum* and *Lactobacillus johnsonii* reveal extensive differences in chromosome organization and gene content. Microbiology 150:3601–3611

Bolotin A, Wincker P, Mauger S, Jaillon O, Malarme K, Weissenbach J, Ehrlich SD, Sorokin A (2001) The complete genome sequence of the lactic acid bacterium *Lactococcus lactis* ssp. *lactis* IL1403. Genome Res 11:731–753

Bolotin A, Quinquis B, Renault P, Sorokin A, Ehrlich SD, Kulakauskas S, Lapidus A, Goltsman E, Mazur M, Pusch GD, Fonstein M, Overbeek R, Kyprides N, Purnelle B, Prozzi D, Ngui K, Masuy D, Hancy F, Burteau S, Boutry M, Delcour J, Goffeau A, Hols P (2004) Complete sequence and comparative genome analysis of the dairy bacterium *Streptococcus thermophilus*. Nat Biotechnol 22:1554–1558

Braat H, Rottiers P, Hommes DW, Huyghebaert N, Remaut E, Remon JP, van Deventer SJ, Neirynck S, Peppelenbosch MP, Steidler L (2006) A phase I trial with transgenic bacteria expressing interleukin-10 in Crohn's disease. Clin Gastroenterol Hepatol 4:754–759

Breiman L (2001) Random forests. Mach Learn 45:5–32

Bron PA, Grangette C, Mercenier A, de Vos WM, Kleerebezem M (2004a) Identification of *Lactobacillus plantarum* genes that are induced in the gastrointestinal tract of mice. J Bacteriol 186:5721–5729

Bron PA, Marco M, Hoffer SM, Van Mullekom E, de Vos WM, Kleerebezem M (2004b) Genetic characterization of the bile salt response in *Lactobacillus plantarum* and analysis of responsive promoters *in vitro* and *in situ* in the gastrointestinal tract. J Bacteriol 186:7829–7835

Bron PA, Molenaar D, de Vos WM, Kleerebezem M (2006) DNA micro-array-based identification of bile-responsive genes in *Lactobacillus plantarum*. J Appl Microbiol 100:728–738

Bron PA, Meijer M, Bongers RS, de Vos WM, Kleerebezem M (2007) Dynamics of competitive population abundance of *Lactobacillus plantarum ivi* gene mutants in faecal samples after passage through the gastrointestinal tract of mice. J Appl Microbiol 103:1424–1434

Brooijmans R, Smit B, Santos F, van Riel J, de Vos WM, Hugenholtz J (2009) Heme and menaquinone induced electron transport in lactic acid bacteria. Microb Cell Fact 8:28

Bruno-Barcena JM, Andrus JM, Libby SL, Klaenhammer TR, Hassan HM (2004) Expression of a heterologous manganese superoxide dismutase gene in intestinal lactobacilli provides protection against hydrogen peroxide toxicity. Appl Environ Microbiol 70:4702–4710

Bruno-Barcena JM, Azcarate-Peril MA, Klaenhammer TR, Hassan HM (2005) Marker-free chromosomal integration of the manganese superoxide dismutase gene (*sodA*) from *Streptococcus thermophilus* into *Lactobacillus gasseri*. FEMS Microbiol Lett 246:91–101

Coconnier MH, Lievin V, Hemery E, Servin AL (1998) Antagonistic activity against *Helicobacter* infection in vitro and in vivo by the human *Lactobacillus acidophilus* strain LB. Appl Environ Microbiol 64:4573–4580

Corcoran BM, Stanton C, Fitzgerald GF, Ross RP (2005) Survival of probiotic lactobacilli in acidic environments is enhanced in the presence of metabolizable sugars. Appl Environ Microbiol 71:3060–3067

Corcoran BM, Stanton C, Fitzgerald GF, Ross RP (2007) Growth of probiotic lactobacilli in the presence of oleic acid enhances subsequent survival in gastric juice. Microbiology 153:291–299

Corcoran BM, Stanton C, Fitzgerald G, Ross RP (2008) Life under stress: the probiotic stress response and how it may be manipulated. Curr Pharm Des 14:1382–1399

De Angelis M, Gobbetti M (2004) Environmental stress responses in *Lactobacillus*: a review. Proteomics 4:106–122

De Angelis M, Bini L, Pallini V, Cocconcelli PS, Gobbetti M (2001) The acid-stress response in *Lactobacillus sanfranciscensis* CB1. Microbiology 147:1863–1873

De Angelis M, Di Cagno R, Huet C, Crecchio C, Fox PF, Gobbetti M (2004) Heat shock response in *Lactobacillus plantarum*. Appl Environ Microbiol 70:1336–1346

De Vries MC (2006) Analyzing global gene expression of *Lactobacillus plantarum* in the human gastrointestinal tract. PhD thesis. Wageningen University, The Netherlands

de Vos WM, Bron PA, Kleerebezem M (2004) Post-genomics of lactic acid bacteria and other food-grade bacteria to discover gut functionality. Curr Opin Biotechnol 15:86–93

Denou E, Berger B, Barretto C, Panoff JM, Arigoni F, Brussow H (2007) Gene expression of commensal *Lactobacillus johnsonii* strain NCC533 during *in vitro* growth and in the murine gut. J Bacteriol 189:8109–8119

Desmond C, Ross RP, O'Callaghan E, Fitzgerald G, Stanton C (2002) Improved survival of *Lactobacillus paracasei* NFBC 338 in spray-dried powders containing gum acacia. J Appl Microbiol 93:1003–1011

Desmond C, Fitzgerald GF, Stanton C, Ross RP (2004) Improved stress tolerance of GroESL-overproducing *Lactococcus lactis* and probiotic *Lactobacillus paracasei* NFBC 338. Appl Environ Microbiol 70:5929–5936

Duwat P, Sourice S, Cesselin B, Lamberet G, Vido K, Gaudu P, Le Loir Y, Violet F, Loubière P, Gruss A (2001) Respiration capacity of the fermenting bacterium *Lactococcus lactis* and its positive effects on growth and survival. J Bacteriol 183:4509–4516

FAO/WHO (2002) Guidelines for the evaluation of probiotics in food. London, Ontario, Canada

Fiocco D, Capozzi V, Goffin P, Hols P, Spano G (2007) Improved adaptation to heat, cold, and solvent tolerance in *Lactobacillus plantarum*. Appl Microbiol Biotechnol 77:909–915

Fiocco D, Capozzi V, Collins M, Gallone A, Hols P, Guzzo J, Weidmann S, Rieu A, Msadek T, Spano G (2010) Characterization of the CtsR stress response regulon in *Lactobacillus plantarum*. J Bacteriol 192:896–900

Fleischmann RD, Adams MD, White O, Clayton RA, Kirkness EF, Kerlavage AR, Bult CJ, Tomb JF, Dougherty BA, Merrick JM, et al. (1995) Whole-genome random sequencing and assembly of *Haemophilus influenzae* Rd. Science 269:496–512

Frees D, Ingmer H (1999) ClpP participates in the degradation of misfolded protein in *Lactococcus lactis*. Mol Microbiol 31:79–87

Gotteland M, Cruchet S (2003) Suppressive effect of frequent ingestion of *Lactobacillus johnsonii* La1 on *Helicobacter pylori* colonization in asymptomatic volunteers. J Antimicrob Chemother 51:1317–1319

Gresham D, Kruglyak L (2008) Rise of the machines. PLoS Genet 4:e1000134

Gross G, van der Meulen J, Snel J, van der Meer R, Kleerebezem M, Niewold TA, Hulst MM, Smits MA (2008) Mannose-specific interaction of *Lactobacillus plantarum* with porcine jejunal epithelium. FEMS Immunol Med Microbiol 54:215–223

Gross G, Snel J, Boekhorst J, Smits MA, Kleerebezem M (2010) Biodiversity of mannose-specific adhesion in *Lactobacillus plantarum* revisited: strain-specific domain composition of the mannose-adhesin. Beneficial Microbes 1:61–66

Guzzo J, Jobin M-P, Delmas F, Fortier L-C, Garmyn D, Tourdot-Maréchal R, Lee B, Diviès C (2000) Regulation of stress response in *Oenococcus oeni* as a function of environmental changes and growth phase. Int J Food Microbiol 55:27–31

Hartke A, Bouche S, Giard JC, Benachour A, Boutibonnes P, Auffray Y (1996) The lactic acid stress response of *Lactococcus lactis* subsp. *lactis*. Curr Microbiol 33:194–199

Hartke A, Frere J, Boutibonnes P, Auffray Y (1997) Differential induction of the chaperonin GroEL and the Co-chaperonin GroES by heat, acid, and UV-irradiation in *Lactococcus lactis* subsp. *lactis*. Curr Microbiol 34:23–26

Herve-Jimenez L, Guillouard I, Guedon E, Gautier C, Boudebbouze S, Hols P, Monnet V, Rul F, Maguin E (2008) Physiology of *Streptococcus thermophilus* during the late stage of milk fermentation with special regard to sulfur amino-acid metabolism. Proteomics 8:4273–4286

Herve-Jimenez L, Guillouard I, Guedon E, Boudebbouze S, Hols P, Monnet V, Maguin E, Rul F (2009) Postgenomic analysis of *Streptococcus thermophilus* cocultivated in milk with *Lactobacillus delbrueckii* subsp. *bulgaricus*: involvement of nitrogen, purine, and iron metabolism. Appl Environ Microbiol 75:2062–2073

Hols P, Hancy F, Fontaine L, Grossiord B, Prozzi D, Leblond-Bourget N, Decaris B, Bolotin A, Delorme C, Dusko Ehrlich S, Guédon E, Monnet V, Renault P, Kleerebezem M (2005) New insights in the molecular biology and physiology of *Streptococcus thermophilus* revealed by comparative genomics. FEMS Microbiol Rev 29:435–463

Hufner E, Markieton T, Chaillou S, Crutz-Le Coq AM, Zagorec M, Hertel C (2007) Identification of *Lactobacillus sakei* genes induced during meat fermentation and their role in survival and growth. Appl Environ Microbiol 73:2522–2531

Kankainen M, Paulin L, Tynkkynen S, von Ossowski I, Reunanen J, Partanen P, Satokari R, Vesterlund S, Hendrickx AP, Lebeer S, De Keersmaecker SC, Vanderleyden J, Hämäläinen T, Laukkanen S, Salovuori N, Ritari J, Alatalo E, Korpela R, Mattila-Sandholm T, Lassig A, Hatakka K, Kinnunen KT, Karjalainen H, Saxelin M, Laakso K, Surakka A, Palva A, Salusjärvi T, Auvinen P, de Vos WM (2009) Comparative genomic analysis of *Lactobacillus rhamnosus* GG reveals pili containing a human-mucus binding protein. Proc Natl Acad Sci USA 106:17193–17198

Kets E, Teunissen P, de Bont J (1996) Effect of compatible solutes on survival of lactic acid bacteria subjected to drying. Appl Environ Microbiol 62:259–261

Kilstrup M, Jacobsen S, Hammer K, Vogensen FK (1997) Induction of heat shock proteins DnaK, GroEL, and GroES by salt stress in *Lactococcus lactis*. Appl Environ Microbiol 63:1826–1837

Kim WS, Perl L, Park JH, Tandianus JE, Dunn NW (2001) Assessment of stress response of the probiotic *Lactobacillus acidophilus*. Curr Microbiol 43:346–350

Klaenhammer TR, Barrangou R, Buck BL, Azcarate-Peril MA, Altermann E (2005) Genomic features of lactic acid bacteria effecting bioprocessing and health. FEMS Microbiol Rev 29:393–409

Kleerebezem M, Boekhorst J, van Kranenburg R, Molenaar D, Kuipers OP, Leer R, Tarchini R, Peters SA, Sandbrink HM, Fiers MW, Stiekema W, Lankhorst RM, Bron PA, Hoffer SM, Groot MN, Kerkhoven R, de Vries M, Ursing B, de Vos WM, Siezen RJ (2003) Complete genome sequence of *Lactobacillus plantarum* WCFS1. Proc Natl Acad Sci USA 100:1990–1995

Kleerebezem M, Hols P, Bernard E, Rolain T, Zhou M, Siezen RJ, Bron PA (2010) The extracellular biology of the lactobacilli. FEMS Microbiol Rev 34:199–230

Kleerebezem M, Vaughan EE (2009) Probiotic and Gut *Lactobacilli* and Bifidobacteria: molecular approaches to study diversity and activity. Annu Rev Microbiol 63:269–290

Lebeer S, Vanderleyden J, De Keersmaecker SC (2008) Genes and molecules of lactobacilli supporting probiotic action. Microbiol Mol Biol Rev 72:728–764, table of contents

Lebeer S, Verhoeven TL, Francius G, Schoofs G, Lambrichts I, Dufrêne Y, Vanderleyden J, De Keersmaecker SC (2009) Identification of a gene cluster for the biosynthesis of a long, galactose-rich exopolysaccharide in *Lactobacillus rhamnosus* GG and functional analysis of the priming glycosyltransferase. Appl Environ Microbiol 75:3554–3563

Leverrier P, Dimova D, Pichereau V, Auffray Y, Boyaval P, Jan G (2003) Susceptibility and adaptive response to bile salts in *Propionibacterium freudenreichii*: physiological and proteomic analysis. Appl Environ Microbiol 69:3809–3818

Lorca GL, Font de Valdez G, Ljungh A (2002) Characterization of the protein-synthesis dependent adaptive acid tolerance response in *Lactobacillus acidophilus*. J Mol Microbiol Biotechnol 4:525–532

Makarova K, Slesarev A, Wolf Y, Sorokin A, Mirkin B, Koonin E, Pavlov A, Pavlova N, Karamychev V, Polouchine N, Shakhova V, Grigoriev I, Lou Y, Rohksar D, Lucas S, Huang K, Goodstein DM, Hawkins T, Plengvidhya V, Welker D, Hughes J, Goh Y, Benson A, Baldwin K, Lee JH, Díaz-Muñiz I, Dosti B, Smeianov V, Wechter W, Barabote R, Lorca G, Altermann E, Barrangou R, Ganesan B, Xie Y, Rawsthorne H, Tamir D, Parker C, Breidt F, Broadbent J, Hutkins R, O'Sullivan D, Steele J, Unlu G, Saier M, Klaenhammer T, Richardson P, Kozyavkin S, Weimer B, Mills D (2006) Comparative genomics of the lactic acid bacteria. Proc Natl Acad Sci USA 103:15611–15616

Marco ML, Peters TH, Bongers RS, Molenaar D, van Hemert S, Sonnenburg JL, Gordon JI, Kleerebezem M (2009) Lifestyle of *Lactobacillus plantarum* in the mouse caecum. Environ Microbiol 11:2747–2757

Macro ML, Wels M, de vos WM, Vaughan EE, Kleerebezem M (2010) Convergence in probiotic *Lactobacillus* gut-adaptive responses in humans and mice. ISME J 11:1481–1484

Martini MC, Lerebours EC, Lin WJ, Harlander SK, Berrada NM, Antoine JM, Savaiano DA (1991) Strains and species of lactic acid bacteria in fermented milks (yogurts): effect on *in vivo* lactose digestion. Am J Clin Nutr 54:1041–1046

McLeod A, Nyquist OL, Snipen L, Naterstad K, Axelsson L (2008) Diversity of *Lactobacillus sakei* strains investigated by phenotypic and genotypic methods. Syst Appl Microbiol 31:393–403

Meijerink M, van Hemert S, Taverne N, Wels M, de vos P, Bron PA, Savelkoul HF, Van Bilsen J, Kleerebezem M, Wells JM (2010) Identification of generic loci in *Lactobacillus plantarum* that modulate the immune response of dendritic cells using comparative genome hybridization. PLos One 5:e10632

Molenaar D, Bringel F, Schuren FH, de Vos WM, Siezen RJ, Kleerebezem M (2005) Exploring *Lactobacillus plantarum* genome diversity by using microarrays. J Bacteriol 187:6119–6127

Noonpakdee W, Sitthimonchai S, Panyim S, Lertsiri S (2004) Expression of the catalase gene *katA* in starter culture *Lactobacillus plantarum* TISTR850 tolerates oxidative stress and reduces lipid oxidation in fermented meat product. Int J Food Microbiol 95:127–135

O'Mahony L, McCarthy J, Kelly P, Hurley G, Luo F, Chen K, O'Sullivan GC, Kiely B, Collins JK, Shanahan F, Quigley EM (2005) *Lactobacillus* and *Bifidobacterium* in irritable bowel syndrome: symptom responses and relationship to cytokine profiles. Gastroenterology 128:541–551

O'Sullivan E, Condon S (1997) Intracellular pH is a major factor in the induction of tolerance to acid and other stresses in *Lactococcus lactis*. Appl Environ Microbiol 63:4210–4215

Perea Velez M, Verhoeven TL, Draing C, Von Aulock S, Pfitzenmaier M, Geyer A, Lambrichts I, Grangette C, Pot B, Vanderleyden J, De Keersmaecker SC (2007) Functional analysis of D-alanylation of lipoteichoic acid in the probiotic strain *Lactobacillus rhamnosus* GG. Appl Environ Microbiol 73:3595–3604

Pfeiler EA, Azcarate-Peril MA, Klaenhammer TR (2007) Characterization of a novel bile-inducible operon encoding a two-component regulatory system in *Lactobacillus acidophilus*. J Bacteriol 189:4624–4634

Pieterse B (2006) Transcriptome analysis of the lactic acid and NaCI-stress response of *Lactobacillus plantarum*. PhD thesis. Wageningen University, The Netherlands

Pieterse B, Leer RJ, Schuren FH, van der Werf MJ (2005) Unravelling the multiple effects of lactic acid stress on *Lactobacillus plantarum* by transcription profiling. Microbiology 151:3881–3894

Prasad J, McJarrow P, Gopal P (2003) Heat and osmotic stress responses of probiotic *Lactobacillus rhamnosus* HN001 (DR20) in relation to viability after drying. Appl Environ Microbiol 69:917–925

Pretzer G, Snel J, Molenaar D, Wiersma A, Bron PA, Lambert J, de Vos WM, van der Meer R, Smits MA, Kleerebezem M (2005) Biodiversity-based identification and functional characterization of the mannose-specific adhesin of *Lactobacillus plantarum*. J Bacteriol 187:6128–6136

Pridmore RD, Berger B, Desiere F, Vilanova D, Barretto C, Pittet AC, Zwahlen MC, Rouvet M, Altermann E, Barrangou R, Mollet B, Mercenier A, Klaenhammer T, Arigoni F, Schell MA (2004) The genome sequence of the probiotic intestinal bacterium *Lactobacillus johnsonii* NCC 533. Proc Natl Acad Sci USA 101:2512–2517

Rainey PB, Preston GM (2000) *In vivo* expression technology strategies: valuable tools for biotechnology. Curr Opin Biotechnol 11:440–444

Rauch PJ, De Vos WM (1992) Characterization of the novel nisin-sucrose conjugative transposon Tn5276 and its insertion in *Lactococcus lactis*. J Bacteriol 174:1280–1287

Rediers H, Rainey PB, Vanderleyden J, De Mot R (2005) Unraveling the secret lives of bacteria: use of *in vivo* expression technology and differential fluorescence induction promoter traps as tools for exploring niche-specific gene expression. Microbiol Mol Biol Rev 69:217–261

Rochat T, Gratadoux JJ, Corthier G, Coqueran B, Nader-Macias ME, Gruss A, Langella P (2005a) *Lactococcus lactis* SpOx spontaneous mutants: a family of oxidative-stress-resistant dairy strains. Appl Environ Microbiol 71:2782–2788

Rochat T, Miyoshi A, Gratadoux JJ, Duwat P, Sourice S, Azevedo V, Langella P (2005b) High-level resistance to oxidative stress in *Lactococcus lactis* conferred by *Bacillus subtilis* catalase KatE. Microbiology 151:3011–3018

Salotra P, Singh DK, Seal KP, Krishna N, Jaffe H, Bhatnagar R (1995) Expression of DnaK and GroEL homologs in *Leuconostoc esenteroides* in response to heat shock, cold shock or chemical stress. FEMS Microbiol Lett 131:57–62

Serrano LM, Molenaar D, Wels M, Teusink B, Bron PA, de Vos WM, Smid EJ (2007) Thioredoxin reductase is a key factor in the oxidative stress response of *Lactobacillus plantarum* WCFS1. Microb Cell Fact 6:29

Sheehan VM, Sleator RD, Fitzgerald GF, Hill C (2006) Heterologous expression of BetL, a betaine uptake system, enhances the stress tolerance of *Lactobacillus salivarius* UCC118. Appl Environ Microbiol 72:2170–2177

Shendure J, Ji H (2008) Next-generation DNA sequencing. Nat Biotechnol 26:1135–1145

Siezen RJ, Wilson G (2010) Probiotic genomics. Microbial Biotechnology 3:1–9

Siezen RJ, Renckens B, van Swam I, Peters S, van Kranenburg R, Kleerebezem M, de Vos WM (2005) Complete sequences of four plasmids of *Lactococcus lactis* subsp. *cremoris* SK11 reveal extensive adaptation to the dairy environment. Appl Environ Microbiol 71:8371–8382

Siezen R, Boekhorst J, Muscariello L, Molenaar D, Renckens B, Kleerebezem M (2006) *Lactobacillus plantarum* gene clusters encoding putative cell-surface protein complexes for carbohydrate utilization are conserved in specific Gram-positive bacteria. BMC Genom 7:126

Siezen RJ, Tzeneva VA, Castioni A, Wels M, Phan HT, Rademaker JL, Starrenburg MJ, Kleerebezem M, Molenaar D, van Hylckama Vlieg JE (2010) Phenotypic and genomic diversity of *Lactobacillus plantarum* strains isolated from various environmental niches. Environ Microbiol 12:758–773

Sleator RD, Hill C (2006) Patho-biotechnology: using bad bugs to do good things. Curr Opin Biotechnol 17:211–216

Smiddy M, Sleator RD, Patterson MF, Hill C, Kelly AL (2004) Role for compatible solutes glycine betaine and L-carnitine in listerial barotolerance. Appl Environ Microbiol 70:7555–7557

Smit G, Smit BA, Engels WJ (2005) Flavour formation by lactic acid bacteria and biochemical flavour profiling of cheese products. FEMS Microbiol Rev 29:591–610

Spano G, Massa S (2006) Environmental stress response in wine lactic acid bacteria: beyond *Bacillus subtilis*. Crit Rev Microbiol 32:77–86

Steidler L, Neirynck S, Huyghebaert N, Snoeck V, Vermeire A, Goddeeris B, Cox E, Remon JP, Remaut E (2003) Biological containment of genetically modified *Lactococcus lactis* for intestinal delivery of human interleukin 10. Nat Biotechnol 21:785–789

van de Guchte M, Serror P, Chervaux C, Smokvina T, Ehrlich SD, Maguin E (2002) Stress responses in lactic acid bacteria. Antonie van Leeuwenhoek 82:187–216

van Hemert S, Meijerink M, Molenaar D, Bron PA, de Vos P, Kleerebezem M, Wells JM, Marco ML (2010) Identification of *Lactobacillus plantarum* genes modulating the cytokine response of human peripheral blood mononuclear cells. BMC Microbiol 10:293

van Kranenburg R, Golic N, Bongers R, Leer RJ, de Vos WM, Siezen RJ, Kleerebezem M (2005) Functional analysis of three plasmids from *Lactobacillus plantarum*. Appl Environ Microbiol 71:1223–1230

Veening JW, Smits WK, Kuipers OP (2008a) Bistability, epigenetics, and bet-hedging in bacteria. Annu Rev Microbiol 62:193–210

Veening JW, Stewart EJ, Berngruber TW, Taddei F, Kuipers OP, Hamoen LW (2008b) Bet-hedging and epigenetic inheritance in bacterial cell development. Proc Natl Acad Sci USA 105:4393–4398

Vogel RF, Pavlovic M, Hormann S, Ehrmann MA (2005) High pressure-sensitive gene expression in *Lactobacillus sanfranciscensis*. Braz J Med Biol Res 38:1247–1252

Wall T, Bath K, Britton RA, Jonsson H, Versalovic J, Roos S (2007) The early response to acid shock in *Lactobacillus reuteri* involves the ClpL chaperone and a putative cell wall-altering esterase. Appl Environ Microbiol 73:3924–3935

Walter J, Heng NC, Hammes WP, Loach DM, Tannock GW, Hertel C (2003) Identification of *Lactobacillus reuteri* genes specifically induced in the mouse gastrointestinal tract. Appl Environ Microbiol 69:2044–2051

Walter J, Loach DM, Alqumber M, Rockel C, Hermann C, Pfitzenmaier M, Tannock GW (2007) D-alanyl ester depletion of teichoic acids in *Lactobacillus reuteri* 100–23 results in impaired colonization of the mouse gastrointestinal tract. Environ Microbiol 9:1750–1760

Wels M (2008) Unraveling the regulatory network of *Lactobacillus plantarum* WCFS1 (Thesis), pp. 1–192, Wageningen University, The Netherlands

Wels M, Francke C, Kerkhoven R, Kleerebezem M, Siezen RJ (2006) Predicting cis-acting elements of *Lactobacillus plantarum* by comparative genomics with different taxonomic subgroups. Nucleic Acids Res 34:1947–1958

Whitehead K, Versalovic J, Roos S, Britton RA (2008) Genomic and genetic characterization of the bile stress response of probiotic *Lactobacillus reuteri* ATCC 55730. Appl Environ Microbiol 74:1812–1819

Xie Y, Chou LS, Cutler A, Weimer B (2004) DNA Macroarray profiling of *Lactococcus lactis* subsp. *lactis* IL1403 gene expression during environmental stresses. Appl Environ Microbiol 70:6738–6747

Zoetendal EG, Booijink CCGM, Klaassens ES, Heilig HGHJ, Kleerebezem M, Smidt H, de Vos WM (2006) Isolation of RNA from bacterial samples of the human gastrointestinal tract. Nat Protoc 1:954–959

Zomer A, Fernandez M, Kearney B, Fitzgerald GF, Ventura M, van Sinderen D (2009) An interactive regulatory network controls stress response in *Bifidobacterium breve* UCC2003. J Bacteriol 191:7039–7049

Zotta T, Asterinou K, Rossano R, Ricciardi A, Varcamonti M, Parente E (2009) Effect of inactivation of stress response regulators on the growth and survival of *Streptococcus thermophilus* Sfi39. Int J Food Microbiol 129:211–220

Chapter 17
Improving the Stress Tolerance of Probiotic Cultures: Recent Trends and Future Directions

Aditya Upadrasta, Catherine Stanton, Colin Hill, Gerald F. Fitzgerald, and R. Paul Ross

17.1 Introduction

The human gastrointestinal tract (GIT) harbors a rich microbiota consisting of up to 1,800 genera and between 15,000 and 36,000 bacterial species, as revealed by culture-independent metagenomic analysis (Frank et al. 2007). The colon is the primary site of microbial colonization in humans and animals, and the majority of the indigenous gut microbes are obligate anaerobes (Berg 1996). Many of these commensals are uncultivatable, exhibiting either poor growth or no growth in synthetic media, and can only be identified using culture-independent methods, such as denatured gradient gel electrophoresis or the recently developed 16S rDNA-based (pyro) sequencing tools. Such compositional sequencing approaches have revealed that more than 98% of the human gut microbiota belong to only four bacterial phyla: *Firmicutes* (64%), *Bacteriodetes* (23%), *Proteobacteria* (8%), and *Actinobacteria* (3%) (Frank et al. 2007).

The gut microbiota have important roles in the physiology of the host's digestive process and the assimilation of nutrients, thereby contributing to host health (Vernazza et al. 2006). Among the most well-studied human gut microbes are *Lactobacillus* spp., which belong to a group of bacteria collectively known as lactic acid bacteria (LAB). These bacteria are ubiquitous and are natural inhabitants of many niches, including milk, meat, plants, grains, as well as the GIT of vertebrates. LAB are a heterogeneous family of Gram-positive, anaerobic, nonsporulating, and acid-tolerant bacteria with low guanine-cytosine (G+C) content in their genomes (Claesson et al. 2007), whereas members of the *Bifidobacterium* genus are rich in G+C (Ventura et al. 2004). The major end product after carbohydrate metabolism

R.P. Ross (✉)
Teagasc, Moorepark Food Research Centre, Fermoy, Co Cork, Ireland

Alimentary Pharmabiotic Centre, Cork, Ireland
e-mail: paul.ross@teagasc.ie

E. Tsakalidou and K. Papadimitriou (eds.), *Stress Responses of Lactic Acid Bacteria*, Food Microbiology and Food Safety, DOI 10.1007/978-0-387-92771-8_17,

in LAB is lactic acid, along with other byproducts, including acetic acid, formic acid, ethanol, and carbon dioxide (Vandenbergh 1993).

Members of LAB have a long traditional history as starter cultures in food and beverage fermentations from ancient times. They contribute to the rapid acidification of food products and also improve the flavor, texture, and nutritional composition of fermented foods (Ross et al. 2002a). It was at the beginning of the twentieth century when Elie Metchnikoff first proposed the scientific rationale that specific bacteria were thought to be beneficial to health (Stanton et al. 2003a). Foods containing probiotics belong to the functional food category, and such foods provide specific health benefits over and above their nutritional value (Stanton et al. 2005). Functional foods include those containing bioactive ingredients such as probiotics; prebiotics; bioactive peptides such as bacteriocins, which are small antimicrobial peptides; bioactive fatty acids such as conjugated linoleic acid (CLA); and organic acids (Stanton et al. 2001). The functional food concept has become popular within the food industry in recent years and is now a global market due to the heightened awareness of consumers of the added value of these foods.

17.1.1 Gut Microbiology, Strains, and Human Health

Members of the *Lactobacillus* and *Bifidobacterium* genera are currently used as probiotics (Stanton et al. 2003b), whereas other groups such as *Enterococcus*, *Oenoccoccus*, *Propionibacterium*, *Bacillus*, *Escherichia coli*, *Clostridium butyricum*, and some yeast strains, such as *Saccharomyces boulardii* (also called *Saccharomyces cerevisiae*), have also been used. In general, *Lactobacillus* spp. and *Bifidobacterium* spp. have been studied extensively as probiotic microorganisms, whereas recent studies have also focused on non-LAB such as yeast, propionibacteria, and other nonpathogenic strains of *E. coli*, *Bacillus*, and *C. butyricum*.

Some examples of currently used probiotic bacteria are listed in Table 17.1. Recent developments in molecular methodologies enable us to understand the diversity and functionality of probiotic intestinal lactobacilli and bifidobacteria in the human GIT and provide a platform to identify effective probiotics as therapeutic agents for humans. A growing body of evidence based on in vitro and in vivo data supports the health benefits associated with the consumption of functional foods containing probiotics. Many of the studies elucidating the health benefits and clinical effects of probiotics have been assessed in various recent reviews (Naidu et al. 1999; Stanton et al. 2003a; Baker et al. 2009; Williams 2010). A selection of probiotic strains with specific proven health effects based on placebo-controlled human clinical trials is listed in Table 17.2. Probiotics have been shown to affect the host immune system positively through immunoglobulin production, antibody response, and other cell-mediated immune responses (Marteau and Rambaud 1993; Matar et al. 2002). The indigenous intestinal microbiota also affords protection of the host from pathogen colonization and invasion into the gut epithelium cell wall by forming

Table 17.1 Examples of currently used probiotic bacteria

Genus	Species – strain	Source
Lactobacillus	*acidophilus* NCFM	Rhodia, Inc. (Madison, WI)
	acidophilus R0052	Institut Rosell (Montreal, Canada)
	acidophilus LA-1	Chr. Hansen (Horsholm, Denmark)
	acidophilus LA-5	Chr. Hansen (Milwaukee, WI)
	acidophilus LB	Lacetol Laboratory (Houdan, France)
	acidophilus SBT-2062	Snow Brand Milk Products Co., Ltd. (Tokyo, Japan)
	casei Immunitas (DN-114 001)	Danone (Paris, France)
	casei strain Shirota	Yakult (Tokyo, Japan)
	fermentum VRI003 (PCC)	Probiomics (Eveleigh, Australia)
	fermentum RC-14	Urex Biotech (London, Ontario, Canada)
	johnsonii La-1 (same as NCC533 and formerly *L. acidophilus*)	Nestlé (Lausanne, Switzerland)
	paracasei 33	GenMont Biotech (Taiwan)
	paracasei GMNL-33	
	paracasei F19	Medipharm (Des Moines, Iowa)
	paracasei CRl 431	Chr. Hansen (Milwaukee, WI)
	plantarum OM	Bio-Energy Systems, Inc. (Kalispell, MT)
	plantarum 299V	Probi AB (Lund, Sweden)
	reuteri RC-14	Chr. Hansens (Milaukee, WI) and Urex Biotech (London, Ontario, Canada)
	reuteri ATCC55730	Probi AB (Lund, Sweden)
	reuteri SD2112	Biogaia (Raleigh, NC)
	rhamnosus R0011	Institut Rosell (Montreal, Canada)
	rhamnosus GG (LGG)	Valio Dairy (Helsinki, Finland)
	rhamnosus GR-1	Urex Biotech (London, Ontario, Canada)
	rhamnosus 271	Probi AB (Lund, Sweden)
	rhamnosus HN001 (DR20)	Fonterra (Wellington, New Zealand)
	rhamnosus GM-020	GenMont Biotech (Taiwan)
	salivarius UCC118	University College (Cork, Ireland)
Lactococcus	*lactis* L1A	Essum AB (Umeå, Sweden)
Bifidobacterium	*animalis* DN 117-001	Danone (Montreal, Canada)
	breve Yakult	Yakut (Tokyo, Japan)
	infantis 35624	Procter & Gamble (Mason, OH)
	longum SBT-2928	Snow Brand Milk Products Co., Ltd. (Tokyo, Japan)
	longum BB536	Morinaga Milk Industry Co., Ltd. (Zama-City, Japan)
	lactis Bb-12	Chr. Hansen (Horsholm, Denmark) Nestlé (Glendale, CA)
	lactis HN019 (DR10)	Fonterra (Wellington, New Zealand)
Streptococcus	*oralis* KJ3	Oragenics Inc. (Alachua, FL)
	uberis KJ2	Oragenics Inc. (Alachua, FL)
	rattus JH145	Oragenics Inc. (Alachua, FL)
Bacillus	*coagulans* BC30	Ganeden Biotech Inc. (Kalispell, MT)
Saccharomyces	*cerevisiae* (*boulardii*)	Biocodex (Creswell, OR)

Table 17.2 Examples of probiotic bacteria and their scientifically proven health effects: based on placebo-controlled human clinical trials

Beneficial effect	Tested strains	Subjects	Dose	Form	References
Diarrhea					
Reduced the frequency of severe diarrhea associated with chemotherapy of colorectal cancer and abdominal discomfort	*L. rhamnosus* GG (ATCC 53103, Gefilus®, Valio Ltd., Helsinki, Finland)	Patients age between 18 to 75 years	Twice a day at a dose of $1–2 \times 10^{10}$	Gelatin capsules	Österlund et al. (2007)
Reduced the severity and duration of acute gastroenteritis (mild diarrheal illness) in young children	*Streptococcus thermophilus*, *B. lactis*, *L. acidophilus* combined with zinc	Children (age between 6 and 12 months)	6×10^9 CFU (2×10^9 of each strain) and 10 mg of zinc/day	Cereals	Shamir et al. (2005)
Alleviating lactose malabsorption symptoms					
	L. bulgaricus *S. thermophilus*	Graduate students	Not specified	Yogurt	Gilliland and Kim (1984)
	L. bulgaricus LA-1 and LA-2 *L. acidophilus* NCFM *L. acidophilus* A-1, B-1, C-1, D-1, and La10	Adults (6 males and 4 females)	$10^7–10^8$ CFU/mL	Nonfermented yogurt and acidophilus milk	Lin et al. (1991)
	L. acidophilus *L. bulgaricus* *Bifidobacterium* spp.	Adult volunteers (age between 20 and 45 years)	Not specified	Traditional yogurt, fermented milk (Ofilus), Bulgofilus	Vesa et al. (1996)
	L. acidophilus NCFM *L. lactis* *S. thermophilus*	20 children (9 boys and 11 girls) age 5–16 years	~10^8 and 10^{10} cells	Low-fat milk	Montes et al. (1995)

Reducing antibiotic-associated diarrhea (AAD)	*L. rhamnosus* GG	16 healthy male volunteers age 18–24 years	Not specified	125 mL of fermented yogurt + erythromycin 400 mg t.i.d.	Siitonen et al. (1990)
	L. rhamnosus GG (*L. casei* subsp. *rhamnosus*)	202 children (age 6 months–10 years)	1×10^{10} (<12 kg) 2×10^{10} (>12 kg) CFU/day	Capsule	Vanderhoof et al. (1999)
	L. casei DN-114 001	541 male soldiers (275 treatment and 266 control)	10^8 CFU/mL	Yogurt	Pereg et al. (2005)
	Lactobacillus spp. (*Weissella. confusa, L. fermentum*)	Children age <5 years	10^8 CFU	Fermented millet product (KSW-koko sour water)	Lei and Jakobsen (2004)
	Saccharomyces boulardii	Hospitalized patients	1 g of lyophilized powder/day	Capsule	Surawicz et al. (1989)
		128 critically ill tube-fed patients	500 mg capsules 4 times/day	Capsule	Bleichner et al. (1997)
		193 in patients age 18–86 years	Lyophilized powder 1 g/day (3×10^{10} CFU)	Capsule	McFarland et al. (1995)
	Clostridium butyricum MIYAIRI	110 children age 1 month–15 years	10^7 CFU/g	Not specified	Seki et al. (2003)
Reduced the risk and occurrence of traveler's diarrhea	*L. rhamnosus* GG	756 persons (different age groups)	2×10^9 CFU/g	Sachets with cold water	Oksanen et al. (1990)
Efficacy against rotavirus diarrhea	*L. sporogenes*	112 newborn infants	10^8 cells/tsp water	Sterile distilled water	Chandra (2002)
Reduced the incidence of *C. difficile*–induced diarrhea and antibiotic-associated diarrhea	*L. bulgaricus* *L. casei* *S. thermophilus*	135 hospitalized patients	100 g (97 mL)	Milk drink	Hickson et al. (2007)

(continued)

Table 17.2 (continued)

Beneficial effect	Tested strains	Subjects	Dose	Form	References
Effective in increasing stool frequency in children with functional constipation	*B. lactis* DN-173010 + starter cultures	160 constipated children (boys and girls) age 3–16 years	6×10^9 CFU/pot and 1.2×10^9 CFU/pot, 2 pots/day (125 g/pot)	Fermented yogurt (Activia®)	Tabbers et al. (2009)
	L. bulgaricus				
	S. thermophilus				
Effective in increasing stool frequency in adult women with constipation	*B. lactis* DN-173010 + starter cultures	135 adult females age 25–65 years	1.25×10^{10} CFU/pot and 1.2×10^9 CFU/pot, 1 pot/day (100 g/pot)	Fermented milk (BIO®, Danone, China)	Yang et al. (2008)
	S. thermophilus				
	L. bulgaricus				
Inflammatory bowel diseases (IBD)					
Crohn's disease: High-dose probiotic and prebiotic (synbiotic) therapy is effective in the treatment of active Crohn's disease	*L. casei*	10 patients age 27 ± 7 years	30×10^9 CFU/day	Oral	Fujimori et al. (2007)
	B. breve (Yakult BL)				
	B. longum (ISAGOL)				
	Prebiotic-psyllium		15×10^9 CFU/day, 3.3 g/day	Dissolved in 100 mL of water	
Effective in the treatment of ulcerative colitis	*Saccharomyces boulardii*	25 patients	250 mg, 3 times/day	Capsule	Guslandi et al. (2003)
Effective in the treatment of ulcerative colitis and remission of the disease equivalent to gold standard therapy with mesalazine	*E. coli* Nissile 1917 (serotype O6:K5:H1)	327 patients age 18–70 years	2.5×10^9 CFU, 200 mg/day	Capsule	Kruis et al. (2004)
	B. longum (Probiotic)	($n = 120$ patients)	2×10^9 CFU	Capsule	Fujimori et al. (2009)

Synbiotic therapy is more effective treatment for ulcerative colitis than probiotic or prebiotic treatments alone	Psyllium (prebiotic) Synbiotic (pro+prebiotic) Synbiotic therapy is more effective in the treatment of ulcerative colitis than the other treatments	40 probiotic therapy 40 prebiotic therapy 40 synbiotic (pro+pre) therapy	8 g of psyllium Both treatments	4 g of psyllium dissolved in water before intake	
	L. casei *L. plantarum* *L. acidophilus* *L. delbreukii* subsp. *bulgaricus* *B. longum* *B. breve* *B. infantis* *Streptococcus salivarius* subsp. *thermophilus* (termed VSL#3)	34 patients age 18–65 years	10^9 CFU/g, 3 g twice a day	Lyophilized powder in sachets	Bibiloni et al. (2005)
Effective in the treatment of recurrent or refractory pouchitis	VSL#3	36 patients	10^9 CFU/g, 6 g/day in a single dose	Lyophilized powder in sachets	Mimura et al. (2004)
Effective in the prevention of the onset of acute pouchitis and improved the quality of life of patients with ileal pouch-anal anastomosis	VSL#3	40 patients age 18–65 years old	10^9 CFU, 3 g of packet/day	Lyophilized powder in packets	Gionchetti et al. (2003, 2007)
Clinical improvement in patients with irritable bowel syndrome (IBS)	VSL#3	10 patients (6 females and 4 males) age 24–58 years	10^9 CFU, 3 g/day	Lyophilized powder	Brigidi et al. (2001)
Alleviated the symptoms of irritable bowel syndrome and stabilizes the gut flora	*L. rhamnosus* GG *L. rhamnosus* Lc705 *Propionibacterium freudenreichii* subsp. *shermanii* JS *B. animalis* subsp. *lactis* Bb12	86 IBS patients age 20–60 years	1×10^7 CFU/mL, 1.2 dL/day	Milk drink (80% lactose free milk and 20% fruit juice)	Kajander et al. (2008)

(continued)

Table 17.2 (continued)

Beneficial effect	Tested strains	Subjects	Dose	Form	References
Efficacy of alleviating IBD symptoms	*L. acidophilus* CUL60 and CUL21 *B. lactis* CUL34 *B. bifidum* CUL20	52 volunteers with IBS age 40 ± years	2.5×10^{10} CFU, 1 capsule/day	Capsule	Williams et al. (2009)
Alleviation of inflammatory bowel syndrome (IBS) symptoms and immune modulation	*L. salivarius* UCC4331 *B. infantis* 35624	77 patients age 18–75 years	1×10^{10} CFU	Malted milk drink	O'Mahony et al. (2005)
Efficacy in relieving many of the symptoms of IBS	Encapsulated *B. infantis* 35624	362 IBS patients	1×10^{8} CFU/mL	Capsule (freeze-dried culture)	Whorwell et al. (2006)
Improved digestive comfort in adults without diagnosed gastric disorders	*B. lactis* DN-173010 + starter cultures *S. thermophilus* *L. bulgaricus*	371 adults age 39.5 ± 12.3 years	1.25×10^{10} CFU/pot and 1.2×10^{9} CFU/pot Group I 1 pot (125 g)/day Group II 2 pots/day	Fermented milk (Activia, Danone)	Guyonnet et al. (2009)
H. pylori–associated diseases					
Effective during therapy against *H. pylori*	Group I *L. rhamnosus* GG Group II *S. boulardii* Group III *Lactobacillus* sp. and Bifidobacteria	85 *H. pylori*-positive asymptomatic patients divided into three groups			Cremonini et al. (2002)
Positive impact on *H. pylori* therapy-associated side effects (antibiotic-associated gastrointestinal side effects)	*L. rhamnosus* GG	Healthy *H. pylori*-positive asymptomatic volunteers, 25 male and 35 female age 40 ± 12 years	6×10^{9} cells	Freeze-dried powder mixed with water	Armuzzi et al. (2001)
Effective in suppressing *H. pylori* and reducing gastric mucosal inflammation	*L. gasseri* OLL2716 (LG21)	*H. pylori* infected humans 31 (29 male and 2 females of age 50 ± 7.4 years)	$1–1.4 \times 10^{7}$ CFU/g, 90 g twice a day	Yogurt	Sakamoto et al. (2001)

Effective in suppressing *H. pylori* infection in humans	*L. acidophilus* La5 *L. bulgaricus* *B. lactis* Bb12 *S. thermophilus*	70 volunteers age 22–59 years	10^7 CFU/mL, 230 mL/day	Yogurt	Wang et al. (2004)
	L. acidophilus (*johnsonii*) La1	20 (8 females and 12 males, mean age 33.1 years)	50 mL of whey-based culture supernatant	Whey	Michetti et al. (1999)
Peptic ulcers, chronic gastritis	*L. casei* strain Shirota	20 *H. pylori*–colonized subjects	10^8 CFU/mL, 65 mL 3 times/day	Yakult milk drink	Cats et al. (2003)
	L. casei DN-114001 + omeprazole, amoxocillin and clarithromycin (OAC)	86 symptomatic *H. pylori*–positive children		Fermented milk (Actimel)	Sýkora et al. (2005)
Protective effect and reduced cancer risk factors in polypectomized colon cancer patients	*L. rhamnosus* GG *B. lactis* Bb12 + BeneoSynergy1 (SYN1 a prebiotic an oligofructose enriched inulin)	80 subjects (37 colon cancer and 43 polypectomized patients)	>$\log_{10}$CFU/g one 12 g sachet/day	Capsugel	Rafter et al. (2007)
Effective in the treatment of postoperative infectious complications in biliary cancer surgery	*L. casei* Shirota (Yakult 400) *B. breve* (Bifel) + galacto oligosaccharides (oligomate 55, a prebiotic)	101 patients with biliary cancer	80 mL of Yakult 400 containing 4×10^{10} CFU and 100 mL of Bifel containing 1×10^{10} CFU oligomate 15 g/day	Yakult 400 Bifel (Yakult Honsha, Japan)	Sugawara et al. (2006)
Reduces the risk of bladder cancer	*L. casei* Shirota	445 subjects	1–4 or more times/week	Fermented milk products	Ohashi et al. (2002)

(continued)

Table 17.2 (continued)

Beneficial effect	Tested strains	Subjects	Dose	Form	References
Reduction in infectious complications in patients with severe acute pancreatitis	*L. acidophilus* *L. casei* *L. salivarius* *L. lactis* *B. bifidum* *B. lactis*	200 adult patients age ≥ 18 years	10^9 CFU twice a day	Through nasojejunal tube	Besselink et al. (2004)
Reducing serum cholesterol levels	*E. faecium* M-74 + 50 μg selenium	43 human volunteers	2×10^9 CFU	Capsule	Hlivak et al. (2005)
Reduces the stress-induced gastrointestinal disorders	*L. acidophilus* Rosell-52 *B. longum* Rosell-175	75 volunteers (male and female) age 18–60 years	3×10^9 CFU/sachet stick once a day	Sachet stick (Probio-Stick)	Diop et al. (2008)
Reduced the cumulative prevalence of eczema	*L. rhamnosus* HN001	Pregnant women and infants	6×10^9 CFU	Capsule (for mothers) and breast milk, water or syringe feeding to infants	Wickens et al. (2008)
Effective in the treatment of early atopic eczema disease in children	*L. rhamnosus* GG	159 mothers	1×10^{10} CFU, 2 capsules/day	Capsule	Kalliomäki et al. (2001)
Immune stimulation during human immunodeficiency virus (HIV) infection and improved the quality of life	*B. bifidum* *S. thermophilus*	77 children age 2–12 years	2.5×10^{10} CFU, 14 g/day	Diluted in milk (NAN2 probiotico or NAN2® Nestlé)	Trois et al. (2008)
Improved the quality of life, reduced diarrheal symptoms, and increased CD4 counts in HIV/AIDS patients	*L. rhamnosus* GR-1 *L. reuteri* RC-14 + starter cultures *L. bulgaricus* *S. thermophilus*	24 female HIV/AIDS patients age 18–44 years	100 mL/day	Yogurt	Anukam et al. (2008)

Increased spermidine levels and reduction of gut mutagenicity	*B. lactis* LKM512 *L. delbreukii* subsp. *bulgaricus* LKM1759 *S. thermophilus* LKM1742	7 healthy adult volunteers (1 male and 6 females), average age 30.5 years	Bifidobacteria 5.2×10^7 Lactobacilli and *Streptococcus* 4.7×10^8 CFU/g 100 g/day	Yogurt	Matsumoto and Benno (2004)
Effective in the reduction of vaginal colonization by pathogenic bacteria and yeast there by offering protection against urogenital disease	*L. rhamnosus* GR-1 *L. fermentum* RC-14	64 healthy women age 19–46 years	$>10^9$ CFU/capsule, 1 capsule/day	Capsule (freeze-dried cultures)	Reid et al. (2003)
Improved well-being in mild rheumatoid arthritis patients	*L. rhamnosus* GG	21 patients age 18–64 years	$\geq 5 \times 10^9$ CFU/capsule twice/day	Capsule (Gefilus® Valio Ltd.)	Hatakka et al. (2003)
Reduction of oral mutans *Streptococci* in orthodontic patients	*B. animalis* subsp. *lactis* DN-173010	24 healthy adolescents age 12–16 years	2×10^8 CFU 200 g/day	Fruit yogurt	Cildir et al. (2009)

a frontline of mucosal defense (Marteau and Rambaud 1993). The mammalian immune system has innate and adaptive components, which cooperate to protect the host against bacterial infections. The innate immune system consists of distinct functional modules that evolved to provide various protective forms against pathogens; nevertheless, it senses the bacteria through pattern-recognition receptors to activate the front-line defense to the host (Medzhitov 2007). Moreover, recently, it was reported that a deficiency of Toll-like receptor 5 (a key component of the innate immune system expressed in the gut mucosa) predisposed mice to developing key features of the metabolic syndrome (Vijay-Kumar et al. 2010). This reveals that changes in the gut microbiome can alter the state of gut health and may promote the development of the metabolic syndrome.

17.1.2 Digestive System and Physiology: An Overview

The digestive system is responsible for the breakdown of food and the assimilation of nutrients that are essential for life. The initial digestive process starts with various secretions, most notably enzymes, first in the mouth, where lysozyme is secreted, and in the upper GIT (stomach, small intestine, and organs such as the liver, pancreas, and gallbladder), where a range of hydrolytic enzymes are secreted, including amylase, mycozyme, lipase, protease, pepsin, pancreatin, bile hydrolase, and cellulase. During an average life span of an individual, 60 tons of food pass through this alimentary channel (Bengmark 1998). Long-term monitoring studies have indicated that intestinal microbiota do not drastically change in adults over periods ranging from several months up to 2 years (Heilig et al. 2002; Ley et al. 2006).

17.1.3 Metagenomics: The Changing Face of Gut Microbiology

The GIT harbors billions of bacteria and provides strong defensive mechanisms of physiological and immune barriers against a variety of bacteria (Galdeano and Perdigon 2006). The bacterial community found at a specific habitat is termed the *microbiome*, and the collective genomes of the total microbiota is termed the *metagenome* (Handelsman et al. 1998). The microbiome was defined as the collective genome of the indigenous microbial population colonizing a person (McAuliffe and Klaenhammer 2002). Recently introduced molecular microbiology methods (16S rRNA and rDNA genes-based determination) facilitate the tracking of uncultivable bacterial species that are residing in these complex ecosystems (Vaughan et al. 2000). Such approaches have facilitated the exploration of the diversity of the human gut microbiota, for example, using cloned libraries of the 16S rRNA gene, followed by Sanger sequencing. Pyrosequencing, a synthesis-based sequencing method, has much higher throughput and much lower cost when compared to Sanger technology (Margulies et al. 2005). Pyrosequencing has been used successfully to study the large microbial communities in the GIT of humans (Bäckhed et al. 2005; Eckburg et al. 2005; Andersson et al. 2008; Dethlefsen et al. 2008), animals (Turnbaugh et al. 2006), soils (Roesch et al. 2007), and oceans (Sogin et al. 2006). Furthermore, emerging genome-based tracking methods such as metagenomics can aid us in exploring the typical composition of different microbial species that are present in the complex gut ecosystem and their entire genetic makeup (Kleerebezem and Vaughan 2009; Schloss and Handelsman 2003; Riesenfeld et al. 2004). The metagenomic analysis of the human GIT microbiota will not only enable us to understand this complex environment, and their pivotal role in host health and disease, but will also provide an insight into the design of strategies to manipulate the proportions of the human microbiota (Turnbaugh et al. 2007). It has been postulated that the human endogenous intestinal microbiota is considered an "organ" of the GIT that plays a role in nourishment, epithelial cell development, and a switch to regulate the innate immunity (Eckburg et al. 2005). The members of *Bacteriodetes* and *Firmicutes*, and one member of the Archaea, called *Methanobrevibacter smithii*, are most predominant in the human GIT (Bäckhed et al. 2005), whereas other prominent anaerobic genera are *Bifidobacterium*, *Eubacterium*, *Fusobacterium*, *Clostridium*, and *Lactobacillus*. Aerobic genera include Gram-negative *E. coli* and *Salmonella* spp., the Gram-positive *Enterococcus*, *Staphylococcus*, and *Streptococcus*, and the fungal species *Candida albicans* (Noverr and Huffnagle 2004). The restricted microbial diversity of the gut is thought to be due to the action of strict adaptive mechanisms necessary for microbial survival (Ley et al. 2006). Moreover, the production of multiple capsular polysaccharides can be an essential requirement for bacterial colonization during host–symbiont mutualism (Liu et al. 2008), and this could be a factor for the limited number of autochthonous microbiota found in the GIT. The endocommensal microbiota are considered resident or autochthonous members of the gut, whereas probiotic microbes are generally considered allochthonous or noncolonizing temporary residents of the GI microbiota (Hord 2008). It is essential to understand the intestinal ecosystem and the changes it encounters during healthy and diseased states to develop effective therapies for specific disease states. The possible role of distal gut microbiota in the developement of obesity was demonstrated by transplanting cecal microbiota from obese mice into lean mice. Intrestingly, when the lean mice were colonized with obese microbiota, they showed significant increases in body fat after 2 weeks of trial (Turnbaugh et al. 2006). These researchers also showed that the development of obesity in leptin-deficient mice coincided with broad, phylum-level changes in the gut microbiota. Among the obese mice, there were reduced numbers of bacteria belonging to *Bacteroidetes* and higher numbers of *Firmicutes* (Ley et al. 2006). Recent studies by Zhang et al. (2009) also supported the role of the gut microbial community in obesity. They studied the genetic diversity of the microbes in the human intestine in relation to obesity and a surgical weight loss procedure using a metagenomics approach. Most strikingly, it was found that obese individuals have distinctly different microbial communities compared with normal-weight individuals; Roux-en-Y gastric bypass alters the intestinal microbial community (Zhang et al. 2009). Furthermore, the *Firmicutes* were selectively dominant in both obese and normal-weight individuals, and their

Table 17.3 Major functions attributed to gut microbiota revealed during recent years

Energy harvest and absorption of inaccessible nutrients and synthesis of vitamins
Xenobiotic metabolism and detoxifying of ingested carcinogens
Regeneration of epithelial cells
Host gut immune (innate and adaptive) system development
Effect on host's organ size
Influence on human behavior

numbers significantly decreased in postgastric-bypassed individuals. A subsequent significant increase in the numbers of *Gammaproteobacteria* was observed. These results suggest that the Bacteria–Archaea synergism may be a novel biomarker of susceptibility to obesity (Zhang et al. 2009). Based on a metagenomics approach, 75% of the genes that are annotated in the metagenomic sequences of 13 human intestinal microbiomes are 40–100% identical at the amino acid level with known genes (Kurokawa et al. 2007). The major functions revealed from studies conducted during recent years that are contributed by gut microbiota (Turnbaugh et al. 2007) are listed in Table 17.3.

The launch of the human microbiome project (Turnbaugh et al. 2007) initiated the exploration of the microbial genes and their possible roles in various biological functions, and this will undoubtedly provide a deep insight leading to new strategies to reveal the causative agents of disease and to maintain human health (Hattori and Taylor 2009). Furthermore, the largest metagenomic survey of human gut microbial genes undertaken by the MetaHIT (Metagenomics of the Human Intestinal Tract) consortium has provided data on the most abundant microbiota in healthy and diseased individuals by cataloging all the possible microbial genes present in the human gut (Qin et al. 2010). The entire cohort harbored between 1,000 and 1,150 bacterial species, and each individual was found to carry at least 160 of such species. The most abundant gut species, as revealed by this study, include members of the *Bacteriodetes* and *Doreal/Eubacterium/Ruminococcus* groups, and also Bifidobacteria, *Proteobacteria*, and streptococci/lactobacilli groups (Qin et al. 2010). Further studies involving the application of metagenomic sequencing using larger subject numbers with varied health status in different population groups are required to fully characterize the host gut–microbiome interactions.

17.2 Factors That Influence Probiotic Strain Viability

Probiotics must survive during food processing and production and retain their viability after reaching the GIT to deliver their therapeutic effect to the host. The survival of probiotic bacteria during food processing and gastric transit depends on a wide range of factors, such as strain selection, culture conditions, growth parameters, choice of fermentation medium, drying conditions, and the selection of delivery vehicles into the GIT. Probiotic bacterial cells encounter various stresses during

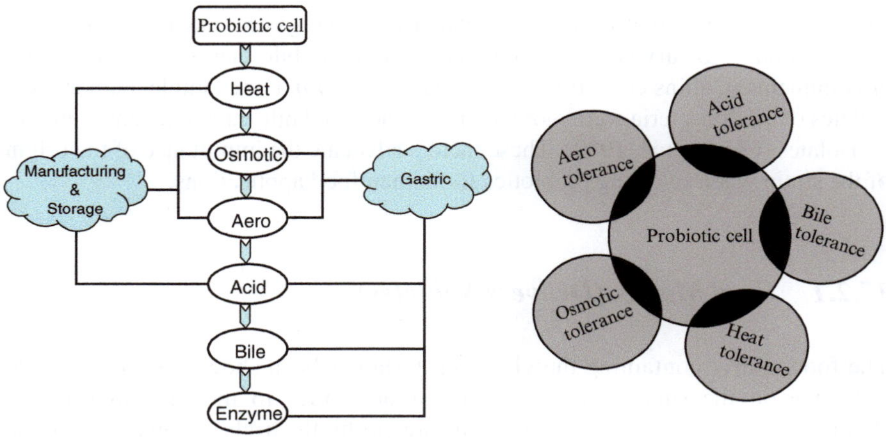

Fig. 17.1 General scheme of probiotic stress resistance required during downstream processing and gastric transit

processing, production, and gastric transit, including environmental stresses such as heat and cold exposure during processing and high hydrostatic pressure during packaging, acid, oxidative, and osmotic stress during production, and acid and bile stress and exposure to digestive enzymes during gastric transit. Thus, the ability of probiotics to remain viable during culture handling, storage, and gastric passage is an important criterion during strain selection. The other major criteria for the strain selection of effective probiotics with improved performance in the gut are shown in Fig. 17.1. The ability to tolerate gastric stresses is considered of prime importance for effective probiotics as suitable adjuncts in functional foods. Acid stress is the primary challenge encountered by probiotics after ingestion, due to the acidic pH (>2) in the stomach (Borgstrom et al. 1957). The tolerance to other stresses, such as thermo, osmotic, and oxygen, may also be pivotal for the successful incorporation of probiotic bacteria into food products (Ross et al. 2005). The initial screening of strains based on various stress- (acid, bile, heat, osmotic, and oxidative) tolerance assays can facilitate the isolation of potential probiotic cultures that can survive during adverse conditions. This approach is useful to estimate and predict their future performance during gastric transit (Ross et al. 2005). The performance and survival of probiotic strains during food processing and gastric transit are strain-dependent and depend on such factors as culturing conditions, fermenting media, drying conditions (either freeze- or spray-drying), and delivery vehicles (food matrices) (Corcoran et al. 2008). Apart from these, reconstitution conditions such as temperature, rehydration rate, and osmolarity of the solution may also play pivotal roles in strain viability (Muller et al. 2009). *Bifidobacterium* strains are less acid-tolerant than *Lactobacillus* strains, and this is reflected by their poor survival in human gastric juice (Dunne et al. 2001). Importantly, bifidobacteria are more resistant to acetate (acetic acid) rather than hydrochloric acid (HCl) in milk-based medium, since acetate is the major byproduct after *Bifidobacterium* fermentation.

In general, cultures that have been isolated from the gastric environment have a superior ability to survive in the presence of acid and bile than strains from other environments (Collins et al. 1998; Godward et al. 2000). For example, human fecal isolates of bifidobacteria were more tolerant to acid and bile than other environmental isolates (Chung et al. 1999). These factors advocate the importance of the origin of the strain when selecting probiotics for human food applications.

17.2.1 Food Matrix (Delivery Vehicles)

The food matrix containing the viable bacteria has been shown to exert a major influence on the survival and delivery of probiotics to the gut environment (Stanton et al. 2005). Furthermore, during probiotic trials, knowledge of the effects of different dietary factors on probiotic survival and efficacy are required (Sanders and Marco 2010). Monitoring the overall host dietary intake may provide a better understanding of probiotic functionality and their response to different dietary patterns. Such data will provide a link between bacteria–food interactions and also the efficacy of probiotics administered along with different food formats.

Following ingestion, probiotic bacteria are exposed to a number of harsh conditions, such as gastric acidity (pH 1.5), bile secretions, and competition with more than 1,000 autochthonous bacterial species (Hooper et al. 2002). It has been shown that most of the Gram-positive anaerobes are sensitive to deconjugated bile salts, whereas the Gram-positive aerobic microorganisms are relatively insensitive (Floch et al. 1972). In general, foods with a high buffering capacity, such as cheese, yogurt, yogurt drinks, and buttermilk, offer protection to probiotics during passage through the GIT (Gardiner et al. 1999; Ross et al. 2002b, 2005). Studies in our laboratory have suggested that probiotics delivered through fermented dairy products such as cheese (Stanton et al. 1998; Gardiner et al. 1999) and yogurt (Kearney et al. 2009) were effectively delivered in viable form to the GIT. Cheddar cheese manufactured with the probiotic *Lactobacillus paracasei* NFBC338 was not altered despite having a high viability of 10^8 log CFU/g after 3 months. However, the suitability of Cheddar cheese as a probiotic delivery vehicle is strain-dependent, and some strains do not survive manufacture and/or ripening (Stanton et al. 1998). Ice cream has been reported as another good matrix to deliver probiotic cultures, as it is composed of milk proteins, fat, and sugar and has a near-neutral pH of up to pH 6.5, in addition to the low storage temperature, which undoubtedly aids probiotic survival in the product (Cruz et al. 2009).

In addition to dairy products, other foods that have been exploited as probiotic delivery matrices include fruit juices, cereal-based products, processed meats, tablets, chocolates, table olives, soy, plant seed extracts, and chewing gums (Champagne et al. 2005). It was reported that certain probiotics can survive and are delivered in an active state to the GIT after storage in commercial fruit juices at 4°C (Champagne and Gardner 2008). Table olives were also identified as a good matrix to deliver probiotic lactobacilli and bifidobacteria to the human GIT, and an improved survival of *Lactobacillus* strains was observed compared with bifidobacteria during a storage period of 90 days at room temperature (Lavermicocca et al. 2005). Similarly, artichokes were found to be an effective carrier matrix to deliver adequate live populations of probiotics to the gut, due to their structure and prebiotic content, particularly the presence of inulin.

17.2.2 Protectants for Enhancing Probiotic Viability

The incorporation of protectants into the growth medium can enhance probiotic cell viability during downstream processing (either spray- or freeze-drying), storage, and gastric transit. A list of commonly used protectants for probiotics bacteria is given in Table 17.4. The addition of Tween-80 into the growth medium was shown to enhance the survival of stationary-phase cultures of *Lactobacillus rhamnosus* GG up to 1,000-fold when compared to controls grown in the absence of Tween-80

Table 17.4 Examples of the most commonly used protectants and claimed beneficial effects in particular to specific probiotic bacteria

Protectant	Strains	Beneficial effect	References
Amino acids (glutamate, aspartate)	*L. delbreukii.* subsp. *bulgaricus*	Increased survival rates during freeze-drying and storage	Martos et al. (2007)
Adonitol	*S. lactis* T164 *S. lactis* T215 *S. cremoris* T162 *S. cremoris* T55 *S. thermophilus* ATCC19258 *L. bulgaricus* T142 *L. cremoris* ATCC19254 *L. casei* ATCC393 *L. fermentum* ATCC9338 *L. leichmanii* ATCC4797 *L. murinus* CNRZ313 *L. plantarum* ATCC8014 *L. leichmanii* ATCC4797 *L. helveticus* ATCC15009	Improved survival rates during freeze-drying	de Valdéz et al. (1983)
Fatty acids (oleic acid)	*L. rhamnosus* GG	Improved survival and acid tolerant	Corcoran et al. (2007)
Soluble starch	*B. infantis* CCRC14633, CCRC14661 *B. longum* ATCC15708, CCRC14634, B6	Improved survival after spray-drying	Lian et al. (2002)

(continued)

Table 17.4 (continued)

Protectant	Strains	Beneficial effect	References
Glucose	*L. rhamnosus* GG, *L. gasseri* ATCC33323 *L. salivarius* UCC500 *L. rhamnosus* E800	Acid tolerance and improved survivability during gastric exposure	Corcoran et al. (2005)
Glucose	*L. delbreukii* sp. *bulgaricus*	Increased survival during storage	Carvalho et al. (2004a)
Fructose Lactose Mannose		In the presence of mannose increased survival rates after freeze-drying	
Gum acacia	*L. paracasei* NFBC338	Improved survival during spray-drying, storage, and gastric exposure; resistance to heat, bile, and H_2O_2	Desmond et al. (2002)
Gum arabic and gelatin	*B. infantis* CCRC14633, CCRC14661 *B. longum* ATCC15708, CCRC14634, B6	Improved survival after spray-drying	Lian et al. (2002)
Monosodium glutamate (MSG)	*L. rhamnosus* GG and E800	Enhanced survival rates during spray-drying and storage at 25°C	Sunny-Roberts and Knorr (2009)
	L. bulgaricus *L. rhamnosus* *L. plantarum* *E. durans* *E. faecalis*	Improved survival rates during storage	Carvalho et al. (2003)
	Oenococcus oeni	Improved survival after freeze-drying	Maicas et al. (2000)
Polydextrose, wheat dextrin, oat flour+β-glucan	*L. rhamnosus*	Enhanced survival rates during freeze-drying and storage	Saarela et al. (2006)
Reconstituted skimmed milk (RSM) nonfat milk solids	*L. paracasei* NFBC 338	Improved survival after spray-drying, storage, and gastric exposure	Desmond et al. (2002)
Sorbitol	*L. bulgaricus* *L. rhamnosus* *L. plantarum* *E. durans* *E. faecalis*	Improved survival rates during storage	Carvalho et al. (2003)
Trehalose	*L. rhamnosus* GG *L. rhamnosus* E800	Improved survival during spray-drying and storage	Sunny-Roberts and Knorr (2009)
Tween-80	*L. rhamnosus* GG	1,000-fold increase in survival rate	Corcoran et al. (2007)

(Corcoran et al. 2007). Furthermore, improved survival rates and acid tolerance were observed in *L. rhamnosus* GG grown in the presence of oleic acid, while the use of other fatty acids, such as linoleic acid and conjugated linoleic acid (CLA), did not result in improved culture viability when exposed to simulated gastric juice (Corcoran et al. 2007). The cultivation of probiotic lactobacilli in reconstituted skim milk (RSM) containing gum acacia led to improved viability of the probiotic culture during drying and storage, and potentially during gastric passage, compared with probiotic cultured in RSM alone. However, gum acacia did not influence the growth rate of *L. paracasei* NFBC338 when used as the sole source of carbohydrate, suggesting that it did not exert growth-stimulating effects (Desmond et al. 2002). Moreover, gum acacia treatment offered 100-fold higher survival during hydrogen peroxide (H_2O_2) stress in a 15-min challenge, compared with control (untreated) cells, and a tenfold improved survival during heat and bile challenges (Desmond et al. 2002). These results suggest that treating cells with gum acacia prior to processing can enhance the survivability of the probiotic cells during manufacturing practices, which aids in the delivery of more effective probiotics in active form to the host GIT.

The acid tolerance and subsequent survival of probiotic *L. rhamnosus* GG, *Lactobacillus gasseri* ATCC33323, *Lactobacillus salivarius* UCC500, *L. rhamnosus* E800, and *L. paracasei* NFBC338 were increased in simulated gastric juice with the addition of glucose (Corcoran et al. 2005). The survivability in gastric juice was strain-dependent, with each of the tested strains showing different survival rates upon exposure to simulated gastric juice (Corcoran et al. 2005). More recently, spray-dried *L. rhamnosus* GG and E800 strains were grown in media containing trehalose and monosodium glutamate (MSG) and showed high survival rates during spray-drying and storage at 25°C over 6 weeks (Sunny-Roberts and Knorr 2009). Carvalho et al. (2003) reported that the addition of sorbitol and MSG into the growth medium did not affect the survival during freeze-drying, but offered subsequent survival stability to the strains *Lactobacillus bulgaricus*, *L. rhamnosus*, *Lactobacillus plantarum*, *Enterococcus durans*, and *Enterococcus faecalis* during longer periods of storage.

17.3 Nutritional Programming of Probiotic Strains to Improve Stress Tolerance

It has been reported that the cultivation of bacteria in the presence of various supplements such as saturated and polyunsaturated fatty acids (PUFAs) offers protection against various environmental stresses (Chang and Cronan 1999; Cronan 2002; Sakamoto and Murata 2002). For example, the survival rates of *L. rhamnosus* GG were enhanced in parallel with the accumulation of fatty acids, which were supplemented as Tween-80 (1 g/L) (Corcoran et al. 2007). In this study, a 55-fold increase in oleic acid was obtained in the total bacterial content when compared to the control cells grown without Tween-80. Further analysis using a neomycin-resistant mutant with 40% F_0F_1-ATPase activity indicated that the effect of Tween-80 was independent of the complex (Corcoran et al. 2007). In the presence of fatty acids and when the pH of the fermentation medium was reduced to pH 3.0, increased levels of unsaturated fatty acids (UFAs), such as palmitoleic, oleic, linoleic, and linolenic acid were found in *Leuconostoc mesenteroides* cell membranes, while the fatty acid composition remained unchanged in the same culture grown at pH 6.5 (Rani and Agrawal 2008). These changes in the fatty acid composition occurred due to altered cell membrane processes. A modification of the fatty acid composition and cell surface physiology was reported in *Bifidobacterium animalis* subsp. *lactis* during bile adaptation (Lorena et al. 2007). In particular, it is the phospholipid bilayer of the bacterial cells that undergoes several modifications in order to withstand adverse environments and facilitate the uptake of exogenous fatty acids, thus maintaining cell membrane homeostasis. Thus, bacterial cells can control the biosynthesis of new fatty acids or can modify the structure of existing fatty acids, which allows bacteria to alter cell membrane viscosity and rapidly adapt to the new environmental conditions (Zhang and Rock 2008). It has been reported that bacteria containing cyclopropane fatty acid (CFA) as a membrane constituent were more acid-tolerant than those containing UFA and that exogenous CFA renders cells more acid-resistant (Chang and Cronan 1999). Bacteria showed high growth rates when the medium was supplemented with fatty acids that were present in their cell membrane lipid bilayer (Rodwell and Peterson 1971). Betaine accumulation was observed to increase when *Lactococcus lactis* NCDO763 cells were subjected to an osmotic stress with NaCl. This exogenous betaine uptake by *L. lactis* was directly proportional to osmotic upshock, growth temperature, and Tween-80 supplementation into the growth medium. With high osmolarity, increased levels of CFA ΔC19:0 were observed in *L. lactis* NCDO763 membrane composition, whereas the unsaturated–saturated fatty acids ratio was remained unchanged (Guillot et al. 2000). The phospholipid bilayer may vary from species to species, but most membrane phospholipids are glycerolipids that contain two fatty acid chains (Zhang and Rock 2008). An acyl-chain modification of membrane phospholipids was proposed as an alternative mechanism in many bacteria in response to environmental stress. The three major acyl-chain modifications were (1) *cis trans* isomerization, (2) cyclopranation of double bonds, and (3) desaturation (Cronan 2002; Sakamoto and Murata 2002). In *E. coli*, *cis trans* isomerization in membrane fatty acids was proposed as a faster heat stress–adaptive mechanism (de Mendoza and Cronan 1983). Certain PUFAs, such as γ-linolenic acid, arachidonic acid, α-linolenic acid, and docosahexanoic acids, in the growth medium exhibited a significant influence on the probiotic adhesion to mucus and epithelial cells. Earlier work by Pasi et al. (2001) reported that the growth and adhesion of the probiotic *Lactobacillus casei* strain Shirota was promoted by γ-linolenic acid and arachidonic acid at a concentration of 5 μg/mL, whereas higher concentrations of PUFAs (10–40 μg PUFA/mL) can be detrimental to microbial growth and mucus adhesion (Pasi et al. 2001). Considering these results, it is essential to evaluate the required optimal concentration of such fatty acids in the growth medium to exert beneficial effects.

The survival rate of probiotic strains during hostile conditions could be enhanced by the incorporation of metabolizable sugars (glucose, sucrose) into the medium prior to drying (Desmond et al. 2001; Carvalho et al. 2003, 2004a; Corcoran et al.

2005). In many studies, disaccharides such as lactose, trehalose, sucrose, and maltose were demonstrated as effective cryoprotectants. For example, the inclusion of disaccharides in the fermentation medium (RSM) offered protection to *L. rhamnosus* GG cells during freeze-drying and subsequent storage under relative vapor pressure (RVP) of 0.0 and 11.4%. Trehalose was the most effective cryoprotectant among all the disaccharides tested (Miao et al. 2008). Improved survival rates of LAB were observed after freeze-drying in sucrose-containing skim milk medium (Otero et al. 2007). Other studies also report skim milk as the most preferable medium for drying LAB (Desmond et al. 2002; Gardiner et al. 2002; Carvalho et al. 2004b; Corcoran et al. 2005). Furthermore, the protective effects of fructan, skim milk, and fructo-oligosaccharide (FOS) were demonstrated in *Lactobacillus reuteri* during freeze-drying (Schwab et al. 2007) and also in *Bifidobacterium* subsp. Bf-1 and Bf-6 (Shin et al. 2000). It is important to note that optimizing the required concentrations of such protectants is crucial to achieve higher viability rates during large-scale production. More studies about the oligosaccharide protection of probiotic cultures are discussed elsewhere in this chapter. These studies concluded that RSM or skim milk is the optimal medium for either culturing or drying the probiotic cultures for preservation. It offers protection to some extent to different stresses, which helps the probiotics to retain their viability during adverse conditions. More studies are required to optimize the fermentation medium and drying matrix to attain greater probiotic viability.

From the above-discussed studies, we can conclude that nutritional modification of the growth medium by providing essential fatty acids and carbohydrates during bacterial cultivation can alter the cell membrane physiology, and subsequently offer protection against different stresses during downstream applications. Whole-genome sequencing and comparative genomics of health-promoting bacteria can be very useful to identify target genes and related proteins playing an important role in stress responses and thus provide an insight to apply genetic or metabolic engineering strategies to create more robust cultures that can adapt to any stressful environment, which enhances the culture performance during pre- and postharvesting times. Techniques like 2D-gel electrophoresis and differential in-gel fluorescence electrophoresis (DIGE) coupled with MALDI-TOF analysis are powerful approaches to exploit the upregulated and downregulated proteins in response to various stresses.

17.3.1 Induction of Stress Responses for Improved Technological Performance

A number of studies have shown that the induction of stress responses in probiotic bacteria led to improved performance during subsequent processing. Adaptive response involves the prior exposure of microbial cells to lesser, or sublethal, stress (such as heat, pH, bile, and osmotic stress), resulting in an enhanced ability of the cells to better resist subsequent damage, thus retaining higher viability rates and greater physiological activity during food processing and gastric transit (Crawford and Davies 1994). For example, prior exposure to sublethal heat and salt stresses led to the improved thermotolerance of probiotic *L. paracasei* NFBC338 during spray-drying (Desmond et al. 2001). It was reported that, preexposure of *L. paracasei* NFBC338 to 0.3 M NaCl offered an enhanced protection during spray-drying at outlet temperatures between 95 and 100°C, when compared to unexposed control cells (Desmond et al. 2001). In another such study, it was reported that prestressed log-phase cells of *Lactobacillus acidophilus* were more tolerant to various stresses applied, whereas the control log-phase cells were killed upon exposure to heat, bile, and NaCl (Kim et al. 2001). In that study, the preexposure of *L. acidophilus* to heat stress did not increase the resistance against bile and NaCl, while the preexposure to either bile or NaCl significantly increased the resistance to heat stress. Stationary-phase cultures of *L. acidophilus* were inherently resistant to these stresses (Kim et al. 2001). In this respect, it is important to emphasize that exposure to one stress can offer the cells cross-protection against other stresses. An example of cross-stress resistance was demonstrated when *L. lactis* subsp. *lactis* cells were exposed to UV light (254 nm), which conferred cross-protection against heat, acid, ethanol, and, to a lesser extent, H_2O_2 stress (Hartke et al. 1995). It has been reported that microbial cells that have entered the stationary phase of growth develop a general stress-resistance mechanism and are more tolerant to various stresses than log-phase cells (Corcoran et al. 2005; Michida et al. 2006).

17.3.2 Fermentation Technology

A minimum level of >10^7 log CFU/mL or per gram product is recommended for probiotic products (Ishibashi and Shimamura 1993; Ouwehand and Salminen 1998). LAB are fastidious microorganisms, as they require complex nutrient media for their growth and metabolism. The large-scale preparation of bacterial cultures is difficult, time-consuming, and expensive. In addition, most of the probiotic strains show poor growth rates in milk-based media. Therefore, it can be difficult to achieve high numbers of viable cells after fermentation (Ross et al. 2005). The media composition and type of substrate used for fermentation can have a great impact on the probiotic strain viability during production and downstream processing (Lacroix and Yildirim 2007). Stress can be applied to bacterial cultures during end-of-fed-batch fermentation to induce a multi-stress response to protect them during subsequent food processing (Muller et al. 2009). During production, further factors such as pH, temperature, and storage duration affect the viability of probiotic bacteria. Under carefully designed conditions, continuous culture fermentations can lead to high cell yield and viability (Doleyres et al. 2002). Bacterial cells produced during continuous-culture fermentation can be maintained in a controlled physiological state that can be manipulated by other parameters, such as growth medium composition and dilution rate. Importantly, continuous fermentations are more difficult to operate under industrial conditions due to the risk of contamination and loss of cell characteristics (Lacroix and Yildirim 2007). A two-stage continuous-fermentation process using stress-adapted bacterial cells could produce a high viable cell yield, with a controlled physiological state (Lacroix and Yildirim 2007). A few studies have reported the use of membrane bioreactors for the large-scale production of probiotic cultures. Using this method with continuous feeding of fresh medium, high cell yields of *Bifidobacterium longum* (Taniguchi et al. 1987) and *Bifidobacterium bifidum* (Corre et al. 1992; Kwon et al. 2006) were achieved. More research is needed to optimize the parameters, such as growth medium, nutrient concentration, and use of stress-adapted cells, in the continuous and free-cell batch fermentations to achieve high cell yields with a good physiological state.

17.3.3 Drying Technology

Drying bacterial culture preparations represents a convenient method for long-term preservation and ease of incorporation of probiotic bacteria into functional foods (Stanton et al. 2005; Meng et al. 2008). Moreover, dried probiotic preparations are readily accessible, easy to handle, and convenient for storage. In this respect, the most commonly used methods are freeze-drying and spray-drying, both of which can be used for large-scale industrial culture preparations.

17.3.3.1 Freeze-Drying Technology

Freeze-drying is the most commonly used method for manufacturing probiotic and dairy starter culture powders (Meng et al. 2008). The exposure of bacterial cells to extremely low freezing temperatures leads to dehydration in the cell, followed by injury and decreased viability (Stanton et al. 2005). Freezing of bacterial cultures can cause damage to cells by the formation of ice crystals and by increasing the internal cell osmolarity due to high solute concentrations (De Angelis and Gobbetti 2004). A wide variety of cryoprotective additives have been employed during the freeze-drying of microorganisms such as monosaccharides (glucose, xylose), disaccharides (sucrose, lactose, maltose, trehalose), tri-saccharides (raffinose), polysaccharides (dextran, mannan, dextrin, hydroxyethyl starch, ficoll, gum acacia), and amino acids (glutamate, aspartate, glycine, proline, aminobutyric acid, glutaric acid) (Hubalek 2003). Saarela et al. (2006) demonstrated that fibers from different sources, such as fruits, oats, flax, wheat dextrin, polydextrose, and inulin, afforded protection to probiotic cells during freeze-drying. Among these, wheat dextrin and polydextrose were promising protectants during freeze-drying and storage (Saarela et al. 2006). Better survival rates of fresh *L. rhamnosus* cells in apple juice (pH 3.5) were achieved during storage at 4 and 20°C when freeze-dried in the presence of oat flour with β-glucan (20% w/v) (Saarela et al. 2006). Freeze-drying bacterial cells in the presence of compatible cryoprotectants resulted in a lower loss of cell viability due to reduced water removal from the bacterial cells (Castro et al. 1997). The preexposure of bacteria cells to other environmental stresses (such as heat, acid, bile, osmotic, oxygen, H_2O_2) have also led to enhanced survival during drying (De Angelis and Gobbetti 2004).

17.3.3.2 Spray-Drying Technology

While freeze-drying of bacterial cultures is "gentler" compared to spray-drying, thus allowing greater probiotic viability rates, it is more expensive and more time-consuming (Stanton et al. 2005). However, spray-drying has not been widely used in the industry to dry probiotic cultures, due to the high losses in cell viability typically found in various studies. Among the main issues associated with spray-drying of probiotics is the long exposure time to high heat, resulting in high losses in culture viability (Stanton et al. 2003b). However, studies have shown that spray-drying can be applicable for the large-scale production of some probiotic *Lactobacillus* strains (Gardiner et al. 2000). A number of factors (inlet and outlet temperatures, drying matrix, and growth patterns) influence the performance and viability of probiotic strains during spray-drying (Santivarangkna et al. 2007). For example, it was shown that the viability of the probiotic *L. paracasei* NFBC338 was inversely related to outlet temperature during spray-drying (Gardiner et al. 2000).

This parameter also influences powder quality, whereas a moisture content of about 3.5% is suggested for shelf-stable products (Zayed and Roos 2004). However, the proper settings of these variables are difficult to estimate in advance (Santivarangkna et al. 2007). The incorporation of the soluble fiber gum acacia in milk-based medium (RSM) prior to spray-drying the probiotic *L. paracasei* NFBC 338 enhanced the viability during storage when compared to the control (Desmond et al. 2002). However, other prebiotics tested, such as inulin and polydextrose, were found not to influence the probiotic viability during spray-drying and storage (Corcoran et al. 2004). Clearly, strain selection is also an important factor, as studies have shown that the technological performance of probiotics during spray-drying is highly strain-dependent. For example, *L. paracasei* NFBC338 showed higher survival rates compared to *L. salivarius* UCC118 when these probiotic cultures were spray-dried in RSM and during storage at 4, 15, and 30°C. A survival rate of >90% was achieved for *L. paracasei* NFBC338 during storage at 4°C over a period of 70 days (Desmond et al. 2002).

17.3.3.3 Optimal Rehydration of Probiotic Powders

During the reconstitution of dried culture powders, bacterial cells encounter various stresses, and employing suitable rehydration conditions for dried probiotic cultures is thus vital to ensure maximum culture viability upon reconstitution. The effects of various rehydration methods and their impact on viability rates of probiotics have been reported. The viability of the rehydrated cultures is strain-dependent and has been shown to be influenced by such parameters as the reconstitution medium, pH, temperature, and osmolarity (Leach and Scott 1959; Ray et al. 1971; de Valdéz et al. 1985; Carvalho et al. 2004b; Muller et al. 2010). For example, temperature was shown not to influence the recovery rate of dried *L. plantarum*, when cells were rehydrated at 30 or 37°C, whereas the viability of *L. bulgaricus* was greatly influenced by temperature (Mille et al. 2004). The pH of the rehydration medium can also greatly influence the viability of the dried probiotic cultures upon reconstitution. Using a multivariate experimental design with a surface-response methodology approach, a high viability of *B. longum* reconstituted in maximum recovery diluent (MRD) at pH 8 with 2% L-arabinose was found, while a lesser effect was observed for *Lactobacillus johnsonii* La1, whereas *L. johnsonii* La1 showed high recovery rates upon rehydration at pH 4 in the presence of maltodextrin (Muller et al. 2010). These studies indicate that the effect of pH and temperature is strain-dependent and may vary among species and strains. Improved survival rates of *Lactobacillus brevis* and *Oenococcus oeni* were observed when reconstituted with 10% (w/v) sucrose and MGY medium, respectively, and this study also reported that different protectants and freeze-drying matrices can greatly influence dried culture-recovery rates (Zhao and Zhang 2005). These studies have demonstrated that the rehydration medium composition and protectants significantly influence cell viability upon reconstitution, and the rehydration conditions (pH, temperature, and osmolarity) impact the culture viability. Furthermore, since the reconstitution rates of probiotics vary among species and strains, the most suitable reconstitution medium and optimal conditions for each bacterium have to be identified separately, thus making it difficult to suggest a universal recovery medium for all bacteria.

17.4 Genes Conferring Stress Resistance

During downstream processing and production, LAB are exposed to a number of environmental stresses, including heat, low pH, osmotic stress, and oxygen. Certain physiological characteristics are desirable for probiotic cells to remain viable during such exposures. In addition, gastric transit also represents a challenge to probiotic viability, given the low acidic pH of the stomach and the exposure to pancreatic secretions such as bile and digestive enzymes in the GIT. Recent developments in molecular techniques (transcriptomics and proteomics) have enabled us to explore and understand the various stresses encountered during processing and gastric transit. The heat-shock genes are classified into six major groups and are listed in Table 17.5.

Table 17.5 Major classes of genes that confer stress resistance in bacteria

Class I (sigma-A-dependent promoter) contains genes encoded on the *dnaK* and *groEL* operons and are most efficiently induced by heat stress (Kilstrup et al. 1997; Desmond et al. 2004)
Class II (sigma-B-dependent promoters) contains *gspA*, *csbA*, *katE*, which are involved in general stress mechanisms
Class III contains *clpB*, *clpC*, *clpE*, *clpL*, and *clpP*
Class IV contains *htpG* and has not yet been reported in LAB
Class V contains *htrA* and *htrB*. These are under the control of CssRS' two-component system, which plays a crucial role in cell wall homeostasis maintenance
Class VI contains a group of genes involved in stress responses whose regulatory mechanisms are unknown and contains the genes *ftsH*, *clpX*, *lon*, and more (Sugimoto and Sonomoto 2008)

The primary mechanism in LAB for the control of intracellular pH involves the F_0F_1-ATPase that translocates protons to the environment by utilizing ATP, while the secondary mechanism for pH homeostasis involves the arginine deiminase (ADI) pathway, which allows microbial cells to neutralize their surrounding environment by NH_3 production. The ADI pathway consists of three cytoplasmic enzymes: arginine deiminase, ornithine carbamoyltransferase, and carbamate kinase (Cotter and Hill 2003). These enzymes are active at a low pH in several *Streptococcus* species (Marquis et al. 1987; Curran et al. 1995). A genetic analysis of the ADI pathway in *Lactobacillus sakei* revealed the presence of five genes (*arcABCTD*) encoding the four components of the pathway and a putative transaminase gene *arcT* (Zuniga et al. 1998). Acid adaptation allows increased microbial resistance to UV radiation and H_2O_2 (Cotter and Hill 2003). It was proposed that the enzyme glutamate decarboxylase, commonly found in bacteria, might strengthen them during gastric transit (Small and Waterman 1998). The different systems that may contribute to acid tolerance in lactobacilli, such as the ADI pathway, urease, glutamine/γ-amino butyric acid (GABA), histamine, gad system, malolactic fermentation, and glutamate decarboxylase, are well studied (De Angelis and Gobbetti 2004). Bile acids are surface-active, water-soluble amphiphatic end products of cholesterol metabolism (Hofmann 1999) and possess potent antimicrobial activity (Begley et al. 2005). They act as detergents that help in the emulsification and solubilization of fats and disrupt cell membranes. They are synthesized from cholesterol and are conjugated to either glycine or taurine in the liver and then pass into the intestine (Begley et al. 2005). Gram-positive anaerobic gut bacteria are more sensitive to unconjugated bile acids; nevertheless, the enteric aerobes are resistant to unconjugated bile acids (Binder et al. 1975). Many of the gut microbiota, including lactobacilli, express bile salt hydrolases (BSHs) (Begley et al. 2006), which function in the hydrolysis of bile salts. These enzymes may play a vital role in the increased resistance of probiotics to the toxic levels of bile salts present in the GIT (Begley et al. 2006). In *L. acidophilus*, an operon containing a two-component regulatory system (2CRS) was upregulated in the presence of bile, and functional analysis revealed that this operon influenced bile tolerance in *L. acidophilus* (Pfeiler et al. 2007). An efflux-based mechanism, that of the ABC-type multidrug-resistance transporter LmrCD, is responsible for bile salt resistance in *L. lactis* (Zaidi et al. 2008). ABC-type multidrug-resistance transporters (ABC-type MDR transporters) may be the crucial factors in the colonization and survival among gut microbiota (Zaidi et al. 2008). The adaptive mechanisms are strain-dependent, and different strains have evolved different mechanisms to combat stressful environments. To elucidate different genes and their expression levels during various stresses, microarray technology can provide a clear insight into the genes responsible for conferring stress resistance. For example, the microarray analysis of bile-induced transcriptome in *L. reuteri* ATCC55730 identified many genes that are affected during bile stress. These include those encoding Ir0004 Clp chaperone (*clpE*), Ir00085 hypothetical protein, Ir1265 multidrug-resistance protein (ABC transporter family), Ir1291 metalloproteinase, Ir1351 conserved membrane protein of unknown function, Ir1516 putative esterase, Ir1584 multidrug-resistance protein, Ir1706 Dps, and lr1864 Clp chaperone (*clpL*) (Whitehead et al. 2008). In *L. reuteri*, the expression levels of 88 genes changed during bile shock (45 genes overexpressed and 43 genes underexpressed), whereas 84 genes changed significantly during bile adaptation (17 genes overexpressed and 67 genes underexpressed) (Whitehead et al. 2008). Genes representing putative functions such as fatty acid and cell wall synthesis have been associated with conferring bile resistance in *E. faecalis* (Breton et al. 2002). When subjected to alkaline stress, proteins induced in *E. faecalis* were mostly similar to the proteins that are expressed during bile stress (Flahaut et al. 1997).

In general, oxidative stress caused by increased levels of reactive oxygen species (ROS) such as the superoxide anion ($O_2^{\bullet-}$), H_2O_2, or hydroxyl radicals ($HO^{\bullet}$) damage biomolecules, including proteins, fatty acids, lipids, and nucleic acids, leading to cell death (Storz and Zheng 2000). The existence of regulatory adaptive mechanisms to oxidative stress was first reported in *E. coli* and *Salmonella enterica* serovar Typhimurium (Tao et al. 1989; Cunningham and Guest 1998) and was subsequently studied in detail in other microorganisms, including *Candida albicans* (Jamieson et al. 1996) and *Clostridium perfringens* (Briolat and Reysset 2002). To counter oxidative stress-mediated cellular damage and to prevent the formation of ROS, bacteria have enzymes such as glutathione reductase, NADH oxidase, catalase, superoxide dismutase, and thioredoxin reductase. They also produce scavengers such as phosphate ABC transporters and FLP-like proteins to obstruct their activity, thereby protecting the cells from cellular damage (van de Guchte et al. 2002). LAB can exhibit an inducible oxidative stress tolerance if they are preexposed to sub-lethal levels of H_2O_2. The major enzymes produced by LAB that are involved in oxygen metabolism include NADH oxidase, pyruvate oxidase, oxidase alpha-glycerophosphate, superoxide dismutase, and NADH peroxidase (Condon 1987). The H_2O-forming NADH oxidase has been proposed to function as a defense mechanism against oxidative stress in *Streptococcus mutans* (Higuchi 1992). Furthermore, the genes encoding NADH peroxidase (*npr*) and NADH oxidase gene (*nox*) have been characterized in detail from a number of LAB, including *E. faecalis* (Ross and Claiborne 1991) and *Streptococcous pneumoniae* (Auzat et al. 1999).

Probiotic bacteria undergo osmotic stress during cultivation and processing and in the GIT. The two general strategies proposed in prokaryotes to combat osmotic stress are (1) "salt in cytoplasm" (this phenomenon is restricted to halophiles) and (2) the accumulation of low-molecular-weight compatible solutes (found in all other bacteria) (Sleator and Hill 2002). The general mechanisms in bacteria to combat osmotic stress include K^+ uptake (genes that are involved in K^+ uptake are *trk* and *kup*, also known as *trkD*) and glycine-betaine accumulation. Indeed, Gram-positive bacteria benefit more from the accumulation of osmoprotectants such as glycine betaine than the electrolyte pair K^+-glutamate (Sleator and Hill 2002).

The term *osmoadaptation* refers to both the physiological and genetic manifestations of adaptation to low- and high-water environments (Galinski 1995). To combat salt stress, halobacterial proteins undergo extensive amino acid substitutions. For example, malate dehydrogenase (*hMDH*) from *Halobacterium marismortui* has a 20% excess of acidic over basic residues (compared to only 61% for its nonhalophilic equivalent) (Mevarech et al. 1977).

Heat-induced thermotolerance has been reported in numerous LAB. The induction of heat-shock proteins and the acquisition of thermotolerance have also been reported to protect the cell from other stresses, that is, cross-protection, as discussed earlier in this chapter. Indeed, general stress mechanisms such as GroESL chaperone proteins, which aid in protein folding, are generally responsible for protection against multiple stresses. The DnaK or HSP70 protein families are the most well-known heat-shock proteins. Numerous genes are involved during stress responses in LAB, but their actual roles, regulation, and expression levels can differ widely among species. Exploiting such genes responsible for stress adaptation can lead to the improved performance of existing probiotics by controlling the expression of those genes, making the strains more robust in adverse environments.

17.5 Recent Trends to Enhance Probiotic Viability in the Gastrointestinal Tract

Many strategies have been investigated to enhance the performance and viability of probiotics during storage and following ingestion, including microencapsulation, genetic modification for development of "designer probiotics," and patho-biotechnology.

17.5.1 Microencapsulation

Microencapsulation technology has been used in the food industry since the 1930s, where the technique has been used in packaging flavors and vitamins (Gouin 2004). Various microencapsulation techniques are available today, with the microencapsulated products widely used in the pharmaceutical, biomedical, agricultural, food, consumer products, and cosmetic industries. In the food industry, microencapsulation is a useful technique to improve the delivery of biogenic metabolites in foods, such as probiotics, minerals, antioxidants, phytosterols, fatty acids, lycopene, and lutein (Champagne and Fustier 2007). A number of different microencapsulation methods are now available, such as spray-drying microencapsulation, spray-cooling, centrifugal coextrusion, extrusion, fluidized bed, and coacervation (Gouin 2004). The most commonly used shell materials for spray-drying microencapsulation include alginate beads, liposomes, gum acacia, maltodextrins, κ-carrageenan, gelatin, starch, and other polysaccharides, such as carboxymethylcellulose and guar gum, and proteins such as whey proteins, soy proteins, and sodium caseinate (Gouin 2004; Stanton et al. 2005). The most commonly reported shell materials for the microencapsulation of probiotic bacteria include calcium alginate, gelatin, whey proteins, starch, and κ-carrageenan (Kailasapathy 2002). Microencapsulation has the potential to protect probiotic cells from the exposure to various environmental stresses in food systems, which allows the controlled release of probiotics at their site of action in the GIT. Exopolysaccharide-producing strains of LAB are naturally entrapped in their own secretions, which function as a capsule, thus affording protection from hostile environments (Ruas-Madiedo et al. 2002; Shah 2002). Recent studies have shown that sodium alginate–coated *L. acidophilus* ATCC43121 exhibited improved survival rates and enhanced performance in simulated gastric juice (pH 1.2 and 1.5) when compared with nonencapsulated cells (Kim et al. 2008). Interestingly, encapsulation did not influence the probiotic adherence to the human HT-29 epithelial cell lines in this study. In another study, the controlled release of *E. coli* K12 was observed in the colon with calcium alginate encapsulation (Iyer et al. 2004). Probiotics have also been encapsulated for incorporation into cheese. For example, *B. bifidum* BB-12 and *L. acidophilus* LA-5 were encapsulated in sodium alginate by either an extrusion technique or an emulsion technique in κ-carrageenan and corn oil as emulsifying agent. The encapsulated strains showed increased viability (of $>10^7$ CFU/g) in white-brined cheese and did not alter the sensory properties when compared with nonencapsulated cells (Özer et al. 2009). Furthermore, medium- and long-chained free fatty acid contents as well as acetaldehyde and acetyl levels were reported to be higher in the cheeses with immobilized probiotics than in control cheese (Özer et al. 2009). *L. rhamnosus* cells were entrapped in the inner aqueous layer of a double water-in-oil-in-water emulsion stabilized with concentrated sweet whey as an emulsifier. This showed an increased survival of 108 and 128% when exposed to low pH and bile salts, respectively (Pimentel-González et al. 2009). In another study, the microencapsulation of the probiotic *Bifidobacterium infantis*, using oil-in-water emulsions (canova vegetable oil, caseinate) and the prebiotic fructo-oligosaccharide Raftilose P95 with either dried glucose syrup or microfluidized starch, followed by spray-drying, improved the performance and viability of the cells during storage at 25°C for 5 weeks. Moreover, it offered protection during exposure to simulated gastric juice (incubated for 2 h at pH 1.2 and adjusted to pH 6.8, then further incubated for 3 h at 37°C) (Crittenden et al. 2006). The immobilization of probiotic cells was reported to yield better survival rates during the frozen storage of food products. For example, Tsen et al. (2007) reported that *L. rhamnosus* cells entrapped in calcium alginate/or κ-carrageenan exhibited better survival during storage at −80°C (Tsen et al. 2007). Similar studies by Homayouni et al. (2008) showed that calcium alginate–encapsulated strains of *L. casei* (Lc-01) and *B. lactis* Bb-12c exhibited significantly improved frozen storage survival in ice cream in the presence of starch as the prebiotic, thus creating a synbiotic product (see ahead). This did not negatively affect the sensory qualities of the synbiotic ice cream product (Homayouni et al. 2008). Sodium alginate–encapsulated probiotic strains of *L. rhamnosus*, *B. longum*, *L. salivarius*, *L. plantarum*, *L. acidophilus*, *L. paracasei*, *B. lactis* type BI-O4, and *B. lactis* type Bi-07 showed better survival at pH 2.0 upon expose to high bile salt concentrations and at moderate heat up to 65°C compared to unencapsulated controls (Ding and Shah 2007). It was recently reported that the microencapsulation of *L. paracasei* subsp. *paracasei* F19 and *B. lactis* BB 12 in the rennet-induced gelation of skim milk protein matrices offered 100-fold survival compared to control cells upon 90-min incubation at pH 2.5 (Heidebach et al. 2009). Furthermore, enhanced survival rates (upon simulated gastric juice exposure) of freeze-dried or fresh cultures of *Bifodobacterium breve* R070 (BB R070) and *B. longum* R023 (BL R023) encapsulated using milk fat and/or denatured whey proteins were observed (Picot and Lacroix 2004). In addition, spray-drying encapsulation was demonstrated as the least destructive immobilization method, yielding a survival rate of 25% for *B. breve* (Picot and Lacroix 2004). Moreover, encapsulated cells exhibited higher viability during 28 days of storage in low-pH yogurts (Picot and Lacroix 2004). Recent studies in our laboratory showed that entrapment of the probiotic strain *L. rhamnosus* GG in heat-treated whey protein gel matrices offered nine log survival when exposed to porcine gastrointestinal contents at pH 1.8 for 3 h, whereas no free cells were detected after 15 min of exposure (Doherty et al. 2010).

17.5.2 Prebiotics (A Synbiotic Approach)

Prebiotics are "nondigestible food ingredients that beneficially affect the host by selectively stimulating the growth and/or activity of one or a limited number of bacterial species" (Gibson and Roberfroid 1995). Synbiotics are generally defined as "a mixture of probiotics and prebiotics that beneficially affect the host by improving the survival and implantation of live microbial dietary supplements in the GIT, by selectively stimulating the growth and/or activating the metabolism of one or a limited number of health promoting bacteria, and thus improving host welfare" (Gibson and Roberfroid 1995). Prebiotics remain undigested in the upper GIT and are fermented by the indigenous anaerobic colonic microbiota to produce lactic acid and short-chain fatty acids (SCFAs; e.g., acetic, butyric, and propionic acids), thus providing metabolic substrates to the colon and stimulating the growth of bifidobacteria (Vernazza et al. 2006). For a food ingredient to be considered prebiotic, it should satisfy the following criteria: (1) neither be hydrolyzed nor absorbed in the upper GIT: (2) be a selective substrate for one or a limited number of potentially beneficial endocommensal bacteria (e.g., bifidobacteria and lactobacilli); and (3) be able to alter the composition of the colonic microbiota into highly potential beneficial bacteria (Gibson et al. 2007). The in vitro and in vivo functional activities (bifidogenic properties) of some prebiotics are well documented (Gibson and Roberfroid 1995; Gibson et al. 2007; Huebner et al. 2007; Su et al. 2007).

When probiotic bacteria are coadministered with prebiotic (synbiotic), the prebiotic oligosaccharides in the product may stimulate the growth and colonization of the added bacteria (Rastall and Maitin 2002). In this respect, the addition of a prebiotic substance during probiotic encapsulation offered very high survival rates of probiotic cultures and successfully demonstrated the advantages of producing synbiotics (Homayouni et al. 2008). Moreover, the therapeutic potential of synbiotics has been evaluated against various GIT diseases using human subjects (Table 17.2), such as Crohn's disease (Fujimori et al. 2007), ulcerative colitis (Fujimori et al. 2009), and colon cancer (Sugawara et al. 2006; Rafter et al. 2007). Furthermore, a number of studies have reported the clinical efficacy of pro-, pre-, and synbiotic applications in critically ill patients (Bengmark and Martindale 2005). Moreover, the incorporation of prebiotics such as fructo-oligosaccharides (Raftilose) and microfluidized starch during the encapsulation of probiotics can improve their technological performance, and, as a result, adequate populations of live cells can be delivered to the GIT (Crittenden et al. 2006; Homayouni et al. 2008).

17.5.3 Genetic Engineering: Designer Probiotics and Patho-Biotechnology

Strain improvement can be achieved either by classical approaches or by using recombinant DNA technology. Through genetic engineering, it is possible not only to improve the performance of the existing probiotic strains but also to design and construct entirely new strains. The developments of new applications such as live vaccines (live antigen carriers), in vivo production of pharmabiotics (Shanahan et al. 2009), and functional foods with probiotics need robust strains that have the ability to survive in the digestive tract and to implement the desired effect at their target site. The genetic manipulation of probiotic strains can be directed either toward their technological improvement during food production or toward their survival and performance in the GIT. For example, the overexpression of heat-shock protein chaperones GroES and GroEL in the probiotic *L. paracasei* NFBC338 improved the thermotolerance of the strain (Desmond et al. 2004). Furthermore, the recombinant probiotic strain exhibited increased resistance to solvent stress (Desmond et al. 2004). Overall, the GroESL-overproducing *Lactobacillus* strain exhibited a tenfold better survival during heat stress when compared with the adapted parent strain and a 54-fold increase over unadapted control cultures (Desmond et al. 2004). Moreover, the GroESL-overproducing strain showed a tenfold increased survival during spray-drying and a 14% increased survival during freeze-drying when compared to the control (Desmond et al. 2004).

The whole-genome sequencing of probiotic strains has yielded increasing numbers of completed genomes of probiotic bacteria, many of which are publicly available on the Internet (NCBI Genomes, Genomes online database [GOLD]). This allows us to characterize their gene expression profiles, and to identify the most adaptable strains in selective environments. Currently, there are 28 genomes of the genus *Lactobacillus* completely sequenced and published, and eight genomes of the genus *Bifidobacteriaceae* are publicly available (NCBI and GOLD 2009). Moreover, many other whole-genome-sequencing projects are currently underway.

Comparative genomics studies of these genomes are already revealing some interesting findings. For example, in a study by O'Sullivan et al. (2009), nine genes were identified as being niche-specific when 11 genomes from either the gut or dairy products, or both, were compared (O'Sullivan et al. 2009). Furthermore, the expression levels of the in vivo inducible (*ivi*) genes in *L. plantarum* WCFS1 are site-specific throughout the mammalian intestine. A 350-fold increase was observed in the relative expression levels of in vivo inducible genes in the mouse intestine when compared to levels in cells grown in vitro on a nutrient-rich medium (Marco et al. 2007).

The term *patho-biotechnology*, first coined by Sleator and Hill (2006), is a recent approach undertaken to improve probiotic strain stress resistance. This approach involves the "exploitation of pathogen stress-survival strategies, and by applying them in the design of more robust probiotic cultures" (Sleator and Hill 2006). Sheehan et al. (2006) demonstrated that the heterologous expression of the listerial betaine-uptake system (BetL) into the probiotic strain *L. salivarius* UCC118, using a nisin-controlled expression system, increased the resistance to several stresses (osmo, cryo, baro, and chill) and to spray- and freeze-drying (Sheehan et al. 2006). Furthermore, it was demonstrated that *B. breve* UCC2003, harboring the betamine-uptake (Betl) gene (from *Listeria monocytogenes*, an intracellular foodborne pathogen), showed an improved tolerance to gastric juice and elevated osmolarity (Sheehan et al. 2007). Similarly, the *L. lactis* strain expressing trehalose synthesis genes of *E. coli* showed an improved tolerance during freeze-drying and an improved tolerance to gastric acid and bile salts, suggesting a protective role for compatible solutes in the gastric environment (Termont et al. 2006). Pathogenic genera represent a useful reservoir of stress-survival mechanisms, which could potentially benefit the physiological and clinical efficacy of probiotics (Sleator and Hill 2006–2008). The ability to confer additional stress tolerance in stress-sensitive cultures is one approach leading to the development and delivery of novel probiotics with maximal therapeutic efficacy. Further scientific assessment is required with rigorous scientific controls to analyze the benefit of using genes from pathogenic species in beneficial bacteria and interpreting by risk–benefit analysis (Sleator and Hill 2008). It is important to emphasize, however, that the genetic modification of probiotics is a very controversial area at present and is not currently allowed for food production in certain countries such as those in the EU. Furthermore, the genetic modification of bacteria for food applications, involving the use of DNA from pathogenic bacteria, is likely to fuel even further controversy.

17.6 Concluding Remarks

Despite being metabolically simple microorganisms, probiotic bacteria employ sophisticated mechanisms to combat stress. In this respect, the expression of a variety of different proteins in a bacterial cell plays a vital role under various stressful conditions, such as those induced by heat, acid, bile, and oxygen. In many cases, similar proteins are expressed during different stressful conditions (heat, acid, and bile), and as such, these mechanisms of resistance are interconnected, which can stabilize the physiology of the cell to survive against the life-threatening consequences of various extrinsic and intrinsic stresses. In this respect, it is important to understand what physiological stresses these bacteria encounter in order to develop more technologically robust probiotic strains and to improve their colonization capabilities in order to reside in the mammalian GIT. Indeed, some lessons might be learned from intestinal pathogens with regard to overcoming the stressful environment of the gut. Probiotics are considered allochthonous microbiota; therefore, frequent consumption is required in order to maintain sufficient population numbers to exert a health benefit. As we learn more about the basic composition and diversity in the human gut microbiota from the rapidly emerging field of metagenomics, new insights will undoubtedly be gleaned on what constitutes a healthy microbiota and what dietary approaches will positively influence this composition. In cases where this involves probiotic intervention, then an understanding of the stresses evoked and how to counteract them should be considered crucial to ensure efficacy.

Acknowledgments AU was supported by a Teagasc Walsh Fellowship. This work was supported by Enterprise Ireland, the Irish Department of Agriculture and Food under the National Development Plan (2000–2006), and Science Foundation Ireland Centre for Science Engineering and Technology (SFI CSET).

References

Andersson AF, Lindberg M, Jakobsson H, Bäckhed F, Nyrén P, Engstrand L (2008) Comparative analysis of human gut microbiota by barcoded pyrosequencing. PLoS One 3:e2836

Anukam KC, Osazuwa EO, Osadolor HB, Bruce AW, Reid G (2008) Yogurt containing probiotic *Lactobacillus rhamnosus* GR-1 and *L. reuteri* RC-14 helps resolve moderate diarrhea and increases CD4 count in HIV/AIDS patients. J Clin Gastroenterol 42:239–243

Armuzzi A, Cremonini F, Bartolozzi F, Canducci F, Candelli M, Ojetti V, Cammarota G, Anti M, De Lorenzo A, Pola P, Gasbarrini G, Gasbarrini A (2001) The effect of oral administration of *Lactobacillus* GG on antibiotic-associated gastrointestinal side-effects during *Helicobacter pylori* eradication therapy. Aliment Pharm Ther 15:163–169

Auzat I, Chapuy-Regaud S, Le Bras G, Dos Santos D, Ogunniyi AD, Le Thomas I, Garel JR, Paton JC, Trombe MC (1999) The NADH oxidase of *Streptococcus pneumoniae*: its involvement in competence and virulence. Mol Microbiol 34:1018–1028

Bäckhed F, Ley RE, Sonnenburg JL, Peterson DA, Gordon JI (2005) Host-bacterial mutualism in the human intestine. American Association for the Advancement of Science, 307:1915–1920

Baker HC, Tran DN, Thomas LV (2009) Health benefits of probiotics for the elderly: a review. J Foodservice 20:250–262

Begley M, Gahan CGM, Hill C (2005) The interaction between bacteria and bile. FEMS Microbiol Rev 29:625–652

Begley M, Hill C, Gahan CGM (2006) Bile salt hydrolase activity in probiotics. Appl Environ Microbiol 72:1729–1738

Bengmark S (1998) Ecological control of the gastrointestinal tract. The role of probiotic flora. Gut 42:2–7

Bengmark S, Martindale R (2005) Prebiotics and synbiotics in clinical medicine. Nutr Clin Pract 20:244–261

Berg RD (1996) The indigenous gastrointestinal microflora. Trends Microbiol 4:430–435

Besselink M, Timmerman H, Buskens E, Nieuwenhuijs V, Akkermans L, Gooszen HG (2004) Probiotic prophylaxis in patients with predicted severe acute pancreatitis (PROPATRIA): design and rationale of a double-blind, placebo-controlled randomised multicenter trial [ISRCTN38327949]. BMC Surgery 4:12

Bibiloni R, Fedorak RN, Tannock GW, Madsen KL, Gionchetti P, Campieri M, De Simone C, Sartor RB (2005) VSL#3 Probiotic-mixture induces remission in patients with active ulcerative colitis. Am J Gastroenterol 100(7):1539–1546

Binder HJ, Filburn B, Floch M (1975) Bile acid inhibition of intestinal anaerobic organisms. Amer J Clin Nutr 28:119

Bleichner G, Bléhaut H, Mentec H, Moyse D (1997) *Saccharomyces boulardii* prevents diarrhea in critically ill tube-fed patients. Intens Care Med 23:517–523

Borgstrom B, Dahlqvist A, Lundh G, Sjovall J (1957) Studies of intestinal digestion and absorption in the human. J Clin Invest 36:1521

Breton YL, Mazé A, Hartke A, Lemarinier S, Auffray Y, Rince A (2002) Isolation and characterization of bile salts-sensitive mutants of *Enterococcus faecalis*. Curr Microbiol 45:434–439

Brigidi P, Vitali B, Swennen E, Bazzocchi G, Matteuzzi D (2001) Effects of probiotic administration upon the composition and enzymatic activity of human fecal microbiota in patients with irritable bowel syndrome or functional diarrhea. Res Microbiol 152:735–741

Briolat V, Reysset G (2002) Identification of the *Clostridium perfringens* genes involved in the adaptive response to oxidative stress. J Bacteriol 184:2333–2343

Carvalho AS, Silva J, Ho P, Teixeira P, Xavier Malcata F, Gibbs P (2003) Protective effect of sorbitol and monosodium glutamate during storage of freeze-dried lactic acid bacteria. Lait 83:203–210

Carvalho AS, Silva J, Ho P, Teixeira P, Malcata FX, Gibbs P (2004a) Effects of various sugars added to growth and drying media upon thermotolerance and survival throughout storage of freeze-dried *Lactobacillus delbrueckii* ssp. *bulgaricus*. Biotechnol Prog 20:248–254

Carvalho AS, Silva J, Ho P, Teixeira P, Malcata FX, Gibbs P (2004b) Relevant factors for the preparation of freeze-dried lactic acid bacteria. Int Dairy J 14:835–847

Castro HP, Teixeira PM, Kirby R (1997) Evidence of membrane damage in *Lactobacillus bulgaricus* following freeze drying. J Appl Microbiol 82:87–94

Cats A, Kuipers EJ, Bosschaert MAR, Pot RGJ, Vandenbroucke-Grauls C, Kusters JG (2003) Effect of frequent consumption of a *Lactobacillus casei*-containing milk drink in *Helicobacter pylori*-colonized subjects. Aliment Pharmacol Therap 17:429–435

Champagne CP, Fustier P (2007) Microencapsulation for the improved delivery of bioactive compounds into foods. Curr Opin Biotech 18(2):184–190

Champagne CP, Gardner NJ (2008) Effect of storage in a fruit drink on subsequent survival of probiotic lactobacilli to gastro-intestinal stresses. Food Res Int 41:539–543

Champagne CP, Gardner NJ, Roy D (2005) Challenges in the addition of probiotic cultures to foods. Crit Rev Food Sci Nutr 45:61–84

Chandra RK (2002) Effect of *Lactobacillus* on the incidence and severity of acute rotavirus diarrhoea in infants. A prospective placebo-controlled double-blind study. Nutr Res 22:65–69

Chang YY, Cronan JE Jr (1999) Membrane cyclopropane fatty acid content is a major factor in acid resistance of *Escherichia coli*. Molec Microbiol 33:249–259

Chung HS, Kim YB, Chun SL, Ji GE (1999) Screening and selection of acid and bile resistant bifidobacteria. Int J Food Microbiol 47:25–32

Cildir SK, Germec D, Sandalli N, Ozdemir FI, Arun T, Twetman S, Caglar E (2009) Reduction of salivary mutans streptococci in orthodontic patients during daily consumption of yoghurt containing probiotic bacteria. Eur J Orthod 31:407–411

Claesson MJ, van Sinderen D, O'Toole PW (2007) The genus *Lactobacillus*—a genomic basis for understanding its diversity. FEMS Microbiol Lett 269:22–28

Collins JK, Thornton G, Sullivan GO (1998) Selection of probiotic strains for human applications. Int Dairy J 8:487–490

Condon S (1987) Responses of lactic acid bacteria to oxygen. FEMS Microbiol Rev 46:269–280

Corcoran BM, Ross RP, Fitzgerald GF, Stanton C (2004) Comparative survival of probiotic lactobacilli spray-dried in the presence of prebiotic substances. J Appl Microbiol 96:1024–1039

Corcoran BM, Stanton C, Fitzgerald GF, Ross RP (2005) Survival of probiotic lactobacilli in acidic environments is enhanced in the presence of metabolizable sugars. Appl Environ Microbiol 71:3060–3067

Corcoran BM, Stanton C, Fitzgerald GF, Ross RP (2007) Growth of probiotic lactobacilli in the presence of oleic acid enhances subsequent survival in gastric juice. Microbiology 153:291

Corcoran BM, Stanton C, Fitzgerald G, Ross RP (2008) Life under stress: the probiotic stress response and how it may be manipulated. Curr Pharm Des 14:1382–1399

Corre C, Madec MN, Boyaval P (1992) Production of concentrated *Bifidobacterium bifidum*. J Chem Tech Biotechnol 53(2):189–194

Cotter PD, Hill C (2003) Surviving the acid test: responses of Gram-positive bacteria to low pH. Microbiol Molec Biol Rev 67:429

Crawford DR, Davies KJ (1994) Adaptive response and oxidative stress. Environ Health Perspect 102(Suppl 10):25

Cremonini F, Di Caro S, Covino M, Armuzzi A, Gabrielli M, Santarelli L, Nista EC, Cammarota G, Gasbarrini G, Gasbarrini A (2002) Effect of different probiotic preparations on anti-*Helicobacter pylori* therapy-related side effects: a parallel group, triple blind, placebo-controlled study. Am J Gastroenterol 97:2744–2749

Crittenden R, Weerakkody R, Sanguansri L, Augustin MA (2006) Synbiotic microcapsules that enhance microbial viability during nonrefrigerated storage and gastrointestinal transit. Appl Environ Microbiol 72:2280–2282

Cronan JE (2002) Phospholipid modifications in bacteria. Curr Opin Microbiol 5:202–205

Cruz AG, Antunes AEC, Sousa A, Faria JAF, Saad SMI (2009) Ice-cream as a probiotic food carrier. Food Res Int 42:1233–1239

Cunningham L, Guest JR (1998) Transcription and transcript processing in the sdhCDAB-sucABCD operon of *Escherichia coli*. Microbiology 144:2113–2123

Curran TM, Lieou J, Marquis RE (1995) Arginine deiminase system and acid adaptation of oral streptococci. Appl Environ Microbiol 61:4494–4496

De Angelis M, Gobbetti M (2004) Environmental stress responses in *Lactobacillus*: a review. Proteomics 4:106–122

de Mendoza D, Cronan JE Jr (1983) Thermal regulation of membrane lipid fluidity in bacteria. Trends Biochem Sci 8:49–52

de Valdéz GF, de Giori GS, de Ruiz Holgado AAP, Oliver G (1983) Protective effect of adonitol on lactic acid bacteria subjected to freeze-drying. Am Soc Microbiol, 45:302–304

de Valdéz GF, de Giori GS, de Ruiz Holgado AP, Oliver G (1985) Rehydration conditions and viability of freeze-dried lactic acid bacteria. Cryobiology 22:574–577

Desmond C, Stanton C, Fitzgerald GF, Collins K, Paul Ross R (2001) Environmental adaptation of probiotic lactobacilli towards improvement of performance during spray drying. Int Dairy J 11:801–808

Desmond C, Ross RP, O'Callaghan E, Fitzgerald G, Stanton C (2002) Improved survival of *Lactobacillus paracasei* NFBC 338 in spray-dried powders containing gum acacia. J Appl Microbiol 93:1003–1011

Desmond C, Fitzgerald GF, Stanton C, Ross RP (2004) Improved stress tolerance of GroESL-overproducing *Lactococcus lactis* and probiotic *Lactobacillus paracasei* NFBC 338. Appl Environ Microbiol 70:5929–5936

Dethlefsen L, Huse S, Sogin ML, Relman DA (2008) The pervasive effects of an antibiotic on the human gut microbiota, as revealed by deep 16S rRNA sequencing. PLoS Biol 6:e280

Ding WK, Shah NP (2007) Acid, bile, and heat tolerance of free and microencapsulated probiotic bacteria. J Food Sci 72:446–450

Diop L, Guillou S, Durand H (2008) Probiotic food supplement reduces stress-induced gastrointestinal symptoms in volunteers: a double-blind, placebo-controlled, randomized trial. Nutr Res 28:1–5

Doherty SB, Gee VL, Ross RP, Stanton C, Fitzgerald GF, Brodkorb A (2010) Efficacy of whey protein gel networks as potential viability-enhancing scaffolds for cell immobilization of *Lactobacillus rhamnosus* GG. J Microbiol Meth 80:231–241

Doleyres Y, Paquin C, LeRoy M, Lacroix C (2002) *Bifidobacterium longum* ATCC 15707 cell production during free-and immobilized-cell cultures in MRS-whey permeate medium. Appl Microbiol Biotechnol 60:168–173

Dunne C, O'Mahony L, Murphy L, Thornton G, Morrissey D, O'Halloran S, Feeney M, Flynn S, Fitzgerald G, Daly C, Kiely B, O'Sullivan GC, Shanahan F, Collins JK (2001) *In vitro* selection criteria for probiotic bacteria of human origin: correlation with *in vivo* findings. Am J Clin Nutr 73:386S–392

Eckburg PB, Bik EM, Bernstein CN, Purdom E, Dethlefsen L, Sargent M, Gill SR, Nelson KE, Relman DA (2005) Diversity of the human intestinal microbial flora. American Association for the Advancement of Science, 308:1635–1638

Flahaut S, Hartke A, Giard JC, Auffray Y (1997) Alkaline stress response in *Enterococcus faecalis*: adaptation, cross-protection, and changes in protein synthesis. Appl Environ Microbiol 63:812–814

Floch MH, Binder HJ, Filburn B, Gershengoren W (1972) The effect of bile acids on intestinal microflora. Amer J Clin Nutr 25:1418–1426

Frank DN, St. Amand AL, Feldman RA, Boedeker EC, Harpaz N, Pace NR (2007) Molecular-phylogenetic characterization of microbial community imbalances in human inflammatory bowel diseases. Proc Natl Acad Sci 104:13780–13785

Fujimori S, Tatsuguchi A, Gudis K, Kishida T, Mitsui K, Ehara A, Kobayashi T, Sekita Y, Seo T, Sakamoto C (2007) High dose probiotic and prebiotic cotherapy for remission induction of active Crohn's disease. J Gastroenterol Hepatol 22:1199–1204

Fujimori S, Gudis K, Mitsui K, Seo T, Yonezawa M, Tanaka S, Tatsuguchi A, Sakamoto C (2009) A randomized controlled trial on the efficacy of synbiotic versus probiotic or prebiotic treatment to improve the quality of life in patients with ulcerative colitis. Nutrition 25:520–525

Galdeano CM, Perdigon G (2006) The probiotic bacterium *Lactobacillus casei* induces activation of the gut mucosal immune system through innate immunity. Clin Vaccine Immunol 13:219–226

Galinski EA (1995) Osmoadaptation in bacteria. Adv Microb Physiol 37:273

Gardiner G, Stanton C, Lynch PB, Collins JK, Fitzgerald G, Ross RP (1999) Evaluation of Cheddar cheese as a food carrier for delivery of a probiotic strain to the gastrointestinal tract. J Dairy Sci 82:1379–1387

Gardiner GE, O'Sullivan E, Kelly J, Auty MAE, Fitzgerald GF, Collins JK, Ross RP, Stanton C (2000) Comparative survival rates of human-derived probiotic *Lactobacillus paracasei* and *L. salivarius* strains during heat treatment and spray drying. Appl Environ Microbiol 66:2605–2612

Gardiner GE, Bouchier P, O'Sullivan E, Kelly J, Collins JK, Fitzgerald G, Ross RP, Stanton C (2002) A spray-dried culture for probiotic Cheddar cheese manufacture. Int Dairy J 12:749–756

Gibson GR, Roberfroid MB (1995) Dietary modulation of the human colonic microbiota: introducing the concept of prebiotics. J Nutr 125:1401

Gibson GR, Willems A, Reading S, Collins MD (2007) Fermentation of non-digestible oligosaccharides by human colonic bacteria. Proc Nutr Soc 55:899–912

Gilliland SE, Kim HS (1984) Effect of viable starter culture bacteria in yogurt on lactose utilization in humans. J Dairy Sci 67:1–6

Gionchetti P, Rizzello F, Helwig U, Venturi A, Lammers KM, Brigidi P, Vitali B, Poggioli G, Miglioli M, Campieri M (2003) Prophylaxis of pouchitis onset with probiotic therapy: a double-blind, placebo-controlled trial. Gastroenterology 124:1202–1209

Gionchetti P, Rizzello F, Morselli C, Poggioli G, Tambasco R, Calabrese C, Brigidi P, Vitali B, Straforini G, Campieri M (2007) High-dose probiotics for the treatment of active pouchitis. Dis Colon Rectum 50:2075–2084

Godward G, Sultana K, Kailasapathy K, Peiris P, Arumugaswamy R, Reynolds N (2000) The importance of strain selection on the viability and survival of probiotic bacteria in dairy foods. Milchwissenschaft 55:441–445

Gouin S (2004) Microencapsulation industrial appraisal of existing technologies and trends. Trends Food Sci Tech 15:330-347

Guillot A, Obis D, Mistou M-Y (2000) Fatty acid membrane composition and activation of glycine-betaine transport in *Lactococcus lactis* subjected to osmotic stress. Int J Food Microbiol 55:47–51

Guslandi M, Giollo P, Testoni PA (2003) A pilot trial of *Saccharomyces boulardii* in ulcerative colitis. Eur J Gastroenterol Hepatol 15:697–698

Guyonnet D, Woodcock A, Stefani B, Trevisan C, Hall C (2009) Fermented milk containing *Bifidobacterium lactis* DN-173 010 improved self-reported digestive comfort amongst a general population of adults. A randomized, open-label, controlled, pilot study. J Digest Disease 10:61–70

Handelsman J, Rondon MR, Brady SF, Clardy J, Goodman RM (1998) Molecular biological access to the chemistry of unknown soil microbes: a new frontier for natural products. Chem Biol 5:245–249

Hartke A, Bouche S, Laplace JM, Benachour A, Boutibonnes P, Auffray Y (1995) UV-inducible proteins and UV-induced cross-protection against acid, ethanol, H_2O_2 or heat treatments in *Lactococcus lactis* subsp. *lactis*. Arch Microbiol 163:329–336

Hatakka K, Martio J, Korpela M, Herranen M, Poussa T, Laasanen T, Saxelin M, Vapaatalo H, Moilanen E, Korpela R (2003) Effects of probiotic therapy on the activity and activation of mild rheumatoid arthritis—a pilot study. Scand J Rheumatol 32:211–215

Hattori M, Taylor TD (2009) The human intestinal microbiome: a new frontier of human biology. DNA Res 16:1

Heidebach T, Först P, Kulozik U (2009) Microencapsulation of probiotic cells by means of rennet-gelation of milk proteins. Food Hydrocoll 23:1670–1677

Heilig H, Zoetendal EG, Vaughan EE, Marteau P, Akkermans ADL, de Vos WM (2002) Molecular diversity of *Lactobacillus* spp. and other lactic acid bacteria in the human intestine as determined by specific amplification of 16S ribosomal DNA. Appl Environ Microbiol 68:114–123

Hickson M, D'Souza AL, Muthu N, Rogers TR, Want S, Rajkumar C, Bulpitt CJ (2007) Use of probiotic *Lactobacillus* preparation to prevent diarrhoea associated with antibiotics: randomised double blind placebo controlled trial. BMJ 335:80

Higuchi M (1992) Reduced nicotinamide adenine dinucleotide oxidase involvement in defense against oxygen toxicity of *Streptococcus mutans*. Oral Microbiol Immunol 7:309–314

Hlivak P, Odraska J, Ferencik M, Ebringer L, Jahnova E, Mikes Z (2005) One-year application of probiotic strain *Enterococcus faecium* M-74 decreases serum cholesterol levels. Bratisl Lek Listy 106:67–72

Hofmann AF (1999) The continuing importance of bile acids in liver and intestinal disease. Arch Intern Med 159:2647–2658

Homayouni A, Azizi A, Ehsani MR, Yarmand MS, Razavi SH (2008) Effect of microencapsulation and resistant starch on the probiotic survival and sensory properties of synbiotic ice cream. Food Chem 111:50–55

Hooper LV, Midtvedt T, Gordon JI (2002) How host-microbial interactions shape the nutrient environment of the mammalian intestine. Annu Rev Nutr 22:283–307

Hord NG (2008) Eukaryotic-microbiota crosstalk: potential mechanisms for health benefits of prebiotics and probiotics. Annu Rev Nutr 28:215–231

Hubalek Z (2003) Protectants used in the cryopreservation of microorganisms. Cryobiology 46:205–229

Huebner J, Wehling RL, Hutkins RW (2007) Functional activity of commercial prebiotics. Int Dairy J 17:770–775

Ishibashi N, Shimamura S (1993) Bifidobacteria: research and development in Japan. Food Tech (Chicago) 47:126–136

Iyer C, Kailasapathy K, Peiris P (2004) Evaluation of survival and release of encapsulated bacteria in *ex vivo* porcine gastrointestinal contents using a green fluorescent protein gene-labelled *E. coli*. LWT—Food Sci Tech 37:639–642

Jamieson DJ, Stephen DWS, Terriere EC (1996) Analysis of the adaptive oxidative stress response of *Candida albicans*. FEMS Microbiol Lett 138:83–88

Kailasapathy K (2002) Microencapsulation of probiotic bacteria: Technology and potential applications. Curr Issues Intestinal Microbiol 3:39–48

Kajander K, Myllyluoma E, Rajilic-Stojanovic M, Kyronpalo S, Rasmussen M, Järvenpää S, Zoetendal EG, de Vos WM, Vapaatalo H, Korpela R (2008) Clinical trial: multispecies probiotic supplementation alleviates the symptoms of irritable bowel syndrome and stabilizes intestinal microbiota. Aliment Pharm Therap 27:48–57

Kalliomäki M, Salminen S, Arvilommi H, Kero P, Koskinen P, Isolauri E (2001) Probiotics in primary prevention of atopic disease: a randomised placebo-controlled trial. Lancet 357:1076–1079

Kearney N, Meng XC, Stanton C, Kelly J, Fitzgerald GF, Ross RP (2009) Development of a spray dried probiotic yoghurt containing *Lactobacillus paracasei* NFBC 338. Int Dairy J 19:684–689

Kilstrup M, Jacobsen S, Hammer K, Vogensen FK (1997) Induction of heat shock proteins DnaK, GroEL, and GroES by salt stress in *Lactococcus lactis*. Appl Environ Microbiol 63:1826–1837

Kim SJ, Cho SY, Kim SH, Song OJ, Shin IS, Cha DS, Park HJ (2008) Effect of microencapsulation on viability and other characteristics in *Lactobacillus acidophilus* ATCC 43121. LWT—Food Sci Tech 41:493–500

Kim WS, Perl L, Park JH, Tandianus JE, Dunn NW (2001) Assessment of stress response of the probiotic *Lactobacillus acidophilus*. Curr Microbiol 43:346–350

Kleerebezem M, Vaughan EE (2009) Probiotic and gut lactobacilli and bifidobacteria: molecular approaches to study diversity and activity. Annu Rev Microbiol 63:269–290

Kruis W, Fric P, Pokrotnieks J, Lukas M, Fixa B, Kascák M, Kamm MA, Weismueller J, Beglinger C, Stolte M, Wolff C, Schulze J (2004) Maintaining remission of ulcerative colitis with the probiotic *Escherichia coli* Nissle 1917 is as effective as with standard mesalazine. Gut 53:1617–1623

Kurokawa K, Itoh T, Kuwahara T, Oshima K, Toh H, Toyoda A, Takami H, Morita H, Sharma VK, Srivastava TP, Taylor TD, Noguchi H, Mori H, Ogura Y, Ehrlich DS, Itoh K, Takagi T, Sakaki Y, Hayashi T, Hattori M (2007) Comparative metagenomics revealed commonly enriched gene sets in human gut microbiomes. DNA Res 14:169

Kwon SG, Son JW, Kim HJ, Park CS, Lee JK, Ji GE, Oh DK (2006) High concentration cultivation of *Bifidobacterium bifidum* in a submerged membrane bioreactor. Biotech Progr 22:1591–1597

Lacroix C, Yildirim S (2007) Fermentation technologies for the production of probiotics with high viability and functionality. Curr Opin Biotech 18:176–183

Lavermicocca P, Valerio F, Lonigro SL, De Angelis M, Morelli L, Callegari ML, Rizzello CG, Visconti A (2005) Study of adhesion and survival of lactobacilli and bifidobacteria on table olives with the aim of formulating a new probiotic food. Appl Environ Microbiol 71:4233–4240

Leach RH, Scott WJ (1959) The influence of rehydration on the viability of dried microorganisms. Microbiology 21:295

Lei V, Jakobsen M (2004) Microbiological characterization and probiotic potential of koko and koko sour water, African spontaneously fermented millet porridge and drink. J Appl Microbiol 96:384–397

Ley RE, Peterson DA, Gordon JI (2006) Ecological and evolutionary forces shaping microbial diversity in the human intestine. Cell 124:837–848

Lian WC, Hsiao HC, Chou CC (2002) Survival of bifidobacteria after spray-drying. Int J Food Microbiol 74:79–86

Lin M-Y, Savaiano D, Harlander S (1991) Influence of nonfermented dairy products containing bacterial starter cultures on lactose maldigestion in humans. J Dairy Sci 74:87–95

Liu CH, Lee SM, VanLare JM, Kasper DL, Mazmanian SK (2008) Regulation of surface architecture by symbiotic bacteria mediates host colonization. Proc Natl Acad Sci 105:3951

Lorena R, Borja S, Patricia R-M, Clara GdlR-G, Abelardo M (2007) Cell envelope changes in *Bifidobacterium animalis* ssp. *lactis* as a response to bile. FEMS Microbiol Lett 274:316–322

Maicas S, Pardo I, Ferrer S (2000) The effects of freezing and freeze-drying of *Oenococcus oeni* upon induction of malolactic fermentation in red wine. Int J Food Sci Tech 35:75–79

Marco ML, Bongers RS, De Vos WM, Kleerebezem M (2007) Spatial and temporal expression of *Lactobacillus plantarum* genes in the gastrointestinal tracts of mice. Appl Environ Microbiol 73:124

Margulies M, Egholm M, Altman WE, Attiya S, Bader JS, Bemben LA, Berka J, Braverman MS, Chen YJ, Chen Z, Dewell SB, Du L, Fierro JM, Gomes XV, Godwin BC, He W, Helgesen S, Ho CH, Irzyk GP, Jando SC, Alenquer ML, Jarvie TP, Jirage KB, Kim JB, Knight JR, Lanza JR, Leamon JH, Lefkowitz SM, Lei M, Li J, Lohman KL, Lu H, Makhijani VB, McDade KE, McKenna MP, Myers EW, Nickerson E, Nobile JR, Plant R, Puc BP, Ronan MT, Roth GT, Sarkis GJ, Simons JF, Simpson JW, Srinivasan M, Tartaro KR, Tomasz A, Vogt KA, Volkmer

GA, Wang SH, Wang Y, Weiner MP, Yu P, Begley RF, Rothberg JM (2005) Genome sequencing in microfabricated high-density picolitre reactors. Nature 437:376–380

Marquis RE, Bender GR, Murray DR, Wong A (1987) Arginine deiminase system and bacterial adaptation to acid environments. Appl Environ Microbiol 53:198–200

Marteau P, Rambaud JC (1993) Potential of using lactic acid bacteria for therapy and immunomodulation in man. FEMS Microbiol Rev 12:207–220

Martos GI, Minahk CJ, Font de Valdez G, Morero R (2007) Effects of protective agents on membrane fluidity of freeze-dried *Lactobacillus delbrueckii* ssp. *bulgaricus*. Lett Appl Microbiol 45:282–288

Matar C, Valdez JC, Medina M, Rachid M, Perdigon G (2002) Immunomodulating effects of milks fermented by *Lactobacillus helveticus* and its non-proteolytic variant. J Dairy Res 68:601–609

Matsumoto M, Benno Y (2004) Consumption of *Bifidobacterium lactis* LKM512 yogurt reduces gut mutagenicity by increasing gut polyamine contents in healthy adult subjects. Mutat Res—Fundamen Molec Mechan Mutagen 568:147–153

McAuliffe OE, Klaenhammer TR (2002) Genomic perspectives on probiotics and the gastrointestinal microflora. In Tannock GW (Ed.), Probiotics and prebiotics: where are we going? Caister Academic Press, Norfolk, UK

McFarland LV, Surawicz C, Greenberg RN, Elmer GW, Moyer KA, Melcher SA, Bowen KE, Cox JL (1995) Prevention of ß-lactam-associated diarrhea by *Saccharomyces boulardii* compared with placebo. Am J Gastroenterol 90:439–448

Medzhitov R (2007) Recognition of microorganisms and activation of the immune response. Nature 449:819–826

Meng XC, Stanton C, Fitzgerald GF, Daly C, Ross RP (2008) Anhydrobiotics: the challenges of drying probiotic cultures. Food Chem 106:1406–1416

Mevarech M, Eisenberg H, Neumann E (1977) Malate dehydrogenase isolated from extremely halophilic bacteria of the Dead Sea. 1. Purification and molecular characterization. Biochemistry 16:3781–3785

Miao S, Mills S, Stanton C, Fitzgerald GF, Roos Y, Ross RP (2008) Effect of disaccharides on survival during storage of freeze dried probiotics. Dairy Sci Technol 88:19–30

Michetti P, Dorta G, Wiesel PH, Brassart D, Verdu E, Herranz M, Felley C, Porta N, Rouvet M, Blum AL, Corthésy-Theulaz I (1999) Effect of whey-based culture supernatant of *Lactobacillus acidophilus* (*johnsonii*) La1 on *Helicobacter pylori* infection in humans. Digestion 60:203–209

Michida H, Tamalampudi S, Pandiella SS, Webb C, Fukuda H, Kondo A (2006) Effect of cereal extracts and cereal fiber on viability of *Lactobacillus plantarum* under gastrointestinal tract conditions. Biochem Eng J 28:73–78

Mille Y, Obert JP, Beney L, Gervais P (2004) New drying process for lactic bacteria based on their dehydration behavior in liquid medium. Biotech Bioeng 88:71–76

Mimura T, Rizzello F, Helwig U, Poggioli G, Schreiber S, Talbot IC, Nicholls RJ, Gionchetti P, Campieri M, Kamm MA (2004) Once daily high dose probiotic therapy (VSL#3) for maintaining remission in recurrent or refractory pouchitis. Gut 53:108–114

Montes RG, Bayless TM, Saavedra JM, Perman JA (1995) Effect of milks inoculated with *Lactobacillus acidophilus* or a yogurt starter culture in lactose-maldigesting children. J Dairy Sci 78:1657–1664

Muller JA, Ross RP, Fitzgerald GF, Stanton C (2009) Manufacture of probiotic bacteria. Prebiotics and probiotics. Science and Technology 1:725–759

Muller JA, Stanton C, Sybesma W, Fitzgerald GF, Ross RP (2010) Reconstitution conditions for dried probiotic powders represent a critical step in determining cell viability. J Appl Microbiol 108:1369–1379

Naidu AS, Bidlack WR, Clemens RA (1999) Probiotic spectra of lactic acid bacteria (LAB). Crit Rev Food Sci Nutr 39:13–126

NCBI, GOLD (2009) http://wwwncbinlmnihgov/genomes/lprokscgi and http://wwwgenomesonlineorg/goldcgi

Noverr MC, Huffnagle GB (2004) Does the microbiota regulate immune responses outside the gut? Trends Microbiol 12:562–568

O'Mahony L, McCarthy J, Kelly P, Hurley G, Luo F, Chen K, O'Sullivan GC, Kiely B, Collins JK, Shanahan F, Quigley EM (2005) *Lactobacillus* and *Bifidobacterium* in irritable bowel syndrome: symptom responses and relationship to cytokine profiles. Gastroenterology 128:541–551

O'Sullivan O, O'Callaghan J, Sangrador-Vegas A, McAuliffe O, Slattery L, Kaleta P, Callanan M, Fitzgerald GF, Ross RP, Beresford T (2009) Comparative genomics of lactic acid bacteria reveals a niche-specific gene set. BMC Microbiol 9:50

Ohashi Y, Nakai S, Tsukamoto T, Masumori N, Akaza H, Miyanaga N, Kitamura T, Kawabe K, Kotake T, Kuroda M, Naito S, Koga H, Saito Y, Nomata K, Kitagawa M, Aso Y (2002) Habitual intake of lactic acid bacteria and risk reduction of bladder cancer. Urol Int 68:273–280

Oksanen PJ, Salminen S, Saxelin M, Hämäläinen P, Ihantola-Vormisto A, Muurasniemi-Isoviita L, Nikkari S, Oksanen T, Pörsti I, Salminen E, et al. (1990) Prevention of travellers diarrhoea by *Lactobacillus* GG. Ann Med 22:53–56

Österlund P, Ruotsalainen T, Korpela R, Saxelin M, Ollus A, Valta P, Kouri M, Elomaa I, Joensuu H (2007) *Lactobacillus* supplementation for diarrhoea related to chemotherapy of colorectal cancer: a randomised study. Br J Cancer 97:1028–1034

Otero MC, Espeche MC, Nader-Macías ME (2007) Optimization of the freeze-drying media and survival throughout storage of freeze-dried *Lactobacillus gasseri* and *Lactobacillus delbrueckii* subsp. *delbrueckii* for veterinarian probiotic applications. Proc Biochem 42:1406–1411

Ouwehand AC, Salminen SJ (1998) The health effects of cultured milk products with viable and non-viable bacteria. Int Dairy J 8:749–758

Özer B, Kirmaci HA, Enel E, Atamer M, Hayalo lu A (2009) Improving the viability of *Bifidobacterium bifidum* BB-12 and *Lactobacillus acidophilus* LA-5 in white-brined cheese by microencapsulation. Int Dairy J 19:22–29

Pasi EK, Seppo JS, Erika I, Yuan Kun L (2001) The influence of polyunsaturated fatty acids on probiotic growth and adhesion. FEMS Microbiol Lett 194:149–153

Pereg D, Kimhi O, Tirosh A, Orr N, Kayouf R, Lishner M (2005) The effect of fermented yogurt on the prevention of diarrhea in a healthy adult population. Am J Infect Control 33:122–125

Pfeiler EA, Azcarate-Peril MA, Klaenhammer TR (2007) Characterization of a novel bile-inducible operon encoding a two-component regulatory system in *Lactobacillus acidophilus*. J Bacteriol 189:4624–4634

Picot A, Lacroix C (2004) Encapsulation of bifidobacteria in whey protein-based microcapsules and survival in simulated gastrointestinal conditions and in yoghurt. Int Dairy J 14:505–515

Pimentel-González DJ, Campos-Montiel RG, Lobato-Calleros C, Pedroza-Islas R, Vernon-Carter EJ (2009) Encapsulation of *Lactobacillus rhamnosus* in double emulsions formulated with sweet whey as emulsifier and survival in simulated gastrointestinal conditions. Food Res Int 42:292–297

Qin J, Li R, Raes J, Arumugam M, Burgdorf KS, Manichanh C, Nielsen T, Pons N, Levenez F, Yamada T, Mende DR, Li J, Xu J, Li S, Li D, Cao J, Wang B, Liang H, Zheng H, Xie Y, Tap J, Lepage P, Bertalan M, Batto JM, Hansen T, Le Paslier D, Linneberg A, Nielsen HB, Pelletier E, Renault P, Sicheritz-Ponten T, Turner K, Zhu H, Yu C, Li S, Jian M, Zhou Y, Li Y, Zhang X, Li S, Qin N, Yang H, Wang J, Brunak S, Doré J, Guarner F, Kristiansen K, Pedersen O, Parkhill J, Weissenbach J (2010) A human gut microbial gene catalogue established by metagenomic sequencing. Nature 464:59–65

Rafter J, Bennett M, Caderni G, Clune Y, Hughes R, Karlsson PC, Klinder A, O'Riordan M, O'Sullivan GC, Pool-Zobel B, Rechkemmer G, Roller M, Rowland I, Salvadori M, Thijs H, Van Loo J, Watzl B, Collins JK (2007) Dietary synbiotics reduce cancer risk factors in polypectomized and colon cancer patients. Am J Clin Nutr 85:488–496

Rani PS, Agrawal R (2008) Effect on cellular membrane fatty acids in the stressed cells of *Leuconostoc mesenteroides*: a native probiotic lactic acid bacteria. Food Biotechnol 22:47–63

Rastall RA, Maitin V (2002) Prebiotics and synbiotics: towards the next generation. Curr Opin Biotechnol 13:490–496

Ray B, Jezeski JJ, Busta FF (1971) Effect of rehydration on recovery, repair, and growth of injured freeze-dried *Salmonella anatum*. Appl Environ Microbiol 22:184

Reid G, Charbonneau D, Erb J, Kochanowski B, Beuerman D, Poehner R, Bruce AW (2003) Oral use of *Lactobacillus rhamnosus* GR-1 and *L. fermentum* RC-14 significantly alters vaginal flora: randomized, placebo-controlled trial in 64 healthy women. FEMS Immunol Med Microbiol 35:131–134

Riesenfeld CS, Schloss PD, Handelsman J (2004) Metagenomics: genomic analysis of microbial communities. Annu Rev Genet 38:525–552

Rodwell AW, Peterson JE (1971) The effect of straight-chain saturated, monoenoic and branched-chain fatty acids on growth and fatty acid composition of *Mycoplasma* strain Y. J Gen Microbiol 68:173–186

Roesch LFW, Fulthorpe RR, Riva A, Casella G, Hadwin AKM, Kent AD, Daroub SH, Camargo FA, Farmerie WG, Triplett EW (2007) Pyrosequencing enumerates and contrasts soil microbial diversity. ISME J 1:283–290

Ross RP, Claiborne A (1991) Cloning, sequence and overexpression of NADH peroxidase from *Streptococcus faecalis* 10 C 1: structural relationship with the flavoprotein disulfide reductases. J Molec Biol 221:857–871

Ross RP, Morgan S, Hill C (2002a) Preservation and fermentation: past, present and future. Int J Food Microbiol 79:3–16

Ross RP, Fitzgerald G, Collins K, Stanton C (2002b) Cheese delivering biocultures-probiotic cheese. Aust J Dairy Technol 57:71–78

Ross RP, Desmond C, Fitzgerald GF, Stanton C (2005) Overcoming the technological hurdles in the development of probiotic foods. J Appl Microbiol 98:1410–1417

Ruas-Madiedo P, Hugenholtz J, Zoon P (2002) An overview of the functionality of exopolysaccharides produced by lactic acid bacteria. Int Dairy J 12:163–171

Saarela M, Virkajärvi I, Nohynek L, Vaari A, Mättö J (2006) Fibres as carriers for *Lactobacillus rhamnosus* during freeze-drying and storage in apple juice and chocolate-coated breakfast cereals. Int J Food Microbiol 112:171–178

Sakamoto I, Igarashi M, Kimura K, Takagi A, Miwa T, Koga Y (2001) Suppressive effect of *Lactobacillus gasseri* OLL 2716 (LG21) on *Helicobacter pylori* infection in humans. Br Soc Antimicrob Chemo, 47:709–710

Sakamoto T, Murata N (2002) Regulation of the desaturation of fatty acids and its role in tolerance to cold and salt stress. Curr Opin Microbiol 5:206–210

Sanders ME, Marco ML (2010) Food formats for effective delivery of probiotics. Annu Rev Food Sci Technol 1:65–85

Santivarangkna C, Kulozik U, Foerst P (2007) Alternative drying processes for the industrial preservation of lactic acid starter cultures. Biotechnol Prog 23:302v315

Schloss PD, Handelsman J (2003) Biotechnological prospects from metagenomics. Curr Opin Biotech 14:303–310

Schwab C, Vogel R, Gänzle MG (2007) Influence of oligosaccharides on the viability and membrane properties of *Lactobacillus reuteri* TMW1.106 during freeze-drying. Cryobiology 55:108–114

Seki H, Shiohara M, Matsumura T, Miyagawa N, Tanaka M, Komiyama A, Kurata S (2003) Prevention of antibiotic-associated diarrhea in children by *Clostridium butyricum* MIYAIRI. Pediatr Int 45:86–90

Shah NP (2002) The exopolysaccharides production by starter cultures and their influence on textural characteristics of fermented milks; June 3, Comwell Scanticon, Kolding, Denmark, p. 5

Shamir R, Makhoul IR, Etzioni A, Shehadeh N (2005) Evaluation of a diet containing probiotics and zinc for the treatment of mild diarrheal illness in children younger than one year of age. J Am Coll Nutr 24:370–375

Shanahan F, Stanton C, Ross P, Hill C (2009) Pharmabiotics: bioactives from mining host–microbe–dietary interactions. Funct Food Rev 1:1–6

Sheehan VM, Sleator RD, Fitzgerald GF, Hill C (2006) Heterologous expression of BetL, a betaine uptake system, enhances the stress tolerance of *Lactobacillus salivarius* UCC118. Appl Environ Microbiol 72:2170–2177

Sheehan VM, Sleator RD, Hill C, Fitzgerald GF (2007) Improving gastric transit, gastrointestinal persistence and therapeutic efficacy of the probiotic strain *Bifidobacterium breve* UCC2003. Microbiology 153:3563

Shin HS, Lee JH, Pestka JJ, Ustunol Z (2000) Growth and viability of commercial *Bifidobacterium* spp in skim milk containing oligosaccharides and inulin. J Food Sci 65:884–887

Siitonen S, Vapaatalo H, Salminen S, Gordin A, Saxelin M, Wikberg R, Kirkkola AL (1990) Effect of *Lactobacillus* GG yoghurt in prevention of antibiotic associated diarrhoea. Ann Med 22:57–59

Sleator RD, Hill C (2002) Bacterial osmoadaptation: the role of osmolytes in bacterial stress and virulence. FEMS Microbiol Rev 26:49–71

Sleator RD, Hill C (2006) Patho-biotechnology: using bad bugs to do good things. Curr Opin Biotech 17:211–216

Sleator RD, Hill C (2007) Patho-biotechnology; using bad bugs to make good bugs better. Sci Prog 90:114

Sleator RD, Hill C (2008) "Bioengineered bugs"—a patho-biotechnology approach to probiotic research and applications. Med Hypoth 70:167–169

Small PLC, Waterman SR (1998) Acid stress, anaerobiosis and gadCB: lessons from *Lactococcus lactis* and *Escherichia coli*. Trends Microbiol 6:214–216

Sogin ML, Morrison HG, Huber JA, Welch DM, Huse SM, Neal PR, Arrieta JM, Herndl GJ (2006) Microbial diversity in the deep sea and the underexplored "rare biosphere." Proc Natl Acad Sci USA 103:12115–12120

Stanton C, Gardiner G, Lynch PB, Collins JK, Fitzgerald G, Ross RP (1998) Probiotic cheese. Int Dairy J 8:491–496

Stanton C, Gardiner G, Meehan H, Collins K, Fitzgerald G, Lynch PB, Ross RP (2001) Market potential for probiotics. Amer J Clin Nutr 73:476S–483S

Stanton C, Desmond C, Fitzgerald G, Ross RP (2003a) Probiotic health benefits: reality or myth? Aust J Dairy Technol 58:107–113

Stanton C, Desmond C, Coakley M, Collins JK, Fitzgerald G, Ross RP (2003b) Two challenges facing development of probiotic-containing functional foods, Farnworth ER (Ed.), Handbook of fermented functional foods. CRC Press, Boca Raton, FL, pp. 27–58

Stanton C, Ross RP, Fitzgerald GF, Sinderen DV (2005) Fermented functional foods based on probiotics and their biogenic metabolites. Curr Opin Biotechnol 16:198–203

Storz G, Zheng M (2000) Oxidative stress. In Storz G, Hengge-Aronis R (Eds.), Bacterial stress responses. ASM Press, Washington, DC, pp. 47–59

Su P, Henriksson A, Mitchell H (2007) Selected prebiotics support the growth of probiotic monocultures *in vitro*. Anaerobe 13:134–139

Sugawara G, Nagino M, Nishio H, Ebata T, Takagi K, Asahara T, Nomoto K, Nimura Y (2006) Perioperative synbiotic treatment to prevent postoperative infectious complications in biliary cancer surgery. Ann Surg 244:706–714

Sugimoto S, Sonomoto K (2008) Molecular chaperones in lactic acid bacteria: physiological consequences and biochemical properties. J Biosci Bioeng 106:324–336

Sunny-Roberts EO, Knorr D (2009) The protective effect of monosodium glutamate on survival of *Lactobacillus rhamnosus* GG and *Lactobacillus rhamnosus* E-97800 (E800) strains during spray-drying and storage in trehalose-containing powders. Int Dairy J 19:209–214

Surawicz CM, Elmer GW, Speelman P, McFarland LV, Chinn J, van Belle G (1989) Prevention of antibiotic-associated diarrhea by *Saccharomyces boulardii*: a prospective study. Gastroenterology 96:981–988

Sýkora J, Valecková K, Amlerová J, Siala K, Dedek P, Watkins S, Varvarovská J, Stozický F, Pazdiora P, Schwarz J (2005) Effects of a specially designed fermented milk product containing probiotic *Lactobacillus casei* DN-114 001 and the eradication of *H. pylori* in children: a prospective randomized double-blind study. J Clin Gastroenterol 39:692–698

Tabbers MM, Chmielewska A, Roseboom MG, Boudet C, Perrin C, Szajewska H, Benninga MA (2009) Effect of the consumption of a fermented dairy product containing *Bifidobacterium lactis* DN-173 010 on constipation in childhood: a multicentre randomised controlled trial (NTRTC:1571). BMC Pediatr 9:22

Taniguchi M, Kotani N, Kobayashi T (1987) High concentration cultivation of *Bifidobacterium longum* in fermenter with cross-flow filtration. Appl Microbiol Biotechnol 25:438–441

Tao K, Makino K, Yonei S, Nakata A, Shinagawa H (1989) Molecular cloning and nucleotide sequencing of oxyR, the positive regulatory gene of a regulon for an adaptive response to oxidative stress in *Escherichia coli*: homologies between OxyR protein and a family of bacterial activator proteins. Molec Gener Genet MGG 218:371–376

Termont S, Vandenbroucke K, Iserentant D, Neirynck S, Steidler L, Remaut E, Rottiers P (2006) Intracellular accumulation of trehalose protects *Lactococcus lactis* from freeze-drying damage and bile toxicity and increases gastric acid resistance. Appl Environ Microbiol 72:7694–7700

Trois L, Cardoso EM, Miura E (2008) Use of probiotics in HIV-infected children: a randomized double-blind controlled study. J Trop Pediatr 54:19–24

Tsen JH, Huang HY, King VAE (2007) Enhancement of freezing-resistance of *Lactobacillus rhamnosus* by the application of cell immobilization. J Gener Appl Microbiol 53:215–219

Turnbaugh PJ, Ley RE, Mahowald MA, Magrini V, Mardis ER, Gordon JI (2006) An obesity-associated gut microbiome with increased capacity for energy harvest. Nature 444:1027–1031

Turnbaugh PJ, Ley RE, Hamady M, Fraser-Liggett CM, Knight R, Gorden JI (2007) The human microbiome project. Nature 449:804–810

van de Guchte M, Serror P, Chervaux C, Smokvina T, Ehrlich SD, Maguin E (2002) Stress responses in lactic acid bacteria. Antonie van Leeuwenhoek 82:187–216

Vandenbergh PA (1993) Lactic acid bacteria, their metabolic products and interference with microbial growth. FEMS Microbiol Rev 12:221–237

Vanderhoof JA, Whitney DB, Antonson DL, Hanner TL, Lupo JV, Young RJ (1999) *Lactobacillus* GG in the prevention of antibiotic-associated diarrhea in children. J Pediatr 135:564–568

Vaughan EE, Schut F, Heilig H, Zoetendal EG, de Vos WM, Akkermans AD (2000) A molecular view of the intestinal ecosystem. Curr Issues Intest Microbiol 1:1–12

Ventura M, van Sinderen D, Fitzgerald GF, Zink R (2004) Insights into the taxonomy, genetics and physiology of bifidobacteria. Antonie van Leeuwenhoek 86:205–223

Vernazza CL, Rabiu BA, Gibson GR (2006) Human colonic microbiology and the role of dietary intervention: introduction to prebiotics. In: Gibson GR, Rastall RA, (Eds.), Prebiotics: development and application, John Wiley and Sons, West Sussex, UK, pp. 1–28

Vesa TH, Marteau P, Zidi S, Briet F, Pochart P, Rambaud JC (1996) Digestion and tolerance of lactose from yoghurt and different semi-solid fermented dairy products containing *Lactobacillus acidophilus* and bifidobacteria in lactose maldigesters: Is bacterial lactase important? Eur J Clin Nutr 50:730–733

Vijay-Kumar M, Aitken JD, Carvalho FA, Cullender TC, Mwangi S, Srinivasan S, Sitaraman SV, Knight R, Ley RE, Gewirtz AT (2010) Metabolic syndrome and altered gut microbiota in mice lacking Toll-like receptor 5. Science 328:228–231

Wang KY, Li SN, Liu CS, Perng DS, Su YC, Wu DC, Jan CM, Lai CH, Wang TN, Wang WM (2004) Effects of ingesting *Lactobacillus*-and *Bifidobacterium*-containing yogurt in subjects with colonized *Helicobacter pylori*. Am J Clin Nutr 80:737–741

Whitehead K, Versalovic J, Roos S, Britton RA (2008) Genomic and genetic characterization of the bile stress response of probiotic *Lactobacillus reuteri* ATCC 55730. Appl Environ Microbiol 74:1812–1819

Whorwell PJ, Altringer L, Morel J, Bond Y, Charbonneau D, O'Mahony L, Kiely B, Shanahan F, Quigley EM (2006) Efficacy of an encapsulated probiotic *Bifidobacterium infantis* 35624 in women with irritable bowel syndrome. Amer J Gastroenterol 101:1581–1590

Wickens K, Black PN, Stanley TV, Mitchell E, Fitzharris P, Tannock GW, Purdie G, Crane J (2008) A differential effect of two probiotics in the prevention of eczema and atopy: a double-blind, randomized, placebo-controlled trial. J Allergy Clin Immunol 122:788–794

Williams EA, Stimpson J, Wang D, Plummer S, Garaiova I, Barker M, Corfe B (2009) Clinical trial: a multistrain probiotic preparation significantly reduces symptoms of irritable bowel syndrome in a double-blind placebo-controlled study. Aliment Pharm Therap 29:97

Williams NT (2010) Probiotics. Amer J Health Syst Pharm 67:449–458

Yang YX, He M, Hu G, Wei J, Pages P, Yang XH, Bourdu-Naturel S (2008) Effect of a fermented milk containing *Bifidobacterium lactis* DN-173010 on Chinese constipated women. World J Gastroenterol 14:6237–6243

Zaidi AH, Bakkes PJ, Lubelski J, Agustiandari H, Kuipers OP, Driessen AJM (2008) The ABC-type multidrug resistance transporter LmrCD is responsible for an extrusion-based mechanism of bile acid resistance in *Lactococcus lactis*. J Bacteriol 190:7357–7366

Zayed G, Roos YH (2004). Influence of trehalose and moisture content on survival of *Lactobacillus salivarius* subjected to freeze-drying and storage. Proc Biochem 39:1081–1086

Zhang H, DiBaise JK, Zuccolo A, Kudrna D, Braidotti M, Yu Y, Parameswaran P, Crowell MD, Wing R, Rittmann BE Krajmalnik-Brown R (2009) Human gut microbiota in obesity and after gastric bypass. Proc Natl Acad Sci USA 106:2365

Zhang Y-M, Rock CO (2008) Membrane lipid homeostasis in bacteria. Nat Rev Micro 6:222–233

Zhao G, Zhang G (2005) Effect of protective agents, freezing temperature, rehydration media on viability of malolactic bacteria subjected to freeze-drying. J Appl Microbiol 99:333–338

Zuniga M, Champomier-Verges M, Zagorec M, Perez-Martinez G (1998) Structural and functional analysis of the gene cluster encoding the enzymes of the arginine deiminase pathway of *Lactobacillus sake*. J Bacteriol 180:4154–4159

Chapter 18
Responses of Lactic Acid Bacteria to Bacteriocins and Other Antimicrobials

Sikder M. Asaduzzaman and Kenji Sonomoto

18.1 Introduction

All organisms have compounds that act as evolutionarily ancient antimicrobial agents. Among these compounds, the diversity of the antimicrobial peptides is so great that more than 1,000 peptides have been included in the antimicrobial sequence database (AMSDb; http://www.bbcm.univ.trieste.it/~tossi/antimic.html). These antibacterial peptides are produced by virtually all forms of life as the first line of defense. Bacteria are remarkable producers of antimicrobial peptides, and these bacterial-derived antimicrobials have a large degree of structural and chemical diversity. Bacteriocins are the well-known subset of antibiotics that have been defined as "ribosomally synthesized proteinaceous molecules released extracellularly by bacteria directed against species which are closely related to the producer bacterium" (Tagg et al. 1976; Nissen-Meyer and Nes 1997). Bacteriocins are generally optimized in the course of evolution toward a narrow spectrum and a specific number of competitors in a certain ecological niche. Specific immunity is usually expressed by the producer strain for self-protection.

K. Sonomoto (✉)
Laboratory of Microbial Technology, Division of Applied Molecular Microbiology and Biomass Chemistry, Department of Bioscience and Biotechnology, Faculty of Agriculture, Graduate School, Kyushu University, 6-10-1 Hakozaki, Higashi-ku, Fukuoka 812-8581, Japan

Laboratory of Functional Food Design, Department of Functional Metabolic Design, Bio-Architecture Center, Kyushu University, 6-10-1 Hakozaki, Higashi-ku, Fukuoka 812-8581, Japan
e-mail: sonomoto@agr.kyushu-u.ac.jp

E. Tsakalidou and K. Papadimitriou (eds.), *Stress Responses of Lactic Acid Bacteria*, Food Microbiology and Food Safety, DOI 10.1007/978-0-387-92771-8_18,

18.2 Bacteriocins of Lactic Acid Bacteria

18.2.1 Background

Because of the intensive investigations on lactic acid bacteria (LAB) bacteriocins, a large number of structurally diverse substances have been isolated, identified, and characterized. The genes involved in bacteriocin biosynthesis are clustered and designated by the generic locus symbol with a more specific genotypic designation for each bacteriocin member (e.g., *nis* for nisin). Bacteriocins producing genes may be found on conjugative transposable elements (e.g., nisin), on the chromosome of the host (e.g., subtilin), or on plasmids (e.g., nukacin ISK-1). The gene clusters for the biosynthesis of representative bacteriocins are depicted in Fig. 18.1. The various gene clusters differ in the order, complexity, and transcriptional organization of the genes that have been identified to be involved in the biosynthesis of bacteriocins' prepeptides, immunity proteins, ABC transporters, and also the other accessory proteins. For example, four genes (*ABCT*) are present in the biosynthesis of the foremost bacteriocin nisin. The structure of nisin was determined by Gross and Morell in an elegant landmark study in 1971 (Gross and Morell 1971).

18.2.2 Classification and Structures of Bacteriocins

Cotter et al. (2005) have classified all bacteriocins into two broad classes. Class I bacteriocins, called *lantibiotics*, are posttranslationally modified and contain lanthionine/3-methyllanthionine. Class II bacteriocins are collectively the *nonlantibiotic bacteriocins*.

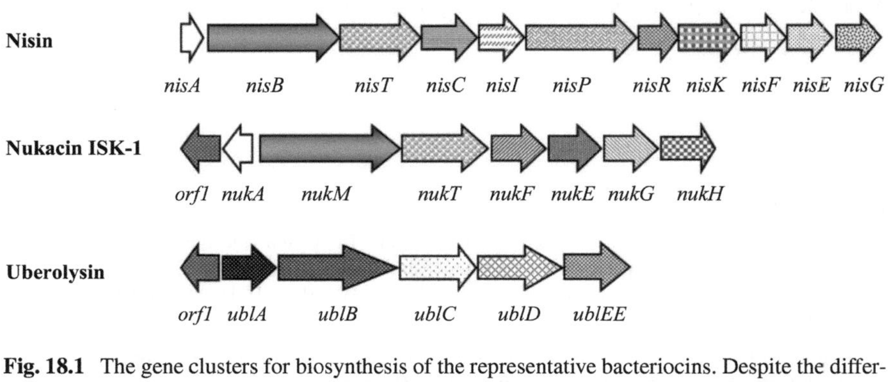

Fig. 18.1 The gene clusters for biosynthesis of the representative bacteriocins. Despite the differences in the gene order, complexity, and transcriptional organization in the biosynthesis of various bacteriocins, comparable gene clusters are usually found in a similar type of bacteriocin. Gene products for nisin are as follows: prepeptide (*nisA*), modification enzyme (*nisB, C*), processing protease (*nisP and nisT*), ABC transporter (*nisT*), immunity protein (*nisFEG, nisI*), and regulatory protein (*nisR, nisK*). *orf* open reading frame

18.2.2.1 Class I Bacteriocins

The class I bacteriocins are small peptides (<5 kDa) called *lantibiotics*. These peptides contain unusual amino acids, such as thioether cross-linked amino acids in lanthionine and 3-methyllanthionine and dehydrated amino acids in 2,3-didehydro-alanine (Dha) and (*Z*)-2,3-didehydrobutyrine (Dhb) (de Vos et al. 1995; Sahl et al. 1995). These unusual amino acids confer high stability and various biological activities to these peptides. Posttranslational modification renders the lantibiotics biologically active. The foremost bacteriocin nisin is the best-studied lantibiotic having a linear structure (Fig. 18.2). Among the subgroups of lantibiotics, the two-peptide lantibiotics have different components for activity. The two components of these post-translationally modified peptides synergistically display strong antibacterial action but individually have little to no activity. Usually, the highest antibacterial efficacy of the two-component bacteriocins is at equimolar concentrations (1:1 stoichiometry).

18.2.2.2 Class II Bacteriocins

The class II bacteriocins are usually heat-stable proteins and are divided into classes IIa, IIb, and IIc. Class IIa (pediocin-like) is the largest subgroup of bacteriocins (Fig. 18.2). They contain an N-terminal consensus sequence YGNGV. Class IIb (circular) (Fig. 18.2)

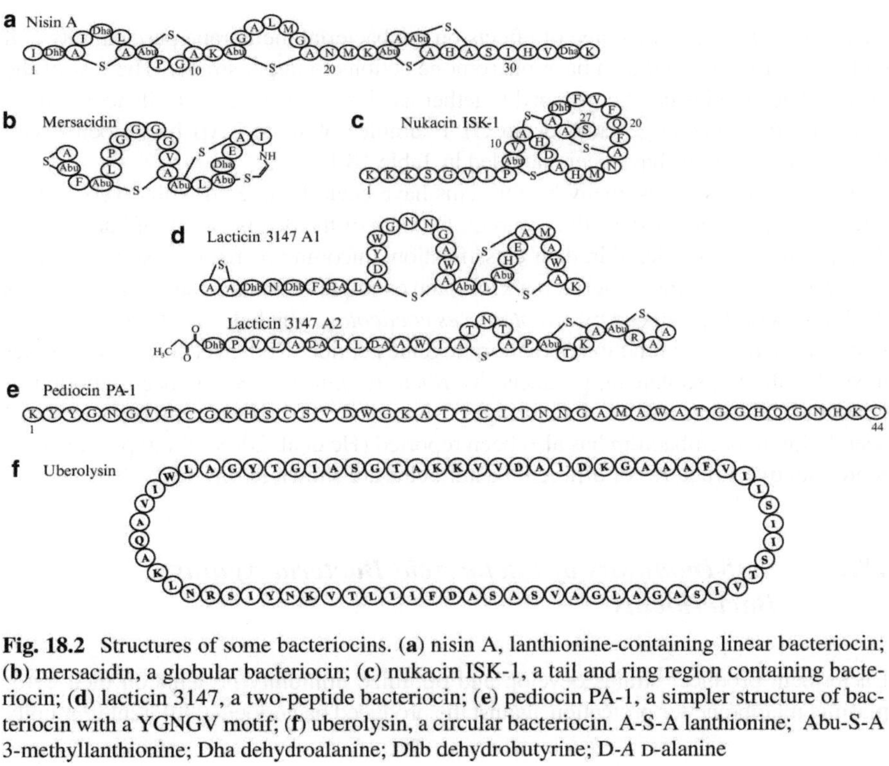

Fig. 18.2 Structures of some bacteriocins. (**a**) nisin A, lanthionine-containing linear bacteriocin; (**b**) mersacidin, a globular bacteriocin; (**c**) nukacin ISK-1, a tail and ring region containing bacteriocin; (**d**) lacticin 3147, a two-peptide bacteriocin; (**e**) pediocin PA-1, a simpler structure of bacteriocin with a YGNGV motif; (**f**) uberolysin, a circular bacteriocin. A-S-A lanthionine; Abu-S-A 3-methyllanthionine; Dha dehydroalanine; Dhb dehydrobutyrine; D-*A* D-alanine

Table 18.1 Some bacteriocins isolated from different species of LAB

Bacteriocin		
Class	Name	Producer strain
I	Nisin	*Lactococcus lactis*
	Mersacidin	*Bacillus* sp.
	Lacticin 481	*Lactococcus lactis*
	Nukacin ISK-1	*Staphylococcus warneri* ISK-1
	Gellidermin	*Staphylococcus gallinarum*
	Actagerdin	*Actinoplanes* sp.
	Lacticin 1347	*Lactococcus lactis*
	Duramycin	*Streptomyces* sp.
	Cinnamycin	*Streptomyces cinnamoneus*
	Epidermin	*Staphylococcus epidermidis*
II	Pediocin PA-1	*Pediococcus acidilactici*
	Mesentericin Y105	*Leuconostoc mesenteroides*
	Sakacin A	*Lactobacillus sakei*
	Enterocin A	*Enterococcus faecium*
	Uberolysin	*Streptococcus uberis*
	Lactacin B	*Lactobacillus acidophilus*
	Lactolin	*Lactobacillus plantarum*
	Acidolin	*Lactobacillus acidophilus*
	Reuterin	*Lactobacillus reuteri*
	Lactobacillin	*Lactobacillus brevis*

bacteriocins have a wide range of effects on the bacterial membrane permeability and cell wall formation and also have pheromone actions to target strains. The rest of the class II bacteriocins can be grouped together as class IIc having miscellaneous structures and functions (e.g., lactococcin G). Examples of some LAB bacteriocins with their producer strains have been included in Table 18.1.

In the past few years, many bacteriocins have been discovered with diverse structures and functions. Besides the common features of the diverse array of bacteriocins that usually are considered in their classification, uncommon characteristics have also been reported for some bacteriocins. One such example of an interesting bacteriocin is the SapB peptide produced by *Streptomyces coelicolor*, which does not show any antimicrobial activity but functions as a morphogenic peptide (Kodani et al. 2004). Another novel lantibiotic, sublancin, produced by *Bacillus subtilis* 168, has been reported; it contains lanthionine and disulfide bridges (Marx et al. 2001). Furthermore, N-terminal acetylation in paenibacillin has also been reported (He et al. 2008). Examples of some representative structures of different bacteriocins are shown in Fig. 18.2.

18.2.3 Self-Immunity of Lactic Acid Bacteria Against Bacteriocins

Bacteriocin producer strains develop and optimize immunity proteins on their cytoplasmic membrane for protection against the produced bacteriocin. Different immunity systems exist. The lantibiotic producer contains an individual small immunity protein LanI (e.g., Pep5), or an ABC transporter LanFEG (e.g., mersacidin), or both LanI/LanFEG (e.g., nisin). The genes responsible for the immunity determinants are found to be colocated with the corresponding biosynthetic genes. The orphan immunity gene is a subject of debate; for instance, the presence of genes encoding the structural and modification genes in the case of lichenicidin is on the DSM 13 chromosome (Veith et al. 2004). In contrast to the colocated association of biosynthetic and immunity genes, in this case, the encoding genes are found within a different region of the chromosome.

The immunity proteins of nisin, subtilin, and other related lantibiotics have been reported for their functional analysis. These lantibiotics have two independent immunity systems (NisFEG/NisI for nisin and SpaFEG/SpaI for subtilin). NisI and SpaI are lipoproteins anchored to the membrane surface via a lipid-modified N-terminal cysteine residue. They intercept by binding to lantibiotics at the surface of the cytoplasmic membrane and are oriented to the outside of the membrane. The heterologous expression of NisI increases the amount of cell-bound nisin and the resistance level (Stein et al. 2003). This indicates that NisI intercepts nisin before it attacks the cell membrane. C-terminal-truncated NisI mutants and a hybrid protein with the immunity protein SpaI show that the C-terminal of NisI confers immunity specifically against nisin (Takala and Saris 2006). The membrane protein complexes (NisFEG and SpaFEG) function as an ABC transporter to transport the lantibiotics from the cytoplasmic membrane into the extracellular space. LanF is an intracellular ATPase subunit, whereas LanE and LanG possess membrane-spanning domains and represent the transmembrane subunits. Another novel type of lantibiotic-binding immunity protein, NukH, is found in nukacin ISK-1 (Aso et al. 2005; Okuda et al. 2005). NukH interacts with nukacin ISK-1 and lacticin 481 but not with nisin. These two immunity systems (NukFEG and NukH) collectively render a higher immunity than that of each individual system. This suggests that NukH provides to the host immunity as a lantibiotic-binding immunity protein in cooperation with NukFEG (Fig. 18.3). Both systems are required for full immunity, and NukFEG and NukH increase the immunity level of nukacin ISK-1-sensitive *Lactococcus lactis* (Aso et al. 2005). Bacteria-associated nukacin ISK-1 is transported by NukFEG, and NukH shows binding activity against nukacin ISK-1, similar to that of NisI, though the features of NukH and NisI are quite different. Truncated and mutated nukacin ISK-1 shows that their binding rates to *nukH*-expressing cells are lower than that of nukacin ISK-1 and also further indicates that unusual amino acids exert an important role in NukH recognition (Okuda et al. 2008). The functional domains of NukH have been characterized by expressing various mutants, proving that there is no correlation between immunity level and NukH-binding activity against nukacin ISK-1. The *lanH* gene is identical to the gene clusters of nukacin ISK-1, ruminococcin, and butyrivibriocin OR79A. The genes associated with lacticin 3147 biosynthesis and immunity are present in two divergent operons, *ltnA1A2M1TM2J* and *ltnRIEF*, respectively. The absence of *ltnFE* in certain circumstances is evident in cells devoid of *ltnI*. The ABC transporter usually functions most efficiently to transport an Ltnα:Ltnβ complex. The presence of an Ltnα:Ltnβ seems to be a prerequisite for complementary functionalities. Both the LtnI and LtnR proteins play a role in lacticin

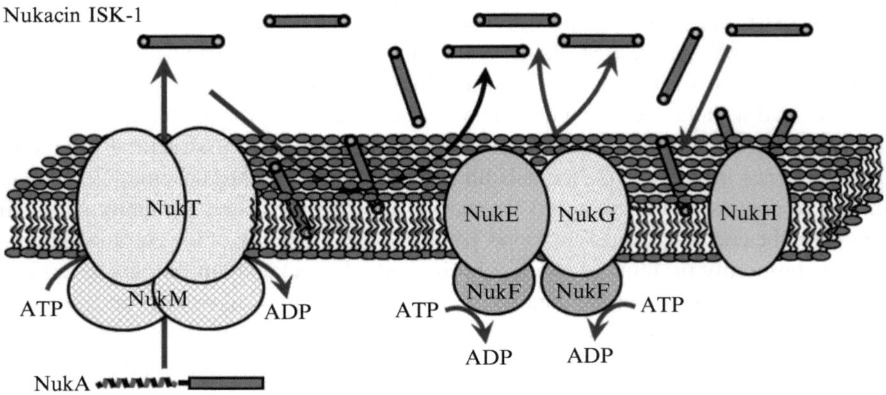

Fig. 18.3 A schematic diagram of nukacin ISK-1 biosynthesis and immunity. The prepeptide NukA undergoes posttranslational modifications, including unusual amino acid formation, cleavage off the leader peptide, and transport by a membrane-located multimeric complex of NukM and NukT for the production of mature nukacin ISK-1. The ABC transporter NukFEG then functions to transport the produced nukacin ISK-1 from the cytoplasmic membrane. The immunity protein NukH gives to the host immunity cooperatively with NukFEG

3147 immunity (McAuliffe et al. 2001). LtnI is similar to the other LanI peptides discussed above, and LtnR works as a regulator of the promoter driving the immunity operon. Generally, immunity to bacteriocins is specific and cross-immunity is rare. However, producers of two-peptide bacteriocins, such as lacticin 3147, face an unusual challenge of exposure to two active peptides (α and β). Staphylococcin C55 producers are cross-immune to lacticin 3147, having a natural repository of *Staphylococcus aureus* strains that are protected against lacticin 3147 (Draper et al. 2009). Functional immunity homologs are also produced by strains of *Bacillus licheniformis* and *Enterococcus faecium.*

18.2.4 Lactic Acid Bacteria Have Diverse Responses to Bacteriocins' Actions

Many bacteriocins (e.g., nisin, pediocin, etc.) bind to the membrane, leading to subsequent actions. In most cases, the activities of bacteriocins are based on different killing mechanisms that are combined in one molecule. The prototypic lantibiotic nisin inhibits peptidoglycan biosynthesis and forms pores through specific interactions with the cell wall precursor lipid II (Wiedemann et al. 2001). Interestingly, the mutant [A12L] gallidermin loses its pore-forming ability but is as potent as wild-type gallidermin. This indicates that pore formation does not contribute to the killing of bacteria for this mutant of gallidermin (Bonelli et al. 2006).

18.2.4.1 Pore Formation into Bacterial Membrane by Bacteriocins

Early findings on bacteriocins' activities relied on membrane permeabilization of the susceptible bacteria, which eventually leads to killing of cells by the leakage of ions and molecules. The binding of lantibiotics to the bacterial membrane followed by insertion into the membrane is a vital step of lantibiotics activities. The phospholipid composition of the target membrane plays an important role in the action of bacteriocins. Electrostatic interactions promote the primary binding of nisin to the membrane. The highly positive-charged C-terminus of nisin interacts initially with the anionic surface of the bacterial membrane (Breukink et al. 1997). Removing the positive charges from the N- or C-terminal region of nisin hampers the initial interactions of the peptide to the membrane (Giffard et al. 1997). The nisin$^{1-12}$ fragment shows a strongly reduced binding affinity to anionic phospholipids, further indicating the importance of the C-terminus of nisin for binding to the membrane (Moll et al. 1997). Although the C-terminus of nisin has been found to be involved in the initial binding to the membrane surface, a variant of nisin Z, in which a short peptide is fused into its C-terminus, showed that the C-terminus translocated across the membrane (van Kraaij et al. 1998). Translocation of the C-terminus is correlated with pore-forming activity and is dependent on anionic lipids of the membrane. Electrostatically bound peptide adopts a membrane-spanning orientation in which a part of the molecules forming the pore is located in the lumen of the vesicle. The N-terminal rings of nisin interact with the disaccharide-pyrophosphate of lipid II, and the positively charged C-terminal primarily binds with the headgroups of the lipids in the membrane bilayer. The lysine-rich, positively charged N-terminal of nukacin ISK-1 binds strongly to the anionic membrane to exert its full antibacterial activity (Asaduzzaman et al. 2006), whereas lacticin 481 (with a net charge of zero) has a higher affinity for the zwitterionic membrane (Demel et al. 1996).

Bacterial cytoplasmic and artificial membranes have been used for the study of the activities of the cationic lantibiotics (Breukink et al. 1999). Early studies prior to the late 1990s focused on the membrane's permeabilization of the bacterial cell as the primary mode of action of bacteriocins. Pore formation causes a leakage of ions and molecules from the bacteria, which eventually leads to cell death (Benz et al. 1991). The lifetimes of the pores formed by lantibiotics may have a few to several 100 ms, and the diameter of nisin's pore is up to 2 nm (Sahl 1991). Nisin's pores are somewhat anion-selective (Breukink et al. 1997), and the pores of nisin and pep5 are rectifying (work only in one direction) (Sahl et al. 1987; Kordel et al. 1988). Nonrectifying and more stable channels are formed by gallidermin and epidermin (Benz et al. 1991). The barrel-stave and wedge models are the best-established mechanisms of pore formation. The cationic lantibiotic monomers bind to the membrane surface through electrostatic interactions in the barrel-stave mechanism. The lantibiotics are assembled into a preaggregate to form the pores at a certain membrane potential, in which the peptides remain perpendicular to the membrane (Sahl 1991). In contrast, the surface-bound lantibiotics bind parallel to the membrane surface in the case of the wedge model. This generates local strain and bends the membrane in such a way that the lipid molecules, together with the lantibiotic, form

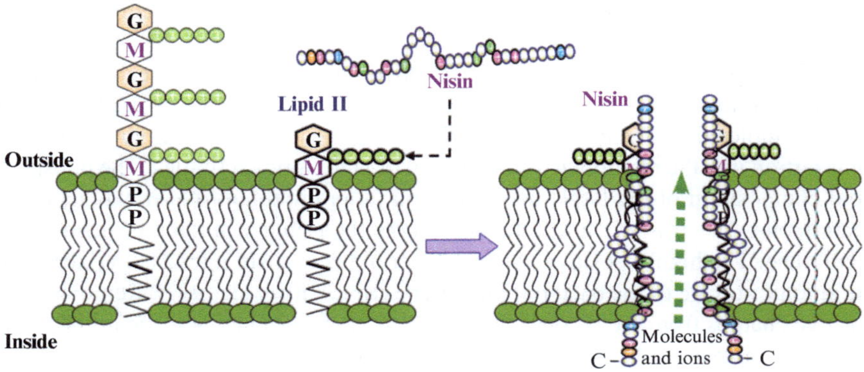

Fig. 18.4 Pore formation by the lantibiotic nisin, using lipid II as docking molecule. Nisin binds to the cell wall precursor lipid II. The N-terminus of nisin binds lipid II, while the C-terminus inserts into the bacterial membrane and subsequently makes a pore to release molecules and ions

a pore (Driessen et al. 1995). Pore formation by the well-studied bacteriocin nisin is presented in Fig. 18.4. Many bacteriocins differ from nisin because they interact with cytoplasmic membranes of sensitive bacteria regardless of their degree of prior energization. This suggests that the loss of permeability of the cytoplasmic membrane occurs in a voltage-independent manner, while nisin acts in a membrane potential–dependent manner. Since LAB vary considerably within each genus and species, the specificity and potency of the activity of bacteriocins are influenced by the lipid composition of the target membranes.

Peptidoglycan is a highly dynamic and continuous-covalent macromolecular structure located on the outside of the cell membrane of almost all eubacteria. The composition of peptidoglycan is quite uniform throughout these organisms and, accordingly, cell wall biosynthesis is an important target for many antibiotics. Lipid I and lipid II are essential intermediates for the biosynthesis of peptidoglycan. Many antibiotics interfere with peptidoglycan biosynthesis through specific binding to these molecules. Transpeptidase and transglycosylase enzymes that are involved in building the cross-linked network of the bacterial cell wall are usually prevented by these molecules. The well-known peptide antibiotic vancomycin kills bacteria by targeting lipid II. Although both vancomycin and nisin bind to lipid II, vancomycin binds to the D-Ala-D-Ala moiety of lipid II, and nisin binds to the disaccharides-pyrophosphate region of lipid II. Therefore, nisin is even effective against vancomycin-resistant strains, though both antibiotics target lipid II (Hsu et al. 2004). Lipid II is the prime target of several other classes of natural products, including many lantibiotics. Lantibiotics interfere with peptidoglycan biosynthesis by binding differently to lipid II; the diverse structures of these compounds help explain the sophisticated activities of lantibiotics to kill bacteria.

Nisin binds with lipid II and uses it as a docking molecule to form pores that are stable and highly efficient (Breukink et al. 2003). This makes nisin's pore-forming activities unique, as compared to those of vancomycin, teicoplanin, and ramoplanin.

The presence of lipid II in the membrane increases the pore-forming efficiency of nisin 1,000 fold as compared to peptides that do not use lipid II. Lipid II increases nisin's activities so greatly that only two lipid II molecules per 10^5 phospholipid molecules dramatically enhance the release of dyes from the vesicles (Breukink et al. 1999; Brötz et al. 1998). The involvement of manifold molecules in the complex of lipid II–nisin is sufficient to form a defined pore of a uniform structure (Breukink et al. 2003). The lipid II–mediated pore complex is highly stable and unique, as other cationic antimicrobial peptides form pores in the membrane that are unstable, transient, and nonuniform in structure (Ottenwälder et al. 1995; Breukink et al. 2003).

As mentioned earlier, the prototypic lantibiotic nisin was initially believed to be involved predominantly in the formation of short-lived pores in bacterial cell membranes. However, a unique mechanism of action has been shown to be exerted by nisin that renders it highly potent against many Gram-positive bacteria, including LAB at nanomolar concentrations (Breukink et al. 1999; Wiedemann et al. 2001). The sensitivities of lipid II–targeting lantibiotics to different bacteria are due to the presence of different lipid II contents among various bacteria (e.g., *Escherichia coli*, 2×10^3 molecules per cell; *Micrococcus lysodeikticus*, 10^5 molecules per cell) (Storm and Strominger 1974; van Heijenoort et al. 1992). Even though the inhibition is caused by binding to lipid I and lipid II, binding to lipid II causes the more predominant effect. Therefore, the interaction of lantibiotics with the cell wall precursor is restricted to lipid II. The pyrophosphate moiety of lipid II interacts with the backbone amides of rings A and B of nisin via six hydrogen bonds (Hsu et al. 2004). Nisin also binds to bactoprenol pyrophosphate, but the affinity is very low compared to the complete lipid II molecule (Bonev et al. 2004), suggesting that the *N*-acetylmuramyl moieties are involved in the high-affinity binding of nisin for additional interactions to take place. Nisin, subtilin, epidermin, gallidermin, plantaricin C, and many other lantibiotics have conserved A and B ring systems responsible for binding to lipid II, in particular to the pyrophosphate moiety. The short-peptides gallidermin and epidermin have a higher affinity to lipid II than nisin. An additional positive charge of these structures at position 4 (lysine) may be the element to enhance binding to the pyrophosphate moiety (Bonelli et al. 2006). Globular lantibiotics (mersacidin, actagardin, cinnamycin, etc.) also bind to lipid II. These lantibiotics form complexes with lipids and disrupt the enzyme function of cell wall biosynthesis (Brötz et al. 1997; Hsu et al. 2003). Mersacidin does not form pores upon binding to lipid II; however, it prevents the activity of transglycosylases (Brötz et al. 1997). LtnA1:LtnA2 peptides of the two-peptide lantibiotic lactiicn 3147 work at nanomolar concentrations with a 1:1 stoichiometry. They work sequentially: The LtnA1 peptide interacts specifically with lipid II; the LtnA1–lipid II complex then recruits the LtnA2 peptide to inhibit cell wall biosynthesis and to form a pore (Wiedemann et al. 2006b).

Lipid II is the target structure in the bacterial cell that nisin recognizes for inhibiting peptidoglycan biosynthesis located near the cell division site (Hyde et al. 2006). Like nisin, epidermin also has the same recognition motif and binds to both lipid I and lipid II (Brötz et al. 1998). The antibacterial activities of plantaricin C are similar to that of nisin. It strongly inhibits in vitro lipid II synthesis and forms a stable complex with lipid II. This indicates that both nisin and plantaricin C may have as a target the same structures in lipid II (Wiedemann et al. 2006a).

The action of the two-peptide bacteriocins is synergistic: The antibacterial activities of the two-peptide lantibiotics are synergistic at optimal concentrations. This emerging subgroup of bacteriocins includes the structurally closely related lacticin 3147, plantaricin W, and staphylococcin C55, and the unrelated streptococcal cytolysin, which combines bacteriocin and cytolytic activity (Gilmore et al. 1994). Lacticin 3147 is a well-studied two-peptide lantibiotic with an exceptional antibiotic efficacy that is achieved when two killing mechanisms (pore formation and inhibition of cell wall biosynthesis) are combined. *L. lactis* is very sensitive to the synergistic effect of the LtnA1 and LtnA2 peptides of lacticin 3147, but *Micrococcus flavus* is relatively insensitive to lacticin 3147. The lactococcal strains are almost equally sensitive to the A1 peptide and mersacidin, but the sensitivity differs by a factor of 30 against *Micrococcus*. The inhibition of cell wall biosynthesis and pore formation needs the sequential actions of the two peptides, in which the interactions of the peptides are stabilized by lipid II.

As compared to nisin, the pore size of lacticin 3147 is smaller (0.6 nm in diameter) (Wiedemann et al. 2006b). While the leakage of metabolites and amino acids from the nisin-treated bacteria is observed, only K^+ leakage is reported from lacticin 3147–sensitive cells. To make a big pore size, several lipid II molecules in the membrane are coupled with nisin molecules to make a complex. Eight nisin and four lipid II molecules are assembled to make a functional pore structure 2 nm in diameter in a sensitive bacterial membrane (Hasper et al. 2006). In contrast, to build a pore size of only 0.6 nm in diameter, such an intricate architecture may not be necessary. It is conceivable that a monomeric or dimeric complex with a 1:1:1 (lipid II:A1:A2) stoichiometry may represent such a pore to build by lacticin 3147. On the contrary, some bacteriocins (e.g., mersacidin) do not form pores. Exploiting the structure-based activities of lantibiotics shows that pore formation in the bacterial membrane is not the main bactericidal mechanism of all bacteriocins.

18.2.4.2 Distinct Responses of Bacteria to Bacteriocins

Abnormal morphogenesis in *B. subtilis* cells is caused by nisin, which is distinctly different from cell wall–inhibiting glycopeptides, β-lactam antibiotics, and the action of other pore-forming peptides (Hyde et al. 2006). Nisin reduces the cross-section of *B. subtilis* cells due to the induced leakage of cytoplasmic contents. The overall aberrations on *B. subtilis* cells for nisin's action are that it permeabilizes cell membranes, accelerates cell division, causes minicell formation, and also deregulates cell envelope formation.

Although the presence of a protein receptor usually augments the antibiotic's action, the sequestration mechanism, in which bacteria are killed by removing lipid II from the cell division site, does not require any receptor molecule for the lantibiotic's activities (Hasper et al. 2006). Here, very clear differences have been shown in comparison with vancomycin. The pyrophosphate of lipid II in bacterial cells is bound by the baseball glove-like structure of the lanthionine rings A and B of nisin. The assembled complex leads only to the formation of a defined pore that kills bacterial cells. However, nisin mutants (e.g., N20P and M21P), while losing their ability to form pores, are still potent antibiotics. Furthermore, mutacin 1140, staphylococcin T, gallidermin, epidermin, and others are unable to form pores but are very efficient at killing bacteria. This observation raises a big question about their activities leading to cell death.

Peptidoglycan synthesis is organized in helical threads along the longitudinal axis of the cell during elongation, and large amounts of peptidoglycan are synthesized at the septum during cell division. A study with fluorescently labeled vancomycin clearly revealed pools of lipid II in the septum along with the helical threads of *B. subtilis* cells (Fig. 18.5). A different lipid II distribution pattern is observed when fluorescently labeled nisin is added to these bacteria. No fluorescence was detected in the septum, and the helical threads were absent. The fluorescence originates from the nisin molecules and clusters in patches on the bacterial cell membrane.

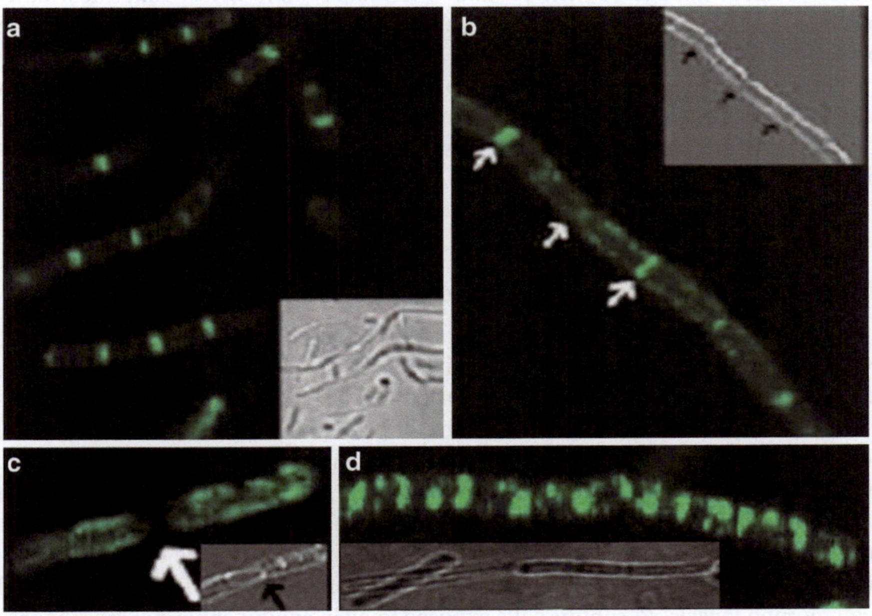

Fig. 18.5 Nisin-induced lipid II segregation observed in bacteria. Nisin segregates lipid II into nonphysiological domains in vivo. (**a**) *B. subtilis* cells stained with 4 μg/mL fluorescent vancomycin. The labeled vancomycin reveals pools of lipid II in the septum and also in helical threads. (**b**) *Bacillus megaterium* cells after incubation for 10 min with 2 μg/mL of labeled vancomycin. The *arrows* indicate the newly formed division sites or older exemplars. (**c**) *B. megaterium* cells incubated with 0.5 μg/mL of fluorescein-labeled nisin. The *arrow* indicates that the bacterium has already divided. (**d**) *B. subtilis* cells incubated with 4 μg/mL of fluorescein-labeled nisin. The fluorescence from nisin appears to be clustered in the patches on the membrane

Interestingly, the nisin mutant (N20PM21P), which lost its pore-forming capacity, also fluoresces similarly to the nisin wild type upon binding to lipid II. Not only *Bacillus* cells but also *L. lactis* fluoresce similarly, suggesting that the sequestration of lipid II is not restricted to rod-shaped bacteria. The presence of such unique structures indicates that peptidoglycan in bacterial cells makes them sensitive regardless of the bacteriocin's capacity to span the lipid bilayer to form pores.

One example of a well-studied class II bacteriocin is pediocin PA-1. It acts on the cytoplasmic membranes of target cells through a multistep process of binding, insertion, and pore formation, which are the common activities of other bacteriocins from LAB. Many studies on the charge distributions of pediocin PA-1 indicated that it functions in the absence of a protein receptor. The primary binding step is mediated by electrostatic interactions between negatively charged phospholipids in the target bacterial membrane and the positively charged amino acid residues in the peptide (Chen et al. 1997).

Circular bacteriocins are another important subgroup of class II bacteriocins. These unusual bacteriocins have diverse characteristics, but have the unifying feature with the head-to-tail circularization of their peptide backbones through the direct covalent linkage of their N- and C-terminal amino acids. Like most other bacteriocins, circular bacteriocins (e.g., enterocin AS-48 and uberolysin) also cause bacterial cell death.

Phages are one of the major causes of fermentation failure. Many *Lactococcus* strains possess one or more prophages integrated in their genome. The bacteriocin Lcn972 and other cell wall antimicrobials induce the prophage in the early exponential growth phase of *L. lactis* (Madera et al. 2009).

18.2.5 Changes in Bacterial Morphology by Bacteriocins

Nisin causes rapid membrane permeabilization of bacteria and subsequent changes in the length, shape, and population distributions (Hyde et al. 2006). The bactericidal action of nisin is due to the intensive effects on membrane permeabilization, followed by cell wall inhibition and metabolic deregulation of bacterial division. Nisin causes partial cell wall detachment from cell membrane in *B. subtilis* cells in comparison to an untreated cross-section (Fig. 18.6). The prime target of action for nisin is located in the region of rapid cell wall growth near the site of septal formation. Significant aberrations in bacterial cells occur during elongation in the cell wall morphogenesis. *B. subtilis* cells exposed to nisin exhibit abnormal morphogenesis during cell division, including multiple septation, cork-screw growth of the cell envelope, and septal malformation (Fig. 18.7). The sequestration mechanisms of lantibiotics, in which the segregation of lipid II from sites of bacterial cell wall synthesis occurs, have also been explained earlier here (Fig. 18.5). Using nisin as an example, the sequestration mechanism indicates a new insight into how short peptides (e.g., gallidermin, epidermin) that are unable to span the membrane may exert their high antibacterial activity.

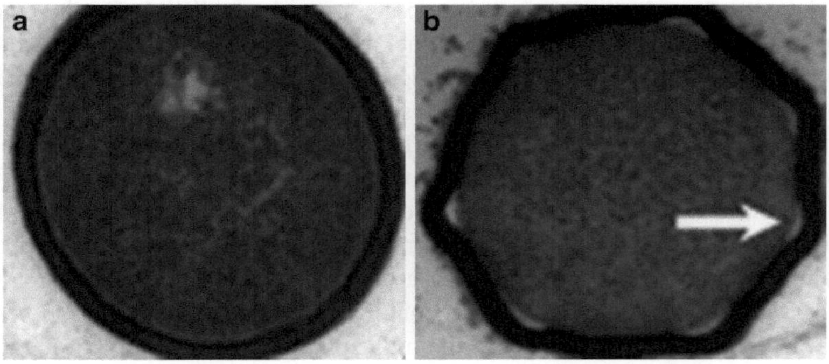

Fig. 18.6 Electron micrograph of bacterial cross-section projections as elucidated by Hyde et al. (2006). (**a**) The cytoplasmic osmotic pressure strengthens the adhesion of the plasma membrane to the peptidoglycan layer in untreated *B. subtilis* cells, resulting in a circular cross-section. (**b**) The detachment of the cell wall from the plasma membrane (*arrow*) is clearly visible after exposure to nisin; pore formation relieves the osmotic stress, leading to an astral cross-section after contraction of the plasma membrane. Scale bar: 100 nm

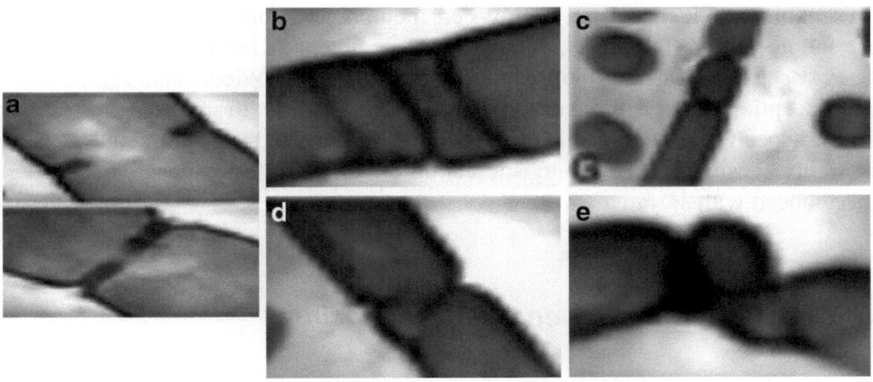

Fig. 18.7 Hyde et al. (2006) observed the morphogenesis of *B. subtilis* cells. (**a**) The normal progression of septal formation in untreated cells; (**b**–**e**) evidence for disturbed bacterial morphogenesis after exposure to nisin that results in morphological aberrations during septation: (**b**) multiseptal divisions; (**c**) one example of division "dead end" that reduces bacteria to produce many nonviable "minicells"; (**d**) disjointed helical septa; and (**e**) corkscrew cell wall morphologies in the growth region near the division site. Scale bar: 200 nm

The action of the major three subgroups of lantibiotics in *B. subtilis* cells was compared using transmission electron micrographs (Asaduzzaman et al. 2009). The well-known lantibiotic nisin (linear) is a lytic-bactericidal agent that causes multiple aberrations, including leakage of cytoplasmic contents, reduction of cell width, acceleration of cell division, minicell formation, abnormal morphogenesis of bacterial cells, and eventual cell death. In contrast, nukacin ISK-1 (a tail and ring region containing lantibiotic) -treated *B. subtilis* cells demonstrate the usual inner-structural arrangement and have good similarity with the internal structure of the untreated cells.

Bacterial cells thus show a static response against nukacin ISK-1. On the contrary, the most widely studied globular lantibiotic, mersacidin, has been reported to cause internal changes in bacterial cells, resulting in the spreading of chromosomes in the cytoplasm and ultimately leading to cell lysis.

Much has been described about the different effects of bacteriocins' actions that are even combined in one molecule to work against bacteria. The efficacy of nisin as an antibiotic is gradually increasing due to the exploitation of its structure-based functions to the target strain. The early findings on nisin were concentrated on the formation of pores leading to the release of molecules and ions (García Garcerá et al. 1993; Driessen et al. 1995). The inhibition of peptidoglycan biosynthesis by nisin was the landmark discovery in the molecular actions of nisin that led to new insights into the molecular mechanisms of bacteriocin modes of action (Breukink et al. 1999). Gallidermin has the lipid II–binding motif of nisin, but the peptide is considerably shorter. In general, it cannot form pores in *Lactococcus*. But, exceptionally, it can form pores in the membrane of only some strains (e.g., *M. flavus* and *Staphylococcus simulans*). Differences in membrane thickness may be the reason behind these dissimilar sensitivities. The average fatty acid acyl chain in the lactococcal membrane contains more than 17 carbon atoms, whereas about 15 atoms are present in staphylococcal and micrococcal membranes (Bonelli et al. 2006). Despite the failure of pore formation, gallidermin is as potent as lacticin 3147 against the *Lactococcus* strains and is 8 times more active than nisin (Wiedemann et al. 2006b). The exceptional antibacterial efficacies of lacticin 3147 are to some extent strain- or species-specific. However, this is a general characteristic for bacteriocins. *L. lactis* is highly sensitive to the synergistic action of A1 and A2 peptides, which is less pronounced with *M. flavus*.

18.2.6 Bacterial Resistance to Bacteriocins

No natural resistance in bacteria has been reported yet, although nisin has been used in the food industry for more than 50 years. However, the development of high-level bacterial resistance to nisin in laboratory has already been reported (Kramer et al. 2006). For instance, the development of bacterial resistance against nisin has been reported due to changes in the composition of the cell membrane; the ratios of phosphatidylglycerol to diphosphatidylglycerol (DPG) increase from 5 to 7. The accumulation of negative charges of DPG probably contributes to stronger binding of cationic nisin. Nisin-resistant variants of *Pediococcus acidilactici* show a remarkable increase in monounsaturated C16:1 and C18:1 fatty acids. Additionally, the resistance of *B. subtilis* to nisin has been suggested by the reduction of the net negative charge of the cell envelope.

The rarity of spontaneous resistance to lacticin 3147 (Guinane et al. 2006) and the very low-level resistance of such mutants are also not big issues for this best-studied two-peptide bacteriocin. However, the cross-immunity shown by the staphylococcin C55 producers to lacticin 3147 (Draper et al. 2009) raises the spectra of resistance through immune mimicry. The resistance of strains due to cross-immunity may therefore emerge by the acquisition of immunity gene homologs. Some LAB strains, including *Lactobacillus casei*, *Lactobacillus rhamnosus*, *Lactobacillus plantarum*, pediococci, and *Leuconostoc* spp., are resistant to vancomycin. This resistance mechanism is usually intrinsic; that is, it is choromosomally encoded and nontransmissible (Klein et al. 1998). In contrast, vancomycin resistance in enterococci is usually plasmid-encoded and transmissible (Quintiliani et al. 1993).

18.3 Other Antimicrobials That Affect Lactic Acid Bacteria

Like bacteriocins, several other antimicrobial compounds are also produced and used by LAB to inhibit the growth of other bacteria present in an ecological niche. Such antagonistic activities caused by LAB include carbon dioxide, which inhibits decarboxylation and also reduces membrane permeability; diacetyl, which interacts with arginine-binding proteins; hydrogen peroxide (H_2O_2)/lactoperoxidase, which cause the oxidation of basic proteins.

Since LAB are industrially important (especially in fermentation), the generation of H_2O_2 inhibits their growth. Because of the sensitivity of wild-type LAB to oxidative stress, more robust strains may provide improvements in the traditional uses of lactococci in industrial fermentations (van de Guchte et al. 2002). Some lactobacilli species grown in the presence of oxygen also generate H_2O_2, which leads to a premature shift from the exponential phase into the stationary phase, and finally produce a reduced biomass (Marty-Teysset et al. 2000). Since dairy fermentations are usually comprised of oxygen-sensitive LAB (van de Guchte et al. 2001), H_2O_2-producing LAB may have negative effects on bacterial yields and the quality of the products. H_2O_2-resistant lactococci gain a survival advantage in the perspective of a mixed-bacterial ecosystem that is usually present in dairy fermentations (Rochat et al. 2005). It was also reported that oxidative stress–resistant *L. lactis* mutants exhibit greater H_2O_2 resistance. On the other hand, oxidative stress increases the sensitivities of 70% of the tested bifidobacteria strains to ampicillin and chloramphenicol, 50% of the strains to cloxacillin and tetracycline, and 40% of the strains to erythromycin, but did not affect sensitivities to vancomycin, kanamycin, and nisin A (Kheadr et al. 2007).

Another study with 10 strains of LAB demonstrated that the strains are sensitive to the Gram-positive spectrum antibiotics erythromycin and novobiocin; the broad spectrum antibiotics rifampicin, spectinomycin, tetracycline, and chloramphenicol; and also β-lactam antibiotics (penicillin, ampicillin, and cephalothin). On the other hand, most of the strains are resistant to the Gram-negative spectrum antibiotics (fusidic acid, nalidixic acid, and polymyxin B and the aminoglycosides neomycin, gentamicin, kanamycin, and streptomycin). Zhou et al. (2005) reported that *Lactobacillus* and *Bifidobacterium* are sensitive to β-lactam antibiotics (penicillin, ampicillin, and cephalothin), Gram-positive spectrum antibiotics (erythromycin and novobiocin), and broad spectrum antibiotics (chloramphenicol, rifampin, spectinomycin, and tetracycline).

18.4 Gene Regulation and Molecular Mechanisms of Lactic Acid Bacterial Response to Bacteriocins and Other Antimicrobials

The transcriptional response of *S. aureus* to subinhibitory concentrations of mersacidin has been explained by gene expression profiling (Sass et al. 2008). Using different sensitivity characteristics of *S. aureus*, they emphasized the important role of lipid II and revealed that the induction of the cell wall stress response is not in every case dependent on inhibitory concentrations of cell wall–active substances. Mersacidin also represents a molecule that seems not to be a substrate for the multidrug resistance transporter VraDE. This can function in the first-line bacterial defense strategy against bacitracin and cationic toxic compounds.

The cell wall stress stimulon is characterized through an inclusive response that involves various cellular processes, which seems to be conserved among Gram-positive bacteria (Jordan et al. 2008). The transcriptional profile of the *S. aureus* response to nisin has become available (Muthaiyan et al. 2008). *S. aureus* responses to nisin are relatively moderate and do not induce the cell wall stress stimulon.

The functional domains of the immunity gene NukH have also been characterized by expressing various mutants (Okuda et al. 2005). No correlation between immunity level and NukH-binding activity against nukacin ISK-1 was found, and thus it was hypothesized that NukH might inactivate nukacin ISK-1 after binding to the cytoplasmic membrane. Furthermore, the *lanH* gene is identical to the gene clusters of nukacin ISK-1, ruminococcin, and butyrivibriocin OR79A.

The responses of a number of Gram-positive bacteria toward cell wall–active antibiotics have been studied by genomewide transcription analysis. The two-component systems (TCSs) and extracellular sigma factors are involved in the intensive action of this complex system. The TCSs are engaged in signal transduction pathways and consist of a sensor or histidine kinase and an effector or response regulator (West and Stock 2001). The LiaSR TCS of *B. subtilis* senses cell envelope stress and is induced by lipid II–interacting bacteriocins (e.g., nisin) and other antibiotics (e.g., bacitracin and ramoplanin) (Mascher et al. 2004). TCS-D is similar to the TCS VraSR of *S. aureus* and is involved in resistance to vancomycin (Kuroda et al. 2003). RmaB belongs to the MarR family and is a transcriptional repressor, which includes the regulators that are usually involved in response to antibiotics. Lactococcal, streptococcal, and staphylococcal CesR/VraR proteins form a group that is distinct from the *Bacillus* LiaR. Further, the LiaR motif is different from CesR by at least three base pairs and is absent in the genome of streptococci and staphylococci (Jordan et al. 2006). The response of *B. subtilis* to the antibiotics like vancomycin, bacitracin, and ramoplanin that bind to lipid II involves a common regulon as well as genes that are drug-specific ones (Mascher et al. 2003). Although theoretically only 46 genes are VraR-dependent, the treatment of *S. aureus* with vancomycin leads to upregulation of more than 100 genes (Kuroda et al. 2003). Martinez et al. (2007) reported that the TCS-D of *L. lactis* is induced upon the inhibition of cell wall biosynthesis by the lipid II–interacting antibiotics, such as the bacteriocin lactococcin 972 (Lcn972) and bacitracin. The synthesis of peptidoglycan at the level of septum formation in *L. lactis* is inhibited by the nonpore-forming bacteriocin Lcn972. The TCS CesSR (previously called TCS-D) orchestrates the response of *L. lactis* to Lcn972. Twenty-six genes are not upregulated by Lcn972 in *L. lactis* Δ*cesR*. Disruption of the CesR response regulator gene results in an increased sensitivity of these bacteria to most of the inducers. Other Gram-positive cocci also have this CesR box. Lipid II–interacting cationic polypeptides strongly induce the CesSR. A clear response of *L. lactis* to cell envelope stress is evident from the response to Lcn972 that shows a distinct transcriptomic signature as the highest upregulated putative operons, which are under the control of CesR.

18.5 Concluding Remarks

In this chapter, responses of LAB to bacteriocins have been discussed concerning the self-immunity of bacteriocin-producing LAB against bacteriocins and the diverse responses to bacteriocin actions. Different immunity systems exist. Bacteriocin producer strains develop and optimize immunity proteins on their cytoplasmic membrane for protection against the produced bacteriocin. The sensitivity of the bacterial strains to bacteriocins is quite diverse. Some LAB strains are killed by some bacteriocins through the inhibition of cell wall biosynthesis and the formation of pores into the bacterial membrane. On the contrary, some bacteria are killed or growth-inhibited by a single mechanism. Several aspects of the LAB responses to bacteriocins and other antimicrobial compounds remain to be elucidated.

References

Asaduzzaman SM, Nagao J, Aso Y, Nakayama J, Sonomoto K (2006) Lysine-oriented charges trigger the membrane binding and activity of nukacin ISK-1. Appl Environ Microbiol 72:6012–6017

Asaduzzaman SM, Nagao J, Iida H, Zendo T, Nakayama J, Sonomoto K (2009) Nukacin ISK-1, a bacteriostatic lantibiotics. Antimicrob Agents Chemother 53:3595–3598

Aso Y, Okuda K, Nagao J, Kanemasa Y, Thi Bich Phuong N, Koga H, Shioya K, Sashihara T, Nakayama J, Sonomoto K (2005) A novel type of immunity protein, NukH, for the lantibiotic nukacin ISK-1 produced by *Staphylococcus warneri* ISK-1. Biosci Biotechnol Biochem 69:1403–1410

Benz R, Jung G, Sahl HG (1991) Mechanism of channel-formation by lantibiotics in black lipid membranes. In: Jung G, Sahl HG (Eds.), Nisin and novel lantibiotics. ESCOM, Leiden, The Netherlands, pp. 359–372

Bonelli RR, Schneider T, Sahl HG, Wiedemann I (2006) Insights into *in vivo* activities of lantibiotics from gallidermin and epidermin mode-of-action studies. Antimicrob Agents Chemother 50:1449–1457

Bonev BB, Breukink E, Swiezewska E, De Kruijff B, Watts A (2004) Targeting extracellular pyrophosphates underpins the high selectivity of nisin. FASEB J 18:1862–1869

Breukink E, van Kraaij CR, Demel A, Siezen RJ, Kuipers OP, de Kruijff B (1997) The C-terminal region of nisin is responsible for the initial interaction of nisin with the target membrane. Biochemistry 36:6968–6976

Breukink E, Wiedemann I, van Kraaij C, Kuipers OP, Sahl H, de Kruijff B (1999) Use of the cell wall precursor lipid II by a pore-forming peptide antibiotic. Science 286:2361–2364

Breukink E, van Heusden HE, Vollmerhaus PJ, Swiezewska E, Brunner L, Walker S, Heck AJ, de Kruijff B (2003) Lipid II is an intrinsic component of the pore induced by nisin in bacterial membranes. J Biol Chem 278:19898–19903

Brötz H, Bierbaum G, Reynolds PE, Sahl HG (1997) The lantibiotic mersacidin inhibits peptidoglycan biosynthesis at the level of transglycosylation. Eur J Biochem 246:193–199

Brötz H, Josten M, Wiedemann I, Schneider U, Götz F, Bierbaum G, Sahl HG (1998) Role of lipid-bound peptidoglycan precursors in the formation of pores by nisin, epidermin and other lantibiotics. Mol Microbiol 30:317–327

Chen Y, Ludescher RD, Montville TJ (1997) Electrostatic interactions, but not the YGNGV consensus motif, govern the binding of pediocin PA-1 and its fragments to phospholipid vesicles. Appl Environ Microbiol 63:4770–4777

Cotter PD, Hill C, Ross RP (2005) Bacteriocins: developing innate immunity for food. Nature Rev Microbiol 3:777–788

de Vos WM, Kuipers OP, van der Meer JR, Siezen RJ (1995) Maturation pathway of nisin and other lantibiotics: post-translationally modified antimicrobial peptides exported by gram-positive bacteria. Mol Microbiol 17:427–437

Demel RA, Peelen T, Siezen RJ, de Kruijff B, Kuipers OP (1996) Nisin Z, mutant nisin Z and lacticin 481 interactions with anionic lipids correlate with antimicrobial activity. A monolayer study. Eur J Biochem 235:267–274

Draper LA, Grainger K, Deegan LH, Cotter PD, Hill C, Ross RP (2009) Cross-immunity and immune mimicry as mechanisms of resistance to the lantibiotic lacticin 3147. Mol Microbiol 71:1043–1054

Driessen AJ, van den Hooven HW, Kuiper W, van de Kamp M, Sahl HG, Konings RN, Konings WN (1995) Mechanistic studies of lantibiotic-induced permeabilization of phospholipid vesicles. Biochemistry 34:1606–1614

García Garcerá MJ, Elferink MG, Driessen AJ, Konings WN (1993) *In vitro* pore-forming activity of the lantibiotic nisin. Role of protonmotive force and lipid composition. Eur J Biochem 212:417–422

Giffard CJ, Dodd HM, Horn N, Ladha S, Mackie AR, Parr A, Gasson MJ, Sanders D (1997) Structure-function relations of variant and fragments of nisin studied with model membrane systems. Biochemistry 36:3802–3810

Gilmore MS, Segarra RA, Booth MC, Bogie CP, Hall LR, Clewell DB (1994) Genetic structure of the *Enterococcus faecalis* plasmid pAD1-encoded cytolytic toxin system and its relationship to lantibiotic determinants. J Bacteriol 176:7335–7344

Gross E, Morell JL (1971) The structure of nisin. J Am Chem Soc 93:4634–4635

Guinane CM, Cotter PD, Hill C, Ross RP (2006) Spontaneous resistance in *Lactococcus lactis* IL1403 to the lantibiotic lacticin 3147. FEMS Microbiol Lett 260:77–83

Hasper HE, Kramer NE, Smith JL, Hillman JD, Zachariah C, Kuipers OP, de Kruijff B, Breukink E (2006) An alternative bactericidal mechanism of action for lantibiotic peptides that target lipid II. Science 313:1636–1637

He Z, Yuan C, Zhang L, Yousef AE (2008) N-terminal acetylation in paenibacillin, a novel lantibiotic. FEBS Lett 582:2787–2792

Hsu STD, Breukink E, Bierbaum G, Sahl HG, de Kruijff B, Kaptein R, van Nuland NA, Bonvin AM (2003) NMR study of mersacidin and lipid II interaction in dodecylphosphocholine micelles. Conformational changes are a key to antimicrobial activity. J Biol Chem 278:13110–13117

Hsu STD, Breukink E, Tischenko E, Lutters MA, de Kruijff B, Kaptein R, Bonvin AM, van Nuland NA (2004) The nisin-lipid II complex reveals a pyrophosphate cage that provides a blueprint for novel antibiotics. Nat Struct Mol Biol 11:963–967

Hyde AJ, Parisot J, McNichol A, Bonev BB (2006) Nisin-induced changes in *Bacillus* morphology suggest a paradigm of antibiotic action. Proc Natl Acad Sci USA 103:19896–19901

Jordan S, Junker A, Helmann JD, Mascher T (2006) Regulation of LiaRS-dependent gene expression in *Bacillus subtilis*: identification of inhibitor proteins, regulator binding site, and target genes of a conserved cell envelope stress-sensing two-component system. J Bacteriol 188:5153–5166

Jordan S, Htchings MI, Mascher T (2008) Cell envelope stress response in Gram-positive bacteria. FEMS Microbiol Rev 32:107–146

Kheadr E, Dabour N, Le Lay C, Lacroix C, Fliss I (2007) Antibiotic susceptibility profile of Bifidobacteria as affected by oxgall, acid and hydrogen peroxide stress. Antimicrob Agents Chemother 51:169–174

Klein G, Pack A, Bonaparte C, Reuter G (1998) Taxonomy and physiology of probiotic lactic acid bacteria. Int J Food Microbiol 41:103–105

Kodani S, Hudson ME, Durrant MC, Buttner MJ, Nodwell JR, Willey JM (2004) The SapB morphogen is a lantibiotic-like peptide derived from the product of the developmental gene *ramS* in *Streptomyces coelicolor*. Proc Natl Acad Sci USA 101:11448–11453

Kordel M, Benz R, Sahl HG (1988) Mode of action of the staphylococcin like peptide Pep 5: voltage-dependent depolarization of bacterial and artificial membranes. J Bacteriol 170:84–88

Kramer NE, van Hijum SA, Knol J, Kok J, Kuipers OP (2006) Transcriptome analysis reveals mechanisms by which *Lactococcus lactis* acquires nisin resistance. Antimicrob Agents Chemother 50:1753–1761

Kuroda M, Kuroda H, Oshima T, Takeuchi F, Mori H, Hiramatsu K (2003) Two-component system VraSR positively modulates the regulation of cell-wall biosynthesis pathway in *Staphylococcus aureus*. Mol Microbiol 49:807–821

Madera C, Garcia P, Rodriguez A, Suarez JE, Martinez B (2009) Prophage induction in *Lactococcus lactis* by the bacteriocin Lactococcin 972. Int J Food Microbiol 129:99–102

Martinez B, Zomer A, Rodriguez A, Kok J, Kuipers OP (2007) Cell envelope stress induced by the bacteriocin Lcn972 is sensed by the lactococcal two-component system CesSR. Mol Microbiol 64:473–486

Marty-Teysset C, de la Torre F, Garel JR (2000) Increased production of hydrogen peroxide by *Lactobacillus delbrueckii* subsp. *Bulgaricus* upon aeration: involvement of an NADH oxidase in oxidative stress. Appl Environ Microbiol 66:262–267

Marx R, Stein T, Entian KD, Glaser SJ (2001) Structure of the *Bacillus subtilis* peptide antibiotic subtilosin A determined by 1 H-NMR and matrix assisted laser desorption/ionization time-of-flight mass spectrometry. J Protein Chem 20:501–506

Mascher T, Margulis NG, Wang T, Ye RW, Helmann JD (2003) Cell wall stress responses in *Bacillus subtilis*: the regulatory network of the bacitracin stimulon. Mol Microbiol 50:1591–1604

Mascher T, Zimmer SL, Smith TA, Helmann JD (2004) Antibiotic-inducible promoter regulated by the cell envelope stress-sensing two-component system LiaRS of *Bacillus subtilis*. Antimicrob Agents Chemother 48:2888–2896

McAuliffe O, O'Keeffe T, Hill C, Ross RP (2001) Regulation of immunity to the two-component lantibiotic, lacticin 3147, by the transcriptional repressor LtnR. Mol Microbiol 39: 982–993

Moll GN, Clark J, Chan WC, Bycroft BW, Roberts GC, Konings WN, Driessen AJ (1997) Role of transmembrane pH gradient and membrane binding in nisin pore formation. J Bacteriol 179:135–140

Muthaiyan A, Silverman JA, Jayaswal RK, Wilkinson BJ (2008) Transcriptional profiling reveals that daptomycin induces the *Staphylococcus aureus* cell wall stress stimulation and genes responsive to membrane depolarization. Antimicrob Agents Chemother 52:980–990

Nissen-Meyer J, Nes IF (1997) Ribosomally synthesized antimicrobial peptides: their function, biogenesis, and mechanism of action. Arch Microbiol 167:67–77

Okuda K, Aso Y, Nagao J, Shioya K, Kanemasa Y, Nakayama J, Sonomoto K (2005) Characterization of functional domains of lantibiotic-binding immunity protein, NukH, from *Staphylococcus warneri* ISK-1. FEMS Microbiol Lett 250:19–25

Okuda K, Yanagihara S, Shioya K, Harada Y, Nagao J, Aso Y, Zendo T, Nakayama J, Sonomoto K (2008) Binding specificity of the lantibiotic-binding immunity protein NukH. Appl Environ Microbiol 74:7613–7619

Ottenwälder B, Kupke T, Brecht S, Gnau V, Metzger J, Jung G, Götz F (1995) Isolation and characterization of genetically engineered gallidermin and epidermin analogs. Appl Environ Microbiol 61:3894–3903

Quintiliani R Jr, Evers S, Courvalin P (1993) The *vanB* gene confers various levels of self-transferable resistance to vancomycin in enterococci. J Infect Dis 167:1220–1223

Rochat T, Gratadoux J, Corthier G, Coqueran B, Nader-Macias ME, Gruss A, Langella P (2005) *Lactococcus lactis* SpOx spontaneous mutants: a family of oxidative-stress-resistant dairy strains. Appl Environ Microbiol 71:2782–2788

Sahl HG (1991) Pore formation in bacterial membranes by cationic lantibiotics. In: Jung G, Sahl HG (Eds.), Nisin and novel lantibiotics. ESCOM, Leiden, The Netherlands, pp. 347–358

Sahl HG, Kordel M, Benz R (1987) Voltage-dependent depolarization of bacterial membranes and artificial lipid bilayers by the peptide antibiotic nisin. Arch Microbiol 149:120–124

Sahl HG, Jack RW, Bierbaum G (1995) Biosynthesis and biological activities of lantibiotics with unique post-translational modifications. Eur J Biochem 230:827–853

Sass P, Jansen A, Szekat C, Sass V, Sahl HG, Bierbaum G (2008) The lantibiotic mersacidin is a strong inducer of the cell wall stress response of *Staphylococcus aureus*. BMC Microbiol 8:186

Stein T, Heinzmann S, Solovieva I, Entian KD (2003) Function of *Lactococcus lactis* nisin immunity genes *nisI* and *nisFEG* after coordinated expression in the surrogate host *Bacillus subtilis*. J Biol Chem 278:89–94

Storm DR, Strominger J (1974) Binding of bacitracin to cells and protoplasts of *Micrococcus lysodeikticus*. J Biol Chem 249:1823–1827

Tagg JR, Dajani AS, Wannamaker LW (1976) Bacteriocins of gram-positive bacteria. Bacteriol Rev 40:722–756

Takala TM, Saris PE (2006) C terminus of NisI provides specificity to nisin. Microbiology 152:3543–3549

van de Guchte M, Ehrlich SD, Maguin E (2001) Production of growth-inhibiting factors by *Lactobacillus delbrueckii*. J Appl Microbiol 91:147–153

van de Guchte M, Serror P, Chervaux C, Smokvina T, Ehrlich SD, Maguin E (2002) Stress responses in lactic acid bacteria. Antonie van Leeuwenhoek 82:187–216

van Heijenoort Y, Gomez M, Derrien M, Ayala J, van Heijenoort J (1992) Membrane intermediates in the peptidoglycan metabolism of *Escherichia coli*: possible roles of PBP 1b and PBP 3. J Bacteriol 174:3549–3557

van Kraaij C, Breukink E, Noordermeer MA, Demel RA, Siezen RJ, Kuipers OP, de Kruijff B (1998) Pore formation by nisin involves translocation of its C-terminal part across the membrane. Biochemistry 37:16033–16040

Veith B, Herzberg C, Steckel S, Feesche J, Maurer KH, Ehrenreich P, Bäumer S, Henne A, Liesegang H, Merkl R, Ehrenreich A, Gottschalk G (2004) The complete genome sequence of *Bacillus licheniformis* DSM13, an organism with great industrial potential. J Mol Microbial Biotechnol 7:204–211

West AH, Stock AM (2001) Histidine kinases and response regulator proteins in two-component signaling systems. Trends Biochem Sci 26:369–376

Wiedemann I, Breukink E, van Kraaij C, Kuipers OP, Bierbaum G, de Kruijff B, Sahl HG (2001) Specific binding of nisin to the peptidoglycan precursor lipid II combines pore formation and inhibition of cell wall biosynthesis for potent antibiotic activity. J Biol Chem 276:1772–1779

Wiedemann I, Bottiger T, Bonelli RR, Schneider T, Sahl HG, Martínez B (2006a) Lipid II-based antimicrobial activity of the lantibiotic plantaricin C. Appl Environ Microbiol 72:2809–2814

Wiedemann I, Böttiger T, Bonelli RR, Wiese A, Hagge SO, Gutsmann T, Seydel U, Deegan L, Hill C, Ross P, Sahl HG (2006b) The mode of action of the lantibiotic lacticin 3147 – a complex mechanism involving specific interaction of two-peptides and the cell wall precursor lipid II. Mol Microbiol 61:285–297

Zhou JS, Pillidge CJ, Gopal PK, Gill HS (2005) Antibiotic susceptibility profiles of new probiotic *Lactobacillus* and *Bifidobacterium* strains. Int J Food Microbiol 98:211–217

Chapter 19
Lactic Acid Bacteria Defenses Against Phages

Rodolphe Barrangou and Philippe Horvath

19.1 Introduction

Historically, starter cultures have been developed to optimize the delivery of fermented products with consistent and desirable attributes. As in most biological processes relying on living organisms, viral predation is an unfortunate yet frequent occurrence that also happens within large-scale industrial fermentation environments. Accordingly, the formulation of starter cultures is based on leveraging various natural resistance mechanisms against viral attack, allowing the sustainable usage of lactic acid bacteria (LAB) for the fermentation of dairy products such as yogurt and cheese. Over time, a significant amount of practical and scientific knowledge has been accumulated, notably about LAB (Nyengaard et al. 1995; Forde and Fitzgerald 1999; Sturino and Klaenhammer 2004a). The food industry heavily relies on a diversity of bacterial species from the *Lactobacillus*, *Lactococcus*, *Leuconostoc*, *Pediococcus*, and *Streptococcus* genera for the manufacture of fermented products. These Gram-positive bacteria readily ferment carbohydrates into lactic acid, primarily, while imparting particular textural and organoleptic properties to the fermentation end product through the generation of desirable compounds derived from the catabolism of lipids and proteins. Factually, the mesophilic *Lactococcus lactis* and the thermophilic *Streptococcus thermophilus* species have been used intensively since the beginning of the twentieth century in commercial dairy fermentations.

Most life forms on Earth are somewhat subjected to viral exposure, and industrial starter cultures provide a consistent prey pool for opportunistic bacteriophages (phages). Actually, viruses are arguably numerically the most dominant biological entity on the planet (Breitbart and Rohwer 2005), and predatory phages are nearly ubiquitous in raw milk, which serves as a continuous viral reservoir and creates a natural challenge for most dairy fermentation (Brüssow 2001). Phages have the

R. Barrangou (✉)
Danisco USA Inc., 3329 Agriculture Drive, Madison, WI 53716, USA
e-mail: rodolphe.barrangou@danisco.com

E. Tsakalidou and K. Papadimitriou (eds.), *Stress Responses of Lactic Acid Bacteria*, Food Microbiology and Food Safety, DOI 10.1007/978-0-387-92771-8_19,

ability to persist in industrial environments due to their resistance to pasteurization, airborne dissemination, and the challenges inherent to implementing effective sanitation strategies in manufacturing facilities (Rousseau and Moineau 2009). A variety of double-stranded DNA phages have the ability to infect lactococci and streptococci. For the former, while *Caudovirales* phages belonging to the *Myoviridae, Siphoviridae,* and *Podoviridae* families are commonly encountered, three main phage groups primarily occur in the industry, namely, 936, c2, and P335 (Deveau et al. 2006; Labrie and Moineau 2007; Rousseau and Moineau 2009). In contrast, streptococcal phages primarily belong to the *Siphoviridae* family and are less diverse, perhaps because widely used strains are more genetically homogeneous (Lévesque et al. 2005; Guglielmotti et al. 2009). The costs inherent to fermentation delays, inconsistencies, and occasional failures have generated interest in understanding and managing phage issues in industrial settings. Concurrently, starter culture manufacturers have a vested interest in ensuring that efficient and high-quality strains have an extended life span. Accordingly, improved manufacturing sanitation, culture rotation strategies, and the use of "phage-resistant" strains have been developed over the years (Moineau 1999). There are several defense mechanisms that LAB may employ to protect themselves against phages (Daly et al. 1996; Josephsen and Neve 1998; Forde and Fitzgerald 1999; Sturino and Klaenhammer 2004a), including primitive defense strategies such as the prevention of adsorption and blocking of DNA injection, as well as elaborate resistance mechanisms such as restriction-modification, abortive infection, and the novel CRISPR/Cas system (Fig. 19.1).

19.2 Native Phage Defense Strategies

19.2.1 Prevention of Adsorption and Blocking of DNA Injection

Adsorption, the initial and obligate step of any phage infection process, is determined by a molecular interaction between the receptor (most often a carbohydrate) present on the surface of the bacterial host cell and a phage structural protein, the antireceptor. The modification or removal of phage receptors is obviously the first barrier that bacteria can implement to thwart the viral threat (Josephsen and Neve 1998; Sturino and Klaenhammer 2004a). As a matter of fact, phage receptor mutation is a frequent outcome of the selection of spontaneous bacteriophage-insensitive mutants (BIMs) following phage challenge (Madera et al. 2003; Dupont et al. 2004). Phage receptor masking, through the production of cell wall–shielding components such as galactosyl-containing lipoteichoic acid or polysaccharides, has also been reported to confer phage resistance in lactococci (Sijtsma et al. 1990; Lucey et al. 1992; Forde et al. 1999; Trotter et al. 2002).

Blocking the injection of phage DNA into the cytoplasm is the next chronological event that can confer phage resistance. Both chromosomal and plasmid-encoded determinants involved in phage DNA injection have been described, including the phage infection protein *pip* gene (Monteville et al. 1994; Garvey et al. 1996; Lucchini et al. 2000).

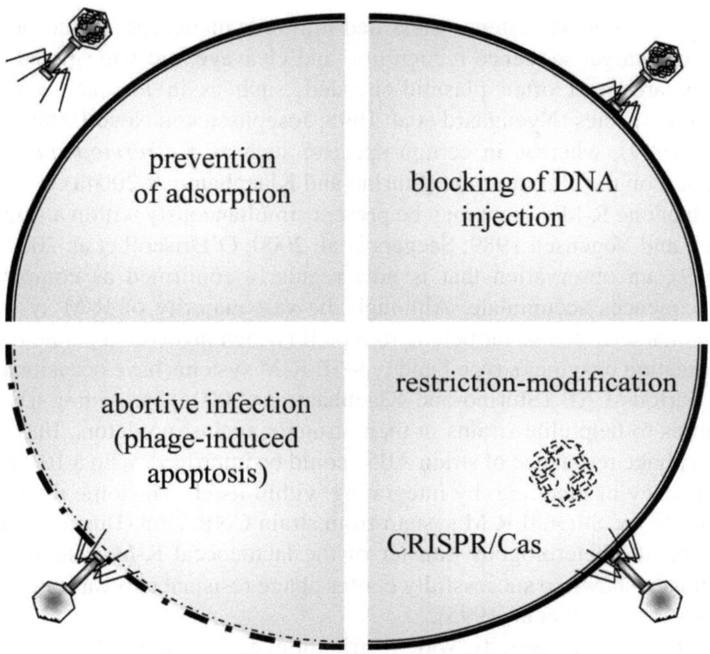

Fig. 19.1 Phage-resistance mechanisms in LAB. The main phage-resistance mechanisms are represented by functional category across interference steps, in their order of occurrence throughout the viral life cycle: prevention of adsorption, blocking of DNA injection, interference with phage DNA by restriction-mofidication and CRISPR/Cas, and abortive infection

Following infection, temperate phages have the possibility to integrate their genome within the host's chromosome, a state called *lysogeny*, loading the host with a significant amount of arguably superfluous sequences (Sturino and Klaenhammer 2004a). This detriment has to be balanced with the "superinfection immunity" conferred by prophage sequences against subsequent infection by identical or similar phages (Alatossava et al. 1995; Bruttin et al. 1997). Superinfection exclusion genes such as *sie* (found in several lactococcal temperate phages) or *ltp* (of the *S. thermophilus* temperate phage TP-J34) also provide resistance against various phages via a DNA injection-blocking mechanism (McGrath et al. 2002a; Sun et al. 2006; Mahony et al. 2008).

19.2.2 Restriction-Modification

Restriction-modification (R-M) systems involve two distinct enzymatic activities: On the one hand, a site-specific restriction endonuclease cleaves the incoming, unprotected phage double-stranded DNA once it has entered the bacterial cytosol; on the other hand, a cognate methylase protects the host genomic DNA by methylation at the same specific sites, allowing a self/nonself-discrimination (Blumenthal and Cheng 2002). R-M systems, classified into four main types based on enzyme structures and target sequence recognition and cleavage, are widespread in LAB, where they are most often plasmid-encoded, such as in *L. lactis* and various *Lactobacillus* species (Nyengaard et al. 1995; Josephsen and Neve 1998; Forde and Fitzgerald 1999), whereas in certain species, such as *S. thermophilus*, they are mainly found on the chromosome (Sturino and Klaenhammer 2004a).

More than one R-M system may be present simultaneously within a single strain (Josephsen and Vogensen 1989; Seegers et al. 2000; O'Driscoll et al. 2004; Suárez et al. 2009), an observation that is now regularly confirmed as complete LAB genome sequences accumulate. Although the vast majority of R-M systems that have been characterized so far belong to type II (which involves separate restriction and modification enzymes), type I and type III R-M systems have occasionally been found in various LAB (Sturino and Klaenhammer 2004a), providing a wealth of opportunities to help elite strains in their struggle against predators. In *S. thermophilus*, the phage resistance of strain A054 could be improved, with a 10^{-4} reduction in the efficiency of plaquing, by integrating within its chromosome the two genes that constitute the Sth368I R-M system from strain CNRZ368 (Burrus et al. 2001). Furthermore, the heterologous transfer of the lactococcal R-M system LlaDCHI was previously shown to successfully confer phage resistance to various *S. thermophilus* hosts (Moineau et al. 1995).

Based on the large diversity, wide distribution, and frequent plasmid linkage of R-M systems in LAB, one can assume that these defense mechanisms are very effective in limiting phage development in natural dairy environments (Forde and Fitzgerald 1999; Sturino and Klaenhammer 2004a). Nevertheless, phages have several escape strategies that allow them to circumvent this first line of intracellular defense, which, for the continuity of the host lineage, has to be secured by other lines of defenses. Moreover, distinct phage-resistance mechanisms may have interconnected functions, as exemplified by the requirement of the LlaKR2I methylase for the expression of the AbiR system in *L. lactis* strain KR2 (Yang et al. 2006).

19.2.3 Abortive Infection

Abortive infection (Abi, also known as phage exclusion) systems have the ability to interfere with several steps of the phage lytic cycle intracellularly, which prevents phage multiplication in the host population, while causing premature bacterial death (Sing and Klaenhammer 1990; Chopin et al. 2005). In LAB, most of the known Abi systems have been described in lactococci, with at least 23 distinct systems. They are mostly plasmid-encoded and documented to interfere with phage genome replication, transcription, translation, encapsidation and particle assembly, and cell lysis (Chopin et al. 2005; Sturino and Klaenhammer 2004a). Functionally, while Abi systems typically result in the death of the host cell, the bacterial population at large is benefiting from the entrapment of the virulent phage in the lost cell before the completion of phage replication. Genetically, most Abi systems are encoded in a single gene, but these various systems (AbiA–AbiZ) do not share significant sequence homology at the gene or protein level. A distinct feature of *abi* sequences found in lactococci is the very low G+C content (25–28%) (Anba et al. 1995), which is notable even for the low G+C genomes of LAB such as lactococci (G+C approx. 35–37%) (Makarova et al. 2006a). Phenotypically, the Abi phage-resistance phenotypes range widely, from partial to complete inhibition of phage plaquing, measured by phage morphology, efficiency of plaquing, efficiency of infective center formation, and burst size (Sing and Klaenhammer 1990). Among the best-characterized Abi systems, AbiA, AbiF, and AbiK prevent DNA replication, while AbiB targets phage mRNA posttranscription, AbiC prevents capsid production, AbiD prevents phage DNA packaging, and AbiZ causes premature lysis (Garvey et al. 1995; Chopin et al. 2005; Fortier et al. 2005; Durmaz and Klaenhammer 2007). These systems affect the phage lytic cycle postadsorption and post-DNA injection into the host, and the interruption of phage development at various stages of the viral lytic cycle either interrupts phage multiplication or results in the release of few or dysfunctional progeny particles.

The presence of Abi systems on plasmids allows for their natural transfer into desirable strains and the development of robust rotation schemes. Also, the association of Abi systems with IS (insertion sequence) elements that include transposases can increase their propensity for dissemination, as shown for *abi-416* and Iso-IS*S1* (Cluzel et al. 1991). Actually, a nonisogenic strain rotation scheme can occasionally be replaced or complemented with isogenic strain rotations that include plasmids with different Abi (and R-M) systems (Durmaz and Klaenhammer 1995). Historically, rotation strategies based on plasmids such as pTR2030, which contains both Abi and R-M systems, have generated more robust starter cultures (Klaenhammer and Sanosky 1985; Hill et al. 1989; Durmaz and Klaenhammer 1995). Similarly, pNP40 (Garvey et al. 1995) carries two Abi determinants that provide resistance against c2 phages.

Abi systems have seldom been described in other LAB, including *Streptococcus thermophilus* (Larbi et al. 1992). In this species, an effective gain of phage resistance could be observed following the transfer of the lactococcal *abiA* gene (Tangney and Fitzgerald 2002).

19.2.4 CRISPR/Cas Systems

Clustered regularly interspaced short palindromic repeats (CRISPR) are a novel DNA repeat family present in the genomes of many bacteria (~40%) and most Archaea (~90%) (Jansen et al. 2002; Horvath and Barrangou 2010). CRISPR loci typically include a series of noncontiguous direct repeats separated by variable sequences called *spacers*, derived from foreign genetic elements. Often times, the repeat-spacer array is adjacent to *cas* (CRISPR-associated) genes, which encode a large and heterogeneous family of proteins that carry functional domains typical of nucleases, helicases, polymerases, and polynucleotide-binding proteins (Haft et al. 2005;

Sorek et al. 2008; Wiedenheft et al. 2009), including the universal markers of CRISPR/Cas systems *cas1* (COG1518) and *cas2* (COG1343, COG3512). In addition to the *cas1–cas6* core genes, subtype-specific genes and genes encoding *repeat-associated mysterious proteins* (*RAMP*) have been classified and further segregated into functional subtypes paired with particular CRISPR repeat sequences (Haft et al. 2005; Makarova et al. 2006b; Sakamoto et al. 2009). CRISPR repeats and spacers vary between 23–47 and 21–72 bp, respectively, with most loci containing fewer than 50 units in LAB (Horvath et al. 2009) (Table 19.1). While CRISPRs are typically located on the chromosome, a few are plasmid-encoded (Lillestøl et al. 2006).

The availability of a plethora of full and draft genome sequences for various LAB provides a basis for a comparative analysis of CRISPR loci in these genomes. We investigated the occurrence of CRISPR loci in 102 genomes across 11 genera and 26 species of LAB belonging to the *Firmicutes* and *Actinobacteria* phyla. A total of 66 CRISPR loci were identified in 47 genomes (46.1%), and 37 sets were classified into eight main families based on CRISPR repeat sequence and *cas* gene content, organization, and sequence (Horvath et al. 2009). Multiple CRISPR loci are widely distributed in lactobacilli and streptococci, but they are notably absent from the genomes of sequenced *Lactococcus*, *Leuconostoc*, and *Oenococcus* species. Interestingly, the occurrence and clustering of CRISPR loci found in LAB do not correlate with the classical phylogenetic tree, which is consistent with their propensity for horizontal gene transfer (Godde and Bickerton 2006; Sorek et al. 2008; Horvath et al. 2009). Actually, two of the eight families were identified in both *Firmicutes* and *Actinobacteria*, indicating that they can be transferred even across phylogenetically distant organisms (Horvath et al. 2009).

Although the CRISPR structure was fortuitously discovered in *Escherichia coli* in 1987 (Ishino et al. 1987) and the term coined in 2002 (Jansen et al. 2002), CRISPR function has remained elusive until recently. While several roles for CRISPR sequences were originally suggested (Mojica et al. 1995; Makarova et al. 2002), the sequence homologies between spacer sequences and extrachromosomal elements such as viruses and plasmids led to the hypothesis that CRISPR may provide adaptive immunity against foreign genetic elements (Mojica et al. 2005; Pourcel et al. 2005; Bolotin et al. 2005; Makarova et al. 2006b). It was established in 2007 in *S. thermophilus* that during the natural generation of phage-resistant variants, bacteria commonly alter their CRISPR loci by the incorporation of novel CRISPR spacers derived from the phages used in the challenge (Barrangou et al. 2007; Deveau et al. 2008; Horvath et al. 2008). Actually, it was shown that spacer content defines immunity against phages and plasmids (Barrangou et al. 2007; Brouns et al. 2008; Marraffini and Sontheimer 2008; van der Ploeg 2009). Several metagenomic studies investigating host/virus population dynamics have confirmed that CRISPR loci evolve in response to viral predation and provide insights both historically and geographically (Andersson and Banfield 2008; Tyson and Banfield 2008; Kunin et al. 2008; Heidelberg et al. 2009; Held and Whitaker 2009; Shah et al. 2009).

Mechanistically, the immune determinants defined by CRISPR spacers become available to the Cas machinery through transcription into a pre-crRNA, which is processed into small RNA molecules (crRNA) that correspond to a spacer flanked

Table 19.1 CRISPR repeat diversity in select strains of the *Enterococcus, Lactobacillus, Pediococcus,* and *Streptococcus* genera

Strain	GenBank accession	Repeat sequence (consensus)	CRISPR family	Repeat number	Repeat size	Spacer size (min–max)
Enterococcus faecium 1,231,408	ACBB00000000	GTTTTARAACCATGTAAA ACAACACAGTTCTCAAAT	Efam1	3	36 bp	30 bp
Lactobacillus acidophilus NCFM	CP000033	RTTTTYTCYAHRWAW RTGGAGSTRMHYYH	Ldbu1	34	29 bp	32 bp (32–34)
Lactobacillus antri DSM16041	ACLL00000000	GTATTCCCCAWRKVYR TRGGGGTGATCCY	Ldbu1	7+7	29 bp	32 bp (32–33)
Lactobacillus casei ATCC334	CP000423	GTTTTCCCCGCACATGC GGGGGTGATCCY	Ldbu1	22	29 bp	32 bp (32–33)
Lactobacillus crispatus JV-V01	ACKR00000000	GTATTCTCCACRHRTGT GGAGGTGATCCW	Ldbu1	5	29 bp	32 bp (32–33)
Lactobacillus delbrueckii subsp. *bulgaricus* ATCC11842	CR954253	GTATTCCCCACGCAAG TGRGGRTGATYC	Ldbu1	41	28 bp	33 bp (33–34)
Lactobacillus fermentum IFO3956	AP008937	GTATTCCCCATGTATG TGGGGGTGATSYY	Ldbu1	21+24	29 bp	32 bp (32–33)
Lactobacillus helveticus DSM20075	ACLM00000000	GTATTCTCCACGTATG TGGAGGTGATCCY	Ldbu1	3	29 bp	32 bp
Streptococcus mutans NN2025	AP010655	ATTTTACCCGCACGAG CGGGGGTGATCCT	Ldbu1	19	29 bp	32 bp (32–33)
Streptococcus thermophilus DGCC7710	Draft genome	GTTTTTCCCGCACACG CGGGGGTGATCCH	Ldbu1	13	29 bp	32 bp (32–33)
Lactobacillus fermentum IFO3956	AP008937	GTCGCACTCCTTCGCGGG GGTGCGTGGATTGAAAT	Lhel1	4	35 bp	34 bp (33–34)
Lactobacillus helveticus DPC4571	CP000517	GTCGCACTCCTTGTGAGT GCGTGGATTGAAAT	Lhel1	23	32 bp	35 bp (34–37)
Streptococcus equi subsp. *zooepidemicus* MGCS10565	CP001129	GTCTCGCCCYWTRCGGGC GAGTGGATTGAAAT	Lhel1	10	32 bp	35 bp (33–35)

(continued)

Table 19.1 (continued)

Strain	GenBank accession	Repeat sequence (consensus)	CRISPR family	Repeat number	Repeat size	Spacer size (min–max)
Streptococcus mutans UA159	AE014133	GTCGCACCCTTCACGGGT GCGTGGATTGAAAT	Lhel1	1	32 bp	Undefined
Streptococcus pyogenes SF370 M1 GAS	AE004092	GTCTCACCCTTCATGGGT GAGTGGATTGAAAT	Lhel1	4	32 bp	34 bp (33–35)
Lactobacillus helveticus DSM20075	ACLM00000000	GTTTTTATTTAACTTA AGAGAAATGTAAAG	Lhel2	44	30 bp	36 bp (35–39)
Lactobacillus brevis subsp. *gravesensis* ATCC27305	ACGG00000000	GCTTTAGTAGGATGTTAAAT CAATGATGTTAAACCC	Lsal1	12	36 bp	30 bp (30–31)
Lactobacillus buchneri ATCC11577	ACGH00000000	GCTTTAGTAGTAGTTYAAARC AATGATGTTTTATYC	Lsal1	29	36 bp	30 bp
Lactobacillus casei BL23	FM177140	GTCTCAGGTAGATGTCGAAT CAATCAGTTCAAGAGC	Lsal1	22	36 bp	30 bp
Lactobacillus crispatus JV-V01	ACKR00000000	GTTTTAGATGATTGTTAGAT CAATGAGGTTTAGATC	Lsal1	5	36 bp	30 bp
Lactobacillus fermentum ATCC14931	ACGI00000000	GTCTTGGATGAGTGTCAGAT CAGTAGTTCCGAGTAC	Lsal1	13	36 bp	30 bp
Lactobacillus gasseri JV-V03	ACGO00000000	GTTTTAGATGGTTGTTAGAT CAATAAGGTTTAGATC	Lsal1	20	36 bp	30 bp
Lactobacillus jensenii 115-3-CHN	ACQN00000000	GTTTTAGAAGGTTGTTAAAT CAGTAAGTTGAAAAAC	Lsal1	33	36 bp	30 bp
Lactobacillus paracasei 8700:2	ABQV00000000	GTCTCAGGTARATGYCGAAT CAATCAGTTCAAGAGC	Lsal1	21	36 bp	30 bp
Lactobacillus rhamnosus GG	FM179322	GTCTCAGGTAGATGTCAGAT CAATCAGTTCAAGAGC	Lsal1	25	36 bp	30 bp
Lactobacillus ruminis ATCC25644	ACGS00000000	GTTTCAGCTGGATGTCATAT CAATGATGTTATGAAC	Lsal1	26	36 bp	30 bp (30–33)
Lactobacillus salivarius subsp. *salivarius* UCC118	CP000233	GTTTCAGAAGTATGTTAAAT CAATAAGGTTAAGACC	Lsal1	30	36 bp	30 bp

Pediococcus acidilactici 7_4	ACXB00000000	GTTTCAGAAGGATGTTAAAT CAATAAGGTTAAGATC	Lsal1	19	36 bp	30 bp
Enterococcus faecalis Fly1	ACAR00000000	GTTTTTGTACTCTCAATAAT TTCTTATCAGTAAAAC	Sthe1	9	36 bp	30 bp (30–31)
Streptococcus gallolyticus UCN34	FN597254	GTTTTTGTACTCTCAAGAT TTAAGTAACCGTAAAAC	Sthe1	16	36 bp	30 bp (29–30)
Streptococcus gordonii Challis substrain CH1	CP000725	GTTTTTGTACTCTCAAGAT TTAAGTAACTGTACAAC	Sthe1	27	36 bp	30 bp (29–31)
Streptococcus infantarius subsp. *infantarius* ATCC BAA-102	ABJK00000000	GTTTTTGTACTCTCAAGAT TTAAGTAACCGTAAAAC	Sthe1	31	36 bp	30 bp (29–30)
Streptococcus oralis ATCC35037	ADMV00000000	GTTTTTRTAGTCTCCAAAAT TGTGACCGATAAAAC	Sthe1	8	35 bp	30 bp
Streptococcus suis 89/1591	AAFA00000000	RTTTTTGTACTCTCAAGAT TTAAGTAACAGTAAAAC	Sthe1	1+7	36 bp	30 bp (29–30)
Streptococcus thermophilus DGCC7710	Draft genome	GTTTTTGTACTCTCAAGAT TTAAGTAACTGTACAAC	Sthe1	33	36 bp	30 bp (29–30)
Streptococcus vestibularis JIM8230	DQ072994	GTTTTTGTACTCTCAAGAT TTAAGTAACTGTACAAC	Sthe1	12	36 bp	30 bp (29–31)
Lactobacillus ruminis ATCC25644	ACGS00000000	GATGATTACCTATCTCCGAG TGAAGGAGACGAAAAC	Sthe2	7	36 bp	40 bp (36–42)
Lactobacillus salivarius ATCC11741	ACGT00000000	RRTAAGAACATATCTCCGAA TRAAGGAGACGAAAAC	Sthe2	12	36 bp	38 bp (35–41)
Streptococcus sanguinis SK36	CP000387	DHDRNHVHYCHNDDHNHBY YBBVDRGRKDVSDRRAMN	Sthe2	43	37 bp	36 bp (32–39)
Streptococcus thermophilus DGCC7710	Draft genome	GATATAAACCTAATTACCT CGAGARGGGACGGAAAC	Sthe2	3	36 bp	39 bp (36–40)
Enterococcus faecalis T1	ACAD00000000	GTTTTAGAGTCATGTTGT TTAGAATGGTACCAAAAC	Sthe3	15	36 bp	30 bp
Enterococcus faecium 1,231,408	ACBB00000000	GTTTTAGAGCTATGCTGAT TTGAATGCTTCCAAAAC	Sthe3	11	36 bp	30 bp

(continued)

Table 19.1 (continued)

Strain	GenBank accession	Repeat sequence (consensus)	CRISPR family	Repeat number	Repeat size	Spacer size (min–max)
Streptococcus agalactiae 2603 V/R	AE009948	GTTTTAGAGCTGTGCTGT TTCGAATGGTTCCAAAAC	Sthe3	26	36 bp	30 bp (29–31)
Streptococcus dysgalactiae subsp. *equisimilis* GGS_124	AP010935	GTTTTAGAGCTATGTTGT TTTGAATGGTCCCAAAAC	Sthe3	19	36 bp	30 bp (30–32)
Streptococcus equi subsp. *zooepidemicus* MGCS10565	CP001129	GTTTTAGAGTTATGCTGT WTTGAATGGTTCCAAAAC	Sthe3	18	36 bp	30 bp
Streptococcus gallolyticus UCN34	FN597254	GTTTTAGAGCTGTGCTGT TTCGAATGGTTCCAAAAC	Sthe3	13	36 bp	30 bp
Streptococcus mutans UA159	AE014133	GTTTTAGAGCTGTGTTGT TTCGAATGGTTCCAAAAC	Sthe3	7	36 bp	30 bp
Streptococcus mutans NN2025	AP010655	GTTTTAGAGCTGTGTTGT TTCGAATGGTTCCAAAAC	Sthe3	70	36 bp	30 bp (30–31)
Streptococcus pyogenes SF370 M1 GAS	AE004092	GTTTTAGAGCTATGCTGT TTTGAATGGTCCCAAAAC	Sthe3	7	36 bp	30 bp
Streptococcus thermophilus DGCC7710	Draft genome	GTTTTAGAGCTGTGTTGT TTCGAATGGTTCCAAAAC	Sthe3	13	36 bp	30 bp (30–32)

Repeat sequences as categorized by families, according to Horvath et al. (2009)

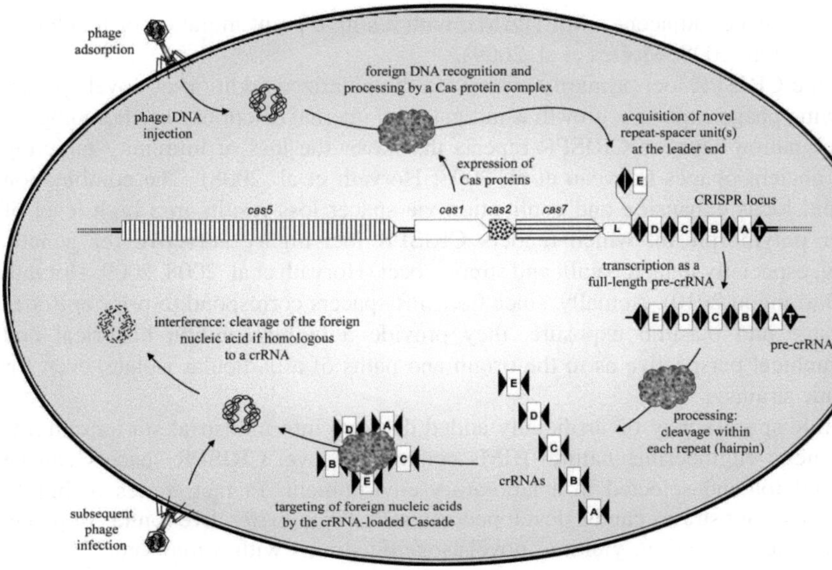

Fig. 19.2 The CRISPR/Cas bacterial immune system. *Top*, immunization step: following phage challenge, *cas* gene products catalyze the acquisition of novel repeat-spacer units within the CRISPR locus, with spacers that are derived from the genome of the invading phage. *Bottom*, interference step: the CRISPR locus is transcribed into a full-length RNA that is processed by Cas proteins into short CRISPR RNAs (crRNAs) that subsequently interfere with the invading nucleic acid bearing an identical sequence

by two partial repeats (Hale et al. 2008, 2009; Brouns et al. 2008; Lillestøl et al. 2009). The crRNAs subsequently and specifically guide the Cas interference machinery toward foreign nucleic acid molecules that match its sequence, ultimately leading to the degradation of invading elements (Hale et al. 2009) (Fig. 19.2).

The polymorphic nature of CRISPR loci, notably that of the Cas machinery, allows them to selectively target either DNA (Marraffini and Sontheimer 2008) or RNA (Hale et al. 2009). While some of the mechanistic idiosyncrasies of the CRISPR system are reminiscent of RNAi, there are several fundamental differences, including the target, the mechanism of action, and the enzymatic machinery involved. Actually, the sequence-specific and adaptive nature of CRISPR/Cas systems shares similarities with the vertebrate adaptive immune system, although CRISPR spacers are DNA-encoded and can be inherited by their progeny.

The CRISPR/Cas system can provide high levels of phage resistance; however, a relatively small proportion of phages (~10^{-6}) retain the ability to infect the "immunized" host (Barrangou et al. 2007; Deveau et al. 2008). These viral particles have specifically mutated either the proto-spacer (sequence within the invading nucleic acid that matches a CRISPR spacer) or the highly conserved adjacent motif, the proto-spacer adjacent motif (PAM), with a single point mutation or a deletion (Deveau et al. 2008; Mojica et al. 2009).

While CRISPR loci primarily evolve via the polarized addition of novel spacers following phage exposure, growth is mitigated by internal deletions via a homologous recombination between CRISPR repeats that favor the loss of immunity targeting more ancient phages (Deveau et al. 2008; Horvath et al. 2008). The combination of rapid locus expansion and contraction via spacer loss results in a high level of spacer polymorphism, which renders CRISPR loci highly attractive for genetic typing, especially in lactobacilli and streptococci (Horvath et al. 2008, 2009; Horvath and Barrangou 2010). Actually, since CRISPR spacers correspond to prior episodes of phage and plasmid exposure, they provide a high-resolution historical and geographical perspective as to the origin and paths of a particular isolate, even for isogenic strains.

While spacers may be artificially added de novo into industrial starter cultures via genetic engineering, natural BIMs containing novel CRISPR spacers can be screened for and selected in a laboratory environment. In fact, series of highly phage-resistant strains can be developed naturally through iterative rounds of phage exposure and selection, yielding novel isogenic strains with a multitude of novel spacers (derived from industrially relevant phage cocktails) that confer broad and deep phage resistance while retaining critical functional traits (Barrangou et al. 2008).

19.3 Engineered Phage Defense Strategies

In addition to these multiple layers of native defense mechanisms, many LAB have been complemented in vitro, with various levels of success, with engineered phage-resistance systems, essentially derived from knowledge on phage genomic sequences (Forde and Fitzgerald 1999; McGrath et al. 2002b; Sturino and Klaenhammer 2004a, 2006).

19.3.1 Phage-Encoded Resistance

Phage-encoded resistance (Per) corresponds to resistance mechanisms that are derived, after cloning into appropriate plasmid vectors, from phage genomic fragments, notably the phage origin (*ori*) of replication (Hill et al. 1990). Cloned LAB phage *ori* have been shown to efficiently compete in *trans* with normal phage replication, especially when they are delivered on high-copy-number vectors (O'Sullivan et al. 1993; Foley et al. 1998; McGrath et al. 1999, 2001; Moscoso and Suárez 2000; Stanley et al. 2000). High levels of resistance against multiple lactococcal phages were also obtained by the cloning and overexpression of a disrupted *c*I repressor gene (Durmaz et al. 2002).

19.3.2 Antisense RNA

Antisense RNAs may be artificially produced by cloning a complete or partial phage gene in the antisense orientation, downstream of a functional promoter (Kim et al. 1992). Following infection, immediately after the transcription of viral genes, the sense target mRNA and antisense RNA form double-stranded molecules that interfere with further phage gene expression and phage proliferation. Such constructions were introduced into various *L. lactis* and *S. thermophilus* hosts, leading to moderate phage insensitivity (Kim et al. 1992; Walker and Klaenhammer 2000; McGrath et al. 2001; Sturino and Klaenhammer 2002b; Sturino and Klaenhammer 2004b).

19.3.3 Phage-Triggered Suicide Systems

Phage-triggered suicide systems can be considered genetically engineered alternatives of Abi systems, where phage infection leads to the host to produce a lethal protein. This kind of phage-induced apoptosis could be obtained in *S. thermophilus* by placing a restriction endonuclease gene under the control of a stringent, phage-inducible promoter into a high-copy-number vector (Djordjevic et al. 1997; Djordjevic and Klaenhammer 1997).

19.3.4 Subunit Poisoning

Within the infected host, phage multimeric proteins can be "poisoned" by the overexpression in *trans* of mutant protein subunits (Sturino and Klaenhammer 2004a, 2006, 2007). To do so, after the identification of critical amino acid residues of the primase protein, site-specific mutations were introduced in vitro into the primase gene of *S. thermophilus* phage κ3. The resulting fragment was then cloned into a high-copy-number vector, which, when introduced into the *S. thermophilus* strain NCK1125, conferred a strong inhibition of phage proliferation (Sturino and Klaenhammer 2007).

19.3.5 Host Factor Elimination

Phage resistance could also be obtained in two LAB species by the suppression of essential host-encoded factors (Sturino and Klaenhammer 2006). The inactivation of bacterial genes leading to phage insensitivity could be achieved either randomly through plasmid-based insertional mutagenesis (Lucchini et al. 2000) or on purpose by the genetic engineering of specific targets that are essential for phage proliferation, such as the disruption of the *thyA* gene in *L. lactis* (Pedersen et al. 2002).

19.4 Industrial Applications and Perspectives

Intuitively, the concurrent presence of multiple phage-resistance systems should provide bacteria used as starter cultures with increased fitness and robustness. Actually, the ability of phages to develop countermeasures and escape resistance systems such as R-M, Abi, and CRISPR (Labrie and Moineau 2007; Deveau et al. 2008; Nechaev and Severinov 2008) by mutations and recombinations requires the stacking of multiple phage-resistance mechanisms for industrial implementation (Coffey and Ross 2002). However, these mutations (especially nonsynonymous mutations, premature stop codons, and deletions) likely have a significant impact on the amino acid sequence and result in a fitness loss for the phage population. Nevertheless, their ability to genetically evolve allows them to persist in industrial settings over time (Labrie and Moineau 2007; Rousseau and Moineau 2009). The presence of Abi and R-M systems on plasmids has allowed the development of rotation schemes based on strains with complementary phage-resistance systems that can be distributed and constructed by the natural transfer of heterologous plasmid DNA through conjugation (Klaenhammer and Sanosky 1985; Sing and Klaenhammer 1990; O'Sullivan et al. 1998). Over time, and by design, strains used in rotation schemes may contain and rely on increasingly complex and advanced combinations of Abi, R-M, and CRISPR systems, although the intrinsic antiplasmid activity of CRISPR/Cas systems may turn out to be incompatible with other plasmid-based defense systems. However, the availability and activity of these various systems vary widely between genera and within species. Accordingly, manufacturers have to leverage the most desirable, synergistic, and effective systems and selectively orchestrate and combine the optimal blend of phage-resistance mechanisms for function consistency and durability. Strategically, the combination and development of these phage-resistance systems will allow us to direct the selective pressure on viral predators, while increasing the chances of survival of the bacterial preys.

The inheritable nature of CRISPR-encoded phage immunity provides the potential for perennial use of industrial microbes. In addition to developing phage resistance, many intrinsic aspects of CRISPR-based immunity have provided avenues of industrial applications, including exploiting the hypervariability for typing purposes, natural genetic tagging of proprietary strains, and engineering of resistance against undesirable foreign genetic elements. Spacer hypervariability has been widely used for typing purposes in a variety of organisms (Pourcel et al. 2005; Vergnaud et al. 2007; Horvath et al. 2008; Cui et al. 2008; Sorek et al. 2008). Actually, exploiting the CRISPR/Cas systems to impede the transfer, and preclude the dissemination, of undesirable genetic elements such as antibiotic-resistance markers has great potential.

Although significant scientific and technological progress has been made in the last few years in our understanding of the host–phage interaction, many efforts are still being applied toward developing and implementing solutions that are applicable and cost-efficient in industrial settings.

Acknowledgments We would like to acknowledge our colleagues and collaborators Patrick Boyaval, Christophe Fremaux, Dennis Romero, Anne-Claire Coûté-Monvoisin, Hélène Deveau, Josiane Garneau, Jessica Labonté, Manuela Villion, and Egon Bech Hansen for their support and many scientific contributions. We would also like to specifically thank Sylvain Moineau, Virgis Siksnys, and Jill Banfield for their insights and expertise on CRISPR.

References

Alatossava T, Forsman P, Ritzenthaler P (1995) Genome homology and superinfection immunity between temperate and virulent *Lactobacillus delbrueckii* bacteriophages. Arch Virol 140:2261–2268

Anba J, Bidnenko E, Hillier A, Ehrlich D, Chopin MC (1995) Characterization of the lactococcal *abiD1* gene coding for phage abortive infection. J Bacteriol 177:3818–3823

Andersson AF, Banfield JF (2008) Virus population dynamics and acquired virus resistance in natural microbial communities. Science 320:1047–1050

Barrangou R, Fremaux C, Deveau H, Richards M, Boyaval P, Moineau S, Romero DA, Horvath P (2007) CRISPR provides acquired resistance against viruses in prokaryotes. Science 315:1709–1712

Barrangou R, Fremaux C, Horvath P, Romero D, Boyaval P (2008) Cultures with improved phage resistance. Patent application WO 2008/108989, 12 Sept 2008

Blumenthal RM, Cheng X (2002) Restriction-modification systems. In: Yasbin RE and Streips UN (Eds.), Modern microbial genetics, 2nd ed. Wiley, New York, pp. 177–225

Bolotin A, Quinquis B, Sorokin A, Ehrlich SD (2005) Clustered regularly interspaced short palindrome repeats (CRISPRs) have spacers of extrachromosomal origin. Microbiology 151:2551–2561

Breitbart M, Rohwer F (2005) Here a virus, there a virus, everywhere the same virus? Trends Microbiol 13:278–284

Brouns SJ, Jore MM, Lundgren M, Westra ER, Slijkhuis RJ, Snijders AP, Dickman MJ, Makarova KS, Koonin EV, van der Oost J (2008) Small CRISPR RNAs guide antiviral defense in prokaryotes. Science 321:960–964

Brüssow H (2001) Phages of dairy bacteria. Annu Rev Microbiol 55:283–303

Bruttin A, Desière F, Lucchini S, Foley S, Brüssow H (1997) Characterization of the lysogeny DNA module from the temperate *Streptococcus thermophilus* bacteriophage ΦSfi21. Virology 233:136–148

Burrus V, Bontemps C, Decaris B, Guédon G (2001) Characterization of a novel type II restriction-modification system, *Sth*368I, encoded by the integrative element ICE*St1* of *Streptococcus thermophilus* CNRZ368. Appl Environ Microbiol 67:1522–1528

Chopin M-C, Chopin A, Bidnenko E (2005) Phage abortive infection in lactococci: variations on a theme. Curr Opin Microbiol 8:473–479

Cluzel P-J, Chopin A, Ehrlich SD, Chopin MC (1991) Phage abortive infection mechanism from *Lactococcus lactis* subsp. *lactis*, expression of which is mediated by an Iso-IS*S1* element. Appl Environ Microbiol 57:3547–3551

Coffey A, Ross P (2002) Bacteriophage-resistance systems in dairy starter strains: molecular analysis to application. Antonie van Leeuwenhoek 82:303–321

Cui Y, Li Y, Gorgé O, Platonov ME, Yan Y, Guo Z, Pourcel C, Dentovskaya SV, Balakhonov SV, Wang X, Song Y, Anisimov AP, Vergnaud G, Yang R (2008) Insight into microevolution of *Yersinia pestis* by clustered regularly interspaced short palindromic repeats. PLoS One 3:e2652

Daly C, Fitzgerald GF, Davis R (1996) Biotechnology of lactic acid bacteria with special reference to bacteriophage resistance. Antonie van Leeuwenhoek 70:99–110

Deveau H, Labrie SJ, Chopin MC, Moineau S (2006) Biodiversity and classification of lactococcal phages. Appl Environ Microbiol 72:4338–4346

Deveau H, Barrangou R, Garneau JE, Labonté J, Fremaux C, Boyaval P, Romero DA, Horvath P, Moineau S (2008) Phage response to CRISPR-encoded resistance in *Streptococcus thermophilus*. J Bacteriol 190:1390–1400

Djordjevic GM, Klaenhammer TR (1997) Bacteriophage-triggered defense systems: phage adaptation and design improvements. Appl Environ Microbiol 63:4370–4376

Djordjevic GM, O'Sullivan DJ, Walker SA, Conkling MA, Klaenhammer TR (1997) A triggered-suicide system designed as a defense against bacteriophages. J Bacteriol 179:6741–6748

Dupont K, Janzen T, Vogensen FK, Josephsen J, Stuer-Lauridsen B (2004) Identification of *Lactococcus lactis* genes required for bacteriophage adsorption. Appl Environ Microbiol 70:5825–5832

Durmaz E, Klaenhammer TR (1995) A starter culture rotation strategy incorporating paired restriction/modification and abortive infection bacteriophage defenses in a single *Lactococcus lactis* strain. Appl Environ Microbiol 61:1266–1273

Durmaz E, Klaenhammer TR (2007) Abortive phage resistance mechanism AbiZ speeds the lysis clock to cause premature lysis of phage-infected *Lactococcus lactis*. Appl Environ Microbiol 189:1417–1425

Durmaz E, Madsen SM, Israelsen H, Klaenhammer TR (2002) *Lactococcus lactis* lytic bacteriophages of the P335 group are inhibited by overexpression of a truncated CI repressor. J Bacteriol 184:6532–6544

Foley S, Lucchini S, Zwahlen MC, Brüssow H (1998) A short noncoding viral DNA element showing characteristics of a replication origin confers bacteriophage resistance to *Streptococcus thermophilus*. Virology 250:377–387

Forde A, Fitzgerald GF (1999) Bacteriophage defence systems in lactic acid bacteria. Antonie van Leeuwenhoek 76:89–113

Forde A, Daly C, Fitzgerald GF (1999) Identification of four phage resistance plasmids from *Lactococcus lactis* ssp. *cremoris* H2O. Appl Environ Microbiol 65:1540–1547

Fortier LC, Bouchard JD, Moineau S (2005) Expression and site-directed mutagenesis of the lactococcal abortive phage infection protein AbiK. J Bacteriol 187:3721–3730

Garvey P, Fitzgerald GF, Hill C (1995) Cloning and DNA sequence analysis of two abortive infection phage resistance determinants from the lactococcal plasmid pNP40. Appl Environ Microbiol 61:4321–4328

Garvey P, Hill C, Fitzgerald GF (1996) The lactococcal plasmid pNP40 encodes a third bacteriophage resistance mechanism, one which affects phage DNA penetration. Appl Environ Microbiol 62:676–679

Godde JS, Bickerton A (2006) The repetitive DNA elements called CRISPRs and their associated genes: evidence of horizontal transfer among prokaryotes. J Mol Evol 62:718–729

Guglielmotti DM, Deveau H, Binetti AG, Reinheimer JA, Moineau S, Quiberoni A (2009) Genome analysis of two virulent *Streptococcus thermophilus* phages isolated in Argentina. Int J Food Microbiol 136:101–109

Haft DH, Selengut J, Mongodin EF, Nelson KE (2005) A guild of 45 CRISPR-associated (Cas) protein families and multiple CRISPR/Cas subtypes exist in prokaryotic genomes. PLoS Comput Biol 1:e60

Hale C, Kleppe K, Terns RM, Terns MP (2008) Prokaryotic silencing (psi)RNAs in *Pyrococcus furiosus*. RNA 14:2572–2579

Hale CR, Zhao P, Olson S, Duff MO, Graveley BR, Wells L, Terns RM, Terns MP (2009) RNA-guided RNA cleavage by a CRISPR RNA-Cas protein complex. Cell 139:945–956

Heidelberg JF, Nelson WC, Schoenfeld T, Bhaya D (2009) Germ warfare in a microbial mat community: CRISPRs provide insights into the co-evolution of host and viral genomes. PLoS One 4:e4169

Held NL, Whitaker RJ (2009) Viral biogeography revealed by signatures in *Sulfolobus islandicus* genomes. Environ Microbiol 11:457–466

Hill C, Pierce K, Klaenhammer TR (1989) The conjugative plasmid pTR2030 encodes two bacteriophage defense mechanisms in lactococci, restriction modification (R^+/M^+) and abortive infection (Hsp^+). Appl Environ Microbiol 55:2416–2419

Hill C, Miller LA, Klaenhammer TR (1990) Cloning, expression, and sequence determination of a bacteriophage fragment encoding bacteriophage resistance in *Lactococcus lactis*. J Bacteriol 172:6419–6426

Horvath P, Barrangou R (2010) CRISPR/Cas, the immune system of Bacteria and Archaea. Science 327:167–170

Horvath P, Romero DA, Coûté-Monvoisin AC, Richards M, Deveau H, Moineau S, Boyaval P, Fremaux C, Barrangou R (2008) Diversity, activity, and evolution of CRISPR loci in *Streptococcus thermophilus*. J Bacteriol 190:1401–1412

Horvath P, Coûté-Monvoisin AC, Romero DA, Boyaval P, Fremaux C, Barrangou R (2009) Comparative analysis of CRISPR loci in lactic acid bacteria genomes. Int J Food Microbiol 131:62–70

Ishino Y, Shinagawa H, Makino K, Amemura M, Nakata A (1987) Nucleotide sequence of the *iap* gene, responsible for alkaline phosphatase isozyme conversion in *Escherichia coli*, and identification of the gene product. J Bacteriol 169:5429–5433

Jansen R, van Embden JD, Gaastra W, Schouls LM (2002) Identification of genes that are associated with DNA repeats in prokaryotes. Mol Microbiol 43:1565–1575

Josephsen J, Neve H (1998) Bacteriophages and lactic acid bacteria. In: Salminen S, von Wright A (Eds.), Lactic acid bacteria: microbiology and functional aspects, 2nd ed. Marcel Dekker, New York, pp. 385–436

Josephsen J, Vogensen FK (1989) Identification of three different plasmid-encoded restriction/modification systems in *Streptococcus lactis* subsp. *cremoris* W56. FEMS Microbiol Lett 59:161–166

Kim JH, Kim SG, Chung DK, Bor YC, Batt CA (1992) Use of antisense RNA to confer bacteriophage resistance in dairy starter cultures. J Ind Microbiol 10:71–78

Klaenhammer TR, Sanosky RB (1985) Conjugal transfer from *Streptococcus lactis* ME2 of plasmids encoding phage resistance, nisin resistance and lactose-fermenting ability: evidence for a high-frequency conjugative plasmid responsible for abortive infection of virulent bacteriophage. J General Microbiol 131:1531–1541

Kunin V, He S, Warnecke F, Peterson SB, Martin HG, Haynes M, Ivanova N, Blackall LL, Breitbart M, Rohwer F, McMahon KD, Hugenholtz P (2008) A bacterial metapopulation adapts locally to phage predation despite global dispersal. Genome Res 18:293–297

Labrie SJ, Moineau S (2007) Abortive infection mechanisms and prophage sequences significantly influence the genetic makeup of emerging lytic lactococcal phages. Appl Environ Microbiol 189:1482–1487

Larbi D, Decaris B, Simonet JM (1992) Different bacteriophage resistance mechanisms in *Streptococcus salivarius* subsp. *thermophilus*. J Dairy Res 59:349–357

Lévesque C, Duplessis M, Labonté J, Labrie S, Fremaux C, Tremblay D, Moineau S (2005) Genomics organization and molecular analysis of virulent bacteriophage 2972 infecting an exopolysaccharide-producing *Streptococcus thermophilus* strain. Appl Environ Microbiol 71:4057–4068

Lillestøl RK, Redder P, Garrett RA, Brügger K (2006) A putative viral defence mechanism in archaeal cells. Archaea 2:59–72

Lillestøl RK, Shah SA, Brügger K, Redder P, Phan H, Christiansen J, Garrett RA (2009) CRISPR families of the crenarchaeal genus *Sulfolobus*: bidirectional transcription and dynamic properties. Mol Microbiol 72:259–272

Lucchini S, Sidoti J, Brüssow H (2000) Broad-range bacteriophage resistance in *Streptococcus thermophilus* by insertional mutagenesis. Virology 275:267–277

Lucey M, Daly C, Fitzgerald GF (1992) Cell surface characteristics of *Lactococcus lactis* harbouring pCI528, a 46 kb plasmid encoding inhibition of bacteriophage adsorption. J Gen Microbiol 138:2137–2143

Madera C, García P, Janzen T, Rodríguez A, Suárez JE (2003) Characterisation of technologically proficient wild *Lactococcus lactis* strains resistant to phage infection. Int J Food Microbiol 86:213–222

Mahony J, McGrath S, Fitzgerald GF, van Sinderen D (2008) Identification and characterization of lactococcal-prophage-carried superinfection exclusion genes. Appl Environ Microbiol 74:6206–6215

Makarova KS, Aravind L, Grishin NV, Rogozin IB, Koonin EV (2002) A DNA repair system specific for thermophilic Archaea and bacteria predicted by genomic context analysis. Nucleic Acids Res 30:482–496

Makarova K, Slesarev A, Wolf Y, Sorokin A, Mirkin B, Koonin E, Pavlov A, Pavlova N, Karamychev V, Polouchine N, Shakhova V, Grigoriev I, Lou Y, Rohksar D, Lucas S, Huang K, Goodstein DM, Hawkins T, Plengvidhya V, Welker D, Hughes J, Goh Y, Benson A, Baldwin K, Lee JH, Díaz-Muñiz I, Dosti B, Smeianov V, Wechter W, Barabote R, Lorca G, Altermann E, Barrangou R, Ganesan B, Xie Y, Rawsthorne H, Tamir D, Parker C, Breidt F, Broadbent J, Hutkins R, O'Sullivan D, Steele J, Unlu G, Saier M, Klaenhammer T, Richardson P, Kozyavkin S, Weimer B, Mills D (2006a) Comparative genomics of the lactic acid bacteria. Proc Natl Acad Sci USA 103:15611–15616

Makarova KS, Grishin NV, Shabalina SA, Wolf YI, Koonin EV (2006b) A putative RNA-interference-based immune system in prokaryotes: computational analysis of the predicted enzymatic machinery, functional analogies with eukaryotic RNAi, and hypothetical mechanisms of action. Biol Direct 1:7

Marraffini LA, Sontheimer EJ (2008) CRISPR interference limits horizontal gene transfer in staphylococci by targeting DNA. Science 322:1843–1845

McGrath S, Seegers JF, Fitzgerald GF, van Sinderen D (1999) Molecular characterization of a phage-encoded resistance system in *Lactococcus lactis*. Appl Environ Microbiol 65:1891–1899

McGrath S, Fitzgerald GF, van Sinderen D (2001) Improvement and optimization of two engineered phage resistance mechanisms in *Lactococcus lactis*. Appl Environ Microbiol 67:608–616

McGrath S, Fitzgerald GF, van Sinderen D (2002a) Identification and characterization of phage-resistance genes in temperate lactococcal bacteriophages. Mol Microbiol 43:509–520

McGrath S, van Sinderen D, Fitzgerald GF (2002b) Bacteriophage-derived genetic tools for use in lactic acid bacteria. Int Dairy J 12:3–15

Moineau S (1999) Applications of phage resistance in lactic acid bacteria. Antonie van Leeuwenhoek 76:377–382

Moineau S, Walker SA, Holler BJ, Vedamuthu ER, Vandenbergh PA (1995) Expression of a *Lactococcus lactis* phage resistance mechanism by *Streptococcus thermophilus*. Appl Environ Microbiol 61:2461–2466

Mojica FJ, Ferrer C, Juez G, Rodríguez-Valera F (1995) Long stretches of short tandem repeats are present in the largest replicons of the Archaea *Haloferax mediterranei* and *Haloferax volcanii* and could be involved in replicon partitioning. Mol Microbiol 17:85–93

Mojica FJ, Díez-Villaseñor C, García-Martínez J, Soria E (2005) Intervening sequences of regularly spaced prokaryotic repeats derive from foreign genetic elements. J Mol Evol 60:174–182

Mojica FJ, Díez-Villaseñor C, García-Martínez J, Almendros C (2009) Short motif sequences determine the targets of the prokaryotic CRISPR defence system. Microbiology 155:733–740

Monteville MR, Ardestani B, Geller BL (1994) Lactococcal bacteriophages require a host cell wall carbohydrate and a plasma membrane protein for adsorption and ejection of DNA. Appl Environ Microbiol 60:3204–3211

Moscoso M, Suárez JE (2000) Characterization of the DNA replication module of bacteriophage A2 and use of its origin of replication as a defense against infection during milk fermentation by *Lactobacillus casei*. Virology 273:101–111

Nechaev S, Severinov K (2008) The elusive object of desire – interactions of bacteriophages and their hosts. Curr Opin Microbiol 11:186–193

Nyengaard N, Vogensen FK, Josephsen J (1995) Restriction-modification systems in *Lactococcus lactis*. Gene 157:13–18

O'Driscoll J, Glynn F, Cahalane O, O'Connell-Motherway M, Fitzgerald GF, Van Sinderen D (2004) Lactococcal plasmid pNP40 encodes a novel, temperature-sensitive restriction-modification system. Appl Environ Microbiol 70:5546–5556

O'Sullivan DJ, Hill C, Klaenhammer TR (1993) Effect of increasing the copy number of bacteriophage origins of replication, in *trans*, on incoming-phage proliferation. Appl Environ Microbiol 59:2449–2456

O'Sullivan D, Coffey A, Fitzgerald GF, Hill C, Ross RP (1998) Design of a phage-insensitive lactococcal dairy starter via sequential transfer of naturally occurring conjugative plasmids. Appl Environ Microbiol 64:4618–4622

Pedersen MB, Jensen PR, Janzen T, Nilsson D (2002) Bacteriophage resistance of a Δ*thyA* mutant of *Lactococcus lactis* blocked in DNA replication. Appl Environ Microbiol 68:3010–3023

Pourcel C, Salvignol G, Vergnaud G (2005) CRISPR elements in *Yersinia pestis* acquire new repeats by preferential uptake of bacteriophage DNA, and provide additional tools for evolutionary studies. Microbiology 151:653–663

Rousseau GM, Moineau S (2009) Evolution of *Lactococcus lactis* phages within a cheese factory. Appl Environ Microbiol 75:5336–5344

Sakamoto K, Agari Y, Agari K, Yokoyama S, Kuramitsu S, Shinkai A (2009) X-ray crystal structure of a CRISPR-associated RAMP superfamily protein, Cmr5, from *Thermus thermophilus* HB8. Proteins 75:528–532

Seegers JF, van Sinderen D, Fitzgerald GF (2000) Molecular characterization of the lactococcal plasmid pCIS3: natural stacking of specificity subunits of a type I restriction/modification system in a single lactococcal strain. Microbiology 146:435–443

Shah SA, Hansen NR, Garrett RA (2009) Distribution of CRISPR spacer matches in viruses and plasmids of crenarchaeal acidothermophiles and implications for their inhibitory mechanism. Biochem Soc Trans 37:23–28

Sijtsma L, Wouters JT, Hellingwerf KJ (1990) Isolation and characterization of lipoteichoic acid, a cell envelope component involved in preventing phage adsorption, from *Lactococcus lactis* subsp. *cremoris* SK110. J Bacteriol 172:7126–7130

Sing WD, Klaenhammer TR (1990) Plasmid-induced abortive infection in lactococci: a review. J Dairy Sci 73:2239–2251

Sorek R, Kunin V, Hugenholtz P (2008) CRISPR-a widespread system that provides acquired resistance against phages in bacteria and archaea. Nat Rev Microbiol 6:181–186

Stanley E, Walsh L, van der Zwet A, Fitzgerald GF, van Sinderen D (2000) Identification of four loci isolated from two *Streptococcus thermophilus* phage genomes responsible for mediating bacteriophage resistance. FEMS Microbiol Lett 182:271–277

Sturino JM, Klaenhammer TR (2002) Expression of antisense RNA targeted against *Streptococcus thermophilus* bacteriophages. Appl Environ Microbiol 68:588–596

Sturino JM, Klaenhammer TR (2004a) Bacteriophage defense systems and strategies for lactic acid bacteria. In: Laskin AI, Bennett JW, Gadd GM (Eds.), Advances in applied microbiology, Vol. 56. Academic, San Diego, pp. 331–378

Sturino JM, Klaenhammer TR (2004b) Antisense RNA targeting of primase interferes with bacteriophage replication in *Streptococcus thermophilus*. Appl Environ Microbiol 70:1735–1743

Sturino JM, Klaenhammer TR (2006) Engineered bacteriophage-defence systems in bioprocessing. Nat Rev Microbiol 4:395–404

Sturino JM, Klaenhammer TR (2007) Inhibition of bacteriophage replication in *Streptococcus thermophilus* by subunit poisoning of primase. Microbiology 153:3295–3302

Suárez V, Zago M, Giraffa G, Reinheimer J, Quiberoni A (2009) Evidence for the presence of restriction/modification systems in *Lactobacillus delbrueckii*. J Dairy Res 76:433–440

Sun X, Göhler A, Heller KJ, Neve H (2006) The *ltp* gene of temperate *Streptococcus thermophilus* phage TP-J34 confers superinfection exclusion to *Streptococcus thermophilus* and *Lactococcus lactis*. Virology 350:146–157

Tangney M, Fitzgerald GF (2002) AbiA, a lactococcal abortive infection mechanism functioning in *Streptococcus thermophilus*. Appl Environ Microbiol 68:6388–6391

Trotter M, Ross RP, Fitzgerald GF, Coffey A (2002) *Lactococcus lactis* DPC5598, a plasmid-free derivative of a commercial starter, provides a valuable alternative host for culture improvement studies. J Appl Microbiol 93:134–143

Tyson GW, Banfield JF (2008) Rapidly evolving CRISPRs implicated in acquired resistance of microorganisms to viruses. Environ Microbiol 10:200–207

van der Ploeg JR (2009) Analysis of CRISPR in *Streptococcus mutans* suggests frequent occurrence of acquired immunity against infection by M102-like bacteriophages. Microbiology 155:1966–1976

Vergnaud G, Li Y, Gorgé O, Cui Y, Song Y, Zhou D, Grissa I, Dentovskaya SV, Platonov ME, Rakin A, Balakhonov SV, Neubauer H, Pourcel C, Anisimov AP, Yang R (2007) Analysis of the three *Yersinia pestis* CRISPR loci provides new tools for phylogenetic studies and possibly for the investigation of ancient DNA. Adv Exp Med Biol 603:327–338

Walker SA, Klaenhammer TR (2000) An explosive antisense RNA strategy for inhibition of a lactococcal bacteriophage. Appl Environ Microbiol 66:310–319

Wiedenheft B, Zhou K, Jinek M, Coyle SM, Ma W, Doudna JA (2009) Structural basis for DNase activity of a conserved protein implicated in CRISPR-mediated genome defense. Structure 17:904–912

Yang JM, Deurraza PJ, Matvienko N, O'Sullivan DJ (2006) Involvement of the LlaKR2I methylase in expression of the AbiR bacteriophage defense system in *Lactococcus lactis* subsp. *lactis* biovar *diacetylactis* KR2. J Bacteriol 188:1920–1928

Chapter 20
Storing Lactic Acid Bacteria: Current Methodologies and Physiological Implications

Chalat Santivarangkna, Ulrich Kulozik, and Petra Foerst

20.1 Introduction

The necessity of maintaining the high viability of lactic acid bacteria (LAB) is increasingly important due to the use of lactic acid starter cultures in concentrated forms and the development of dried probiotic products. Conventionally, cultures of LAB are freshly prepared by the successive propagation of cells from a stock culture until a bulk inoculum with approximately 10^{8-9} viable cells has been obtained. The inoculum will be added to food matrices to get approximately 10^{6-7} viable cells/g of raw material. In this practice, a successful preservation process merely relies on the ability to maintain a sufficient number of cells to continue producing culture with the same characteristics as those of the original culture. The viability for a sufficient number of cells may be as low as 0.1% (Heckly 1961). In contrast, nowadays LAB cultures are increasingly produced by starter culture suppliers in frozen and dried concentrated forms (10^{11-12} cells/g). It is a prerequisite that the concentrated cultures are stable during storage and possess a high enough viability to resume fermentation activities immediately after the direct inoculation to food matrices (Direct Vat Set [DVS]; Direct Vat Inoculation [DVI]). In contrast to starter cultures, the probiotic products rely on specific functions of LAB, such as improving lactose tolerance, reducing cholesterol level, normalizing the intestinal microbial composition, and immunomodulizing. These properties are dose-dependent (e.g., a daily dose of 10^9 CFU) (Knorr 1998). Therefore, a high level of viable cells is required in the dried products.

C. Santivarangkna (✉)
Chair for Food Process Engineering and Dairy Technology, Centre of Life and Food Sciences, Technische Universität München, Weihenstephaner Berg 1, 85354 Freising-Weihenstephan, Germany
e-mail: santivar@wzw.tum.de

E. Tsakalidou and K. Papadimitriou (eds.), *Stress Responses of Lactic Acid Bacteria*, Food Microbiology and Food Safety, DOI 10.1007/978-0-387-92771-8_20,

In this chapter, the storage of LAB in frozen and dried forms – which are the most widespread techniques for the long-term storage of LAB – will be described together with the role of sublethal stress treatments in the preservation of the cells.

20.2 Storing Lactic Acid Bacteria in Frozen Form

20.2.1 Freezing

Freezing converts a liquid cell suspension into a mixture of ice and solute concentrate. It is similar to drying in the sense that it removes water from a solution in the form of ice crystals. The cytoplasm of a cell generally remains unfrozen down to −10 to −15°C. Thus, extracellular ice crystals usually form first during freezing. Consequently, cellular water will migrate outward due to the higher extracellular osmotic pressure. The inactivation mechanisms during freezing depend on the cooling rate (Mazur 1970). If the rate of cooling is very low, cells will lose water rapidly enough to maintain the osmotic equilibrium between their inside and outside, resulting in cell dehydration. If the cooling rate is so high that cells are not able to lose water rapidly enough, the osmotic equilibrium is maintained by frozen water inside the cells, resulting in intracellular ice formation (IIF). It was suggested that a cooling rate of 10°C/min is generally suitable for most living cells (Nakamura 1996). In LAB, the exact optimal cooling rate is species-dependent (Champagne et al. 1991; Sanders et al. 1999). Some studies reported that streptococci generally survive better than lactobacilli (Tsvetkov and Shishkova 1982; Fonseca et al. 2000). Interestingly, *Lactobacillus plantarum* is relatively unaffected by freezing. The viability is quite high over the wide range of the cooling rates (100 and 87% for a cooling rate of 5 and 250°C/min, respectively), and there is no significant change in its viability between a low (5°C/min) and a very high freezing rate (30,000°C/min) (Dumont et al. 2004).

Although theoretically an optimal cooling rate can be determined for each strain of LAB, this approach often takes too much time and is very laborious. In practice, the slow and rapid freezing of concentrated cell suspension is often realized by placing the suspension in a freezer (−30 or −40°C) or a deep freezer (−80°C) and by immersing it in a dry ice–alcohol mixture (−78°C) or liquid nitrogen (−196°C). To avoid IIF, a two-step method was recommended for the preservation in liquid nitrogen. LAB cells should be cooled to a temperature between −20 and −40°C to allow sufficient efflux of water before the subsequent rapid freezing and maintaining of cells in liquid nitrogen (De Valdez 2001). Instead of the immersion in liquid nitrogen, the maintenance of cells in the vapor phase of liquid nitrogen (−130°C) is also used, because the contamination from leaking samples and the container surface is minimized. Nevertheless, the contamination in the vapor phase is still possible and was reported in some studies (Morris 2005; Grout and Morris 2008, 2009).

Commercial frozen cultures are produced by dispensing the cell concentrate in containers such as metal cans and plastic pouches. These starter cultures are rapidly frozen by liquid nitrogen. Another form of frozen cell concentrate is the granule form, which is relatively more convenient for weighing and dispensing. The frozen granules can be produced by distributing the cell suspension through a droplet disc with bores or through a nozzle into liquid nitrogen. Liquid nitrogen may not be completely sterile, but the sterilization would probably be too costly. Furthermore, the large number of revived cells in the concentrated cultures would be expected to suppress the trivial number of contaminants (Keogh 1970). The frozen cell concentrate granules are commonly dispensed in laminated cardboard containers or plastic pouches. In addition to direct freezing in liquid nitrogen, innovative rapid freezing methods, namely, spray-freezing and pressure shift-freezing, were explored by Volkert et al. (2008) in *Lactobacillus rhamnosus*. A high viability (90.5%) of *L. rhamnosus* after storage for a day at −20°C was obtained by spray-freezing 20% skim milk.

In freezing formulations, cryoprotectants are efficiently used to forestall the injuries of cells. For instance, in a study with *Lactobacillus delbrueckii* subsp. *bulgaricus* (Fonseca et al. 2006), the water outflow was limited during slow freezing (5°C/min) due to the high viscosity of the residual unfrozen solution in the presence of glycerol, and IIF was not observed during the rapid freezing (2,500°C/min). Common cryoprotectants used for LAB are glycerol (Thunell et al. 1984; Beal and Corrieu 1994; Beal et al. 2001), sugars such as lactose, sucrose, and trehalose (Carcoba and Rodriguez 2000), and skim milk. Skim milk (10–20% total solid) is a favorite suspending medium/protectant, either used alone or in combination with the previously mentioned protectants (Carcoba and Rodriguez 2000; Juarez et al. 2004; Modesto et al. 2004; Volkert et al. 2008). Growth medium M17 (Kim and Dunn 1997) and nutrients used in growth media such as yeast extract (Baumann and Reinbold 1966) and meat extract (Johannsen 1972; Carcoba and Rodriguez 2000) were also reported for their cryoprotective effects. Dimethyl sulfoxide (DMSO), which is a common cryoprotectant for microorganisms, is used less often with LAB. Moreover, DMSO may be prohibited for applications related to foods because of food regulations and its unpleasant garlic flavor imparted to the starter cultures. It is advisable to allow the protectants to equilibrate the intracellular solutes before freezing. The equilibration time depends on the type of the protectants. This is typically 10–60 min at 0–10°C (Hubalek 2003). The incubation of cells with cryoprotectants at a higher temperature should be avoided. It was found that the incubation of *L. delbrueckii* subsp. *bulgaricus* with glycerol at 42°C for 1 h ascribes to the higher susceptibility to freezing and thawing (Panoff et al. 2000).

20.2.2 Storage and Thawing of Frozen Cells

The viability during frozen storage inversely correlates to the storage temperature. Only a small reduction in the viability of LAB was observed during storage at −70 and −80°C (Foschino et al. 1996; Juarez et al. 2004; Fonseca et al. 2006). Generally, commercial frozen starter cultures can be stored at a temperature lower than −45°C for 12 months. A very low storage temperature (e.g., −80 or −196°C) and rapid thawing are recommended when frozen cells are obtained by a high cooling rate, such as with liquid nitrogen (−196°C). This is due to the fact that the freezing of cells at a high cooling rate to a very low final freezing temperature does not allow sufficient time for ice crystal growth. Subsequently, during storage at an increased temperature, fluctuation of the storage temperature, and thawing, the small ice crystals will reorganize into lethally large crystals (Mazur 1984).

In laboratories where LAB are stored in small cryogenic ampules in liquid nitrogen, rapid thawing can be carried out by plunging it into a water bath at 37–40°C. With moderate agitation, ice will be melted in 40–60 s (Nakamura 1996). For commercial frozen DVS cultures, thawing and inoculation are accomplished simultaneously. The thawing occurs rapidly due to the small pellet size and the comparatively small amount of the cultures employed, typically 500 g of frozen cultures/2,500–10,000 L of milk. The frozen cultures filled in a can may need to be detached from the container by placing it in water containing 25–50 ppm available choline at 20–22°C for 20 min. The complete thawing in the fluid food matrix can be accelerated by agitation (Sandine 1996).

20.3 Storing Lactic Acid Bacteria in Dried Form

20.3.1 Drying

Generally, bacterial cells contain between approximately 70–95% water. On the one hand, the removal of water during drying inactivates cells. On the other hand, LAB should be dried to a water content low enough to slow down metabolic processes and preserve cellular structures and functions during storage. In addition to heat or cryo-injuries due to harsh temperatures from drying processes like spray- and freeze-drying, when cells are dried to low water content, a number of cell components such as DNA or RNA and proteins will be inflicted. However, it is quite clear that the cell membrane is the principal site of inactivation (Brennan et al. 1986; Selmer-Olsen et al. 1999a; Santivarangkna et al. 2007). In membrane phases where the relative water content is less than 0.25 g H_2O/g dry weight, phospholipids are likely to undergo phase transition (Wolfe 1987), for example, from the liquid crystalline (L_α) to gel phase (L_β). Consequently, at a physiological temperature where lipids are usually fluid and molten, the lipids may begin changing to the gel phase during drying. Vice versa, the phase transition is repeated again during rehydration, leading to membrane leakage. A common means to protect cells during drying is mixing protectants to the drying media. Sugars are preferable protectants because of their relatively low prices and chemically innocuous nature. It is believed that protectants protect cells by replacing water, interacting with membrane phospholipids, and hence prevailing during the phase transition (Crowe et al. 1983; Leslie et al. 1995). Effective protectants for LAB are, for example, trehalose (Leslie et al. 1995;

Conrad et al. 2000), sorbitol (Carvalho et al. 2002; Santivarangkna et al. 2006), sucrose (Leslie et al. 1995; Carvalho et al. 2003a), skim milk (Corcoran et al. 2004; Ananta et al. 2005), and gum acacia (Desmond et al. 2002a).

The preservation of LAB in dried forms can be carried out with freeze-drying or lyophilization, which is the established process for both stock cultures and commercial starter cultures. However, due to the high costs of freeze-drying, alternative drying processes are now of increasing interest. The alternative processes have different advantages over freeze-drying. Aside from the lower costs, they have shorter processing time, more reliable process monitoring, enhanced stability at room temperature, and a possible continuous-processing mode. Furthermore, they may allow a better viability for LAB that are sensitive to freezing, such as *L. bulgaricus* (Beal and Corrieu 1994; Fonseca et al. 2000) and *Lactobacillus sanfranciscensis* (De Angelis and Gobbetti 2004).

20.3.1.1 Freeze-Drying

The typical freeze-drying process consists of three steps: freezing, primary drying, and secondary drying. The freezing stages can be conducted with techniques already mentioned. The inactivation of the freeze-dried LAB is mostly attendant on the freezing step (Tsvetkov and Brankova 1983; To and Etzel 1997). In a work of Gehrke (1991), the viability of microorganisms, including *L. plantarum,* decreased only slightly during the drying stage of freeze-drying. However, a drastic decrease occurred again when the moisture content of samples was lower than 4%. The primary drying stage represents most of the freeze-drying time. At this stage, ice crystals are removed by sublimation. In addition to free water, which is removed during primary drying, cells contain about 0.25 g water/g dry weight of bound water, which is not frozen. This bound water is partly removed by desorption during secondary drying. At the end of the secondary drying, the sample should retain a water content that is optimal for storage. This is generally at 1% or less for long-term shelf life (Nakamura 1996). However, the moisture content of 4% was also suggested to be enough in practice (Gardiner et al. 2000). In principle, at low-water-content levels, the suspending medium is in the glassy state, which elicits a high storage stability. The glassy state of cells is likely to occur as well when a high concentration of solutes is accumulated inside. However, it is not clear whether cells per se can vitrify and whether all parts of the cell vitrify or not.

20.3.1.2 Spray-Drying

Spray-drying produces a dry powder by atomizing the liquid at high velocity and directing the spray of droplets (10–200 μL) into a flow of hot air, about 150–200°C. The process requires a very short drying time. The outlet air temperature, or the temperature at which the product leaves the drying chamber, is believed to be the major processing parameter affecting the viability of spray-dried starter cultures. However, the outlet air temperature is not easy to control and depends on many parameters. Studies reported that the lower the outlet air temperature was, the higher was the viability of LAB (Kim and Bhowmik 1990; Bielecka and Majkowska 2000; Desmond et al. 2002a; Ananta et al. 2005). Nevertheless, too low of an outlet air temperature may cause high residual moisture contents that exceed the level required for long-term storage. Reduction in the viability caused by increased outlet air temperature varies with the carrier used. For example, the greatest reduction was observed in bifidobacteria dried in soluble starch, and a lesser reduction was seen in other carriers, such as gelatin gum arabic and skim milk (Lian et al. 2002).

20.3.1.3 Fluidized Bed-Drying

A fluidized bed is a bed of solid particles with a stream of air or gas blowing upward through the particles at a rate high enough to keep them in motion. Like spray-drying, the process is also suitable for large-scale continuous production. Fluidized bed-drying time ranges from 1 min to 2 h. Heat inactivation can be minimized and controlled by using relatively low air temperatures. The use of fluidized bed-dryers may be limited because of the irregular particle sizes and sticky nature of the granulated materials, which can lead to an inhomogeneous bed, agglomerated particles, and a decreased drying rate. The drawback for its use with LAB is that only granulatable materials can be dried. Cells must therefore be entrapped or encapsulated in support materials such as skim milk (Roelans and Taeymans 1990), potato starch (Linders et al. 1997b), alginate (Selmer-Olsen et al. 1999b), and casein (Mille et al. 2004). Notably, an extremely desiccated support material (e.g., $a_w \leq 0.1$) may pose an osmotic shock, leading to a reduction in viability after mixing LAB (Mille et al. 2004).

20.3.1.4 Vacuum-Drying

Under vacuum conditions, moisture can be removed at a low temperature, and the oxidation reactions can be minimized for oxygen-sensitive LAB. The drying time of the process varies from 20 to over 100 h. The vacuum-drying system is conventionally a batch process consisting of a chamber with heated shelves. Trays containing the wet materials are placed on the shelves, and water is removed in a vacuum pump and condensed in a condenser. Studies on the use of vacuum-drying in LAB are rare, although comparable survival rates between LAB dried by controlled low-temperature vacuum-drying and freeze-drying were reported in *Lactobacillus acidophilus* (King and Su 1993) and in *Lactobacillus paracasei* subsp. *paracasei* (Higl et al. 2008).

20.3.1.5 Mixed Drying Systems

The combination of spray- and fluidized bed-drying was studied in *Lactobacillus casei* (Zimmermann 1987) and in *L. plantarum* and *Enterococcus faecium* (Strasser et al. 2009) using lactose and NaCl particles, and powdered cellulose, as carriers, respectively. The concentrated cells were sprayed by a nozzle over carrier materials, which were fluidized in the dryer. The granules with the culture on the surface were dried in a fluidized bed chamber; therefore, the drying can be better controlled. However, the viabilities of LAB from these studies were rather low (*L. casei*, 20%; *L. plantarum*, 16.3%; *E. faecium*, 11.0%). The combination of fluidized bed-drying and freeze-drying was studied in *Streptococcus thermophilus* by Wolff et al. (1990) in a so-called atmospheric pressure freeze-drying process (Wolff et al. 1986). Instead of working under vacuum, this process is based on a decrease in the partial pressure of water vapor by adsorption in a fluidized adsorbent bed. Although the process saves 35% of energy costs, it requires approximately twice the drying time than that of vacuum freeze-drying and still needs improvement to obtain comparable viabilities after drying (33% compared to 55% from conventional freeze-drying).

20.3.2 Storage and Rehydration of Dried Cells

In the dry state, cells are more susceptible to oxygen. The major damage during storage of LAB is ascribed to the lipid oxidation of membranes (Castro et al. 1995, 1996; Teixeira et al. 1996; Andersen et al. 1999). Linoleic ($C_{18:2}$) and linolenic ($C_{18:3}$) acids of freeze-dried LAB were reported as polyunsaturated fatty acids that are sensitive to oxidation during storage (Yao et al. 2008; Coulibaly et al. 2009). It was suggested also that various types of nonenzymatic browning reactions play an important role in the inactivation, and these reactions and the oxidation may indeed be interrelated processes (Kurtmann et al. 2009c). To protect LAB from the oxidation reaction, antioxidants such as ascorbate (Kurtmann et al. 2009a) and glutamate (Sunny-Roberts and Knorr 2009) are often added to drying matrices. Despite its possible role as a prooxidant, ascorbate was reported in some studies (Teixeira et al. 1995b ; Melin et al. 1999) to lead to an improved stability in *L. acidophilus* stored in the presence of 10% (w/w) ascorbate under atmospheric oxygen level (Kurtmann et al. 2009a). Some sugars, such as sorbitol (Linders et al. 1997b; Carvalho et al. 2003b) and mannitol (Efiuvwevwere et al. 1999), were also suggested as effective antioxidants.

Although the stability of cells during storage is believed to be species-dependent, Miyamoto-Shinohara et al. (2008) attempted to determine some tentative correlation between the morphology of LAB and their storage stability. For *Enterococcus*, *Lactococcus*, and *Streptococcus,* species not producing exopolysaccharides (EPS) (*Enterococcus avium*, *Lactococcus lactis*, *E. faecium*, *Streptococcus equi*, and *Enterococcus faecalis*) had higher viabilities than many of the species producing EPS (*Streptococcus mutans*, *Streptococcus gordonii*, *Streptococcus salivarius*, *Streptococcus sanguinis*, and *Streptococcus mitis*). For *Lactobacillus,* species with glycerol teichoic acid (GTA) in both the cell wall and cell membrane (*Lactobacillus helveticus*) and in just the cell membrane (*Lactobacillus fermentum*) had higher viability than those having GTA solely in the cell wall (*Lactobacillus brevis* and *L. delbrueckii*). Furthermore, with regard to antigens in the cell wall, species with ribitol teichoic acid (RTA) (*L. plantarum*) or polysaccharides (*L. casei* and *Lactobacillus salivarius*) had a higher viability than species with GTA.

The stability of starter cultures strongly corresponds to their storage temperatures and moisture content (King et al. 1998; Andersen et al. 1999; Gardiner et al. 2000; Silva et al. 2002; Higl et al. 2007). In general, freeze-dried products have vastly expanded dry surface area and are therefore more susceptible to oxygen and moisture. In contrast, fluidized bed-dried cells have the advantage that the encapsulation provides a microenvironment that protects cells from oxygen and light during storage (Selmer-Olsen et al. 1999b). Principally, dried cells should be kept under low relative humidity (e.g., $a_w \leq 0.1$) and in a package that is moisture-, light-, and air-impermissible to prevent a regain of moisture and an accelerated oxidation reaction. It was shown that an aluminum laminated pouch is superior for the storage of LAB and bifidobacteria compared to glass and polyester (PET) bottles (Wang et al. 2004). A reduced oxygen level is very important when storage of cells at $a_w \leq 0.1$ cannot be achieved. However, the exclusion of oxygen may not assist in overcoming the accelerated reduction in the viability when LAB are stored at too high an $a_{w,}$ that is, >0.32 (Kurtmann et al. 2009a). For stock cultures, the viability of cells dried in ampules is strongly influenced by the degree of vacuum under which the ampules are sealed. Miyamoto-Shinohara et al. (2006) suggested that the viability of cultures in ampules sealed under vacuum <1 Pa survived better than those sealed under 7 Pa.

The storage temperature of dried LAB should be lower than their glass transition temperature (T_g) at the corresponding moisture content (Higl et al. 2007). T_g is the temperature below which a material vitrifies into a so-called amorphous glass. An amorphous glass has a very high viscosity (e.g., 10^{13} Pa s) and exhibits solid-like behavior (Aguilera and Karel 1997). The high viscosity immobilizes cellular constituents, thus inhibiting diffusion and slowing down deleterious reactions or changes in the structures and chemical composition. A storage temperature of 4°C is recommended to be suitable for long-term storage of freeze-dried LAB (Wang et al. 2004; Coulibaly et al. 2009). Commercial freeze-dried starter cultures can be stored at this temperature for longer than 12 months. Solutes with a high T_g can improve the stability of the dried cells by increasing the T_g of the whole drying matrix. However, it is noteworthy that the nature of the solutes is greatly important for storage stability. The viability obtained from freeze-dried *L. acidophilus* stored in the glassy matrix of a reducing sugar lactose was lower than that from the nonglassy matrix of a nonreducing sugar sucrose (Kurtmann et al. 2009b). Recently, there was a report on a possible decrease in the amount of bacteriocin produced by LAB during storage. It was shown that the amount of bacteriocin production in freeze-dried vaginal *Lactobacillus* strains was reduced after storage for 12 months, whereas the production of other antimicrobial substances (lactic acid and hydrogen peroxide) and the autoaggregation property were still retained (Juarez et al. 2009). However, the results contradict those from previous studies with relatively shorter storage times. In studies by Mauriello et al. (1999)

in spray-dried *L. lactis* subsp. *lactis* and *Lactobacillus curvatus*, and by Silva et al. (2002) in spray-dried *Carnobacterium divergens*, *L. salivarius*, and *Lactobacillus sakei*, the bacteriocin production property was retained after storing for 2 and 3 months, respectively.

Studies showed that the recovery of dried cells depends on the rehydration medium, which is species-dependent (Sinha et al. 1974; De Valdez et al. 1985; Zhao and Zhang 2005) and correlates directly to the rehydration temperature. A higher viability is obtained when cells are dried with the warm media (Teixeira et al. 1995a; Mille et al. 2004; Wang et al. 2004). LAB dried with some drying processes may require a long rehydration time. The Ca-alginate gel beads with entrapped *L. helveticus* needed a lag time of almost 3 h to reach the original bead volume during rehydration (Selmer-Olsen et al. 1999b). The gel beads are inclined to retard the fermentation process or cause an uneven distribution of cells in the nonliquid food matrix. Thus, this drying process may be more suitable for time-consuming fermentation processes, such as malolactic fermentation (Clementi and Rossi 1984).

20.4 Physiological Implications

As shown above, LAB have to cope with many harsh conditions and lethal stresses during preservation processes, and their physiological state greatly affects the resistance to such stresses. Generally, exposing the cells to a mild stress triggers the cells' protective mechanisms to subsequent lethal stresses. Therefore, the induction of these stresses and modification or transfer of genes related to them can be used as strategies to improve the tolerance of LAB during the preservation. The influence of sublethal stress treatments on the tolerance of LAB to the preservation processes is summarized in Table 20.1.

20.4.1 Osmotic Stress

It is well known that some organisms confronted with a decreasing water activity accumulate compatible solutes. Compatible solutes, or osmolytes, are small organic compounds that do not interfere with cell functions and are used for osmotic adjustment (Brown 1976) as well as for protection of cells from freezing (De Antoni et al. 1989; Duong et al. 2006), drying (Welsh and Herbert 1999; Desmond et al. 2002b), and high temperature (Silva et al. 2004; Kilimann et al. 2006). There is a direct relationship between the accumulation of the compatible solute betaine in osmotically stressed LAB and the better ability of the cells to survive drying (Kets and De Bont 1994; Kets et al. 1996). It is important that the accumulation of compatible solutes is induced during the growing of cells or before drying and freezing because the accumulation is energy-dependent, and an energy source such as glucose must be available (Hutkins et al. 1987). Moreover, LAB have limited or no possibilities

Table 20.1 Summary on the influence of sublethal stress treatments on the viabilities of lactic acid bacteria subjected to different preservation methods

Lactic acid bacteria	Preservation/challenges	Sublethal stress treatments		Viability (control/stressed)	References
L. acidophilus CRL639	Freezing (−20°C)	Cold	Grown at 25 vs. 37°C	7/77%	Lorca and De Valdez (1998)
L. acidophilus ATCC4356	Freezing (−80°C)	Cold	Grown at 22 vs. 37°C	79.2/89.2%	Baati et al. (2000)
		Cold	22°C, 6 h	75/85%	
L. acidophilus RD758	Freezing (−20°C)	Cold	Grown at 30 vs. 37°C	(100/0 min)[a, b]	Wang et al. (2005)
		Heat	Grown at 42 vs. 37°C	(100/170 min)[a, b]	
		Acid	Grown at pH 5 vs. pH 6	(100/0 min)[a, b]	
L. bulgaricus LBB	Freezing (−20°C)	Cold shock	30°C, 60 min	(10/27%)[b]	de Urraza and De Antoni (1997)
L. bulgaricus CFL1	Freezing (−20°C)	Acid/stationary phase	pH 5.25, 30 min	4.7/42.9%	Streit et al. (2008)
L. bulgaricus CIP101027 T	Freezing-thawing (−20/+37°C, 12 cycles)	Cold	28°C, 24 h	(0.04/0.2)[b, c]	Panoff et al. (2000)
L. bulgaricus NCFB1489	Spray-drying	Heat/log phase	50°C, 30 min	5.87/6.05 log CFU/mL	Teixeira et al. (1995a)
		Heat/stationary phase	50°C, 30 min	7.37/7.14 log CFU/mL	
L. bulgaricus ESB285	Spray-drying	Acid	pH 3.4 vs. pH (6.5)[b]	(6.5/8.0)[b] log CFU/mL	Silva et al. (2005)
L. bulgaricus URL-LB1	Air-drying	Hyperosmotic	0.15 M NaCl	0.05/0.4%	Kets et al. (1996)
L. coryniformis Si3	Freeze-drying	Acid and cold	Grown at pH 4.5, 30°C vs. pH 5.5, 34°C	67/38%	Schoug et al. (2008)
		Heat, cold, acid, high pH	42°C, 26°C, pH 4.5, pH 6.5 vs. pH 5.5, 34°C	67/72, 18, 40, 28%	
L. helveticus LB1	Freezing (−20°C)	Cold	10°C; 2, 5 h	6/(6, 15)[b]%	Kim and Dunn (1997)
L. johnsonii NCK88	Freezing (−20°C)	Heat	55°C, 45 min	26/45%	Walker et al. (1999)
L. paracasei NFBC338	Spray-drying (outlet temperature: 95–100°C, 100–105°C)	Heat	52°C, 15 min	4.3, 0.5/8.0, 9.0%	Desmond et al. (2002b)
		Hyperosmotic	0.5 M NaCl, 30 min	8.27, 1.15/33.46, 18.4%	

L. paraplantarum C7	Freezing-thawing (−70/+25°C, 4 cycles)	Cold	5°C, 2, 4, 6 h	22/35.4, 37.8, 46.8%	Kim et al. (2004)
			10°C, 2, 4, 6 h	22/28.3, 30.3, 34.1%	
L. plantarum P743	Air-drying	Hyperosmotic	1 M NaCl; without and with 2 mM betaine	4.3/11.7 and 26.0%	Kets et al. (1996)
L. rhamnosus HN001	Fluidized bed-drying	Heat/stationary phase	50°C, 10 min	(7.3/1.6)[d]	Prasad et al. (2003)
		Heat/log phase	50°C, 10 min	(7.3/4.2)[d]	
		Hyperosmotic/log phase	0.6 M NaCl	(7.3/2.0)[d]	
L. lactis subsp. *cremoris* M126, M149, M179	Freezing (−20°C)	Cold shock	10°C; 2 and 5 h	15, 26, 1/(15, 30, 1 and 15, 40, 1)[b]%	Kim and Dunn (1997)
L. lactis subsp. *cremoris* LC10-1, LC11-1, LC12-1	Freezing (−20°C)	Cold/log phase	10°C, 2 h	(20, 21, 20/19, 25, 20)[b]%	Kim et al. (1999)
		Stationary phase		(20, 21, 20/45, 65, 20)[b]%	
L. lactis subsp. *cremoris* MM160, MM310	Freezing (−60°C)	Heat	39°C, 25 min	(0.54, 0.63/0.76, 0.72)[e]	Broadbent and Lin (1999)
		Cold	10°C, 1–4 h	(0.54, 0.61/0.60–0.90, 0.81–0.89)[e]	
L. lactis subsp. *cremoris* MM160	Freeze-drying	Heat	42°C, 25 min	(0.24/0.33)[e]	Broadbent and Lin (1999)
L. lactis subsp. *cremoris* MM160, MM310	Freeze-drying	Cold	10°C, 2 h	(0.23, 0.23/0.34, 0.49)[e]	Broadbent and Lin (1999)
		Cold	10°C, 2 h	(0.23, 0.25/0.35, 0.24)[e]	
L. lactis subsp. *diacetylactis* DRC-2	Freezing-thawing (−20°C/ambient, 3 cycles)	Stationary phase		6.1/23.7%	Lee (2004)

(continued)

Table 20.1 (continued)

Lactic acid bacteria	Preservation/challenges	Sublethal stress treatments		Viability (control/stressed)	References
		Cold/log, stationary phase	10°C, 4 h	6.1, 23.7/28.5, 60%	
		Cold/stationary phase	4, 10, 16°C, 4 h	23.7/61.6,60.6, 46.6%	
L. lactis subsp. lactis LL40-1, LL41-1, LL43-1	Freezing (−20°C)	Cold/log phase	10°C, 2 h	(60, 20, 13/93, 60, 45)[b]%	Kim et al. (1999)
		Stationary phase		(60, 20, 13/75, 60, 67)[b]%	
L. lactis subsp. *lactis* MG1363	Freezing-thawing (−20/+30°C, 4 cycles)	Cold/log phase	10°C; 2, 4 h	(0.1/3, 7%)[b]	Wouters et al. (1999a)
		Heat/log phase	42°C, 10 min	(0.1/0.3%)[b]	
		Stationary phase		(0.1/2%)[b]	
L. lactis subsp. *lactis* IL1403	Freezing-thawing (−20/+20°C, 7 cycles)	Chemicals/log phase; β-mercaptoethanol, cadmium chloride, mercury(II) chloride, sodium azide	35.8 mg/mL, 35 μg/mL, 5 μg/mL, 37 μg/mL,	7/(1, 10, 0.7, 7%)[b]	Panoff et al. (1995)
		Cold-/log phase	8°C, 48 h	7/100%[b]	
		Heat-/log phase	42°C, 30 min	7/3%[b]	
L. lactis subsp. *lactis* MM210, FG2	Freeze-drying	Heat	42°C, 25 min	(0.20, 0.25/0.48, 0.63)[e]	Broadbent and Lin (1999)
S. thermophilus CNRZ302	Freezing-thawing (−20°C/+30°C, 4 cycles)	Cold	20°C; 2, 4 h	(0.006/0.5, 5%)[b]	Wouters et al. (1999b)
			10°C; 2, 4 h	(0.006/0.02, 0.06%)[b]	
S. thermophilus SFi39	Freezing-thawing (−20/+30°C, 3 cycles)	Heat	50°C, 30 min	1/5%	Varcamonti et al. (2006)

		Cold	20°C, 4 h	1/25%	
Enterococcus faecium URL-EF1	Air-drying	Hyperosmotic	1 M NaCl; without and with 2 mM betaine	17.1/40.6 and 66.1%	Kets et al. (1996)
O. oeni	Freeze-drying	Acid	ATB medium, pH 4.0, 3.5, 3.2 vs. pH 4.8	67.5/54.6, 90.7, 76.8%	Hua et al. (2009)
Pediococcus pentosaceus PO_2	Freezing (−20°C)	Cold	10°C; 2, 5 h	27/45, (52)[b]%	Kim and Dunn (1997)

[a]Difference in acidification activity (time in minutes necessary for the inoculated milk to reach pH 5.5) before and after freezing
[b]Approximate values read from figure
[c]Relative survival: survival of cells before freezing was considered as 1.0
[d]Log-unit reduction
[e]Fraction of original cell population (before sublethal stress treatment) that could form colonies after freezing and freeze-drying

to synthesize compatible solutes de novo and are dependent on their uptake or the uptake of their precursors from the growth medium. For example, LAB are not able to synthesize proline, ectoine, and betaine, which are the principal compatible solutes identified in many studies (Le et al. 2007). In addition, during preservation processes such as spray-drying, freeze-drying, and freezing, a response of cells to the stress is not likely due to the very short drying time and the very low temperature. It should be noted that in case a washing step was applied after the harvesting of cells, the loss of accumulated compatible solutes may occur due to the osmotic downshift. In *L. casei* subsp. *rhamnosus*, an efflux of 90% of the accumulated proline occurred within 1 min upon osmotic downshock (Jewell and Kashket 1991). Similarly, *L. plantarum* cells rapidly released glycine betaine, proline, and some glutamates (Glaasker et al. 1996).

On the one hand, the osmotic stress promotes the assimilation of compatible solutes into the cells (Glaasker et al. 1996; Kets et al. 1997). On the other hand, too high of a salt concentration (e.g., 1 M NaCl) may be detrimental to cells during drying (Linders et al. 1997a), and a lower NaCl concentration may compromise the growth rate of LAB (Kilstrup et al. 1997; Prasad et al. 2003). In addition to NaCl, hyperosmotic conditions can be induced by sugars (Sunny-Roberts and Knorr 2007, 2008). It was reported that the preincubation of *L. delbrueckii* subsp. *bulgaricus* with lactose, trehalose, and sucrose induces the cryotolerance of the cells. The increased cryotolerance is due to their roles as adaptive agents rather than cryoprotectants (Panoff et al. 2000). Although the osmotic stress induction by sugars seems to be less effective (Carvalho et al. 2003a; Glaasker et al. 1998), sugars have an additional advantage in that they are also known to be dehydration protectants (Santivarangkna et al. 2008). However, it should be noted that under the osmotic stress of a sugar or salt, the intracellular accumulation of compatible solutes and their protective effects are probably different in each LAB strain and depend on the dedicated transport systems of the solutes and the permeability of salts and sugars (Obis et al. 2001; Le et al. 2007).

20.4.2 Cold Stress

Since freezing and freeze-drying are established methods for the preservation of LAB, the cold-shock adaptation has been studied intensively. Cold shock can improve the cryotolerance of LAB. It can be realized by growing cells at a temperature that is lower than the optimal temperature. For instance, an improved cryotolerance was observed in *L. acidophilus* grown at 30°C (Wang et al. 2005), 25°C (Lorca and De Valdez 1998), and 22°C (Baati et al. 2000). However, a drawback is that the low temperature often compromises the growth and yield of LAB (Baati et al. 2000; Kim et al. 2004; Derzelle et al. 2003). Alternatively, a cold shock can be carried out by chilling or preincubating cells at low temperatures after the cultivation. The preincubation of cells before freezing was shown to improve the viability of *L. acidophilus* (Baati et al. 2000), *L. bulgaricus* (De Urraza and De Antoni 1997; Panoff et al. 2000), *L. helveticus, L. lactis* subsp. *lactis* (Wouters et al. 1999a), *L. lactis* subsp. *cremoris* and *Pediococcus pentosaceus* (Kim and Dunn 1997) after freezing, as well as the viability of *L. lactis* subsp. *cremoris* and *L. lactis* subsp. *lactis* after freeze-drying (Broadbent and Lin 1999). Interestingly, in a study with *L. acidophilus*, the viabilities after freezing of the cells cold-shocked by growing at 22°C and by preincubating at the same temperature for 6 h were almost the same. This may imply a possible sublethal treatment of cells without compromising the growth and yield of cells, which is more economical in practice.

Notably, the reduction in viability may occur during cold shock at a very low temperature. The viability of log-phase cells of *L. lactis* subsp. *diacetylactis* was reduced when being held in iced water for an hour (Lee 2004). Furthermore, physiological adaptations at a very low temperature are unlikely due to the inhibited metabolisms and stress proteins' synthesis (Panoff et al. 1995). The induction of a cold-shock gene (*cspA*) of *L. bulgaricus* was delayed at 15°C compared to 25°C (Serror et al. 2003), and the cold-shock proteins of *S. thermophilus* were hardly induced upon cold shock at 10°C (Wouters et al. 1999b). The viability of *L. bulgaricus* cold-shocked at 0 and 7°C was even decreased after frozen storage (De Urraza and De Antoni 1997).

The effectiveness of cold shock also depends on the suspending media. For instance, the cryotolerance in cold-shocked *L. acidophilus* suspended in phosphate buffer was higher than that in phosphate buffer saline (Baati et al. 2000), and the enhanced cryotolerance of cold-shocked *L. bulgaricus* was achieved only in the presence of fermentable sugars (De Urraza and De Antoni 1997). The advantage from cold shock over the unshocked cells was reported to be reduced over the storage time. In a study with *L. lactis*, the viability was improved only for a short period of frozen storage, and it had less significant effects after storage for a year (Kim et al. 1998).

20.4.3 Heat Stress

A high temperature damages almost all cellular components. However, sublethal heat treatments can increase the tolerance of LAB to a subsequent threat, for example, during spray-drying where a heat inactivation is possible. The adaptation to heat improved the viability after spray-drying of log-phase cells of *L. delbrueckii* subsp. *bulgaricus* (Teixeira et al. 1995a) and *L. paracasei* dried at high outlet air temperatures, namely, 100–105°C (Desmond et al. 2002b). For spray-drying, it is likely that the adaptation does not confer further advantages if considerable viability is already obtained from other factors. Heat shock increased the viability of log-phase cells of *L. delbrueckii* subsp. *bulgaricus* as compared to its control but did not boost the viability over normal levels found in unstressed stationary-phase cells. Similarly, when *L. paracasei* was spray-dried at lower outlet air temperatures (85–90 and 90–95°C), where high viabilities of unadapted cells are obtained, the protective effect from heat shock could not be observed (Desmond et al. 2002b).

For spray-drying, skim milk is often used as the suspending medium of LAB. The sublethal heat treatment of the cells in this matrix also rendered a better heat tolerance compared to the use of plain buffer (Teixeira et al. 1993).

Heat-shock treatment was also reported to infer a cross-protection against other stresses. In a study with fluidized bed-drying, where dehydration is the major inactivation mechanism, the viabilities of *L. rhamnosus* were reduced by 7.3 log units after 14 weeks' storage for the unstressed log-phase cells, whereas the reduction in viabilities for heat-shocked stationary and mid-log-phase cells was 1.6 and 4.16 log units, respectively (Prasad et al. 2003). Similarly, improved cryotolerance was also reported during freezing of *L. lactis* subsp. *lactis* (Broadbent and Lin 1999; Wouters et al. 1999a) and *S. thermophilus* (Varcamonti et al. 2006), and during freeze-drying of *L. lactis* subsp. *cremoris* (Broadbent and Lin 1999).

Due to the negative effect of heat, the number of cells before heat shock should be taken into consideration for the viability improvement. In a study by Walker et al. (1999) with *Lactobacillus johnsonii* heat-shocked at 55°C for different time intervals, the significant improvement in viability after freezing was found only for 45 min (increased by 19%). However, the number of cells was reduced by 1.5 log units for this heating duration; therefore, the heat shock actually has a negative impact on the net viability after the preservation.

20.4.4 Stationary-Phase-Associated Stress Responses

Generally, the exponential growth of cells is limited by the depletion of essential nutrients, the accumulation of fermentation products, and nonoptimal growth conditions. In industry, LAB are harvested either in the late log phase or in the early stationary phase. Besides maximal yield, the entry into the stationary phase also increases the tolerance of the cells to several stresses. This is due to the production of general stress proteins, whose expression is controlled by the response to starvation and by the stringent response (van de Guchte et al. 2002). For instance, the cryotolerance of stationary-phase cells of *L. lactis* subsp. *lactis* (Kim et al. 1999; Wouters et al. 1999a), *L. lactis* subsp. *cremoris* (Kim et al. 1999), *L. bugaricus* and *S. thermophilus* (Fonseca et al. 2001), and *L. lactis* subsp. *diacetylactis* (Lee 2004) was significantly higher than in the control log-phase cells, except in a strain of *L. lactis* subsp. *cremoris*, in which a comparable viability was observed. The cryotolerance of stationary-phase cells was also reported to depend on the growth medium. A higher cryotolerance of *S. thermophilus* was observed only in cells grown in the modified TPPY broth, whereas no difference was found for cells grown in skim milk (Morice et al. 1992). Similarly, a higher viability was found also in stationary-phase cells of *L. rhamnosus* (Corcoran et al. 2004) and *L. delbrueckii* subsp. *bulgaricus* (Teixeira et al. 1995a) after spray-drying, compared to log-phase cells.

The greater tolerance of stationary-phase cells to cryo- and dehydration stresses is probably due to the depletion of nutrients. Glucose starvation is a known trigger for stress response, leading to the resistance to many stresses, such as osmotic and heat stress (van de Guchte et al. 2002). The starvation-induced *E. faecalis* cells show multiresistance against stresses; particularly, heat and oxidative tolerances increase progressively with the duration of starvation (Giard et al. 1996). These tolerances may be beneficial to cells dried by spray-drying and fluidized bed-drying, where cells are subjected to either oxidative or heat and oxidative stresses.

20.4.5 Acid Stress

Acid stress in LAB is of increasing interest not only because it enables an improved viability of the frozen and dried cells, but also because of the necessity for probiotic LAB to survive exposure to gastric acid. The influence of pH on the viability of cells after drying was reported to vary with the drying process applied. According to a study by Schweigart (1971), the pH optima before drying of LAB cultures are 4.6–4.8, 5.1–5.3, and 4.9–5.2 for freeze-drying, spray-drying, and oven-drying, respectively. In a study by Hua et al. (2009) with *Oenococcus oeni*, different media with different pH were used for the cultivation. The decrease in medium pH increased the viability after freeze-drying. However, at the lowest pH studied (3.2), the viability was slightly declined. Similar results were also found in *L. acidophilus* grown at pH 5.0 The strain showed no loss in acidification activity during freezing and a low rate of loss in acidification activity during frozen storage. However, the cryotolerance was decreased when a more acidic medium (pH 4.5) was used for growth of the cells (Wang et al. 2005).

In addition to growing cells at a low pH, the acid-stress induction can be conducted by the cell cultivation without a pH control and by exposing cells to low pH at the end of the cultivation. Cells of *L. delbrueckii* subsp. *bulgaricus* grown under uncontrolled pH (end pH of ca. 4.0) had higher viabilities during spray-drying (Silva et al. 2005). However, the uncontrolled pH did not influence stability during storage of the dried cells. After a pH-controlled fermentation (pH 6.0), *L. delbrueckii* subsp. *bulgaricus* was acidified at pH 5.25, and a higher viability was observed in acid-shocked cells after freezing (42.9%) compared to control cells (4.7%) (Streit et al. 2008). The cryotolerance also depends on the type of acid used; H_2SO_4 was shown to be superior to HCl (Streit et al. 2007). Similarly, the heat resistance of *E. faecium* varied with the acid used, resulting in the following order: citric ≥ acetic > malic ≥ lactic > hydrochloric ≥ ascorbic acid-adapted cells (Fernandez et al. 2009).

20.4.6 Genetic Modification

It was reported that some strains do not respond to, or do not have a higher tolerance upon, sublethal stress treatments (Kim and Dunn 1997; Zotta et al. 2008). Although a higher viability of cells after preservation can still be obtained by the selection of robust strains (Chou and Weimer 1999; Monnet et al. 2003) or some specific mutants

(Rivals et al. 2007), studies have shown that the stress tolerance of LAB can also be specifically modified by a genetic manipulation. In a study by Desmond et al. (2004), a *L. paracasei* strain was genetically modified to overproduce the chaperone protein GroESL, which refolds denatured proteins during heat stress. The GroESL-overproducing strain exhibits a tenfold increase in heat tolerance when compared to a control parent strain, while the heat-adapted cells exhibit a higher level of an approximately 50-fold increase. This may be expected when considering that the heat-tolerance mechanisms also associate with other heat-shock proteins (e.g., DnaK, DnaJ, Clp, HtrA, and FtsH), glycolysis machinery, and other stationary-phase–related factors (Prasad et al. 2003). Hence, the induction of the whole battery of heat-stress tolerance mechanisms by sublethal heat treatment may give higher viability. In another study with this GroESL-overproducing *L. paracasei,* the strain was dried by spray-drying and freeze-drying and its viability was about ten and twofold higher than that of controls. Furthermore, as betaine was reported to protect LAB from desiccation, the nisin-controlled expression system was used to direct the heterologous expression of the listerial betaine-uptake system, BetL, in *L. salivarius* UCC118 (Sheehan et al. 2006). Following nisin induction, it was found that the UCC118-BetL$^+$ survives freeze-drying and spray-drying significantly better than UCC118-BetL$^-$. Similarly, the nisin-controlled expression system was used in studies with *L. lactis* that are related to the overproduction of different cold-shock proteins and to the increased accumulation of trehalose. The overproduction of CspD (Wouters et al. 1999a) and CspB (Wouters et al. 2000) resulted in about a tenfold and a five to tenfold increase in the viabilities of the cells after successive freeze–thaw cycles, compared to their respective control strains. There was only small improvements in the freeze-protective effect from the overproduction of CspE, and no such improvements from the overproduction of CspA or CspC and CspA (Wouters et al. 2000). Trehalose is a well-known and commonly used dehydration and cryoprotective excipient. However, its biosynthesis pathway was not evidenced in *L. lactis;* therefore, the expression of the trehalose biosynthesis genes *ots BA* from *Escherichia coli* was investigated in *L. lactis* NZ9000 (Termont et al. 2006). The genes were able to express in the strain NZ9000, and the intracellular accumulation of trehalose reached a maximum at approximately 50 mg/g wet cell weight after 2–3 h of nisin induction. The viability of the induced NZ9000 was markedly higher than that of the noninduced NZ9000 after freeze-drying with skim milk (94.0 vs. 56.7%) and without skim milk (ca. 100 vs. 20%). The induced cells also had higher storage stability and retained almost 100% viability for at least 1 month when being stored at 8°C and 10% relative humidity.

The overproduction of cold-shock proteins was also investigated in *L. plantarum.* Among its three identified cold-shock proteins, CspC, CspL, and CspP (Derzelle et al. 2000), the cryotolerance of the CspC- and CspL-overproducing strains did not differ from their controls, whereas an increase in cryotolerance was found from CspP overproduction only during four to six freeze–thaw cycles (Derzelle et al. 2003). In addition to cold-shock genes, in a study with *L. casei* by Beaufils et al. (2007), it was shown that several mutations affecting carbon catabolite repression (CCR) also alter the cells' sensitivity to freezing. The glucose-grown *ptsI* mutant had an almost twofold higher viability (75%) after four freeze–thaw cycles compared to the wild-type strain (45%). The wild-type strain transports glucose via the phosphotransferase system (PTS), whereas the *ptsI* mutant transports glucose via an unknown non-PTS permease. However, when growing in ribose and maltose (non-PTS sugars), the viabilities of this mutant were comparable to the wild-type strain. It was suggested that the connection between CCR and cold-shock response is due to the fact that a cold shock drastically disturbs the metabolism of bacterial cells and thereby probably also alters the phophorylation state of PTS proteins.

20.5 Concluding Remarks

LAB can be stored long term with high viability in frozen and dried forms by freezing and freeze-drying processes. Some alternative drying processes are promising for the preparation of dried cells, but they still require further improvement to resemble freeze-drying. A high viability of LAB after long-term storage is obtained only when each process step, that is, the freezing and drying, storage, and thawing/rehydration, is carried out under optimal conditions. It should be kept in mind that neither the drying nor the freezing process may be applicable universally to every LAB. It is advisable that where the loss of culture is vitally important, storage of the culture by both methods should be employed in parallel. The viability of LAB also depends on the physiology of the cells. Predominantly, sublethal stress treatments can improve viability of LAB. These treatments are osmotic, cold, heat, and acid shock, as well as the employment of stationary-phase cells and a genetic modification technique. Indeed, sublethal stress treatment is a process per se, where the response also depends on the conditions used for the treatments, such as salt concentration, temperature, pH, time, and suspending media. In most cases, the stress treatments were carried out at specific conditions, and this may explain some controversial results in some studies (Panoff et al. 1995; Linders et al. 1997a; Succi et al. 2007), besides the intrinsic nature of the LAB.

Furthermore, there are some points that still require more studies because the information is still limited. They are, for example, the changes in important traits of LAB like probiotic properties after long-term storage, and the sublethal stress treatments of the LAB after the harvesting of cells. The former is of importance for probiotic product development. The latter allows the stress treatment of LAB without compromising the cells' yield or prolonging the cultivation time.

References

Aguilera JM, Karel M (1997) Preservation of biological materials under desiccation. Crit Rev Food Sci Nutr 37:287–309

Ananta E, Volkert M, Knorr D (2005) Cellular injuries and storage stability of spray-dried *Lactobacillus rhamnosus* GG. Int Dairy J 15:399–409

Andersen AB, Fog-Peterson MS, Larsen H, Skibsted LH (1999) Storage stability of freeze-dried starter cultures (*Streptococcus thermophilus*) as related to physical state of freezing matrix. Food Sci Technol 32:540–547

Baati L, Fabre-Gea C, Auriol D, Blanc PJ (2000) Study of the cryotolerance of *Lactobacillus acidophilus*: effect of culture and freezing conditions on the viability and cellular protein levels. Int J Food Microbiol 59:241–247

Baumann DP, Reinbold GW (1966) Freezing of lactic cultures. J Dairy Sci 49:259–264

Beal C, Corrieu G (1994) Viability and acidification activity of pure and mixed starters of *Streptococcus salivarius* ssp *thermophilus*-404 and *Lactobacillus delbrueckii* ssp *bulgaricus*-398 at the different steps of their production. Food Sci Technol 27:86–92

Beal C, Fonseca F, Corrieu G (2001) Resistance to freezing and frozen storage of *Streptococcus thermophilus* is related to membrane fatty acid composition. J Dairy Sci 84:2347–2356

Beaufils S, Sauvageot N, Maze A, Laplace JM, Auffray Y, Deutscher J, Hartke A (2007) The cold shock response of *Lactobacillus casei*: relation between HPr phosphorylation and resistance to freeze/thaw cycles. J Mol Microbiol Biotechnol 13:65–75

Bielecka M, Majkowska A (2000) Effect of spray drying temperature of yoghurt on the survival of starter cultures, moisture content and sensoric properties of yoghurt powder. Nahrung-Food 44:257–260

Brennan M, Wanismail B, Johnson MC, Ray B (1986) Cellular-damage in dried *Lactobacillus acidophilus*. J Food Prot 49:47–53

Broadbent JR, Lin C (1999) Effect of heat shock or cold shock treatment on the resistance of *Lactococcus lactis* to freezing and lyophilization. Cryobiology 39:88–102

Brown AD (1976) Microbial water stress. Bacteriol Rev 40:803–846

Carcoba R, Rodriguez A (2000) Influence of cryoprotectants on the viability and acidifying activity of frozen and freeze-dried cells of the novel starter strain *Lactococcus lactis* ssp *lactis* CECT 5180. Eur Food Res Technol 211:433–437

Carvalho AS, Silva J, Ho P, Teixeira P, Malcata FX, Gibbs P (2002) Survival of freeze-dried *Lactobacillus plantarum* and *Lactobacillus rhamnosus* during storage in the presence of protectants. Biotechnol Lett 24:1587–1591

Carvalho AS, Silva J, Ho P, Teixeira P, Malcata FX, Gibbs P (2003a) Effects of addition of sucrose and salt, and of starvation upon thermotolerance and survival during storage of freeze-dried *Lactobacillus delbrueckii* ssp. *bulgaricus*. J Food Sci 68:2538–2541

Carvalho AS, Silva J, Ho P, Teixeira P, Malcata FX, Gibbs P (2003b) Protective effect of sorbitol and monosodium glutamate during storage of freeze-dried lactic acid bacteria. Lait 83:203–210

Castro HP, Teixeira P, Kirby R (1995) Storage of lyophilized cultures of *Lactobacillus bulgaricus* under different relative humidities and atmospheres. Appl Microbiol Biotechnol 44:172–176

Castro HP, Teixeira PM, Kirby R (1996) Changes in the cell membrane of *Lactobacillus bulgaricus* during storage following freeze-drying. Biotechnol Lett 18:99–104

Champagne CP, Gardner N, Brochu E, Beaulieu Y (1991) The freeze drying of lactic acid bacteria – a review. Can Inst Food Sci Technol J 24:118–128

Chou LS, Weimer B (1999) Isolation and characterization of acid- and bile-tolerant isolates from strains of *Lactobacillus acidophilus*. J Dairy Sci 82:23–31

Clementi F, Rossi J (1984) Effect of drying and storage conditions on survival of *Leuconostoc oenos*. Am J Enol Vitic 35:183–186

Conrad PB, Miller DP, Cielenski PR, de Pablo JJ (2000) Stabilization and preservation of *Lactobacillus acidophilus* in saccharide matrices. Cryobiology 41:17–24

Corcoran BM, Ross RP, Fitzgerald GF, Stanton C (2004) Comparative survival of probiotic lactobacilli spray-dried in the presence of prebiotic substances. J Appl Microbiol 96:1024–1039

Coulibaly I, Amenan AY, Lognay G, Fauconnier ML, Thonart P (2009) Survival of freeze-dried *Leuconostoc mesenteroides* and *Lactobacillus plantarum* related to their cellular fatty acids composition during storage. Appl Biochem Biotechnol 157:70–84

Crowe JH, Crowe LM, Mouradian R (1983) Stabilization of biological membranes at low water activities. Cryobiology 20:346–356

De Angelis M, Gobbetti M (2004) Environmental stress responses in *Lactobacillus*: a review. Proteomics 4:106–122

De Antoni GL, Perez P, Abraham A, Anon MC (1989) Trehalose, a cryoprotectant for *Lactobacillus bulgaricus*. Cryobiology 26:149–153

De Urraza P, De Antoni G (1997) Induced cryotolerance of *Lactobacillus delbrueckii* supsp. *bulgaricus*b LBB by preincubation at suboptimal temperatures with a fermentable sugar. Cryobiology 35:159–164

De Valdez GF (2001) Maintenance of lactic acid bacteria. In: Spencer JFT, De Spencer ALG (Eds.), Food microbiology protocols. Humana Press, Totawa, pp. 163–172

De Valdez GF, Degiori GS, Holgado APD, Oliver G (1985) Effect of the rehydration medium on the recovery of freeze dried lactic acid bacteria. Appl Environ Microbiol 50:1339–1341

Derzelle S, Hallet B, Francis KP, Ferain T, Delcour J, Hols P (2000) Changes in *cspL*, *cspP*, and *cspC* mRNA abundance as a function of cold shock and growth phase in *Lactobacillus plantarum*. J Bacteriol 182:5105–5113

Derzelle S, Hallet B, Ferain T, Delcour J, Hols P (2003) Improved adaptation to cold-shock, stationary-phase, and freezing stresses in *Lactobacillus plantarum* overproducing cold-shock proteins. Appl Environ Microbiol 69:4285–4290

Desmond C, Ross RP, O'Callaghan E, Fitzgerald G, Stanton C (2002a) Improved survival of *Lactobacillus paracasei* NFBC 338 in spray-dried powders containing gum acacia. J Appl Microbiol 93:1003–1011

Desmond C, Stanton C, Fitzgerald GF, Collins K, Ross RP (2002b) Environmental adaptation of probiotic lactobacilli towards improvement of performance during spray drying. Int Dairy J 12:183–190

Desmond C, Fitzgerald GF, Stanton C, Ross RP (2004) Improved stress tolerance of GroESL-overproducing *Lactococcus lactis* and probiotic *Lactobacillus paracasei* NFBC 338. Appl Environ Microbiol 70:5929–5936

Dumont F, Marechal PA, Gervais P (2004) Cell size and water permeability as determining factors for cell viability after freezing at different cooling rates. Appl Environ Microbiol 70:268–272

Duong T, Barrangou R, Russell WM, Klaenhammer TR (2006) Characterization of the tre locus and analysis of trehalose cryoprotection in *Lactobacillus acidophilus* NCFM. Appl Environ Microbiol 72:1218–1225

Efiuvwevwere BJO, Gorris LGM, Smid EJ, Kets EPW (1999) Mannitol-enhanced survival of *Lactococcus lactis* subjected to drying. Appl Microbiol Biotechnol 51:100–104

Fernandez A, Varez-Ordonez A, Lopez M, Bernardo A (2009) Effects of organic acids on thermal inactivation of acid and cold stressed *Enterococcus faecium*. Food Microbiol 26:497–503

Fonseca F, Beal C, Corrieu G (2000) Method of quantifying the loss of acidification activity of lactic acid starters during freezing and frozen storage. J Dairy Res 67:83–90

Fonseca F, Beal C, Corrieu G (2001) Operating conditions that affect the resistance of lactic acid bacteria to freezing and frozen storage. Cryobiology 43:189–198

Fonseca F, Marin M, Morris GJ (2006) Stabilization of frozen *Lactobacillus delbrueckii* subsp. *bulgaricus* in glycerol suspensions: freezing kinetics and storage temperature effects. Appl Environ Microbiol 72:6474–6482

Foschino R, Fiori E, Galli A (1996) Survival and residual activity of *Lactobacillus acidophilus* frozen cultures under different conditions. J Dairy Res 63:295–303

Gardiner GE, O'Sullivan E, Kelly J, Auty MA, Fitzgerald GF, Collins JK, Ross RP, Stanton C (2000) Comparative survival rates of human-derived probiotic *Lactobacillus paracasei* and *L. salivarius* strains during heat treatment and spray drying. Appl Environ Microbiol 66:2605–2612

Gehrke H (1991) Untersuchungen zur Gefriertrocknung von Mikroorganismen [in German]. Dissertation, Universität Braunschweig

Giard JC, Hartke A, Flahaut S, Benachour A, Boutibonnes P, Auffray Y (1996) Starvation-induced multiresistance in *Enterococcus faecalis* JH2–2. Curr Microbiol 32:264–271

Glaasker E, Konings WN, Poolman B (1996) Osmotic regulation of intracellular solute pools in *Lactobacillus plantarum*. J Bacteriol 178:575–582

Glaasker E, Tjan FSB, Ter Steeg PF, Konings WN, Poolman B (1998) Physiological response of *Lactobacillus plantarum* to salt and nonelectrolyte stress. J Bacteriol 180:4718–4723

Grout BWW, Morris GJ (2008) Contaminated liquid nitrogen storage vessels as potential vectors for pathogens. Cryo-Lett 29:74–75

Grout BWW, Morris GJ (2009) Contaminated liquid nitrogen vapour as a risk factor in pathogen transfer. Theriogenology 71:1079–1082

Heckly RJ (1961) Preservation of bacteria by lyophilization. Adv Appl Microbiol 3:1–76

Higl B, Kurtmann L, Carlsen CU, Ratjen J, Forst P, Skibsted LH, Kulozik U, Risbo J (2007) Impact of water activity, temperature, and physical state on the storage stability of *Lactobacillus paracasei* ssp. *paracasei* freeze-dried in a lactose matrix. Biotechnol Prog 23:794–800

Higl B, Santivarangkna C, Foerst P (2008) Bewertung und Optimierung von Gefrier-und Vakuumtrocknungsverfahren in der Herstellung von mikrobiellen Starterkulturen [in German]. Chemie Inginieur Technik 80: 1157–1164

Hua L, Zhao WY, Wang H, Li ZC, Wang AL (2009) Influence of culture pH on freeze-drying viability of *Oenococcus oeni* and its relationship with fatty acid composition. Food Bioprod Process 87:56–61

Hubalek Z (2003) Protectants used in the cryopreservation of microorganisms. Cryobiology 46:205–229

Hutkins RW, Ellefson WL, Kashket ER (1987) Betaine transport imparts osmotolerance on a strain of *Lactobacillus acidophilus*. Appl Environ Microbiol 53:2275–2281

Jewell JB, Kashket ER (1991) Osmotically regulated transport of proline by *Lactobacillus acidophilus* IFO 3532. Appl Environ Microbiol 57:2829–2833

Johannsen E (1972) Influence of various factors on the survival of *Lactobacillus leichmannii* during freezing and thawing. J Appl Bacteriol 35:415–421

Juarez TMS, Ocana VS, Nader-Macias ME (2004) Viability of vaginal probiotic lactobacilli during refrigerated and frozen storage. Anaerobe 10:1–5

Juarez TMS, Bru E, Martos G, Nader-Macias ME (2009) Stability of freeze-dried vaginal *Lactobacillus* strains in the presence of different lyoprotectors. Can J Microbiol 55:544–552

Keogh BP (1970) Survival and activity of frozen starter cultures for cheese manufacture. Appl Microbiol 19:928–931

Kets EPW, De Bont JAM (1994) Protective effect of betaine on survival of *Lactobacillus plantarum* subjected to drying. FEMS Microbiol Lett 116:251–255

Kets EPW, Teunissen PJM, De Bont JAM (1996) Effect of compatible solutes on survival of lactic acid bacteria subjected to drying. Appl Environ Microbiol 62:259–261

Kets EPW, Groot MN, Galinski EA, De Bont JAM (1997) Choline and acetylcholine: novel cationic osmolytes in *Lactobacillus plantarum*. Appl Microbiol Biotechnol 48:94–98

Kilimann KV, Doster W, Vogel RF, Hartmann C, Ganzle MG (2006) Protection by sucrose against heat-induced lethal and sublethal injury of *Lactococcus lactis*: an FT-IR study. Biochim Biophys Acta 1764:1188–1197

Kilstrup M, Jacobsen S, Hammer K, Vogensen FK (1997) Induction of heat shock proteins DnaK, GroEL, and GroES by salt stress is *Lactococcus lactis*. Appl Environ Microbiol 63:1826–1837

Kim SJ, Kim JH, Park JY, Kim HT, Jeong SJ, Ha YL, Yun HD, Kim JH (2004) Cold adaptation of *Lactobacillus paraplantarum* C7 isolated from kimchi. J Microbiol Biotechnol 14:1071–1074

Kim SS, Bhowmik SR (1990) Survival of lactic acid bacteria during spray drying of plain yogurt. J Food Sci 55:1008–1011

Kim WS, Dunn NW (1997) Identification of a cold shock gene in lactic acid bacteria and the effect of cold shock on cryotolerance. Curr Microbiol 35:59–63

Kim WS, Khunajakr N, Dunn NW (1998) Effect of cold shock on protein synthesis and on cryotolerance of cells frozen for long periods in *Lactococcus lactis*. Cryobiology 37:86–91

Kim WS, Ren J, Dunn NW (1999) Differentiation of *Lactococcus lactis* subspecies *lactis* and subspecies *cremoris* strains by their adaptive response to stresses. FEMS Microbiol Lett 171:57–65

King VAE, Su JT (1993) Dehydration of *Lactobacillus acidophilus*. Proc Biochem 28:47–52

King VAE, Lin HJ, Liu CF (1998) Accelerated storage testing of freeze-dried and controlled low-temperature vacuum dehydrated *Lactobacillus acidophilus*. J Gen Appl Microbiol 44:161–165

Knorr D (1998) Technology aspects related to microorganisms in functional foods. Trends Food Sci Technol 9:295–306

Kurtmann L, Carlsen CU, Risbo J, Skibsted LH (2009a) Storage stability of freeze-dried *Lactobacillus acidophilus* (La-5) in relation to water activity and presence of oxygen and ascorbate. Cryobiology 58:175–180

Kurtmann L, Carlsen CU, Skibsted LH, Risbo J (2009b) Water activity-temperature state diagrams of freeze-dried *Lactobacillus acidophilus* (La-5): influence of physical state on bacterial survival during storage. Biotechnol Prog 25:265–270

Kurtmann L, Skibsted LH, Carlsen CU (2009c) Browning of freeze-dried probiotic bacteria cultures in relation to loss of viability during storage. J Agric Food Chem 57:6736–6741

Le MC, Bon E, Lonvaud-Funel A (2007) Tolerance to high osmolality of the lactic acid bacterium *Oenococcus oeni* and identification of potential osmoprotectants. Int J Food Microbiol 115:335–342

Lee K (2004) Cold shock response in *Lactococcus lactis* ssp. *diacetylactis*: a comparison of the protection generated by brief pre-treatment at less severe temperatures. Proc Biochem 39:2233–2239

Leslie SB, Israeli E, Lighthart B, Crowe JH, Crowe LM (1995) Trehalose and sucrose protect both membranes and proteins in intact bacteria during drying. Appl Environ Microbiol 61:3592–3597

Lian WC, Hsiao HC, Chou CC (2002) Survival of bifidobacteria after spray-drying. Int J Food Microbiol 74:79–86

Linders LJM, Meerdink G, Vantriet K (1997a) Effect of growth parameters on the residual activity of *Lactobacillus plantarum* after drying. J Appl Microbiol 82:683–688

Linders LJM, Wolkers WF, Hoekstra FA, Vantriet K (1997b) Effect of added carbohydrates on membrane phase behavior and survival of dried *Lactobacillus plantarum*. Cryobiology 35:31–40

Lorca GL, De Valdez GF (1998) Temperature adaptation and cryotolerance in *Lactobacillus acidophilus*. Biotechnol Lett 20:847–849

Mauriello G, Aponte M, Andolfi R, Moschetti G, Villani F (1999) Spray-drying of bacteriocin-producing lactic acid bacteria. J Food Prot 62:773–777

Mazur P (1970) Cryobiology – freezing of biological systems. Science 168:939–949

Mazur P (1984) Freezing of living cells: mechanisms and implications. Am J Physiol 247:C125-C142

Melin AM, Perromat A, Deleris G (1999) Pharmacologic application of FTIR spectroscopy: effect of ascorbic acid-induced free radicals on *Deinococcus radiodurans*. Biospectroscopy 5:229–236

Mille Y, Obert JP, Beney L, Gervais P (2004) New drying process for lactic bacteria based on their dehydration behavior in liquid medium. Biotechnol Bioeng 88:71–76

Miyamoto-Shinohara Y, Sukenobe J, Imaizumi T, Nakahara T (2006) Survival curves for microbial species stored by freeze-drying. Cryobiology 52:27–32

Miyamoto-Shinohara Y, Sukenobe J, Imaizumi T, Nakahara T (2008) Survival of freeze-dried bacteria. J Gen Appl Microbiol 54:9–24

Modesto M, Mattarelli P, Biavati B (2004) Resistance to freezing and freeze-drying storage processes of potential probiotic bifidobacteria. Ann Microbiol 54:43–48

Monnet C, Beal C, Corrieu G (2003) Improvement of the resistance of *Lactobacillus delbrueckii* ssp. *bulgaricus* to freezing by natural selection. J Dairy Sci 86:3048–3053

Morice M, Bracquart P, Linden G (1992) Colonial variation and freeze-thaw resistance of *Streptococcus thermophilus*. J Dairy Sci 75:1197–1203

Morris GJ (2005) The origin, ultrastructure, and microbiology of the sediment accumulating in liquid nitrogen storage vessels. Cryobiology 50:231–238

Nakamura LK (1996) Preservation and maintenance of eubacteria. In: Hunter-Cevera JC, Belt A (Eds.), Maintaining cultures for biotechnology and industry. Academic, London, pp. 65–84

Obis D, Guillot A, Mistou MY (2001) Tolerance to high osmolality of *Lactococcus lactis* subsp. *lactis* and cremoris is related to the activity of a betaine transport system. FEMS Microbiol Lett 202:39–44

Panoff JM, Thammavongs B, Laplace JM, Hartke A, Boutibonnes P, Auffray Y (1995) Cryotolerance and cold adaptation in *Lactococcus lactis* subsp *lactis* IL1403. Cryobiology 32:516–520

Panoff JM, Thammavongs B, Gueguen M (2000) Cryoprotectants lead to phenotypic adaptation to freeze-thaw stress in *Lactobacillus delbrueckii* ssp. *bulgaricus* CIP 101027T. Cryobiology 40:264–269

Prasad J, McJarrow P, Gopal P (2003) Heat and osmotic stress responses of probiotic *Lactobacillus rhamnosus* HN001 (DR20) in relation to viability after drying. Appl Environ Microbiol 69:917–925

Rivals JP, Beal C, Thammavongs B, Gueguen M, Panoff JM (2007) Cryotolerance of *Lactobacillus delbrueckii* subsp. *bulgaricus* CFL1 is modified by acquisition of antibiotic resistance. Cryobiology 55:19–26

Roelans E, Taeymans D (1990) Effect of drying conditions on survival and enzyme activity of microorganims. In: Spiess WEL, Schubert H (Eds.), Engineering and food, Vol. 3: advanced processes. Elsevier Applied Science, London, pp. 559–569

Sanders JW, Venema G, Kok J (1999) Environmental stress responses in *Lactococcus lactis*. FEMS Microbiol Rev 23:483–501

Sandine WE (1996) Commercial production of dairy starter cultures. In: Cogan TM, Accolas JP (Eds.), Dairy starter cultures. VCH, New York, pp. 191–206

Santivarangkna C, Kulozik U, Foerst P (2006) Effect of carbohydrates on the survival of *Lactobacillus helveticus* during vacuum drying. Lett Appl Microbiol 42:271–276

Santivarangkna C, Wenning M, Foerst P, Kulozik U (2007) Damage of cell envelope of *Lactobacillus helveticus* during vacuum drying. J Appl Microbiol 102:748–756

Santivarangkna C, Higl B, Foerst P (2008) Protection mechanisms of sugars during different stages of preparation process of dried lactic acid starter cultures. Food Microbiol 25:429–441

Schoug A, Fischer J, Heipieper HJ, Schnurer J, Hakansson S (2008) Impact of fermentation pH and temperature on freeze-drying survival and membrane lipid composition of *Lactobacillus coryniformis* Si3. J Ind Microbiol Biotechnol 35:175–181

Schweigart F (1971) The drying of lactic acid bacteria cultures for Mahewu production. Lebensmittel-Wissenschaft und -Technologie 4:20–23

Selmer-Olsen E, Birkeland SE, Sorhaug T (1999a) Effect of protective solutes on leakage from and survival of immobilized *Lactobacillus* subjected to drying, storage and rehydration. J Appl Microbiol 87:429–437

Selmer-Olsen E, Sorhaug T, Birkeland SE, Pehrson R (1999b) Survival of *Lactobacillus helveticus* entrapped in Ca-alginate in relation to water content, storage and rehydration. J Ind Microbiol Biotechnol 23:79–85

Serror P, Dervyn R, Ehrlich SD, Maguin E (2003) *csp*-like genes of *Lactobacillus delbrueckii* ssp. *bulgaricus* and their response to cold shock. FEMS Microbiol Lett 226:323–330

Sheehan VM, Sleator RD, Fitzgerald GF, Hill C (2006) Heterologous expression of BetL, a betaine uptake system, enhances the stress tolerance of *Lactobacillus salivarius* UCC118. Appl Environ Microbiol 72:2170–2177

Silva J, Carvalho AS, Teixeira P, Gibbs PA (2002) Bacteriocin production by spray-dried lactic acid bacteria. Lett Appl Microbiol 34:77–81

Silva J, Carvalho AS, Pereira H, Teixeira P, Gibbs PA (2004) Induction of stress tolerance in *Lactobacillus delbrueckii* ssp. *bulgaricus* by the addition of sucrose to the growth medium. J Dairy Res 71:121–125

Silva J, Carvalho AS, Ferreira R, Vitorino R, Amado F, Domingues P, Teixeira P, Gibbs PA (2005) Effect of the pH of growth on the survival of *Lactobacillus delbrueckii* subsp. *bulgaricus* to stress conditions during spray-drying. J Appl Microbiol 98:775–782

Sinha RN, Dudani AT, Ranganathan B (1974) Protective effect of fortified skim milk as suspending medium for freeze drying of different lactic acid bacteria. J Food Sci 39:641–642

Strasser S, Neureiter M, Geppl M, Braun R, Danner H (2009) Influence of lyophilization, fluidized bed drying, addition of protectants, and storage on the viability of lactic acid bacteria. J Appl Microbiol 107:167–177

Streit F, Corrieu G, Beal C (2007) Acidification improves cryotolerance of *Lactobacillus delbrueckii* subsp. *bulgaricus* CFL1. J Biotechnol 128:659–667

Streit F, Delettre J, Corrieu G, Beal C (2008) Acid adaptation of *Lactobacillus delbrueckii* subsp. *bulgaricus* induces physiological responses at membrane and cytosolic levels that improves cryotolerance. J Appl Microbiol 105:1071–1080
Succi M, Tremonte P, Reale A, Sorrentino E, Coppola R (2007) Preservation by freezing of potentially probiotic strains of *Lactobacillus rhamnosus*. Ann Microbiol 57:537–544
Sunny-Roberts EO, Knorr D (2007) Physiological analysis of *Lactobacillus rhamnosus* VTT E-97800 – adaptive response to osmotic stress induced by trehalose. Br Food J 109:735–748
Sunny-Roberts EO, Knorr D (2008) Evaluation of the response of *Lactobacillus rhamnosus* VTT E-97800 to sucrose-induced osmotic stress. Food Microbiol 25:183–189
Sunny-Roberts EO, Knorr D (2009) The protective effect of monosodium glutamate on survival of *Lactobacillus rhamnosus* GG and *Lactobacillus rhamnosus* E-97800 (E800) strains during spray-drying and storage in trehalose-containing powders. Int Dairy J 19:209–214
Teixeira P, Castro H, Kirby R (1993) Inducible thermotolerance in *Lactobacillus bulgaricus*. Lett Appl Microbiol 18:218–221
Teixeira P, Castro H, Kirby R (1995a) Spray drying as a method for preparing concentrated cultures of *Lactobacillus bulgaricus*. J Appl Bacteriol 78:456–462
Teixeira PC, Castro MH, Malcata FX, Kirby RM (1995b) Survival of *Lactobacillus delbrueckii* ssp. *bulgaricus* following spray-drying. J Dairy Sci 78:1025–1031
Teixeira P, Castro H, Kirby R (1996) Evidence of membrane lipid oxidation of spray-dried *Lactobacillus bulgaricus* during storage. Lett Appl Microbiol 22:34–38
Termont S, Vandenbroucke K, Iserentant D, Neirynck S, Steidler L, Remaut E, Rottiers P (2006) Intracellular accumulation of trehalose protects *Lactococcus lactis* from freeze-drying damage and bile toxicity and increases gastric acid resistance. Appl Environ Microbiol 72:7694–7700
Thunell RK, Sandine WE, Bodyfelt FW (1984) Frozen starters from internal pH control grown cultures. J Dairy Sci 67:24–36
To BCS, Etzel MR (1997) Spray drying, freeze drying, or freezing of three different lactic acid bacteria species. J Food Sci 62:576–585
Tsvetkov T, Brankova R (1983) Viability of micrococci and lactobacilli upon freezing and freeze drying in the presence of different cryoprotectants. Cryobiology 20:318–323
Tsvetkov T, Shishkova I (1982) Studies on the effects of low temperatures on lactic acid bacteria. Cryobiology 19:211–214
van de Guchte M, Serror P, Chervaux C, Smokvina T, Ehrlich SD, Maguin E (2002) Stress responses in lactic acid bacteria. Antonie von Leeuwenhoek 82:187–216
Varcamonti M, Arsenijevic S, Martirani L, Fusco D, Naclerio G, De Felice M (2006) Expression of the heat shock gene ClpL of *Streptococcus thermophilus* is induced by both heat and cold shock. Microb Cell Fact. doi: 10.1186/1475–2859–5-6
Volkert M, Ananta E, Luscher C, Knorr D (2008) Effect of air freezing, spray freezing, and pressure shift freezing on membrane integrity and viability of *Lactobacillus rhamnosus* GG. J Food Eng 87:532–540
Walker DC, Girgis HS, Klaenhammer TR (1999) The groESL chaperone operon of *Lactobacillus johnsonii*. Appl Environ Microbiol 65:3033–3041
Wang YC, Yu RC, Chou CC (2004) Viability of lactic acid bacteria and bifidobacteria in fermented soymilk after drying, subsequent rehydration and storage. Int J Food Microbiol 93:209–217
Wang Y, Corrieu G, Beal C (2005) Fermentation pH and temperature influence the cryotolerance of *Lactobacillus acidophilus* RD758. J Dairy Sci 88:21–29
Welsh DT, Herbert RA (1999) Osmotically induced intracellular trehalose, but not glycine betaine accumulation promotes desiccation tolerance in *Escherichia coli*. FEMS Microbiol Lett 174:57–63
Wolfe J (1987) Lateral stresses in membranes at low water potential. Aust J Plant Physiol 14:311–318
Wolff E, Louvet P, Gilbert H (1986) Atmospheric freeze drying: design and energy considerations. In: Arun S, Mujumdar (Eds.), Drying'86. 5th International Symposium on Drying, Cambridge, MA, August. Hemisphere Publishing, Washington, pp. 432–439

Wolff E, Delisle B, Corrieu G, Gibert H (1990) Freeze-drying of *Streptococcus thermophilus*: a comparison between the vacuum and the atmospheric method. Cryobiology 27:569–575

Wouters JA, Jeynov B, Rombouts FM, De Vos WM, Kuipers OP, Abee T (1999a) Analysis of the role of 7 kDa cold-shock proteins of *Lactococcus lactis* MG1363 in cryoprotection. Microbiology 145:3185–3194

Wouters JA, Rombouts FM, De Vos WM, Kuipers OP, Abee T (1999b) Cold shock proteins and low-temperature response of *Streptococcus thermophilus* CNRZ302. Appl Environ Microbiol 65:4436–4442

Wouters JA, Mailhes M, Rombouts FM, De Vos WM, Kuipers OP, Abee T (2000) Physiological and regulatory effects of controlled overproduction of five cold shock proteins of *Lactococcus lactis* MG1363. Appl Environ Microbiol 66:3756–3763

Yao AA, Coulibaly I, Lognay G, Fauconnier ML, Thonart P (2008) Impact of polyunsaturated fatty acid degradation on survival and acidification activity of freeze-dried *Weissella paramesenteroides* LC11 during storage. Appl Microbiol Biotechnol 79:1045–1052

Zhao G, Zhang G (2005) Effect of protective agents, freezing temperature, rehydration media on viability of malolactic bacteria subjected to freeze-drying. J Appl Microbiol 99:333–338

Zimmermann K (1987) Einflussparameter und mathematische Modellierung der schonenden Trocknung von Starterkulturen [in German]. Dissertation, Technische Universität München

Zotta T, Ricciardi A, Ciocia F, Rossano R, Parente E (2008) Diversity of stress responses in dairy thermophilic streptococci. Int J Food Microbiol 124:34–42

Part V
Conclusion

Chapter 21
Future Challenges in Lactic Acid Bacteria Stress Physiology Research

Konstantinos Papadimitriou and Jan Kok

21.1 Introduction

Bacteria have been among the most important models for studying the stress physiology of cellular organisms. This can be partially explained by the ease with which one can handle these organisms in the laboratory; however, it also reflects their stunning inherent ability to occupy highly divergent ecological niches. It is well established today that bacteria are equipped with a remarkable arsenal toward the alleviation of perturbations caused by stressful conditions, allowing them to adapt and survive even in extreme environments that would be prohibitive for most other forms of life (Pikuta et al. 2007). In the landmark book *Bacterial Stress Responses* published 10 years ago, the information presented mainly referred to the stress behavior of *Escherichia coli* and *Bacillus subtilis* (Storz and Hengge-Aronis 2000), which is not surprising given that these two bacteria have served as the model organisms for almost all areas of bacteriology.

Over the past decade, a significant amount of knowledge concerning the stress physiology of lactic acid bacteria (LAB) has also accumulated. The use of LAB in food fermentations, their probiotic potential, and in some instances their severe pathogenicity have been the key reasons fueling stress research in this group of bacteria. From a historical perspective, the topic started to systematically appear in

J. Kok (✉)
Department of Molecular Genetics, Groningen Biomolecular Sciences and Biotechnology Institute, University of Groningen, Nijenborgh 7, 9747 AG Groningen, The Netherlands

Top Institute Food and Nutrition, Nieuwe Kanaal 9A, 6709 PA Wageningen, The Netherlands

The Netherlands Kluyver Centre for Genomics of Industrial Fermentations, NCSB, Julianalaan 67, 2628 BC Delft, The Netherlands
e-mail: jan.kok@rug.nl

E. Tsakalidou and K. Papadimitriou (eds.), *Stress Responses of Lactic Acid Bacteria*, Food Microbiology and Food Safety, DOI 10.1007/978-0-387-92771-8_21,

the literature by the late 1970s to early 1980s. Nevertheless, it was not until the 1990s that the development of tools for the elaborate genetic manipulation of several LAB strains (Delcour et al. 2000; Mills 2001) laid the background for the first detailed studies of the stress responses in these economically and medically important bacteria. The publication of the *Lactococcus lactis* genome sequence back in 2001 marked the shift of LAB research (including stress physiology) from the classical genetic approaches to those of modern genomics (Bolotin et al. 2001). Since then, the field has greatly benefited from advances in sequencing technologies, bioinformatics, functional genomics, and proteomics/metabolomics. Today more than 25 complete genome sequences of LAB have been deposited in public databases, while the number is expected to exceed 100 in the near future (Siezen and Wilson 2010). This wealth of information will soon completely change our understanding of LAB, since we will be able to better appreciate their evolution, physiology, and metabolism, as well as their technological and/or probiotic properties and pathogenicity.

The available data clearly indicate that during their domestication in nutrient-rich environments, LAB adapted by excessive gene loss and, to a lesser extent, by gene gain events (Klaenhammer et al. 2005; Makarova et al. 2006). The net effect was a reductive evolutionary mechanism that decreased the size of their genomes and simplified their metabolism. The presence of a variable number of pseudogenes in several LAB genomes is also indicative of ongoing genome-decay processes (Makarova et al. 2006). The adaptation of LAB to their new niches might have affected stress-related genes. In favor of this notion, the limited number of oxidative stress-resistance genes of the dairy starter *Streptococcus thermophilus*, in comparison to the repertoire of such genes in pathogenic streptococci, has been suggested to contribute to its diminished pathogenicity (Hols et al. 2005).

LAB, like other bacteria, have developed sophisticated regulatory mechanisms for controlling and altering gene expression that clearly diverge from those of the Gram-positive paradigm *B. subtilis* in order to persist under stress conditions. In *B. subtilis* and several other Gram-positive bacteria, including *Listeria monocytogenes* and *Staphylococcus aureus*, the alternative sigma factor σ^B modulates their responses to various stresses (van Schaik and Abee 2005). The expression of the σ^B regulon, also known as the general stress or class II heat-shock regulon, confers resistance to the cells. The analysis of the completed genome sequences of LAB has shown that a σ^B ortholog is not present in these bacteria (van de Guchte et al. 2002; Yother et al. 2002; Sugimoto et al. 2008). This is considered a major evolutionary deviation of LAB. While several studies have demonstrated a species- or even subspecies-dependent overlapping core of genes or proteins that are upregulated in LAB under different stress conditions, a well-defined *Bacillus*-like general stress regulon (>100 genes) that is under the control of a single σ factor does not exist. Most LAB are equipped with the negative response regulators (RRs) HrcA and CtsR, present in *B. subtilis* as well, that regulate the expression of class I and class III heat-shock genes, respectively (Sugimoto et al. 2008). Even so, differences have been reported for LAB, like the dual regulation of certain operons by both HrcA and CtsR or the absence of *hrcA* from the genome of *Oenococcus oeni* (Grandvalet et al. 2005). As a result, a number of LAB species are emerging as distinct models for stress research in their own right, since the control of the expression of stress genes in Gram-positive bacteria seems not to be well conserved and findings in *B. subtilis* are not always applicable to other members of this large group of bacteria.

In the preceding parts of the book, the state of the art of LAB stress physiology has been presented. In this concluding chapter, we will attempt to summarize the most important areas that could be the focus of research in the field and that we consider will significantly improve our conception of stress behavior in LAB.

21.2 The Requirement for Functional Analysis of the Stress Responses of More Species and Strains

LAB constitute a large group of bacteria, and thus research about their stress physiology has been uneven among different species and/or subspecies. *L. lactis*, *Lactobacillus plantarum*, *Streptococcus mutans*, *Enterococcus faecalis*, and *O. oeni* have been used as models of their genera. These choices were made mostly on the basis of the strains' economic importance (i.e., in the case of starters) or of their pathogenicity. Interestingly, *L. lactis* can be considered the best-studied LAB in terms of its stress physiology, although in the case of acid stress, available information about *S. mutans* is comparable. The exploitation of specific strains has assisted the detailed analysis of their stress responses, while, conversely, this strategy has hindered the identification of mechanisms that are species- or even subspecies-dependent. The situation seems to be settled for lactobacilli, in which, even though *L. plantarum* is the leading species in the stress physiology research of the genus, several other species (e.g., *Lactobacillus acidophilus*, *Lactobacillus casei*, and *Lactobacillus bulgaricus*) have also been adequately investigated. Surprisingly, there are important LAB species for which the key aspects of their physiology under stress conditions remain understudied. For example, little is known about *S. thermophilus* and its acid, starvation, and osmotic stress responses. This is also true for a number of other streptococci, such as *Streptococcus pyogenes* and *Streptococcus agalactiae*, for which there are limited data despite their medical impact. Additionally, strains belonging to the genera *Leuconostoc*, *Weissella*, *Tetragenococcus*, and *Carnobacterium* have been scarcely examined. Consequently, it would be critical for the progress of LAB stress physiology research to expand the investigation to other species and strains than those traditionally employed. Fortunately, current technologies in genomics and proteomics offer the unique opportunity for in silico and functional analysis of the stress responses of a bacterium in a rapid and cost-effective manner, even in the absence of prior knowledge of its biology.

21.3 Stress and the Cell Cycle

As is clear from the chapters in this book, LAB are oftentimes confronted with various stressful conditions, be it the environmental changes encountered by industrial LAB throughout fermentation or during the beneficial or sometimes detrimental interactions of probiotic or pathogenic LAB, respectively, with the host. Conditions in which resources are plentiful can gradually, due to growth and nutrient use, or quickly change to circumstances where nutrients are scarce. With abundant sources of nutrients, cell division rate and cell sizes are large, whereas in nutrient-poor situations, growth rate and cell size are reduced. One of the consequences of stressful situations is that they very often interfere with cell proliferation. Changes in cell growth rate should lead to adaptations in the cell's division cycle so that it is coordinated with the replication and segregation of the chromosome. Multiple signaling pathways relay nutritional and growth-rate information to the cell cycle machinery in the Gram-negative and Gram-positive paradigms *E. coli* and *B. subtilis* (for recent reviews, see Mott and Berger 2007; Wang and Levin 2009; Katayama et al. 2010). Thus, the cells are able to continuously fine-tune their division cycle with the state of the environment. Almost nothing is known about the cell cycle machinery and its regulation under normal or stressful conditions in LAB, an omission that needs to be corrected if one wants to really understand the intricacies in the response of LAB to stress.

21.4 Stress and DNA Damage Control

Many of the stresses that bacteria encounter lead to DNA damage, and LAB are no different in this respect. Notorious for their genotoxic effects are acid and oxidative stresses, which increase the formation of abasic sites in DNA, and in several instances described in this book, LAB have been shown to induce genes for DNA repair and recombination enzymes in response to these and other stressful conditions. A wealth of information is available on the inducible repair of DNA damage, especially in *E. coli* (see, for instance, Walker et al. 2000; Schlacher and Goodman 2007).

Any condition that leads to a transition from the active growth of bacterial cells to the stationary phase entails the risk of accumulating mutations in the genome of the organism. Safeguards to DNA damage in exponentially growing cells, in which the DNA is in a dispersed conformation, are different from those in the stationary phase, at least in the organisms in which this has been extensively studied (for a review, see Frenkiel-Krispin and Minsky 2006). A drastic reorganization of bacterial nucleoids into ordered, tightly packed structures takes place in cells in the stationary state, promoting DNA repair on the one hand and offering protection by restricting the accessibility of the genomic material to detrimental factors, on the other hand. Increased resistance of bacteria in the stationary phase is often seen to multiple stresses, including those that cause DNA damage, such as various antibiotics (see Sect. 21.5). Virtually nothing is known about the machinery responsible for sensing genome integrity in the various species of LAB and for DNA repair in these organisms, a deficiency that should be the focus of future research considering their significance in stress control and stress cross-resistance. In particular, the apparent link between DNA damage repair and the development of antibiotic resistance is of eminent importance for the medically relevant LAB.

21.5 Stress-Induced Mutagenesis

Stress-induced, or stationary-phase, mutagenesis refers to the mutations generated under a hypermutable state that the cells may enter during stress (Gonzalez et al. 2008). The phenomenon has been verified across different cell types, from bacterial to yeast and cancer cells, and has changed our notion of a static genome to a genome that is plastic and responds to stress through mutations (Galhardo et al. 2007). The induction of mutations as a stress response can potentially contribute to adaptive evolution by enhancing genetic diversity and increasing the likelihood for the generation of mutants that are able to survive and persist under the specific environmental stressful condition (Foster 2007; Galhardo et al. 2007). From a mechanistic perspective, adaptive mutations can arise from cellular stress responses, error-prone DNA repair and erroneous transcription machineries, and/or from the movement of mobile genetic elements (for reviews, refer to Foster 2007; Galhardo et al. 2007; Robleto et al. 2007). In addition, such mutations appear when cells are maladapted to their environment and are thus nondividing, and they are not associated with the replication errors occurring during cell proliferation (Robleto et al. 2007; Gonzalez et al. 2008). Stress-induced mutagenesis has been shown to drive bacterial pathogenesis and antibiotic resistance. This distinct type of stress response has been very rarely investigated in LAB. It has been demonstrated that the exposure of *Streptococcus uberis* cells to UV irradiation caused a significant increase in the proportion of both rifampin- and ciprofloxacin-resistant mutants (Varhimo et al. 2007). This increase in antibiotic-resistant mutants was dependent on the induction of a small DNA damage-inducible gene cassette that contained a Y-family error-prone DNA polymerase. Furthermore, during industrial processes, starter LAB strains reside under adverse environmental conditions designed to be bacteriostatic or bactericidal for food spoilage and pathogenic bacteria. This situation could lead to the accumulation of stress-induced mutants that may display altered functional properties. Indeed, using appropriate mutator reporter strains of *L. plantarum*, Machielsen et al. demonstrated that even though mutation frequency was unaffected during high temperature, low pH, osmotic, or starvation stress, it was increased by a factor of 100 after exposure to sublethal levels of hydrogen peroxide (H_2O_2) (Machielsen et al. 2010). Interestingly, the preadaptation of *L. plantarum* cells at 42°C, a nonmutagenic condition, attenuated the mutagenic effect of oxidative stress. As the authors of this study note, this finding may be of technological importance, since it offers a methodology to protect cells against stress-induced mutations and to enhance the genetic stability of the starter culture. Irrespective of the specific niche they occupy, LAB spend most of their lifetime under stressful conditions that hinder cell proliferation, and thus stress-induced mutagenesis as a response may have been overlooked. Accordingly, further investigation is needed, since the available literature indicates that stress-induced mutagenesis may have important implications for the stress physiology of LAB.

21.6 Sensing and Signaling Stress

Bacteria are able to sense environmental changes and to subsequently transmit the perceived stimulus through signal transduction processes to the cellular machineries necessary to mount an adaptive response. A significant amount of work has been performed to identify stress sensors in *E. coli* and *B. subtilis*, and some of the mechanisms by which these organisms sense stresses are well elucidated (for reviews, refer to Schumann 2007; Kramer 2010). In most cases, stress sensors are of a macromolecular nature (i.e., DNA, RNA, and proteins), and bacterial sensing relies on conformational changes in the macromolecules due to the stress (Schumann 2007). The hallmark apparatus bacteria use for sensing and signaling stress is the two-component system (TCS). In response to signal binding, the sensor histidine kinase (HK) of the TCS activates its RR, which in turn induces or represses gene expression (Hoch and Varughese 2001). Signaling between the HK and the RR is achieved through His-to-Asp phosphoryl-group transfer. The same type of phosphorylation events is also seen in phosphorelay systems, where the TCS is involved in a more complex signaling cascade, like the sporulation phosphorelay of *B. subtilis* (Hoch and Varughese 2001). For many years, signal transduction in bacteria was thought to be mediated solely by HKs and the only residues considered to be phosphorylated were His and Asp (Soufi et al. 2008b; Majakovic 2010). It was not until recently that eukaryotic-type Ser/Thr kinases as well as bacterial Tyr kinases without counterparts in *Eukarya* were discovered. Such kinases have already been implicated in the regulation of different cellular processes, like antibiotic resistance, pathogenesis, and stress responses through protein phosphorylation (Klein et al. 2003; Rajagopal et al. 2003; Lacour et al. 2006). The appreciation of the multiplicity of bacterial kinases and phosphoprotein phosphatases after the analysis of the available complete microbial genomes also raised interest in the determination of the phosphoproteome of bacteria. Original attempts based on 2D-gel electrophoresis (2-DE) coupled with mass spectrometry (MS) identification revealed several phosphorylated proteins, though characterization of the phosphorylation sites was not possible (Soufi et al. 2008b). It also became clear that the phosphoproteome is dynamic, responding to different stimuli, and phosphorylated proteins seemed to accumulate under stress conditions (Rosen et al. 2004; Levine et al. 2006). Advances in automated liquid chromatography and high-resolution MS have made possible the site-specific gel-free approaches of examining phosphoproteomes that allow detection of low-abundant phosphoproteins, as well as phosphorylation sites (Soufi et al. 2008b). The detailed phosphoproteomes of a number of bacteria are now available, including those of *B. subtilis* (Macek et al. 2007) and of two LAB representatives, *L. lactis* and *S. pneumoniae* (Soufi et al. 2008a; Sun et al. 2010). From the available data, the majority of the phosphorylated proteins are involved in carbon metabolism, while others participate in DNA and protein metabolism and stress responses. New techniques, like SILAC (stable isotope labeling by amino acids in cell culture), hold promise for the quantitative study of phosphoproteome dynamics (Zhong et al. 2007; Soufi et al. 2008b).

It is obvious that research on bacterial protein phosphorylation is entering a new era and the tools to answer several questions now exist. Determining the substrate specificity of different kinases, dissecting protein phosphorylation cascades, and assessing the actual role of each phosphorylation event will be the next challenges.

Unfortunately, as several authors highlight in Parts I and II of this book, our understanding of stress sensing and signaling by LAB is still rather basic. Most studies are confined to the role of TCSs under different stress conditions. The existence of non-TCS stress sensors has been rarely experimentally supported (e.g., the BusA/OpuA osmosensor in *L. lactis*), and most proposed mechanisms are the result of extending findings from other bacteria to LAB. Another example that reflects our refractory view of the LAB stress-sensing capabilities is the yet-unappreciated role of one-component systems. One-component systems are transcriptional regulators that contain the sensing and response domains found separately in HKs and RRs of TCSs in one fused molecule. Furthermore, it has been suggested that one-component systems might dominate signal transduction in bacteria, since they are more abundant and more widely distributed than TCSs (Ulrich et al. 2005). Throughout this book, only three examples of one-component systems have been presented: the BcrR membrane-bound bacitracin sensor in *E. faecalis*; the PrkC modulator of antimicrobial resistance in the same organism; and the PknB regulator of biofilm formation in *S. mutans* (for details, refer to Chap. 8). It is evident that the systematic study of how LAB sense and signal stress is necessary. The identification and characterization of novel stress sensors, as well as the detailed charting of signal transduction pathways and their dynamics under environmental changes, will be very important. Ultimately, this knowledge will allow us to manipulate these pathways in order to increase robustness in the case of food-related and probiotic LAB or to design novel antimicrobials in the case of pathogenic LAB.

21.7 Stress in the Context of the Single Cell

Although bacteria are clonal organisms, cultures of a single bacterial strain are not as homogeneous as one would expect. Recent work in many different microbial species has demonstrated that pure cultures of most strains contain diverse subpopulations that are not genetically different, for example, through the loss or acquisition of plasmids, or the movement of transposable elements, but diverge phenotypically. Phenotypic heterogeneity can occur in phenotypes that are crucial for the fitness and development of the organism and results in a dynamic source of diversity. It creates variant subpopulations that are potentially better equipped to persist or survive during environmental perturbations or to exploit new niches (Davidson and Surette 2008; Losick and Desplan 2008; Veening et al. 2008; Eldar and Elowitz 2010). Genetic changes are not involved in the processes that lead to phenotypic heterogeneity, allowing rapid reversion to the original phenotype, if required. The known causes of heterogeneity are, for example, stochasticity in gene expression, individual rates of cell growth, meta-stable inheritance of epigenetic modifications of the chromosome, and the physiological state of the cell. It is becoming clearer and clearer that the resistance of microorganisms to many stressors, such as acid, osmotic pressure, heat, oxygen, and DNA damage, is a heterogeneous noninheritable trait that is believed to strengthen the population fitness under stress (Booth 2002; Thattai and van Oudenaarden 2004; Aertsen and Michiels 2005; Lidstrom and Konopka 2010).

The responses of LAB to stressful conditions, as we have seen in the various chapters in this book, have been addressed by thorough genetic, physiological, and molecular biology approaches, and increasingly also using the various genomic technologies that are now at hand (transcriptomics, proteomics, metabolomics studies, metabolic and gene regulatory network modeling). Phenotypic heterogeneity at the single-cell level, however, is masked in the global analyses, as the data are averaged across millions of cells. Current developments in single-cell analyses techniques (Brehm-Stecher and Johnson 2004; Ducret et al. 2009; Muller and Nebe-von-Caron 2010) allow a detailed description of the extent of cell individuality in bacteria. The application of these technologies to the study of the response of LAB to stress should soon provide a wealth of new information on the ways these organisms cope with stressful changes in their environments. Single-cell labeling and (flow cytometric) sorting of subpopulations, followed by "omics" analyses of the sorted subpopulations, should allow quantitative information on cellular behavior (and that of metabolism) under stress conditions to be obtained.

21.8 Stress in the Context of Microbial Communities

LAB used in the fermentation industry are, more often than not, part of mixed microbial communities: Strains and species are combined for optimal industrial performance to produce foods with the best possible attributes. On the other hand, probiotic strains as well as pathogenic LAB are either part of existing competing populations or have to try and get a foothold in already residing communities in, for example, the human gastrointestinal (GI) tract. Interactions in these various niches involve fierce competition for various nutrients, but also cell-to-cell communication, mutual cross-feeding, the production of antimicrobial substances, and so forth. Recently, several papers have appeared on the microbial ecology of LAB using (post) genomic approaches. Thus, novel insight has been gained in the proto-cooperation of *S. thermophilus* and *Lactobacillus delbrueckii* subsp. *bulgaricus*, which involves not only nutrient exchange but also a stress response on the part of *S. thermophilus* to cope with the H_2O_2 produced by *L. bulgaricus* (Herve-Jimenez et al. 2009). The growth of *L. lactis* together with *S. aureus* led to a response in the transcriptome of the former that was suggestive of nutritional competition taking place. Interestingly, the presence of *L. lactis* hardly influenced the growth of *S. aureus*, but enterotoxin production by *S. aureus* was greatly modulated, with a downregulation of expression seen for two of the six enterotoxin genes (Even et al. 2009). The knowledge gained from studies like that of Even and coworkers might ultimately lead to the novel use of certain species of LAB for the biocontrol of (foodborne) pathogens.

Bacteriophages have been described for all LAB under study, and these bacterial predators add another and very important level of complexity on top of the phenotypic heterogeneity that may also exist at this level in the ecosystems in which LAB subsist. In these mixed communities, global but also very local stresses will be encountered. The latter is especially true in the GI tract, but also if one considers that even in milk fermentations, where the starter LAB are evenly dispersed at the start of the fermentation, the bacteria ultimately are randomly distributed and constrained in a coagulated matrix, where they are faced with very local circumstances such as restricted nutrient availability (Jeanson et al. 2011). There are, however, also indications that localization in the cheese matrix could be strain-dependent; for instance, higher numbers of cells were observed in the curd for some strains when a lower amount of fat was present in the cheese milk, while for other strains, the number of cells in the whey increased (Rehman et al. 2003). The location of starter cells in dairy products has been extensively studied (Parker et al. 1998; Fitzsimons et al. 2001; Ercolini et al. 2003; Ly-Chatain et al. 2010). Cells were often present around fatty globules, particularly in whey pockets (Marcellino and Benson 1992; Hassan et al. 2003; Lopez et al. 2006). Depending on the conditions, cells could grow out to mixed (micro-)colonies or to (contesting) neighbors. Tools and techniques to examine such microenvironments and to visualize bacterial interactions and the responses of the microbes to (adverse) environmental conditions in situ will no doubt soon lead to fundamentally new insights in how LAB cope with the various stresses they might encounter at the cellular level.

21.9 Concluding Remarks

Much still remains to be learned with respect to the mechanisms that LAB use to cope with the various stresses they encounter during their lives. Prospects for near-future major advances in our understanding thereof are good, as novel tools for the analyses of the responses at the cellular, ensemble, mixed-culture, and biofilm levels are rapidly advancing. Combining molecular analyses with global genomics approaches and fostering the cooperation of researchers in the field of LAB stress research with bioinformaticians and modelers will prove to be critical. For one thing, small RNAs (sRNAs) and small ORFs (sORFs) are currently being implicated in the genetic regulation of many important traits (among them, specifically, stress resistance) in an increasing number of eukaryotes and in bacteria (Waters and Storz 2009; Gripenland et al. 2010; Hemm et al. 2010; Liu and Camilli 2010; Romby and Charpentier 2010). The presence and roles of sRNAs and sORFs are practically unexplored in LAB. Powerful next-generation sequencing allows for the rapid and in-depth quantitative analysis of, among others, (meta-)transcriptomes (RNAseq; see, e.g., MacLean et al. 2009), and should be applied in LAB stress research to examine the role of these small-molecule regulators in the regulation of LAB stress responses. The gene regulatory network models that are available for some of the LAB paradigms are in dire need of the incorporation of this information.

The challenge and goal will ultimately be to use the obtained knowledge for biotechnology applications, to improve the robustness and functionality of probiotic strains as well as the strains and species used in food fermentations, or to employ the information in a medical setting, to successfully fight the pathogenic LAB.

References

Aertsen A, Michiels CW (2005) Diversify or die: generation of diversity in response to stress. Crit Rev Microbiol 31:69–78

Bolotin A, Wincker P, Mauger S, Jaillon O, Malarme K, Weissenbach J, Ehrlich SD, Sorokin A (2001) The complete genome sequence of the lactic acid bacterium *Lactococcus lactis* ssp. *lactis* IL1403. Genome Res 11:731–753

Booth IR (2002) Stress and the single cell: intrapopulation diversity is a mechanism to ensure survival upon exposure to stress. Int J Food Microbiol 78:19–30

Brehm-Stecher BF, Johnson EA (2004) Single-cell microbiology: tools, technologies, and applications. Microbiol Mol Biol Rev 68:538–559

Davidson CJ, Surette MG (2008) Individuality in bacteria. Annu Rev Genet 42:253–268

Delcour J, Ferain T, Hols P (2000) Advances in the genetics of thermophilic lactic acid bacteria. Curr Opin Biotechnol 11:497–504

Ducret A, Maisonneuve E, Notareschi P, Grossi A, Mignot T, Dukan S (2009) A microscope automated fluidic system to study bacterial processes in real time. PLoS One 4:e7282

Eldar A, Elowitz MB (2010) Functional roles for noise in genetic circuits. Nature 467:167–173

Ercolini D, Hill PJ, Dodd CE (2003) Bacterial community structure and location in Stilton cheese. Appl Environ Microbiol 69:3540–3548

Even S, Charlier C, Nouaille S, Ben Zakour NL, Cretenet M, Cousin FJ, Gautier M, Cocaign-Bousquet M, Loubière P, Le Loir Y (2009) *Staphylococcus aureus* virulence expression is impaired by *Lactococcus lactis* in mixed cultures. Appl Environ Microbiol 75:4459–4472

Fitzsimons NA, Cogan TM, Condon S, Beresford T (2001) Spatial and temporal distribution of non-starter lactic acid bacteria in Cheddar cheese. J Appl Microbiol 90:600–608

Foster PL (2007) Stress-induced mutagenesis in bacteria. Crit Rev Biochem Mol Biol 42:373–397

Frenkiel-Krispin D, Minsky A (2006) Nucleoid organization and the maintenance of DNA integrity in *E. coli*, *B. subtilis* and *D. radiodurans*. J Struct Biol 156:311–319

Galhardo RS, Hastings PJ, Rosenberg SM (2007) Mutation as a stress response and the regulation of evolvability. Crit Rev Biochem Mol Biol 42:399–435

Gonzalez C, Hadany L, Ponder RG, Price M, Hastings PJ, Rosenberg SM (2008) Mutability and importance of a hypermutable cell subpopulation that produces stress-induced mutants in *Escherichia coli*. PLoS Genet 4:e1000208

Grandvalet C, Coucheney F, Beltramo C, Guzzo J (2005) CtsR is the master regulator of stress response gene expression in *Oenococcus oeni*. J Bacteriol 187:5614–5623

Gripenland J, Netterling S, Loh E, Tiensuu T, Toledo-Arana A, Johansson J (2010) RNAs: regulators of bacterial virulence. Nat Rev Microbiol 8:857–866

Hassan AN, Ipsen R, Janzen T, Qvist KB (2003) Microstructure and rheology of yogurt made with cultures differing only in their ability to produce exopolysaccharides. J Dairy Sci 86:1632–1638

Hemm MR, Paul BJ, Miranda-Rios J, Zhang A, Soltanzad N, Storz G (2010) Small stress response proteins in *Escherichia coli*: proteins missed by classical proteomic studies. J Bacteriol 192:46–58

Herve-Jimenez L, Guillouard I, Guedon E, Boudebbouze S, Hols P, Monnet V, Maguin E, Rul F (2009) Postgenomic analysis of *Streptococcus thermophilus* cocultivated in milk with *Lactobacillus delbrueckii* subsp. *bulgaricus*: involvement of nitrogen, purine, and iron metabolism. Appl Environ Microbiol 75:2062–2073

Hoch JA, Varughese KI (2001) Keeping signals straight in phosphorelay signal transduction. J Bacteriol 183:4941–4949

Hols P, Hancy F, Fontaine L, Grossiord B, Prozzi D, Leblond-Bourget N, Decaris B, Bolotin A, Delorme C, Dusko Ehrlich S, Guédon E, Monnet V, Renault P, Kleerebezem M (2005) New insights in the molecular biology and physiology of *Streptococcus thermophilus* revealed by comparative genomics. FEMS Microbiol Rev 29:435–463

Jeanson S, Chadoeuf J, Madec MN, Aly S, Floury J, Brocklehurst TF, Lortal S (2011) Spatial distribution of bacterial colonies in a model cheese. Appl Environ Microbiol. 77:1493–1500

Katayama T, Ozaki S, Keyamura K, Fujimitsu K (2010) Regulation of the replication cycle: conserved and diverse regulatory systems for DnaA and *oriC*. Nat Rev Microbiol 8:163–170

Klaenhammer TR, Barrangou R, Buck BL, Azcarate-Peril MA, Altermann E (2005) Genomic features of lactic acid bacteria affecting bioprocessing and health. FEMS Microbiol Rev 29:393–409

Klein G, Dartigalongue C, Raina S (2003) Phosphorylation-mediated regulation of heat shock response in *Escherichia coli*. Mol Microbiol 48:269–285

Kramer R (2010) Bacterial stimulus perception and signal transduction: response to osmotic stress. Chem Rec 10:217–229

Lacour S, Doublet P, Obadia B, Cozzone AJ, Grangeasse C (2006) A novel role for protein-tyrosine kinase Etk from *Escherichia coli* K-12 related to polymyxin resistance. Res Microbiol 157:637–641

Levine A, Vannier F, Absalon C, Kuhn L, Jackson P, Scrivener E, Labas V, Vinh J, Courtney P, Garin J, Séror SJ (2006) Analysis of the dynamic *Bacillus subtilis* Ser/Thr/Tyr phosphoproteome implicated in a wide variety of cellular processes. Proteomics 6:2157–2173

Lidstrom ME, Konopka MC (2010) The role of physiological heterogeneity in microbial population behavior. Nat Chem Biol 6:705–712

Liu JM, Camilli A (2010) A broadening world of bacterial small RNAs. Curr Opin Microbiol 13:18–23

Lopez C, Maillard MB, Briard-Bion V, Camier B, Hannon JA (2006) Lipolysis during ripening of Emmental cheese considering organization of fat and preferential localization of bacteria. J Agric Food Chem 54:5855–5867

Losick R, Desplan C (2008) Stochasticity and cell fate. Science 320:65–68

Ly-Chatain MH, Le ML, Thanh ML, Belin J-M, Wachı Y (2010) Cell surface properties affect colonisation of raw milk by lactic acid bacteria at the microstructure level. Food Res Int 43:1594–1602

Macek B, Mijakovic I, Olsen JV, Gnad F, Kumar C, Jensen PR, Mann M (2007) The serine/threonine/tyrosine phosphoproteome of the model bacterium *Bacillus subtilis*. Mol Cell Proteomics 6:697–707

Machielsen R, van Alen-Boerrigter IJ, Koole LA, Bongers RS, Kleerebezem M, Van Hylckama Vlieg JE (2010) Indigenous and environmental modulation of frequencies of mutation in *Lactobacillus plantarum*. Appl Environ Microbiol 76:1587–1595

MacLean D, Jones JD, Studholme DJ (2009) Application of "next-generation" sequencing technologies to microbial genetics. Nat Rev Microbiol 7:287–296

Majakovic I (2010) Protein phosphorylation in bacteria. Microbe 5:21–25

Makarova K, Slesarev A, Wolf Y, Sorokin A, Mirkin B, Koonin E, Pavlov A, Pavlova N, Karamychev V, Polouchine N, Shakhova V, Grigoriev I, Lou Y, Rohksar D, Lucas S, Huang K, Goodstein DM, Hawkins T, Plengvidhya V, Welker D, Hughes J, Goh Y, Benson A, Baldwin K, Lee JH, Díaz-Muñiz I, Dosti B, Smeianov V, Wechter W, Barabote R, Lorca G, Altermann E, Barrangou R, Ganesan B, Xie Y, Rawsthorne H, Tamir D, Parker C, Breidt F, Broadbent J, Hutkins R, O'Sullivan D, Steele J, Unlu G, Saier M, Klaenhammer T, Richardson P, Kozyavkin S, Weimer B, Mills D (2006) Comparative genomics of the lactic acid bacteria. Proc Natl Acad Sci USA 103:15611–15616

Marcellino SN, Benson DR (1992) Scanning electron and light microscopic study of microbial succession on Bethlehem St. Nectaire cheese. Appl Environ Microbiol 58:3448–3454

Mills DA (2001) Mutagenesis in the post genomics era: tools for generating insertional mutations in the lactic acid bacteria. Curr Opin Biotechnol 12:503–509

Mott ML, Berger JM (2007) DNA replication initiation: mechanisms and regulation in bacteria. Nat Rev Microbiol 5:343–354

Muller S, Nebe-von-Caron G (2010) Functional single-cell analyses: flow cytometry and cell sorting of microbial populations and communities. FEMS Microbiol Rev 34:554–587
Parker ML, Gunning PA, Macedo AC, Malcata FX, Brocklehurst TF (1998) The microstructure and distribution of micro-organisms within mature Serra cheese. J Appl Microbiol 84:523–530
Pikuta EV, Hoover RB, Tang J (2007) Microbial extremophiles at the limits of life. Crit Rev Microbiol 33:183–209
Rajagopal L, Clancy A, Rubens CE (2003) A eukaryotic type serine/threonine kinase and phosphatase in *Streptococcus agalactiae* reversibly phosphorylate an inorganic pyrophosphatase and affect growth, cell segregation, and virulence. J Biol Chem 278:14429–14441
Rehman S-U, Farkye NY, Drake M (2003) Reduced-fat Cheddar cheese from a mixture of cream and liquid milk protein concentrate. Int J Dairy Technol 56:94–98
Robleto EA, Yasbin R, Ross C, Pedraza-Reyes M (2007) Stationary phase mutagenesis in *B. subtilis*: a paradigm to study genetic diversity programs in cells under stress. Crit Rev Biochem Mol Biol 42:327–339
Romby P, Charpentier E (2010) An overview of RNAs with regulatory functions in Gram-positive bacteria. Cell Mol Life Sci 67:217–237
Rosen R, Becher D, Buttner K, Biran D, Hecker M, Ron EZ (2004) Highly phosphorylated bacterial proteins. Proteomics 4:3068–3077
Schlacher K, Goodman MF (2007) Lessons from 50 years of SOS DNA-damage-induced mutagenesis. Nat Rev Mol Cell Biol 8:587–594
Schumann W (2007) Bacterial stress sensors. In: Atassi MZ (Ed.), Protein reviews. Springer, New York
Siezen RJ, Wilson G (2010) Probiotics genomics. Microb Biotechnol 3:1–9
Soufi B, Gnad F, Jensen PR, Petranovic D, Mann M, Mijakovic I, Macek B (2008a) The Ser/Thr/Tyr phosphoproteome of *Lactococcus lactis* IL1403 reveals multiply phosphorylated proteins. Proteomics 8:3486–3493
Soufi B, Jers C, Hansen ME, Petranovic D, Mijakovic I (2008b) Insights from site-specific phosphoproteomics in bacteria. Biochim Biophys Acta 1784:186–192
Storz G, Hengge-Aronis R (2000) Bacterial stress responses. ASM Press, Washington, DC
Sugimoto S, Abdullah Al M, Sonomoto K (2008) Molecular chaperones in lactic acid bacteria: physiological consequences and biochemical properties. J Biosci Bioeng 106:324–336
Sun X, Ge F, Xiao CL, Yin XF, Ge R, Zhang LH, He QY (2010) Phosphoproteomic analysis reveals the multiple roles of phosphorylation in pathogenic bacterium *Streptococcus pneumoniae*. J Proteome Res 9:275–282
Thattai M, van Oudenaarden A (2004) Stochastic gene expression in fluctuating environments. Genetics 167:523–530
Ulrich LE, Koonin EV, Zhulin IB (2005) One-component systems dominate signal transduction in prokaryotes. Trends Microbiol 13:52–56
van de Guchte M, Serror P, Chervaux C, Smokvina T, Ehrlich SD, Maguin E (2002) Stress responses in lactic acid bacteria. Antonie van Leeuwenhoek 82:187–216
van Schaik W, Abee T (2005) The role of sigmaB in the stress response of Gram-positive bacteria—targets for food preservation and safety. Curr Opin Biotechnol 16:218–224
Varhimo E, Savijoki K, Jalava J, Kuipers OP, Varmanen P (2007) Identification of a novel streptococcal gene cassette mediating SOS mutagenesis in *Streptococcus uberis*. J Bacteriol 189:5210–5222
Veening JW, Smits WK, Kuipers OP (2008) Bistability, epigenetics, and bet-hedging in bacteria. Annu Rev Microbiol 62:193–210
Walker GC, Smith BT, Sutton M (2000) The SOS response to DNA damage. In: Storz G, Hengge-Aronis R (Eds.), Bacterial stress responses. ASM Press, Washington, DC, pp. 131–144
Wang JD, Levin PA (2009) Metabolism, cell growth and the bacterial cell cycle. Nat Rev Microbiol 7:822–827
Waters LS, Storz G (2009) Regulatory RNAs in bacteria. Cell 136:615–628
Yother J, Trieu-Cuot P, Klaenhammer TR, De Vos WM (2002) Genetics of streptococci, lactococci, and enterococci: review of the Sixth International Conference. J Bacteriol 184:6085–6092
Zhong J, Molina H, Pandey A (2007) Phosphoproteomics. Curr Protoc Protein Sci, Chapter 24: Unit 24

Index

E. Tsakalidou and K. Papadimitriou (eds.), *Stress Responses of Lactic Acid Bacteria*, Food Microbiology and Food Safety, DOI 10.1007/978-0-387-92771-8,

O

P

MIX
Papier aus verantwortungsvollen Quellen
Paper from responsible sources
FSC® C105338

If you have any concerns about our products,
you can contact us on
ProductSafety@springernature.com

In case Publisher is established outside the EU,
the EU authorized representative is:
Springer Nature Customer Service Center GmbH
Europaplatz 3, 69115 Heidelberg, Germany

Printed by Libri Plureos GmbH
in Hamburg, Germany